Engineering Design for the Control of Workplace Hazards

Engineering Design for the Control of Workplace Hazards

Richard A. Wadden
Environmental and Occupational Sciences
School of Public Health
University of Illinois at Chicago

Peter A. Scheff
Pritzker Department of Environmental Engineering
Illinois Institute of Technology

With contributions by

Robert J. Allen *University of Illinois at Chicago*
John E. Franke *Peterson Associates, Milwaukee, Wisconsin*
Edward R. Hermann *University of Illinois at Chicago*
Floyd G. Miller *University of Illinois at Chicago*

McGraw-Hill Book Company

New York St. Louis San Francisco Auckland Bogotá
Hamburg Johannesburg London Madrid Mexico
Milan Montreal New Delhi Panama
Paris São Paulo Singapore
Sydney Tokyo Toronto

Library of Congress Cataloging-in-Publication Data

Wadden, R. A.
Engineering design for the control of workplace hazards.

Includes index.
1. Industrial safety. 2. Engineering design.
I. Scheff, Peter A. II. Title.
T55.W34 1987 620.8′6 86-27803
ISBN 0-07-067664-X

1234567890 DOC/DOC 89210987

ISBN 0-07-067664-X

The editors for this book were Betty Sun and Lynne Lackenbach, the designer was Naomi Auerbach, and the production supervisor was Annette Mayeski. It was set in Century Schoolbook by Techna Type, Inc.

Printed and bound by R. R. Donnelley & Sons Inc.

To Angela, Terese, Noah, and Nuiko

Contents

Preface

Design of control systems to enhance occupational safety and health is becoming a basic necessity in most industries. The requirements of OSHA, increased employer liability, and heightened worker awareness of the implications of unsafe and unhealthy work environments are all contributing to recognition of this educational need. Although a small cadre of ventilation and safety engineers and industrial hygienists are available to deal with design and control of occupational hazards, their area of application is often limited to large installations or companies. And their contributions are often only belatedly incorporated, after equipment or process design and installation have been completed. The vast majority of engineers concerned with design and operation are not trained to recognize or control existing or potential safety and health hazards. In addition, while many of the principles of engineering control technology are well known, these ordinarily are not collected in one volume convenient for use by the practitioner in the field.

The major thrust of our effort is to reach those engineers who probably will not become specialists in the occupational safety and health field. However, this large group is knowingly or unknowingly often called upon to make significant decisions that will affect the workers' well-being. Our aim is to provide these engineers with guidelines with which to evaluate the problem, and design tools with which to make at least initial estimates of the control technology required. Our text encourages incorporation of control concepts at an early stage of design, and will supply practicing engineers with information to support bid specifications and to ask "the right questions" of contractors and consultants carrying out the actual installations.

We have structured our presentation into four general areas: control rationale, control specification, system design, and control requirements for specific hazards. Chapter 2, "Occupational Injury and Disease," supplies the control rationale, which is ultimately protection of worker health. In this chapter we summarize in some detail injury statistics and the findings of occupational epidemiology in a variety of hazardous industries. We have not attempted to present an exhaustive review, but we have tried to present enough information to give those who are not specialists in epidemiology a reasonable appreciation of the data that do exist.

Under control specification we have included Chapter 3, "Hazard Evaluation and Control," and Chapter 4, "Design Criteria." Chapter 3 is intended as a brief summary of approaches that are used for workplace evaluation. Standard references such as *Patty's Industrial Hygiene and Toxicology* (Clayton and Clayton, Wiley, 1978), NIOSH's *The Industrial Environment* (NIOSH, 1973 and soon to be revised), the National Safety Council's *Fundamentals of Industrial Hygiene* (Olishifski, 1979), and an ever-expanding literature are available for the details of industrial hygiene practice. Likewise, although we have left specific discussion of toxicological testing to more specialized texts, we have presented a variety of relationships based on such testing, which may be used to predict worker exposure. Every design procedure requires a criterion on which to judge success or failure. Appropriate criteria for the workplace, both as performance and specification standards and guidelines, are summarized in Chapter 4, "Design Criteria."

System design includes "Ventilation" (Chapter 5), "Air Cleaning" (Chapter 6), "Process Alteration" (Chapter 7), and "Worker-Machine Interactions" (Chapter 8). Each of these presentations emphasizes procedures to implement the first level of a design solution. The nature of the problems dealt with in these areas is somewhat broader than those summarized under control requirements for specific hazards. "Guarding" (Chapter 9), "Ionizing Radiation" (Chapter 10), "Nonionizing Radiation" (Chapter 11), "Heat" (Chapter 12), and "Noise and Vibration" (Chapter 13) fall under the latter category.

A common format has been used for each design chapter. This includes a description of the particular problems associated with each area (e.g., heat), presentation of the appropriate design equations, and worked-out solutions to realistic design problems. In general, we have included enough data and other reference material so that the reader need not search elsewhere. The choice of design procedures was based on existing practice, with preference given to those methods that have been thoroughly validated in field and laboratory. We have not discussed the theoretical basis for such procedures in depth, nor have we presented approaches that are presently in the speculative stage, although numerous references are cited for the interested practitioner.

A panel of chemical engineers in industry recently advised the AIChE that chemical engineering design classes should present the simplest valid design procedure and the most complex (e.g., numerical integration); and dispense with teaching any of the intermediate "short-cut" approaches of varying complexity, which were formerly so useful before general access to computers became available. We subscribe to this philosophy and have presented the first-level procedures in our text. Although at first glance these may not appear to be "back-of-the-envelope" estimators, all of the calculations can be carried out efficiently with a hand-held calculator.

As with any book of this size, many persons contributed to its evolution. In order to implement the intended scope of design, the need soon became evident for insightful contributions from others knowledgeable in the field. It has been our good fortune to be able to call on just such a competent group at the University of Illinois in Chicago who generously made their expertise available. Five of six authors are engineers (chemical, civil, environmental, industrial), four are certified industrial hygienists, and three have professional engineering registration. Dr. Edward Hermann, School of Public Health, University of Illinois, wrote Chapters 12 and 13, "Heat" and "Noise and Vibration"; Dr. Robert Allen of the same school contributed Chapter 5, "Ventilation," and collaborated with us on Chapter 6, "Air Cleaning"; Dr. John Franke, formerly of the University of Illinois and now with Peterson Associates, contributed to Chapters 3 and 4, "Hazard Evaluation" and "Design Criteria," and just as importantly, critiqued the whole book from the vantage point of his extensive experience as a practicing industrial hygienist. Dr. Floyd Miller of the Mechanical Engineering Department of the University co-authored Chapter 9, "Guarding."

John Talty of the National Institute for Occupational Safety and Health (NIOSH) and William Martin, Consulting Environmental Engineer and formerly with NIOSH, were most helpful in supplying information, documents, and review comments. Our thanks also go to Barbara Bates for developing the tables in Chapter 2; Dr. Chung Fung-Li and Thomas Galassi for reference checking; Dr. Daniel Hryhorczuk for his review of Chapter 2; Ms. Dianne Martia Salaty for the original artwork; and the students who participated in our courses in engineering control for occupational safety and health at the Illinois Institute of Technology (ENVE 532) and the University of Illinois at Chicago (CEMM 393). We also offer our greatest appreciation to Frederica Davis, Denise Umstead, and Rosita Plaza for all the typing that was required. Most particularly, we would like to thank Mrs. Davis, who labored full-time for almost seven months through a bewildering morass of scrawled drafts, always with the greatest equanimity.

The choice and presentation of materials represents the authors' pooled academic and workplace experience, and the editor's judgments. Our coverage has been of necessity selective, although we hope that most topics will be useful. We, of course, stand responsible for all errors, omissions, and oversights.

R. A. Wadden
P.A. Scheff

Illustration Credits

Figures 5.1, 5.2, 5.4, 5.6, 5.10, 5.18, 5.20, 5.22, and 6.24 are reprinted by permission of the ACGIH and the Committee on Industrial Ventilation.

Figure 5.3 is reprinted by permission of the ACGIH and the Carrier Corporation.

Figures 5.14, 5.15, 5.16, 6.7, and 6.8 are reprinted by permission of the American Institute of Chemical Engineers from the *American Institute of Chemical Engineers Journal.*

Figures 5.21 and 5.23 are reprinted by permission of the *ASHRAE Handbook, 1980 Systems.*

Figures and tables in Appendix B, Chapter 5, are reprinted by permission of the *ASHRAE Handbook, 1985 Fundamentals.*

Figures 6.3 and 6.6 are reprinted by permission of McGraw-Hill from *Chemical Engineering.*

Figures 6.5, 6.23, 8.9, 8.10, 12.1, 13.12, 13.13, 13.14, 13.15, 13.16, 13.17, 13.18, 13.19, 13.20, 13.22, 13.24, 13.25, and 13.29 are reprinted by permission of McGraw-Hill.

Figure 6.9 is reprinted by permission of the *ASHRAE Handbook and Product Directory, 1983 Equipment.*

Figures 6.12, 6.17, and 6.18 are reprinted by permission of the *Journal of Air Pollution Control Association.*

Figure 6.13 is reprinted by permission of Harper & Row.

Figures 7.8 and 7.9 are reprinted by permission of *Hydrocarbon Processing.*

Figure 8.3 is reprinted by permission of the Institute of Industrial Engineers from *Industrial Engineering.*

Figure 8.4 is reprinted by permission of Harvard University Press.

Figures 8.5 and 8.12 are reprinted by permission of Taylor and Francis, Ltd.

Figure 8.6 is reprinted by permission of the *British Journal of Psychology.*

Figure 8.7 is reprinted by permission of the *American Industrial Hygiene Association Journal.*

Figure 9.7 is reprinted by permission of the Alliance of American Insurers.

Figures 9.9 and 9.34 are reprinted by permission of the Liberty Mutual Insurance Group.

Figures 10.2, 10.6, 10.11, 10.12, 10.13, 10.14, 10.15, and 10.16 are reprinted by permission of the National Council on Radiation Protection.

Figure 10.9 is reprinted by permission of Pergamon Press from *Health Physics*.

Figures 10.10, 11.4, 11.5, 11.6, 11.11, and 11.16 are reprinted by permission of John Wiley & Sons, Inc.

Figures 11.1 and 11.10 are reprinted by permission of the National Safety Council.

Figures 11.2 and 11.9 are reprinted by permission of the *Journal of Applied Optics*.

Figure 11.3 is reprinted by permission of the *American Welding Society from Welding Journal*.

Figures 11.7, 11.13, and 11.14 are reprinted by permission of the American National Standards Institute.

Figure 11.12 is reprinted by permission of the ACGIH.

Figure 11.15 is reprinted by permission of the New York Academy of Science.

Figure 13.1 is reprinted by permission of Peterson and Gross.

Chapter

1

Workplace Hazard Control

R. A. Wadden

The underlying rationale for workplace control is protection of worker and community health. The actual response to a control need may be stimulated by the presence of a regulation or guideline, the desire to maintain good employer-employee relationships, an interest in obtaining reasonable insurance rates, or the threat or implementation of a law suit. Regulations, such as the Occupational Safety and Health Administration's *General Industry Standards* (OSHA, 1983), and guidelines, such as the American Conference of Governmental Industrial Hygienists' *TLV's, Threshold Limit Values for Chemical Substances and Physical Agents in the Work Environment* (ACGIH, 1985), provide specifications for developing appropriate design solutions. These are often in the form of numerical goals related to worker exposure. For instance, the OSHA noise regulation specifies 90 dBA as an acceptable 8-hr exposure; and the TLV of 0.05 mg/m^3 of air for chromic acid (expressed as Cr) is suggested as an appropriate average 8-hr air exposure that ordinarily is not expected to cause adverse health effects. Other types of specification detail the actual design configuration that must be used to protect workers from illness and injury, e.g., point of operation guarding for a mechanical power press.

Less easily formulated into design specifications, but just as important a stimulus for control in our present society, is the possibility of legal action. Probably the most notable hazard in this regard, at least up to the present, has been asbestos production and use. Table 1.1 indicates the number of lawsuits that have been filed against several U.S. companies engaged in asbestos work. It has been estimated that asbestos-caused diseases could cost the industry and its insurers $200 billion or more by the end of the century (*Economist,* 1982). Manville Corporation, the largest U.S. asbestos producer, requested relief from asbestos litigation by filing for reorganization under federal bankruptcy laws (NYT, 1982); and additionally sued several of its insurers for $5 billion, alleging that they had wrongfully refused to pay appropriate claims (Brody, 1982).

When numerical guidelines exist for worker exposure, it is usually necessary to determine existing levels for comparison purposes. Careful workplace monitoring of noise and vibration, chemicals in air, temperature (heat exposure), and ionizing and nonionizing radiation will indicate: (1) whether an acceptable level is being met; (2) which machines or processes are responsible for acceptable levels not being met; and (3) what reduction in exposure and discharge is required to reach and maintain acceptable levels. Some approaches to interpretation of monitoring data in order to answer these questions are given in Chapter 3.

Given an acceptable level to design to, it is then necessary to choose an appropriate mode of control. This may be through engineering design, which includes local and area ventilation; process and equipment modification; substitution of less hazardous materials; pollutant removal, isolation, and shielding; use of devices for personal protection; and administrative controls. As an illustration of each, let us consider

TABLE 1.1 Asbestos-Related Lawsuits

Company	Suits outstanding
Manville	16,500
Amatex	9,300
GAF	10,000
Celotex	11,000
Owens-Illinois	"Substantial number"
Eagle-Picher	11,500
Pittsburgh Corning	8,000
Raymark	10,443
Rock Wool Mfg.	3,000–7,000
Bairnes	11,000
H. K. Porter	Not available
UNR Industries	17,000

SOURCE: *Economist,* 1982.

the options available for minimizing noise exposure in the engine room of a large ocean-going merchant ship, *The Hawaiian Queen* (Jensen et al., 1978). Measurements in the engine room indicated average levels of 120 dBA over 8 hrs, well above the applicable OSHA standard of 90 dBA. In addition, the primary stage of a nested-type double-reduction gear unit of the 9000-hp steam turbine was identified as the major noise source. Requiring protective earmuffs for all engine room workers is one possible approach (although not a complete solution, since earmuffs alone would probably not reduce exposures by 30 dBA). Administrative solutions might be to minimize worker entry into the engine room; or to specify (possibly illegally) that only those sailors who already have significant hearing impairment be employed at this worksite. An engineering solution, which was actually implemented, was to enclose the gear train casing with modular acoustic panels. A split commercial silencer was also installed at the point of propeller shaft penetration into the enclosure to attenuate sounds that would otherwise escape around the shaft. These steps reduced the noise to engine room ambient levels. Before and after sound pressure levels are given in Figure 1.1. However, the ambient noise level is 91 dBA, which is still above the OSHA standard. Consequently, further control is required. Monitoring of personal exposures would still be advisable to determine whether hearing protection is required, since workers will spend different amounts of time near the equipment and elsewhere in the engine room.

The thrust of this book is toward engineering design to solve such problems. We have attempted to explain and demonstrate ways to control workplace hazards at their source. We have suggested methods of control, tested by ourselves and others, which are generally agreed upon in the environmental and occupational field. Procedures to predict

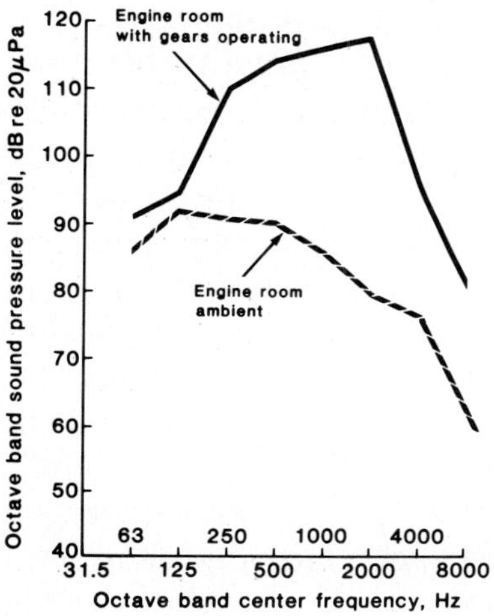

Figure 1.1 Engine room sound pressure levels (*Jensen et al., 1978*).

control efficacy are recommended whenever appropriate. The goal of these calculations is to make a reasonable first estimate of the important design parameters for a particular problem. For example, the appropriate thickness and material for shielding from a neutron source are determined in Chapter 10; the acoustical treatment required for an enclosure to control a 500-Hz noise source is specified in Chapter 13; the design requirements for ventilating an open-surface dip tank are developed in Chapter 5. These estimates can subsequently be used by the practitioner for planning purposes, initial cost estimates, and control equipment and process specification and evaluation. They also serve as a basis for informed discussion and consultation with design specialists and contractors, and provide a format for field evaluation of the effectiveness of the ultimate design solution.

The suggested design methods will often not supply a final answer to a workplace problem. This will ordinarily require a more intensive investigation of the technical and economic feasibility of various available alternatives, followed by a specific design in more depth at the nuts-and-bolts level. The details of final design solutions for such diverse problems as shielding of ruby lasers, local ventilation of a chromic acid electroplating bath for bicycle parts, or reduction in the vibrational energy of a metal grinder probably will come from specialists in these areas.

Nor are our recommended procedures meant to be all-inclusive or to produce an "instant expert." Most design solutions require several rounds of "cut-and-try." However, the following chapters do supply an approach to problem definition leading to an initial evaluation of control alternatives. Application of the design methods will allow the practitioner to intelligently and systematically initiate and participate in the design process.

Of necessity, many of the problems faced by plant engineers are due to existing situations. Processes may be unsafe because of poor original design, or they may have become so because of change in function. Sometimes the health implications of unsolved or unrecognized safety hazards may not be realized for long periods, even years. It is hoped that this text will be helpful in responding to such immediate concerns. But the underlying motive of our approach is to encourage consideration and inclusion of safe design at the early, conceptual stages of project planning. Making hazard control an integral part of the design process provides a powerful method of attaining a safe workplace.

A case in point is a recent report called *Computerized Manufacturing Automation* (OTA, 1984). Although it discusses the economic implications of robotics in terms of both projected lost jobs and increased plant investment, and also the potential development of a "dehumanizing" work environment, there is only one mention of the possible workplace hazards associated with such devices. Unfortunately, an equally current NIOSH report (NIOSH, 1984) details the death of a trained, experienced die-cast worker after being pinned between the back end of an industrial robot and a steel safety pole. The hydraulic robot had been installed to remove and transfer die-cast parts from a die-cast machine. The worker entered the robot's work envelope to clean up scrap metal accumulated on the floor. To do so, he evaded the safety rail interlock system and evidently did not recognize the hazards associated with the movement of parts of the robotic assembly other than the active arm. The need for more thorough worker training and supervision, adequate clearance for robot operation, and the incorporation of motion sensors or light curtains to inactivate the robot when a person crosses the work envelope boundaries are among the recommendations for future robot designs.

Retrofit controls are by their nature often expensive, and frequently are considered as incremental "extra" production and investment costs due to the necessity to fulfill recently recognized regulatory or legal requirements. However, many times good engineering design for hazard control at the planning stage, and even retroactively, will result in economically attractive solutions. An example is given in Chapter 7 of an oxychlorination process in which a process change increased yield as well as reducing pollutant production and discharge. The control of leaks in chemical plants and petroleum refineries (discussed in Chapters 3 and 7) is also a procedure that has the potential for yielding financial dividends.

Better employee morale is another likely outcome. The finishing of stereo cabinets in one spray-painting plant offers an example (O'Brien and Hurley, 1981). In this process the wooden cabinets are finished with a base coat, a glaze, and a clear lacquer in a multistep, conveyorized operation. At one of the steps a robot was installed to apply a heavy coat of glaze to cabinet exteriors and lid undersides because of high employee turnover in this messy operation. Besides eliminating the operator from an unpleasant function, using the robot reduced waste from overspray and increased material utilization by 15%, and the company expects a 5-year payout.

In another example, from a corrugated box factory, slit side trim was blown through a conveyor duct to collection bins (Jensen et al., 1978). The 12-in. duct was suspended 10 ft from the floor, extended across the 40-ft work room, and contained blower fans with heavy blades, which chopped up and conveyed the trim. The trim hitting the walls caused ear-level sound levels of 93 dBA (Figure 1.2). Noise-level control was sought to improve worker communication as well as to meet the OSHA 90-dBA requirement. The solution was to wrap the problem duct locations (mainly at the bends) with 2 in. of mineral-wool building

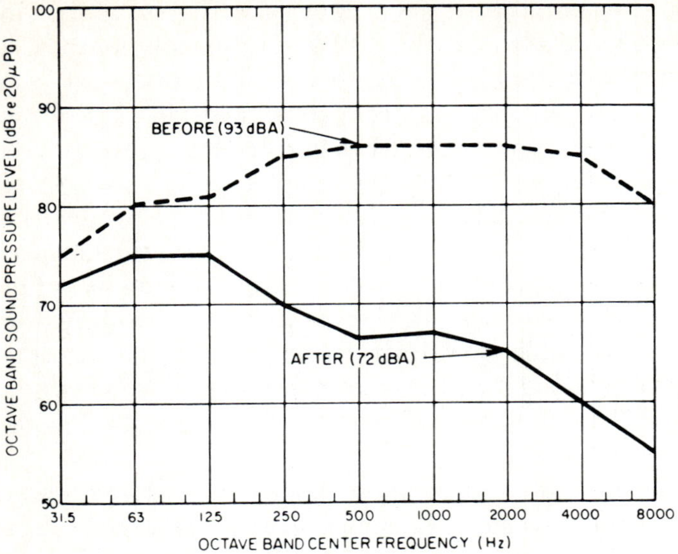

Figure 1.2 Noise levels in scrap duct for corrugated box industry, before and after covering (*Jensen et al., 1978*).

insulation to furnish an absorbing and resilient layer. Two impervious layers of heavy tar paper, spirally wrapped with 50% overlay (to prevent noise leakage) were placed over the insulation. The resulting noise level was 72 dBA (Figure 1.2). This reduction was more than enough to meet the OSHA standards, and the conveyor noise could hardly be heard above the ambient noise in the building. Plant safety and employee morale improved because speech interference was greatly reduced.

Every design that is implemented to control workplace hazards, whether retroactive or original, needs to be followed up with field measurement to check the efficacy of control. There is no substitute for this step. Of the unsuccessful design solutions we have observed, most occurred because the designer and client assumed that specification and installation were equivalent to effective operation. Prediction methods will not be 100% accurate, and actual operation of a plant may differ considerably from design specifications. Maintenance of ventilation and control systems is as important as initial design.

Figure 1.3 shows the effect of engineering controls on vinyl chloride monomer (VCM) air concentrations at a number of sites in a polyvinylchloride (PVC) dispersion polymerization process (Enviro, 1978). The controls, instituted in 1975, included improved seal-leak detection, maintenance, and venting; individual permanent local exhaust systems for batch mix tanks, orifices, pumps and compressors, reactors, blend tanks, filter enclosure, and PVC cake breaker; and enclosed

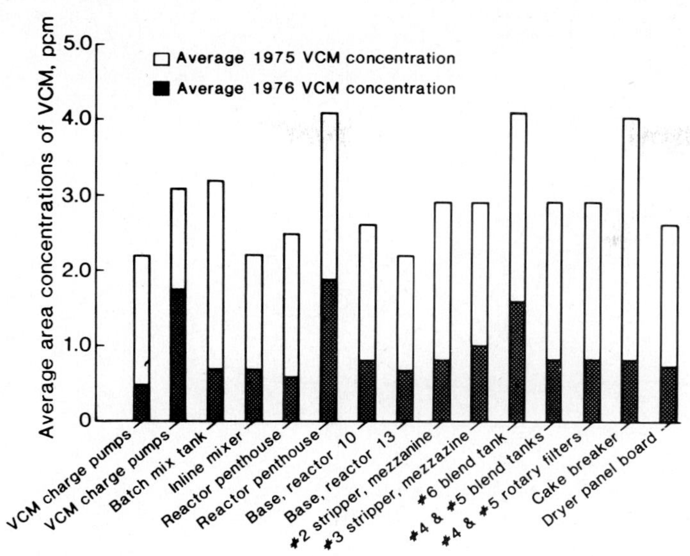

Figure 1.3 Average area VCM concentration at work stations (*Enviro, 1978*). (*Note:* VCM charge pumps and blend tanks were in areas that do not require regular operator attendance; and the reactor penthouse was posted to require personal protective devices.)

waste-treating facilities for process liquids. Although improvement is still necessary [the OSHA permissible exposure limit for VCM is 1 ppm (OSHA, 1983), and the material is a recognized human carcinogen (ACGIH, 1985)], the impact of engineering controls is evident.

The noise control design of a 70,000 bbl/day catalytic hydrosulfurizing facility (CHD) also illustrates such design follow-up (Jensen et al., 1978). The design objective was 85 dBA at normal work stations and 87 dBA maximum in passageways and normal maintenance areas. Figure 1.4 shows the noise contours at the plant site generated from vendor-guaranteed noise emissions for 78 pieces of as-purchased equipment, and simplified assumptions about source location and noise propagation. The problem areas are clearly delineated. Figure 1.5 shows the predicted noise levels that would result from optimum placement of equipment and off-the-shelf noise controls. After the plant was actually built to these specifications, a field test resulted in the contours of Figure 1.6. The general pattern was the same as predicted, particularly close to the plant boundary, but significant levels still occurred at several locations. These were traced to an unexpectedly noisy stripper bottoms pump, two improperly insulated valves, and an unpredicted noisy coupling. The solutions to these problems were relatively straightforward.

This project pointed up the difficulty of gathering good source-noise emission data, and this lack is a problem for other types of emissions

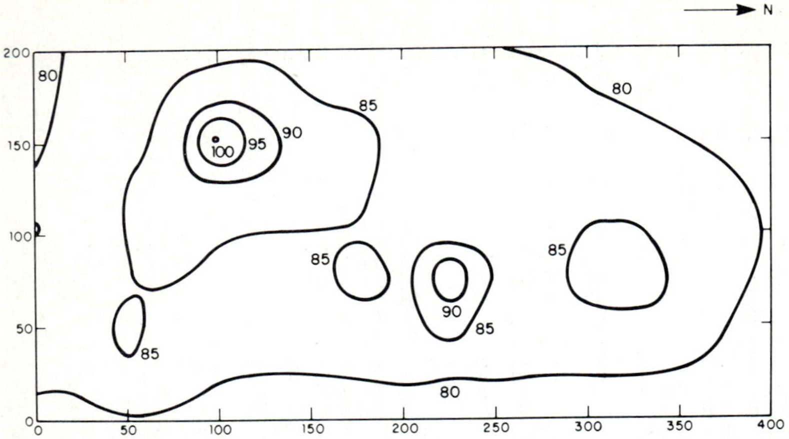

Figure 1.4 Uncontrolled A-weighted sound-level contours (dBA) generated for 70,000 bbl/day catalytic hydrosulfurization unit (CHD) site (*Jensen et al., 1978*). (*Note:* Contour lines are labeled on the decreasing side. Axes are in units of feet.)

as well. It also demonstrated the fact that equipment vendors will often respond to hazard control considerations only if their customers specify or demand them. Continuous tube coupling guards are now available as a standard item because of owner-vendor resolution of the noisy coupling problem (Jensen et al., 1978).

Another major consideration in implementing controls is the education of workers, management, and designers to the nature and hazard of the workplace problem; and the care and considerations required for its solution. For instance, a movable ventilation hood was recently specified to remove metal dust from a grinding operation to finish

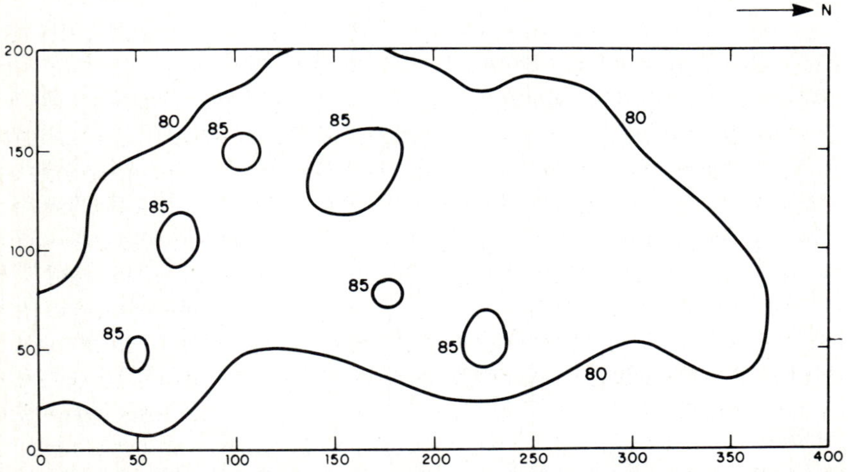

Figure 1.5 A-weighted sound-level contours (dBA) generated for CHD site using off-the-shelf controls (*Jensen et al., 1978*). (*Note:* Contour lines are labeled on the decreasing side. Axes are in units of feet.)

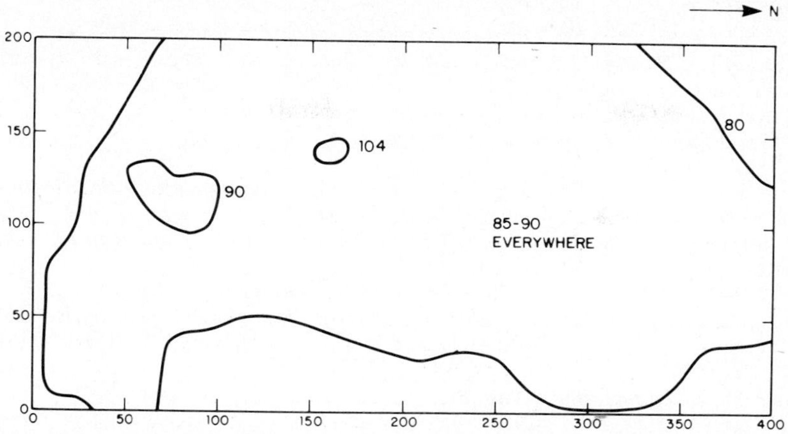

Figure 1.6 A-weighted sound-level contours actually measured at CHD site (*Jensen et al., 1978*). Axes are in units of feet.

foundry castings. After work on one casting was completed, the hood was rolled out of the way and the casting was removed and replaced by crane. The hood was then pushed back into place in order to remove dust during grinding on the new piece. However, upon return to the worksite after several months, the hood was discovered essentially abandoned. Because of the small wheels and the considerable amount of sand on the floor, the workers found the device very difficult to push back and forth. They and their immediate management were also not sufficiently convinced of the hazard that existed to request or devise alternative solutions. Nor were they confident of the efficacy of the ventilation control, which, indeed, had not been tested.

The need for communication to foster proper maintenance and implementation is ongoing. It is implicit in all of our design chapters. In addition, we have suggested some specific concerns and control considerations in Chapter 8.

Plant and equipment design are usually complex processes, involving many individuals and often conflicting criteria. In the past, many hazard considerations were not included. But the human and economic implications of neglecting such requirements have become too severe to overlook. The tragedies of the Bhopal methylisocyanate exposure and the Willow Island, West Virginia, cooling tower collapse in 1978 [which claimed 51 lives (OSHA, 1985)], as well as many lesser-known occurrences (Kletz, 1985), stand as reminders for our continual concern. The advent of new technologies involving different, perhaps more toxic chemicals; increased use of robotics, computer control, and laser operations; and the potential advent of more stringent environmental controls will continue to call upon the resourcefulness of the design and operating engineer to meet the requirements of "loss management."

REFERENCES

ACGIH (1985). *TLV's Threshold Limit Values for Chemical Substances and Physical Agents in the Work Environment Adapted by ACGIH for 1985–1986*. American Conference of Governmental Industrial Hygienists, Cincinnati, Ohio.

Brody, M. (1982). Fallout from Manville. *Barrons, 62,* 8, 10, Aug. 30.

Economist (1982). *284,* 79–80, Sept. 4.

Enviro Control, Inc. (1978). *Engineering Control Technology Assessment for the Plastics and Resins Industry*. NIOSH Report No. 78-159, March.

Jensen, P., Jokel, C. R., and Miller, L. N. (1978). *Industrial Noise Control Manual,* rev. ed. NIOSH Report No. 79-117, December.

Kletz, T. (1985). *What Went Wrong*? Gulf Publishing Co., Houston, Texas.

NIOSH (1984). *Request for Assistance in Preventing the Injury of Workers by Robots*. National Institute for Occupational Safety and Health, DHHS (NIOSH) Publication No. 85-103, December.

NYT (*The New York Times*) (1982). Aug. 27.

O'Brien, D. M., and Hurley, D. E. (1981). *An Evaluation of Engineering Control Technology for Spray Painting*. NIOSH Publication 81-121, June.

OSHA (1983). *General Industry Standards Part 1910*. Occupational Safety and Health Administration 2206, Revised, March 11.

OSHA (1985). Concrete and masonry construction safety standards. *Fed. Reg. 50,* 37543–37554, September 16.

OTA (1984). *Computerized Manufacturing Automation*. Office of Technology Assessment, Number OTA-CIT-236, April.

Chapter

2

Occupational Injury and Disease

R. A. Wadden

I. Methods of Evaluation

The need to protect worker health is basic to developing effective engineering controls. Understanding the effects that various occupational hazards induce is useful in supplying a rationale for control and for developing appropriate solutions. Some effects have been observed in systematic ways. Others are measurable but are not ordinarily reported. Still others are reported but not in ways that will show definite work-related patterns.

The National Institute for Occupational Safety and Health (NIOSH) recently developed a reference list of the 10 leading categories of work-related disease and injury (Table 2.1). The criteria used to choose the categories were the frequency of occurrence of the disease or injury; the severity of its manifestations in individual cases; and the potential for prevention.

Work-related injury and fatality statistics are one type of information that describes worker health (BLS, 1984; NSC, 1983). These data are

TABLE 2.1 The 10 Leading Work-Related Diseases and Injuries—United States, 1982[a]

1. Occupational lung diseases: asbestosis, byssinosis, silicosis, coal workers' pneumoconiosis, lung cancer, occupational asthma
2. Musculoskeletal injuries: disorders of the back, trunk, upper extremity, neck, lower extremity; traumatically induced Raynaud's phenomenon
3. Occupational cancers (other than lung): leukemia; mesothelioma; cancers of the bladder, nose, and liver
4. Amputations, fractures, eye loss, lacerations, and traumatic deaths
5. Cardiovascular diseases: hypertension, coronary artery disease, acute myocardial infarction
6. Disorders of reproduction: infertility, spontaneous abortion, teratogenesis
7. Neurotoxic disorders: peripheral neuropathy, toxic encephalitis, psychoses, extreme personality changes (exposure-related)
8. Noise-induced loss of hearing
9. Dermatologic conditions: dermatoses, burns (scaldings), chemical burns, contusions (abrasions)
10. Psychologic disorders: neuroses, personality disorders, alcoholism, drug dependency

[a]The conditions listed under each category are to be viewed as selected examples, not comprehensive definitions of the category.

SOURCE: NIOSH, 1983a.

reported as a rate or incidence, I, which is the number of deaths, injuries, or lost workdays occurring in a working population during a specified time divided by the total working population. The dimensions are events/person·time. For convenience, fatality incidence is usually reported per 100,000 workers; and lost workdays or injuries or illnesses are reported per 100 equivalent full-time workers:

$$I_{\text{lost workdays}} = \frac{\text{number of lost workdays/time interval}}{\left[\dfrac{\text{(total hours worked by all employees/time interval)}}{\text{(200,000 work hours/100 employee·years)}}\right]} = \frac{\text{lost workdays}}{\text{100 employee·years}} \tag{2.1}$$

The value of 200,000 is based on 40 hr/week, 50 weeks/year. The usual reporting period is 1 year. Incidences are also determined for more specific work classifications and work populations, e.g., anthracite mining. Such statistics allow us to judge the effects of hazards in various large industrial classifications and to make some evaluation about whether work conditions are improving from year to year.

Clinical case histories of work-related injuries or illnesses also serve to document worker health. For example, a recent NIOSH report presents the details of two fatalities resulting from occupational heat stroke (NIOSH, 1984c). Although such reports highlight the particular occupational problem and effects (e.g., one of the workers was obese and the other had a history of alcoholism, both predisposing factors that rendered them susceptible to heat stroke), they do not tell us anything about how frequently such problems occur, and often only a limited amount of information about what caused the effect. Epidemiology is the study of the distribution and determinants of disease or injury occurrence in a group of people. It is a method of systematically determining patterns and causes of ill health. Besides incidence, occupational epidemiology also uses measures of prevalence. The point prevalence, P, is the proportion of persons in a group who are affected by a disease or injury at a particular point in time (Schlesselman, 1982). The point prevalence is related to the person·time incidence rate (Eq. 2.1) by

$$P = I \cdot \Delta t \tag{2.2}$$

where Δt is the duration of the period being considered. For most occupational studies relative risk, RR, is the ratio of the incidence of a disease in an exposed population to incidence of the disease in a nonexposed group. Ideally the two groups differ only in terms of exposure, although corrections can be made for age, race, and sex. The standardized mortality ratio, SMR, is the ratio of observed deaths in an exposed population to the expected deaths in the same group multiplied by 100 (Mausner and Kramer, 1985). Often the expected number of deaths will be estimated using mortality rates of the general population (e.g., the U.S. population) adjusted to reflect the same age, sex, and race characteristics as those of the study group.

Some data are also available from controlled studies of human and animal exposure to hazardous materials. However, ethical considerations require human exposures to be quite low, and the measure of effect must be reversible and usually subclinical, i.e., one that does not require medical intervention. Animal testing offers more flexibility but suffers from the problem of extrapolating effects to humans.

As an example of how some of these measures are used, let us consider several epidemiological studies from the rubber industry. The reported *incidence* of occupational injury and disease for 1983 was 13.0 cases/100 worker·years, and the lost workday incidence was 101.4 lost workdays/100 worker·years (shown later in Table 2.4). These incidences are in contrast with those for general manufacturing of 10 cases/100 worker·years and 73.5 lost workdays/100 worker·years. McMichael et al. (1974) reported an SMR of 315 ($p < 0.001$) for leukemia for 40- to 64-year-old men in a cohort of 6678 rubber workers. This means that about three times as many deaths occurred due to leukemia in this group of workers as would have been predicted to occur in a group of the same age and race based on mortality statistics of the U.S. male population. Monson and Fine (1978) reported an RR = 3.6 for leukemia for rubber workers who had worked for more than 5 years in the calendering area. This relative risk (no measures of statistical significance were reported) indicates that the incidence of leukemia cases in workers with more than 5 years of calendering experience was three times greater (57 cases of cancer/100,000 person·years) than the incidence of leukemia cases for rubber workers in the 13,570-men cohort who had not worked in the calendering area or had not worked in the calendering area for longer than 5 years (16 cases of cancer/100,000 person·years). The control group also did not include other subgroups that had been identified as having elevated leukemia risks, e.g., those in tire curing for >15 years. Both incidence rates were standardized to the same age-specific person-year distribution of all 13,570 men in the study. Presumably, since both the study and control groups were drawn from the same working cohort, the effect of leaving confounding exposure variables in the evaluation is lessened. For example, one would not expect calendering workers to have smoking or recreational habits significantly different from other rubber workers.

RR and SMR are epidemiological measures that are often (but not necessarily) based on mortality. Some epidemiology studies examine changes in morbidity (number and intensity of illnesses) or changes in physiological or biochemical parameters that might rationally be related to exposure. Since the initial organ of exposure to air pollutants is the lung, lung disease prevalence and severity are common measures of occupational morbidity. Bronchitis and emphysema are generally recognized conditions of lung dysfunction with medical definitions. Certain lung diseases are associated with particular occupational exposures. Coal workers' pneumoconiosis (CWP) and byssinosis in textile workers are two examples. Each of these diseases has its definition and grading system(s) of severity of effect.

Determination of lung function is another method of characterizing the effect of air pollutants. Typical measures are forced vital capacity (FVC), forced expiratory volume at 1 sec ($FEV_{1.0}$), forced expiratory flow rate at 50% and 75% of forced vital capacity (FEF_{50} and FEF_{75}), and forced expiratory flow rate between 25% and 75% of the 1-sec forced expiratory volume ($FEF_{25-75\%}$). All of these variables can be extracted from a test where the patient breathes hard into a testing device that records the total volume and the volume/time exhaled. The various measures can be compared against expected values for the healthy population or those for some other control group. Sometimes the study population is also used as the control group—for instance, when comparing measurements of lung function in a group of workers before and after a shift. Depending on extent, incremental decreases in lung function parameters may or may not reflect a particular lung disease.

That is, lung function may decrease due to an air pollutant, but the decrease may not constitute or reflect an illness condition (morbidity).

Appropriate measures of morbidity for physical agents include hypertension (elevation of blood pressure) for heat exposures, increase in hearing threshold levels for noise exposure, and severity of vibration white finger disease for segmental vibration.

II. Injury and Disease Statistics

Table 2.2 shows several estimates of occupational fatalities for 1982. The two sets of information are drawn from different sources, the Bureau of Labor Statistics (BLS, 1984) and the National Safety Council (NSC, 1983), which are not strictly comparable. The BLS data are extracted from the reports required of private-sector employers under the Occupational and Safety Health Act of 1970. OSHA excludes unpaid family workers, some self-employed, and farms with fewer than 11 employees. The BLS survey includes only occupationally caused illnesses in the year reported, with no consideration of continuing conditions. In addition, since occupationally related illness may be unrecognized and therefore unreported, the survey values may be underestimates. The NSC values are based on data from the National Center for Health Statistics, state departments of health, state industrial commissions, and government employers. The NSC definition of worker includes all persons gainfully employed, including owners, managers, other paid employees, the self-employed, and unpaid family workers, but excluding domestic servants.

Also shown in Table 2.2 are some age-adjusted death rates for various causes of death. The crude occupationally related death rates would have to be adjusted for age, sex, and race to be strictly comparable. Working groups also are often found to be more healthy than the general population, since those with health dysfunction tend to self-select out of physiologically stressful work environments (Wen et al., 1983; McMichael, 1976). Consequently, some real, but difficult to quantitate, adjustment might also be appropriate for the "healthy worker" effect.

III. Job-Related Injuries

Severe job-related traumatic injuries include deaths, amputations, fractures, and lacerations. It has been estimated that at least 10 million traumatic injuries occur in occupational settings each year, with 30% being severe (NIOSH, 1983b). In 1982, an estimated 21,000 workers suffered amputations of fingers (93%), hands (1.6%), and toes (2.9%) (NIOSH, 1984b). Fifty-four percent of amputations involve machines,

TABLE 2.2 Occupational Fatalities for 1982

Industry division	Employers in the private sector (BLS, 1984; Cotter, 1984)			All industries (NSC, 1983)		
		Fatalities			Fatalities	
	Average annual employment (thousands)	Numbers	Incidence Deaths/ 10^5 workers·year	Average annual employment (thousands)	Numbers	Incidence Deaths/ 10^5 workers·year
Private sector (total)	62,646	4,090	7.4	98,800[a]	11,200	11
Agriculture, forestry, and fishing	729	180	28.4	3,400	1,800	52
Mining	1,056	440	44.3	1,100	600	55
Construction	2,894	720	28.7	5,200	2,100	40
Manufacturing	197	770	4.5	19,000	1,000	5
Transportation and public utilities	4,630	970	21.9	5,300	1,400	26
Wholesale and retail trade	15,646	490	3.8	22,500	1,100	5
Finance, insurance, and real estate	4,253	100	2.5	26,700	1,700	6
Services	14,242	420	3.5			
Government	—	—	—	15,600	1,500	10

Some Annual Age-Adjusted Death Rates/10^5 Persons for Total U.S. Population—1980 (NCHS, 1983)[b]

Moter vehicle accidents	23.7
Other accidents	19.7
Malignant neoplasms of respiratory system	36.7
Homicide and legal intervention	11.4

[a]Includes government workers as well as those from the private sector.

[b]Crude occupational incidence rates would have to be adjusted for age, sex, race, and the healthy worker effect to be directly comparable with these values. Occupational fatalities are incremental to deaths occurring in private-sector employed population in nonoccupational settings.

including power presses (9.3%), saws (8.3%), and shears, slitters, and slicers (7.3%) (McCaffrey, 1981). Understandably, machine operators have the highest proportion of occupational amputations (8%) when compared to other job categories. An estimated 400,000 work-related fractures occurred in 1982. Fall to a working surface accounted for 15% of fractures, with blows from unspecified or falling objects being listed as the cause of another 31%. Truck drivers (5% of all reported fractures), miscellaneous laborers (4%), and construction laborers (3%) were most at risk. The major sources of an estimated 2,250,000 lacerations in 1982 were knives (13%), other sharp metal items (13%), saws (6%), glass items (5%), nails (5%), and other machines (3%). Workers in eating and drinking establishments (7% of all lacerations), grocery stores (4%), general building construction (2%), and meat packing (2%) were at greatest risk (NIOSH 1984b).

Musculoskeletal injuries accounted for an estimated 580,000 emergency room-treated occupational injuries in 1982 (NIOSH, 1983b). Causes of such injuries included manual handling of materials (i.e., unaided lifting and lowering), repetitive motion, and vibration. About 1 million workers sustain back injuries every year, which accounts for one of every five reported injuries in the workplace. Repeated twisting or awkward postures, particularly when combined with high force, has been associated with approximately 20,000 injuries per year. Most at risk are workers in the construction, food preparation, chemical, product fabrication, and mining industries. Whole-body vibration (such as is endured by truck drivers) may contribute to low back pain, vertebragenic pain, and degenerative disk disease. Segmental vibration, the effects of which are discussed in more detail later in this chapter, is a well-documented cause of "vibration syndrome," which reduces the sensitivity of fingers to heat, cold, and pain (NIOSH, 1983b).

Table 2.3 shows the distribution of occupational fatalities by cause and major industrial category. Although the data were collected from the private sector with 11 or more employees, the distribution is likely to hold for all of industry. As is indicated by the NSC data in Table 2.2, it has been estimated that ~10,000 persons are killed on the job each year (NIOSH, 1984b; NSC, 1983). By far the largest fraction of fatalities are caused by motor vehicle incidents.

Table 2.4 summarizes total reported cases and lost workdays due to occupational injury and illness. The low reported number of illness cases relative to those for injury is evident. The semiconductor industry provides an example of some of the problems involved in illness reporting. A high rate of illness associated with nervous system disorder was reported for California workers in this industry in 1977 and 1978 (La Dou, 1984). The rate of occupational illness of all types in this industry for this period was 1.3 illness/100 workers compared to 0.6

TABLE 2.3 Distribution of Fatalities by Cause: Occupational Injury and Illness Fatalities in Units with 11 Employees or More, Private Sector, 1982 and 1983[a]

Cause[b]	Total private sector[c]	Agriculture, forestry, and fishing	Mining—oil and gas extraction only	Construction	Manufacturing	Transportation and public utilities[d]	Wholesale and retail trade	Finance, insurance, and real estate	Services
Number of fatalities for 1982 and 1983	7190	260	680	1390	1500	1540	930	170	730
Incidence[e] for 1983	5.6	12.7	27.6	26.3	4.3	13.3	3.3	1.7	2.5
					Percent				
Total—all causes	100	100	100	100	100	100	100	100	100
Highway vehicles	28	24	18	16	20	47	28	51	32
Falls	12	13	11	26	9	5	7	13	11
Heart attacks	11	3	10	13	11	7	9	13	20
Industrial vehicles or equipment	10	25	14	15	12	4	11	3	4
Assaults	7	1	1	1	2	3	28	10	15
Electrocutions	7	12	2	13	6	5	6	2	4
Caught in, under, or between objects other than vehicles or equipment	5	[f]	5	3	5	11	4	0	1

Struck by objects other than vehicles or equipment	4	3	13	3	7	3	2	0	1
Aircraft crashes	4	2	3	[f]	4	7	1	10	5
Explosions	3	1	5	2	5	2	1	0	1
Fires	2	11	6	2	4	1	[f]	0	[f]
Plant machinery operations	2	0	1	1	7	[f]	2	0	0
Gas inhalation	2	1	8	1	3	2	0	0	[f]
All other	3	3	3	4	5	2	[f]	0	4

[a]It is impossible to estimate year-to-year changes precisely, because sampling errors are large at the industry division level. Therefore, the results are for both years rather than a comparison between them.

[b]Cause is defined as the object or event associated with the fatality.

[c]Excludes coal, metal and nonmetal mining, and railroads, for which data are not available.

[d]Excludes railroads.

[e]The incidence rates I represent the number of fatalities per 100,000 full-time workers and were calculated as

$$I = (N/EH) \cdot 200{,}000{,}000$$

where N = number of fatalities
EH = total hours worked by all employees during calendar year
200,000,000 = base for 100,000 full-time equivalent workers (working 40 hr/week, 50 weeks/year)

[f]less than 1%.

NOTE: Because of rounding, percentages may not add to 100.

SOURCE: BLS, 1985a; Cotter, 1985.

TABLE 2.4 Occupational Injury and Illness Cases and Lost Workday Statistics for 1983

		Total cases			Injury and illness, lost workdays	
				Incidence[b]		Incidence[b]
Industry[a]	1983 annual average employment (thousands)	Total injury cases (thousands)	Total illness cases (thousands)	Injury and illness cases per 100 full-time workers	Average lost workdays per lost workday case	Lost workdays per 100 full-time workers
Private sector[c]	74,749.9	4,748.0	106.1	7.6	17	58.5
Agriculture, forestry, and fishing[c]	893.2	84.6	3.5	11.9	15	90.8
Agricultural production[c]	NA	53.5	2.0	13.6	14	101.1
Agricultural services	NA	29.2	1.3	10.0	15	77.2
Forestry	NA	1.3	0.1	10.7	18	101.5
Fishing, hunting, and trapping	NA	0.6	[c]	4.9	24	73.1
Mining	957.0	77.9	1.0	8.4	28	125.1
Metal mining[e]	56.9	2.6	0.1	6.0	27	92.5
Anthracite mining[e]	2.9	0.1	[c]	6.7	77	470.4
Bituminous coal and lignite mining[e]	191.8	11.6	0.2	7.1	34	188.5
Oil and gas extraction	599.9	58.9	0.7	9.8	26	120.5
Nonmetallic minerals, except fuels[e]	105.6	4.6	0.1	3.9	25	66.1
Construction	3,940.0	489.2	5.4	14.8	19	118.2
General building contractors	1,014.7	119.5	1.2	14.4	18	113.0

Heavy construction contractors	787.4	110.2	1.6	15.4	20	122.4
Special trade contractors	2,137.3	259.5	2.6	14.8	19	119.0
Manufacturing	18,497.0	1,712.9	59.8	10.0	17	73.5
Durable goods	10,774.0	1,042.0	36.0	10.3	17	73.4
Lumber and wood products	657.5	111.1	1.7	18.3	18	163.5
Furniture and fixtures	446.9	57.6	1.2	14.1	14	83.0
Stone, clay, and glass products	572.6	71.4	1.8	13.1	19	112.0
Primary metal industries	838.0	98.0	2.4	12.4	19	103.4
Fabricated metal products	1,373.8	192.4	6.0	15.1	16	96.5
Machinery, except electrical	2,038.3	186.8	5.7	9.8	16	58.1
Electrical and electronic equipment	2,023.8	117.6	7.3	6.3	16	41.4
Transportation equipment	1,756.3	141.5	6.4	8.4	18	64.5
Instrument and related products	694.9	32.8	2.1	5.2	17	35.6
Miscellaneous manufacturing industries	371.3	32.7	1.3	9.9	17	66.3
Nondurable goods	7,724.0	670.9	23.9	9.6	17	73.6
Food and kindred products	1,622.0	242.4	11.9	16.5	17	131.2
Tobacco manufacturers	68.5	3.8	[d]	6.5	14	42.8
Textile mill products	743.5	52.5	1.0	7.4	18	51.4
Apparel and other textile products	1,1164.1	64.0	1.4	6.4	17	40.6
Paper and allied products	661.6	64.3	1.2	10.0	20	90.3

TABLE 2.4 Occupational Injury and Illness Cases and Lost Workday Statistics for 1983 *(Continued)*

		Total cases			Injury and illness, lost workdays	
				Incidence[b]		Incidence[b]
Industry[a]	1983 annual average employment (thousands)	Total injury cases (thousands)	Total illness cases (thousands)	Injury and illness cases per 100 full-time workers	Average lost workdays per lost workday case	Lost workdays per 100 full-time workers
Nondurable goods (Continued)						
Printing and publishing	1,295.8	73.6	1.5	6.6	16	44.6
Chemicals and allied products	1,046.8	53.9	2.5	5.5	17	42.3
Petroleum and coal products	195.4	10.3	0.4	5.5	19	46.8
Rubber and miscellaneous plastics products	718.1	89.1	2.5	13.0	16	101.4
Leather and leather products	207.9	17.0	1.4	10.0	20	87.3
Transportation and public utilities	4,958.0	374.0	5.0	8.2	20	94.9
Railroad transportation[e]	376.6	28.7	0.6	8.0	15	76.9
Local and interurban passenger transit	254.1	18.0	0.2	9.7	21	114.3
Trucking and warehousing	1,214.3	152.7	0.7	13.3	23	185.7
Water transportation	194.1	19.2	0.3	10.8	36	229.9
Transportation by air	451.7	51.1	0.7	12.7	13	97.7
Pipelines, except natural gas	20.7	0.8	—	4.2	16	24.7

Transportation services	227.3	6.9	—	3.4	21	37.9
Communication	1,337.0	35.8	1.0	2.9	18	27.1
Electric, gas, and sanitary services	881.9	60.8	1.4	7.2	17	61.4
Wholesale and retail trade	20,805.0	1,177.1	9.3	7.2	16	47.8
Wholesale trade	5,259.0	342.5	3.7	7.0	16	50.6
Wholesale trade—durable goods	3,064.0	184.5	2.0	6.4	15	39.7
Wholesale trade—nondurable goods	2,195.0	158.0	1.7	7.9	16	66.3
Retail trade	15,545.0	834.6	5.6	7.3	16	46.7
Building materials and garden supplies	613.0	46.7	0.3	8.5	16	57.9
General merchandise stores	2,161.3	145.5	—	9.7	14	55.8
Food stores	2,560.3	196.4	0.9	10.4	18	79.8
Automotive dealers and service stations	1,666.7	105.1	0.9	6.8	16	39.7
Apparel and accessory stores	955.6	16.6	—	2.4	18	17.9
Furniture and home furnishing stores	602.8	19.0	0.2	3.7	17	30.4
Eating and drinking places	5,006.7	249.8	2.3	7.8	13	40.5
Miscellaneous retail	1,978.5	55.4	0.3	3.6	18	31.4
Finance, insurance, and real estate	5,467.0	93.6	1.6	2.0	15	12.8
Banking	1,663.4	24.4	0.4	1.6	13	8.2
Credit agencies other than banks	630.8	7.4	0.1	1.3	12	6.3

TABLE 2.4 Occupational Injury and Illness Cases and Lost Workday Statistics for 1983 *(Continued)*

		Total cases			Injury and illness, lost workdays	
				Incidence[b]		Incidence[b]
Industry[a]	1983 annual average employment (thousands)	Total injury cases (thousands)	Total illness cases (thousands)	Injury and illness cases per 100 full-time workers	Average lost workdays per lost workday case	Lost workdays per 100 full-time workers
Finance, insurance, and real estate (Continued)						
Security, commodity brokers and service	309.0	2.5	[d]	0.8	13	3.8
Insurance carriers	1,233.7	19.3	0.4	1.8	16	12.5
Insurance agents, brokers and service	487.4	3.1	0.1	0.8	17	4.4
Real estate	989.8	34.8	0.6	4.4	16	35.0
Holding and other investment offices	136.8	1.9	[d]	1.7	13	9.3
Services	19,232.7	738.6	20.5	5.1	16	37.0
Hotels and other lodging places	1,166.9	80.5	1.3	9.2	14	54.5
Personal services	954.2	20.0	0.5	2.9	15	20.8
Business services	3,539.4	128.4	2.7	4.7	16	33.5
Auto repair, services and garages	615.7	38.5	0.4	7.1	16	52.3
Miscellaneous repair services	284.2	19.4	0.5	7.9	17	61.8

Motion pictures	216.0	6.3	0.1	4.1	18	29.0
Amusement and recreation services	830.5	43.7	0.9	8.2	15	53.9
Health services	5,973.4	285.9	10.7	6.3	16	50.3
Legal services	601.1	2.3	0.2	0.5	24	4.8
Educational services	1,216.8	27.8	0.8	3.4	13	18.3
Social services	1.221.4	45.6	1.2	5.3	15	37.6
Museums, botanical, zoological gardens	37.6	1.7	0.1	6.4	13	33.5
Membership organizations	1,511.2	26.5	0.7	2.6	18	18.7
Miscellaneous services	1,064.3	11.9	0.4	1.3	14	7.7

[a]Industry division totals include data for industries not shown separately.

[b]Incidence rates represent the number of injuries and illnesses per 100 full-time workers and were calculated as

$$I = (N/EH) \cdot 200{,}000$$

where N = number of injuries and illnesses or lost workdays
EH = total hours worked by all employees during the calendar year
200,000 = base for 100 full-time workers (working 40 hr/week, 50 weeks/year)

[c]Excludes farms with fewer than 11 employees.

[d]Estimates of fewer than 50 cases.

[e]Data conforming to OSHA definitions for coal and lignite mining (SIC 11 and 12), and metal and nonmetal mining (SIC 10 and 14), and railroads (SIC 40) were provided by the Mine Safety and Health Administration, U.S. Department of Labor, and by the Federal Railroad Administration, U.S. Department of Transportation. Independent mining contractors are excluded from the coal, metal, and nonmetal mining industries.

NOTE: NA = data not available. Dashes indicated that data do not meet publication guidelines. Because of rounding, components may not add to totals.

The number of lost workdays for the industry division levels shown in this table can be approximated by multiplying the number of lost workday cases by the average lost workdays per lost workday case.

SOURCE: BLS, 1985a.

illnesses/100 workers in general manufacturing (for 1977). In 1979 the California rate dropped precipitously to 0.8 illness/100 workers. A state investigation revealed that reporting methods had changed. Short-term, accidental exposures to the toxic gases used in production, such as arsine, phosphine, and diborane, which did not result in hospitalization or absence of more than one day from work, were being reported as occupational injuries rather than illnesses. This change in reporting specification may also have confused evaluation of the health effects of toxic gases in the industry (Bass, 1984).

Tables 2.2–2.4 identify mining, construction, and agriculture as the most hazardous areas, although certain manufacturing occupations (lumber; stone, clay, and glass; primary metals) individually have high lost workday incidences. Previous to 1984, the industries surveyed by the BLS had shown four straight years of improvement in injury and illness rates. However, in 1984 there was an increase of 12% in total cases and 15% in lost workday cases, the largest single-year increases since record keeping was initiated in 1973 (BLS, 1985b). Injury rates rose for all industry divisions except for agriculture, and finance, insurance, and real estate.

IV. Specific Injury and Disease Effects

The following tables and discussion summarize recent epidemiologic disease and illness information for some specific industries and job classifications. The emphasis is on occupational diseases since, as has been noted, these tend to be underreported in annual morbidity and mortality statistics. In particular, occupational cancer often becomes evident only after a latent period extending from 5 to 40 years from initial exposure to the carcinogen (NIOSH, 1984a). Of course, many hazardous occupations are not covered. But the summaries illustrate the kind of information available from epidemiological studies, and the form in which it is reported. In general, judgments about whether a particular type of work is hazardous rely on the general consistency of results from several studies rather than on the specific RR or other findings from a single study.

A. Rubber manufacture

Epidemiological studies on workers in the rubber industry are summarized in Table 2.6. Although there are differences between studies, it is evident that exposure to air pollutants in some work areas (e.g., curing) can cause short-term and permanent decreases in lung function

($FEV_{1.0}$ and FVC); and elevated risks of lung cancer, particularly for those employed longer than 5 or 10 years. In addition, elevated risks are often found for lymphatic leukemia, malignant neoplasm of the gastrointestinal tract, diabetes, and bladder and pancreatic cancers.

B. Textile production

Byssinosis, characterized by acute and chronic lung dysfunction, is associated with exposure to cotton, hemp or flax fibers. Table 2.7 summarizes some of the more recent studies on U.S. and British textile workers. Industrial bronchitis is also a commonly encountered disease in such industries. An estimated 35,000 current and retired textile workers have been disabled by severely impaired lung function (OSHA, 1978).

Although there is some variation in findings, occupational studies have generally shown that smoking textile workers have a higher prevalence of byssinosis than nonsmoking workers, and that prevalence increases with dust exposure level, although these may have to be time-weighted (Figure 2.1 and Table 2.5).

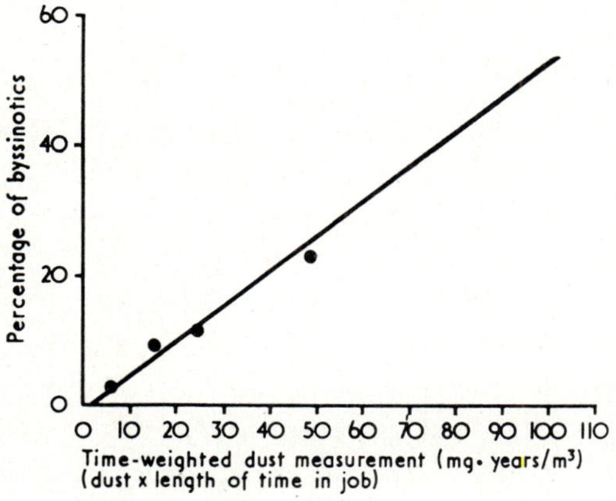

Figure 2.1 Percentage of byssinotics regressed against time-weighted dust measurement (1140 males and females; % = −0.62 + 0.53 · (dust concentration · years on job); r = +0.64) (*Fox et al., 1973; reprinted by permission of the British Journal of Industrial Medicine, R. I. McCallum, ed.*).

TABLE 2.5 Prevalence of Byssinosis in Time-Weighted Dust Exposure Groups

Time-weighted dust group (mg·years/m³)	Number of subjects examined	Mean time-weighted dust measurement level (mg·years/m³)	Precent prevalence of byssinois (grade ½ and over)
0.0–10.0	350	5.75	3.63
10.1–20.0	257	15.34	9.73
20.1–30.0	206	24.06	12.31
>30.0	347	48.50	22.19

SOURCE: Fox et al., 1973.

TABLE 2.6 Epidemiologic Studies of Diseases of Rubber Workers

Reference	Period of study	Number in study	Number of cases or deaths	Summary
Andjelkovic et al., 1976	1964–1973	8,418	2,373	Considered overall and age-specific death rates for all causes of death and for deaths due to specific causes. Examined mortality among active and retired workers separately. Death rate due to all causes did not indicate an excess. Death rate due to all causes among retired workers, aged 40–64 years, were clearly in excess, with an SMR[a] of 202 ($p < 0.001$). For several causes of death, an excess rate was indicated though not always with statistical significance. Among causes of death for which an excess rate was indicated for one or both age groups (aged 40–64 years and aged 65–84 years) were malignant neoplasm (M.N.) of the digestive organs and peritoneum, M.N. of genitourinary organs, M.N. of other and unspecified sites, and neoplasms of the lymphatic and hematopoietic tissue. An excess was also found for deaths due to diabetes.
Baxter and Werner, 19980	1967–1976	40,867	5,773	1. Explored the relationship between excess bladder cancer and exposure to antioxidants based on 1- and 2-naphthylamine, which was withdrawn from use in 1949. Workers were divided into three exposure groups: *a.* Men who started work before January 1, 1950, in a factory that had used the suspect antioxidants (12,779 men) *b.* Men who started work on or after January 1, 1950, in a factory that used the suspect antioxidants (18,118 men) *c.* Men who worked in factories that had never used the suspect antioxidants (9,970 men)

				For the whole population, a significant excess death rate due to bladder cancer was found with a SMR[a] of 127 ($p < 0.05$) based on 73 deaths. Group *a* experienced an SMR[a] of 144 ($0.025 < p < 0.05$) based on 36 deaths. Groups *b* and *c* had SMRs[a] of 108 (24 deaths) and 97 (13 deaths), respectively. 2. Excess mortality also was found for all malignant neoplasms (SMR = 109; 1,748 deaths), stomach cancer (SMR = 122; 216 deaths), and cancer of the trachea, bronchus, and lung (SMR = 115; 822 deaths). Three sectors of the rubber industry in particular had excess mortality due to lung cancer. Workers in the Ebonite and Vulcanite sector experienced an SMR[a] of 186 (18 deaths) due to lung cancer. Workers in the adhesives, rubber solutions, and sealing compounds sector had an SMR of 176 (21 deaths); and those in the belting, hose, rubber with asbestos flooring sector had an SMR[a] of 139 (99 deaths). In comparison, the tire sector of the rubber industry experienced an SMR of 109 (326 deaths) of lung cancer.
Delzell and Monson, 1981	1940–1978	15,643	6,797	Explored the relationship between mortality and (1) duration of employment, (2) years since starting work, (3) age, (4) active vs. terminated employment. No excess mortality overall or due to specific causes was found for active employees. Among terminated employees, there was excess overall mortality during the first 5 years after termination. Among terminated employees, there was excess mortality due to bladder cancer and leukemia during the 5-year interval after termination.
Delzell et al., 1982	1964–1973	Cases and controls are members of a cohort of 8,418	121 deaths due to lung cancer (cases); 448 deaths due to other causes (controls)	Determined lung cancer risk associated with 19 production work areas at a rubber plant. Men who worked >5 years in rubber reclaim operations experienced a relative risk of 2.2 ($p = 0.03$) based on 11 lung cancer deaths. Men who worked in special products (primarily fuel cell manufacturing) experienced a relative risk of 1.7 ($p = 0.03$) based on 31 lung cancer deaths.

TABLE 2.6 Epidemiologic Studies of Diseases of Rubber Workers *(Continued)*

Reference	Period of study	Number in study	Number of cases or deaths	Summary
Fine and Peters, 1976a	1972–1973		121 cases (curing workers); 189 controls (rubber workers in jobs other than curing)	1. To determine acute effects of curing fumes, 29 of the 121 curing workers were divided into high- and low-exposure groups. $FEV_{1.0}$[b] and FVC[c] were measured before and after a work shift. The eight high-exposure workers had a significant decrease in $FEV_{1.0}$ (-115 ml, $p = 0.05$) and FVC (-164 ml, $p < 0.05$). 2. 92 white curing workers over the age of 24 were divided into high- and low-exposure groups. Cross-sectionally, the residual $FEV_{1.0}$[d] and the residual FVC[d] of the heavy-exposure group ($N = 30$) was significantly decreased. The average residual $FEV_{1.0}$ among the heavy-exposure group was -169 ± 471 ml ($p = 0.05$), and the average residual FVC was -266 ± 675 ml ($p = 0.01$). 3. 98 of the 121 curing workers were retested 1 year later. The mean 1-year loss of $FEV_{1.0}$ was -173 ± 228 ml for curing workers who were exposed for >10 years. This was significantly greater ($p < 0.01$) than the control group.
Fine and Peters, 1976b	1972–1973		65 cases, (processing area workers); 189 controls (rubber workers in jobs other than processing)	Analyzed (1) prevalence of respiratory symptoms in processing area workers, (2) chronic effect of exposure in processing on pulmonary function. Based on questionnaire responses, processing workers, who are exposed to a variety of respirable particulate materials, had a statistically significant increased prevalence of two symptoms (cough and phlegm) associated with mucous hypersecretion and bronchitis. No significant increase in the symptoms of recurrent infections of chronic obstructive lung disease. The only significant difference between processing workers and controls was a decrease in the ratio of $FEV_{1.0}$[b] to FVC[c]. But in workers with >10 years' ex-

				posure, there was a statistically significant decrease in $FEV_{1.0}$, residual $FEV_{1.0}{}^{d}$, FEV/FVC ratio, and expiratory flow rates at 50% FVC and 25% FVC. Years of exposure in the processing area was a significant predictor of $FEV_{1.0}$ ($p = 0.003$) and FVC ($p = 0.009$).
Fine et al., 1976	1972–1974		80 cases (cases were workers exposed to talc below TLV); 189 controls (rubber workers not exposed to talc)	1. Respiratory questionnaires were administered to determine prevalence of respiratory symptoms in talc workers. No significant increase of symptoms associated with recurrent pulmonary infections was found. A high prevalence was found of two symptoms associated with mucous hypersecretion and chronic bronchitis. A high prevalence of one symptom (wheezing) associated with obstructive respiratory disease was also found. 2. 28 workers were studied for single-shift reduction in $FEV_{1.0}{}^{b}$ and FVC[c]. No significant reduction was found. 3. Looked at the chronic effect of talc exposure on $FEV_{1.0}$, FVC, FEV/FVC, residual $FEV_{1.0}{}^{d}$, residual FVC, and flow rates at 50% FVC, 25% FVC, and 12.5% FVC. Significant reduction in residual $FEV_{1.0}$ found only for workers exposed more than 10 years. Years of exposure to talc was found to be a significant ($p = 0.015$) predictor of $FEV_{1.0}$. Researchers estimated a 26-ml loss of $FEV_{1.0}$ each year in excess of expected value.
Fox et al., 1974	1967–1971	40,867	2,029	Surveyed occupational cancer in the rubber and cablemaking industries and attempted to relate biological cancer sites to various work areas. Also looked at patterns of mortality in three different exposure groups: 1. Men employed in factories that used known carcinogenic compounds and men who entered those factories before January 1, 1950 2. Men employed in factories that used known carcinogenic compounds and men who entered those factories after December 31, 1949

TABLE 2.6 Epidemiologic Studies of Diseases of Rubber Workers *(Continued)*

Reference	Period of study	Number in study	Number of cases or deaths	Summary
				3. Men employed in factories that had never used known carcinogenic compounds For the whole population, a significant excess death rate due to all neoplasms was found with an SMR[a] of 115 ($p < 0.001$) based on 659 deaths. Exposure groups 1 and 3 did not have a significant excess death rate due to all neoplasms, but exposure group 2 experienced an SMR of 129 ($p < 0.001$) based on 274 deaths. In the tire sector of the rubber and cablemaking industry, an excess death rate due to bronchial carcinoma was found with an SMR of 134 ($p < 0.001$) based on 131 deaths.
Lednar et al., 1977	1964–1973	4,302	73 cases (workers terminated with pulmonary disability retirement); 146 controls (rubber workers who terminated without disability retirement)	Evaluated exposure and duration of exposure of 73 terminated workers. 53.4% terminated due to emphysema, 13.7% due to cancer of the lung and bronchus, 11% due to asthma, 21.9% due to other pulmonary conditions. Significant relative risk[e] (RR) of pulmonary disability associated with extrusion (RR = 2.44, $p < 0.01$), curing preparation (RR = 3.88, $p < 0.01$), curing (RR = 2.47, $p < 0.01$), finishing and inspection (RR = 2.05, $p < 0.05$), and service (RR = 2.19, $p < 0.05$). Smoking and exposure were found to interact in their association with pulmonary disability.
McMichael et al., 1974	1964–1972	6,678	1,783	Analyzed age-specific and cause-specific death rates. In the 40–84 age group:

				1. An SMR[a] of 187 ($p < 0.001$) for stomach cancer (39 deaths) 2. An SMR of 142 ($p < 0.05$) for prostate cancer (49 deaths) 3. An SMR of 226 ($p < 0.01$) for lymphosarcoma (14 deaths) 4. An SMR of 143 ($p < 0.05$) for diabetes (43 deaths) 5. An SMR of 154 ($p < 0.05$) for arteriosclerosis (34 deaths) In the 40–64 age group, an SMR of 315 ($P < 0.001$) for leukemia (11 deaths).
Monson and Nakano, 1976	1940–1974	13,571	5,079	Looked at cause-specific death rates for workers in specific work areas. For all workers, an SMR[a] of 128 (55 deaths) was found for leukemia and an SMR of 122 (48 deaths) was found for bladder cancer; in processing, an excess of cancer of the gastrointestinal tract (SMR = 140; 52 deaths), stomach cancer (SMR = 190; 18 deaths), and large intestinal cancer (SMR = 130; 14 deaths); in tire curing, an excess of lung cancer (SMR = 160; 20 deaths), and pneumonia (SMR = 160; 12 deaths); in tire building, an excess of brain cancer (SMR = 190; 7 deaths), bladder cancer (SMR = 190; 9 deaths), and lymphatic cancer and myeloma (SMR = 160; 11 deaths); in warehouse and shipping, an excess of diabetes (SMR = 190; 11 deaths) and bladder cancer (SMR = 310; 8 deaths).
Monson and Fine, 1978	1940–1976	13,570	1,129 deaths due to cancer; 230 incident cases of cancer; 1,359 total cases	Evaluated cancer mortality and morbidity. If an external comparison of mortality rates versus mortality rates for U.S. white males indicated an excess, an internal comparison of cancer morbidity rates among workers in various work areas was made. Excess cases of specific cancers were found for employees who worked 5+ years in rubber making, a relative risk[e] (RR) of 2.2 (13 cases) was found for stomach cancer and an RR = 2.0 (17 cases) was found for intestinal cancer. For employees who worked 5+ years in tire curing, an RR of 2.5 (8 cases) was found for pancreatic cancer and an RR = 2.2

TABLE 2.6 Epidemiologic Studies of Diseases of Rubber Workers *(Continued)*

Reference	Period of study	Number in study	Number of cases or deaths	Summary
				(31 cases) was found for lung cancer. For those in tire building for 5+ years, an RR = 2.5 (8 cases) was found for lymphatic cancer. Employees in tire assembly for 5+ years had an RR = 6.5 (12 cases) for skin cancer and an RR = 4.1 (8 cases) for brain cancer. An excess rate of leukemia was found in several work areas, including: 1. Calendering, 5+ years, RR = 3.6 (8 cases) 2. Tire curing, 15+ years, RR = 3.1 (8 cases) 3. Elevators, 5+ years, RR = 2.9 (4 cases) 4. Rubber fabrics 0+ years, RR = 3.5 (4 cases)
Wolf et al., 1981	1964–1973		72 leukemia cases; 288 controls	Evaluated leukemia cases at four rubber and tire manufacturing companies. Each case was given an occupational title group designation. In one company a weak association ($p = 0.07$) between lymphatic leukemia and work histories of potential high solvent exposure was found. A weak association was found between lymphatic leukemia and general service jobs for all companies combined ($p = 0.07$). Also a weak association ($p = 0.07$) was found between lymphatic leukemia and compounding and mixing jobs. No elevated risk was found for myeloid leukemia.

[a]Standardized mortality ratio:

$$\text{SMR} = \frac{\text{observed deaths}}{\text{expected deaths}} \times 100$$

[b]Forced expiratory volume in 1 sec.

[c]Forced vital capacity.

[d]Residual $FEV_{1.0}$ and residual FVC provide a method of comparing the $FEV_{1.0}$ and FVC between groups. The residual $FEV_{1.0}$ and FVC is the difference between a predicted value and the observed value determined from pulmonary function tests. Various methods are used to calculate the predicted values.

[e]$$\text{Relative risk} = \frac{\text{incidence rate among those exposed}}{\text{incidence rate among unexposed}}$$

TABLE 2.7 Textile Worker Lung Disease

Reference	Period of study	Number in study	Byssinosis cases (% prevalence)	Summary
Bouhuys et al., 1969	—	214	61 (29%)	Study conducted on male workers exposed to dust in carding and spinning areas of a cotton textile mill.
Zuskin et al., 1969	—	59 carders 99 spinners	15 (25%) 12 (12%)	Significant decrease ($p < 0.01$) in $FEV_{1.0}$[a] during a Monday work shift.
Schrag and Gullett, 1970	—	509	63 (12%)	Highest prevalence of byssinosis was found among carders (28 cases, 29% prevalence).
Merchant et al., 1972	—	441	25 (6%) (preparation area, 20%; yarn processing, 2%)	1. A significant effect ($p = 0.01$) of smoking on byssinosis index[b] among men, with smokers having a higher index (more severe) than nonsmokers. No significant effect of exposure-risk group or age on byssinosis index for men or women. No effect due to smoking among women. 2. Among men, there were significant effects due to smoking ($p = 0.0012$) and exposure-risk group ($p = 0.0003$) on the severity of bronchitis. Among women, smoking had a significant effect on the bronchitis index ($p = 0.0162$). 3. 76% of predicted $FEV_{1.0}$[e] and 90% of predicted FVC[c] was found for male smokers in high-exposure areas ($N = 66$). Nonsmokers had consistently better pulmonary function than smokers. 4. 27 employees in the high-exposure area were studied before the start of a Monday shift and after 6 hr of dust exposure to determine any change in flow at 50% of vital capacity. Of the 27, 5 were byssinotics and 4 were reactors.[d] Byssinotics had a 29% decrease, reactors had a 39% decrease and the 18 other employees had an 8% increase in flow at 50% of the volume after 6 hr of dust exposure.

TABLE 2.7 Textile Worker Lung Disease *(Continued)*

Reference	Period of study	Number in study	Byssinosis cases (% prevalence)	Summary
Imbus and Suh, 1973	—	10,133	466 (4.5%) (preparation area, 26%) Bronchitis—993 cases (9.8% prevalence)	1. Strong association found between byssinosis and bronchitis, with higher (more severe) byssinosis indices[b] associated with higher incidence of bronchitis. Male smokers had a significantly higher incidence ($p < 0.05$) of byssinosis than did nonsmokers. No difference was found in women. 2. $FEV_{1.0}$[a] was obtained for all workers previous to the start of a Monday work shift and percent of predicted $FEV_{1.0}$[e] was calculated. Nonbyssinotic, nonbronchitic nonsmokers had the highest average percent of predicted $FEV_{1.0}$ (males, 89.4%; females, 85.7%). Male nonsmoking bronchitics had a slightly lower average percent of predicted $FEV_{1.0}$ (88.2%). Male nonsmoking byssinotics had an even lower average percent of predicted $FEV_{1.0}$ (84.5%). Among nonsmoking males, the lowest average percent of predicted $FEV_{1.0}$ was found among bronchitic byssinotics (80.1%). Similar differences between categories were found in smokers. Percent of predicted $FEV_{1.0}$ did not decrease with increased length of employment in nonsmoking, nonbyssinotic males, but in nonsmoking byssinotic males, percent of predicted $FEV_{1.0}$ dropped sharply after 18 years of employment. Employees experience an increasing risk of developing respiratory impairment in the following order: smoking employees; byssinotic employees; smoking byssinotics. 3. FVC[c] and $FEV_{1.0}$ were obtained for 8,025 employees before the start of a Monday shift and after 6 hr of cotton

			dust exposure. Byssinosis produced more of a decrement during the shift than did bronchitis (male nonsmokers, 5.2% decrease based on 28 byssinotics; male nonsmokers, 3.8% decrease based on 59 bronchitics). Byssinosis and bronchitis together produced the highest decrement (male, nonsmokers, 11.1% decrease based on 16 bronchitic byssinotics) during the shift.
Braun et al., 1973	995		Data on 611 carders were compared to data on 384 noncarders. 1. 21.7% of carders (133) reported subjective symptoms (chest tightness) compared to 5.4% (20) noncarders. 2. 28.3% (173) carders had <80% of the normal predicted $FEV_{1.0}$[e] compared to 21.6% (83) noncarders. 3. 22.9% (140) of carders had 10% decline in $FEV_{1.0}$[a] during the Monday work shift as compared to 11.4% (43) noncarders. 4. A direct relationship was found between length of employment and both Monday decline in $FEV_{1.0}$ and an $FEV_{1.0}$ below 80% of predicted. 5. There was very little effect due to smoking on prevalence of symptoms.
Merchant et al., 1973	1,772	262	Mathematical dose-response curves were determined for data on byssinosis prevalence by median dust level (<1.5 mg/m^3). Data were categorized by sex, smoking group, work area, and mill type. For all subgroups a strong linear association between dust level and byssinosis prevalence was found. In almost all groups (except where the subpopulation was too small), the correlation coefficient was greater than 0.90. A linear association between percent change in $FEV_{1.0}$[a] (over a Monday work shift) and median dust levels was also found and had a correlation coefficient of 0.82.

TABLE 2.7 Textile Worker Lung Disease *(Continued)*

Reference	Period of study	Number in study	Byssinosis cases (% prevalence)	Summary
Fox et al., 1973		1,140		1. Four groups of workers categorized by time-weighted dust concentrations that were then compared to prevalence of byssinosis. % prevalence = $-0.62 + 0.53 \cdot$ (dust concentration, mg/m^3·years on job); $r = 0.64$. (See also Figure 2.1.) 2. Found no contribution of cotton dust exposure to the prevalence of bronchitis. 3. Regression lines for time-weighted dust exposure against percent predicted $FEV_{1.0}{}^e$ were determined separately for smokers and nonsmokers. Correlations were low; smokers, $r = -0.13$; nonsmokers, $r = -0.08$. Slopes of the two lines were significantly different from zero ($p < 0.001$).
Berry et al., 1974	1970–1973	1,359 cotton workers, 227 artificial-fiber workers		*Bronchitis* 1. Higher prevalence of bronchitis in cotton mills than in artificial-fiber mills. 2. No effect due to dust exposure on the prevalence of bronchitis. 3. Significant positive trend ($p < 0.05$) in bronchitis prevalence with increasing length of exposure. *Byssinosis* 1. Significant effect ($p < 0.05$) due to smoking after allowing for dust concentration and length of exposure, smokers having 40% more byssinosis then nonsmokers. 2. Significantly higher byssinosis prevalence in men than in women ($p < 0.01$). The effect due to sex was larger than the effect due to smoking. 3. Highest prevalence occurred for strippers and grinders (all males). The second highest prevalence occurred for drawframe tenters (all females)

			Change of Symptoms Of 1,586 subjects, 796 workers completed a second questionnaire 2 years after the first. Findings include: 1. An overall test showed cotton mills to be significantly ($p < 0.01$) more conducive to the development and retention of symptoms of bronchitis than the artificial-fiber mill environment. 2. A significant ($p < 0.001$) increase in incidence of progression of byssinosis to more serious levels with increase in dust concentration in women. 3. For byssinosis, a significant ($p < 0.01$) decrease in remission with increasing length of exposure. 4. Higher incidence of byssinosis in smokers than in non-/ex-smokers, but number too small to permit significance judgment.
Zuskin et al., 1975	61 carding machine operators	55 (90%)	During Monday shift dust exposures: 1. Workers experienced a significant decrease in $FEV_{1.0}$[a] (males $p < 0.01$; females $p < 0.025$) 2. FVC[c] decreased slightly but was significant only for male cotton workers ($p < 0.01$). 3. $FEV_{1.0}$/FVC decreased significantly only for hemp workers ($p < 0.01$) and flax workers ($p < 0.05$). 4. Residual volume increased significantly for hemp workers ($p < 0.01$), flax workers ($p < 0.025$), and male cotton workers ($p < 0.025$). 5. In most cases, FRC[f] and TLC[g] increased slightly but not significantly. 6. Decrease in MEF50%[h] was significant for all groups (males, $p < 0.01$; females, $p < 0.025$). 7. Workers with a byssinosis index[b] of 1 or 2 had higher decreases in $FEV_{1.0}$ and $MEF_{50\%}$, but differences were not significant.

TABLE 2.7 Textile Worker Lung Disease (*Continued*)

Reference	Period of study	Number in study	Byssinosis cases (% prevalence)	Summary
Bouhuys et al., 1979		691 cotton textile workers (age ≥ 45); 7,984 community residents (age ≥7 years)		1. The average residual $FEV_{1.0}$[i] of 467 cotton mill workers (age, 45–64 years) was compared to the average residual $FEV_{1.0}$ of 988 community residents (age, 45–64 years). In all subgroups by age, smoking, and sex, cotton mill workers had more negative values than comparable community residents. 2. Prevalence of disabling lung function loss[j] among cotton textile workers was 6.7% compared to a prevalence of 2.9% in community residents ≥ 45 years, with the difference significant at $p < 0.01$.

[a]1-sec forced expiratory volume.

[b]Byssinosis indices vary somewhat from study to study. A typical grading system is: 0 = no symptoms on Mondays; $\frac{1}{2}$ = occasional symptoms (chest tightness, dyspnea, or cough) on Mondays; 1 = symptoms regularly on most Mondays; 2 = symptoms on Mondays and other workdays.

[c]Forced vital capacity.

[d]Reactors have no history of byssinosis but have decrements in flow of 15% or more at 50% of vital capacity.

[e]Percent predicted of $FEV_{1.0} = (FEV_{1.0}\text{ observed}/FEV_{1.0}\text{ predicted}) \times 100\%$ (various methods used to determine predicted $FEV_{1.0}$).

[f]Functional residual capacity.

[g]Total lung capacity [calculated from residual volume (RV) and forced vital capacity (FVC)].

[h]Maximum flow rates at 50% of the control vital capacity.

[i]Residual $FEV_{1.0}$ = observed $FEV_{1.0}$ − predicted $FEV_{1.0}$.

[j]Disabling lung function loss defined as: $FEV_{1.0} \leq 1.50$ liters for males 45–64 years, $FEV_{1.0} \simeq 1.20$ liters for males ≤ 65 years; $FEV_{1.0} \leq 1.00$ liters for females.

C. Coal mining

Coal workers' pneumoconiosis (simple CWP) is a disease that destroys lung function by deposition and retention of coal dust. Various categories can be discriminated based on the size and number of opacities occurring in chest x-rays. Progressive massive fibrosis (called PMF or complicated CWP) is an extreme form of lung dysfunction that may occur as a consequence of various lung diseases including CWP. It is characterized by larger areas of opacity in chest x-rays than simple CWP and severe restriction of the lung's capacity to transfer oxygen.

The estimated prevalence of CWP among coal miners currently employed is about 4.5%; and 0.2% have been diagnosed as having PMF (Morgan and Lapp, 1976). Approximately 4000 deaths/year are associated with legislatively defined "black lung disease," and in 1974 there were an estimated 19,400 cases of CWP (NIOSH, 1983a). Other medical conditions associated with coal dust exposures, such as industrial bronchitis, are also common to coal workers and lead to decreased lung function.

Table 2.8 summarizes much of the available information on coal miners' lung diseases. Although not all the findings are consistent, it is apparent that coal mine exposures are detrimental to lung function and contribute to CWP and PMF. Figures 2.2–2.4 illustrate some of these relationships. Although there is a predictable decrease with age in forced expiratory volume in 1 sec. ($FEV_{1.0}$), it is evident that dust exposure (in g · hr/m^3) is significant for both smokers (Figure 2.2*a*) and nonsmokers (Figure 2.2*b*). This effect is emphasized by the data in Table 2.9 and Figure 2.4; and by the regression coefficients reported

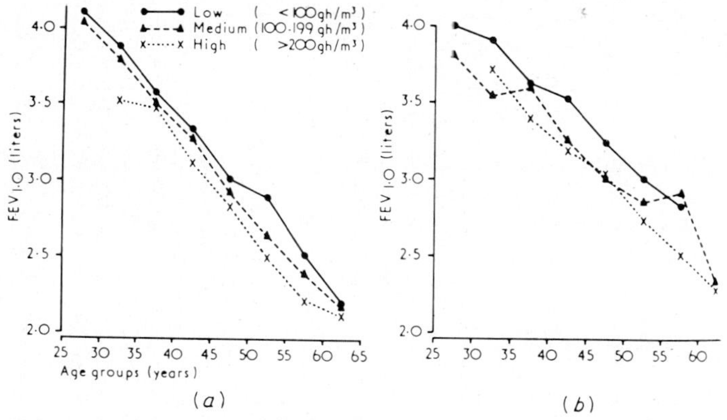

Figure 2.2 Relationship between mean $FEV_{1.0}$ and age among (*a*) smokers and (*b*) nonsmokers, grouped according to cumulative dust exposures (age groups with less than five men not included). Work carried out as part of the British National Coal Board's pneumoconiosis field research (*Rogen et al., 1973; reprinted by permission of the British Journal of Industrial Medicine, R. I. McCallum, ed.*).

in Table 2.10, where the contribution of dust exposure to decreased $FEV_{1.0}$ is significant at the $p < 0.001$ level. Figure 2.3 uses "years underground" as a surrogate for work exposures. The relationship between years underground and severity of CWP is essentially the same for anthracite and bituminous miners. Figure 2.4 illustrates the difference that can occur in CWP prevalence due to different exposure conditions.

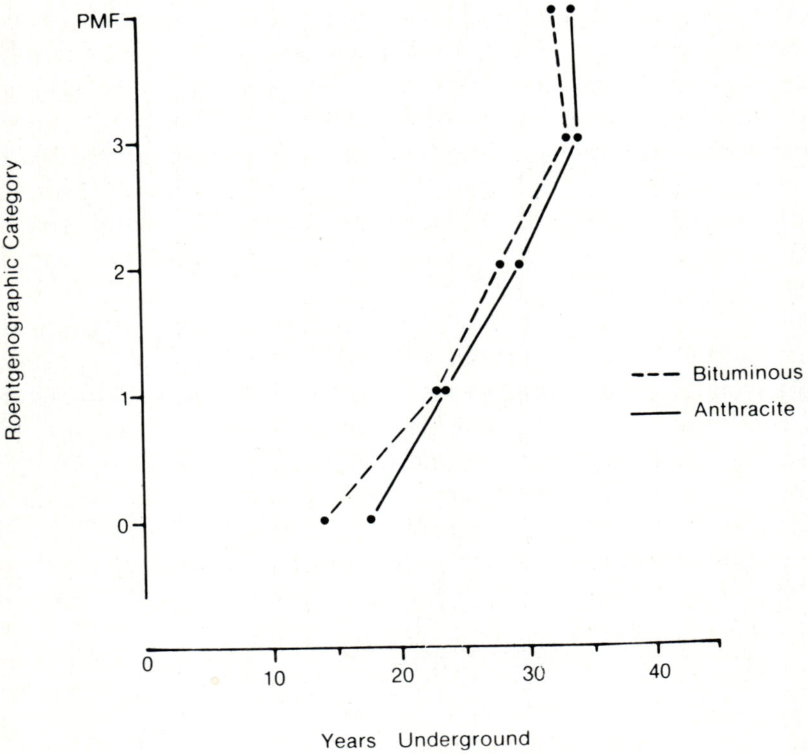

Figure 2.3 Relationship of CWP to years underground (*Morgan et al., 1973; reprinted by permission of the Archives of Environmental Health, a publication of the Helen Dwight Reid Educational Foundation*).

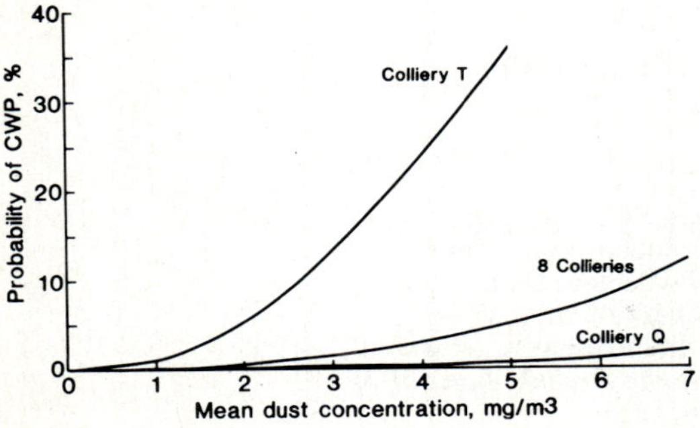

Figure 2.4 Estimates of probabilities of developing Category 2/1 or more simple pneumoconiosis over roughly 35 years (*Hurley et al., 1982; reprinted by permission of the British Journal of Industrial Medicine, H. A. Waldron, ed.*).

TABLE 2.8 Coal Workers' Lung Diseases

Reference	Study period	Study population	Prevalence	Summary
Rogan et al., 1973	1971	3,581 British coal face workers		Statistically significant reduction in $FEV_{1.0}$[a] with increasing dust exposure ($p < 0.001$) in nearly all age categories for smoking and nonsmoking groups. Developed regression equation for $FEV_{1.0}$ with respect to dust exposure, smoking, age, height, and weight. (See also Table 2.10 and Figure 2.2.)
Morgan et al., 1973	Oct. 1969– July 1971	9,076 U.S. miners 8,553 bituminous 523 anthracite	30% with CWP[b]; 2.5% with PMF[c]	1. Roentgenographic category of simple pneumoconiosis increases with the number of years worked underground. (See Figure 2.3.) 2. Although miners with PMF showed no increase in the amount of time spent underground over Category 3 CWP, this finding may have occurred because the development of PMF leads to premature disabilities and sometimes death.
Attfield and Hudak, 1980	Aug. 1972– Feb. 1975	9,343	8% with CWP; 1% with PMF	3. Second round of national coal study confirmed relationship between time underground and CWP.
Pemberton, 1956		242 working U.S. coal miners		Mean $FEV_{1.0}$[a] of the miners was compared to that of two control groups (rural workers, and an urban group of factory workers). The miners had a slightly lower $FEV_{1.0}$ than did either of the control groups, despite the fact that miners smoked somewhat less.
Hyatt et al., 1964	June 1960– July 1962	267 random samples of West Virginia bituminous miners	39% had simple CWP[b]; 7% had compli-	Relationships were found between increasing impairment of ventilatory function ($FEF_{25-75\%}$)[d], and increasing prevalence of bronchitis, with the number of years worked underground. Values not corrected

TABLE 2.8 Coal Workers' Lung Diseases *(Continued)*

Reference	Study period	Study population	Prevalence	Summary
		and ex-miners; 134 (50%) in mining industry; 36 (14%) in other occupations; 80 (30%) unemployed; 17 (6%) disabled	cated CWP or PMF[c]	for age. Only for Category 3 simple CWP and complicated CWP was it demonstrated that CWP per se led to a reduction of ventilatory capacity.
Higgins et al., 1968		5,540 working and former miners, and nonminers in five communities in West Virginia; 2,675 male, 2,865 female		Demonstrated that only in miners and ex-miners with 30 years or more of underground experience was the $FEV_{1.0}$[a] significantly decreased. In miners with <30 years of underground exposure, the $FEV_{1.0}$ did not differ significantly from that of nonminers from the same communities.
Lainhart et al., 1969	1962–1963	3,500 working and nonworking miners in Appalachia		Because adequate control groups were not available in each of the communities in which the study was carried out, the results were compared with predicted values. The FVC[e] of the miners did not differ from predicted values. Among the older miners, the $FEV_{1.0}$[a] was slightly lower than the predicted values. Most of the data were not corrected for age.
Morgan et al., 1974		9,076 U.S. working miners		Because adequate control groups were not available in each of the communities, the results were compared to predicted values. The $FEV_{1.0}$[a] and FVC[e] of anthracite miners were lower than those of bitumi-

		nous miners. Both the anthracite and bituminous miners had a mean $FEV_{1.0}$ that was lower than predicted, but the difference between observed and predicted values was greater among the anthracite miners.
Lapp et al. 1972	93 U.S. underground working miners; 42 nonminers as controls	1. Found small, but significant decreases in $FEV_{1.0}$[a] FVC[e] and (FEF_{75})[f] during the course of the work shift in the miners. When viewed according to smoking status, all of these indices were decreased after the shift in the smoking miners, whereas among the nonsmoking miners, only the FEF_{75} decreased. 2. When the miners were divided into groups according to high, medium, and low exposure to dust on the basis of time spent at the working face, there was a general tendency for the high and medium groups to show decreases after the work shift, whereas the low groups did not.
Lapp et al., 1976	82 U.S. miners with various categories of CWP[b] and 58 age-matched control subjects	1. The two groups did not differ with regard to the standard spirometric indices, such as $FEV_{1.0}$,[a] FVC,[e] or $FEV_{1.0}$/FVC. 2. The FEF_{50},[g] adjusted for age, height, and smoking, was significantly lower among the nonanthracite miners than the nonbronchitic control subjects but it did not distinguish between miners and control subjects when grouped according to smoking habits. 3. In contrast, the age- and height-adjusted closing capacities were consistently higher in the miners than in the control group, regardless of smoking status. 4. Only in the nonsmoking group were there significant differences between miners and control subjects with regard to closing volume.

TABLE 2.8 Coal Workers' Lung Diseases *(Continued)*

Reference	Study period	Study population	Prevalence	Summary
				5. Neither closing capacity nor closing volume was significantly different between miners with and without simple pneumoconiosis.[h]
Hurley et al., 1982		2,600 British coal miners aged 35–65, all working in the industry for at least 20 years	13.5% in Category 0/1 CWP; 3.1% in Category 2/1 CWP	1. The mean cumulative dust exposure of these 2,600 coal miners was 183 $g \cdot hr/m^3$ (14% had greater than 300 $g \cdot hr/m^3$). 2. The probability P of showing Category 2/1 or more CWP among working miners in 10 collieries for t thousand hours at dust exposure d (mg/m^3) was expressed by the regression equation: $\ln\left(\frac{P}{1-P}\right) = -24.60 + 2.435\ln(d) + 4.322\ln(t) - 1.906Q + 2.342T$ where Q and T are dummy variables with the values of 0 or 1 depending on whether colliery Q and T are included. The prevalence of 2/1 CWP was much different in collieries Q and T than in the other 8. (See also Figure 2.4 and Table 2.9.)
Fairman et al., 1977	1972–1973	1,438 surface coal miners	4% with some CWP	1. Of the 1,171 miners who had never previously worked underground, only 0.17% showed evidence of CWP.[b] 2. Multiple regression analysis for miners who had never worked underground indicated significant effect of years mining on FVC[e] only for smokers ($p = 0.0001$).

Amandus et al., 1984	1972–1978	Radiographs of 1,438 strip-mine workers reevaluated		The prevalence of Category 2 or higher pneumoconiosis in miners without underground coal mine experience was 2.1% among anthracite strip miners, but was not found in bituminous strip miners.
Reger and Hancock, 1980		722 pairs; controls were coal miners not exposed to diesel emissions		Diesel-exposed group reported higher prevalence of respiratory symptoms (cough, phlegm production) and had lower measures of pulmonary function than controls.
Rom et al., 1980		242 Utah miners	25% with CWP[b]; 1 case of PWF[c]; 57% with chronic bronchitis	Significant associations between coal dust exposure index and $FEF_{25-75\%}$,[d] CWP, and chronic bronchitis, the latter two associations also being significant for nonsmokers when evaluated separately.

[a]$FEV_{1.0}$ = forced expiratory volume in 1 sec.
[b]CWP = coal workers' pneumoconiosis.
[c]PMF = progressive massive fibrosis.
[d]$FEF_{25-75\%}$ = forced expiratory flow between 25% and 75% of the forced expiratory volume.
[e]FVC = forced vital capacity.
[f]FEF_{75} = forced expiratory flow at 75% of exhaled vital capacity.
[g]FEF_{50} = forced expiratory flow at midvital capacity.
[h]Closing volume and closing capacity are other measures of lung function.

TABLE 2.9 Number of Men Studied, and (in Parentheses) Percentage Classified as Category 2/1 or more CWP, in Relation to Dust and Time

Cumulative hours worked (in 1000s)	Cumulative dust exposure (g·hr/m³)								
	≤79	80–119	120–159	160–199	200–239	240–279	280–359	≥360	All
0–39	73 (0)	72 (0)	84 (0)	42 (0)	19 (0)	10 (0)	3 (0)	3 (0)	306 (0)
40–48	97 (0)	81 (0)	87 (2.3)	106 (2.3)	45 (0.9)	26 (0)	20 (3.0)	3 (0)	465 (1.2)
49–56	91 (0)	65 (0)	68 (2.4)	60 (0)	79 (3.3)	45 (2.7)	41 (1.0)	7 (0)	456 (1.3)
57–64	70 (0)	51 (0)	53 (1.5)	58 (1.4)	67 (3.9)	81 (8.9)	64 (6.3)	37 (11.4)	481 (4.1)
65–72	75 (0)	48 (0)	36 (0.6)	36 (6.7)	42 (6.2)	46 (4.8)	70 (10.6)	33 (7.3)	386 (4.5)
73–80	42 (0)	35 (0)	29 (0)	35 (5.1)	29 (4.1)	38 (13.2)	68 (11.8)	64 (13.8)	340 (7.3)
>80	33 (0.6)	22 (0)	15 (0)	18 (0)	17 (7.1)	15 (6.7)	25 (8.8)	21 (9.5)	166 (4.0)
All	481 (0.0)	374 (0)	372 (1.2)	355 (2.1)	298 (3.6)	261 (6.4)	291 (7.8)	168 (10.4)	2600 (3.1)

SOURCE: Hurley et al., 1982.

TABLE 2.10 Multiple Regression Analysis of $FEV_{1.0}$ Results from 3581 Male Miners Aged 25 to 64[a]

Variable	Units	Mean	Regression coefficient	Standard error	Significance, $p <$
Dust exposure	$g \cdot hr/m^3$	174.8	−0.0006	0.0001	0.001
Smoking	cigs/day	13.1	−0.0046	0.0011	0.001
Age	years	47.0	−0.0471	0.0012	0.001
Height	cm	169.6	0.0306	0.0025	0.001
Sitting height	cm	90.1	0.0086	0.0048	0.1
Weight	kg	72.2	0.0034	0.0010	0.001

[a]Constant = −0.84 liter, residual standard deviation = 0.32 liter, degrees of freedom = 3574.

SOURCE: Rogan et al., 1973.

D. Elevated heat exposures

Occupational heat-related health conditions include heat cramps, heat exhaustion, dehydration, and skin disorders in addition to the increased risk of unintentional injuries. The most severe condition, heat stroke, may lead to death. An estimated 6 million U.S. workers are exposed to occupational heat stress (NIOSH, 1984c). However, because worksite conditions and occupation are usually not listed on hospital records or death certificates, estimates of deaths and heat-related illnesses are difficult to obtain. Analysis of California Department of Health Services data covering 1973–1976 indicated 7 fatalities out of 1128 acute occupational heat-related illnesses, with 10–15% of the surviving patients requiring hospitalization and an additional 40% being absent from work for varying periods after their illnesses (NIOSH, 1984b).

Table 2.11 lists presently available epidemiologic information on the effects of heat stress. Elevated blood pressure and hypertension have been observed in metallurgical workers (Kloetzel et al., 1973); and the relative risk of death from diseases of the digestive system, excluding cirrhosis of the liver, has been reported for steel workers (Redmond et al., 1979). Figure 2.5 illustrates the employment history of 38 metallurgical workers with hypertension (as characterized by blood pressures > 160/95). One difficulty with epidemiological evaluation of jobs involving elevated heat exposures is that those least able to adjust to heat stress tend to self-select out of such settings. Another problem is that heat exposures are often accompanied by exposures to high concentration of dust or pollutant gases (e.g., in a foundry).

E. Steel making

Table 2.12 summarizes the major findings of a retrospective study of 58,821 Allegheny County (Pennsylvania) male steelworkers employed in 1953 and having worked for at least one continuous month. The mortality experience of this cohort has been observed from 1953 to 1975, and has been analyzed with respect to 57 work areas (Lloyd and Ciocco, 1969; Redmond et al., 1969; Lloyd et al., 1970; Redmond et al., 1981). The study population has been subdivided into three employment cohorts (Redmond et al., 1981):

1. Employment in a particular work area based on the first job held in 1953. Men so employed are compared to all other steelworkers.

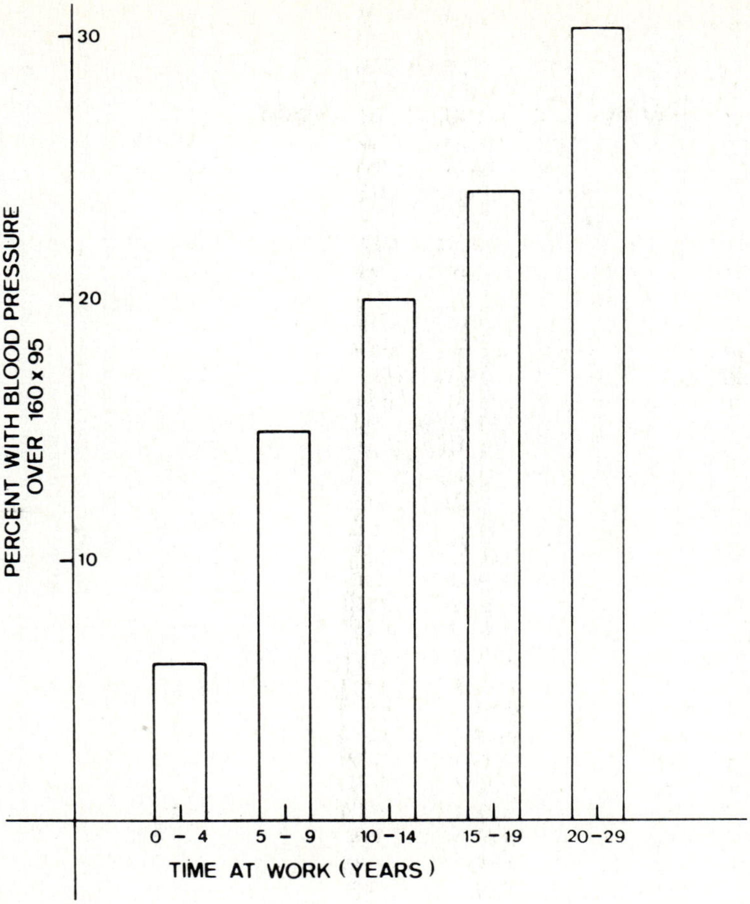

Figure 2.5 Distribution of work duration for metallurgical workers with hypertension (*Kloetzel et al., 1973; reprinted by permission of the American Occupational Medicine Association from the Journal of Occupational Medicine*).

2. Employment in a particular work area based on having worked in that area for at least 1 month prior to or during 1953. Men so employed in a given work area are compared to steelworkers never employed in this work area through 1953.
3. Employment in a particular work area based on having worked in that area for at least 5 years prior to or during 1953. Men so employed in a given work area are compared to all other steelworkers having accumulated 5 or more years experience in the steel industry prior to or during 1953.

By 1975 about 31% of the original population had died, 68% were still alive, and 1% were lost to follow-up.

Use of the steel population not employed in a specific area eliminates

TABLE 2.11 Health Effects of Elevated Heat Exposures

Reference	Study period	Study population	Number of cases	Summary
Kloetzel et al., 1973		330 Brazilian metallurgical workers		1. Workers grouped according to heat exposure categories: *a.* "Extreme heat"—temperatures of 40–45 °C; most workers are at furnaces or rolling mill lines—median age 36.8 years. *b.* "Some heat"—conditions were uncomfortable only for those individuals working close to the furnace doors—median age 27.1 years. *c.* "No heat"—median age 25.5 years. Group *a* was significantly older than groups *b* and *c*, but race and salt intake differences were not significant.

Blood Pressure and (Number of Workers Evaluated)

	% > 140/90 mm Hg (elevated level)		% > 160/95 mm Hg (hypertension level)
	All workers	Workers < 35 years	
a.	42 (186)	27 (80)	15 (186)
b.	11 (54)	15 (34)	4 (54)
c.	10 (90)	3 (61)	8 (90)

Blood pressure differences were significantly elevated ($p < 0.05$) for group *a* even when the analysis was restricted to workers < 35 years.

2. Found increasing incidence of hypertension with duration of time at work (although this relationship was not adjusted for age) (see also Figure 2.5). These findings support the observation that prolonged exposure to high levels of environmental heat has to be considered as an additional factor of risk in the development of hypertension.

Redmond et al., 1979	1953–1970	59,414 U.S. steel workers	12,818 deaths	1. Relative risk (RR[a]) of death from cardiovascular-renal disease for workers in hot jobs compared to other steel workers not in hot jobs was not elevated; indeed, a significant deficit was found primarily because of lower-than-expected rates of arteriosclerotic heart disease. There is a slight suggestion that workers with <1 year in hot jobs may have elevated risk of death from this disease. 2. RR was elevated for digestive diseases, excluding cirrhosis of the liver, for worksites with estimated average WBGT[b] in the summer season > 27.50°C (2.22; observed deaths = 17, expected deaths = 10.1, $p < 0.05$); and RR increased, although without statistical significance, with increasing WBGT for lower temperatures. The same trend was observed when job categories were discriminated by average kcal/min activity levels as well as WBGT. RR = 5.37 was also significant ($p < 0.05$) for those who had worked more than 15 years at the highest temperature category.
Kenedi, 1948				Pathological electrocardiograms occurred four times as frequently among Hungarian foundry workers exposed to intense heat than among those exposed to moderate heat.
Mayer-Theveniaud, 1977				Found a higher incidence of cardiac conduction anomalies among French workers in hot jobs.
Kiev, 1979	1979	217 Russian foundry workers		217 workers who performed the hottest jobs in the Lenin steel works had the worst hemodynamic parameters as determined by polycardiography

[a]RR = relative risk = ratio of the incidence in the heat exposed group to the incidence in an unexposed group with an adjustment for age and race differences between the two groups. The control group was composed of steel workers employed in 1953 who never worked in any of the high-heat-exposure areas.

[b]WBGT = wet-bulb globe temperature for evaluation of heat exposure in working environment. WBGT = 0.7 (NWB) + 0.3(GT); NWB = natural wet-bulb temperature; GT = globe temperature.

TABLE 2.12 Relative Mortality Risks for Steelworkers[a]

Work area	Cause of death (race)	Employment cohort: First job	Ever work	>5 years
Coke plant				
	All cancers (nonwhite)	1.65*	1.52	1.88*
	Cancer of lungs, bronchus, and trachea (nonwhite)	2.92*	2.75*	3.98*
	Cancer of kidney (white)	7.60*	5.16*	11.99*
	Nonmalignant respiratory diseases	1.49		
Iron and steel				
Open hearth	Cancer of kidney (nonwhite)		5.54	
Foundry	All cancers (white)	1.27		1.27
	Genitourinary cancer (white)	1.99	1.81*	2.40*
	Cancer of prostate	2.16	1.91	2.20
	Cancer of kidney		2.55	3.68*
	Arteriosclerotic heart disease (white)	0.76		
Blast furnace	Other nonmalignant digestive disease (white)	1.62	1.40	1.60
Electric furnace	Cancer of bladder and other urinary organs	5.42*	2.56	
Billet, bloom, and slab mill				
	Vascular lesions of the central nervous system (nonwhite)		1.80	
	Hypertensive heart disease, (nonwhite)		2.17	
	Tuberculosis of the respiratory system (white)	2.84*		
	Ill-defined causes (white)	1.78		1.96*
Secondary mills				
Structural mill	All causes (white)	1.23*	1.19*	1.22*
	Cancer of lymph and hematopoietic tissues (white)	2.22		
	Cardiovascular-renal disease (white)		1.18*	1.17
	Arteriosclerotic heart disease (white)	1.20	1.24	1.24
	Accidents (white)	1.76	1.63	2.03
Plate mill	Cancer of the kidney (white)		4.11	
Axle mills	Cancer of the lungs, bronchus, and trachea (white)		2.04*	
General finishing and shipping	Digestive and peritoneal cancer (nonwhite)		2.28*	
Hot strip mill	Diabetes (white)	5.94*		
Rail mill finishing	Nonmalignant digestive disease (white)		1.76	

TABLE 2.12 Relative Mortality Risks for Steelworkers[a] *(Continued)*

Work area	Cause of death (race)	Employment cohort: First job	Ever work	>5 years
Sheet and tin mill				
Batch pickling and sheet dryers	All causes (white)	1.77	1.41	1.63
	Arteriosclerotic heart disease (white)		1.57	2.03
Coating	Hypertensive heart disease (white)	6.36*	4.28*	7.47*
Hot pack mills	Accidents (white)		1.68	
Tin finishing and shipping	Lymphatic and hematopoietic cancer (white)	3.30	3.95	3.14
	All other cancers (white)		2.48	
Stainless annealing, pickling, processing	All cancers (white)	3.08*		1.65
	Digestive and peritoneal cancer (white)		2.04	2.52
Annealing and normalizing	Arteriosclerotic heart disease (nonwhite)			2.79
Cold reducing mills	Respiratory cancer (white)	1.52		
Continuous pickling and electric cleaning	Causes not otherwise listed (white)			3.80*
Sheet finishing and shipping	Accidents (white)	1.59*		
Tube mill				
	Nonmalignant respiratory disease (white)	2.64		
Maintenance				
Mason department	Cancer of lymphatic and hematopoietic tissues (nonwhite)	2.97		
	Nonmalignant respiratory disease (white)	2.44*	1.74*	2.69*
	All causes (nonwhite)	1.22	1.24	1.34
	Cardiovascular-renal disease (nonwhite)		1.25	
	Homicides and suicides (nonwhite)		2.27	8.60*
Blacksmith shop	Respiratory cancer (white)	1.89	1.68	2.19*
	Cancer of the lungs, bronchus, and trachea		1.64	2.01
Electric maintenance assigned	Cancer of the kidney (white)	3.51*		
Machine shop	Digestive and peritoneal cancer (white)			1.49*

TABLE 2.12 Relative Mortality Risks for Steelworkers[a] *(Continued)*

		Employment cohort		
Work area	Cause of death (race)	First job	Ever work	>5 years
All other maintenance	Respiratory cancer (white)	2.50		
Miscellaneous				
Janitors	All causes (white)	1.36*	1.31*	1.31
	Cardiovascular-renal disease (white)	1.32	1.23	
	Arteriosclerotic heart disease	1.32		
	Nonmalignant digestive disease (white)	2.26*	1.88*	
	Genitourinary cancer (nonwhite)		2.23*	
	Cancer of the prostate	2.32	2.90*	2.79*
Plant protection	All causes (white)	1.28*	1.20*	1.23*
	Cardiovascular-renal disease (white)	1.47*	1.32*	1.46*
	Cardiovascular-renal disease (white) (cont.)			
	Arteriosclerotic heart disease	1.46*	1.35*	1.40*
	Diabetes (white)	2.74*	2.05*	2.23
Power and fuel	All other cancers (white)			1.73
	Hypertensive heart disease (white)			1.82
	All other causes			1.74
Transportation and yard	Digestive and peritoneal cancer (nonwhite)		1.61	
	(white)	1.72*		
	Cirrhosis of the liver (white)			1.85
	Respiratory cancer			
	(white)			0.50
	(nonwhite)		1.66	
White river or marine ways	Nonmalignant digestive disease			2.46
	Cirrhosis of the liver			3.51

[a]All values significant for $p < 0.05$; *indicates $p < 0.01$. Relative risk determined from the ratio of the incidence in an exposed group to the incidence in an appropriate unexposed group with an adjustment for age and race differences between the two groups.

SOURCE: Redmond et al., 1981.

a possible bias related to employability. However, there still may be selection based on health concerns within specific work categories. For instance, elevated risks for janitors and plant protection personnel are thought to be related to preemployment or subsequent selective assignment due to reasons of health (Redmond et al., 1981). Other risk factors outside of occupational exposures, such as significant differences in smoking habits between those in a particular area and the comparison population, may contribute to apparent associations. The areas of employment used for each worker only apply to his experience prior to or during 1953, and consequently do not evaluate exposures during interviewing years. Nonetheless, the findings of this study have been consistent when mortality was evaluated to 1966, to 1970, and to 1975. Table 2.12 includes only relative risks that were significant at the $p < 0.05$ or $p < 0.01$ level.

Some of the most notable findings include the following:

1. Nonwhite coke oven workers had significant excess in mortality to all cancers, and particularly respiratory cancer. White workers show the same mortality trend, but relatively few were employed for long periods in coke oven areas. Kidney cancer, however, was a significant disease for white workers in this job category (Redmond et al., 1972; Redmond et al., 1981).
2. White foundry workers incurred excess risk of cancer to genitourinary organs, especially of the prostate and kidney (Redmond et al., 1981).
3. Masons (white) experienced increased risk to nonmalignant respiratory disease, possibly from exposure to dusty and abrasive materials (Rockette and Redmond, 1976; Redmond et al., 1981).
4. White workers in various sheet and tin mill locations had excess mortality from arteriosclerotic heart disease, hypertensive heart diseases, lymphatic and hematopoietic cancers, and digestive and peritoneal cancer (Mazumdar et al., 1975; Redmond et al., 1981). The use of pickling materials and of nickel may have contributed to these rates.

F. Occupational noise

There are considerable data relating occupational noise exposure with hearing losses (OSHA, 1981). Elevated noise levels also can cause other adverse effects such as degraded job performance, increased accident and absenteeism rates, and stress-related illnesses, although these are less easily quantified relative to workplace exposures. Figure 2.6 summarizes permanent median hearing losses that would be predicted to occur in a population exposed 8 hr per day to noise of various frequencies over 10 and 40 years. The plots, used by OSHA to develop a hearing conservation regulation, are based on a variety of British, Dutch, Swedish, and U.S. studies (OSHA, 1981; Passchier-Vermeer, 1968; Burns and Robinson, 1970). Other studies, though differing in detail, are in general agreement with these data (Berger et al., 1978; Yerg et al., 1978; Baughn, 1973). Figure 2.7 shows results of a NIOSH

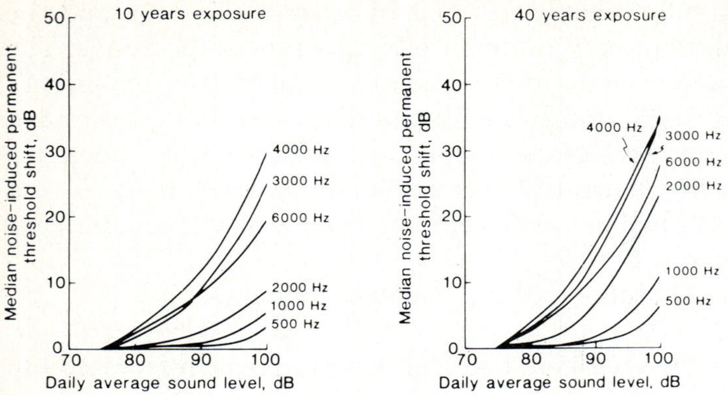

Figure 2.6 Noise-induced permanent threshold shift from 8-hr exposures over 10 and 40 years (*OSHA, 1981*).

survey on 792 noise-exposed workers compared with 380 nonexposed workers as controls (OSHA, 1981). The hearing level differences between median subject and control values are somewhat greater than those indicated by Figure 2.6.

Several problems persist in characterizing permanent noise-induced hearing losses that can be ascribed to occupational settings. One of these is the normal hearing loss (presbycusis) due to aging. The values in Figure 2.6 are in addition to presbycusis. In most situations, individuals will not be significantly impaired due to presbycusis alone unless they live to be very old (OSHA, 1981). Another problem lies in the definition of what constitutes hearing impairment. Prior to 1979, hearing impairment was defined as an average hearing threshold level in excess of 25 decibels (dB) at frequencies of 500, 1000, and 2000 Hz (ACGIH, 1984, pp. 466–470). These were considered to be the frequencies that were most related to the ability to hear and understand speech. As is evident from Figures 2.6 and 2.7, much larger reductions in hearing acuity will occur from 3000 to 6000 Hz than at lower frequencies. However, no consensus has been reached on appropriate inclusion of higher frequencies into the impairment formula (Kryter, 1973; EPA, 1974; Ward, 1975), although both the American Academy of Otolaryngology and OSHA have specified inclusion of 3000 Hz (AAOO, 1979; OSHA, 1981).

Table 2.13 summarizes the cumulative percentage of workers exposed to 8-hr time-weighted average noise levels > 80 dBA in a variety of industrial categories (OSHA, 1981). The determination and interpretation of dBA are given in Chapter 13. The industries selected for survey were thought to be those that contained the noisiest workplaces. Accordingly, almost 2.9 million U.S. workers are exposed to noise levels ≥ 90 dBA, the present OSHA standard (see Chapter 4); and an

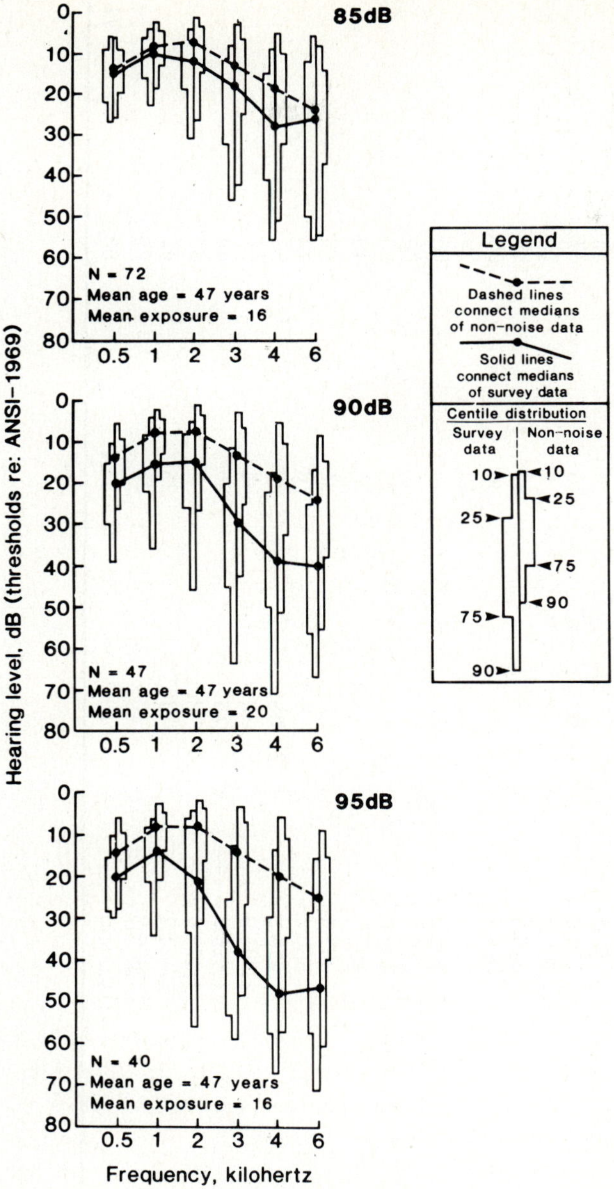

Figure 2.7 Hearing-level distribution for workers aged 43 to 51 years exposed to daily average noise levels of 85 dB, 90 dB, and 95 dB, and for non-noise-exposed workers (*OSHA, 1981*): N = number of workers; mean exposure = mean value for years of exposure.

TABLE 2.13 U.S. Industrial Noise Exposure Levels

Industry title	Production workers (thousands)	Cumulative percentage of production workers exposed to noise at various levels			
		≥90 dB	≥85 dB	≥80 dB	≥75 dB
Food	1,176.2	16	28	47	—
Tobacco	52.5	6.6	9.7	28	—
Textiles	770.0	52	75	87	—
Apparel	1,122.2	0	1	20	—
Lumber and wood	646.3	72	94	97	—
Furniture and fixtures	398.0	12	30	53	—
Paper	541.5	21	40	59	—
Printing and publishing	702.2	19	45	66	—
Chemicals	636.9	20	37	55	—
Petroleum and coal	139.7	52	76	82	—
Rubber and plastics	601.1	8.9	20	40	—
Leather	207.4	0	1	20	—
Stone, clay, and glass	560.5	4.8	16	42	—
Primary metals	978.3	38	63	81	—
Fabricated metals	1,305.9	19	34	56	—
Machinery, except electrical	1,616.2	13	26	48	—
Electrical machinery	1,378.6	2.5	7	27	—
Transportation equipment	1,404.2	13	23	42	—
Utilities	659.3	30	74	89	—
Total	14,904.0	19.3	34.4	53.1	71.5

SOURCE: OSHA, 1981.

additional 2.2 million endure levels between 85 and 90 dBA. OSHA has estimated, based on this survey, representative distributions of age and presbycusis, and the predicted hearing loss from Figure 2.6, that 1.06 million workers currently have hearing impairment due to occupational noise in excess of 25 dB based on 1000, 2000, and 3000 Hz (OSHA, 1981).

G. Exposure to asbestos

Asbestosis, lung cancer, and mesothelial tumors of the pleura and peritoneum are the diseases commonly associated with asbestos exposure. All are characterized by latent periods from first exposure to development of disease symptoms of 10–40 years. Asbestosis involves irreversible diffuse scarring of the lung and progressive shortness of breath. It has been estimated that 10–18% of asbestos insulation workers will die of asbestosis; and that risk of death from asbestosis is increased by a factor of 2 or 3 by smoking (NIOSH, 1983a). It has also been proposed that up to 11% of asbestos-exposed workers may develop mesothelioma (NIOSH, 1984a).

Table 2.14 summarizes the results of six epidemiological studies on asbestos miners and millers, insulation and shipyard workers, and workers in cement pipe, textiles, sheet and tile manufacture. For most settings it appears that elevated mortality risks exist and that development of one or more of the diseases is a function of cumulative exposure. Table 2.15 shows the mortality experience for a large cohort of U.S. and Canadian insulation workers. Both mesothelioma and asbestosis are rare in the general population, and the large number of occupationally related deaths in this cohort from these diseases is notable. Table 2.16 gives a comparison for several measures of mortality for different workplace categories. The differences reflect types of asbestos exposure (e.g., crysotile, crocidolite, or amosite) as well as worksite conditions of ventilation and particle control.

H. Segmental vibration

An estimated 1.2 million U.S. workers are exposed to segmental vibration (i.e., that which affects only a part of the body) (Wasserman et al., 1974). Most such exposures occur from the use of power tools such as jackhammers, chain saws, and grinders. Vibration white finger syndrome (VWF), sometimes referred to as Raynaud's phenomenon, can often arise from segmental vibration. The condition is characterized by intermittent numbness and blanching of the fingers; experience of pain with return of circulation; loss of finger dexterity; and, in advanced stages, irreversible changes in digital arteries with reduction in blood supply.

TABLE 2.14 Epidemiological Studies of Asbestos-Exposed Workers

Reference	Period of study	Number in study	Deaths	Summary
Dement et al., 1983	1940–1975	1,261 asbestos textile workers (chrysotile)	308 deaths	1. Statistically significant ($p < 0.05$) excess mortality was observed for all causes combined (SMR[a] = 150; 308 deaths); all malignant neoplasms (SMR = 168; 59 deaths); malignant neoplasms of the trachea, bronchus, and lung (SMR = 315; 35 deaths); diseases of the circulatory system (SMR = 125; 105 deaths); and all nonmalignant respiratory diseases (SMR = 294; 28 deaths). An increasing trend in the lung cancer SMR with increased employment was observed with SMR of 185 ($p < 0.05$) for the 15 deaths occurring in workers with less then 10 years of employment and a SMR of 976 ($p < 0.05$) for the 12 deaths occurring among workers with 20–29 years of employment. 2. A strong exposure-response relationship was found for lung cancer for workers with 15 or more years since initial employment.

Cumulative dust exposure[b,c] (fiber·day/cm^3)	Significant SMR ($p < 0.05$) (no. of deaths)
$<10^3$	not significant
$10^3 - 10^4$	279 (9)
$10^4 - 4 \cdot 10^4$	352 (7)
$4 \cdot 10^4 - 10^5$	1099 (10)
$>10^5$	1818 (2)

				For asbestosis and pulmonary fibrosis a strong increasing trend in incidence density (cases/1000 person·years) with exposure was observed, with 0.32 cases/1000 person·years for <1000 fiber·day/cm^3 and 15.87 cases/1000 person·years for >100,000 fibers·day/cm^3.
Henderson and Enterline, 1979	1941–1973	1,075 asbestos, manufacturing workers (asbestos cement pipes, shingles, and sheets)	781	1. Excess mortality was observed for all cancers (SMR[a] = 159.0; 173 deaths); for cancer of the digestive system (SMR = 137.8; 55 deaths); cancer of the respiratory system (SMR = 270.4; 63 deaths); and for respiratory diseases (SMR = 173.0; 68 deaths). 2. A strong relationship was found between total dose expressed as mppcf·years[d] and mortality due to respiratory cancer expressed by: $SMR_{resp.\ cancer}$ = 100.0 + 0.658 (mppcf·years). Higher risk found for manufacture of cement pipe (crocidolite and chrysotile; SMR = 521.7; 12 deaths) than for sheet and shingle workers (chrysotile only; SMR = 230.8; 15 deaths). Maintenance workers, who tend to be mobile throughout the plant, had an SMR[a] of 300.0 (15 deaths). Pneumoconiosis-pulmonary fibrosis death rates[e] were found to be related to cumulative dust exposure at the time of retirement: <250 mppcf·years, 151.6 deaths/100,000; >750 mppcf·years, 936.1 deaths/100,000.

TABLE 2.14 Epidemiological Studies of Asbestos-Exposed Workers (*Continued*)

Reference	Period of study	Number in study	Deaths	Summary
Nicholson et al., 1979	At least 20 years' seniority prior to 1961 was required for entry in cohort. Cohort followed until Aug. 1977.	544 chrysotile asbestos miners and millers	178	Excess mortality found for all cancers (SMR = 134; 49 deaths); lung cancer (SMR = 252; 28 deaths); and noninfectious pulmonary diseases (SMR = 448; 30 deaths). Details of the mortality experience according to time from onset of exposure indicate an increase in mortality between 30 and 50 years from first exposure and a decline in excess mortality after 50 or more years from first exposure.
Berry and Newhouse, 1983	1942–1980	13,460 manufacturing workers (chrysotile and crocidolite)		1. No overall or site-specific excess of mortality for any diseases except mesothelioma. 2. 8 of 10 pleural mesotheliomas associated with elevated crocidolite exposures. 3. For lung cancer a linear relationship between relative risk and cumulative exposure was fitted using matched case-controls: $RR_{\text{lung cancer}} = 1.00 + 0.00058$ (fiber·years/cm^3) At 100 fiber·4 years/cm^3, the upper 90% confidence limit was 1.80. 4. Reported that only 5% of men employed after 1941 experienced >100 fiber·years/cm^3, which was interpreted as historically low.

Finkelstein, 1983		328 asbestos cement manufacturing workers: 186 production 55 maintenance 87 controls (Ontario, Canada)	58	Control group had mortality rate similar to those of the general Ontario population. Exposed group had all-cause mortality rate that was twice that of controls; malignancy mortality 5 times expected; lung cancer mortality 8 times expected. 10 of the 58 deaths among the production workers were due to malignant mesothelioma and 20 due to lung cancer.
Selikoff et al., 1979	1943–1976	17,800 U.S. and Canadian insulation workers	2271	Observed overall mortality was elevated above that for the age-adjusted white male population as well as for total cancer, noninfections pulmonary disease, and many cancer sites (see Table 2.15).

[a]SMR (standardized mortality ratio) $= \dfrac{\text{number of deaths observed}}{\text{number of deaths expected}} \times 100$

[b]A worker's cumulative dust exposure is equal to the cumulative product of the estimated average dust concentrations for particular jobs held by the worker and the time spent in those jobs.

[c]Concentrations are expressed as fibers > 5 μm in length/cm^3 of air.

[d]Concentrations are expressed as million particles per cubic foot (mppcf).

[e]Death rates were calculated in the following manner:

$$\frac{\text{number of deaths}}{\text{person·years lived}} \times 100{,}000$$

TABLE 2.15 Deaths among 17,800 Asbestos Insulation Workers in the United States and Canada, January 1, 1967–December 31, 1976

Number of Men 17,800; Man-Years of Observation 166,853

Underlying cause of death	Number of deaths			Observed/expected deaths	
	Expected[a]	Observed (BE)	Observed (DC)	(BE)	(DC)
Total deaths, all causes	1658.9	2271	2271	1.37	1.37
Total cancer, all sites	319.7	995	922	3.11	2.88
Cancer of lung	105.6	486	429	4.60	4.06
Pleural mesothelioma	†	63	25	—	—
Peritoneal mesothelioma	†	112	24	—	—
Mesothelioma[b]	†	0	55	—	—
Cancer of esophagus	7.1	18	18	2.53	2.53
Cancer of stomach	14.2	22	18	1.54	1.26
Cancer of colon-rectum	38.1	59	58	1.55	1.52
Cancer of larynx	4.7	11	9	2.34	1.91
Cancer of pharynx, buccal	10.1	21	16	2.08	1.59
Cancer of kidney	8.1	19	18	2.36	2.23
All other cancer	131.8	184	252	1.40	1.91
Noninfectious pulmonary disease, total	59.0	212	188	3.59	3.19
Asbestosis	†	168	78	—	—
All other causes	1280.2	1064	1161	0.83	0.91

[a]Expected deaths are based on white male age-specific U.S. death rates of the U.S. National Center for Health Statistics, 1967–1976.

[b]Not otherwise specified.

†Rates are not available, but these have been rare causes of death in the general population.

(BE): Best evidence. Number of deaths categorized after review of best available information (autopsy, surgical, clinical).

(DC): Number of deaths as recorded from death certificate information only.

SOURCE: Selikoff et al., 1979.

TABLE 2.16 Comparison of Mortality Observations for Jobs Involving Asbestos

Asbestos job category	Lung cancer mortality SMR	Mesothelioma mortality incidence, deaths/10^3 person·years
Mining and milling[a]	252	0.13
Manufacturing[a]	270	2.92
Insulating[b]	480	2.20

[a]Nicholson et al., 1979.
[b]Selikoff et al., 1979.

Tables 2.17 and 2.18 summarize some of the epidemiological evidence that establishes power-tool use as an occupational hazard. The prevalence values are considerably elevated over the 6–8% normally expected in the community from hereditary influences, other disease conditions, and exposure to some pollutants (such as vinyl chloride) (ACGIH, 1984, pp. 493–497). Also included in Table 2.17 are several predictive equations based on the epidemiological data. These relate the exposure duration and intensity, as expressed by the dominant frequency-weighted root-mean-square acceleration, to the onset of VWF symptoms. Prevention of such symptoms over a working lifetime of 25–30 years' exposure appears to require 8-hr, 40-hr/week acceleration levels < 1–4.7 m/sec^2. Figure 2.8 shows a relationship between prevalence and exposure based on 21 of the epidemiological studies.

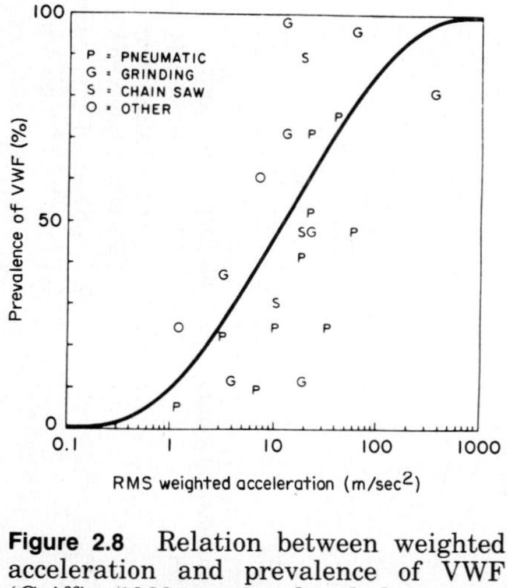

Figure 2.8 Relation between weighted acceleration and prevalence of VWF (*Griffin, 1982; reprinted with the permission of John Wiley and Sons, Inc.*).

TABLE 2.17 Segmental Vibration Effects

Reference	Study population	Percent prevalence of vibration syndrome[a]	Summary
Behrens et al., 1982	205 vibration-exposed workers: 147 gray iron foundry workers; 58 shipyard foundry workers; 63 nonexposed foundry workers (controls)	 83 64 0	Vibration-induced white finger (VWF) from pneumatic chipping and grinding tools was graded according to stage.[a] In the iron foundries, the median latency[b] of tingling was 0.6 year, numbness was 0.8 year, and blanching was 1.4 years. In the shipyard foundry, tingling had a median latency of 4.2 years, numbness 9.5 years, and blanching 16.5 years. There was a statistically significant ($p < 0.001$) association between exposure time using a pneumatic chipping hammer and the severity (stage) of VWF.
Pyykko et al., 1982	Year and number of forest workers: 1972, 118 1975, 187 1976, 181 1977, 189 1978, 203 1979, 210 1980, 197	 40 23 19 11 10 10 7	Longitudinal study based on workers who used chain saw ≥ 500 hr/year during three consecutive years: 1. Since 1972 only three new subjects have been afflicted with VFW. 2. Retrospective analysis of the history of attacks of VWF among 187 subjects studied in 1975 indicated a mean latency[b] for development of VWF of 4–5 years (annual work rate of 1300 hr/year).
Riddle and Taylor, 1982	89 forest workers: 43 who previously worked with chain saws; 28 who have worked with unprotected chain saws in the past and now work with chain saws incorporated with		Longitudinal study 9 years after the introduction of anti-vibration devices. Stage[a] of VWF was assessed for each: 1. Among the 43 ex-chain sawyers, whose mean saw usage time was 11.7 ± 5 years, termination of vibration exposure resulted in a decline in prevalence of VWF (stage 01, 02, and 03) from 65% (worst stage of VWF recorded for this group) to 30% (1981). In terms

	antivibration devices (AV saws); 18 who have worked with AV saws only		of the current status of change in their condition, 4 report some improvement in their condition, and 4 report some deterioration in their condition. 2. Among the 28 active sawyers (mean saw usage time of 16 ± 5 years) with previous experience using non-AV saws, a decline in prevalence of VWF (stages 01, 02, and 03) from 61% (worst stage of VWF recorded for this group) to 46% (in 1981) was observed. As of 1981, 10 of the 28 sawyers never had VWF, 9 report the status of their condition to be stationary, 6 report an improvement in their condition, and 3 report their condition to be deteriorating. 3. Among the 18 active sawyers whose experience is limited to AV saws, an increase in prevalence of VWF (stages 01, 02, 03) from 11% (worst stage of VWF previously recorded for this group) to 17% (1981) was observed. The mean saw usage time for this group was 5 ± 3 years.
Pelnar et al., 1982	566 forest workers; 323 power saw operators; 243 other forestry workers	21.2 27.9 12.2	1. Among power saw operators, a correlation was found between period of work with the power saw and prevalence of finger blanching ($r = 0.83$). 2. Prevalence of VWF was twice as high in those who had a history of hand injury (i.e., cuts, lacerations, fractures, or frostbite).
Miyashita et al., 1982	312 forestry workers; 266 chain saw operators; 46 controls		Total chain saw operating time (TOT)[c] was calculated for each subject with classification into five groups of N subjects: Prevalence 1. Never exposed to vibration (N = 46); P = 0% 2. TOT < 2000 hr (N = 39); P = 7.9% 3. 2000 ≤ TOT < 5000 hr (N = 76); P = 29% 4. 5000 ≤ TOT < 8000 hr (N = 51); P = 33.3% 5. TOT ≥ 8000 hr (N = 100); P = 47%

TABLE 2.17 Segmental Vibration Effects *(Continued)*

Reference	Study population	Percent prevalence of vibration syndrome[a]	Summary
Taylor and Pelmear, 1976	1029 sawyers, chippers, grinders, and swagers; 722 vibration-exposed 307 controls		1. The prevalence of VWF[a] varied between 8% and 95% depending on intensity and duration of exposure. Majority of pedestal grinding and swaging cases were in class 03 VWF. 2. See Table 2.18 for summary of exposures and effects.
NIOSH, 1983c; Wasserman et al., 1982	268 foundry and shipyard workers; vibration-exposed (chipping hammers): 147 foundry workers, 58 shipyard workers; controls: 63 workers	 47 19 0	1. Found a direct relationship between years exposed and severity of vibration syndrome.[a] Vibration syndrome[a] of stage 1 or greater severity was found in 31% of the workers exposed 1.5 years or less, in 41% of the workers exposed 1.5 to 3 years, and in 71% of the workers exposed more than 3 years. A similar relationship was found among shipyard workers. 2. The average latency[b] for the appearance of blanching for those with advanced vibration syndrome of stage 1 or greater severity was 2 years for foundry workers and 17 years for shipyard workers.
Brubaker et al., 1983	Lumberjacks: 146 treefellers; 142 controls;	 51.4 4.9	1. The mean latency[b] was 8.6 years (S.D. 6.5), whereas the median (a better indicator of latency due to extreme variability of the data) was 7.3 years. 2. Effect of chain saw exposure:

Chain saw exposure (years)	VWF[a] ≥stage 01: Number	VWF[a] ≥stage 01: Percent exposure group
0	2	2.4
1–5	1	11.1

	6–10	13	48.1
	11–15	14	70.0
	16–20	8	47.0
	21–25	9	81.8
	>26	3	60.0

Griffin, 1982

1. Review of various studies to determine dose-effect regression relationship between frequency-weighted acceleration[d] (a_k) and the latent period[b] before onset of VWF resulted in

$$L = 9.34a_k^{-0.36}$$

with a correlation coefficient of 0.539. But a large amount of scatter in the data creates some doubt about accuracy.

2. A regression between the mean weighted acceleration and prevalence of VWF reported in 21 studies gave

$$P = 0.5[1 + \sin(22 \ln (a_k) - 55)]$$

with correlation coefficient of 0.56.

Brammer, 1982a

1. Dose-effect relationships based on reported findings from 40 population groups consisting of workers who operate a power tool or are involved in an industrial process that results in vibration exposure throughout the workday. Based on data from seven studies, the average latent period of VWF,[b] L (in years), in population groups was related to the frequency-weighted acceleration,[d] a_k (in m/sec^2), of vibration entering the hand. A least-squares best fit of the data ($r^2 = 0.82$) resulted in

$$L = 78.7a_k^{-1.07}$$

2. Data reported in studies on chain saw operators indicate that the range of exposure durations prior to the

TABLE 2.17 Segmental Vibration Effects *(Continued)*

Reference	Study population	Percent prevalence of vibration syndrome[a]	Summary
			development of VWF by a population group follows a normal distribution. Based on data from 14 studies, the relationship between t_{LI}, the average latent period in years, and S, the standard deviation or range of latent periods (in years) for persons with different susceptibility to VWF was developed. A least-squares best fit of the data resulted in $S = 0.01 + 0.46t_{LI}$ 3. The progression of VWF in a population group to stage 3 VWF[a] can be predicted using the relationship $L_{03/L} = 3.2$ (where L_{03} is the average duration of exposure necessary for development of stage 3 VWF). This relationship was determined using data from studies where population groups consisted of at least 30 workers and the stage 3 subgroup consisted of at least 6 persons.
Brammer, 1982b			Three methods were used to predict the threshold limit[e] of vibration exposure: 1. Extrapolation of $L = 78.7a_k^{-1.07}$ 2. Analysis of data from a comparative occupational survey of the onset of VWF (Miura et al., 1959) 3. Analysis of data from a prospective study of the development of VWF (Taylor et al., 1977)

Based on these predictions, the threshold limit for workers whose hands are exposed to vibration throughout each workday was estimated to lie in the range $1 < a_k < 2\ m/sec^2$.

Threshold for the onset of disorders derived by three methods (expressed as a frequency-weighted, rms, component acceleration in m/sec^2).

Brammer,[1]	Miura et al.,[2]	Taylor et al.,
1–2.9	1–2.1	<4.7

[1]Onset of episodes of finger blanching delayed for the duration of a working lifetime of 25–30 years.
[2]Value depends on the dominant vibration frequency.

[a]Stage 00—no tingling, numbness, or blanching of digits.
Stage 0T—intermittent tingling.
Stage 0N—intermittent numbness.
Stage 0TN—intermittent tingling and numbness.
Stage 01—blanching of one or more fingertips with or without tingling and numbness.
Stage 02—blanching of fingers beyond tips, usually confined to winter.
Stage 03—extensive blanching of digits. Frequent episodes summer and winter.
Stage 04—extensive blanching, most fingers. Frequent episodes summer and winter.
Prevalence values based on symptoms of stage 01 or greater.

[b]Latency is the period from the first occupational exposure to hand-arm vibration to the onset of symptoms of VWF.

[c]Total operating time (TOT) = chain saw operating hours per day × days per year × years.

[d]Frequency weighting used is from the Draft International Standard, ISO 5349 (ISO, 1982).

[e]Threshold limit is the vibration level at which the duration of exposure prior to the onset of finger blanching corresponds to a working lifetime.

TABLE 2.18 Summary of Some VMF Prevalence and Latent Periods

Population and process		Latent interval (years)	Percent prevalence: VWF group $N = 722$	Percent prevalence: Control $N = 307$	Frequency-weighted rms acceleration (m/sec^2)
1. Chain sawyers	Thetford	2.8	90	6	25
	Kielder	3.6	70	8	
	Dumfries	3.0	40	4	
2. Grinding	Pedestal	1.8			50
	Pedestal	4.5			12.2
	Hand	8.7	8–96	4–17	
	Hand	13.7			
	Pedestal	14.0			3
3. Chipping,	Birkenhead	5.3	54–85	19	
cauking,	Rosyth	14.0			
nobbling	Rosyth	16.5	35–55	—	
4. Swaging	Liverpool	0.6	95	6	70

SOURCE: Taylor and Pelmear, 1976.

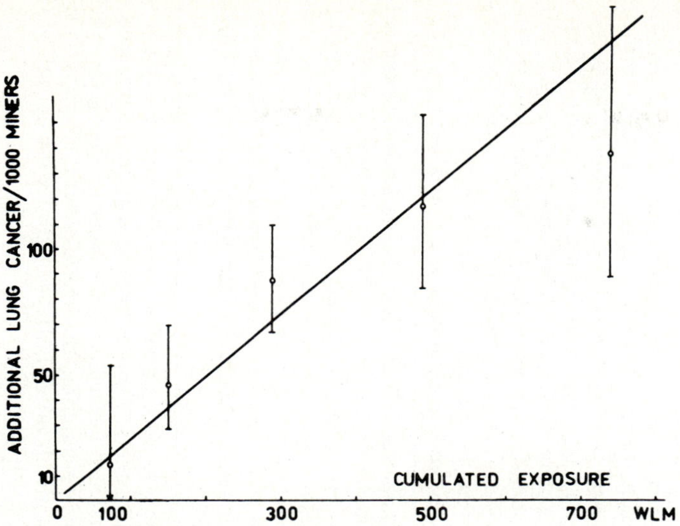

Figure 2.9 Additional lung cancer frequency and cumulative exposure of Czech uranium miners to radon daughters (*Kunz et al., 1979; reprinted by permission of Pergamon Press, Ltd. from Health Physics*). 95% confidence limits are indicated. All excess frequencies except that at 99 WLM are significantly elevated ($p < 0.01$).

I. Lung cancer in metal mines

Table 2.19 summarizes epidemiological findings for metal miners from Sweden, Cornwall, New Mexico, and Czechoslovakia. Although different metals were extracted, the mining environment in all cases was characterized by elevated levels of short-lived decay products of radon-222 (radon daughters). SMRs for lung cancer vary between 211 (Fox et al., 1981) and 1630 (Axelson and Sundell, 1978); and relative risks and observed/expected lung cancer death ratios are correspondingly elevated. The risk appears to apply to nonsmokers as well as smokers.

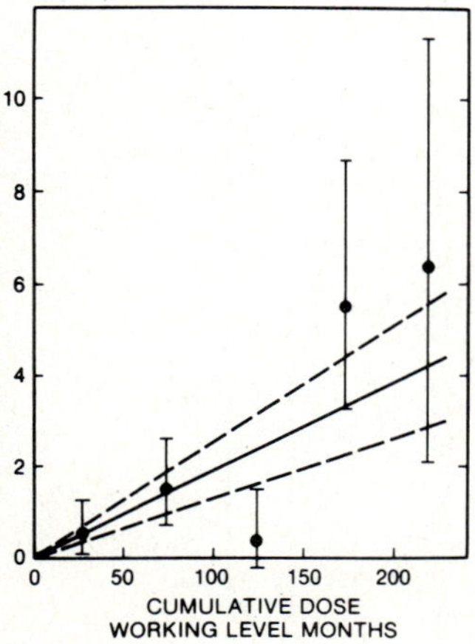

Figure 2.10 Absolute risk of lung cancer from exposure to radon daughters in 1294 underground iron miners studied from January 1, 1951, to December 31, 1976 (*Radford and St. Clair Renard, 1984; reprinted by permission of The New England Journal of Medicine*). Solid line joins origin and mean value of exposure and risk based on the whole group of miners. The line also is approximately a least-squares fit constrained through the origin ($r = 0.86$). Dotted lines are ± 90% confidence limits.

TABLE 2.19 Some Recent Epidemiologic Studies of Lung Cancer in Uranium and Metal Miners

References	Period of study	Study population	Cases or deaths	Summary
Axelson and Sundell, 1978	1956–1976	Swedish zinc-lead miners; 29 lung cancer cases (21 miners); 174 controls, subjects who died at same time from other causes but came from same area (19 miners)		1. Expected lung cancer deaths were determined from national Swedish mortality rates and the nonmining subjects. The ratio of observed lung cancer in miners to expected lung cancer in nonminers resulted in an SMR[a] = 1,630 ($p < 0.0001$). Differences were ascribed to exposure to radon daughters. 2. Risk of lung cancer in miners only was higher among nonsmokers than among smokers. But the induction-latency time for lung cancer to develop was shorter for smoking miners (mean = 34 years) than for nonsmoking miners (mean = 43.4 years) (difference significant at $p = 0.016$).
Kunz et al., 1979	1948–1975	Czech uranium miners (>3,300)		Miners were divided into five categories based on their cumulative radiation exposure, expressed as working level months (WLM)[b]; and these levels were used to evaluate each group's lung cancer experience. Additional lung cancer frequency (the difference between the observed and expected frequencies) increased with cumulative exposure in the whole group (see Figure 2.9).
Fox et al., 1981	1939–1976	1,333 Cornish tin miners (27,631 person-years at risk from two mines)	936 deaths	Subjects were classified as either underground or surface miners. Those whose occupation could not be ascertained formed a third category. Among the whole study population, the overall mortality rate was significantly elevated (SMR[a] = 122). Excess mortality was

			found for respiratory diseases (SMR = 160; 185 deaths), lung cancer (SMR = 117; 61 deaths), and stomach cancer (SMR = 141; 33 deaths). Of the deaths due to respiratory diseases, 93 were due to silicosis and 56 due to silicotuberculosis. In relation to occupation, underground miners experienced an SMR of 473 (104 deaths), due to respiratory diseases and silicotuberculosis combined, as compared to an SMR for the same cause of 76 (34 deaths) among surface miners. Similarly, among underground workers the SMR for lung cancer was 211 (28 deaths) and for stomach cancer 200 (10 deaths) compared to 74 (13 deaths) and 106 (9 deaths), respectively, among surface workers. Annual exposure rates to radon daughters was estimated at 25 WLM[b] and 15 WLM for the two mines.
Edling, 1982	1957–1977	Swedish iron miners; 47 lung cancer cases (42 miners); 897 controls, subjects who died from other causes at same time but who were from the same area (331 miners) matched by year of birth, sex, and age at death	1. The SMR[a] for miners was 1,620 ($p < 0.0001$) with the lung cancer mortality among nonminers serving as the expected mortality rate. Typical mine exposures ranged from 0.3 to 1.0 WLM[b]. 2. Forty-four miner cases were matched to 44 miner controls to assess the relationship among mining, lung cancer, and smoking habits. The relative risk[c] of lung cancer for smoking miners vs. nonsmoking miners was 2.0, with a 95% confidence interval of 0.7–5.7. The mean induction-latency time for lung cancer to develop was shorter for heavy smokers (mean = 36.9 years) than for nonsmoking miners (mean = 39.6 years), but the difference was not significant.

TABLE 2.19 Some Recent Epidemiologic Studies of Lung Cancer in Uranium and Metal Miners *(Continued)*

References	Period of study	Study population	Cases or deaths	Summary
Edling and Axelson, 1983	1966–1977	Swedish iron ore miners; 38 lung cancer deaths (33 miners); 503 controls matched by age at birth, age at death, and sex who died of nonmalignant causes (183 miners)		The SMR[a] among miners was 1,150 with the unexposed used as the standard. Twenty-eight miner cases were matched with 28 miner controls to assess the effect of smoking. The relative risk[c] for lung cancer among smoking miners vs. nonsmoking miners was 1.5. Estimates of lifetime risk of lung cancer due to radon and its daughters among miners were calculated from the data. Among the group that died of lung cancer between the ages of 50 and 65 years, an excess risk was estimated of 25.9 lung cancer cases per 10^6 person-years · WLM[b]; among the group that died of lung cancer after the age of 65, the estimated excess risk was 54.3 lung cancer cases for 10^6 person-years · WLM.
Samet et al., 1984	1969–1982	Navajo men in a uranium mining community; 32 lung cancer cases (23 uranium miners); nonlung cancer deaths matched by age and diagnosis date (no identified uranium miners)		The association between lung cancer and uranium mining was statistically significant at $p = 1.1 \times 10^{-11}$. An estimate of the relative risk for lung cancer among uranium miners was determined by using the binomial distribution. The lower limit of the 95% confidence limit for relative risk was 14.4. The median induction-latency period for lung cancer to develop for the 23 cases with uranium mining experience was 23.5 years with a range from 12 to 38 years. Smoking did not appear to be a factor, as 62% of the cases who were miners never smoked; and smoking miners smoked a median value of only 4 cigarettes/day.
Radford and St. Clair Renard, 1984	1951–1976	1,415 Swedish iron miners	532 deaths	For all miners combined, the average exposure to radon daughters was 4.8 WLM[b] per year. The average total cumulative dose was 93.7 WLM with a range of 2 to 300 WLM.

During the study period there were 139 deaths due to all types of cancer, compared with 89.4 expected based on national mortality rates for Sweden. The observed/expected deaths (O/E) was 1.56 ($p < 0.01$). The excess was mostly due to 50 deaths from lung cancer (O/E = 3.42; $p < 0.01$) and 28 deaths from stomach cancer (O/E = 1.89; $p < 0.01$). The O/E for all other types of cancer was 1.02 based on 61 deaths (not significant). For all causes except cancer, miners experienced an O/E of 1.26 based on 393 deaths ($p < 0.01$).

No evidence that silicosis was a contributory factor was found. A smoking-specific analysis was used to determine expected lung cancer rates among smoking and nonsmoking miners. Among nonsmoking miners, O/E = 10 (18 deaths); among smoking miners O/E = 2.9 (32 deaths). The absolute risk of lung cancer due to exposure to radon daughters for smokers was 21.8 excess cases of lung cancer per 10^6 person-years per WLM. The risk for nonsmokers was 16.3 excess cases of lung cancer per 10^6 person-years per WLM. The latency period from the beginning of underground work to death from lung cancer was 41.3 years for the 8 cases who never smoked, 39.7 years for the 10 long-term ex-smokers (stopped more than 10 years previously), and 38.8 years for the 32 smokers.

---

[a]SMR = standardized mortality ratio:

$$\text{SMR} = \frac{\text{observed deaths}}{\text{expected deaths}} \times 100$$

[b]WLM = working-level month. One WLM is defined as exposure for 1 working month of 170 hr to a concentration of one working level, where one working level is any combination of short-lived decay products of radon-222 per liter of air that will result in the emission of 1.3×10^5 MeV of alpha energy during complete decay.

$$^{c}\text{Relative risk} = \frac{\text{incidence among those exposed}}{\text{incidence among those unexposed}}$$

TABLE 2.20 Risk Estimates of Lung Cancer in Underground Mining from Radon-Daughter Exposure

Miner group and period studied	Number	Absolute risk coefficient: Excess lung cancers/ 10^6 person-years · WLM	Relative risk coefficient: (Observed/expected lung cancer cases)/WLM
Swedish iron[a] 1951–1976	1,294	19	0.036
Smokers		22	0.024
Nonsmokers		16	0.107
U.S. uranium[b] 1950–1974	3,356	6–9	0.008–0.011
Czech uranium[c] 1950–1975	2,530	18–21	0.023
Ontario uranium[d] 1959–1977	14,558	6–15	0.017–0.038
Newfoundland fluorspar[e]	2,414	18	0.024

[a]Radford and St. Clair Renard, 1984.
[b]BEIR, 1980.
[c]Sevc et al., 1976; Kunz et al., 1979.
[d]Thomas and McNeill, 1982; Muller et al., 1983.
[e]Thomas and McNeill, 1982.
SOURCE: Radford and St. Clair Renard, 1984.

Figures 2.9 and 2.10 show exposure-effect relationships for Czech uranium miners (Kunz et al., 1979) and Swedish iron miners (Radford and St. Clair Renard, 1984), respectively, the latter plot describing an effect even at relatively low levels. Table 2.20 summarizes various estimates of absolute and relative risk of exposure. Exposure is usually expressed as working-level months (WLM), the definition of which is given in Table 2.19. The number of excess cancer cases due to radon-daughter exposure range from 6 to 22 cancers/10^6 person-years · WLM.

REFERENCES

AAOO (American Academy of Otolarngology) (1979). Guide for the evaluation of hearing handicap. *Otolaryngol. Head Neck Surgery., 87,* 539–551.

ACGIH (American Conference of Governmental Industrial Hygienists) (1984). *Documentation of the Threshold Values.*

Amandus, H. E., Hanke, W., Kullman, G., and Reger, R. B. (1984). A re-evaluation of radiological evidence from a study of U.S. strip coal miners. *Arch. Environ. Health, 39,* 346–351.

Andjelkovic, D., Taulbee, J., and Symons, M. (1976). Mortality experience of a cohort of rubber workers, 1964–1973. *J. Occup. Med., 18,* 387–394.

Attfield, M., and Hudak, J. (1980). National coal study, prevalence of coal workers' pneumoconiosis: Comparison of first and second rounds. In: *Health Implications of*

New Energy Technologies (Rom, W. N., and Archer, V. E., eds.). Ann Arbor Science, Ann Arbor, Mich., pp. 203–212.
Axelson, O., and Sundell, L. (1978). Mining, lung cancer and smoking. *Scand. J. Work Environ. Health, 4,* 46–52.
Bass, A. (1984). Defining toxic exposure: A battle of semantics. *Technol. Review, 25,* 31, May–June.
Baughn, W. L. (1973). *Relation between Daily Noise Exposure and Hearing Loss.* Aerospace Medical Research Lab. AMRL-TR-73-53, June.
Baxter, P. J., and Werner, J. B. (1980). *Mortality in the British Rubber Industries 1967–76.* Her Majesty's Stationary Office, London.
Behrens, V., Taylor, W., and Wasserman, D. E. (1982). Vibration syndrome in workers using pneumatic chipping and grinding tools. In: *Vibration Effects on the Hand and Arm in Industry* (Brammer, A. J., and Taylor, W., eds.). Wiley, New York, pp. 147–156.
BEIR (Committee on Biological Effects of Ionizing Radiation) (1980). *The Effects on Populations of Exposures to Low Levels of Ionizing Radiation: 1980.* National Academy Press, Washington, D.C.
Berger, E. H., Register, L. H., and Thomas, W. G. (1978). Presumed noise-induced permanent threshold shift resulting from an exposure to an A-weighted L_{eq} of 89 dB. *J. Acoust. Soc. Amer., 64,* 192–197.
Berry, G., Molyneux, M. K. B., and Tombleson, J. B. L. (1974). Relationships between dust levels and byssinosis and bronchitis in Lancashire cotton mills. *Brit. J. Ind. Med., 31,* 18–27.
Berry, G., and Newhouse, M. L. (1983). Mortality of workers manufacturing friction materials using asbestos. *Brit. J. Ind. Med., 40,* 1–7.
BLS (1984). *Occupational Injuries and Illnesses in the United States by Industry, 1982.* Bureau of Labor Statistics, U.S. Dept. of Labor, Bulletin 2196, April.
BLS (1985a). *Occupational Injuries and Illnesses in the United States by Industry, 1983.* Bureau of Labor Statistics, U.S. Dept. of Labor, Bulletin 2236, June.
BLS (1985b). Occupational injuries and illnesses in 1984. *U.S. Dept. of Labor News,* 85-483, Bureau of Labor Statistics, November 13.
Bonhuys, A., Beck, G. J., and Schoenberg, J. B. (1979). Priorities in prevention of chronic lung diseases. *Lung, 156,* 129–148.
Bouhuys, A., Wolfson, R. L., Horner, D. W., Brain, J. D., and Zuskin, E. (1969). Byssinosis in cotton textile workers: Respiratory survey of a mill with rapid labor turnover. *Ann. Inter. Med., 71,* 257–269.
Brammer, A. J. (1982a). Relations between vibration exposure and the development of the vibration syndrome. In: *Vibration Effects on the Hand and Arm in Industry* (Brammer, A. J, and Taylor, W., eds.). Wiley, New York, pp. 283–290.
Brammer, A. J. (1982b). Threshold limit for hand-arm vibration exposure throughout the workday. In: *Vibration Effects on the Hand and Arm in Industry* (Brammer, A. J., and Taylor, W., eds.). Wiley, New York, pp. 291–301.
Braun, D. C., Jurgiel, J. A., Kaschak, M. C., and Babyak, M. A. (1973). Prevalence of respiratory signs and symptoms among U.S. cotton textile workers. *J. Occup. Med., 15,* 414–419.
Brubaker, R. L., Machenzie, C. J. G., Eng., P. R., and Bates, D. V. (1983). Vibration white finger disease among tree fellers in British Columbia. *J. Occup. Med., 25,* 403–408.
Burns, W., and Robinson, D. W. (1970). *Hearing and Loss in Industry.* Her Majesty's Stationery Office, London.
Cotter, D. M. (1985). Work-related deaths dropped sharply during 1983, BLS survey finds. *Monthly Labor Rev.,* September, pp. 41–44.
Delzell, E., Andjelkovich, D., and Tyroler, H. A. (1982). A case-control study of employment experience and lung cancer among rubber workers. *Amer. J. Ind. Med., 3,* 393–404.
Delzell, E., and Monson, R. R. (1981). Mortality among rubber workers: IV. General mortality patterns. *J. Occup. Med., 23,* 850–856.
Dement, J. M., Harris, R. L., Symons, M. J., and Shy, C. M. (1983). Exposure and

mortality among chrysotile asbestos workers. Part II: Mortality. *Amer. J. Ind. Med., 4,* 421–433.

Edling, C. (1982). Lung cancer and smoking in a group of iron ore miners. *Amer. J. Ind. Med., 3,* 191–199.

Edling, C., and Axelson, O. (1983). Quantitative aspects of radon daughter exposure and lung cancer in underground miners. *Brit. J. Ind. Med., 40,* 182–187.

EPA (1974). *Information on Levels of Environmental Noise Requisite to Protect Public Health and Welfare with an Adequate Margin of Safety.* U.S. EPA Report No. 550/9-74-004.

Fairman, R. P., O'Brien, R. J., Swecker, S., Amandus, H. E., and Shoub, E. P. (1977). Respiratory status of surface coal miners in the United States. *Arch. Environ. Health, 32,* 211–215.

Fine, L. J., and Peters, J. M. (1976a). Respiratory morbidity in rubber workers: II. Pulmonary function in curing workers. *Arch. Environ. Health, 31,* 10–14.

Fine, L. J., and Peters, J. M. (1976b). Studies of respiratory morbidity in rubber workers: III. Respiratory morbidity in processing workers. *Arch. Environ. Health, 31,* 136–140.

Fine, F. J., Peters, J. M., Burgess, W. A., and Berardinis, L. J. (1976). Studies of respiratory morbidity in rubber workers: IV. Respiratory morbidity in talc workers. *Arch. Environ. Health, 31,* 195–200.

Finkelstein, M. M. (1983). Mortality among long-term employees of an Ontario asbestos-cement factory. *Brit. J. Ind. Med., 40,* 138–144.

Fox, A. J., Goldblatt, P., and Kinlen, L. J. (1981). A study of the mortality of Cornish tin miners. *Brit. J. Ind. Med., 38,* 378–380.

Fox, A. J., Lindars, D. C., and Owen, R. (1974). A survey of occupational cancer in the rubber and cablemaking industries: Results of five-year analysis, 1967–71. *Brit. J. Ind. Med., 31,* 140–151.

Fox, A. J., Tombleson, B. L., Watt, A., and Wilkie, A. G. (1973). A survey of respiratory disease in cotton operatives. Part II. Symptom, dust estimation, and the effect of smoking habit. *Brit. J. Ind. Med., 30,* 48–53.

Griffin, M. J. (1982). Hand-arm vibration standards and dose-effect relationships. In: *Vibration Effects on the Hand and Arm in Industry* (Brammer, A. J., Taylor, W., eds.). Wiley, New York, pp. 259–268.

Henderson, V. L., and Enterline, P. E. (1979). Asbestos exposure: Factors associated with excess cancer and respiratory disease mortality. *Ann. N. Y. Acad. Sci., 330,* 117–126.

Higgins, I. T. T., Higgins, M. W., Lockshin, M. D., and Canale, N. (1968). Chronic respiratory disease in mining communities in Marion County, West Virginia. *Brit. J. Ind. Med., 25,* 165–175.

Hurley, J. F., Burns, J., Copland, L. Z., Dodgson, J., and Jacobsen, M. (1982). Coalworker's simple pneumoconiosis and exposure to dust at 10 British coal mines. *Brit. J. Ind. Med., 39,* 120–127.

Hyatt, R. E., Kistin, A. D., and Mahan, T. K. (1964). Respiratory disease in southern West Virginia coal miners. *Amer. Rev. Respir. Dis., 89,* 387–401.

Imbus, H. R., and Suh, M. W. (1973). Byssinosis: A study of 10,133 textile workers. *Arch. Environ. Health, 26,* 184–191.

ISO (International Organization for Standardization) (1979). *Principles for the Measurement and the Evaluation of Human Exposure to Vibration Transmitted to the Hand.* Draft International Standard ISO/DIS 5349.

Kenedi, I. A. (1948). Hoemunkasok szive. *Iparegeszseguegyi koezlemenyek, 1,* 125–145 (as referenced in F. N. Dukes-Dobos, Hazards of heat exposure, *Scand. J. Work Environ. Health, 7,* 73–83, 1981).

Kiev report on 7th Congress of International Ergonomics Association, Warsaw, Poland (1979) (as referenced in F. N. Dukes-Dobos, Hazards of heat exposure, *Scand. J. Work Environ. Health, 7,* 73–83, 1981).

Kloetzel, K., Etelvino de Andrade, A., Falleiros, J., and Pacheco, J. C. (1973). Relationship between hypertension and prolonged exposure to heat. *J. Occup. Med., 15,* 878–880.

Kryter, K. D. (1973). Impairment to hearing from exposure to noise. *J. Accoust. Soc. Amer., 53,* 1211–1234.

Kunz, E., Sevc, J., Placek, V., and Horacek, J. (1979). Lung cancer in man in relation to different time distribution of radiation exposure. *Health Phys., 36,* 699–706.

LaDou, J. (1984). The not-so-clean business of making chips. *Technol. Rev.* May–June, pp. 23, 24, 26–30, 32–36.

Lainhart, W. S., Doyle, H. M., Enterline, P. E., Henschel, A., and Kendrick, M. A. (1969). *Pneumoconiosis in Appalachian Bituminous Coal Miners,* pp. 77–111. U.S. Public Health Service, U.S. Dept. of Health, Education and Welfare, U. S. Government Printing Office, Washington, D.C.

Lapp, N. L., Block, J., Boehlecke, B., Lippmann, M., Morgan, W. K. C. and Reger, R. B. (1976). Closing volume in coal miners. *Am. Rev. Respir. Dis., 113,* 155–161.

Lapp, N. L., Hankinson, J. L., Burgess, D. B., and O'Brien, R. (1972). Changes in ventilatory function in coal miners after a work shift. *Arch. Environ. Health, 24,* 204–208.

Lednar, W. M., Tyroler, H. A., McMichael, A. J., and Shy, C. M. (1977). The occupational determinants of chronic disabling pulmonary disease in rubber workers. *J. Occup. Med., 19,* 263–268.

Lloyd, J. W., and Ciocco, A. (1969). Long-term mortality study of steelworkers. I. Methodology. *J. Occup. Med., 11,* 299–310.

Lloyd, J. W., Lundin, F. E., Jr., Redmond, C. K., and Geiser, P. B. (1970). Long-term mortality study of steelworkers. IV. Mortality by work area. *J. Occup. Med., 12,* 151–157.

Mausner, J. S., and Kramer, S. (1985). *Mausner and Bahn Epidemiology—An Introductory Text,* 2d ed. W. B. Saunders, Philadelphia.

Mayer-Theveniaud, N. (1977). Les trauvaux en temperture elevee. Doctoral Dissertation, Faculty of Medicine of Dijon, France (as referenced in F. N. Dukes-Dobos, Hazards of heat exposure, *Scand. J. Work Environ. Health, 7,* 73–83, 1981).

Mazumdar, S., Lerer, T., and Redmond, C. K. (1975). Long-term mortality study of steelworkers. IX. Mortality patterns among sheet and tin mill workers. *J. Occup. Med., 17,* 751–755.

McCaffrey, D. P. (1981). Work-related amputations by type and prevalence. *Monthly Labor Rev. 104,* 35–41.

McMichael, A. J. (1976). Standardized mortality ratios and the "healthy worker effect": Scratching beneath the surface. *J. Occup. Med., 18,* 165–168.

McMichael, A. J., Spirtas, R., and Kupper, L. L. (1974). An epidemiologic study of mortality within a cohort of rubber workers, 1964–72. *J. Occup. Med., 16,* 458–464.

Merchant, J. A., Kilburn, J. H., O'Fallon, W. M., Hamilton, J. D., and Lumsden, J. C. (1972). Byssinosis and chronic bronchitis among cotton textile workers. *Ann. Intern. Med., 76,* 423–433.

Merchant, J. A., Lumsden, J. C., Kilburn, K. H., O'Fallon, W. M., Ujda, J. R., Germino, V. H., and Hamilton, J. D. (1973). Dose response studies in cotton textile workers. *J. Occup. Med., 15,* 222–230.

Miura, T., Morioka, M., Kimura, K., and Akutu, A. (1959). On the occupational hazards by vibrating tools. Report 4. *J. Sci. Labour (Japan), 35,* 760–767.

Miyashita, K., Shiomi, S., Itoh, N., Kasamatsu, T., and Iwata, H. (1982). Development of the vibration syndrome among chain sawyers in relation to their total operation time. In: *Vibration Effects on the Hand and Arm in Industry* (Brammer, A. J., and Taylor, W., eds.). Wiley, New York, pp. 269–276.

Monson, R. R., and Fine, L. J. (1978). Cancer mortality and morbidity among rubber workers. *J. Natl. Cancer Inst., 61,* 1047–1053.

Monson, R. R., and Nakano, K. K. (1976). Mortality among rubber workers: I. White male union employees in Akron, Ohio. *Amer. J. Epidem., 103,* 284–296.

Morgan, W. K. C., Burgess, D. B., Jacobson, G., O'Brien, R. J., Pendegrass, E. P., Reger, R. B., and Shoub, E. P. (1973). The prevalence of coal workers' pneumoconiosis in U.S. coal miners. *Arch. Environ. Health, 27,* 221–226.

Morgan, W. K. C., Handelsman, L., Kibelstis, J., Lapp, N. L., and Reger, R. B. (1974). Ventilatory capacity and lung volumes of U.S. coal miners. *Arch. Environ. Health, 28,* 182–189.

Morgan, W. K. C., and Lapp, N. L. (1976). Respiratory disease in coal miners. *Amer. Rev. Respir. Des., 113,* 531–559.

Muller, J., Wheeler, W. C., Gentleman, J. F., Suranyi, G., and Kusiak, R. A. (1983). *Study of Mortality of Ontario Miners, 1955–1977.* Part I. Ontario Ministry of Labor, Toronto, May.

NCHS (1983). *Health, United States, 1983.* National Center for Health Statistics, Public Health Service DHHS Pub. No. (PHS) 84-1232, December.

Nicholson, W. J., Selikoff, I. J., Seidman, H., Lilis, R., and Formby, P. (1979). Long-term mortality experience of chrysotile miners and millers in Thetford mines, Quebec. *Ann. N. Y. Acad. Sci., 330,* 11–31.

NIOSH (1983a). Leading work related diseases and injuries. *Morbidity and Mortality Weekly Report, 32,* 24–26, 32, January 21.

NIOSH (1983b). Musculoskeletal injuries. *Morbidity and Mortality Weekly Report, 32,* 189–191, April 15.

NIOSH (1983c). Vibration syndrome. *Current Intelligence Bulletin 38.* U.S. Dept. of Health and Human Services, Public Health Service, Centers for Disease Control, National Institute for Occupational Safety and Health, March 29.

NIOSH (1984a). Occupational cancers (other than lung). *Morbidity and Mortality Weekly Report, 33,* 125–128, March 9.

NIOSH (1984b). Severe occupational traumatic injuries. *Morbidity and Mortality Weekly Report, 33,* 213–215, April 27.

NIOSH (1984c). Fatalities from occupational heat exposure. *Morbidity and Mortality Weekly Report, 33,* 410–412, July 20.

NSC (1983). *Accident Facts, 1983 ed.* National Safety Council, Chicago.

OSHA (1978). Occupational exposure to cotton dust. *Fed. Reg., 43,* 27350–27463, June 23.

OSHA (1981). Occupational noise exposure; hearing conservation amendment. *Fed. Reg., 46,* 4078–4179, January 16.

Passchier-Vermeer, W. (1968). *Hearing Loss Due to Exposure to Steady-State Broadband Noise.* Report No. 235, Institute for Public Health Engineering, The Netherlands.

Pelnar, P. V., Gibbs, G. W., and Pathak, B. P. (1982). A pilot investigation of the vibration syndrome in forestry workers of eastern Canada. In: *Vibration Effects on the Hand and Arm in Industry* (Brammer, A. J., and Taylor, W., eds.). Wiley, New York, pp. 173–188.

Pemberton, J. (1956). Chronic bronchitis, emphysema, and bronchial spasm in bituminous coal workers. *Arch. Indust. Health, 13,* 529–544.

Pyykko, I., Korhonen, O. S., Farkkila, M. A., Starck, J. P., and Aatola, S. A. (1982). A longitudinal study of the vibration syndrome in Finnish forestry workers. In: *Vibration Effects on the Hand and Arm in Industry* (Brammer, A. J., and Taylor, W. eds.). Wiley, New York, pp. 157–168.

Radford, E. P., and St. Clair Renard, K. G. (1984). Lung cancer in Swedish iron miners exposed to low doses of radon daughters. *N. Engl. J. Med., 310,* 1485–1494.

Redmond, C. K., Ciocco, A., Lloyd, J. W., and Rush, H. W. (1972). Long-term mortality study of steelworkers. VI. Mortality from malignant neoplasms among coke oven workers. *J. Occup. Med., 14,* 621–629.

Redmond, C. K., Emes, J. J., Mazumdar, S., Magee, P. C., and Kamon, E. (1979). Mortality of steelworkers employed in hot jobs. *J. Environ. Pathol. Toxic., 2,* 75–96.

Redmond, C. K., Smith, E. M., Lloyd, J. W., and Rush, H. W. (1969). Long-term mortality study of steelworkers. III. Follow-up *J. Occup. Med., 11,* 513–521.

Redmond, C. K., Wieand, H. S., Rockette, H. E., Sass, R., and Weinberg, G. (1981). *Long-Term Mortality of Steelworkers,* NIOSH Report No. 81–120, June.

Reger, R., and Hancock, J. (1980). Coal miners exposed to diesel-exhaust emissions. In: *Health Implications of New Energy Technologies* (Rom, W. N., and Archer, V. E., eds.). Ann Arbor Science, Ann Arbor, Mich., pp. 213–231.

Riddle, H. F. V., and Taylor, W. (1982). Vibration-induced white finger among chain sawyers nine years after the introduction of anti-vibration measures. In: *Vibration Effects on the Hand and Arm in Industry* (Brammer, A. J., and Taylor, W., eds). Wiley, New York, pp. 169–172.

Rockette, H. E., and Redmond, C. K. (1976). Long-term mortality study of steelworkers. X. Mortality patterns among masons. *J. Occup. Med., 18,* 541–545.

Rogan, J. M., Ahfield, M., Jacobsen, M., Rae, S., Walker, D. D., and Walton, W. H. (1973). Role of dust in the working environment in development of chronic bronchitis in British coal miners. *Brit. J. Ind. Med., 30,* 217–226.

Rom, W., Barkman, H., Turner, W., Kanner, R., Wright, W., Nichols, M., and Renzetti, A. (1980). Coal workers respiratory disease in Utah: A preliminary report. In: *Health Implications of New Energy Technologies* (Rom, W. N., and Archer, V. E., eds.). Ann Arbor Science, Ann Arbor, Mich., pp. 247–256.

Samet, J. M., Kutvirt, D. M., Waxweiler, R. J., and Key, C. R. (1984). Uranium mining and lung cancer in Navajo men. *N. Engl. J . Med., 310,* 1481–1484.

Schlesselman, J. J. (1982). *Case-Control Studies.* Oxford University Press, New York.

Schrag, P. E., and Gullett, A. D. (1970). Byssinosis in cotton textile mills. *Amer. Rev. Respir. Dis., 101,* 497–503.

Selikoff, I. J., Hammond, E. C., and Seidman, H. (1979). Mortality experience of insulation workers in the United States and Canada, 1943–1976. *Ann. N. Y. Acad. Sci., 330,* 91–116.

Sevc, J., Kunz, E., and Placek, V. (1976). Lung cancer in uranium miners and long-term exposure to radon daughter products. *Health Phys. 30,* 433–437.

Taylor, W., and Pelmear, P. L. (1976). Raynaud's phenomenon of occupational origin; an epidemiological survey. *Acta Chir. Scand.* (Suppl.), *465,* 27–32.

Taylor, W., Pearson, J. C. G., and Keighley, G. D. (1977). A longitudinal study of Raynaud's phenomenon in chain saw operations. In: *Proceedings of the International Occupations Hand-arm Vibration Conference* (Wasserman, D. E., Taylor, W., and Curry, M. G., eds.) Dept. of Health and Human Services, NIOSH publication 77–170, Cincinnati, Ohio, pp. 69–76.

Thomas, D. C., and McNeill, K. G. (1982). *Risk Estimates for the Health Effects of Alpha Radiation.* Report to the Candian Atomic Energy Control Board, Ottawa.

Ward, W. D. (1975). Acoustic trauma and noise induced hearing loss. In: *The Nervous System,* vol. 3, *Human Communication and Its Disorders* (Tower, D. B., ed.). Raven Press, New York.

Wasserman, E. E., Badger, D. W., Doyle, T. E., and Margolies, L. (1974). Industrial vibration—an overview. *Amer. Soc. Safety Eng. J., 19,* 38–43.

Wasserman, D. E., Behrens, V., Taylor, W., Samueloff, S., and Reynolds, D. (1982). *Vibration White Finger Disease in U.S. Workers Using Pneumatic Chipping and Grinding Handtools, Volume I: Epidemiology Results,* DHHS (NIOSH) Publication No. 82–118. U.S. Dept. of Health and Human Services, Public Health Service, Centers for Disease Control, National Institute for Occupational Safety and Health, Cincinnati, Ohio.

Wen, C. P., Tsai, S. P., and Gibson, R. L. (1983). Anatomy of the healthy worker effect: A critical review. *J. Occup. Med., 25,* 283–289.

Wolf, P. H., Andjelkovich, D., Smith, A., and Tyroler, H. (1981). A case-control study of leukemia in the U.S. rubber industry. *J. Occup. Med., 23,* 103–108.

Yerg, R. A., Sataloff, J., Glorig, A,. and Menduke, H. (1978). Inter-industrial noise study. *J. Occup. Med., 20,* 351–358.

Zuskin, E., Valic, F., Butkovic, D., and Bouhuys, A. (1975). Lung function in textile workers. *Brit. J. Ind. Med., 32,* 283–288.

Zuskin, E., Wolfson, R. L., Harpel, G., Welborn, J. W., and Bouhuys, A. (1969). Byssinosis in carding and spinning workers. *Arch. Environ. Health, 19,* 666–675.

Chapter

3

Hazard Evaluation and Control

R. A. Wadden

J. E. Franke

I. Evaluation of Risk

The first step in determining appropriate engineering design to prevent workplace hazards is the evaluation of the health or safety risk. This ordinarily will be carried out with reference to a standard or guideline, or relative to some other motivating factor such as the possibility of legal action. Often the specific health risk is difficult (or impossible) to quantitate. In some cases, however, change in biological function can be related to workplace exposure, and exposure is used in general practice as a surrogate for risk. Once an acceptable level of exposure has been determined, the required level of control can be specified to restrict release of hazardous material or energy from a source. Appropriate design solutions can then be developed to meet this control stipulation. The following discussion outlines some of the pertinent considerations for evaluation of risk.

A. Dose-effect relationships

Although health-related standards or guidelines exist for many materials (see Chapter 4), it is not unusual to encounter a substance that is suspected to be hazardous but for which no design criteria are available. Under these conditions it is necessary to estimate a safe environmental concentration, either by chemical or biochemical analogy to similar compounds or by carrying out or evaluating studies that compare dosage to health end point. For some materials, where the effects on health are known to be completely reversible, controlled exposure of human subjects may be possible. Ordinarily, however, evaluation of toxicity will involve exposure of test animals to controlled conditions. The results of animal studies must then be extrapolated to humans, a procedure fraught with uncertainty.

Even for materials with standards, it is useful to have dose-effect information for design of controls. Such relationships give the designer a basis for judging the flexibility of a control solution. A case in point is the hearing loss curves given in Chapter 2 for exposure to various frequencies and noise intensities. The curves provide a basis for an alternative that meets a 90-dB guideline (and also an 80-dB guideline if this level becomes a control requirement in the future).

Studies that involve workplace exposures and attendent health effects are closer to reflecting actual conditions. Epidemiological studies of such data, of the type summarized in Chapter 2, are a way to determine possible cause-effect relationships. These may be retrospective or prospective in nature. In retrospective studies, persons are chosen with and without a particular disease condition, and are compared on the bases of exposure history. The prospective approach involves comparison of all individuals without the disease but with known exposures. This cohort is then followed until illness does occur. Retrospective analyses are usually cheaper to carry out and deal with a fixed population, but they ordinarily suffer from inadequate exposure information, biased recall, and the difficulty of defining the whole cohort that has undergone exposure. Prospective studies do not have exposure bias but are more expensive, must deal with attrition of the subject population, and have long follow-up times.

Often only a few cases of severe ill health will occur in a given setting, and extensive epidemiological evaluation is not possible. The recourse is to reconstruct the exposure conditions that apparently led to the health effects.

B. Measures of effect on people

A variety of measures can be used to describe the health effect of the dose-effect relationship. Most pollutants will cause a continuum of biological effects, starting with slight disruption of biochemical cycles at low concentrations and proceeding to morbidity and ultimately, at high enough exposures, to death. Mortality and morbidity constitute clinical measures of effect. Other effects may be subclinical but still represent incorporation of a particular chemical into the body. Blood concentrations of 4% carboxyhemoglobin and 25 μg of lead/100 ml are ordinarily not considered to be health-threatening, but nonetheless do reflect environmental exposures.

In general, one or more specific organ systems will be affected by exposure to a particular material. Consequently, liver dysfunction will initially be of most interest with respect to solvent exposures; and sulfur dioxide will primarily attack the lung. Measures of effect will ordinarily be related to the target organ that is most likely to be affected.

C. Characterization of exposure

The characterization of dose in the ideal dose-effect relationship requires a measurement or estimation of worker exposure to a pollutant or other environmental stressor. Some of the questions that need to be answered to supply exposure information include: (1) What materials are being discharged? (2) At what environmental levels do they exist and for how long? (3) Who is likely to be most exposed? The first question can be answered through a process analysis, the second through personal or area sampling and biological monitoring, and the third through observation of worker activity patterns, job definitions, and human-machine interactions.

1. Process analysis. The first step in evaluating exposure is a process evaluation. This may be carried out from process design and operating data, preferably in conjunction with a process walk-through. Some important considerations include: (1) Is process operation batch or continuous? (2) What are the raw materials, products, and by-products? (3) What chemical and physical reactions are involved? (4) What materials storage and handling procedures are used? (5) How is energy generated, used, and transferred? (6) How is the process actually operated, as opposed to how it was designed? (7) What changes in operation occur during start-up, shut-down, and upset? (8) How are process wastes disposed of?

2. Environmental measurement. A specific measure of individual worker exposure can be gained from personal sampling. Although either instantaneous or integrated measurements are possible, most personal devices are of the integrative type. The film badge for ionizing radiation, passive dosimeter for NO_2 and formaldehyde determinations, and some noise meters all supply integrated exposures. Other devices that contain small pumps are also used for collection of some chemicals and airborne particulate matter. Most personal air sampling procedures attempt to collect air that is representative of the breathing zone.

Table 3.1 is an introduction to sampling characteristics and availability of some commercial passive sampling systems. Because these samplers have no pumps, they are very compact and lightweight. Mass collection sensitivity and accuracy were determined in laboratory experiments where interfering substances and sampling variables were minimized. When using passive dosimeters in the field, side-by-side sampling with a standard method at one or more locations should be included to ensure that interferences and sampling variables on-site are not significant. The technology of many passive sampling systems is new and untested. Products have been marketed and then withdrawn, sometimes for technical reasons such as an unanticipated interference, a short shelf life, or a short storage life after use.

One of the requirements of a personal monitor is that it be small and lightweight so as not to interfere with worker activity. Since many available devices do not conform to this requirement, personal exposure detection is ordinarily carried on over limited periods rather than as a routine procedure.

A complementary approach to workplace monitoring is area sampling. More sophisticated and bulkier devices can be used, and monitoring is more easily carried out over extended time intervals. However, in order to extend these monitoring data to estimate personal exposures, it is necessary to characterize worker activity patterns accurately with respect to the sources of pollutants and other workplace stressors.

Commonly used general area survey meters are described in Table 3.2. Many models of these meters are small enough, or are equipped with sampling probes, so that a worker's breathing zone may be sampled for short time periods. Meters frequently have more than one concentration range. Response times and sensitivities vary with manufacturer. Nevertheless, Table 3.2 provides a general introduction to what the common survey meter classes can and cannot do. Interferences may alter the results of most survey meters. For example, a carbon monoxide monitor with electrochemical sensor reads 1 ppm too high when methane is present at a 5 ppm concentration (Nader and Lauderdale, 1983). A formaldehyde meter reads higher when exposed to substances such as methanol, ethanol, phenol, and formic acid. Some meters, such as detector tubes or piezoelectric meters, are really short-term batch samplers. Other devices (infared photometers, electrochemical sensors) operate continuously but must be zeroed, spanned, and calibrated periodically. Meters calibrated at the factory with a standard vapor or aerosol (aerosol photometers, photoionization detectors) should also be calibrated in the field by the users.

Several examples illustrate the differences between the results of personal and area sampling. Figure 3.1 shows the average air exposure of various workers to vinyl chloride monomer in a polyvinyl chloride (PVC) polymerization process (Enviro Control, 1978). The concentrations are based on 8-hr collection with charcoal tubes and personnel samples taken once per month. The effect of engineering controls instituted in late 1975 is reflected in most of the exposures. However, the dispersion area foreman and reactor cleaner still have levels exceeding the PEL of 1 ppm (see Table 4.2). These values are in contrast to the area samples shown in Figure 1.3 for the same plant. The Figure 1.3 levels are averages of gas chromatograph analyses taken every 15 min. None of the area locations monitored, which were reported as being ordinarily accessible, showed concentrations that would reflect the foreman or reactor cleaner exposures.

Figure 3.2 shows the operator noise exposure (97 dBA) resulting from a bank of 15 centrifuges in a blood plasma production room (Jensen et al., 1978). Close-in monitoring at various locations was also carried out to determine which components were the major noise sources. The motor exhaust and pulley guard surface contributed the maximum sound pressure levels in the 500–4000 Hz octave bands. And sound pressures near the pulley guard surface, which was highly reflective, were actually also caused by the motor exhaust. (The exhaust was subsequently fitted with a muffler.) The levels shown in Figure 3.2

(*Text continues on page 96.*)

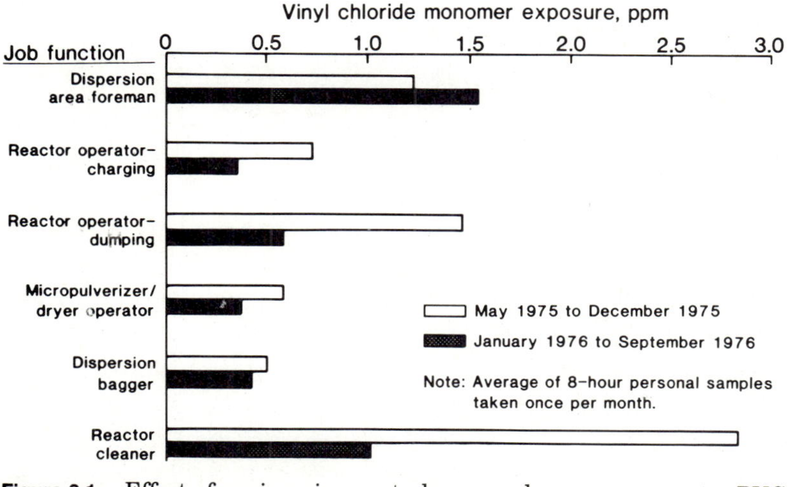

Figure 3.1 Effect of engineering controls on employee exposure at a PVC dispersion polymerization process (*Enviro Control, 1978*).

TABLE 3.1 Some Commercially Available Passive Sampling Systems

Components measured	Mass collection sensitivity	Accuracy	References
Organic vapors such as:			Lautenberger et al., 1980; EPA, 1981; Bamberger et al., 1978; Mazur et al., 1981; Gillespie and Daniel, 1979; Hickey and Bishop, 1981
Alkyl lead	5 $\mu g/m^3 \cdot$ 8 hr		
Benzene	0.02–0.2 ppm · 8 hr		
Carbon tetrachloride	0.1 ppm · 8 hr	9.2%	
Chlorobenzene	0.15 ppm · 8 hr	9.2%	
Chloroform	22.5 ppb · 24 hr	9.2%	
1,2-Dichloroethane	1.7 ppm · 24 hr	9.2%	
Methylchloroform	21 ppm · hr		
Perchloroethylene	0.1 ppm · 8 hr	9.2%	
Trichloroethylene	11 ppm · hr		
Vinyl chloride	0.02 ppm · 8 hr		
Formaldehyde	0.77–2 ppm · hr	13%	Balmat and Meadows, 1985; Kring et al., 1982; Wallace and Ott, 1982
Ethylene oxide	2 ppm · hr	4.3–36%	Kring et al., 1985; Mullins, 1985
Hydrogen cyanide	0.01–10 ppm · 8 hr		MDA, 1981
Phosgene	0.01–0.1 ppm · 8 hr		EPA, 1981
Ammonia	0.4–10 ppm · 8 hr	23%	American Gas and Chemical, 1981; McKee and McConnaughey, 1985; West, 1980
Chlorine	0.013–2 ppm · 8 hr		EPA, 1981; Hardy et al., 1979
Carbon monoxide	50 ppm · 8 hr	21%	American Gas and Chemical, 1981; McConnaughey et al., 1985
Hydrazine	>5 ppm · 8 hr		EPA, 1981

Hydrogen fluoride	0.1 ppm · 4 hr	17.4%	Young and Monat, 1982
Hydrogen sulfide	0.01–10 ppm · 8 hr	16%	Hardy et al., 1981 Kring et al., 1984
Nitrogen dioxide	1–40 ppm · hr		E. I. DuPont, 1981; EPA, 1981; Palmes et al., 1976
Nitrous oxide	<1.3 ppm · 1.4 hr	30.5%	Bishop and Houssain, 1984
Ozone	1 ppm · 8 hr		EPA, 1981
Sulfur dioxide	0.01–0.1 ppm · 8 hr	7–60%	Cassinelli et al., 1985; West and Reiszner, 1979
Gamma-, x-, and beta-radiation:			Wilkening, 1978; Cember, 1983
Pocket dosimeter	Varies; detection ranges are 0–200 mR and 0–200 R	15% for 0.05–2 MeV energies	
Film badge	20 keV (x-rays); 200 keV (beta)	Varies with film quality, radiation energy, environmental factors	
Thermoluminescent device	Varies; detection ranges from 10 mrad to 100,000 rad for 0.1- to 1.3-MeV energies	Better than film badge for >10 R exposure	
Nonionizing radiation	None currently available		
Heat stress	None currently available		
Continuous noise	1 dBA	±3 dB	

TABLE 3.2 General Area Survey Meters[a]

Meter	Typical contaminants measured	Selected range	Response time	Sensitivity
Catalytic combustion	Combustible gases and vapors	25–1000 ppm	2 sec or more	25 ppm (methane)
Flame ionization	Total hydrocarbons	1–100,000 ppm	2–5 sec	1–20 ppm (methane)
Photoionization detector	Organic vapors (nonspecific)	0.1–20 ppm	2 sec	0.1 ppm (benzene)
Halide meter	Trichloroethylene	10–500 ppm		
Detector tubes	Ammonia	25–700 ppm	4 min	5 ppm
	Benzene	5–40 ppm	1–6 min	1 ppm
	Carbon monoxide	5–150 ppm	4 min	10 ppm
	Chlorine	0.3–5 ppm	3 min	0.3 ppm
	n-Hexane	100–3000 ppm	3 min	100 ppm
	Hydrogen sulfide	1–20 ppm	4 min	1 ppm
	Nitrogen dioxide	0.5–25 ppm	2 min	0.5 ppm
	Trichloroethylene	10–500 ppm	2 min	10 ppm
Portable gas chromatograph[b]	Ammonia	170 ppm/ml	—	
	Benzene	89 ppm/ml	—	
	Carbon monoxide	22 ppm/ml	—	
	Chlorine	20 ppm/ml	—	1 ppm
	Hydrogen sulfide	20 ppm/ml	—	
	Nitrogen dioxide	76 ppm/ml	—	
Infrared spectrometer[b]	Ammonia	0.2 ppm		
	Benzene			
	Carbon monoxide	0.2 ppm		
	n-Hexane	0.2 ppm		
	Sulfur dioxide	0.5 ppm	1 min	
	Trichloroethylene	0.2 ppm		
	Carbon dioxide	5 ppm		

Ultraviolet photometer	Mercury	0.02–0.2 mg/m^3		
	Trichloroethylene			
Electrochemical oxidizing sensor	Carbon monoxide	1–100 ppm	25 sec	1 ppm
	Chlorine	0.05–2 ppm	1 min	0.04 ppm
	Hydrogen sulfide	0.5–20 ppm	1 min	0.4 ppm
	Nitrogen dioxide	0.04–2 ppm	30 sec	0.04 ppm
	Formaldehyde	0.3–10 ppm	20 sec	0.1 ppm
Colorimetric analyzer	Ammonia, chlorine, hydrogen sulfide, nitrogen dioxide, sulfur dioxide	0.025–0.25 ppm (SO_2)	4 min	0.025 ppm (SO_2)
Coulometry	Total oxidants	0.02–1 ppm	1 min	0.06 ppm
Chemiluminescense	Ozone	0.01–1 ppm	<5 sec	0.02 ppm
Aerosol photometer	Respirable dust	0.01–20 mg/m^3	6 sec	0.01 mg/m^3
	Fibrous dust	0.1–30 fibers/cm^3	1 min	0.1 fiber/cm^3
Beta-attenuation	Respirable dust	0.2–50 mg/m^3	4 min	0.02 mg/m^3
Piezoelectric meter	Respirable dust	0.1–10 mg/m^3	2 min	0.1 mg/m^3
Geiger-Muller survey meter	x-, gamma-, and beta-rays	0.2–2 mR/hr	2–10 sec	5000 counts per mR (gamma)
Cutie pie	x-, gamma-, and beta-rays	0.1–3 mR/hr for 7 keV–1.3 MeV (gamma)	<1 sec	0.1 mR/hr
Neutron survey meter	Thermal neutrons	0–12,000 neutrons/cm^2	15 sec	100 counts
Photovoltaic meter	Ultraviolet radiation	20–6000 μW/cm^2	<1 sec	20 μW/cm^2
Electromagnetic leakage detector	Microwave radiation	100–20,000 μW/cm^2	0.5–3 sec	Difficult calibration
Sound-level meter	Noise	35–150 dBA	$\frac{1}{8}$–1 sec	±1 dB (type 1 meter) ±3 dB (type 2 meter)

[a]Information extracted from equipment manuals, advertisements, Wilkening, 1978, and Nader and Lauderdale, 1983.

[b]Only the lower detectable limit is listed in the selected range.

indicate that worker exposure will be a function of how much time is spent at various locations in the vicinity of a pollution source.

3. Biochemical indices. Although environmental monitoring is the major method of characterizing exposures, biological measurements can serve as useful concomitants. While measurement of dysfunction of target organs is the most direct reflection of exposure conditions, this is usually not feasible, and biological measurements are carried out on more accessible body systems and materials. The most useful measures from a monitoring standpoint are those that are noninvasive: exhaled air and urine. Blood samples are of use in evaluation but frequent collection is not practical. Collection of composite urine samples over 24-hr periods is also difficult in the industrial setting. Analysis of perspir-

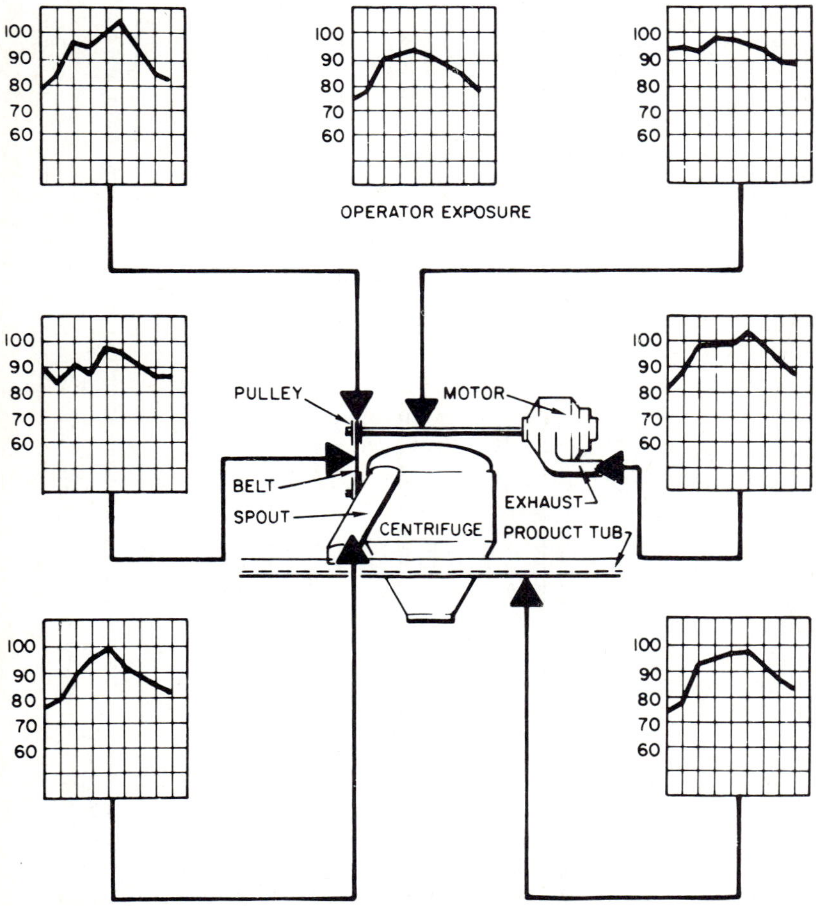

Figure 3.2 Operator noise exposure (dB) and close-in diagnostic levels from blood plasma centrifuges at various locations (*Jensen et al., 1978*). (Divisions on the horizontal scale correspond to octave band center frequencies of 31.5, 63, 125, 250, 500, 1000, 2000, 4000, and 8000 Hz.)

ation, tears, nails, milk, saliva, and feces have also been suggested for biological monitoring but have been of limited use.

Some materials change identity upon incorporation into the body. Their detection in body fluids then requires use of a surrogate measure such as a metabolite. Table 3.3 shows a number of air exposure-urine concentration relationships for organic chemicals. The predictions from such equations need to be interpreted in terms of typical background levels and intersubject variability of the metabolites, the analytical method used, and the specificity of the air pollutant-metabolite relationship (Waritz, 1979).

The relationship between CO and carboxyhemoglobin (COHb) is much better defined (Coburn et al., 1965) and is given by the equation:

$$\frac{\alpha[\mathrm{COHb}]_t - \beta\dot{V}_{\mathrm{CO}} - P_1(\mathrm{CO})}{\alpha[\mathrm{COHb}]_0 - \beta\dot{V}_{\mathrm{CO}} - P_1(\mathrm{CO})} = \exp(-\alpha t/\beta V_B) \qquad (3.1)$$

where

t = duration of exposure, min
$\alpha = \overline{p}c(O_2)/M[O_2Hb]'$
$\beta = 1/D_L + P_L/\dot{V}_A$
V_B = blood volume, ml
$[\mathrm{COHb}]_t$ = concentration of COHb in blood at time t, ml/ml
(COHb%) = concentration of COHb expressed as percent of saturation = $[\mathrm{COHb}]_t \cdot 100/[(1.38\ \mathrm{ml}\ O_2/\mathrm{g\ Hb}) \cdot m_{\mathrm{Hb}}]$
$[\mathrm{COHb}]_0$ = initial concentration of COHb in blood, ml/ml
M = Haldane constant, ratio of the affinity of Hb for CO to that for O_2 (220 for pH 7.4)
m_{Hb} = hemoglobin in blood concentration, g Hb/ml blood
D_L = diffusion capacity of the lungs for CO, ml/min · mmHg
P_b = barometric pressure, mmHg
$\dot{V}_A$ = alveolar ventilation rate, ml/min
$P_1(\mathrm{CO})$ = inspired carbon monoxide partial pressure, mmHg
$\dot{V}_{\mathrm{CO}}$ = endogenous carbon monoxide production rate, ml/min
$\overline{p}c(O_2)$ = mean pulmonary capillary oxygen pressure, mmHg
$[O_2Hb]'$ = concentration of oxyhemoglobin, ml/ml = $(1.38)(m_{\mathrm{Hb}}) - [\mathrm{COHb}]_t$, ml/ml
$P_L = P_b$ − partial pressure of water vapor at body temperature mmHg, (47 mmHg at 37°C)

Equation (3.1) is applicable to both air exposures and exhaled air.

Example Electrical parts are manufactured in a tight building that has no mechanical ventilation. Propane-fueled forklift trucks routinely drive from the loading docks, through a punch press area, to a storage room. The lift trucks are infrequently maintained and consequently emit large amounts of carbon monoxide. Over 4 hr, a punch press operator's average exposure is 70 ppm CO. What is his carboxyhemoglobin level at this time? (*Note:* Assume that P_b = 750 mmHg.)

Solution appears on pages 100–101.

TABLE 3.3 Relationships between Index Chemicals in Urine and Air Exposures to Various Organics

Exposure chemical	Index chemical	Equation[a]
Methyl chloroform (1,1,1-trichloroethane)	Total trichloro compounds	$Y = 0.27X + 0.54$
		$= 0.18X + 0.29$[h]
	1,1,1-Trichloroethanol	$= 0.19X + 0.27$
	Trichloroacetic acid	$= 0.073X + 0.21$
Trichloroethylene	Total trichloro compounds	$Y = 8.37X + 17.12$[i]
	1,1,1-Trichloroethanol	$= 5.19X + 12.28$
	Trichloroacetic acid	$= 3.17X + 4.84$
Trichloroethylene	Total trichloro compounds	$Y = 7.25X + 5.5$
		$= 4.97 + 1.9$[h]
		$Y = 6X(10^{-kt})$[j]
	1,1,1-Trichloroethanol	$= 5.57X + 4.4$
	Trichloracetic acid	$= 2.74X + 0.7$
Benzene	Phenol	$Y = 19X + 30$[b,h]
Phenol	Phenol	$Y = 108X + 28$[b]
		$= 80X + 40$[b,h]
Toluene	Hippuric acid	$Y = 23X + 800$[b]
		$= 16X + 600$[b,h]
		$Z = 18X + 350$
Toluene	Hippuric acid	$Y = 40X + 550$[b,i]
		$Z = 24X + 500$
Toluene	Hippuric acid	$Y = 9.8X + 400$[b]
Toluene	Hippuric acid	$Y = 31X + 400$[b]
		$= 25X + 300$[b,i]
m-Xylene	*m*-Methylhippuric acid	$Y = 27X$[b]
		$= 26X$[b,i]
Styrene	Mandelic acid	$Y = 18.4X + 149$[d,i]
	Mandelic acid	$= -14.0X + 6125$[e,i]
	Phenylglyoxylic acid	$= 2.7X + 79$[d,i]
	Phenylglyoxylic acid	$= -1.0X + 540$[e,i]
	Phenylglyoxylic acid	$Y = 12.5X + 101$[d,h]
	Mandelic acid	$Y = 4X + 80$[d,h]
	Styrene	$= 0.0124X$

[a] Y = concentration of index chemical in urine, mg/liter of urine; X = air concentration of exposure chemical, ppm; Z = concentration of index chemical in urine in mg/liter of creatinine.

[b] Calculated by R. S. Waritz from investigators' data.

[c] Calculated by investigators.

[d] Exposures <150 ppm.

[e] Exposures >150 ppm.

Maximum atmospheric concentration evaluated	Urinary concentration (mg/liter) calculated at given exposure ppm	Standard deviation (ppm) at given exposure ppm	Urine sample type	Reference
50 ppm · 8 hr/day · 3 days	14 at 50	~20 at 50[b]	Spot; after 3 days of exposure	Seki et al., 1975
	9 at 50	~15 at 50[b]		
	10 at 50	~20 at 50[b]		
	4 at 50	~25 at 50[b]		
40 ppm · 8 hr/day · $2\frac{1}{2}$ days	854 at 100[c]	—	Spot; after $2\frac{1}{2}$ days of exposure	Ogata et al., 1971
	531 at 100[c]			
	322 at 100[c]			
175 ppm · 8 hr/day · 3 days	730 at 100	~40 at 100[b]	Spot; after 3 days of exposure	Ikeda et al., 1972
	500 at 100	~20 at 100[b]		Nomiyama, 1971
	575 at 100	~40 at 100[b]		
	210 at 100	—		
150 ppm · 8 hr	120 at 50[b]		Spot; at end of workday	Rainsford and Davies, 1965
3.5 ppm · 8 hr	568 at 5[b]	~1 at 3.5[b]	Spot; at end of workday	Ohtsuji and Ikeda, 1972
	440 at 5[b]	~1 at 3.5[b]		
240 ppm · 6 hr	3100 at 100[b]	~45 at 100[b]	Spot; near end of workday	Ikeda and Ohtsuji, 1969
	2200 at 100[b]	~45 at 100[b]		
170 ppm · 8 hr	4550 at 100[b]	—	Spot; at end of workday	Pagnotto and Lieberman, 1967[g]
600 ppm · 8 hr	1380 at 100[b]	—	24 hr and end of day	Van Oettingen et al., 1942
800 ppm · 6 hr		—		
200 ppm · 7 hr	3100 at 100[b]	~10 at 100[b]	Spot; at end of exposure	Ogata et al., 1970
	2500 at 100[b]	~15 at 100[b]		
200 ppm · 7 hr	2700 at 100	~55 at 100[b]		
	2600 at 100	~30 at 100[b]		
290 ppm · 8 hr with occasional excursions of 1500 ppm · 10 min	1000 at 50[c]	—	Spot; at end of workday	Gotell et al., 1972
	2000 at 100[c]			
	—			
	—			
—	—	—	—	Ikeda et al., 1982
—	—	—	—	
80 ppm · 6 hr	1.0 at 80			Ramsey et al., 1980

[g]Includes female workers, but results for females not separated by investigators.
[h]Corrected for specific gravity = 1.016.
[i]Corrected for specific gravity = 1.024.
[j]k = 0.23 day^{-1} for males, 0.20 day^{-1} for females; t = time after last 8-hr exposure, days.

SOURCE: Revised from Waritz, 1979.

TABLE 3.4 Typical Values of Some Physiological Parameters

	Typical values (extreme values) [95%]	
	Men	Women
Blood volume, ml/kg body weight (V_B)	69	62
Diffusion capacity of the lung, ml min^{-1} mmHg^{-1} (D_L)	22.4(10.5–45)	
Alveolar ventilation rate, liters/min ($\dot{V}_A$)	6.4[1.9–10.9]	
Mean pulmonary capillary oxygen pressure, mmHg [$pc(O_2)$]	100(95–105)	
Hemoglobin in blood concentration, g Hb/100 ml blood ($m_{Hb} \times 100$)	15.85[13.4–17.3]	
Endogenous CO production rate, ml/min ($\dot{V}_{CO}$)	0.007	
Endogenous COHb% levels, normal	0.4–0.7	
hemolytic anemia	4–6	
Smoker COHb% levels	5–20	
Creatinine excretion, mg/kg body weight/day	1.8[1.1–2.5]	1.2[1.1–1.3]
Urine excretion, liters/day	1.02(0.51–2.0)	0.99(0.50–1.88)
specific gravity	(1.010–1.025)	
Diffusion capacity of the lung for CO ml/min · mmHg (D_L)	22.4(10.5–45)	

SOURCES: Diem and Lentner, 1972; ACGIH, 1984.

solution Typical values of the physiological parameters are (Table 3.4):

V_B = 5500 ml for a 175-lb man

$\dot{V}_A$ = 6000 ml/min

$\bar{p}c(O_2)$ = 100 mmHg

m_{Hb} = 0.16 g Hb/ml blood

D_L = 30 ml/min · mmHg

$\dot{V}_{CO}$ = 0.007 ml/min

$(COHb\%)_0$ = 0.8% (nonsmoker) and $[COHb]_0 = 1.77 \times 10^{-3}$ ml/ml

Referring to the above definitions,

$$[O_2Hb]' = (1.38)(0.16) - [COHb]_t$$

$$M = 218$$

$$\alpha = \frac{\bar{p}c(O_2)}{M[O_2Hb]'} = \frac{0.4587}{0.2208 - [COHb]_t} \text{ mm/ml/ml}$$

$$P_L = 750 - 47 = 703 \text{ mmHg}$$

$$\beta = \frac{1}{D_L} + \frac{P_L}{\dot{V}_A} = \frac{1}{30} + \frac{703}{6000} = 0.150 \text{ min} \cdot \text{mm/ml}$$

$$P_1(CO) = (70 \times 10^{-6} \text{ mmHg CO/mmHg air})(750 \text{ mmHg})$$
$$= 0.0525 \text{ mmHg CO}$$

Substituting in Eq. (3.1) gives:

$$\frac{[(0.4587)[\text{COHb}]_t/(0.2208 - [\text{COHb}]_t)] - 0.0011 - 5.25 \times 10^{-2}}{[(0.4587)[\text{COHb}]_0/(0.2208 - [\text{COHb}]_t)] - 0.0011 - 5.25 \times 10^{-2}}$$
$$= \exp\{-(5.54 \times 10^{-4})(t)/(0.2208 - [\text{COHb}]_t)\}$$

For $[\text{COHb}]_0 = 1.77 \times 10^{-3}$ ml/ml and $t = 240$ min, the trial-and-error solution will be:

$$[\text{COHb}]_t = 0.0124 \text{ ml/ml} \quad \text{or} \quad 5.62 \text{ COHb\%}$$

Table 3.5 lists tentative empirical equations for methylene chloride, styrene, tetrachloroethylene, and trichloroethylene that describe the concentration in breath air, C_t, as a function of time after termination of exposure, t, given an initial exposure to concentration C_0, for a given time. The equation has the general form:

$$\frac{C}{C_0} = K_A\exp(-\alpha t) + K_B\exp(-\beta t) \tag{3.2}$$

where C/C_0 is the ratio of concentration in exhaled air at time t, in hours, after exposure to C_0; and K_A, K_B, α, and β are empirically determined constants. The coefficients and exponents are specific for the given exposure times; and the equations should be used only to predict uptake within these limits.

Blood lead level, (PbB), resulting from lead in air exposure, (PbA), is an example of an air concentration-blood relationship. Although the relationship may not be linear over all air concentration ranges, it is typically represented by:

$$\text{PbB} = \beta(\text{PbA}) + \alpha \tag{3.3}$$

where PbB is in μg Pb/100 ml blood; PbA is expressed as a concentration in μg Pb/m^3; β is the change of blood lead/unit change in air lead, m^3/100 ml; and α is an intercept value, μg Pb/100 ml. Table 3.6 summarizes values of β inferred from some recent occupational studies. The slope is typically 0.05–0.1 m^3/100 ml. For lower levels of exposure in nonoccupational settings, the values of β are, on the average, somewhat higher than those of Table 3.6: 1.64 ± 0.22 m^3/100 ml for adults with exposures ≤ 3.2 μg/m^3 and PbB ≤ 30 μg/100 ml; and ~0.5 m^3/100 ml for adults exposed to 9–36 μg/m^3 and PbB = 30–40 μg/100 ml (EPA, 1984).

Equation (3.3) may be used to predict blood lead level increase or decrease due to changes in occupational exposure directly from air. Indirect exposure, e.g., ingestion, would increase the blood lead concentrations, so the corresponding value of β would be increased, perhaps by a factor of 2 (EPA, 1984). The value of α from mathematical fitting

TABLE 3.5 Decay of Organics in Human Exhaled Air

Chemical	Hours of exposure	Hours studied postexposure	Concentrations evaluated (ppm)	Air sample[a]	Number of subjects[b]
Methylene chloride	2	3	100, 200	E	11 M
Styrene	8	5	25, 115, 260	E	15 M
	7	6	99	A	6 M
	6	47	80	E	4 M
Tetrachloro-ethylene	7	110	101	A	16 M
	2	4	100, 150, 200	A	23 M, 1 F
	4	4			
	8	4			
Trichloro-ethylene	1	22	20, 100, 200	A	3–9 (M&F)
	3				
	$7\frac{1}{2}$				
	8		50	A	
Toluene[e]	8		100		

[a]E = normal exhaled air; A = alveolar air.
[b]M = male, F = female.
[c]C/C_0 = ratio of concentration in exhaled air at time t (hr) after exposure to C_0.
[d]Calculations made by R. S. Waritz using data obtained from graphs presented in the authors' original scientific papers and without access to the original raw data. Therefore they are subject to errors introduced by the original translation to graph form and by printing. They should be considered illustrative only.

of the data is a calculational convenience without physical meaning, because exposure to zero lead concentration is unlikely. However, some general background values that might be used for estimating the consequences of occupational exposures in the 18- to 74-year age group are 16.6 ± 0.29 μg/100 ml for white males, 19.1 ± 0.70 μg/100 ml for black males, 11.7 ± 0.23 μg/100 for white females, and 12.7 ± 0.44 μg/100 for black females. These are average blood lead concentrations from a recent U.S. national survey (NCHS, 1982). Since the half-life of lead in the blood and soft tissue is typically 30 days (EPA, 1984), Eq. (3.3) is applicable only to average exposure conditions over relatively long periods.

4. Job definition. The interaction of the worker with the equipment or materials he or she works with needs to be well defined. Source type and location, and the activity patterns that surround the process, are

Coefficients and exponents[c] $[C/C_0 = K_A \exp(-\alpha t) + K_B \exp(-\beta t)]$				
K_A	K_B	α	β	Reference
1.8×10^{-2}	10^{-1}	1.1×10^{-1}	10^{-2}	DiVencenzo et al., 1972[d]
1.5×10^{-2}	5×10^{-3}	1.01	1.3×10^{-1}	Gotell et al., 1972[d]
1.5×10^{-2}	5×10^{-3}	1.01	1.3×10^{-1}	Stewart et al., 1968[d]
9.3×10^{-3}	1.4×10^{-3}	0.80	3.6×10^{-2}	Ramsey et al., 1980 ACGIH, 1984[c]
3.5×10^{-2}	3.5×10^{-2}	8.3×10^{-2}	8.8×10^{-3}	Stewart et al., 1970[d]
2.9×10^{-1}	1.6×10^{-1}	1.2×10^{-2}	8.0×10^{-3}	Fernandez et al., 1976
2.1×10^{-1}	2.0×10^{-1}	9.9×10^{-2}	7.5×10^{-3}	
10^{-1}	2.2×10^{-1}	4.9×10^{-2}	4.4×10^{-3}	
4.9×10^{-2}	4.8×10^{-2}	4.6×10^{-1}	5.1×10^{-2}	Stewart et al., 1974[d]
5×10^{-2}	1.3×10^{-2}	1.6	8.1×10^{-2}	
9×10^{-2}	2.6×10^{-2}	1.8	6.9×10^{-2}	
5.2×10^{-2}	3.2×10^{-2}	2.1	0.23	ACGIH, 1984[f] Fernandez et al., 1975
8.5×10^{-2}	5.22×10^{-2}	26.2	1.57	ACGIH, 1984[c]

[e] C/C_0 equation applies to end-exhaled air and has additional $K_c \exp(-\delta t)$, where $K_c = 1.37 \times 10^{-2}$ and $\delta = 0.189$.

[f] C/C_0 equation applies to alveolar air and has additional $K_c \exp(-\delta t)$, where $K_c = 0.5 \times 10^{-2}$ and $\delta = 0.024$.

SOURCE: Revised from Waritz, 1979.

important factors in determining "dose." As an example, Figure 3.3 shows the average concentration change in a welding shop in condensation nuclei concentration (CNC) with change in number of welding or grinding stations in use (Franke and Wadden, 1985).

Retrospective epidemiological studies often require a measure of dose, or at least exposure. In the absence of actual measured values, job classifications are often used as a surrogate. An indicated in Chapter 2, this approach was used successfully in an evaluation of coke oven workers (Redmond et al., 1972; Redmond et al., 1981). Full-time "topside" coke oven workers have been found to have 4.9–6.9 times the incidence of lung cancer as other steelworkers. "Side-oven" workers incurred 1.6–1.9 times the lung cancer risk, depending on length of employment (Redmond et al., 1981). However, such classifications are often not so specific, or the duties and exposure change over time, and consequently should be used with caution. If at all possible it is desirable to compare job classifications with actual activity observations, or, for retrospective studies, with interviews of long-term workers.

TABLE 3.6 Some Recent Analyses on Occupational 8-hr Exposures to High Air Lead Levels

Analysis	Study		Air lead[a] (μg/m³)	Blood lead (μg/100 ml)	β, slope (m³/100 ml)
Ashford et al., 1977	Williams et al., 1969		50–300	40–90	0.19
	Globe Union				0.10
	Delco-Remy				
King et al., 1979	Factory 1, 1975		35–1200	25–90	0.032
	Factory 2a, 1975				0.07
	Factory 3a, 1975				0.068
Gartside et al., 1982	Delco-Remy, 1974–1976		10–350	22–72	0.0514
					Nonlinear: at 50 μg/m³
Bishop and Hill, 1983	Battery plants 1975–1981	A	20–170	12–50	0.081
		B	2–200	18–72	0.045
		C	7–195	22–60	0.048
		D	20–140	24–75	0.022
		E	4–140	18–60	0.045
		F		15–53	0.101
Chavalitnitikul et al., 1984	Battery plant 1981		2–165	12–51	0.165
[Better fit of data $r^2 = 0.786$ with			log (PbB) = 1.0425 + 0.2738 log (PbA)]		

[a]Assumed 8-hr exposure; divide by 3 for 24-hr equivalent.
SOURCE: Revised from EPA, 1984.

5. Estimation methods. When actual measures of dose or uptake are not available, it becomes necessary to estimate the probable dose. Environmental modeling is often used to interpret or extend limited measurement data or source emission information from the literature.

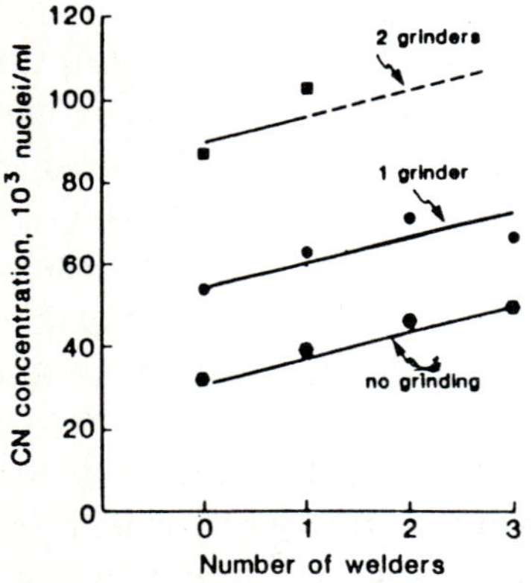

Figure 3.3 Covariance plot of condensation nuclei concentration and number of welders showing lines of common slope (*Franke and Wadden, 1987*).

The most common approach is to consider the workplace as a space that approaches the conditions of a completely mixed reactor. An equation describing the pollutant mass balance around such as space is (Wadden and Scheff, 1983):

$$C_i = \frac{k[q_0(1 - F_0) + q_2]C_o + S - R}{k(q_0 + q_1F_1 + q_2)} \times [1 - \exp\{-(k/V)(q_0 + q_1F_1 + q_2)t\}] + C_s\exp\{-(k/V)(q_0 + q_1F_1 + q_2)t\} \tag{3.4}$$

where C is concentration indoors (C_i) and outdoors (C_o), and $C_i = C_s$ at $t = 0$; t is time; q is volumetric flow rate for make-up air (q_0), recirculation (q_1), infiltration (q_2), exfiltration (q_3), and exhaust (q_4); F is filter efficiency for make-up (F_0) and recirculation air (F_1) (often the same); V is room volume; S is indoor source emission rate; R is indoor sink removal rate; and k, a factor that accounts for inefficiency of mixing, is the fraction of incoming air that mixes completely within the room volume. For the case where R is a first-order function of C_i, the solution will have the form:

$$C_i = \frac{k[q_0(1 - F_0) + q_2]C_o + S}{k(q_0 + q_1F_1 + q_2) + E} \times (1 - \exp\{-[(k/V)(q_0 + q_1F_1 + q_2) + E/V]t\}) + C_s\exp\{-[(k/V)(q_0 + q_1F_1 + q_2) + E/V]t\} \tag{3.5}$$

where E is a proportionality constant for the particular pollutant of interest, such that $R = EC_i$. Steady-state values of indoor concentrations, $C_{i,ss}$, will result by letting t approach ∞, which, for Eq. (3.5), results in:

$$C_{i,ss} = \frac{k[q_0(1 - F_0) + q_2]C_o + S}{k(q_0 + q_1F_1 + q_2) + E} \tag{3.6}$$

These equations have been used successfully to predict concentration of NO_2, CO, respirable particles (RP), and condensation nuclei (CN) in a welding shop and lunchroom (Franke and Wadden, 1987); ozone in offices and laboratories (Shair and Heitner, 1974); RP and sulfates in homes (Dockery and Spengler, 1981); RP in meeting rooms (Ishizu, 1980); and a variety of other applications (Wadden and Scheff, 1983). One of the major advantages of this approach is that it allows emission rates from various types of pollution sources to be used in different settings to estimate workplace concentrations.

Few situations actually have ideal mixing ($k = 1.0$). More common conditions are $0.1 \leq k \leq 0.3$. Often, the decay or removal term, R, is not evaluated explicitly, but rather S is considered a net emission rate with an implicit sink term. An estimate of C_o is often available from environmental monitoring data. A variety of methods exist for estimating q_2 and q_3, depending on the type of building, wind speed, and indoor-outdoor temperature difference (Wadden and Scheff, 1983). In general, the natural infiltration rate into structures of ordinary construction (e.g., not designed to be very airtight for energy efficiency) is about 1 air change/hr. If forced ventilation is used, the actual volumetric flow rates should be used.

Equations (3.4)–(3.6) may be used with appropriate emission factors to make initial or retrospective estimates of area exposures. They can also be used to interpret area monitoring data to determine appropriate emission factors for a particular workplace; and these factors can later be used with the equations to predict exposures. A typical example for RP in a welding shop with five operating welding stations for 1 hr, and appropriate emission factors from Table 3.7, is as follows:

Given:

$$V = 7000 \text{ m}^3$$

$$R = 0$$

$$S = 5 \times 23.6 \text{ mg/min} = 118 \text{ mg/min}$$

$$t = 1 \text{ hr}$$

$$C_o = 0$$

$$C_s = 0.5 \text{ mg/m}^3$$

$$q_0 = q_1 = 0$$

$$kq_2/V = 1.1 \text{ ach}$$

$$C_i = \frac{S}{k(q_2/V)V}[1 - \exp\{-k(q_2/V)t\}] + C_s\exp\{-k(q_2/V)t\}$$

$$= \frac{(118 \text{ mg/min})(60 \text{ min/hr})}{(1.1 \text{ ach})(7000 \text{ m}^3)} \times [1 - \exp\{-(1.1 \text{ ach})(1 \text{ hr})\}] + (0.5 \text{ mg/m}^3)[\exp\{-(1.1 \text{ ach})(1 \text{ hr})\}]$$

$$C_i = 0.78 \text{ mg/m}^3$$

At steady state:

$$C_{i,ss} = \frac{S}{k(q_2/V)V} = 0.92 \text{ mg/m}^3$$

and this concentration will occur in about 4 hr if emissions and infiltration remain constant.

TABLE 3.7 Welding Shop Source Emission Rates

	Activity factor		Emission rate/activity level	
Contaminant	Activity	Levels	mg/min/level	10^{12} nuclei/min/level
CN	Welding	0, 1, 2, 3	30.9	2.9
RP	Welding	0, 1, 2, 3	23.6	
NO_2	Heaters	0, 1	1.5	
CN	Grinding	0, 1, 2		3.3–6.4
CN	Spray painting	0, 1		1.8

SOURCE: Franke and Wadden, 1987.

II. Control Alternatives

A. Control

Workers can be protected from hazardous exposures through isolation, dilution or removal of the pollutant, process modification, substitution of a less hazardous material, shielding, and personal protection. Administrative controls, such as work practice rules that minimize exposure to high-heat environments, are methods of ameliorating the effect of exposures. The thrust of our discussion here and in succeeding chapters will focus on engineering controls as opposed to personal protective devices.

Isolation can either be of the worker from the hazard, as would be an overhead crane operator in an air-conditioned cab in a steel rolling mill; or the process from the worker as in local ventilation systems (Chapter 5). *Dilution* of pollutant concentrations relies on the forced or natural ventilation rate being high enough to maintain personal exposures to a level less than applicable health-related standards. *Removal-control* devices, such as electrostatic precipitators for particulate matter, collect hazardous substances before they can be released into the workspace (Chapter 6).

Process modification can take several paths. One of these is equipment modification. As an example, in pneumatic spraying of automobiles, up to 50% of the paint does not coat the workpiece; and 20% of the particles $<12\ \mu m$ do not deposit (O'Brien and Hurley, 1981). Alternatives to pneumatic spraying are use of airless sprayers, where the paint is dispersed by special nozzles under high pressure; or electrostatic sprayers, where particles adhere because of their charge. Both of these devices waste much less paint than pneumatic equipment (O'Brien and Hurley, 1981). In foundries, production cleaning of castings by hand grinding has always been a source of dust exposure. The installation of industrial robots to carry out this activity in an isolated space removes the human being from the worker-workpiece interface where highest concentrations are encountered (Svensson, 1981).

An alternative control approach is *substitution* of a less toxic, or physiologically inert, material for the hazardous one. Traditionally, ship and automotive painting called for a zinc chromate primer, the chromate being the source of potential health problems. However, the trend is now toward a primer containing zinc dust, a material that achieves better bonding as well as having much lower toxicity. Similarly, less hazardous organic coatings have been substituted for lead- and chromium-containing paint pigments (O'Brien and Hurley, 1981).

Process redesign so that toxic by-products are never produced is another type of modification. The production of *o*-chlorophenol has been associated with the formation of tetrachlorodibenzodioxin, TCDD, some of the isomers of which are extremely toxic (Esposito et al., 1980). Recycled phenol, mixed with various chlorinated phenolic reaction products, was used as one of the reactants. Since recycle streams may concentrate by-product materials not specifically removed, use of virgin phenol instead of the recycled mixture was suggested as a solution. This example is discussed in more detail in Chapter 7.

Changes in conditions for *equipment* and *process operation* may also be useful in minimizing hazardous release. For instance, in sulfuric acid production, maintaining inert gas atmospheres on process vessels until all equipment has reached design operating temperatures will greatly reduce emissions of gaseous sulfur oxides and acid mist. Tungsten inert gas welding (TIG) is usually considered to produce lower amounts of fume than other methods. However, if TIG welding is used without filler wire, more amperage is required to produce the seams, and more fume is produced than expected.

TABLE 3.8 Electrostatic Precipitator Specifications

Input voltage	120 V, 60 Hz, single phase
Input current	1.0 A nominal
Power consumption	100 W/0.94 m^3/sec (100 W/2000 ACFM)
Weight	138 kg (320 lb)
Maximum recommended air capacity	0.94 m^3/sec (2000 ACFM)
Ionizer section	
Voltage	8,600–11,000 V dc, positive corona
Current	2.0 mA nominal
Collector section	
Plate material	Aluminum
Plate voltage	4500–5000 V dc

SOURCE: Holcomb and Scholz, 1981.

Shielding, or placing a barrier between worker and hazard source, is a common mode of control, particularly for physical hazards. Machine guarding (Chapter 9) prevents contact with pinch points, rotating, stamping, cutting, or shearing equipment. Suppression barriers are frequently used for control of heat exposure (Chapter 12), ionizing radiation (Chapter 10), nonionizing radiation (Chapter 11), and noise (Chapter 13).

B. Maintenance

Even though a design change has been implemented, there continues to be a need for maintenance of the device, and monitoring of the device and workplace. The electrostatic precipitator described in Table 3.8 provides an example. The precipitator was used to clean recycled air in a welding shop. Figure 3.4 illustrates that a significant change in efficiency occurred after 4 days of operation at an inlet gas concentration of 23 mg RP/m^3 (Holcomb and Scholz, 1981). A maintenance schedule that cleaned out the precipitator at 3-day intervals would prevent this breakthrough.

The EPA has proposed regulations for control of fugitive volatile organic compounds (VOC) in petroleum refineries (EPA, 1983). These emissions include those from pressure-relief devices, compressors, open-ended valves, sampling connection systems, valves, and pumps. Labor

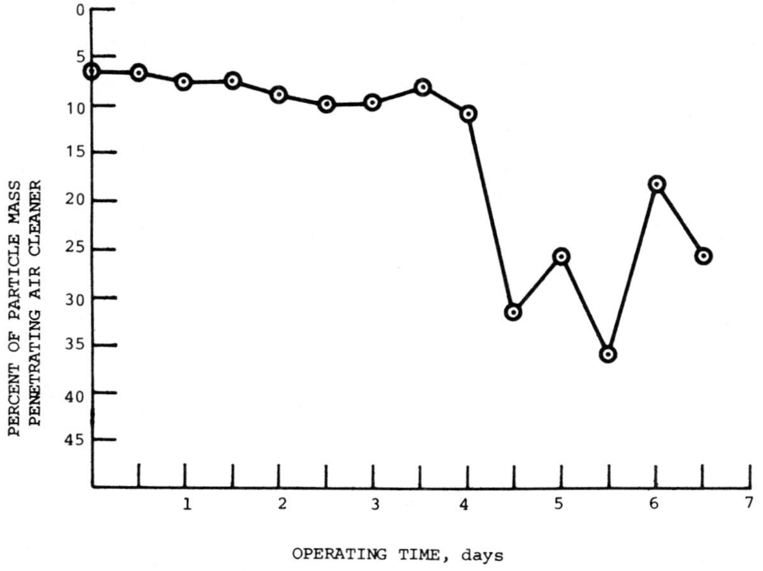

Figure 3.4 Electrostatic precipitator operation preceding and during a breakthrough failure; inlet RP concentration = 23 mg/m^3 (*Holcomb and Scholz, 1981*).

TABLE 3.9 Refinery Monitoring and Maintenance Labor-Hour Requirement for Fugitive VOC Control[a]

		Leak detection			Leak repair		
Source type	Number of components	Type of monitoring[b]	Times monitored per year	Monitoring labor-hours required[c,d]	Percent of sources leaking	Estimated number of leaks[e]	Maintenance labor-hours[f]
Valves							
Gas vapor	260	Instrument	4	35	10	11	12
		Instrument	12	104		16	18
Light liquid[g]	500	Instrument	1	17	11	11	12
		Instrument	4	67		22	25
		Instrument	12	200		33	37
Pump seals							
Light liquid	14	Instrument	1	2.3	24	1	80
		Instrument	12	28		2	160
		Visual	52	6.1			
Relief valves							
Gas vapor	7	Instrument	4	7.5	7	0	0
Compressor seals							
Gas vapor	3	Instrument	4	2	35	1	40

[a]EPA, 1982; model refinery contains units for alkylation, thermal cracking, solvent extraction, reforming, and vacuum distillation.

[b]Assumes that instrument monitoring requires a two-person team and visual monitoring one person.

[c]Monitoring time per person: pumps-instrument 5 min, visual $\frac{1}{2}$ min; compressors 5 min; valves 1 min, and safety/relief valves 8 min.

[d]Monitoring labor-hours = number of workers × number of components × time to monitor × times monitored per year.

[e]Annual percent recurrence factors have been applied for monthly, quarterly, and annual instrument inspections. It is assumed that 5% of leaks initially detected are found with monthly monitoring ($0.05 \times 12 = 0.6$), that 10% of leaks initially detected are found with quarterly monitoring ($0.1 \times 4 = 0.4$), and that 20% of leaks initially detected are found by annual monitoring ($0.2 \times 1 = 0.2$). Number of leaks = number of components × % sources leaking × annual % recurrence factor.

[f]Leak repair = number of leaks × repair time. Labor-hours: repair time per component: pumps, 80 hr; compressors, 40 hr; valves, 1.13 hr. (Basis: weighted average on 75% of the leaks repaired on-line requiring 10 min per repair, and on 25% of the leaks repaired off-line requiring 4 hr per repair. Safety relief valves, 0 hr. (It is assumed that these leaks are corrected by routine maintenance at no additional labor requirement.)

[g]Materials lighter than kerosene.

TABLE 3.10 Refinery Valve Maintenance Study Results

	Repaired valves with initial screening values ≥10,000 ppmv	
	Directed maintenance[a]	Undirected maintenance[b]
Number of valves repaired	9	23
Number of successful repairs (<10,000 ppmv after maintenance) (%)	8 (88.9)	13 (56.5)
Measured VOC emissions before maintenance, kg/hr	0.107	1.809
Measured VOC emissions after maintenance, kg/hr (% reduction)	0.0139 (87.0)	0.318 (82.4)
Number of valves with decreased emissions (%)	9 (100.0)	21 (91.3)
Number of valves with increased emissions (%)	0 (0.0)	2 (8.7)

[a]Directed maintenance consists of performing valve repairs while simultaneously monitoring emissions. Repair procedures such as tightening the packing gland or adding grease are continued until VOC concentration drops to specified level or further reductions are not possible.

[b]Undirected maintenance involves repair without simultaneous monitoring.

SOURCE: EPA, 1982.

requirements for leak detection and repair are given in Table 3.9 for a refinery containing units for alkylation, thermal cracking, solvent extraction, re-forming, and vacuum distillation (EPA, 1982). Some typical measured emission reductions for valves are given in Table 3.10. While valves and pipe fittings are designed not to leak, they frequently do. Effective control requires recognition of this fact and appropriate repair or replacement.

REFERENCES

ACGIH (1984). *Documentation of the Threshold Limit Values,* 4th ed. and Supplemental Documentations for 1981–1984. American Conference of Governmental Industrial Hygienists, Cincinnati, Ohio.

American Gas and Chemical Co. (1981). *Leak-Tec Personal Protection Indicators*. Northvale, N. J.

Ashford, W. A., Gecht, R. D., Hattis, D. B., and Katz, J. I. (1977). *The Effects of OSHA Medical Removal Protection on Labor Costs of Selected Lead Industries.* Center for Policy Alternatives Report No. CPA 77/11, Massachusetts Institute of Technology, Cambridge, Mass.

Balmat, J. L., and Meadows, G. W. (1985). Monitoring of formaldehyde in air. *Amer. Ind. Hyg. Assoc. J., 46,* 578–584.

Bamberger, R. L., Esposito, G. G., Jacobs, B. W., Podalak, G. E., and Mazur, J. F. (1978). A new personal sampler for organic vapors. *Amer. Ind. Hyg. Assoc. J., 39,* 701–708.

Bishop, E. C., and Hossain, M. A. (1984). Field comparison between two nitrous oxide (N_2O) passive monitors and conventional sampling methods. *Amer. Ind. Hyg. Assoc. J., 45,* 812–816.

Bishop, R. L., and Hill, W. J. (1983). A study of the relationship between blood lead levels and occupational air lead levels. *Amer. Stat. 37,* 471–475.

Cassinelli, M. R., Hull, D. R., and Cuendet, P. A. (1985). Performance of sulfur dioxide passive monitors. *Amer. Ind. Hyg. Assoc. J., 46,* 599–608.
Cember, H. (1983). *Introduction to Health Physics,* 2d ed. Pergamon Press, New York.
Chavalitnitikul, C., Levin, L., and Lung-Chi, C. (1984). Study and models of total lead exposures of battery workers. *Amer. Ind. Hyg. Assoc. J., 45,* 802–808.
Coburn, R. F., Forester, R. E., and Kane, P. B. (1965). Consideration of the physiological variables that determine the blood carboxyhemoglobin concentration in man. *J. Clin. Invest. 44,* 1899–1910.
Diem, K., and Lentner, C. (1972). *Documenta Geigy Scientific Tables.* Ciba-Geigy Ltd., Basle, Switzerland.
DiVincenzo, G. D., Yanno, P. F., and Astill, B. D. (1972). Human and canine exposures to methylene chloride vapor. *J. Amer. Ind. Hyg. Assoc., 33,* 125–135.
Dockery, D. W., and Spengler, J. D. (1981). Indoor-outdoor relationships of respirable sulfates and particles. *Atmos. Environ., 15,* 335–343.
Dupont, E. I. 1981. *High Quality Air and Noise Monitoring Instruments.* DuPont Occupational and Environmental Health Products, Applied Technology Division, Wilmington, Del.
Enviro Control, Inc. (1978). *Engineering Control Technology Assessment for the Plastics and Resin Industry.* NIOSH Report No. 78-159, March.
EPA (1981). *Workshop on Indoor Air Quality Research Needs.* Environmental Protection Agency, Department of Energy, Interagency Research Group on Indoor Air Quality, Leesburg, Va., April.
EPA (1982). *VOC Fugitive Emissions in Petroleum Refining Industry.* EPA Report No. EPA-450/3-81-015a, November.
EPA (1983). Standards of performance for new stationary sources; VOC fugitive emission sources; petroleum refineries. *Fed. Reg. 48,* 279–299, January 4.
EPA (1984). *Air Quality Criteria for Lead.* Vol. III, EPA Report No. EPA 600/8-83-028B, August.
Esposito, M. P., Tiernan, T. O., and Dryden, F. E. (1980). *Dioxins.* EPA Report No. EPA-600/2-80-197, November.
Fernandez, J., Guberan, E., and Caperos, J. (1976). Experimental human exposures to tetrachlorethylene vapor and elimination in breath after inhalation. *Amer. Ind. Hyg. Assoc. J., 37,* 143–150.
Fernandez, J., Humbert, B. E., Droz, P. O., and Caperos, J. (1975). Trichloroethylene exposure. Percentage studies of absorption, excretion and metabolism by human subjects. *Arch. Mal. Prof., 36,* 397–407.
Franke, J., and Wadden, R. A. (1987). Indoor contaminant emission rates characterized by source activity factors. *Environ. Sci. Technol.,* (in press).
Gartside, P. S., Bucher, C. R., and Lerner, S. (1982). Relationship of air lead and blood lead for workers at an automobile battery factory. *Int. Arch. Occup. Environ. Health, 50,* 1–10.
Gillespie, J. C., and Daniel, L. B. (1979). A new sampling tool for monitoring exposure to toxic gases and vapors. *Proceedings, Symposium on Development and Usage of Personal Monitors for Exposure and Health Effect Studies.* U.S. Environmental Protection Agency, EPA-600/9-79-32, June.
Gotell, P., Axelson, O., and Lindelof, B. (1972). Field studies on human styrene exposure. *Work Environ. Health, 9,* 76–83.
Hardy, J. K., Dasguptas, P. K., Reiszner, K. D., and West, P. W. (1979). A personal chlorine monitor utilizing permeation sampling. *Environ. Sci. Technol., 13,* 1090–1093.
Hardy, J. K., Strecker, D. T., Savariar, C. P., and West, P. W. (1981). A method for the personal monitoring of hydrogen sulfide utilizing permeation sampling. *Amer. Ind. Hyg. Assoc. J., 42,* 283–286.
Hickey, J. L. S., and Bishop, C. C. (1981). Field comparison of charcoal tubes and passive vapor monitors with mixed organic vapors. *Amer. Ind. Hyg. Assoc. J., 42,* 264–267.
Holcomb, M. L., and Scholz, R. C. (1981). *Evaluation of Air Cleaning and Monitoring Equipment Used in Recirculation Systems.* NIOSH Report No. 81-113, April.
Ikeda, M., Koizumi, A., Miyasaka, M., and Watanabe, T. (1982). Styrene exposure and

biologic monitoring in F.R.P. boat production plants. *Int. Arch. Occup. Environ. Health, 2,* 21–26.

Ikeda, M., and Ohtsuji, H. (1969). Significance of urinary hippuric acid determination as an index of toluene exposure. *Brit. J. Ind. Med., 26,* 244–246.

Ikeda, M., Ohtsuji, H., Imamura, T., and Komoike, Y. (1972). Urinary excretion of total trichloro-compounds, trichloroethanol and trichloroacetic acid as a measure of exposure to trichloroethylene and tetrachloroethylene. *Brit. J. Ind. Med., 29,* 328–333.

Ishizu, Y. (1980). General equation for the estimation of indoor pollution. *Environ. Sci. Technol., 14,* 1254–1257.

Jensen, P., Jokel, C. R., and Miller, L. N. (1978). *Industrial Noise Control Manual,* rev. ed. NIOSH Report No. 79-117, December.

King, E., Conchie, A., Hiett, D., and Milligan, B. (1979). Industrial lead absorption. *Ann. Occup. Hyg., 22,* 213–239.

Kring, E. V., Damrell, D. J., Henry, T. J., DeMoor, H. M., Basilio, A. N., and Simon, C. E. (1984). Laboratory validation and field verification of a new passive colorimetric air monitoring badge for sampling hydrogen sulfide in air. *Amer. Ind. Hyg. Assoc. J., 45,* 1–9.

Kring, E. V., McGibney, P. D., and Thornley, G. D. (1985). Laboratory validation of five commercially available methods for sampling ethylene oxide in air. *Amer. Ind. Hyg. Assoc. J., 46,* 620–624.

Kring, E. V., Thornley, G. D., Dessenberger, C., Lautenberger, W. J., and Ansol, G. R. (1982). A new passive colorimetric air monitoring badge for sampling formaldehyde in air. *Amer. Ind. Hyg. Assoc. J., 43,* 786–795.

Lautenberger, W. J., Kring, E. V., and Morello, J. A. (1980). A new personal badge monitor for organic vapors. *Amer. Ind. Hyg. Assoc. J., 41,* 737–747.

Mazur, J. F., Rinehart, D. S., Esposito, G. G., and Podolak, G. E. (1981). Evaluation of passive monitors for assessing vapor degreaser emissions. *Amer. Ind. Hyg. Assoc. J., 42,* 752–756.

McConnaughey, P. W., McKee, E. S., and Pritts, L. M. (1985). Passive colorimetric dosimeter tubes for ammonia, carbon monoxide, carbon dioxide, hydrogen sulfide, nitrogen dioxide, and sulfur dioxide. *Amer. Ind. Hyg. Assoc. J., 45,* 357–362.

McKee, E. S., and McConnaughey, P. W. (1985). A passive, direct reading, length of stain dosimeter for ammonia. *Amer. Ind. Hyg. Assoc. J., 46,* 407–410.

MDA (1981). The Monitox personal gas detection alarm system. Environmental Division, MDA Scientific, Inc., Glenview, Ill.

Mullins, H. E. (1985). Sub-part-per-million diffusional sampling for ethylene oxide with the 3M #3550 ethylene oxide monitor. *Amer. Ind. Hyg. Assoc. J., 46,* 625–631, 1985.

Nader, J. S., and Lauderdale, J. F. (1983). Direct reading instruments for analyzing airborne gases and vapors. In: *Air Sampling Instruments for Evaluation of Atmospheric Contaminants,* 6th ed. American Conference of Governmental Industrial Hygienists, Cincinnati, Ohio.

NCHS (1982). (Annest, J. L., Mahaffey, K. R., Cox, D. H., and Roberts, J.). Blood lead levels for persons 6 months–74 years of age: United States, 1976–1980. *Advance Data from Vital and Health Statistics* No. 79, DHHS Pub. No. (PHS) 82-1250, Public Health Service, Hyattsville, Md., May 12.

Nomiyama, K. (1971). Estimation of trichloroethylene exposure by biological materials. *Int. Arch. Arbeitsmed., 27,* 281–292.

O'Brien, D. M., and Hurley, D. E. (1981). *An Evaluation of Engineering Control Technology for Spray Painting.* NIOSH Report No. 81-121, June.

Ogata, M., Takatsuka, Y., and Tomokuni, K. (1971). Excretion of organic chlorine compounds in the urine of person exposed to vapors of trichloroethylene and tetrachloroethylene. *Brit. J. Ind. Med., 28,* 386–391.

Ogata, M., Tomokuni, K., and Takasuka, Y. (1970). Urinary excretion of hippuric acid and *m*- or *p*-methylhippuric acid in the urine of persons exposed to vapors of toluene and *m*- or *p*-xylene as a test of exposure. *Brit. J. Ind. Health, 27,* 43–50.

Ohtsuji, H., and Ikeda, M. (1972). Quantitative relationship between atmospheric phenol vapour and phenol in the urine of workers in Bakelite factories. *Brit. J. Ind. Med., 29,* 70–73.

Pagnotto, L. D., and Lieberman, L. M. (1967). Urinary hippuric acid excretion as an index of toluene exposure. *Amer. Ind. Hyg. Assoc. J., 28,* 129–134.

Palmes, E. D., Gunnison, A. F., DiMattio, J., and Tomczyk, C. (1976). Personal sampler for nitrogen dioxide. *Amer. Ind. Hyg. Assoc. J., 37,* 570–577.

Rainsford, S. G., and Davies, T. A. L. (1965). Urinary excretion of phenol by men exposed to vapour of benzene: A screening test. *Brit. J. Ind. Med., 22,* 21–26.

Ramsey, J. C., Young, J. D., Karbowski, R. J., Chenoweth, M. B., McCarty, L. P., and Braun, W. H. (1980). Pharmacokinetics of inhaled styrene in human volunteers. *Toxicol. Appl. Pharmacol., 53,* 54–63.

Redmond, C. K., Ciocco, A., Lloyd, J. W., and Rush, H. W. (1972). Long-term mortality study of steelworkers. VI. Mortality from malignant neoplasms among coke oven workers. *J. Occup. Med., 14,* 621–629.

Redmond, C. K., Wieand, H. S., Rockette, H. E., Sass, R., and Weinberg, G. (1981). *Long-Term Mortality of Steelworkers.* NIOSH Report No. 81–120, June.

Seki, Y., Urashima, Y., Aikawa, H., Matsumura, H., Ichikawa, Y., Hiratsuka, F., Yoshioka, Y., Shimbo, S., and Ikeda, M. (1975). Trichloro-compounds in the urine of humans exposed to methyl chloroform at sub-threshold levels. *Int. Arch. Arbeitsmed., 34,* 39–49.

Shair, F. H., and Heitner, K. L. (1974). Theoretical model for relating indoor pollutant concentrations to those outside. *Environ. Sci. Technol., 8,* 444–451.

Stewart, R. D., Baretta, E. D., Dodd, H. C., and Torkelson, T. R. (1970). Experimental human exposure to tetrachloroethylene. *Arch. Environ. Health, 20,* 224–229.

Stewart, R. D., Dodd, H. C., Baretta, E. D., and Schaffer, E. D. (1968). Human exposure to styrene vapor. *Arch. Environ. Health, 16,* 656–662.

Stewart, R. D., Hake, C. L., and Peterson, J. E. (1974). Use of breath analysis to monitor trichloroethylene exposures. *Arch. Environ. Health, 29,* 6–13.

Svensson, I. (1981). Swedish improvements in cleaning of castings with regard to environmental factors. In: *Proceedings, Symposium on Occupational Health Hazard Control Technology in the Foundry and Secondary Non-Ferrous Smelting Industries.* NIOSH Report No 81-114, August.

Von Oettingen, W. F., Neal, P. A., and Donahue, D. D. (1942). The toxicity and potential dangers of toluene. *J. Amer. Med. Assoc., 118,* 579–584.

Wadden, R. A., and Scheff, P. A. (1983). *Indoor Air Pollution.* Wiley, New York.

Wallace, L. A., and Ott, W. R. (1982). Personal monitors: A state-of-the-art survey. *J. Air Pollution Control Assoc., 32,* 601–610.

Waritz, R. S. (1979). Biological indicators of chemical dosage and burden. In: *Patty's Industrial Hygiene and Toxicology,* Vol. III, Cralley, L. V., and Cralley, L. J., eds. Wiley, New York.

West, P. W. (1980). Passive monitoring of personal exposures to gaseous toxins. *Amer. Lab.,* July, pp. 35–46.

West, P. W., and Reiszner, K. D. (1979). Personal monitoring by means of gas permeation. *Proceedings, Symposium on Development and Usage of Personal Monitors for Exposure and Health Effects Studies.* U.S. Environmental Protection Agency, EPA-600/9-79-032, June.

Wilkening, G. M. (1978). Nonionizing radiation; and Ionizing radiation. In: *Patty's Industrial Hygiene and Toxicology,* Vol. I, Clayton, G. D., and Clayton, F. E., Wiley, New York, pp. 359–489.

Williams, M. K., King, E., and Walford, J. (1969). An investigation of lead absorption in an electric accumulator factory with the use of personal samplers. *Brit. J. Ind. Med., 26,* 202–216.

Young, M. S., and Monat, J. P. (1982). Development of a passive dosimeter for hydrogen fluoride monitoring. *Amer. Ind. Hyg. Assoc. J., 43,* 890–896.

Chapter

4

Design Criteria

R. A. Wadden

J. E. Franke

I. Standards, Codes and Other Control Benchmarks

In order to design a device or system that will solve a hazard problem, it is necessary to have a specification of the condition the design is expected to meet. This specification may be in the form of a standard promulgated by a national or state governmental agency, a code required by a local governmental body, an external guideline suggested by a professional or standard-setting organization, or an internal guideline based on current "good practice" in a particular industry. The specification should be based on criteria that relate to minimizing or eliminating the effect of the hazard, but as we shall discuss, this is not always the case.

The purpose of health and safety standards should be to prevent illness or injury to the working population. For example, 8-hr, time-weighted threshold limit values (TLVs) are concentrations of potentially hazardous air pollutants at which most workers are not expected to show adverse health effects, based on a 40-hr work week. The TLVs are developed from workplace and laboratory studies of human health effects and animal exposures. On the other hand, some design standards are not adequately related to safety and health criteria, and may merely represent the least expensive (most profitable) equipment design.

The Occupational Safety and Health Act specifies duties of the employer for the benefit of employees. However, only the government is authorized to enforce these responsibilities. Most private suits are grounded on allegations of negligence, i.e., that everyone has a legal duty to act reasonably and prudently. But even for private suits the existence of standards or guidelines will be important. Determination of what is "reasonable and prudent" depends on the situation, and is the responsibility of court and jury. OSHA regulations, voluntary consensus standards, or industry or corporate guidelines help to define what would have been appropriate conduct. As a general rule, greater evidentiary weight will be given to governmental standards than to those that are voluntary (Johnson, 1985).

An indirect factor that may force standards to be related more specifically to health and safety criteria is the increase in legal actions concerning process and product liability. A case in point was the action of the Manville Corporation and UNR Industries in filing for reorganization under Chapter 11 of the U.S. Bankruptcy Act because of the large number of lawsuits and judgments against the companies resulting from asbestos exposures and effects (NYT, 1982a, 1982b). About 16,500 asbestos-related lawsuits have already been filed against Manville Corporation and another 32,000 new suits have been forecast, with a total estimated liability of more than $2 billion (NYT, 1982b). The reduction of such liability may encourage more pertinent design requirements.

Design criteria are formed or specified in several ways. The OSHA standards (OSHA, 1983b) and the EPA Air Quality Standards (EPA, 1971) are examples of design criteria prescribed by a government authority. The ASHRAE ventilation standard for indoor air quality (ASHRAE, 1986) is an example of a consensus standard recommended by a professional organization involved in the field. Consensus standards are desirable from the standpoint that they represent generally agreed-upon design conditions or specifications. However, they also often represent only the lowest common denominator of workplace protection. A third category of formation is "recommended practices." These usually are related to the economics of design and only occasionally reflect health and safety requirements; they are not ordinarily presented in as organized a form as prescribed or consensus standards. For example, removing lead-based paint from steel plate by grinding before welding was specified in the maritime industry to improve weld quality and reduce a fire hazard, but the practice also reduced the lead fume exposure hazard.

Design criteria may be of the "specified" or "performance" type. Specification standards detail how a particular control is to be accomplished. An example is the OSHA standard for bench and floor stand abrasive grinding wheels, which states:

> The angular exposure of the grinding wheel periphery and sides for safety guards used on machines known as bench and floor stands should not exceed 90° or one-fourth of the periphery. This exposure shall begin at a point not more than 65° above the horizontal plane of the wheel spindle [OSHA, 1983b, 1910.215(3)].

An example of a performance standard is the OSHA standard for ethylene oxide (OSHA, 1984):

> . . . [OSHA] establishes a permissible exposure limit for occupational exposure to ethylene oxide (EtO) of 1 part EtO per million parts of air (1 ppm) determined as an 8-hour time-weighted average concentration.

This concentration level can be achieved through appropriate ventilation or source control or replacement. The particular approach to be used is left up to the designer.

Table 4.1 shows some of the groups that promulgate or develop design criteria and standards that influence workplace design. Many times one or more of these organizations will develop consensus standards that at some later period are adopted into a government regulation.

Compliance with a standard is a function of the designating authority. OSHA standards and local fire protection ordinances are regulations supported by the force of law, and they are mandatory for the organizations and individuals affected. Voluntary standards are merely recommendations and are more likely to be followed if there is little cost associated with conformance.

Knowledge of current "good practice" may generate internal recommended practices. This knowledge can be gained informally when company representatives communicate through professional meetings, seminars, journals, and trade shows, and in the daily conduction of business. For example, good practice may dictate that small foundry castings be allowed to cool before knocking off gating systems with hammers to reduce heat stress; that pressure gauges not be in direct contact with the ethylene oxide gas inside hospital sterilizers; that powdered materials be added mechanically rather than manually to batch reactors containing hazardous liquids and their vapors; or that ultrafine synthetic fibrous materials of undefined chronic toxicity be worked with as carefully as asbestos would be. It is good practice to

TABLE 4.1 Some Organizations That Set or Recommend Standards for the Workplace

Government:	
Bureau of Mines	(BM)
Consumer Product Safety Commission	(CPSC)
Department of Defense	(DOD)
Department of Transportation	(DOT)
Environmental Protection Agency	(EPA)
National Bureau of Standards	(NBS)
National Institute for Occupational Safety and Health	(NIOSH)
Nuclear Regulatory Commission	(NRC)
Occupational Safety and Health Administration	(OSHA)
Others:	
American Conference of Governmental Industrial Hygienists	(ACGIH)
American Industrial Hygiene Association	(AIHA)
American National Standards Institute	(ANSI)
American Society of Heating, Refrigerating and Air Conditioning Engineers	(ASHRAE)
American Society for Testing and Materials	(ASTM)
Factory Mutual	(FM)
Illumination Engineers Society	(IES)
Institute of Electrical and Electronic Engineers	(IEEE)
International Commission for Radiation-Protection	(ICRP)
International Standards Organization	(ISO)
National Commission for Radiation-Protection	(NCRP)
National Fire Protection Association	(NFPA)
National Safety Council	(NSC)
Underwriters Laboratory	(UL)
World Health Organization	(WHO)

specify maximum emission rates for noise and air contaminants when purchasing equipment. Good practice is more likely to be followed and updated if a reduction in liability is perceived by management.

II. Chemical Hazards

The primary organs of contact for workplace chemical hazards are the skin and lung. Acceptable airborne concentrations are designated as threshold limit values (TLVs), ordinarily expressed as time-weighted average concentrations (TWA) for a normal 8-hr workday and a 40-hr work week. The TLVs are to be used as guides for health hazard control, and not as a definition of safe concentrations. However, they are ordinarily considered to be levels to which nearly all workers may be repeatedly exposed without adverse effect (ACGIH, 1984a). Provided the 8-hr TWA is not exceeded, some substances also have designated short-term exposure limits (STELs). These are higher concentrations to which workers can be continuously exposed for 15 min no more than four times a day without irritation, chronic or irreversible tissue damage, or narcosis that would interfere with normal functioning (ACGIH, 1984a). Tables 4.2 and 4.3 list current TLV-TWA values. The notation "skin" indicates that additional absorption can occur through the skin. However, the TLVs are structured under the assumption that only air exposure occurs. Exposure through other paths would lower the acceptable exposure from air. The letter "C" designates that the TLV is a ceiling value and should never be exceeded at any time during the work period. The concentration for other TLVs may exceed the stated value for limited periods, provided that concentrations also occur below the standard. However,

> . . . short-term exposures should exceed three times the TLV-TWA for no more than a total of 30 minutes during a work day and under no circumstances should they exceed five times the TLV, provided the TLV-TWA is not exceeded (ACGIH, 1984a).

Recognized and suspected carcinogens are also identified, although guidelines for these materials, when given, are not necessarily related to carcinogenic potential.

The TLVs are guidelines for design and are based on the health implications of exposure, but they have no force of law. Some of the TLVs have been incorporated into OSHA regulations as permissible exposure limits (PELs), and these are also shown in Tables 4.2 and 4.3 (OSHA, 1983b). Many more substances are covered by the TLV list, and the levels are generally the same or more stringent than the PELs, since the latter have not been updated for many years. In addition, NIOSH has recommended standards for a limited number of airborne substances and published these individually in criteria documents.

(Text continues on page 138.)

TABLE 4.2 Design Criteria for Air Contaminants

	Threshold limit values (ACGIH, 1984a)[a]		OSHA standards (OSHA, 1983b)	
Substance[b]	ppm	mg/m³, 25°C	ppm	mg/m³, 25°C
Acetaldehyde	100	180	200	360
Acetic acid	10	25	10	25
Acetic anhydride	C 5	C 20	5	20
Acetone	750	1780	1000	2400
Acetonitrile—Skin	40	70	40	70
2-Acetylaminofluorene	—	—	*	*
Acetylene	E	—	—	—
Acetylene dichloride: See 1,2-Dichloroethylene				
Acetylene tetrabromide	1	15	1	14
Acetylsalicylic acid (aspirin)	—	5	—	—
Acrolein	0.1	0.25	0.1	0.25
Acrylamide—Skin	—	0.3	—	0.3
Acrylic acid	10	30	—	—
Acrylonitrile—Skin	2*	4.5*	2**	4.5**
Aldrin—Skin	—	0.25	—	0.25
Allyl alcohol—Skin	2	5	2	5
Allyl chloride	1	3	1	3
Allyl glycidyl ether (AGE)—Skin	5	22	C 10	C 45
Allyl propyl disulfide	2	12	2	12
α-Alumina	—	D	—	—
Aluminum				
Metal and oxide	—	10	—	—
Pyro powders	—	5	—	—
Welding fumes	—	5	—	—
Soluble salts	—	2	—	—
Alkyls (not otherwise classified)	—	2	—	—
4-Aminodiphenyl—Skin	—	**	—	*
2-Aminoethanol: See Ethanolamine				
2-Aminopyridine	0.5	2	0.5	2
3-Amino 1,2,4-triazole: See Amitrole			—	—
Amitrole	—	(0.2)*	—	—
Ammonia	25	18	50	35
Ammonium chloride fume	—	10	—	—
Ammonium persulfate, as S_2O_8	—	(5)	—	—
Ammonium sulfamate	—	10	—	15
n-Amyl acetate	100	530	100	525
sec-Amyl acetate	125	670	125	650
Aniline and homologs—Skin	2	10	5	19
Anisidine (*o,p* isomers)—Skin	0.1	0.5	—	0.5
Antimony and compounds, as Sb	—	0.5	—	0.5
Antimony trioxide				
Handling and use, as Sb	—	0.5	—	—
Production	—	*	—	—
ANTU (α-naphthyl thiourea)	—	0.3	—	0.3
Argon	E	—	—	—
Arsenic and soluble compounds, as As	—	0.2	—	—

TABLE 4.2 Design Criteria for Air Contaminants (*Continued*)

Substance[b]	Threshold limit values (ACGIH, 1984a)[a]		OSHA standards (OSHA, 1983b)	
	ppm	mg/m³, 25°C	ppm	mg/m³, 25°C
Arsenic, inorganic	—	—	—	0.01**
Arsenic organic compounds, as As	—	—	—	0.5
Arsenic trioxide production	—	*	—	—
Arsine	0.05	0.2	0.05	0.2
Asbestos (see also Table 4.3)	—	**	—	**
Asphalt (petroleum) fumes	—	5	—	—
Atrazine	—	5	—	—
Azinphos-methyl—Skin	—	0.2	—	0.2
Barium soluble compounds, as Ba	—	0.5	—	0.5
Benomyl	0.8	10	—	—
Benzene	10*	30*	10	30
Benzidine—Skin	—	**	—	*
p-Benzoquinone: See Quinone				
Benzoyl peroxide	—	5	—	5
Benzo(a)pyrene	—	*	—	—
Benzyl chloride	1	5	1	5
Beryllium	—	0.002*	—	0.002
Biphenyl	0.2	1.5	0.2	1
Bismuth telluride	—	10	—	—
Se-doped	—	5	—	—
Borates, tetra, sodium salts				
Anhydrous	—	1	—	—
Decahydrate	—	5	—	—
Pentahydrate	—	1	—	—
Boron oxide	—	10	—	15
Boron tribromide	(C 1)	(C 10)	—	—
Boron trifluoride	C 1	C 3	C 1	C 3
Bromacil	1	10	—	—
Bromine	0.1	0.7	0.1	0.7
Bromine pentafluoride	0.1	0.7	—	—
Bromochloromethane: See Chlorobromomethane			—	—
Bromoform—Skin	0.5	5	0.5	5
1,3-Butadiene	(10*)	(22*)	1000	2200
Butane	800	1900	—	—
Butanethiol: See Butyl mercaptan				
2-Butanone: See Methyl ethyl ketone (MEK)				
2-Butoxyethanol—Skin	25	120	50	240
n-Butyl acetate	150	710	150	710
sec-Butyl acetate	200	950	200	950
tert-Butyl acetate	200	950	200	950
Butyl acrylate	10	55	—	—
n-Butyl alcohol—Skin	C 50	C 150	100	300
sec-Butyl alcohol	100	305	150	450
tert-Butyl alcohol	100	300	100	300
Butylamine—Skin	C 5	C 15	C 5	C 15
tert-Butyl chromate, as CrO_3—Skin	—	C 0.1	—	C 0.1
n-Butyl glycidyl ether (BGE)	25	135	50	270

TABLE 4.2 Design Criteria for Air Contaminants *(Continued)*

Substance[b]	Threshold limit values (ACGIH, 1984a)[a]		OSHA standards (OSHA, 1983b)	
	ppm	mg/m³, 25°C	ppm	mg/m³, 25°C
n-Butyl lactate	5	25	—	—
Butyl mercaptan	0.5	1.5	10	35
o-sec-Butylphenol—Skin	5	30	—	—
p-tert-Butyltoluene	10	60	10	60
Cadmium dust and salts, as Cd	—	0.05	—	0.1 (fume) 0.2 (dust)
Cadmium oxide				
Fume, as Cd	—	C 0.05	—	5
Production	—	0.05	—	—
Calcium carbonate/marble	—	D	—	—
Calcium cyanamide	—	0.5	—	—
Calcium hydroxide	—	5	—	—
Calcium oxide	—	2	—	5
Calcium silicate	—	D	—	—
Camphor, synthetic	2	12	—	2
Caprolactam				
Dust	—	1	—	—
Vapor	5	20	—	—
Captafol—Skin	—	0.1	—	—
Captan	—	5	—	—
Carbaryl (Sevin)	—	5	—	5
Carbofuran	—	0.1	—	—
Carbon black	—	3.5	—	3.5
Carbon dioxide	5000	9000	5000	9000
Carbon disulfide—Skin	10	30	20	62.3
Carbon monoxide	50	55	50	55
Carbon tetrabromide	0.1	1.4	—	—
Carbon tetrachloride—Skin	5*	30*	10	63
Carbonyl chloride: See Phosgene				
Carbonyl fluoride	2	5	—	—
Catechol (Pyrocatechol)	5	20	—	—
Cellulose (paper fiber)	—	D	—	—
Cesium hydroxide	—	2	—	—
Chlordane—Skin	—	0.5	—	0.5
Chlorinated camphene—Skin	—	0.5	—	0.5
Chlorinated diphenyl oxide	—	0.5	—	0.5
Chlorine	1	3	C 1	C 3
Chlorine dioxide	0.1	0.3	0.1	0.3
Chlorine trifluoride	C 0.1	C 0.4	C 0.1	C 0.4
Chloroacetaldehyde	C 1	C 3	C 1	C 3
α-Chloroacetophenone (phenacyl chloride)	0.05	0.3	0.05	0.3
Chloroacetyl chloride	0.05	0.2	—	—
Chlorobenzene (monochlorobenzene)	75	350	75	350
o-Chlorobenzylidene malononitrile—Skin	C 0.05	C 0.4	0.05	0.4
Chlorobromomethane	200	1050	200	1050
2-Chloro-1,3-butadiene: See β-Chloroprene				

TABLE 4.2 Design Criteria for Air Contaminants *(Continued)*

Substance[b]	Threshold limit values (ACGIH, 1984a)[a]		OSHA standards (OSHA, 1983b)	
	ppm	mg/m³, 25°C	ppm	mg/m³, 25°C
Chlorodifluoromethane	1000	3500	—	—
Chlorodiphenyl (42% chlorine)—Skin	—	1	—	1
Chlorodiphenyl (54% chlorine)—Skin	—	0.5	—	0.5
1-Chloro, 2,3-epoxy-propane: See Epichlorohydrin				
2-Chloroethanol: See Ethylene chlorohydrin				
Chloroethylene: See Vinyl chloride				
Chloroform	10*	50*	C 50	C 240
bis-Chloromethyl ether	0.001**	0.005**	*	*
Chloromethyl methyl ether	*	*	*	*
1-Chloro-1-nitropropane	2	10	20	100
Chloropentafluoroethane	1000	6320	—	—
Chloropicrin	0.1	0.7	0.1	0.7
β-Chloroprene—Skin	10	45	25	90
o-Chlorostyrene	50	285	—	—
o-Chlorotoluene—Skin	50	250	—	—
2-Chloro-6-(trichloromethyl) pyridine: See Nitrapyrin				
Chlorpyrifos—Skin	—	0.2	—	—
Chromite ore processing (chromate), as Cr	—	0.05**	—	—
Chromium metal	—	0.5	—	1.0
Chromium				
Sol. chromic or Chromous salts, as Cr	—	—	—	0.5
Chromic acid and chromates	—	—	—	0.1
Chromium(II) compounds, as Cr	—	0.5	—	—
Chromium(III) compounds, as Cr	—	0.5	—	—
Chromium(VI) compounds, as Cr				
Water-soluble	—	0.05	—	0.5
Certain water-insoluble	—	0.05**	—	1
Chromyl chloride	0.025	0.15	—	—
Chrysene	*	*	—	—
Clopidol	—	10	—	—
Coal tar pitch volatiles, as benzene solubles	—	0.2**	—	0.2[f]
Cobalt, as Co metal, dust and fume	—	(0.05)	—	0.1
Cobalt carbonyl, as Co	—	0.1	—	—
Cobalt hydrocarbonyl, as Co	—	0.1	—	—
Coke oven emissions	—	—	—	0.150**
Copper				
Fume	—	0.2	—	0.1
Dusts and mists, as Cu	—	1	—	1
Cotton dust, raw (respirable)	—	0.2	—	1[h]
Crag herbicide	—	—	—	15
Cresol, all isomers—Skin	5	22	5	22
Crotonaldehyde	2	6	2	6
Crufomate	—	5	—	—

TABLE 4.2 Design Criteria for Air Contaminants (*Continued*)

	Threshold limit values (ACGIH, 1984a)[a]		OSHA standards (OSHA, 1983b)	
Substance[b]	ppm	mg/m³, 25°C	ppm	mg/m³, 25°C
Cumene—Skin	50	245	50	245
Cyanamide	—	2	—	—
Cyanides, as CN—Skin	—	5	—	5
Cyanogen	10	20	—	—
Cyanogen chloride	C 0.3	C 0.6	—	—
Cyclohexane	300	1050	300	1050
Cyclohexanol	50	200	50	200
Cyclohexanone	25	100	50	200
Cyclohexene	300	1015	300	1015
Cyclohexylamine—Skin	10	40	—	—
Cyclonite—Skin	—	1.5	—	—
Cyclopentadiene	75	200	75	200
Cyclopentane	600	1720	—	—
Cyhexatin	—	5	—	—
2,4-D	—	10	—	10
DDT (Dichlorodiphenyl-trichloroethane)—Skin	—	1	—	1
Decaborane—Skin	0.05	0.3	0.05	0.3
Demeton—Skin	0.01	0.1	—	0.1
Diacetone alcohol	50	240	50	240
1,2-Diaminoethane: See Ethylenediamine				
Diazinon—Skin	—	0.1	—	—
Diazomethane	0.2	0.4	0.2	0.4
Diborane	0.1	0.1	0.1	0.1
1,2-Dibromo-3-chloropropane	—	—	0.001**	0.0097**
1,2-Dibromoethane: See Ethylene dibromide				
2-N-Dibutylaminoethanol—Skin	2	14	—	—
Dibutyl phosphate	1	5	1	5
Dibutyl phthalate	—	5	—	5
Dichloroacetylene	C 0.1	C 0.4	—	—
o-Dichlorobenzene	C 50	C 300	C 50	C 300
p-Dichlorobenzene	75	450	75	450
3,3′-Dichlorobenzidine—Skin	—	*	—	*
Dichlorodifluoromethane	1000	4950	1000	4950
1,3-Dichloro-5,5-dimethyl hydantoin	—	0.2	—	0.2
1,1-Dichloroethane	200	810	100	400
1,2-Dichloroethane: See Ethylene dichloride				
1,1-Dichloroethylene: See Vinylidene chloride				
1,2-Dichloroethylene	200	790	200	790
Dichloroethyl ether—Skin	5	30	C 15	C 90
Dichlorofluoromethane	10	40	1000	4200
Dichloromethane: See Methylene chloride				
1,1-Dichloro-1-nitroethane	2	10	C 10	C 60
1,2-Dichloropropane: See Propylene dichloride				
Dichloropropene—Skin	1	5	—	—
2,2-Dichloropropionic acid	1	6	—	—
Dichlorotetrafluoroethane	1000	7000	1000	7000

TABLE 4.2 Design Criteria for Air Contaminants *(Continued)*

Substance[b]	Threshold limit values (ACGIH, 1984a)[a]		OSHA standards (OSHA, 1983b)	
	ppm	mg/m³, 25°C	ppm	mg/m³, 25°C
Dichlorvos (DDVP)—Skin	0.1	1	—	1
Dicrotophos—Skin	—	0.25	—	—
Dicyclopentadiene	5	30	—	—
Dicyclopentadienyl iron	—	10	—	—
Dieldrin—Skin	—	0.25	—	0.25
Diethanolamine	3	15	—	—
Diethylamine	10	30	25	75
Diethylaminoethanol—Skin	10	50	10	50
Diethylene triamine—Skin	1	4	—	—
Diethyl ether: See Ethyl ether				
Diethyl ketone	200	705	—	—
Diethyl phthalate	—	5	—	—
Difluorodibromomethane	100	860	100	860
Diglycidyl ether (DGE)	0.1	0.5	C 0.5	C 2.8
Dihydroxybenzene: See Hydroquinone				
Diisobutyl ketone	25	150	50	290
Diisopropylamine—Skin	5	20	5	20
Dimethoxymethane: See Methylal				
Dimethyl acetamide—Skin	10	35	10	35
Dimethylamine	10	18	10	18
4-Dimethylaminoazobenzene	—	—	*	*
Dimethylaminobenzene: See Xylidene				
Dimethylaniline (N,N-dimethylaniline)—Skin	5	25	5	25
Dimethylbenzene: See Xylene				
Dimethyl carbamyl chloride	*	*	—	—
Dimethyl-1,2-dibromo-2,2-dichloroethyl phosphate (Dibrom): See Naled				
Dimethylformamide—Skin	10	30	10	30
2,6-Dimethyl-4-heptanone: See Diisobutyl ketone				
1,1-Dimethylhydrazine—Skin	0.5*	1*	0.5	1
Dimethylphthalate	—	5	—	5
Dimethyl sulfate—Skin	0.1*	0.5*	1	5
Dinitolmide	—	5	—	—
Dinitrobenzene (all isomers)—Skin	0.15	1	—	1
Dinitro-*o*-cresol—Skin	—	0.2	—	0.2
3,5-Dinitro-*o*-toluamide: See Dinitolmide				
Dinitrotoluene—Skin	—	1.5	—	1.5
Dioxane, tech. grade—Skin	25	90	100	360
Dioxathion—Skin	—	0.2	—	—
Diphenyl: See Biphenyl				
Diphenylamine	—	10	—	—
Diphenylmethane diisocyanate: See Methylene bisphenyl isocyanate				
Dipropylene glycol methyl ether—Skin	100	600	100	600
Dipropyl ketone	50	235	—	—
Diquat	—	0.5	—	—
di-*sec*-Octyl phthalate (di-2-ethylhexylphthalate)	—	5	—	5

TABLE 4.2 Design Criteria for Air Contaminants (*Continued*)

Substance[b]	Threshold limit values (ACGIH, 1984a)[a] ppm	Threshold limit values (ACGIH, 1984a)[a] mg/m³, 25°C	OSHA standards (OSHA, 1983b) ppm	OSHA standards (OSHA, 1983b) mg/m³, 25°C
Disulfiram	—	2	—	—
Disulfoton	—	0.1	—	—
2,6-*ditert*-butyl-*p*-cresol	—	10	—	—
Diuron	—	10	—	—
Divinyl benzene	10	50	—	—
Emery	—	D	—	—
Endosulfan—Skin	—	0.1	—	—
Endrin—Skin	—	0.1	—	0.1
Enflurane	(75)	(575)	—	—
Epichlorohydrin—Skin	2	10	5	19
EPN—Skin	—	0.5	—	0.5
1,2-Epoxypropane: See Propylene oxide				
2,3-Epoxy-1-propanol: See Glycidol				
Ethane	E	—	—	—
Ethanethiol: See Ethyl mercaptan				
Ethanolamine	3	8	3	6
Ethion—Skin	—	0.4	—	—
2-Ethoxyethanol—Skin	5	19	200	740
2-Ethoxyethyl acetate—Skin	5	27	100	540
Ethyl acetate	400	1400	400	1400
Ethyl acrylate—Skin	5	20	25	100
Ethyl alcohol (ethanol)	1000	1900	1000	1900
Ethylamine	10	18	10	18
Ethyl amyl ketone	25	130	25	130
Ethyl benzene	100	435	100	435
Ethyl bromide	200	890	200	890
Ethyl butyl ketone	50	230	50	230
Ethyl chloride	1000	2600	1000	2600
Ethylene	E	—	—	—
Ethylene chlorohydrin—Skin	C 1	C 3	5	16
Ethylenediamine	10	25	10	25
Ethylene dibromide—Skin	*	*	20	153.7
Ethylene dichloride	10	40	50	202.4
Ethylene glycol vapor	C 50	C 125	—	—
Ethylene glycol dinitrate—Skin	0.05	0.3	0.2	1
Ethylene glycol methyl ether acetate: See 2-Methoxyethyl acetate				
Ethylene oxide	1*	2*	1[g]	2[g]
Ethyleneimine—Skin	0.5	1	0.5	1*
Ethyl ether	400	1200	400	1200
Ethyl formate	100	300	100	300
Ethylidene chloride: See 1,1-Dichloroethane				
Ethylidene norbornene	C 5	C 25	—	—
Ethyl mercaptan	0.5	1	C 10	C 25
N-Ethylmorpholine—Skin	5	23	20	94
Ethyl silicate	10	85	100	850
Fenamiphos—Skin	—	0.1	—	—
Fensulfothion	—	0.1	—	—
Fenthion—Skin	—	0.2	—	—

TABLE 4.2 Design Criteria for Air Contaminants (*Continued*)

Substance[b]	Threshold limit values (ACGIH, 1984a)[a] ppm	Threshold limit values (ACGIH, 1984a)[a] mg/m³, 25°C	OSHA standards (OSHA, 1983b) ppm	OSHA standards (OSHA, 1983b) mg/m³, 25°C
Ferbam	—	10	—	15
Ferrovanadium dust	—	1	—	1
Fibrous glass dust	—	10	—	—
Fluorides, as F	—	2.5	—	2.5
Fluorine	1	2	0.1	0.2
Fluorotrichloromethane: See Trichlorofluoromethane				
Fonofos—Skin	—	0.1	—	—
Formaldehyde	(1*)	(1.5*)	3	3.7
Formamide	20	30	—	—
Formic acid	5	9	5	9
Furfural—Skin	2	8	5	20
Furfuryl alcohol—Skin	10	40	50	200
Gasoline	300	900	—	—
Germanium tetrahydride	0.2	0.6	—	—
Glass, fibrous or dust: See Fibrous glass dust				
Glutaraldehyde	C 0.2	C 0.7	—	—
Glycerin mist	—	D	—	—
Glycidol (2,3-epoxy-1-propanol)	25	75	50	150
Glycol monoethyl ether: See 2-Ethoxyethanol				
Grain dust	—	(4)	—	—
Graphite (natural): See Table 4.3				
Graphite (synthetic)	—	D	—	—
Gypsum	—	D	—	—
Hafnium	—	0.5	—	0.5
Halothane	(50)	(400)	—	—
Helium	E	—	—	—
Heptachlor—Skin	—	0.5	—	0.5
Heptane (*n*-heptane)	400	1600	500	2000
2-Heptanone: See Methyl *n*-amyl ketone				
3-Heptanone: See Ethyl butyl ketone				
Hexachlorobutadiene	0.02*	0.24*	—	—
Hexachlorocyclopentadiene	0.01	0.1	—	—
Hexachloroethane	10	100	1	10
Hexachloronaphthalene—Skin	—	0.2	—	0.2
Hexafluoroacetone—Skin	0.1	0.7	—	—
Hexamethylphosphoramide—Skin	*	*	—	—
Hexane (*n*-hexane)	50	180	500	1800
Other isomers	500	1800	—	—
2-Hexanone: See Methyl *n*-butyl ketone				
Hexone: See Methyl isobutyl ketone				
sec-Hexyl acetate	50	300	50	300
Hexylene glycol	C 25	C 125	—	—
Hydrazine—Skin	0.1*	0.1*	1	1.3
Hydrogen	E	—	—	—
Hydrogenated terphenyls	0.5	5	—	—
Hydrogen bromide	(C 3)	(C 10)	3	10

TABLE 4.2 Design Criteria for Air Contaminants *(Continued)*

	Threshold limit values (ACGIH, 1984a)[a]		OSHA standards (OSHA, 1983b)	
Substance[b]	ppm	mg/m^3, 25°C	ppm	mg/m^3, 25°C
Hydrogen chloride	C 5	C 7	C 5	C 7
Hydrogen cyanide—Skin	C 10	C 10	10	11
Hydrogen fluoride, as F	(C 3)	(C 2.5)	3	2.5
Hydrogen peroxide	1	1.5	1 (90%)	1.4 (90%)
Hydrogen selenide, as Se	0.05	0.2	0.05	0.2
Hydrogen sulfide	10	14	C 20	C 28
Hydroquinone	—	2	—	2
4-Hydroxy-4-methyl-2-pentanone: See Diacetone alcohol				
2-Hydroxypropyl acrylate—Skin	0.5	3	—	—
Indene	10	45	—	—
Indium and compounds, as In	—	0.1	—	—
Iodine	C 0.1	C 1	C 0.1	C 1
Iodoform	0.6	10	—	—
Iron oxide fume (Fe_2O_3), as Fe	—	5	—	10
Iron pentacarbonyl, as Fe	0.1	0.8	—	—
Iron salts, soluble, as Fe	—	1	—	—
Isoamyl acetate	100	525	100	525
Isoamyl alcohol	100	360	100	360
Isobutyl acetate	150	700	150	700
Isobutyl alcohol	50	150	100	300
Isooctyl alcohol	50	270	—	—
Isophorone	C 5	C 25	25	140
Isophorone diisocyanate—Skin	0.01	0.09	—	—
Isopropoxyethanol	25	105	—	—
Isopropyl acetate	250	950	250	950
Isopropyl alcohol	400	980	400	980
Isopropylamine	5	12	5	12
N-Isopropylaniline—Skin	2	10	—	—
Isopropyl ether	250	1050	500	2100
Isopropyl glycidyl ether (IGE)	50	240	50	240
Kaolin	—	D	—	—
Ketene	0.5	0.9	0.5	0.9
Lead, inorg., dusts and fumes, as Pb	—	0.15	—	0.05
Lead aresnate, as $Pb_3(AsO_4)_2$	—	0.15	—	—
Lead chromate, as Cr	—	0.05*	—	—
Limestone	—	D	—	—
Lindane—Skin	—	0.5	—	0.5
Lithium hydride	—	0.025	—	0.025
LPG (liquified petroleum gas)	1000	1800	1000	1800
Magnesite	—	D	—	—
Magnesium oxide fume	—	10	—	15
Malathion—Skin	—	10	—	15
Maleic anhydride	0.25	1	0.25	1
Manganese, as Mn				
Dust and compounds	—	C 5	—	C 5
Fume	—	1	—	—

TABLE 4.2 Design Criteria for Air Contaminants *(Continued)*

Substance[b]	Threshold limit values (ACGIH, 1984a)[a] ppm	mg/m³, 25°C	OSHA standards (OSHA, 1983b) ppm	mg/m³, 25°C
Manganese cyclopentadienyl tricarbonyl, as Mn—Skin	—	0.1	—	—
Manganese tetroxide	—	1	—	—
Marble/calcium carbonate	—	D	—	—
Mercury, as Hg—Skin				
Alkyl compounds	—	0.01	—	0.01
All forms except alkyl				
Vapor	—	0.05	—	C 0.1
Aryl and inorganic compounds	—	0.1	—	—
Mesityl oxide	15	60	25	100
Methacrylic acid	20	70	—	—
Methane	E	—	—	—
Methanethiol: See Methyl mercaptan				
Methomyl—Skin	—	2.5	—	—
Methoxychlor	—	10	—	15
2-Methoxyethanol—Skin	5	16	25	80
2-Methoxyethyl acetate—Skin	5	24	25	120
4-Methoxyphenol	—	5	—	—
Methyl acetate	200	610	200	610
Methyl acetylene	1000	1650	1000	1650
Methyl acetylene-propadiene mixture (MAPP)	1000	1800	1000	1800
Methyl acrylate—Skin	10	35	10	35
Methylacrylonitrile—Skin	1	3	—	—
Methylal	1000	3100	1000	3100
Methyl alcohol (methanol)—Skin	200	260	200	260
Methylamine	10	12	10	12
Methyl amyl alcohol: See Methyl isobutyl carbinol				
Methyl *n*-amyl ketone	50	235	100	465
N-Methyl aniline—Skin	0.5	2	2	9
Methyl bromide—Skin	5	20	C 20	C 80
Methyl *n*-butyl ketone	5	20	100	410
Methyl chloride	50	105	100	206
Methyl chloroform	350	1900	350	1900
Methyl chloromethyl ether	—	—	—	*
Methyl 2-cyanoacrylate	2	8	—	—
Methylcyclohexane	400	1600	500	2000
Methylcyclohexanol	50	235	100	470
o-Methylcyclohexanone—Skin	50	230	100	460
Methylcyclopentadienyl manganese tricarbonyl, as Mn—Skin	—	0.2	—	—
Methyl demeton—Skin	—	0.5	—	—
Methylene bisphenyl isocyanate (MDI)	C 0.02	C 0.2	C 0.02	C 0.2
Methylene chloride	100	350	500	1737
4,4′-Methylene bis(2-chloroaniline)—Skin	0.02*	0.22*	—	—

TABLE 4.2 Design Criteria for Air Contaminants (*Continued*)

	Threshold limit values (ACGIH, 1984a)[a]		OSHA standards (OSHA, 1983b)	
Substance[b]	ppm	mg/m³, 25°C	ppm	mg/m³, 25°C
Methylene bis(4-cyclohexylisocyanate)	C 0.01	C 0.11	—	—
4,4-Methylene dianiline—Skin	(0.1*)	(0.8*)	—	—
Methyl ethyl ketone (MEK)	200	590	200	590
Methyl ethyl ketone peroxide	C 0.2	C 1.5	—	—
Methyl formate	100	250	100	250
5-Methyl-3-heptanone: See Ethyl amyl ketone				
Methyl hydrazine—Skin	C 0.2*	C 0.35*	C 0.2	C 0.35
Methyl iodide—Skin	2*	10*	5	28
Methyl isoamyl ketone	50	240	—	—
Methyl isobutyl carbinol—Skin	25	100	25	100
Methyl isobutyl ketone	50	205	100	410
Methyl isocyanate—Skin	0.02	0.05	0.02	0.05
Methyl isopropyl ketone	200	705	—	—
Methyl mercaptan	0.5	1	C 10	C 20
Methyl methacrylate	100	410	100	410
Methyl parathion—Skin	—	0.2	—	—
Methyl propyl ketone	200	700	200	700
Methyl silicate	1	6	—	—
α-Methyl styrene	50	240	C 100	C 480
Metribuzin	—	5	—	—
Mevinphos—Skin	0.01	0.1	—	—
Molybdenum, as Mo				
Soluble compounds	—	5	—	5
Insoluble compounds	—	10	—	15
Monocrotophos	—	0.25	—	—
Morpholine—Skin	20	70	20	70
Naled	—	3	—	—
Naphthalene	10	50	10	50
α-Naphthylamine	—	—	—	*
β-Naphthylamine	—	**	—	*
Neon	E	—	—	—
Nickel				
Metal	—	1	—	1
Soluble compounds, as Ni	—	0.1	—	1
Nickel carbonyl, as Ni	0.05	0.35	0.001	0.007
Nickel sulfide roasting, fume and dust, as Ni	—	1**	—	—
Nicotine—Skin	—	0.5	—	0.5
Nitrapyrin	—	10	—	—
Nitric acid	2	5	2	5
Nitric oxide	25	30	25	30
p-Nitroaniline—Skin	—	3	1	6
Nitrobenzene—Skin	1	5	1	5
p-Nitrochlorobenzene—Skin	(0.5)	(3)	—	1
4-Nitrodiphenyl	—	**	—	*
Nitroethane	100	310	100	310

TABLE 4.2 Design Criteria for Air Contaminants *(Continued)*

Substance[b]	Threshold limit values (ACGIH, 1984a)[a]		OSHA standards (OSHA, 1983b)	
	ppm	mg/m³, 25°C	ppm	mg/m³, 25°C
Nitrogen dioxide	3	6	C 5	C 9
Nitrogen trifluoride	10	30	10	29
Nitroglycerin (NG)—Skin	0.05	0.5	C 0.2	C 2
Nitromethane	100	250	100	250
1-Nitropropane	25	90	25	90
2-Nitropropane	(10*)	(35*)	25	90
N-Nitrosodimethylamine (dimethylnitrosoamine)—Skin	—	*	—	*
Nitrotoluene—Skin	2	11	5	30
Nitrotrichloromethane: See Chloropicrin				
Nonane	200	1050	—	—
Octachloronaphthalene—Skin	—	0.1	—	0.1
Octane	300	1450	500	2350
Oil mist, mineral	—	5	—	5
Osmium tetroxide, as Os	0.0002	0.002	—	0.002
Oxalic acid	—	1	—	1
Oxzygen difluoride	(C 0.05)	(C 0.1)	0.05	0.1
Ozone	0.1	0.2	0.1	0.2
Paraffin wax fume	—	2	—	—
Paraquat, respirable sizes—Skin	—	0.1	—	0.5
Parathion—Skin	—	0.1	—	0.1
Particulate polycyclic aromatic hydrocarbons (PPAH): See Coal tar pitch volatiles				
Pentaborane	0.005	0.01	0.005	0.01
Pentachloronaphthalene	—	0.5	—	0.5
Pentachlorophenol—Skin	—	0.5	—	0.5
Pentaerythritol	—	D	—	—
Pentane	600	1800	1000	2950
2-Pentanone: See Methyl propyl ketone				
Perchloroethylene	50	325	100	678
Perchloromethyl mercaptan	0.1	0.8	0.1	0.8
Perchloryl fluoride	3	14	3	13.5
Persulfates, alkali metal, as S_2O_8	—	(5)	—	—
Petroleum distillates (naphtha)	—	—	500	2000
Phenol—Skin	5	19	5	19
Phenothiazine—Skin	—	5	—	—
N-Phenyl-beta-naphthylamine	*	*	—	—
p-Phenylene diamine—Skin	—	0.1	—	0.1
Phenyl ether, vapor	1	7	1	7
Phenyl ether—biphenyl mixture, vapor	—	—	1	7
Phenylethylene: See Styrene, monomer				
Phenyl glycidyl ether (PGE)	1	6	10	60
Phenylhydrazine—Skin	(5*)	(20*)	5	22
Phenyl mercaptan	0.5	2	—	—
Phenylphosphine	C 0.05	C 0.25	—	—

TABLE 4.2 Design Criteria for Air Contaminants (*Continued*)

Substance[b]	Threshold limit values (ACGIH, 1984a)[a] ppm	Threshold limit values (ACGIH, 1984a)[a] mg/m^3, 25°C	OSHA standards (OSHA, 1983b) ppm	OSHA standards (OSHA, 1983b) mg/m^3, 25°C
Phorate—Skin	—	0.05	—	—
Phosdrin: See Mevinphos				
Phosgene	0.1	0.4	0.1	0.4
Phosphine	0.3	0.4	0.3	0.4
Phosphoric acid	—	1	—	1
Phosphorus (yellow)	—	0.1	—	0.1
Phosphorus oxychloride	0.1	0.6	—	—
Phosphorus pentachloride	0.1	1	—	1
Phosphorus pentasulfide	—	1	—	1
Phosphorus trichloride	0.2	1.5	0.5	3
Phthalic anhydride	1	6	2	12
m-Phthalodinitrile	—	5	—	—
Picloram	—	10	—	—
Picric acid—Skin	—	0.1	—	0.1
Pindone	—	0.1	—	—
Piperazine dihydrochloride	—	5	—	—
2-Pivalyl-1,3-indandione: See Pindone				
Plaster of Paris	—	D	—	—
Platinum				
Metal	—	1	—	—
Soluble salts, as Pt	—	0.002	—	0.002
Polychlorobiphenyls: See Chlorodiphenyls				
Polytetrafluoroethylene decomposition products	—	Minimal	—	—
Potassium hydroxide	—	C 2	—	—
Potassium persulfate, as S_2O_8	—	(5)	—	—
Propane	E	—	1000	1800
Propane sultone	*	*	—	—
Propargyl alcohol—Skin	1	2	—	—
β-Propiolactone	0.5*	1.5*	*	*
Propionic acid	10	30	—	—
Propoxur	—	0.5	—	—
n-Propyl acetate	200	840	200	840
Propyl alcohol—Skin	200	500	200	500
n-Propyl nitrate	25	105	25	110
Propylene	E	—	—	—
Propylene dichloride	75	350	75	350
Propylene glycol dinitrate—Skin	0.05	0.3	—	—
Propylene glycol monomethyl ether	100	360	—	—
Propyleneimine—Skin	2*	5*	2	5
Propylene oxide	20	50	100	240
Propyne: See Methyl acetylene				
Pyrethrum	—	5	—	5
Pyridine	5	15	5	15
Quinone	0.1	0.4	0.1	0.4
RDX: See Cyclonite				

TABLE 4.2 Design Criteria for Air Contaminants *(Continued)*

Substance[b]	Threshold limit values (ACGIH, 1984a)[a] ppm	Threshold limit values (ACGIH, 1984a)[a] mg/m³, 25°C	OSHA standards (OSHA, 1983b) ppm	OSHA standards (OSHA, 1983b) mg/m³, 25°C
Resorcinol	10	45	—	—
Rhodium				
Metal	—	1	—	0.1 (fume and dusts)
Insoluble compounds, as Rh	—	1	—	—
Soluble compounds, as Rh	—	0.01	—	0.001 (soluble salts)
Ronnel	—	10	—	15
Rosin core solder pyrolysis products, as formaldehyde	—	0.1	—	—
Rotenone (commercial)	—	5	—	5
Rouge	—	D	—	—
Rubber solvent (Naphtha)	400	1600	100	400 (Naphtha)
Selenium compounds, as Se	—	0.2	—	0.2
Selenium hexafluoride, as Se	0.05	0.2	0.05	0.4
Sesone	—	10	—	—
Silane: See Silicon tetrahydride				
Silicon	—	D	—	—
Silicon carbide	—	D	—	—
Silicon tetrahydride (silane)	5	7	—	—
Silver, as Ag				
Metal	—	0.1	—	0.01
Soluble compounds	—	0.01	—	0.01
Sodium azide	C 0.1	C 0.3	—	—
Sodium bisulfite	—	5	—	—
Sodium 2,4-dichloro-phenoxyethyl sulfate: See Sesone				
Sodium fluoroacetate—Skin	—	0.05	—	0.05
Sodium hydroxide	—	C 2	—	2
Sodium metabisulfite	—	5	—	—
Sodium persulfate, as S_2O_8	—	(5)	—	—
Starch	—	D	—	—
Stibine	0.1	0.5	0.1	0.5
Stoddard solvent	100	525	500	2900
Strychnine	—	0.15	—	0.15
Styrene, monomer	50	215	100	—
Subtilisins (proteolytic enzymes as 100% pure crystalline enzyme)	—	C 0.00006	—	—
Sucrose	—	D	—	—
Sulfotep—Skin	—	0.2	—	0.2
Sulfur dioxide	2	5	5	13
Sulfur hexafluoride	1000	6000	1000	6000
Sulfuric acid	—	1	—	1
Sulfur monochloride	(C 1)	(C 6)	1	6
Sulfur pentafluoride	(C 0.01)	(C 0.1)	0.025	0.25
Sulfur tetrafluoride	(C 0.1)	(C 0.4)	—	—
Sulfuryl fluoride	5	20	5	20
Sulprofos	—	1	—	—

TABLE 4.2 Design Criteria for Air Contaminants (*Continued*)

Substance[b]	Threshold limit values (ACGIH, 1984a)[a] ppm	Threshold limit values (ACGIH, 1984a)[a] mg/m³, 25°C	OSHA standards (OSHA, 1983b) ppm	OSHA standards (OSHA, 1983b) mg/m³, 25°C
Systox: See Demeton				
2,4,5-T	—	10	—	10
Tantalum	—	5	—	5
TEDP: See Sulfotep				
Tellurium and compounds, as Te	—	0.1	—	0.1
Tellurium hexafluoride, as Te	0.02	0.2	0.02	0.2
Temephos	—	10	—	—
TEPP—Skin	0.004	0.05	—	0.05
Terphenyls	C 0.5	C 5	C 1	C 9
1,1,1,2-Tetrachloro-2,2-difluoroethane	500	4170	500	4170
1,1,2,2-Tetrachloro-1,2-difluoroethane	500	4170	500	4170
1,1,2,2-Tetrachloroethane—Skin	1	7	5	35
Tetrachloroethylene: See Perchloroethylene				
Tetrachloromethane: See Carbon tetrachloride				
Tetrachloronaphthalene—Skin	—	2	—	2
Tetraethyl lead, as Pb—Skin	—	0.1	—	0.075
Tetrahydrofuran	200	590	200	590
Tetramethyl lead, as Pb—Skin	—	0.15	—	0.075
Tetramethyl succinonitrile—Skin	0.5	3	0.5	3
Tetranitromethane	1	8	1	8
Tetrasodium pyrophosphate	—	5	—	—
Tetryl (2,4,6-trinitrophenylmethyl-nitramine)—Skin	—	1.5	—	1.5
Thallium				
Soluble compounds, as Tl—Skin	—	0.1	—	0.1
4,4′-Thiobis(6-*tert*-butyl-*m*-cresol)	—	10	—	—
Thioglycolic acid	1	5	—	—
Thionyl chloride	(C 1)	(C 5)	—	—
Thiram	—	5	—	5
Tin				
Metal	—	2	—	—
Oxide and inorganic compounds except SnH_4, as Sn	—	2	—	0.1
Organic compounds, as Sn—Skin	—	0.1	—	0.1
Inorganic compounds, except oxides	—	—	—	2
Titanium dioxide, as Ti	—	D	—	15
o-Tolidine—Skin	*	*	5	22
Toluene (toluol)—Skin	100	375	200	754
Toluene-2,4-diisocyanate (TDI)	0.005	0.04	C 0.02	C 0.14
m-Toluidine—Skin	(2)	(9)	—	—

TABLE 4.2 Design Criteria for Air Contaminants *(Continued)*

Substance[b]	Threshold limit values (ACGIH, 1984a)[a]		OSHA standards (OSHA, 1983b)	
	ppm	mg/m^3, 25°C	ppm	mg/m^3, 25°C
o-Toluidine—Skin	2*	9*	5	22
p-Toluidine—Skin	(2*)	(9*)	—	—
Toxaphene: See Chlorinated camphene				
Tributyl phosphate	0.2	2.5	—	5
Trichloroacetic acid	1	5	—	—
1,2,4-Trichlorobenzene	C 5	C 40	—	—
1,1,1-Trichloroethane: See Methyl chloroform				
1,1,2-Trichloroethane—Skin	10	45	10	45
Trichloroethylene	50	270	100	546
Trichlorofluoromethane	C 1000	C 5600	1000	5600
Trichloromethane: See Chloroform				
Trichloronaphthalene	—	5	—	5
Trichloronitromethane: See Chloropicrin				
1,2,3-Trichloropropane	50	300	50	300
1,1,2-Trichloro-1,2,2-trifluoroethane	1000	7600	1000	7600
Tricyclohexyltin hydroxide: See Cyhexatin				
Triethylamine	10	40	25	100
Trifluorobromomethane	1000	6100	1000	6100
Trimelitic anhydride	0.005	0.04	—	—
Trimethylamine	10	24	—	—
Trimethyl benzene	25	125	—	—
Trimethyl phosphite	2	10	—	—
2,4,6-Trinitrophenol: See Picric acid				
2,4,6-Trinitrophenyl-methylnitramine: See Tetryl				
2,4,6-Trinitrotoluene (TNT)—Skin	—	0.5	—	1.5
Triorthocresyl phosphate	—	0.1	—	0.1
Triphenyl amine	—	5	—	—
Triphenyl phosphate—Skin	—	3	—	3
Tungsten, as W				
Insoluble compounds	—	5	—	—
Soluble compounds	—	1	—	—
Turpentine	100	560	100	560
Uranium (natural), soluble and insoluble compounds, as U	—	0.2	—	0.05 (sol.) 0.25 (insol.)
Valeraldehyde	50	175	—	—
Vanadium, as V_2O_5, respirable dust and fume	—	0.05	—	(dust) C 0.5 (fume) C 0.1
Vegetable oil mists	—	D	—	—
Vinyl acetate	10	30	—	—
Vinyl benzene: See Styrene				
Vinyl bromide	5*	20*	—	—
Vinyl chloride	5**	10**	1*	2.6*
Vinyl cyanide: See Acrylonitrile	—	—	—	—
Vinyl cyclohexene dioxide	10*	60*	—	—
Vinylidene chloride	5	20	—	—
Vinyl toluene	50	240	100	480
VM & P Naphtha	300	1350	—	—
Warfarin	—	0.1	—	0.1
Welding fumes	—	5	—	—
Wood dust (certain hardwoods as beech and oak)	—	1	—	—
Soft wood	—	5	—	—

TABLE 4.2 Design Criteria for Air Contaminants *(Continued)*

Substance[b]	Threshold limit values (ACGIH, 1984a)[a] ppm	Threshold limit values (ACGIH, 1984a)[a] mg/m^3, 25°C	OSHA standards (OSHA, 1983b) ppm	OSHA standards (OSHA, 1983b) mg/m^3, 25°C
Xylene (*o*, *m*, *p* isomers)	100	435	100	435
m-Xylene α,α'-diamine—Skin	—	C 0.1	—	—
Xylidine—Skin	2	10	5	25
Yttrium	—	1	—	1
Zinc chloride fume	—	1	—	1
Zinc chromate	—	0.05*	—	—
Zinc oxide				
Fume	—	5	—	5
Dust	—	D	—	—
Zinc stearate	—	D	—	—
Zirconium compounds, as Zr	—	5	—	5

[a]Values listed in parentheses are intended changes (ACGIH, 1984).
*Possible carcinogen.
**Human carcinogen.
[b]Skin: Indicates that skin absorption may also occur. TLVs are based only on airborne exposures.
C: Indicated value is ceiling limit as well as 8-hr, time-weighted TLV.
D: Nuisance dust; TLV is 10 mg/m^3 total dust with <1% quartz or 5 mg/m^3 respirable dust.
E: Acts primarily as simple asphyxiant without other significant physiologic effects.
[f]Includes anthracene, BaP, phenanthrene, acridine, chrysene, and pyrene.
[g]Promulgated August 21, 1984 (OSHA, 1984), but struck down by the 4th U.S. Circuit Court of Appeals in 2–1 decision November 7, 1984.
[h]See Table 4.3 as well.

TABLE 4.3 Design Criteria for Mineral Dusts, Cotton Dusts, and Nuisance Particulates

Substance	Threshold limit values, 8-hr, time-weighted (ACGIH, 1984a) mg/m^3: Respirable dust[b]	Threshold limit values, 8-hr, time-weighted (ACGIH, 1984a) mg/m^3: Total dust[b]	OSHA standards, 8-hr, time-weighted (OSHA, 1983b) mg/m^3: Respirable dust	OSHA standards, 8-hr, time-weighted (OSHA, 1983b) mg/m^3: Total dust
Silica, SiO_2				
Crystalline				
Quartz	(0.1)	(0.3)	$\frac{10}{\%\ SiO_2 + 2}$	$\frac{30}{\%\ SiO_2 + 2}$
Cristobalite	(0.05)	(0.15)	$\frac{5}{\%\ SiO_2 + 2}$	$\frac{15}{\%\ SiO_2 + 2}$
Tridymite	(0.05)	(0.15)	$\frac{5}{\%\ SiO_2 + 2}$	$\frac{15}{\%\ SiO_2 + 2}$
Silica, fused	(0.1)	(0.3)	—	—
Tripoli	(0.1)	(0.3)	—	—
Amorphous		—	$\frac{80}{\%\ SiO_2}$ (includes natural diatomaceous earth)	
Diatomaceous earth (uncalcined)	—	(10)	—	—
Precipitated silica and silica gel	(5)	(10)	—	—

TABLE 4.3 Design Criteria for Mineral Dusts, Cotton Dusts, and Nuisance Particulates *(Continued)*

Substance	Threshold limit values, 8-hr, time-weighted (ACGIH, 1984a) mg/m³: Respirable dust[b]	Threshold limit values: Total dust[b]	OSHA standards, 8-hr, time-weighted (OSHA, 1983b) mg/m³: Respirable dust	OSHA standards: Total dust
Silicates and other dusts (<1% quartz)				
Asbestos	—		0.2 fibers/cm³ > 5 μm[g]	
Amosite[a]	0.5 fibers/cm³ > 5 μm[g]		—	—
Chrysotile[a]	2 fibers/cm³ > 5 μm		—	—
Crocidolite[a]	0.2 fibers/cm³ > 5 μm		—	—
Other forms[a]	2 fibers/cm³ > 5 μm		—	—
Barium sulfate	—	(10)	—	—
Graphite (natural)	(2.5)	(5)	15 Mppcf[c]	
Graphite (synthetic)	—	(10)	—	—
Mica	(0.3)	—	20 Mppcf[c]	
Mineral wool fiber	—	10	—	—
Perlite	5[c]	(10)	—	—
Portland cement	5[c]	(10)	50 Mppcf[c]	
Soapstone	(3)	(6)	20 Mppcf[c]	
Talc (containing no asbestos fibers)	2	—	20 Mppcf[c]	
Talc (containing asbestos fibers)	[d]	—	Use asbestos limit	
Tremolite	—	—	See talc, containing asbestos fibers	
Coal dust				
Respirable dust (>5% quartz)	(0.1)	—	$\frac{10}{\% SiO_2 + 2}$	—
Respirable dust (<5% quartz)	2	—	2.4	—
Nuisance dust[e]				
Total dust <1% quartz	5	10	5	15
Cotton dust				
Respirable (<15 μm), lint-free	0.2			
Yarn manufacture			0.2[f]	
Slashing and weaving			0.75[f]	
Cotton waste			1.0	
Processing operations of waste recycling (sorting, blending, cleaning, and willowing) and garnetting				

[a]Human carcinogen.

[b]Values listed in parentheses are intended changes (ACGIH, 1984a).

[c]Mppcf = million particles per cubic foot of air based on impinger samples counted by light field technics. For silicates, when mass mean diameter (mmd) and particle density are not available, an approximate conversion factor, assuming mmd = 1.5 μm and particle density = 2.5 g/cm³, is 6 Mppcf ≃ 1 mg/m³ of respirable dust (ACGIH, 1984a).

[d]Use asbestos TLV for specific type of asbestos, but talc concentration should not exceed 2 mg/m³.

[e]Nuisance dusts include α-alumina, calcium carbonate, calcium silicate, cellulose, emery, glycerin mist, graphite (synthetic), gypsum, kaolin, limestone, magnesite, marble, mineral wool fiber, pentaerythritol, plaster of Paris, Portland cement, rouge, silicon, silicon carbide, starch, sucrose, titanium dioxide, vegetable oil mists (except castor, cashew nut, or similar irritant oils), zinc stearate, and zinc oxide dust.

[f]OSHA (1985).

[g]OSHA (1986a); "asbestos" includes chrysotile, amosite, crocidolite, tremolite asbestos, anthophyllite asbestos, and actinolite asbestos; "fiber" includes all particulate forms >5μm with a length-to-diameter ratio of at least 3:1.

Biological exposure indices (BEIs), when available, are useful supplementary indicators in determining whether the workplace is healthful. The BEI is a measure of how much of a pollutant is actually absorbed. A number of predictive equations for the effect of air exposure on biological uptake were given in Chapter 3. Some of these were used to establish the BEI guidelines for 8-hr/day, 40-hr/week exposures given in Table 4.4. When the BEI levels are not exceeded, most workers are not expected to incur ill effects. However, for some of the chemicals there are identifiable population groups that may be at risk at values less than the BEI. For instance, individuals with heart problems are likely to continue to be at risk at the BEI COHb level of 8%.

TABLE 4.4 Suggested Biological Exposure Indices for 8-hr Days and 40-hr Weeks

Chemical	Timing of sample	BEI
Carbon monoxide		
Carboxyhemoglobin in blood, COHb[a]	End of shift	<8%
CO in end-exhaled air[a]	End of shift	<40 ppm
Ethyl benzene		
Mandelic acid in urine[a]	End of shift	2 g/liter[c]
	End of work week	1.5 g/g creatinine[c]
Ethyl benzene in end-exhaled air[b]	Prior to shift	2 ppm
Styrene monomer		
Mandelic acid in urine[a]	End of shift	1.0 g/liter[c], 0.8 g/g creatinine[c]
Styrene in mixed exhaled air	Prior to shift	40 ppb
Phenylglyoxylic acid in urine[a]	End of shift	250 mg/liter[c], 240 mg/g creatinine[c]
Styrene in mixed exhaled air[b]	During shift	18 ppm
Styrene in venous blood[b]	End of shift	0.55 mg/liter
	Prior to shift	0.02 mg/liter
Toluene		
Hippuric acid in urine[a]	End of shift	2.5 g/g creatinine[c]
	Last 4 hr of shift (timed sample)	3 mg/min[c]
Toluene in venous blood[b]	End of shift	1.0 mg/liter
Toluene in end-exhaled air[b]	During shift	20 ppm
Trichloroethylene		
Trichloroacetic acid in urine[a]	End of work week	100 mg/liter[c]
Total trichloroacetic acid and trichloroethanol in urine[a]	End of shift and end of work week	300 mg/liter[c], 320 mg/g creatinine[c]
Free trichloroethanol in blood[a]	End of shift and end of work week	4 mg/liter
Trichloroethylene in end-exhaled air	Prior to shift and end of work week	0.5 ppm
Xylene		
Methyl hippuric acids in urine	End of shift	1.5 g/g creatinine
	Last 4 hr of shift (timed sample)	2 mg/min

[a]Quantitative but not specific test for chemical.
[b]Should be used only as a confirmatory test, since quantitative interpretation is ambiguous.
[c]Because of individual variation, these values should be considered a mean value of a group of workers rather than an index for an individual.
SOURCE: ACGIH, 1984a.

General ventilation guidelines also exist for indoor air (ASHRAE, 1986). However, these are based on maintaining CO_2 exposures at less than 1000 ppm given human habitation. Usually, TLVs for specific materials will result in greater ventilation rates than those determined using the CO_2 level in indoor air.

III. Physical Agents

A. Heat stress

The ACGIH TLV guideline for limiting heat exposure is based on preventing deep-body temperatures from exceeding 38°C (ACGIH, 1984a). Since measurement of this parameter is not practical in work situations, the wet-bulb globe temperature index (WBGT) has been recommended as the most practical substitute. The WBGT requires three different temperature measurements and is calculated from:

$$\text{WBGT} = 0.7\ \text{NWB} + 0.2\ \text{GT} + 0.1\ \text{DB} \quad \text{(outdoors with solar load)} \tag{4.1a}$$

$$\text{WBGT} = 0.7\ \text{NWB} + 0.3\ \text{GT} \quad \text{(indoors or outdoors with no solar load)} \tag{4.1b}$$

where NWB is the natural wet-bulb temperature, DB is the dry-bulb temperature, GT is the globe temperature, and all measurements are in degrees Celsius. Further discussion on the meaning and application of this and other heat stress parameters is given in Chapter 12. Table 4.5 shows recommended maximum values of WBGT. Higher exposures are permissible under special conditions of close medical surveillance, but in no case is the deep-body temperature to exceed 38°C. The total heat load is the sum of environmental heat and the workload. Workload is estimated from the nature of the work activity and the basal metabolic rate. Table 4.6 shows typical values for different activities.

The workload is then the sum of appropriate values from each of the three categories in Table 4.6. The WBGT guideline is determined from Table 4.5 using the criteria that <200 kcal/hr corresponds to light work, 200–350 kcal/hr corresponds to moderate work, and >350 kcal/hr corresponds to heavy work. When rest times are taken under quite different conditions than exist at the worksite, e.g., in an air conditioned room, appropriate weighting is suggested to determine the workload (ACGIH, 1984a). At present there are no OSHA standards specifying heat stress restrictions.

TABLE 4.5 Permissible Heat Exposure, Threshold Limit Values, °C, WBGT

Work-rest regimen	Workload: Light	Workload: Moderate	Workload: Heavy
Continuous work	30.0	26.7	25.0
75% work, 25% rest each hour	30.6	28.0	25.9
50% work, 50% rest each hour	31.4	29.4	27.9
25% work, 75% rest each hour	32.2	31.1	30.0

SOURCE: ACGIH, 1984a.

TABLE 4.6 Heat Generation for Different Activities

Body position and movement		Type of work		Basal metabolism
Activity	kcal/min	Activity	kcal/min	kcal/min
Sitting	0.3	Hand		1.0
Standing	0.6	Light	0.4	
Walking	2.0–3.0	Heavy	0.9	
Walking uphill	Add 0.8/m of rise	One arm		
		Light	1.0	
		Heavy	1.8	
		Two arms		
		Light	1.8	
		Heavy	2.8	
		Whole body		
		Light	3.5	
		Moderate	5.0	
		Heavy	7.0	
		Very heavy	9.0	

SOURCE: ACGIH, 1984a.

B. Noise

As is discussed in more detail in Chapter 13, sound levels are often expressed in decibels on the A-weighted scale of sound-level meters (dBA). TLVs and PELs in dBA are given in Table 4.7. These are based generally on the type of hearing loss data presented in Chapter 2, and consequently neither of these guidelines will ensure that no hearing loss will take place in exposed workers. An alternative design approach involves control of noise at various frequencies to levels that do not cause hearing loss in 50% or 95% of the working population (see Chapter 2). Another OSHA requirement (1981, 1983a, 1983b) is the institution of a hearing conservation program whenever workplace exposures are greater than 85 dBA.

TABLE 4.7 TLVs and PELs for Noise Exposures

Duration, hr/day	Sound levels, dBA TLV[a]	Sound levels, dBA PEl[b]
Continuous or intermittent		
16	80	—
8	85	90
6		92
4	90	95
3		97
2	95	100
1	100	105
$\frac{1}{2}$	105	110
$\frac{1}{4}$	110	115 (maximum)
$\frac{1}{8}$	115 (maximum)	115
Impact[c]		
Impulses/day		
10,000	120[d]	140[d] (maximum)
1,000	130	140
100	140 (maximum)	140

[a]ACGIH, 1984a.
[b]OSHA, 1983b.
[c]Impact noise are those variations in noise level that involve maxima at intervals of more than 1 sec; intervals of less than 1 sec are considered continuous noise.
[d]Decibel peak sound pressure, not dBA, re 20 μPa.

C. Mechanical vibration

Some of the health implications of hand-arm segmental vibration have been reviewed in Chapter 2. Table 4.8 shows suggested TLVs for various exposures to frequency-weighted root-mean-square (rms) acceleration. The vibration can occur in any of three directions, X_h, Y_h, or Z_h. The basis of the recommendation is to prevent vibration white finger disease after a working lifetime exposure. But because of individual susceptibilities, the TLV is not considered to define the boundary between safe and dangerous levels (ACGIH, 1984b).

Whole-body vibration (head to toe), such as is experienced by drivers of large vehicles, may be associated with low back and vertebrogenic

TABLE 4.8 TLVs for Segmental Vibration

Daily exposure, t	Values of the dominant,[a] frequency-weighted, rms component acceleration not to be exceeded, a_k	
hr	m/sec^2	g^b
$4 < t < 8$	4	0.40
$2 < t < 4$	6	0.61
$1 < t < 2$	8	0.81
$t < 1$	12	1.22

[a]Usually one axis of vibration is dominant over the remaining two axes. Frequency weighting carried out with a network specified in ISO (1984).
[b]g = 9.81 m/sec^2.
SOURCE: ACGIH, 1984a.

pain, and degenerative disk disease (NIOSH, 1983). At very low frequencies resonance may take place, i.e., the frequency at which the body and organs are most susceptible to mechanical vibration. Whole-body resonance occurs between 4 and 8 Hz, nominally 5 Hz; head and shoulders, 20–30 Hz; and eyeballs, 60–90 Hz. Manual tracking is affected at 5 Hz and visual acuity temporarily impaired from 1 to 25 Hz (ACGIH, 1984b). At this point no satisfactory guideline exists for industrial settings, although one has been proposed (ISO, 1978). No OSHA PELs exist for any type of vibration.

IV. Equipment Specification

A. Machine guarding

The general OSHA requirements for machine guarding and anchoring are (OSHA, 1983b, §1910.212):

> (a) Machine guarding—(1) Types of guarding. One or more methods of machine guarding shall be provided to protect the operator and other employees in the machine area from hazards such as those created by point of operation, ingoing nip points, rotating parts, flying chips and sparks. Examples of guarding methods are—barrier guards, two-hand tripping devices, electronic safety devices, etc.
>
> (2) General requirements for machine guards. Guards shall be affixed to the machine where possible and secured elsewhere if for any reason attachment to the machine is not possible. The guard shall be such that it does not offer an accident hazard in itself.
>
> (3) Point of operation guarding. (i) Point of operation is the area on a machine where work is actually performed upon the material being processed.

(ii) The point of operation of machines whose operation exposes an employee to injury, shall be guarded. The guarding device shall be in conformity with any appropriate standards therefor, or, in the absence of applicable specific standards, shall be so designed and constructed as to prevent the operator from having any part of his body in the danger zone during the operating cycle.

(iii) Special handtools for placing and removing material shall be such as to permit easy handling of material without the operator placing a hand in the danger zone. Such tools shall not be in lieu of other guarding required by this section, but can only be used to supplement protection provided.

(iv) The following are some of the machines which usually require point of operation guarding:

(a) Guillotine cutters.
(b) Shears.
(c) Alligator shears.
(d) Power presses.
(e) Milling machines.
(f) Power saws.
(g) Jointers.
(h) Portable power tools.
(i) Forming rolls and calenders.

(4) Barrels, containers, and drums. Revolving drums, barrels, and containers shall be guarded by an enclosure which is interlocked with the drive mechanism, so that the barrel, drum, or container cannot revolve unless the guard enclosure is in place.

(5) Exposure of blades. When the periphery of the blades of a fan is less than seven (7) feet above the floor or working level, the blades shall be guarded. The guard shall have openings no larger than one-half ($\frac{1}{2}$) inch.

(b) Anchoring fixed machinery. Machines designed for a fixed location shall be securely anchored to prevent walking or moving.

The principles of appropriate guarding design are given in Chapter 9. OSHA has also designated guarding and control requirements for woodworking machinery (§1910.213), abrasive wheel machinery (§1910.215), mills and calenders in the rubber and plastics industry (§1910.216), mechanical power presses (§1910.217), forging machines (§1910.218), and hand and portable powered tools (§1910.241–244). Mechanical power-transmission apparatus is covered in §1910.219, and the specifications for prime-mover guards are [OSHA, 1983b, §1910.219(b)]:

(b) Prime-mover guards—(1) Flywheels. Flywheels located so that any part is seven (7) feet or less above floor or platform shall be guarded in accordance with the requirements of this subparagraph:

(i) With an enclosure of sheet, perforated, or expanded metal, or woven wire:

(ii) With guard rails placed not less than fifteen (15) inches nor more than twenty (20) inches from rim. When flywheel extends into pit or is within 12 inches of floor, a standard toeboard shall also be provided;

(iii) When the upper rim of flywheel protrudes through a working floor, it shall be entirely enclosed or surrounded by a guardrail and toeboard.

(iv) For flywheels with smooth rims five (5) feet or less in diameter, where the preceding methods cannot be applied, the following may be used: A disk attached to the flywheel in such manner as to cover the spokes of the wheel on the exposed side and present a smooth surface and edge, at the same time providing means for periodic inspection. An open space, not exceeding four (4) inches in width, may be left between the outside edge of the disk and the rim of the wheel if desired, to facilitate turning the wheel over. Where a disk is used, the keys or other dangerous projections not covered by disk shall be cut off or covered. This subdivision does not apply to flywheels with solid web centers.

(v) Adjustable guard to be used for starting engine or for running adjustment may be provided at the flywheel of gas or oil engines. A slot opening for jack bar will be permitted.

(vi) Wherever flywheels are above working areas, guards shall be installed having sufficient strength to hold the weight of the flywheel in the event of a shaft or wheel mounting failure.

(2) Cranks and connecting rods. Cranks and connecting rods, when exposed to contact, shall be guarded in accordance with paragraphs (m) and (n) of this section, or by a guardrail as described in paragraph (o)(5) of this section.

(3) Tail rods or extension piston rods. Tail rods or extension piston rods shall be guarded in accordance with paragraphs (m) and (o) of this section, or by a guardrail on sides and end, with a clearance of not less than fifteen (15) nor more than twenty (20) inches when rod is fully extended.

•
•
•

(m) Standard guards—general requirements—(1) Materials. (i) Standard conditions shall be secured by the use of the following materials. Expanded metal, perforated or solid sheet metal, wire mesh on a frame of angle iron, or iron pipe securely fastened to floor or to frame of machine.

(ii) All metal should be free from burrs and sharp edges.

(2) Methods of manufacture. (i) Expanded metal, sheet or perforated metal, and wire mesh shall be securely fastened to frame.

(n) [Reserved]

(o) Approved materials—

•
•
•

(5) Guardrails and toeboards. (i) Guardrail shall be forty-two (42) inches in height, with midrail between top rail and floor.

(ii) Posts shall be not more than eight (8) feet apart; they are to be permanent and substantial, smooth, and free from protruding nails, bolts, and splinters. If made of pipe, the post shall be one and one-fourth ($1\frac{1}{4}$) inches inside diameter, or larger. If made of metal shapes or bars, their section shall be equal in strength to that of one and one-half ($1\frac{1}{2}$) by one and one-half ($1\frac{1}{2}$) by three-sixteenths ($\frac{3}{16}$) inch angle iron. If made of wood, the posts shall be two by four (2 × 4) inches or larger. The upper rail shall be two by four (2 × 4) inches, or two one by four (1 × 4) strips, one at the top and one at the side of posts. The midrail may be one by four (1 × 4) inches or more. The rails (metal shapes, metal bars, or wood), should be on that side of the posts which gives the best protection and support. Where panels are fitted with expanded metal or wire mesh as noted in Table 0–12 the middle rails may be omitted. Where guard is exposed to contact with moving equipment, additional strength may be necessary.

(iii) Toeboards shall be four (4) inches or more in height, of wood, metal, or of metal grill not exceeding one (1) inch mesh. Toeboards at flywheel pits should preferably be placed as close to edge of the pit as possible.

B. Fire protection

Fire and explosion hazards are associated with many airborne gases, vapors, and dusts. The list of flammable gases and vapors presented in Table 4.9 includes substances from the TLV list (ACGIH, 1984a) and commonly used industrial solvents. The table gives available data on boiling point, flash point, lower and upper flammable limits, and dry time relation. The flashpoint of a material is the minimum temperature at which vapor escapes from a liquid or solid in a test apparatus such that an ignitable mixture is formed with air near the surface. The minimum concentration of vapor in air below which propagation of flame does not occur on contact with an ignition source is called the lower flammable limit (LFL). The maximum concentration above which propagation of flame does not occur is called the upper flammable limit (UFL). The upper flammable limit can be deceptive. For example, an empty gasoline tank may contain vapors above the UFL of 7.6%. As a combustible gas meter probe is inserted into the tank, it will pass through a vapor interface that is within the narrow flammable range. The meter indicator will jump quickly and then settle back to zero once the UFL concentration is exceeded in the meter. An unwary tester might certify that the tank is free of residual gasoline vapor when it is not. Dry time relation compares the evaporation rate of a substance to that of ethyl ether. For example, acetaldehyde evaporates three times slower than ethyl ether from a nonporous surface.

(*Text continues on page 157.*)

TABLE 4.9 Characteristics of Flammable Gases and Vapors

Substance	Boiling point, °C	Flash point, °C	Flammable limits, percent by volume		Dry time relation
			Lower	Upper	
Acetaldehyde	21	−39	4.0	60	3
Acetic acid (glacial)	118	39	4.0	20.0	11.0
Acetic anhydride	140	49	2.7	10.3	17.4
Acetone	56	−20	2.5	13	1.9
Acetonitrile	82	6	3.0	16.0	5.2
Acetylene	−83	—	2.5	100	Gas
Acetylene tetrabromide	243	—			>100
Acrolein	52	−26	2.8	31	
Acrylic acid (glacial)	142	50	2.4	8.0	
Acrylonitrile	77	0	3.0	17	
Allyl alcohol	97	21	2.5	18.0	8.5
Allyl chloride	45	−32	2.9	11.1	2.0
Allyl glycidyl ether	154	57			
2-Aminopyridine	211	68			
Ammonia, anhydrous	−33	—	15	28	Gas
n-Amyl acetate	149	16–21	1.1	7.5	11.6
sec-Amyl acetate	121	32	1		15
n-Amyl alcohol	138	33	1.2	10.0 (100)	32.1
Aniline	184	70	1.3	11	200
o-Anisidine	224	118			
Arsine	−63	230a			
Asphalt (petroleum)		>205			
Benzene	80	−11	1.3	7.9	2.8
Benzoyl peroxide	a	80			
Benzyl chloride	179	67			
Biphenyl	254	113	0.6 (111)	5.8 (155)	

1,3-Butadiene	−4	—	2.0	12.0	Gas
Butane	−1	—	1.6	8.4	Gas
2-Butoxyethanol	171	62	1.1	12.7	85
			(93)	(135)	
n-Butyl acetate	127	22	1.7	7.6	7.8
sec-Butyl acetate	112	31	1.7	9.8	5.2
tert-Butyl acetate	96	17–22	1.5		7.8
Butyl acrylate	145	48	1.5	9.9	
n-Butyl alcohol	117	37	1.4	11.2	19.6
sec-Butyl alcohol	94	24	1.7	9.8	12.3
			(100)	(100)	
tert-Butyl alcohol	83	11	2.4	8.0	11.9
Butylamine	78	−12	1.7	9.8	5.1
n-Butyl glycidyl ether	164	58			
n-Butyl lactate	160	71			>145
Butyl mercaptan	98	2			
n-Butyl propionate	145	32			16.5
p-*tert*-Butyltoluene	193	68			
Camphor	204	66	0.6	3.5	
Carbon disulfide	46	−30	1.3	50.0	1.6
Carbon monoxide	−192	—	12.5	74	Gas
Carbon tetrachloride	77	—			2.6
Chlorinated camphene	a	135			
Chlorine dioxide	10	100a	10		
Chloroacetaldehyde	90	88			
α-Chloroacetophenone	247	118			
Chlorobenzene	132	28	1.3	9.6	8.2
Chlorodiphenyl (42% Cl)	325–366	176			
Chlorodiphenyl (54% Cl)	365–390	222			
Chloroform	61	—			2.2
bis-Chloromethyl ether	106				
Chloromethyl methyl ether	59	−9			
1-Chloro-1-nitropropane	141	62			17.9
β-Chloroprene	59	−20	4	20	

TABLE 4.9 Characteristics of Flammable Gases and Vapors *(Continued)*

Substance	Boiling point, °C	Flash point, °C	Flammable limits, percent by volume		Dry time relation
			Lower	Upper	
p-Chlorotoluene	179	67	1.1		48
Cresol	191–201	81–86	1.1–1.4		>400
			(150)		
Crotonaldehyde	102	13	2.1	15.5	4.9
Cumene	152	36	0.9	6.5	13.9
Cyanamide	260	141			
Cyanogen	−21	—	6.6	32	Gas
Cyclohexane	82	−20	1.3	8	2.6
Cyclohexanol	161	68	2.4		150
			(68)		
Cyclohexanone	156	44	1.1	9.4	22.2
			(100)		
Cyclohexene	83	< −7	1.2	4.8	
			(100)	(100)	
Cyclohexylamine	134	31			82.9
Cyclopentadiene	42	<32			
Cyclopentane	49	< −7	1.5		
Decaborane	213	<80			
Diacetone alcohol	164	64	1.8	6.9	60
Diborane	−92	—	0.8	88	Gas
2-N-dibutylaminoethanol	224	93			
Dibutyl phthalate	340	157	0.5		200
			(235)		
o-Dichlorobenzene	180	66	2.2	9.2	39.7
p-Dichlorobenzene	174	66	2.5		
			(66)		
1,1-Dichloroethane	57–59	−6	5.6	15.9	1.9
1,2-Dichloroethylene	48	2	5.6	12.8	2.5
Dichloroethyl ether	178	55	2.7		200

1,1-Dichloro-1-nitroethane	124	76			12.7
1,3-Dichloropropene	104	35	5.3	14.5	
2,3-Dichloropropene	94	15	2.6	7.8	
Dicyclopentadiene	172	32			
Diethanolamine	268	172			>5000
Diethylamine	57	−23	1.8	10.1	2.2
Diethylaminoethanol	162	60	6.7	11.7	
Diethyl carbonate	126	25			9.0
Diethylene triamine	207	98	2	6.7	>400
Diethyl ketone	103	13	1.6		
Diethyl phthalate	296	161	0.7 (186)		>300
Diisobutyl ketone	168	49	0.8 (93)	7.1 (93)	30.8
Diisopropylamine	84	−1	1.1	7.1	3.1
Dimethyl acetamide	165	70	1.8 (100)	11.5 (160)	>100
Dimethylamine	7	—	2.8	14.4	Gas
Dimethylaniline	224	97	1.0		90
Dimethylformamide	153	58	2.2 (100)	15.2	45
1,1-Dimethylhydrazine	63	−15	2	95	
Dimethylphthalate	282	146	0.9 (180)		>200
Dimethyl sulfate	188a	83			
Dinitrobenzene	318	150			
Dinitrotoluene	300	207			
Dioxane	101	12	2.0	22	5.8
Dipentene	170	45	0.7 (150)	6.1 (150)	89
Diphenylamine	302	153			
Dipropylene glycol methyl ether	190	85			
Dipropyl ketone	143	49			34
Di-*sec*-octyl phthalate	231	218	0.3 (245)		

TABLE 4.9 Characteristics of Flammable Gases and Vapors *(Continued)*

Substance	Boiling point, °C	Flash point, °C	Flammable limits, percent by volume		Dry time relation
			Lower	Upper	
Distilled water	100	—			60
2,6-*ditert*-Butyl-*p*-cresol	265	127			
Divinyl benzene	200	76	1.1	6.2	
Epichlorohydrin	115	31	3.8	21.0	17.0
Ethane	−89	Gas	3.0	12.5	
Ethanolamine	172	85	5.5	17	>5000
2-Ethoxyethanol	135	43	1.7	15.6	28.1
			(93)	(93)	
2-Ethoxyethyl acetate	156	47	1.7	12.7	32.4
Ethyl acetate	77	−4	2.0	11.5	2.7
Ethyl acrylate	99	10	1.4	14	
Ethyl alcohol	78	13	3.3	19	7
Ethylamine	17	<−18	3.5	14	Gas
Ethyl amyl ketone	168	57			37.2
Ethyl benzene	136	21	1.0	6.7	9.4
Ethyl bromide	38	—	6.8	8.0	
Ethyl butyl ketone	148	46	1.4	8.8	
Ethyl chloride	12	−50	3.8	15.4	Gas
Ethylene	−104	—	2.7	36.0	Gas
Ethylene chlorohydrin	130	60	4.9	15.9	23.9
Ethylenediamine	116	34	4.2	14.4	>5000
Ethylene dibromide	132	—			12.4
Ethylene dichloride	84	13	6.2	16	3.3
Ethylene glycol	197	111	3.2	15.3	
Ethylene oxide	11	−20	3.0	100	
Ethyleneimine	56	−11	3.6	46	
Ethyl ether	35	−45	1.9	36.0	1.0
Ethyl formate	54	−20	2.8	16.0	1.8

2-Ethyl hexyl acetate	199	71	0.8	8.1	152
Ethylidene norbornene	67 @ 50 mmHg	38			
Ethyl mercaptan	35	< −18	2.8	18.0	40
N-ethylmorpholine	138	32	1.0	9.8	
Ethyl silicate	168	52	1.3	23	
Formaldehyde	−19	—	7.0	73	Gas
Formamide	210	154			<300
Formic acid	101	69	18	57	
Fuel oil No. 1 (kerosene)	151–301	43–72	0.7	5	627
Furfural	161	60	2.1	19.3	60
Furfuryl alcohol	171	75	1.8	16.3	
Gasoline	38–204	−43	1.4	7.6	2.5
Glycerin	290a	160			
Glycidol	167a	72			
Heptane	98	−4	1.1	6.7	2.7
Hexachloronaphthalene	343–387	180			
Hexane	69	−22	1.1	7.5	1.9
sec-Hexyl acetate	141	45	0.9 (45)		14
sec-Hexyl alcohol	140	58			42
Hexylene glycol	197	102	1.2	8.1	>300
Hydrazine	113	38	2.9	98	
Hydrogen	−252	Gas	4.0	75	
Hydrogen cyanide	26	−18	5.6	40.0	
Hydrogen sulfide	−60	—	4.0	44.0	Gas
Hydroquinone	286	165			
2-Hydroxypropyl acrylate	191	65	1.8		
Indene	182	100			
Iron pentacarbonyl	105	−15			
Isoamyl acetate	143	25	1.0 (100)	7.5	12.2
Isoamyl alcohol	132	43	1.2	9.0 (100)	30.7
Isobutyl acetate	118	18	1.3	10.5	5.8

TABLE 4.9 Characteristics of Flammable Gases and Vapors *(Continued)*

Substance	Boiling point, °C	Flash point, °C	Flammable limits, percent by volume		Dry time relation
			Lower	Upper	
Isobutyl alcohol	107	28	1.7	10.6	16.3
			(51)	(94)	
Isooctyl alcohol	182–195	82			300
Isophorone	215	84	0.8	3.8	200
Isophorone diisocyanate	158 @ 10 mmHg				
Isopropoxyethanol	140	49			
Isopropyl acetate	90	2	1.8	8	3.0
Isopropyl alcohol	83	12	2.0	12	7.7
Isopropylamine	32	−37	2.0	10.4	1.6
Isopropyl ether	69	−28	1.4	7.9	1.4
Isopropyl glycidyl ether	127	33			
Liquefied petroleum gas (LPG)	0	−75	1.9	9.5	Gas
Maleic anhydride	202	102	1.4	7.1	
Mesityl oxide	130	31	1.4	7.2	8.4
Methacrylic acid	158	77			
Methane	−162	Gas	5.0	15.0	
2-Methoxyethanol	124	39	1.8	14	21
2-Methoxyethyl acetate	145	49	1.5	12.3	22
			(93)	(93)	
Methyl acetate	60	−10	3.1	16	2.2
Methyl acetylene	−23	—	1.7	11.7	Gas
Methyl acetylene-propadiene mixture	−34	—	3.4	10.8	Gas
Methyl acrylate	80	−3	2.8	25	
Methylacrylonitrile	90	13	2	6.8	
Methylal	44	−32	2.2	13.8	1.5
Methyl alcohol	64	11	6.0	36	5
Methylamine	−6	—	4.9	20.7	Gas

Methyl *n*-amyl ketone	150	39	1.1	7.9	17.4
			(66)	(121)	
N-methyl aniline	194	79			
Methyl bromide	4	537	10	16.0	
Methyl *n*-butyl ketone	128	25	1.2	8	8.1
Methyl chloride	−24	−50	8.1	17.4	
Methyl chloroform	74	—	7.5	12.5	2.7
Methyl cyclohexane	101	−4	1.2	6.7	
Methylcyclohexanol	165	65			140
o-Methylcyclohexanone	163	48			38
Methylcyclopentadienyl manganese tricarbonyl	232	110			
Methylene bisphenyl isocyanate	172	202			
Methylene chloride	40	—	14	22	1.8
4,4′-Methylene dianiline	398 @ 78 mmHg	220			
Methyl ethyl ketone	80	−9	1.4	11.4	2.7
			(93)	(93)	
Methyl formate	−19	32	4.5	23	1.6
Methyl hydrazine	88	−8	2.5	92	
Methyl isoamyl ketone	146	36	1.0	8.2	
			(93)	(93)	
Methyl isobutyl carbinol	130	41	1.0	5.5	36
Methyl isobutyl ketone	118	18	1.2	8.0	5.6
			(93)	(93)	
Methyl isocyanate	39	−7	5.3	26	
Methyl mercaptan	6	<0	3.9	21.8	
Methyl methacrylate	100	10	1.7	8.2	
Methyl propyl ketone	102	7	1.5	8.2	6.9
α-Methyl styrene	165	54	1.9	6.1	
Mineral spirits No. 360	149	40	0.8	6.0	47
			(100)	(100)	
Mineral spirts, odorless	354–388	54	1.0	6.0	39
			(100)	(100)	
Morpholine	129	38	1.8	11	
Naphtha, coal tar	138–215	38–43			>29

TABLE 4.9 Characteristics of Flammable Gases and Vapors *(Continued)*

Substance	Boiling point, °C	Flash point, °C	Flammable limits, percent by volume		Dry time relation
			Lower	Upper	
Naphtha, hi-flash	150–204	29–43	0.8	6.0	4.4
Naphtha, V.M.&P.	118–179	14	1.2	6.0	7.1
Naphthalene	218	79	0.9	5.9	
β-Naphthylamine	300	157			
Nickel carbonyl	43	<−24	2		
Nicotine	246	95	0.7	4.0	
p-Nitroaniline	336	199			
Nitrobenzene	211	88	1.8 (93)		160
p-Nitrochlorobenzene	242	127			
Nitroethane	114	28	3.4		
Nitromethane	101	35	7.3		6.6
1-Nitropropane	131	36	2.2		10.9
2-Nitropropane	120	24	2.6	11.0	7.7
Nitrotoluene	238	106	1.6–2.2		
Nonane	151	31	0.8		2.9
Octane	126	13	1.0	6.5	5.9
Oil mist, mineral	68	135			
Pentaborane	60	36	0.42		
pent-Acetate	127	37			20
Pentachloronaphthalene	330	180			
Pentane	36	<−40	1.51	7.8	1
Perchloroethylene	121	—			6.6
Phenol	181	79	1.8	8.6	
p-Phenylene diamine	267	156	0.6		
Phenyl ether	257	96	0.8	1.5	
Phenylhydrazine	244a	88			
Phenyl mercaptan	168	56			
Phosphine	−88	Gas			

Phthalic anhydride	284	152	1.7	10.5	
Picric acid	>300a	150			
Propane	−42	—	2.1	9.5	Gas
Propargyl alcohol	114	36			
β-Propiolactone	155	74	2.9		
Propionic acid	147	52	2.9	12.1	
n-Propyl acetate	102	13	1.7 (38)	8	4.8
Propyl alcohol	97	23	2.2	13.7	7.8
n-Propyl nitrate	111	20	2	100	
Propylene	−47	—	2.0	11.1	Gas
Propylene dichloride	96	16	3.4	14.5	4
Propylene glycol monomethyl ether	120	38			
Propyleneimine	66	−4			
Propylene oxide	35	−37	2.3	36	
Pyridene	115	20	1.8	12.4	8.2
Quinone	—	38–93			
Resorcinol	277	127	1.4 (200)		
Styrene	146	31	1.1	7.0	12.4
Sulfur monochloride	138	118			
o-Terphenyl	332	163			
m-Terphenyl	363	191			
Tetrabutyl tin	145 @ 10 mmHg	124			
1,1,2,2-Tetrachloroethane	146	—			19
Tetrachloronaphthalene	311–360	210			
Tetraethyl lead	110a	93	1.8		
Tetrahydrofuran	66	−14	2	11.8	2.0
Tetramethyl lead	100a	38	1.8		
Toluene	111	4	1.2	7.1	4.5
Toluene-2,4-diisocyanate	251	127	0.9	9.5	
o-Toluidine	200	85	1.5 (85)		140
Tributyl phosphate	293	146			>100
1,2,4-Trichlorobenzene	213	105	2.5 (150)	6.6 (150)	

TABLE 4.9 Characteristics of Flammable Gases and Vapors *(Continued)*

Substance	Boiling point, °C	Flash point, °C	Flammable limits, percent by volume		Dry time relation
			Lower	Upper	
1,1,2-Trichloroethane	114	—			12.6
Trichloroethylene	87	—	8	10.5	3.1
Trichloronaphthalene	302–354	200			
1,2,3-Trichloropropane	156	82	3.2	12.6	28
Triethylamine	89	−7	1.2	8.0	2.7
Trimethylamine	3	—	2.0	11.6	Gas
1,2,4-Trimethyl benzene	165	44	0.9	6.4	
Trimethyl phosphite	111	54			
2,4,6-Trinitrotoluene	240	246a			
Triorthocresyl phosphate	410a	225			<200
Turpentine	153–175	32–46	0.8		375
Valeraldehyde	103	12			
Vinyl acetate	72	−8	2.6	13.4	3.3
Vinyl bromide	16	—	9	15	
Vinyl chloride	−14	−78	3.6	33.0	Gas
Vinyl cyclohexene dioxide	220	110			
Vinylidene chloride	32	−28	6.5	15.5	
Vinyl toluene	168	58	0.8	11.0	
Xylene	138–144	27–32	1.0	7.0	9.2
m-Xylene α,α′-diamine	247	134			
Xylidine	224	97	1.0		
Zinc stearate		277			

NOTES: (1) Listing an "a" with a boiling or flash point means that the substance decomposes or explodes at that temperature.
(2) Listing a number in parentheses below a flammable limit means that the limit was determined at that temperature (in °C).

REFERENCES: ACGIH, 1984b; AAI, 1981; NFPA, 1984; NIOSH, 1981.

Table 4.9 is not a comprehensive list of all flammable materials and their properties. Flash point and the other properties vary with factors such as substance purity, process temperature and pressure, and test apparatus. The data in Table 4.9 should be regarded as approximations, and the cited references should be consulted if more detail is required. Nonlisting of a material property in the table means that either the data were not available or the property does not apply to that substance. Other properties may be important to the fire hazard associated with a substance. For example, ignition temperature is the minimum temperature required to initiate combustion of a solid, liquid, or gas. The ignition temperature of carbon disulfide is so low (90°C) that the surface of a lightbulb could ignite vapors of this substance that are within the flammable range.

Other materials on the TLV list have physical or chemical properties that contribute to fire hazard. For example, oxidizing materials may greatly intensify a fire. Some oxidizing materials are methyl ethyl ketone peroxide, chlorine, chlorine triflouride, nitric acid, and perchloryl fluoride. Other materials, such as ethylene glycol dinitrate, picric acid, dichloroacetylene, and trinitrotoluene, are highly explosive. Pyrophoric materials ignite spontaneously in air and include silicon tetrahydride and triethylaluminum. Reactive combinations of materials may also increase a fire hazard. For example, a violent polymerization of acetaldehyde may result when reacted with sodium hydroxide (NFPA, 1984).

Flammable liquids are those that have flash points less than 37.8°C, and are designated as Class I liquids (OSHA, 1983, §1910.106). Flammable liquids are further categorized as Class IA (flash point $<$ 22.8°C, boiling point $<$ 37.8°C), Class IB (flash point $<$ 22.8°C, boiling point $\geq$ 37.8°C), and Class IC (flash point $\geq$ 22.8°C, boiling point $<$ 37.8°C). Combustible liquids are defined as being Class II with flash points between 37.8°C and 60°C, and Class III with flash points above 60°C. One use of flash points is specification of solvent storage and handling procedure according to flammable and combustible liquid classifications. For example, Table 4.10 lists allowable container sizes for flammable and combustible liquids. Products and processes to which this table does not apply are given in OSHA 1910.106(d)(1). To illustrate the use of flash point, V.M.&P. naphtha [boiling point 118–179°C; flash point 14°C (Table 4.9)] is a Class IB flammable liquid. The material could be stored in a 5-gal safety can according to Table 4.10.

Enclosed processing buildings for industrial plants such as those compounding cosmetics, pharmaceuticals, solvents, cleaning fluids, and insecticides, and chemical operations such as oxidation, reduction, halogenation, hydrogenation, alkylation, and polymerization (but not these processes in refineries, chemical plants, or distilleries), have the

TABLE 4.10 Maximum Allowable Size of Containers and Portable Tanks

	Flammable liquids			Combustible liquids	
Container type	Class IA	Class IB	Class IC	Class II	Class III
Glass or approved plastic	1 pt	1 qt	1 gal	1 gal	1 gal
Metal (other than DOT drums)	1 gal	5 gal	5 gal	5 gal	5 gal
Safety cans	2 gal	5 gal	5 gal	5 gal	5 gal
Metal drums (DOT specifications)	60 gal	60 gal	60 gal	60 gal	60 gal
Approved portable tanks	660 gal	660 gal	660 gal	660 gal	660 gal

SOURCE: OSHA, 1983b, §1910.106(d), Table H-12.

following ventilation requirements [OSHA, 1983b, §1910.106(e)(3)(v) and (h)(3)(iii)]:

> (iii) Ventilation. (a) Enclosed processing buildings shall be ventilated at a rate of not less than 1 cubic foot per minute per square foot of solid floor area. This shall be accomplished by natural or mechanical ventilation with discharge or exhaust to a safe location outside of the building. Provisions shall be made for introduction of makeup air in such a manner as not to short circuit the ventilation. Ventilation shall be arranged to include all floor areas or pits where flammable vapors may collect.
>
> (b) Equipment used in a building and the ventilation of the building shall be designed so as to limit flammable vapor-air mixtures under normal operating conditions to the interior of equipment, and to not more than 5 feet from equipment which exposes Class I liquids to the air. Examples of such equipment are dispensing stations, open centrifuges, plate and frame filters, open vacuum filters, and surfaces of open equipment.

These regulations typify ventilation specifications for fire hazard control. Ventilation specified in this manner does not necessarily control air contaminant health hazards. Furthermore, a flammable mixture of a vapor and air is not necessarily much heavier than air. Ventilation system design should take this fact into account.

Example A paint storage room is 15 ft high, wide, and deep. Natural ventilation is provided through a gravity exhaust duct using the stack effect (see Chapter 12). Specifically, warm supply air flows from the adjacent work area through a grill in the door, mixes with air in the storage room, flows into the exhaust duct opening near the floor, and discharges above roof level into the cooler outdoor air. Stored materials stacked against the exhaust duct opening have reduced the ventilation by 80%. V.M.&P. naphtha leaks from a stored container at a rate of 0.5 gal/hr. What is the average solvent concentration in the room?

Solution
Given:

Steady-state conditions
Mixing factor = 0.3
Indoor temperature = 20°C
All of the naphtha evaporates
Design ventilation rate = 1 cfm/ft^2
Molecular weight = 112 g/g-mole
Liquid density = 0.86 gm/ml

Using the notation of Chapter 3, the actual ventilation rate will be

$$q_2 = (1 - 0.8)(225 \text{ ft}^2)(1 \text{ cfm/ft}^2 \text{ floor area}) = 45 \text{ cfm}$$

$$\frac{\text{Vapor volume}}{\text{hr}} = \left(\frac{0.5 \text{ gal}}{\text{hr}}\right)\left(\frac{3785 \text{ m}l}{\text{gal}}\right)\left(\frac{0.86 \text{ g}}{\text{m}l}\right)\left(\frac{22.4 \text{ liters}}{\text{g-mole}}\right)\left(\frac{\text{g-mole}}{112 \text{ g}}\right)\left(\frac{293 \text{ K}}{273 \text{ K}}\right)$$

$$= 349.4 \text{ liters/hr}$$

From Eq. (3.6):

$$C_{i,ss} = \frac{S}{kq_2} = \frac{(349.4 \text{ liters/hr})(\text{hr}/60 \text{ min})(100\%)}{(0.3)(45 \text{ cfm})(28.32 \text{ liters/ft}^3)} = 1.52\%$$

which is within the flammable range of 1.2–6% for V.M.&P. naphtha.
If the exhaust duct opening is cleared,

$$q_2 = 225 \text{ cfm} \quad \text{and} \quad C_{i,ss} = 0.3\% \text{ or } 3000 \text{ ppm}$$

Although the concentration is reduced to 25% of the lower flammable limit (the maximum acceptable for flammable vapors), a health hazard still exists because the level is 10 times greater than the TLV for V.M.&P. naphtha (Table 4.2).

Materials in the form of dusts also contribute to fire hazard. Although explosions of dispersed dusts can be spectacular, dust fires caused by ignition of dust layers and deposits are more common (Palmer, 1973). Table 4.11 summarizes some fire and explosion properties of dusts (Palmer, 1973). Presented are data from the United States, United Kingdom, and the German Federal Republic that represent the most hazardous of the published values. Since each country uses different test methods, properties such as minimum ignition temperature of dust clouds may differ somewhat. These properties also vary with particle size, dust moisture content, and dust layer thickness.

Table 4.11 lists the minimum ignition temperatures by a heated surface of dusts in either dispersed or deposited forms. Minimum explosible concentration for dust clouds is the concentration below which propagation of flame does not occur upon contact with an ignition source. Minimum ignition energy is a measure of the energy needed to ignite a dust cloud with a spark. Maximum explosion pressure is an indicator of the severity of an explosion on buildings and structures.

(*Text continues on page 174.*)

TABLE 4.11 Explosion Characteristics of Various Dusts

Dust	Minimum ignition temperature, °C		Minimum explosible concentration, g/liter	Minimum ignition energy, mJ	Maximum explosion pressure, lb/in.² gauge	Maximum rate of pressure rise, lb/in.²·sec	Maximum oxygen concentration to prevent ignition, percent by volume	References[a]
	Cloud	Layer						
Acetamide[b]	560	—	—	—	—	—	—	8
Aceto acetanilide	560	—	0.030	20	90	4,800	—	4
Acetoacet-*p*-phenetedide	560	—	0.030	10	87	>10,000	—	1
Acetoacet-*o*-toluidine	710	—	—	—	—	—	—	1
2-Acetylamino-5-nitrothiazole	450	450	0.160	40	137	9,000	—	1
Acetyl-*p*-nitro-*o*-toluidine	450	—	—	—	—	—	—	8
Adipic acid	550	—	0.035	60	95	4,000	—	4,8
Alfalfa	460	200	0.100	320	88	1,100	—	5
Almond shell	440	200	0.065	80	101	1,400	—	5
Aluminium, atomized	650	760	0.045	50	84	>20,000	—	3
Aluminium, flake	610	320	0.045	10	127	>20,000	—	3
Aluminium-cobalt alloy	950	570	0.180	100	92	11,000	—	3
Aluminium-copper alloy	—	830	0.100	100	95	4,000	—	3
Aluminium-iron alloy	550	450	—	—	36	300	—	3
Aluminium-lithium alloy	470	400	<0.1	140	96	6,000	—	3
Aluminium-magnesium alloy	430	480	0.020	80	86	10,000	—	3
Aluminium-nickel alloy	950	540	0.190	80	96	10,000	—	3
Aluminium-silicon alloy	670	—	0.040	60	85	7,500	—	3
Aluminium acetate[c]	560	640	—	—	59	950	—	3
Aluminium octoate	460	—	—	—	—	—	—	8
Aluminium stearate	400	380	0.015	10	86	>10,000	—	3
2-Amino-5-nitrothiazole	460	460	0.075	30	110	5,600	—	3
Anthracene	505	Melts	—	—	68	700	—	2,8
Anthranilic acid	580	—	0.030	35	84	6,500	—	1
Anthraquinone	670	—	—	—	—	—	—	8
Antimony	420	330	0.420	1,920	28	300	—	3
Antipyrin	405	Melts	—	—	53	—	—	2
Asphalt	510	500	0.025	25	94	4,800	—	7

Aspirin	550	Melts	0.015	16	87	7,700	—	8
Azelaic acid	610	—	0.025	25	76	4,700	—	4
αα′-Azo-isobutyronitrile	430	350	0.015	25	134	8,000	—	4
Barley	370	—	—	—	—	—	—	8
Benzethonium chloride	380	410	0.020	60	91	6,700	—	1
Benzoic acid	600	Melts	0.011	12	95	10,300	—	8
Benzotriazole	440	—	0.030	30	103	9,200	—	1
Benzoyl peroxide	—	—	—	21	—	—	—	8
Beryllium[d]	910	540	—	—	Did not ignite		—	3
Beryllium acetate, basic[e]	620	—	0.080	100	87	2,200	15	1
Bis(2-hydroxy-5-chloro-phenyl)-methane[e]	570	—	0.040	60	70	2,000	13	1
Bis(2-hydroxy-3,5,6-trichloro-phenyl)-methane	Did not ignite	450	—	—	—	—	—	1
Bone meal[c]	490	230	—	—	11	100	—	6
Boron[c]	730	390	Did not ignite	—	41	200	—	3
Bread	450	—	—	—	—	—	—	8
Brunswick green	360	—	—	—	—	—	—	8
p-t-Butyl benzoic acid	560	—	0.020	25	88	6,500	—	4
Cadmium	570	250	—	4,000	7	100	—	3
Cadmium yellow	390	—	—	—	—	—	—	8
Calcium carbide	555	325	—	—	13	—	—	2
Calcium citrate[b]	470	—	—	—	—	—	—	8
Calcium gluconate[b]	550	—	—	—	—	—	—	8
Calcium DL pantothenate	520	—	0.050	80	105	4,600	—	1
Calcium propionate	530	—	—	—	90	1,900	—	8
Calcium silicide	540	540	0.060	150	86	20,000	—	3
Calcium stearate	400	—	0.025	15	97	>10,000	—	1
Caprolactam	430	—	0.07	60	79	1,700	8	8
Carbon, activated[f]	660	270	0.100	—	92	1,700	—	7
Carbon, black	510	—	—	—	—	—	—	7
Carboxy methylcellulose	460	310	0.060	140	130	5,000	—	4
Carboxy methylhydroxy ethylcellulose	380	—	0.200	960	83	800	—	4
Carboxy polymethylene	520	—	0.115	640	76	1,200	—	4
Casein	460	—	—	—	89	1,200	—	8
Cellulose	410	300	0.045	40	117	8,000	—	4
Cellulose acetate[g]	340	—	0.035	20	114	6,500	5	4,8

TABLE 4.11 Explosion Characteristics of Various Dusts *(Continued)*

Dust	Minimum ignition temperature, °C		Minimum explosible concentration, g/liter	Minimum ignition energy, mJ	Maximum explosion pressure, lb/in.² gauge	Maximum rate of pressure rise, lb/in.²·sec	Maximum oxygen concentration to prevent ignition, percent by volume	References[a]
	Cloud	Layer						
Cellulose acetate butyrate	370	—	0.025	30	81	2,700	7	4
Cellulose proprionate	460	—	0.025	60	105	4,700	—	4
Cellulose triacetate	390	—	0.035	30	107	4,300	—	4,8
Cellulose tripropionate	460	—	0.025	45	88	4,000	—	4
Charcoal	530	180	0.140	20	100	1,800	—	7
Chloramine-T[c]	540	150	—	—	7	150	—	1
o-Chlorobenzmalono nitrile	—	—	0.025	—	90	>10,000	—	1
o-Chloroaceto acetanilide	640	—	0.035	30	94	3,900	—	1
p-Chloroaceto acetanilide	650	—	0.035	20	85	5,500	—	1
Chloro amino toluene sulfonic acid	650	—	—	—	—	—	—	8
4-Chloro-2-nitroaniline	590	120	<0.750	140	123	3,500	—	1
p-Chloro *o*-toluidine hydrochloride	650	—	—	—	—	—	—	8
Chocolate crumb	340	—	—	—	—	—	—	8
Chromium	580	400	0.230	140	56	5,000	—	3
Cinnamon	440	230	0.060	30	121	3,900	—	5
Citrus peel	500	330	0.060	100	51	1,200	—	5
Coal, brown[h]	485	230	—	—	—	—	—	2
Coal, 8% volatiles	730	—	—	—	—	—	—	7
Coal, 12% volatiles	670	240	—	—	—	—	—	7
Coal, 25% volatiles	605	210	0.120	120	62	400	—	7
Coal, 37% volatiles[i]	610	170	0.055	60	90	2,300	—	7
Coal, 43% volatiles	575	180	0.050	50	92	2,000	—	7
Cobalt	760	370	—	—	—	—	—	3
Cocoa	500	200	0.065	120	69	1,200	—	5
Coconut	450	280	—	—	—	—	—	2
Coconut shell	470	220	0.035	60	115	4,200	—	5

Coffee[e]	360	270	0.085	160	38	150	10	5,8
Coffee, extract	600	—	—	—	47	—	—	2
Coffee, instant	410	350	0.280	Did not ignite	68	500	—	5
Coke	>750	430	—	—	—	—	—	2
Coke, petroleum, 13% volatiles[f]	670	—	1.00	—	36	200	—	7
Colophony	325	Melts	—	—	—	—	—	2
Copal[j]	330	Melts	—	—	68	—	—	2
Copper	700	—	—	Did not ignite	Did not ignite	Did not ignite	—	3
Copper-zinc, gold bronze	370	190	1.00	—	44	1,300	—	3
Cork	460	210	0.035	35	96	7,500	—	6
Corn cob	450	240	0.045	45	127	3,700	—	5
Corn dextrine	410	390	0.040	40	124	7,000	—	5
Cornflour	390	—	—	—	—	—	—	8
Cornstarch	390	—	0.040	30	145	9,500	—	5
Cotton flock	470	—	0.050	25	94	6,000	—	4
Cotton linters	520	—	0.50	1,920	73	400	5	5
Cottonseed meal	530	200	0.055	80	89	2,200	—	5
Coumarone-indene resin	550	—	0.015	10	93	11,000	11	4
Crystal violet	475	Melts	—	—	—	—	—	2
Cyclohexanone peroxide	—	—	—	21	84	5,600	—	8
Dehydroacetic acid	430	—	0.030	15	87	8,000	—	1
Dextrin	410	440	0.050	40	99	9,000	—	6
Dextrose monohydrate	350	—	—	—	—	—	—	8
Diallyl phthalate	480	—	0.030	20	90	8,500	—	1
Diamino stilbene disulfonic acid[b]	550	—	—	—	—	—	—	8
Diazo aminobenzene	550	—	0.015	20	114	>10,000	—	1
Di-*t*-butyl-*p*-cresol	420	—	0.015	15	79	13,000	9	4
Dibutyl tin maleate	600	—	—	—	—	—	—	8
Dibutyl tin oxide	530	—	—	—	—	—	—	8
Dichlorophene	770	—	—	—	72	3,000	—	1
2,4-Dichlorophenoxy ethyl benzoate	540	—	0.045	60	84	2,200	—	1
Dicyclopentadiene dioxide	420	—	0.015	30	89	9,500	—	4
Dihydrostreptomycin sulfate	600	230	0.520	—	42	200	7	1

TABLE 4.11 Explosion Characteristics of Various Dusts *(Continued)*

Dust	Minimum ignition temperature, °C		Minimum explosible concentration, g/liter	Minimum ignition energy, mJ	Maximum explosion pressure, lb/in.^{2} gauge	Maximum rate of pressure rise, lb/in.^{2}·sec	Maximum oxygen concentration to prevent ignition, percent by volume	References[a]
	Cloud	Layer						
3,3′-Dimethoxy-4,4′-diaminodiphenyl	—	—	0.030	—	82	>10,000	—	1
Dimethylacridan	540	—	—	—	—	—	—	8
Dimethyl diphenyl urea	490	—	—	—	—	—	—	8
Dimethyl isophthalate	580	—	0.025	15	84	8,000	—	4
Dimethyl terephthalate	570	—	0.030	20	105	12,000	6	4
S-S′-Dimethyl xanthogen-ethylene bis dithiocarbamate	400	—	0.300	3,200	84	1,500	—	1
Dinitroaniline	470	—	—	—	—	—	—	8
3,5-Dinitrobenzamide	500	Melts	0.040	45	163	6,500	—	1
3,5-Dinitrobenzoic acid	460	—	0.050	45	139	4,300	—	1
Dinitrobenzoyl chloride	380	—	—	—	—	—	—	8
Dinitrocresol	340	Melts	0.030	—	—	—	—	1,2
4,4′-Dinitro-sym-diphenyl urea	550	—	0.095	60	102	2,500	—	1
Dinitro stillbene disulfonic acid	450	—	—	—	—	—	—	8
Dinitrotoluamide	500	—	0.050	15	153	>10,000	—	6
Diphenyl	630	—	0.015	20	82	3,700	—	1
4,4′-Diphenyl disulfonylazide	590	140	0.065	30	143	5,500	—	1
Diphenylol propane (Bis-phenol-A)[g]	570	—	0.012	11	81	11,800	5	4,8
Egg white	610	—	0.14	640	58	500	—	5
Epoxy resin	490	—	0.015	9	94	8,500	—	4,8
Esparto grass	—	—	—	—	94	7,300	—	8
Ethylcellulose	340	330	0.025	15	112	7,000	—	4
Ethylene diamine tetraacetic acid	450	—	0.075	50	106	3,000	—	1

Ethyl hydroxyethylcellulose	390	—	0.020	30	94	2,200	—	6
Ferric ammonium ferrocyanide	390	210	1.500	—	17	100	—	1
Ferric dimethyl dithiocarbamate	280	150	0.055	25	86	6,300	—	1
Ferric ferrocyanide	370	—	—	—	82	1,000	—	1
Ferrochromium	790	670	2.00	—	—	—	—	3
Ferromanganese	450	290	0.130	80	62	5,000	—	3
Ferrosilicon (45% Si)	640	—	—	—	—	—	—	2
Ferrosilicon (90% Si)	Did not ignite	980	0.240	1,280	113	3,500	—	3
Ferrotitanium	370	400	0.140	80	55	9,500	—	3
Ferrous ferrocyanide	380	190	0.400	—	—	—	—	1
Ferrovanadium	440	400	1.300	400	—	—	—	3
Fish meal	485	—	—	—	—	—	—	2
Fumaric acid	520	—	0.085	35	103	3,000	—	4
Garlic	360	—	0.10	240	57	1,300	—	5
Gelatin, dried	620	480	<0.5	—	78	1,200	—	1
Gilsonite	580	500	0.020	25	78	4,500	—	7
Graphite	730	580	—	—	—	—	—	7
Grass	—	—	—	—	56	400	—	8
Gum arabic	500	260	0.060	100	117	3,000	—	4
Gum Karaya	520	240	0.100	180	116	2,500	—	4
Gum manila (copal)	360	390	0.030	30	89	6,000	—	4
Gum tragacanth	490	260	0.040	45	123	5,000	—	4
Hexamethylene tetramine	410	—	0.015	10	98	11,000	11	4
Horseradish	—	—	<0.100	—	96	1,600	—	6
Hydrazine acid tartrate	570	—	0.175	460	30	200	—	1
p-Hydroxybenzoic acid	620	—	0.040	—	37	—	—	1
Hydroxyethylcellulose	410	—	0.025	40	106	2,600	—	6
Hydroxyethylmethylcellulose	410	—	—	—	—	—	—	8
Hydroxypropylcellulose	400	—	0.020	30	96	2,900	—	6
Iron	430	240	—	—	—	—	—	2
Iron, carbonyl	420	230	0.105	100	47	8,000	—	3
Iron pyrites	380	280	1.00	8,200	5	100	—	3
Isatoic anhydride	700	—	0.035	25	80	4,900	—	1
Isinglass	520	—	—	—	None	None	—	8
Isophthalic acid	700	—	0.035	25	78	3,100	—	4

TABLE 4.11 Explosion Characteristics of Various Dusts *(Continued)*

Dust	Minimum ignition temperature, °C		Minimum explosible concentration, g/liter	Minimum ignition energy, mJ	Maximum explosion pressure, lb/in.^{2} gauge	Maximum rate of pressure rise, lb/in.^{2}·sec	Maximum oxygen concentration to prevent ignition, percent by volume	References[a]
	Cloud	Layer						
Kelp	570	220	Did not ignite		19	200	—	5
Lactalbumin	570	240	0.040	50	97	3,500	—	4
Lampblack	730	—	—	—	—	—	—	7
Lauryl peroxide	—	—	—	12	90	6,400	—	8
Lead[k]	790	290	—	Did not ignite	3	100	—	3
Leather	390	—	—	—	—	—	—	8
Lignin	450	—	0.040	20	102	5,000	7	4
Lignite	450	200	0.030	30	94	8,000	—	7
Lycopodium	480	310	0.025	40	75	3,100	9	5
Magnesium	560	430	0.030	40	116	15,000	—	3
Maize husk	430	—	—	—	75	700	—	8
Maize starch	410	—	—	—	—	—	—	2
Maleic anhydride	500	Melts	—	—	—	—	—	2
Malt barley	400	250	0.055	35	95	4,400	—	5
Manganese	460	240	0.125	305	53	4,900	—	3
Manganese ethylene bis dithiocarbamate	270	—	0.07	35	—	—	—	8
Manioc	430	—	—	—	—	—	—	8
Mannitol	460	—	0.065	40	97	2,800	—	1
Melamine formaldehyde resin	410	—	0.02	50	93	1,800	—	4,8
DL Methionine	370	360	0.025	35	119	5,700	7	1
1-Methylaminoanthraquinone	830	Melts	0.055	50	71	3,300	—	1
Methylcellulose	360	340	0.030	20	133	6,000	—	4
2,2-Methylene bis-4-ethyl-6-*t*-butyl phenol	310	—	—	—	76	7,300	—	8
Milk	440	—	—	—	—	—	—	8
Milk, skimmed	490	200	0.050	50	95	2,300	—	5

Milk sugar	450	Melts	—	—	31	—	—	2
Molybdenum	720	360	—	—	—	—	—	3
Molybdenum disulfide	570	290	—	—	—	—	—	6
Monochloracetic acid	620	—	—	—	—	—	—	8
Monosodium salt of trichloro ethylphosphate[b]	540	—	—	—	—	—	—	8
Moss, Irish	530	230	Did not ignite		21	300	—	5
Naphthalene	575	Melts	—	—	87	—	—	2
β-Naphthalene-azo-dimeth-ylaniline	510	Melts	0.020	50	70	2,300	—	1
β-Naphthol	670	—	—	—	—	—	—	8
Naphthol yellow	415	395	—	—	—	—	—	2
Nigrosine hydrochloride	630	—	—	—	—	—	—	8
p-Nitro-*o*-anisidene	400	—	—	—	—	—	—	8
p-Nitro-benzene arsonic acid	360	280	0.195	480	77	900	—	1
Nitrocellulose	—	—	—	30	>256	>20,900	—	8
Nitro diphenylamine	480	—	—	—	—	—	—	8
Nitro furfural semicarbazone	240	—	—	—	>143	8,600	—	8
Nitropyridone	430	Melts	0.045	35	111	>10,000	—	1
p-Nitro-*o*-toluidine	470	—	—	—	—	—	—	8
m-Nitro-*p*-toluidine	470	—	—	—	—	—	—	8
Nylon	500	430	0.030	20	95	4,000	6	4
Oilcake meal	470	285	—	—	—	—	—	2
Onion, dehydrated	410	—	0.130	Did not ignite	35	500	—	5
Paper	440	270	0.055	60	96	3,600	—	6
Paraformaldehyde	410	—	0.040	20	133	13,000	—	4
Peanut hull	460	210	0.045	50	116	8,000	—	5
Peat	420	295	—	—	—	—	—	2
Peat, sphagnum	460	240	0.045	50	104	2,200	—	5
Pectin	410	200	0.075	35	132	8,000	—	5
Penicillin, N-ethyl piperidine salt of	310	—	—	—	—	—	—	8
Pentaerythritol	450	—	0.030	10	90	9,500	7	4
Phenol formaldehyde	450	—	0.015	10	107	6,500	—	4,8
Phenol furfural resin	530	—	0.025	10	88	8,500	—	4
Phenothiazine	540	—	0.030	—	56	3,000	—	1
p-Phenylenediamine	620	—	0.025	30	94	11,000	—	4

TABLE 4.11 Explosion Characteristics of Various Dusts *(Continued)*

Dust	Minimum ignition temperature, °C		Minimum explosible concentration, g/liter	Minimum ignition energy, mJ	Maximum explosion pressure, lb/in.² gauge	Maximum rate of pressure rise, lb/in.²·sec	Maximum oxygen concentration to prevent ignition, percent by volume	References[a]
	Cloud	Layer						
Phosphorus, red	360	305	—	—	—	—	—	2
Phosphorus pentasulfide	280	270	0.050	15	64	>10,000	—	1
Phthalic acid	650	Melts	—	—	62	—	—	2
Phthalic anhydride	605	Melts	0.015	15	72	4,200	11	2,4
Phthalimide	630	—	0.030	50	89	4,800	—	1
Phthalodinitrile	>700	Melts	—	—	43	—	—	2
Phytosterol	330	Melts	0.025	10	76	>10,000	—	1
Piperazine	480	—	—	—	72	1,400	—	1
Pitch	710	—	0.035	20	88	6,000	—	7
Polyacetal	440	—	0.035	20	113	4,100	—	4
Polyacrylamide	410	240	0.040	30	85	2,500	—	4
Polyacrylonitrile	500	460	0.025	20	89	11,000	—	4
Polycarbonate	710	—	0.025	25	96	4,700	—	4
Polyethylene	390	—	0.020	10	80	7,500	—	4,8
Polyethylene oxide	350	—	0.030	30	106	2,100	5	4
Polyethylene terephthalate	500	—	0.040	35	98	5,500	—	4
Polyisobutyl methacrylate	500	280	0.020	40	74	2,800	—	4
Polymethacrylic acid	450	290	0.045	100	97	1,800	—	4
Polymethyl methacrylate	440	—	0.020	15	101	1,800	7	4
Polymonochlorotrifluoro-ethylene	600	720	Did not ignite		—	—	—	4
Polypropylene	420	—	0.020	30	76	5,500	—	4
Polystyrene	500	500	0.020	15	100	7,000	—	4
Polytetrafluoroethylene	670	570	Did not ignite		—	—	—	4
Polyurethane foam	510	440	0.030	20	87	3,700	—	4
Polyurethane foam, fire retardant	550	390	0.025	15	96	3,700	—	4
Polyvinyl acetate[e]	450	—	0.040	160	69	1,000	11	4,8

Polyvinyl alcohol	450	Melts	—	—	78	—	—	2
Polyvinyl butyral	390	—	0.020	10	84	2,000	5	4
Polyvinyl chloride[k]	670	—	Did not ignite		38	500	—	4
Polyvinylidene chloride[b]	670	—	—	—	—	—	—	8
Polyvinyl pyrrolidone	465	Melts	—	—	15	—	—	2
Potassium hydrogen tartrate	520	—	—	—	—	—	—	1
Potassium sorbate	380	180	0.120	60	79	9,500	—	1
Potato, dried	450	—	—	—	97	1,000	—	8
Potato starch	430	—	—	—	—	—	—	2
Provender	370	—	—	—	93	1,400	—	8
Pyrethrum	460	210	0.100	80	95	1,500	—	1
Quillaia bark	450	—	—	—	—	—	—	8
Rape seed meal	465	—	—	—	—	—	—	2
Rayon, viscose	420	—	—	—	—	—	—	8
Rayon, flock	—	—	0.03	—	—	—	—	8
Rice	440	240	0.050	50	105	2,700	—	5
Rosin	390	—	0.015	10	87	12,000	—	4
Rubber	380	—	—	—	—	—	—	8
Rubber, crude, hard	350	—	0.025	50	80	3,800	13	4
Rubber, crumb	440	—	—	—	84	3,300	—	8
Rubber, vulcanized	360	—	—	—	40	—	—	2
Rye flour	415	325	—	—	35	—	—	2
Saccharin	690	—	—	—	—	—	—	1
Salicylanilide	610	Melts	0.040	20	73	4,800	—	1
Salicylic acid	590	—	0.025	—	84	6,800	—	4,8
Sawdust	430	—	—	—	97	2,000	—	8
Sebacic acid	—	—	—	—	74	400	—	8
Senna	440	—	0.010	105	49	300	—	8
Shellac	400	—	0.020	10	73	3,600	9	4
Silicon	Did not ignite	760	<0.10	80	94	13,000	—	3
Soap	430	600	0.085	100	77	2,800	—	6
Sodium acetate	590	—	0.030	35	90	4,600	—	1
Sodium amatol	580	Melts	0.140	—	65	800	—	1
Sodium benzoate	560	680	0.050	80	91	3,700	—	1
Sodium carboxymethylcellulose	320	—	1.10	440	49	400	5	8
Sodium 2-chloro-5-nitrobenzene sulfonate	550	440	—	—	—	—	—	1

TABLE 4.11 Explosion Characteristics of Various Dusts *(Continued)*

Dust	Minimum ignition temperature, °C		Minimum explosible concentration, g/liter	Minimum ignition energy, mJ	Maximum explosion pressure, lb/in.² gauge	Maximum rate of pressure rise, lb/in.²·sec	Maximum oxygen concentration to prevent ignition, percent by volume	References[a]
	Cloud	Layer						
Sodium 2,2-dichloropropionate	500	—	0.260	220	68	500	—	1
Sodium dihydroxy naphthalene disulfonate[b]	510	—	—	—	—	—	—	8
Sodium glucaspaldrate	600	—	—	—	—	—	—	8
Sodium glucoheptonate	600	—	—	—	—	—	—	8
Sodium monochloracetate	550	—	—	—	—	—	—	8
Sodium *m*-nitrobenzene sulfonate	—	—	—	—	92	400	—	1
Sodium *m*-nitrobenzoate	—	—	—	—	87	2,900	—	1
Sodium pentachlorophenate	Did not ignite	360	—	—	Did not ignite		—	1
Sodium propionate	479	—	—	—	70	700	—	8
Sodium secobarbital	520	—	0.100	960	76	800	—	1
Sodium sorbate	400	140	0.050	30	87	6,500	—	1
Sodium thiosulfate[c]	510	330	—	—	11	<100	—	1
Sodium toluene sulfonate	530	—	—	—	—	—	—	8
Sodium xylene sulfonate	490	—	—	—	—	—	—	8
Soot	>690	535	—	—	Did not ignite		—	2
Sorbic acid[g]	440	460	0.020	15	106	>10,000	5	1,8
L-Sorbose	370	—	0.065	80	76	4,700	—	1
Soya flour	550	340	0.060	100	94	800	9	5
Soya protein	540	—	0.050	60	98	6,500	9	5
Starch	470	—	—	—	—	—	—	8
Starch, cold water	490	—	—	—	—	—	—	8
Stearic acid	290	—	—	25	80	8,500	—	1

Steel	450	—	—	—	—	—	—	8
Streptomycin sulfate	700	—	—	—	—	—	—	8
Sucrose	420	Melts	0.045	40	86	5,500	—	1
Sugar	370	400	0.045	30	109	5,000	—	5
Sulfur	190	220	0.035	15	78	4,700	—	1
Tantalum	630	300	<0.20	120	55	4,400	—	3
Tartaric acid	350	—	—	—	—	—	—	8
Tea	500	—	—	—	93	1,700	—	8
Tea, instant	580	340	Did not ignite		48	400	—	5
Tellurium	550	340	—	—	—	—	—	3
Terephthalic acid	680	—	0.050	20	84	8,000	—	4
Tetranitro carbazole	395	Melts	—	—	—	—	—	2
Thiourea	420	Melts	—	—	29	100	—	1
Thorium	270	280	0.075	5	79	5,500	—	3
Thorium hydride	260	20	0.080	3	81	12,000	—	3
Tin	630	430	0.190	80	48	1,700	—	3
Titanium	375	290	0.045	15	85	11,000	Ignites in carbon dioxide	2,3
Titanium hydride	480	540	0.070	60	121	12,000	3	3
Tobacco	485	290	—	—	—	—	—	2
Tobacco, dried	320	—	—	—	85	1,000	—	8
Tobacco, stem	420	230	Did not ignite		53	400	—	5
Tribromosalicyl anilide	880	Melts	—	—	—	—	—	1
Trinitro toluene	—	—	0.070	75	63	2,100	—	6
s-Trioxane	480	—	0.143	—	85	600	—	1
α,α′-Trithiobis (N,N-dimethyl-thioformamide)	280	230	0.060	35	96	6,000	—	1
Tung	540	240	0.070	240	74	1,900	—	5
Tungsten	730	470	—	—	Did not ignite		—	3
Uranium	20	100	0.060	45	69	5,000	—	3
Uranium hydride	20	20	0.060	5	74	9,000	—	3
Urea[b]	900	—	Did not ignite		—	—	—	4,8
Urea formaldehyde molding powder	460	—	0.085	80	89	3,600	9	4
Urea formaldehyde resin	430	—	0.02	34	110	1,600	—	8

TABLE 4.11 Explosion Characteristics of Various Dusts *(Continued)*

Dust	Minimum ignition temperature, °C		Minimum explosible concentration, g/liter	Minimum ignition energy, mJ	Maximum explosion pressure, lb/in.^{2} gauge	Maximum rate of pressure rise, lb/in.^{2}·sec	Maximum oxygen concentration to prevent ignition, percent by volume	References[a]
	Cloud	Layer						
Vanadium	500	490	0.220	60	57	1,000	10	3
Vitamin B_1 mononitrate	380	190	0.035	35	120	9,000	—	1
Vitamin C	460	280	0.070	60	88	4,800	—	1
Walnut shell	420	210	0.035	60	121	5,500	—	5
Wax, accra	260	—	—	—	—	—	—	8
Wax, carnauba	340	—	—	—	—	—	—	8
Wax, paraffin	340	—	—	—	—	—	—	8
Wheat, flour	380	360	0.050	50	109	3,700	—	5
Wheat, grain dust	420	290	—	—	43	—	—	2
Wheat starch	430	—	0.045	25	100	6,500	—	5
Wood	360	—	—	—	90	5,700	5	8
Wood, bark	450	250	0.020	60	103	7,500	—	4
Wood, flour	430	—	0.050	20	94	8,500	7	4
Wood, hard	420	315	—	—	66	—	—	2
Wood, soft	440	325	—	—	63	—	—	2
Yeast	520	260	0.050	50	123	3,500	—	5
Zinc	680	460	0.500	960	70	1,800	—	3
Zinc ethylene dithiocarbamate	480	180	—	—	45	300	—	1
Zinc stearate	315	Melts	0.020	10	80	>10,000	—	1,2
Zirconium	20	220	0.045	5	75	11,000	Ignites in carbon dioxide	3
Zirconium hydride	350	270	0.085	60	90	9,500	3	3

—Indicates no value available, but not that it was unmeasurable.

[a]References:

1. Dorsett and Nagy, 1968.

2. Freytag, 1965.
3. Jacobson et al., 1964.
4. Jacobson et al., 1962.
5. Jacobson et al., 1961.
6. Nagy et al., 1968.
7. Nagy et al., 1965.
8. Raftery, 1968.

[b]Group (b) dust, U.K. designation of dusts that did not propagate flame in test apparatus.

[c]Guncotton ignition source in pressure test.

[d]Contained 8% oxide.

[e]Inert gas carbon dioxide.

[f]Guncotton ignition source in minimum explosible concentration and maximum explosion pressure tests.

[g]Inert gas nitrogen.

[h]See also Lignite.

[i]Standard Pittsburgh coal.

[j]See also Gum manila.

[k]Flame ignition source in pressure test.

SOURCE: Palmer, 1973.

TABLE 4.12 Hazard Classification for Dust Deflagrations

Hazard class	K_{st} (bar·m·sec^{-1}) for strong ignition source (~10,000 W·sec)	Maximum rate of pressure rise in Hartmann apparatus (psi/sec)
St-1	≤200	≤7,300
St-2	201–300	7,300–22,000
St-3	>300	>22,000

SOURCE: NFPA, 1979.

Maximum rate of pressure rise is used when designing explosion relief venting. Maximum oxygen concentration to prevent ignition is used when protecting against dust cloud explosions with an inert atmosphere. For example, displacing oxygen in air with nitrogen will prevent ignition of a cellulose acetate dust cloud if oxygen content is no more than 5%.

Because of differences in measurement techniques, the maximum explosion pressure and maximum rate of pressure rise given in Table 4.11 should be considered in a relative sense rather than as absolute values for equipment design. In particular, dust explosion measurements made in test vessels with volumes < 20 liters tend to give rate of pressure rise values that are significantly too low (Bartknecht, 1981; NFPA, 1979). Most of the U.S. Bureau of Mines data given in Table 4.11 are in this category, since they were collected with the Hartmann apparatus (1.23 liters) (Nagy and Verakis, 1983). For test volumes $>$ 20 liters, the product of the maximum pressure rise, $(dP/dt)_{max}$, and the cube root of the test vessel volume, V, is essentially constant:

$$\left(\frac{dP}{dt}\right)_{max} \cdot V^{1/3} = K_{st} \tag{4.2}$$

where K_{st} is the "explosion violence" constant (Bartknecht, 1981). A number of nomographs have been developed that relate vent area to vessel volume for various dust hazard classifications based on K_{st} (NFPA, 1979; Bartknecht, 1981).

These hazard classifications are shown in Table 4.12. Table 4.13 gives a comparison between U.S. measurements taken with the Hartmann device and German observations from a 1-m^3 test vessel. The data from the 1-m^3 apparatus is considered to be more applicable for equipment of commercial size (NFPA, 1979). The details of vent design are covered in NFPA (1979). An instructive comparison of various vent area estimation procedures is given in Nagy and Verakis (1983); and some of the limitations of testing and design are discussed by Bartknecht (1981) and by Britton and Chippett (1985). Equation (4.2) also holds for gases, with K_{st} being replaced by K_G, provided that significant autoturbulent

TABLE 4.13 Explosibility of Various Dusts in 1.23-liter and 1-m^3 Test Devices

	U.S., Hartmann apparatus, 1.23 liters			German, 1 m^3			
Type of dust	Maximum pressure[a] (psia)	Maximum rate (psi/sec)	Hazard class	Maximum pressure[a] (psia)	Maximum rate (psi/sec)	K_{st} (bar·m·sec^{-1})	Hazard class
Aluminum (atomized)	100	>20,000	St-3	200	10,700	730	St-3
Magnesium	120	15,000	St-2	272	7,500	510	St-3
Cornstarch	150	10,000	St-2	165	2,940	320	St-3
Cork	110	7,500	St-2	156	2,970	200	St-2
Phenol formaldehyde	120	7,000	St-1	150	1,900	130	St-1
Wheat starch	115	6,500	St-1	159	1,940	130	St-1
Soybean flour	115	6,500	St-1	150	1,610	110	St-1
Wood flour	115	5,500	St-2	152	2,460	170	St-1
				113–154[b]		83–211[b]	St-1–St-2[b]
Sugar	100	5,500	St-1	140	2,000	140	St-1
Coal, Pittsburgh	105	2,300	St-1	152	1,370	90	St-1
Cocoa	85	1,200	St-1	124	630	40	St-1

[a]Note that maximum pressures given in Table 4.11 are gauge pressures, psig.
[b]Bartknecht, 1981.

SOURCES: Nagy and Verakis, 1983; NFPA, 1979.

effects are not present and the test container is >26 liters (Britton and Chippett, 1985).

C. Electrical hazards

Electrical system design specifications are contained in Subpart S of the OSHA regulations (§1910.310–1910.399). The requirements for grounding are [OSHA, 1983b, §1910.304(f)]:

> (1) Systems to be grounded. The following systems which supply premises wiring shall be grounded:
>
> (i) All 3-wire DC systems shall have their neutral conductor grounded.
>
> (ii) Two-wire DC systems operating at over 50 volts through 300 volts between conductors shall be grounded unless:
>
> (A) They supply only industrial equipment in limited areas and are equipped with a ground detector; or
>
> (B) They are rectifier-derived from an AC system complying with paragraphs (f)(1)(iii), (f)(1)(iv), and (f)(1)(v) of this section; or
>
> (C) They are fire-protective signaling circuits having a maximum current of 0.030 amperes.
>
> (iii) AC circuits of less than 50 volts shall be grounded if they are installed as overhead conductors outside of buildings or if they are supplied by transformers and the transformer primary supply system is ungrounded or exceeds 150 volts to ground.
>
> (iv) AC systems of 50 volts to 1000 volts shall be grounded under any of the following conditions, unless exempted by paragraph (f)(1)(v) of this section:
>
> (A) If the system can be so grounded that the maximum voltage to ground on the ungrounded conductors does not exceed 150 volts;
>
> (B) If the system is nominally rated 480Y/277 volt, 3-phase, 4-wire in which the neutral is used as a circuit conductor;
>
> (C) If the system is nominally rated 240/120 volt, 3-phase, 4-wire in which the midpoint of one phase is used as a circuit conductor; or
>
> (D) If a service conductor is uninsulated.
>
> (v) AC systems of 50 volts to 1000 volts are not required to be grounded under any of the following conditions:
>
> (A) If the system is used exclusively to supply industrial electric furnaces for melting, refining, tempering, and the like.
>
> (B) If the system is separately derived and is used exclusively for rectifiers supplying only adjustable speed industrial drives.
>
> (C) If the system is separately derived and is supplied by a transformer that has a primary voltage rating less than 1000 volts, provided all of the following conditions are met:
>
> (1) The system is used exclusively for control circuits,
>
> (2) The conditions of maintenance and supervision assure that only qualified persons will service the installation,
>
> (3) Continuity of control power is required, and
>
> (4) Ground detectors are installed on the control system.

(D) If the system is an isolated power system that supplies circuits in health care facilities.

(2) Conductors to be grounded. For AC premises wiring systems the identified conductor shall be grounded.

(3) Grounding connections. (i) For a grounded system, a grounding electrode conductor shall be used to connect both the equipment grounding conductor and the grounded circuit conductor to the grounding electrode. Both the equipment grounding conductor and the grounding electrode conductor shall be connected to the grounded circuit conductor on the supply side of the service disconnecting means, or on the supply side of the system disconnecting means or overcurrent devices if the system is separately derived.

(ii) For an ungrounded service-supplied system, the equipment grounding conductor shall be connected to the grounding electrode conductor at the service equipment. For an ungrounded separately derived system, the equipment grounding conductor shall be connected to the grounding electrode conductor at, or ahead of, the system disconnecting means or overcurrent devices.

(iii) On extensions of existing branch circuits which do not have an equipment grounding conductor, grounding-type receptacles may be grounded to a grounded cold water pipe near the equipment.

(4) Grounding path. The path to ground from circuits, equipment, and enclosures shall be permanent and continuous.

(5) Supports, enclosures, and equipment to be grounded.—(i) Supports and enclosures for conductors. Metal cable trays, metal raceways, and metal enclosures for conductors shall be grounded, except that:

(A) Metal enclosures such as sleeves that are used to protect cable assemblies from physical damage need not be grounded; or

(B) Metal enclosures for conductors added to existing installations of open wire, knob-and-tube wiring, and nonmetallic-sheathed cable need not be grounded if all of the following conditions are met: (1) runs are less than 25 feet; (2) enclosures are free from probable contact with ground, grounded metal, metal laths, or other conductive materials; and (3) enclosures are guarded against employee contact.

(ii) Service equipment enclosures. Metal enclosures for service equipment shall be grounded.

(iii) Frames of ranges and clothes dryers. Frames of electric ranges, wallmounted ovens, counter-mounted cooking units, clothes dryers, and metal outlet or junction boxes which are part of the circuit for these appliances shall be grounded.

(iv) Fixed equipment. Exposed non-current-carrying metal parts of fixed equipment which may become energized shall be grounded under any of the following conditions:

(A) If within 8 feet vertically or 5 feet horizontally of ground or grounded metal objects and subject to employee contact.

(B) If located in a wet or damp location and not isolated.

(C) If in electrical contact with metal.

(D) If in a hazardous (classified) location.

(E) If supplied by a metal-clad, metal-sheathed, or grounded metal raceway wiring method.

(F) If equipment operates with any terminal at over 150 volts to ground; however, the following need not be grounded:

(1) Enclosures for switches or circuit breakers used for other than service equipment and accessible to qualified persons only;

(2) Metal frames of electrically heated appliances which are permanently and effectively insulated from ground; and

(3) The cases of distribution apparatus such as transformers and capacitors mounted on wooden poles at a height exceeding 8 feet above ground or grade level.

(v) Equipment connected by cord and plug. Under any of the conditions described in paragraphs (f)(5)(v)(A) through (f)(5)(v)(C) of this section, exposed non-current-carrying metal parts of cord- and plug-connected equipment which may become energized shall be grounded.

(A) If in hazardous (classified) locations (see §1910.307).

(B) If operated at over 150 volts to ground, except for guarded motors and metal frames of electrically heated appliances if the appliance frames are permanently and effectively insulated from ground.

(C) If the equipment is of the following types:

(1) Refrigerators, freezers, and air conditioners;

(2) Clothes-washing, clothes-drying and dishwashing machines, sump pumps, and electrical aquarium equipment;

(3) Hand-held motor-operated tools;

(4) Motor-operated appliances of the following types; hedge clippers, lawn mowers, snow blowers, and wet scrubbers;

(5) Cord- and plug-connected appliances used in damp or wet locations or by employees standing on the ground or on metal floors or working inside of metal tanks or boilers;

(6) Portable and mobile X-ray and associated equipment;

(7) Tools likely to be used in wet and conductive locations; and

(8) Portable hand lamps.

Tools likely to be used in wet and conductive locations need not be grounded if supplied through an isolating transformer with an ungrounded secondary of not over 50 volts. Listed or labeled portable tools and appliances protected by an approved system of double insulation, or its equivalent, need not be grounded. If such a system is employed, the equipment shall be distinctively marked to indicate that the tool or appliance utilizes an approved system of double insulation.

(iv) Nonelectrical equipment. The metal parts of the following nonelectrical equipment shall be grounded: frames and tracks of electrically operated cranes; frames of nonelectrically driven elevator cars to which electric conductors are attached; hand operated metal shifting ropes or cables of electric elevators, and metal partitions, grill work, and similar metal enclosures around equipment of over 750 volts between conductors.

(6) Methods of grounding fixed equipment. (i) Non-current-carrying metal parts of fixed equipment, if required to be grounded by this subpart,

shall be grounded by an equipment grounding conductor which is contained within the same raceway, cable, or cord, or runs with or encloses the circuit conductors. For DC circuits only, the equipment grounding conductor may be run separately from the circuit conductors.

(ii) Electric equipment is considered to be effectively grounded if it is secured to, and in electrical contact with, a metal rack or structure that is provided for its support and the metal rack or structure is grounded by the method specified for the non-current-carrying metal parts of fixed equipment in paragraph (f)(6)(i) of this section. For installations made before April 16, 1981, only, electric equipment is also considered to be effectively grounded if it is secured to, and in metallic contact with, the grounded structural metal frame of a building. Metal car frames supported by metal hoisting cables attached to or running over metal sheaves or drums of grounded elevator machines are also considered to be effectively grounded.

(7) Grounding of systems and circuits of 1000 volts and over (high voltage).—(i) General. If high voltage systems are grounded, they shall comply with all applicable provisions of paragraphs (f)(1) through (f)(6) of this section as supplemented and modified by this paragraph (f)(7).

(ii) Grounding of systems supplying portable or mobile equipment. (See §1910.302(b)(3).) Systems supplying portable or mobile high voltage equipment, other than substations installed on a temporary basis, shall comply with the following:

(A) Portable and mobile high voltage equipment shall be supplied from a system having its neutral grounded through an impedance. If a delta-connected high voltage system is used to supply the equipment, a system neutral shall be derived.

(B) Exposed non-current-carrying metal parts of portable and mobile equipment shall be connected by an equipment grounding conductor to the point at which the system neutral impedance is grounded.

(C) Ground-fault detection and relaying shall be provided to automatically de-energize any high voltage system component which has developed a ground fault. The continuity of the equipment grounding conductor shall be continuously monitored so as to de-energize automatically the high voltage feeder to the portable equipment upon loss of continuity of the equipment grounding conductor.

(D) The grounding electrode to which the portable or mobile equipment system neutral impedance is connected shall be isolated from and separated in the ground by at least 20 feet from any other system or equipment grounding electrode, and there shall be no direct connection between the grounding electrodes, such as buried pipe, fence, etc.

(iii) Grounding of equipment. All non-current-carrying metal parts of portable equipment and fixed equipment including their associated fences, housings, enclosures, and supporting structures shall be grounded. However, equipment which is guarded by location and isolated from ground need not be grounded. Additionally, pole-mounted distribution apparatus at a height exceeding 8 feet above ground or grade level need not be grounded.

Unlike some other segments of the OSHA regulations, the electrical safety requirements are regularly updated. More specific information may be obtained from NFPA (1983) and OSHA (1986b).

REFERENCES

AAI (1981). *Handbook of Organic Industrial Solvents,* 5th ed. Alliance of American Insurers, Chicago.

ACGIH (1984a). *Threshold Limit Values for Chemical Substances and Physical Agents in the Work Environment Adopted by ACGIH for 1984–1985.* American Conference of Governmental Industrial Hygienists, Cincinnati, Ohio.

ACGIH (1984b). *Documentation of the Threshold Limit Values,* 4th ed. and Supplemental Documentations for 1981–1984. American Conference of Governmental Industrial Hygienists, Cincinnati, Ohio.

ASHRAE (1986). *A Proposed American National Standard for Ventilation for Acceptable Indoor Air Quality.* American Society of Heating, Refrigerating and Air-Conditioning Engineers, New York.

Bartknecht, W. (1981). *Explosions.* Springer-Verlag, Berlin.

Britton, L. G., and Chippett, S. (1985). Practical aspects of dust deflagration testing. Paper 58d, AIChE 1985 Spring National Meeting, Houston, March 24–28.

Dorsett, H. G., and Nagy, J. (1968). *Explosibility of Chemical Drugs, Dyes and Pesticides.* U.S. Bureau of Mines RI 7132.

EPA (1971). National primary and secondary ambient air quality standards. *Fed. Reg., 36,* 8186, November 25.

Freytag, H. H., ed. (1965). *Handbuch der Raumexplosionen.* Verlag Chemie, Weinheim.

ISO (1978). *Guide for the Evaluation of Human Exposure to Whole-Body Vibration.* International Standards Organization Pub. No. ISO 2631-1978E.

ISO (1984). *Draft Proposal for the Measurement and the Evaluation of Human Exposure to Vibration Transmitted to the Hand.* International Standards Institute Pub. No. ISO-5349 (Second DIS).

Jacobson, M., Cooper, A. R., and Nagy, J. (1964). *Explosibility of Metal Powders.* U.S. Bureau of Mines RI 6516.

Jacobson, M., Nagy, J., and Cooper, A. R. (1962). *Explosibility of Dusts Used in the Plastics Industry.* U.S. Bureau of Mines RI 5971.

Jacobson, M., Nagy, J., Cooper, A. R., and Ball, F. J. (1961). *The Explosibility of Agricultural Dusts.* U.S. Bureau of Mines RI 5753.

Johnson, R. K. (1985). Legal implications of compliance and noncompliance with standards and internal corporate guidelines. *Chem. Eng. Prog. 81*(8), 9–14, August.

Nagy, J., Cooper, A. R., and Dorset, H. G. (1968). *Explosibility of Miscellaneous Dusts.* U.S. Bureau of Mines RI 7208.

Nagy, J., Dorsett, H. G., and Cooper, A. R. (1965). *Explosibility of Carbonaceous Dusts.* U.S. Bureau of Mines RI 6597.

Nagy, J., and Verakis, H. C. (1983). *Development and Control of Dust Explosions.* Marcel Dekker, New York.

NFPA (1979). *Guide for Explosion Venting, NFPA 68-1978.* National Fire Protection Association, Quincy, Mass., February 23.

NFPA (1983). *The National Electrical Code Handbook,* 3d ed., Schram, P. J., Caloggero, J. M., and Tedesco, J. A., eds. National Fire Protection Association, Quincy, Mass.

NFPA (1984). *Fire Protection Guide on Hazardous Materials,* 8th ed. National Fire Protection Association, Quincy, Mass.

NIOSH (1981). *Occupational Health Guidelines for Chemical Hazards.* National Institute for Occupational Safety and Health, DHHS Pub. No. 81-123.

NIOSH (1983). Musculoskeletal injuries. *Morbid. Mortal. Week. Rep. 32,* 189–191.

NYT (*The New York Times*) (1982a). August 27.

NYT (*The New York Times*) (1982b). August 28.

OSHA (1981). Occupational noise exposure; hearing conservation amendment. *Fed. Reg., 46,* 4078–4179, January 16.
OSHA (1983a). Occupational noise exposure; hearing conservation amendment (Final rule). *Fed. Reg., 48,* 9737–9785, March 8.
OSHA (1983b). *General Industry Standards Part 1910.* Occupational Safety and Health Administration 2206, Revised March 11.
OSHA (1984). Occupational exposure to ethylene oxide; final standard. *Fed. Reg., 49,* 25733–25809, June 22.
OSHA (1985). Occupational exposure to cotton dust; final rule. *Fed. Reg., 50,* 51119–51179, December 13.
OSHA (1986a). Occupational exposure to asbestos, tremolite, anthophyllite, and actinolite (Final rules). *Fed. Reg., 51,* 22612–22790, June 20.
OSHA (1986b). Electrical standards for construction (Final rule). *Fed. Reg., 51,* 25294–25335, July 11.
Palmer, K. N. (1973). *Dust Explosions and Fires.* Chapman & Hall, London.
Raftery, M. M. (1968). Explosibility tests for industrial dusts. Ministry of Technology and Fire Offices' Committee. *Fire Research Technical Paper No. 21,* Her Majesty's Stationery Office, London.

Chapter 5

Ventilation

R. J. Allen

Replacement of workroom air, already affected by material or energy release from process sources, with air of different characteristics is one of the most important means of hazard control. The replacement air may be from outdoors (make-up air) introduced with or without cleaning and/or conditioning; or from the workspace itself after appropriate treatment (recirculated air). Adequate ventilation is required for (NIOSH, 1973):

1. Controlling contaminant exposure for health
2. Controlling heat and humidity for comfort
3. Preventing fires and explosions

The first of these goals will be the primary subject of this chapter. The control of air temperature and humidity to maintain worker health is discussed in Chapter 12, and the concentration requirements to minimize fire and explosion hazards are covered in Chapter 4.

Two general approaches to minimizing exposure to hazardous materials are local and general (or dilution) ventilation. In Chapter 3 it was pointed out that personal monitoring is carried out in order to quantitate the actual breathing zone exposure of a particular worker.

On the other hand, area monitoring, while often more practical to implement, characterizes only the general exposure that may be incurred by persons within the workspace. This general exposure may or may not be representative of personal exposure, depending on the characteristics of pollutant release from process sources, the nature of the human-process interface, and diluting effects of general ventilation. Although the ultimate goal of ventilation is to minimize exposure to air pollutants, local and general ventilation accomplish this in quite different ways. The aim of local ventilation is to restrict personal (breathing zone) exposures at the human-machine (or process) interface. The purpose of general ventilation is to dilute the concentration of hazardous materials that have already been released into the workspace, i.e., to minimize area or general exposure.

Local ventilation removes contaminants from the workspace before such materials can cross the human-process interface and enter the worker's breathing zone. The system used to bring this about includes an air mover to supply, or draw, air with enough velocity to overcome the tendency of process pollutants to disperse into the workroom. These velocities are called control or capture velocities, depending on system design, and are discussed further in Section VI. (Occasionally, air is also supplied in such a manner as to form an isolation barrier.) Exhaust air from a local ventilation system will ordinarily have a pollutant concentration in excess of occupational or environmental health standards. Consequently, such systems will also require appropriate ducting to contain the exhaust as well as a device to clean the air before release to the outdoor environment or recirculation to the workspace.

General ventilation involves supply and exhaust of air from a room, a number of rooms, or an entire building. Make-up air may enter the workspace by natural ventilation through designed openings, by infiltration through leaks in the building envelope, or by forced ventilation with an air mover. In contrast to local exhaust, the concentration of pollutants in general exhaust air is essentially the same as that in the work area. Ducting may or may not be required, and an exhaust air cleaner is usually not required. In general, dealing with the contaminant after it is mixed into the workspace requires handling a much larger volume of air.

There are a number of factors to keep in mind when selecting an appropriate ventilation system. These include:

1. General ventilation is used primarily for comfort control, with contaminant control usually a secondary consideration. Dilution ventilation controls area exposure to contaminants.
2. Local exhaust ventilation is more commonly used to control personal exposure to airborne contaminants.

3. Local exhaust systems will provide general ventilation for the work area.
4. A balance of supply and exhaust systems is required for either system to function as designed.
5. To provide effective dilution ventilation, the air must be delivered in the working zone (below 3 m).
6. In many cases natural ventilation will not be sufficient, and mechanical supply and exhaust will be required for adequate performance of a general ventilation system. If source emission factors are available, Eqs. (3.4)–(3.6) can be used to estimate the necessary area ventilation air rate (see Chapter 3).
7. General building exhaust may be required in addition to local exhaust.
8. The exhaust discharge, whether local or general, should be directed so as to reduce the possibility of reentry (exhausted air being drawn back into the working zones). Exhaust openings should not be located near windows, roof vents, or air intakes.
9. Care should be taken to avoid unintentional short-circuiting (supply air going straight to exhaust without mixing into the space).
10. Local exhaust ventilation should not direct contaminated air through the worker's breathing zone, as shown on Figure 5.1. Rather the air should be directed away from the breathing zone.
11. Enclosing a contaminant source will ordinarily reduce the amount of air to be exhausted to control contaminant release into the workspace, as is illustrated on Figure 5.2. Reducing the amount of air exhausted will reduce costs.
12. Exhausting "dirty" air outdoors only transfers an air pollution problem from the workspace to the community. It may be necessary to clean the contaminant from the exhausted air before it is released outdoors.

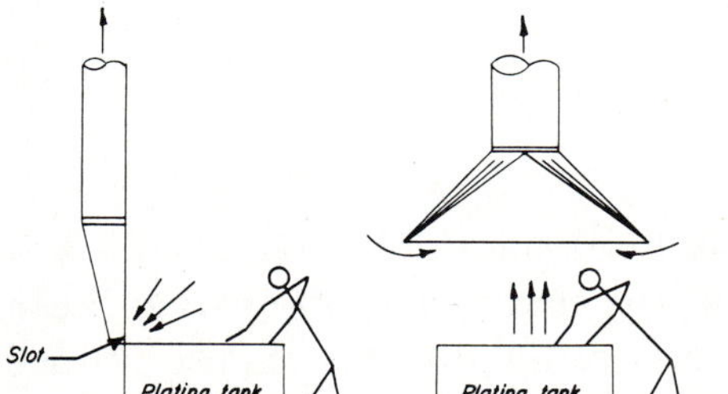

Figure 5.1 Direction of air flow (*ACGIH, 1984*). Locate the hood so the contaminant is removed away from the breathing zone of the worker.

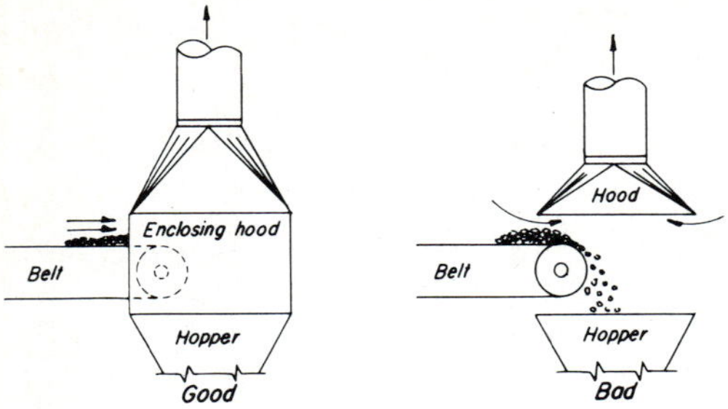

Figure 5.2 Enclose the source (*ACGIH, 1984*). Enclose the operation as much as possible. The more completely the source is enclosed, the less air is required for control.

13. All ventilation systems require periodic testing to determine how well they operate.
14. All ventilation systems require regular maintenance to ensure optimum operation.

I. Properties of Air

The molecular weight of clean, dry air is 28.96 g/g-mole. The density at 70°F and 1 atm can be determined from the molecular weight and the ideal gas law [recognizing that the volume, V_n, of 1 g-mole of any gas is 22.4 liters at 32°F (492°R) and 1 atm]:

$$\rho_{\text{air}} = \frac{\text{MW}_{\text{air}}}{V_n}$$

$$\rho_{\text{air}},\ 70°\text{F, 1 atm} = \left(\frac{28.96\ \text{g/g-mole}}{22.4\ \text{liters/g-mole}}\right)\left(\frac{492°\text{R}}{530°\text{R}}\right)\left(\frac{\text{lb}_\text{m}}{453.6\ \text{g}}\right)\left(\frac{10^3\ \text{liters}}{35.314\ \text{ft}^3}\right)$$
$$= 0.075\ \text{lb}_\text{m}/\text{ft}^3 \qquad (5.1)$$

Typically, if air temperature is $40°\text{F} < T < 100°\text{F}$ and altitude is within ±1000 ft of sea level, no correction is made for temperature and pressure. Correction factors for pressure and temperature are shown in Table 5.1. The correction factor (d) is:

$$d = \frac{\rho_{\text{actual conditions}}}{0.075\ \text{lb}_\text{m}/\text{ft}^3} \qquad (5.2)$$

Note that English units are used in the calculation of air density. Most ventilation calculations and handbooks in the United States still use English rather than metric units. To remain consistent with these other works, English units will be used throughout this chapter. Since the English system has units of pounds mass (lb_m) and pounds force (lb_f), these will be used for clarity rather than lb. Using the English

TABLE 5.1 Air Density Correction Factor, *d*

Altitude, ft		−1,000	Sea level	1,000	2,000	3,000	4,000	5,000	6,000	7,000	8,000	9,000	10,000
Barometer	in.Hg	31.02	29.92	28.86	27.82	26.52	25.84	24.90	23.98	23.09	22.22	21.39	20.58
	in.H_2O	422.2	407.5	392.8	378.6	365.0	351.7	338.9	326.4	314.3	302.1	291.1	280.1
Air temp., °F	−40	1.31	1.26	1.22	1.17	1.13	1.09	1.05	1.01	0.97	0.93	0.90	0.87
	0	1.19	1.15	1.11	1.07	1.03	0.99	0.95	0.91	0.89	0.85	0.82	0.79
	40	1.10	1.06	1.02	0.99	0.95	0.92	0.88	0.85	0.82	0.79	0.76	0.73
	70	1.04	1.00[a]	0.96	0.93	0.89	0.86	0.83	0.80	0.77	0.74	0.71	0.69
	100	0.98	0.95	0.92	0.88	0.85	0.81	0.78	0.75	0.73	0.70	0.68	0.65
	150	0.90	0.87	0.84	0.81	0.78	0.75	0.72	0.69	0.67	0.65	0.62	0.60
	200	0.83	0.80	0.77	0.74	0.71	0.69	0.66	0.64	0.62	0.60	0.57	0.55
	250	0.77	0.75	0.72	0.70	0.67	0.64	0.62	0.60	0.58	0.56	0.53	0.51
	300	0.72	0.70	0.67	0.65	0.62	0.60	0.58	0.56	0.54	0.52	0.50	0.48
	350	0.68	0.65	0.62	0.60	0.58	0.56	0.54	0.52	0.51	0.49	0.47	0.45
	400	0.64	0.62	0.60	0.57	0.55	0.53	0.51	0.49	0.48	0.46	0.44	0.42
	450	0.60	0.58	0.56	0.54	0.52	0.50	0.48	0.46	0.45	0.43	0.42	0.40
	500	0.57	0.55	0.53	0.51	0.49	0.47	0.45	0.44	0.43	0.41	0.39	0.38
	550	0.54	0.53	0.51	0.49	0.47	0.45	0.44	0.42	0.41	0.39	0.38	0.36
	600	0.52	0.50	0.48	0.46	0.45	0.43	0.41	0.40	0.39	0.37	0.35	0.34
	700	0.47	0.46	0.44	0.43	0.41	0.39	0.38	0.37	0.35	0.34	0.33	0.32
	800	0.44	0.42	0.40	0.39	0.37	0.36	0.35	0.33	0.32	0.31	0.30	0.29
	900	0.40	0.39	0.37	0.36	0.35	0.33	0.32	0.31	0.30	0.29	0.28	0.27
	1,000	0.37	0.36	0.35	0.33	0.32	0.31	0.30	0.29	0.28	0.27	0.26	0.25

[a] Standard air density, sea level, 70°F = 0.075 lb_m/ft^3.

SOURCE: ACGIH, 1984.

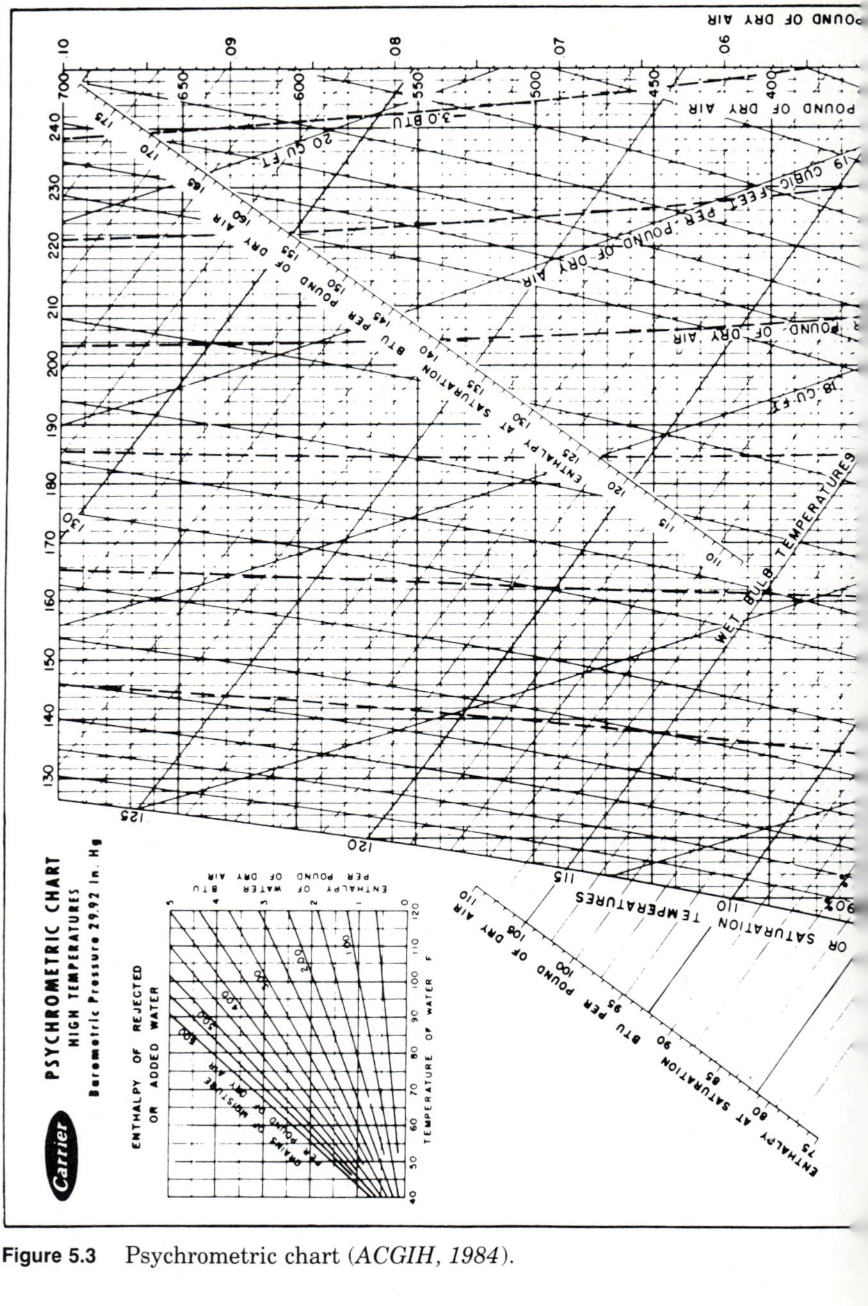

Figure 5.3 Psychrometric chart (*ACGIH, 1984*).

system of units often requires use of the conversion factor g_c, which is equal to:

$$g_c = \frac{32.2 \text{ lb}_m \cdot \text{ft}}{\text{lb}_f \cdot \text{sec}^2} \tag{5.3}$$

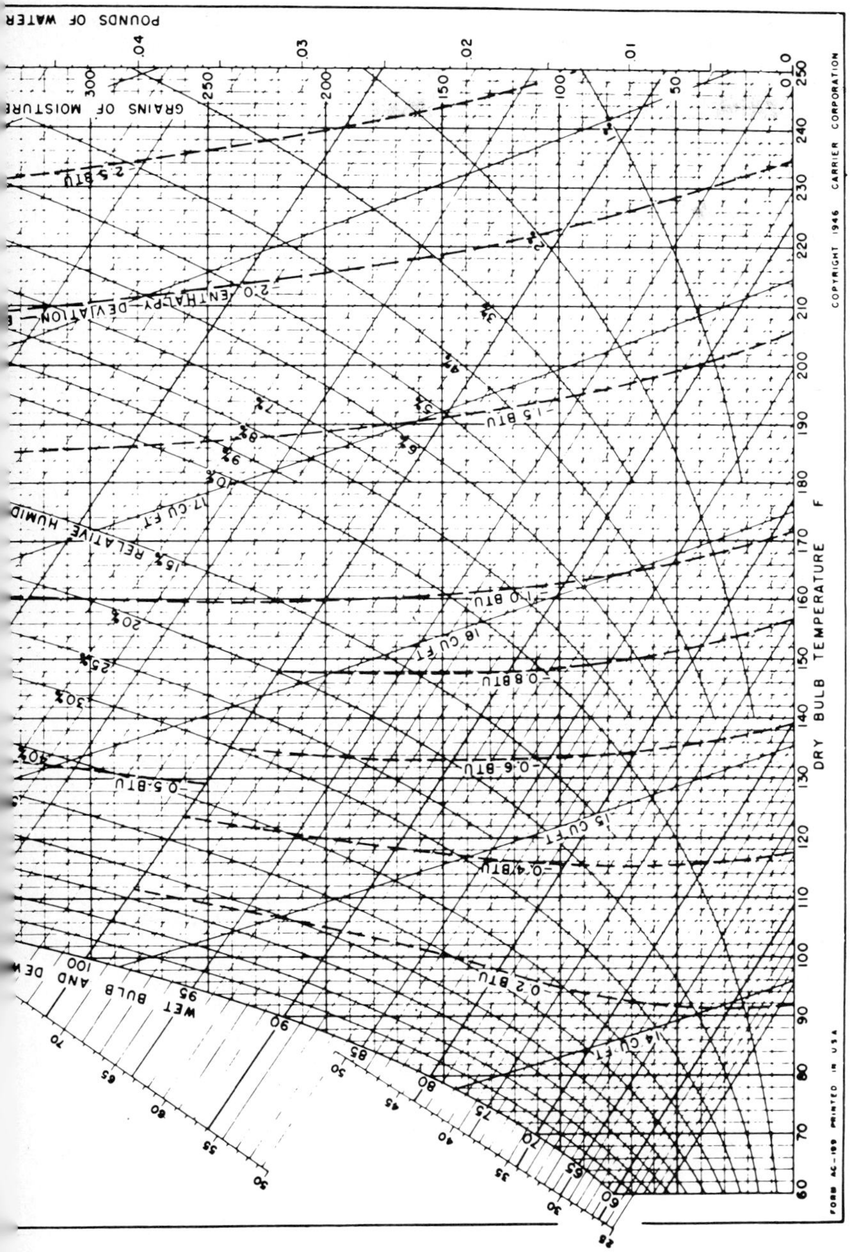

The addition of water vapor to dry air will change the density of the air. In general, a density correction should be applied when the water vapor content of air is greater than 0.02 lb_m water vapor per 1.0 lb_m of dry air. The density of moist air is:

$$\rho_{\text{moist air}} = \frac{\text{mass moist air}}{\text{volume moist air}} \tag{5.4}$$

and the mass of moist air equals:

$$\text{Mass of moist air} = \text{mass of dry air} + \text{mass of water vapor} \tag{5.5}$$

The mass of dry air, mass of water vapor, and volume of moist air can be obtained from the psychrometric chart shown on Figure 5.3 by entering wet- and dry-bulb temperatures.

Example 5.1 Find the density of air having a dry-bulb temperature of 180°F and a wet-bulb temperature of 110°F.

solution From Figure 5.3, the intersection of the wet- and dry-bulb temperatures is at a point where the ratio of mass of water vapor to mass of dry air is 0.042 and the moist volume is 17.2 ft^3 per pound of dry air. Using 1.0 lb$_m$ of dry air as a basis:

$$\rho_{\text{moist air}} = \frac{1.0\ \text{lb}_m\ \text{dry air} + [(0.042\ \text{lb}_m\ \text{water vapor})/(1.0\ \text{lb}_m\ \text{dry air})](1.0\ \text{lb}_m\ \text{dry air})}{[(17.2\ \text{ft}^3)/(1.0\ \text{lb}_m\ \text{dry air})](1.0\ \text{lb}_m\ \text{dry air})}$$

$$= \frac{1.042\ \text{lb}_m}{17.2\ \text{ft}^3} = 0.061\ \text{lb}_m/\text{ft}^3$$

It is not uncommon to find the density of air expressed as specific weight, γ (force per unit volume), rather than mass per unit volume. Specific weight is:

$$\gamma = \frac{\rho g}{g_c} \tag{5.6}$$

where g = gravitational acceleration = 32.2 ft/sec^2

Therefore:

$$\gamma = \frac{(\rho,\ \text{lb}_m/\text{ft}^3)(32.2\ \text{ft/sec}^2)}{32.2\ \text{lb}_m \cdot \text{ft/lb}_f \cdot \text{sec}^2} = \frac{\text{lb}_f}{\text{ft}^3}$$

The specific weight of clean, dry air is 0.075 lb$_f$/ft^3.

The density of air also changes with the addition of contaminants. Unless the contaminant concentration is very large, density changes very little, as is illustrated in the following example.

Example 5.2 What is the percent change in air density if the concentration of ethyl alcohol increases from zero to 1000 ppm?

solution

Clean air concentration = 999,000 ppm
(free of ethyl alcohol)

Ethyl alcohol concentration = 1,000 ppm

Molecular weight of mixture = (0.999)(28.96 g/g-mole) + (0.001)(46.07 g/g-mole)
= 28.98 g/g-mole

$$\frac{\text{Change in molecular weight}}{\text{Original molecular weight}} = \frac{\text{change in density}}{\text{original density}} = \frac{(28.98 - 28.96)\ \text{g/g-mole}}{28.96\ \text{g/g-mole}}$$

$$= 0.0006$$

$$= 0.06\%$$

Ethyl alcohol was chosen for this example because it has a relatively high TLV (1000 ppm) and a relatively high molecular weight for a gas.

Pressure has the dimensions of force per unit area. Pressure measurements in air can be expressed in several different units, including atm, in. or mmHg, in. or mmH_2O, Pa, and in. water gauge. For conversion purposes:

$$1\ \text{atm} = 29.92\ \text{in.Hg} = 760\ \text{mmHg} = 14.7\ \text{lb}_f/\text{in.}^2$$

$$= 1.01 \times 10^2\ \text{kPa} = 1.01 \times 10^5\ \text{N/m}^2 = 407.5\ \text{in.H}_2\text{O}$$

$$= 407.5\ \text{in. water gauge (W.G.)}$$

Because pressure changes encountered in ventilation are usually small and because of the preference for English units, units in terms of inches of water or inches water gauge are most commonly used.

Pressure measurements can also be expressed as absolute or gauge. The relationship between the two is:

$$\text{Gauge pressure} = (\text{absolute pressure}) - (\text{atmospheric pressure}) \quad (5.7)$$

Ventilation pressure measurements are usually expressed as gauge pressures.

II. Air Movement through Ducts

Fluid flow can be classified as being either laminar or turbulent. Laminar flow is characterized by the smooth motion of one layer of fluid past another. Turbulent flow is characterized by irregular and nearly random motion superimposed on the main motion of the fluid. The kind of fluid motion can be determined through use of the Reynolds number (N_{Re}), defined as (Whitaker, 1968):

$$N_{Re} = \frac{\rho v d}{\mu} \quad (5.8)$$

where ρ = fluid density (mass/volume)
v = fluid velocity (length/time)
d = characteristic length (length), usually duct diameter in ventilation work
μ = fluid viscosity (mass/length · time)

Inserting values having consistent units gives the following values of

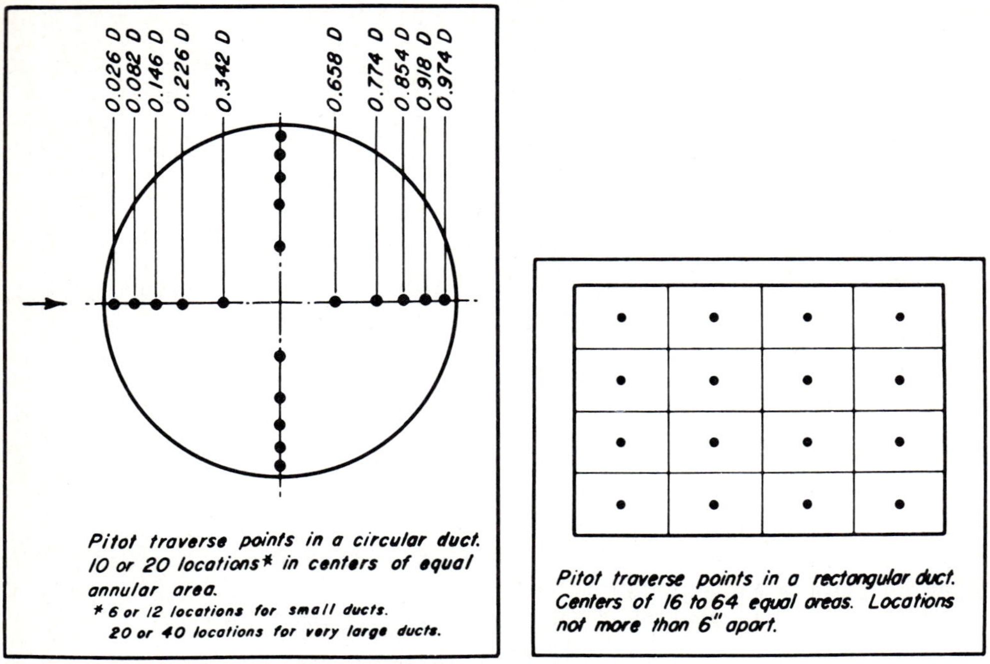

Figure 5.4 Traversing locations (*ACGIH, 1984*).

N_{Re}, which characterize pipe flow:

$N_{Re} \leq 2100$:	laminar flow
$N_{Re} \geq 4000$:	turbulent flow
$2100 < N_{Re} < 4000$:	transition flow (may be laminar or turbulent)

In ventilation work it is very unusual to find a value for N_{Re} less than 4000, as is illustrated in the following example.

Example 5.3 Is flow laminar or turbulent in a circular duct, 6 in. in diameter, with air moving at 100 ft/min? (Note that a 6-in.-diameter duct and air velocity of 100 ft/min are small values for a ventilation system.)

solution

$$N_{Re} = \frac{(0.075 \text{ lb}_m/\text{ft}^3)(100 \text{ ft/min})(6 \text{ in.})(1 \text{ ft}/12 \text{ in.})}{(1.2 \times 10^{-5} \text{ lb}_m/\text{ft} \cdot \text{sec})(60 \text{ sec/min})}$$

$$= 5208$$

The flow is turbulent.

Velocity measurements are often required to characterize the operation of ventilation systems. For stable turbulent flow, the velocity profile will be symmetrical. Such a profile requires an upstream straight run of duct equivalent to more than 10 diameters. If only one measurement can be made (at the duct centerline), the average velocity is approximately 0.9 times the centerline velocity. Since sufficient straight runs of duct are sometimes not available, a more accurate measurement of average velocity can be obtained by traversing the duct. Figure 5.4 shows where to make velocity measurements in circular and rectangular ducts by dividing the duct into subareas of equal area.

To describe flow it is necessary to perform an accounting of mass (continuity equation) and an accounting of energy (Bernoulli's equation). The continuity equation states that the mass of fluid entering a duct segment is equal to the mass of fluid leaving the duct segment plus the mass of fluid accumulated in the duct segment. Referring to Figure 5.5,

$$\text{Mass entering} = \frac{\rho_1 v_1 \pi d_1^2}{4}$$

and

$$\text{Mass leaving} = \frac{\rho_2 v_2 \pi d_2^2}{4}$$

If $\rho_1 = \rho_2$ (which is almost always the case), and the mass accumulated is zero, then:

$$\frac{v_1 \pi d_1^2}{4} = \frac{v_2 \pi d_2^2}{4}$$

or

$$v_1 A_1 = v_2 A_2$$

where A is the cross-sectional area.

Since the product of velocity and area equals flow rate, $Q_1 = Q_2$. If there are several inlets and outlets to a duct segment, this relationship can be generalized to:

$$\Sigma Q_{in} = \Sigma Q_{out} \tag{5.9}$$

Bernoulli's equation for an incompressible, frictionless fluid and steady flow (no time variation) states that the sum of external energy, kinetic energy, and potential energy is a constant along a streamline. (A streamline is a line drawn through the fluid such that the velocity vector is tangent to the line.) Mathematically:

$$\frac{P}{\gamma} + \frac{v^2}{2g} + z = \text{constant} \tag{5.10}$$

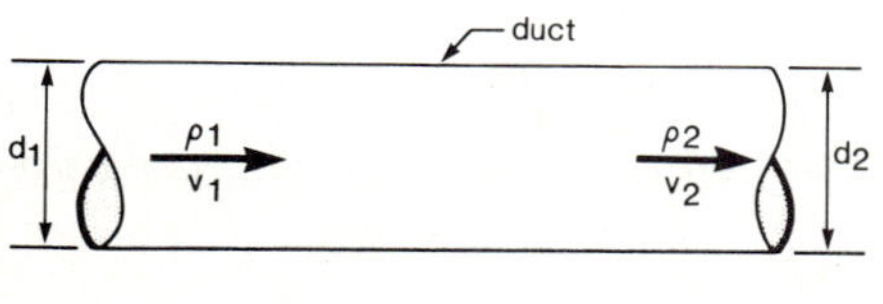

Figure 5.5 Flow schematic.

where P = pressure (force/area)
γ = specific weight (force/volume)
v = velocity (length/time)
g = gravitational acceleration (length/time^2)
z = height above some reference (length)

Note that each term has dimensions of length.

In ventilation work, special names are attached to the external and kinetic energy terms in Bernoulli's equation:

$$\frac{P}{\gamma} = \text{static head}$$

and $$\frac{v^2}{2g} = \text{velocity head}$$

Equation (5.10) can be rearranged by multiplying through by γ:

$$P + \frac{\rho v^2}{2g_c} + \gamma z = (\gamma)\,(\text{constant})$$
$$= \text{constant} \left(\text{since for an incompressible fluid } \gamma = \frac{\rho g}{g_c}\right) \tag{5.11}$$

Examination of Eq. (5.11) shows that all terms now have dimensions of pressure. The first term (P) is called the static pressure (SP); and the second term, $\rho v^2/2g_c$, is called the velocity pressure (VP). The constant is known as total pressure (TP). For ventilation applications it is rare that air is either raised or lowered a vertical distance sufficient to cause the potential energy of the air to be significantly different from one point to another. Therefore, total pressure is defined as:

$$\text{TP} = \text{SP} + \text{VP} \tag{5.12}$$

Since TP is constant along a streamline, TP at point 1 is equal to TP at point 2, or:

$$\text{SP}_1 + \text{VP}_1 = \text{SP}_2 + \text{VP}_2 \tag{5.13}$$

Since real fluid flow is not frictionless, fluid moving from point 1 to point 2 will lose energy to friction. Energy is also lost when flow changes direction. These losses of energy will be manifested as a loss in pressure. Similarly, to get the fluid to move from point 1 to point 2, energy must be added to the system. Modifying Eq. (5.13) to account for the pressure loss and energy addition going from point 1 to point 2 gives:

$$\text{SP}_1 + \text{VP}_1 + \text{pressure (energy) input}_{1-2}$$
$$= \text{SP}_2 + \text{VP}_2 + \text{pressure loss}_{1-2} \tag{5.14}$$

Measurement of static, velocity, and total pressures is illustrated in

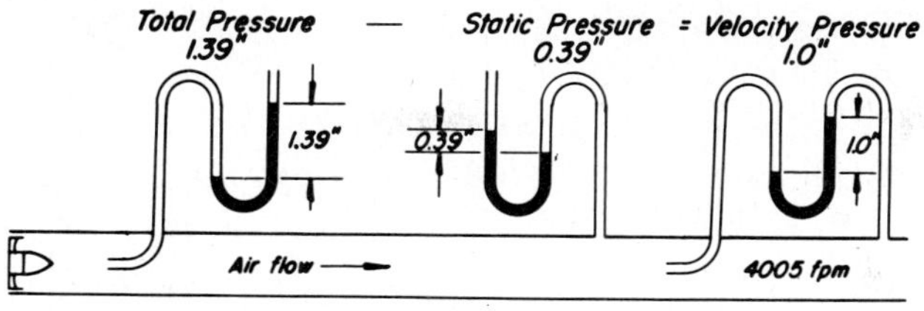

Figure 5.6 Measurement of total, static, and velocity pressures (*ACGIH, 1984*).

Figure 5.6. Note that the value of velocity pressure is always positive; the values of the static and total pressures may be positive or negative (gauge pressure). In studying the operation of a ventilation system, it is more common to make measurements of SP and VP than it is to make measurements of velocity alone. If velocity is needed, it can be calculated from VP by using

$$v = 4005(\mathrm{VP})^{0.5} \tag{5.15}$$

where v = velocity (ft/min)
VP = velocity pressure (in.H_2O)

To obtain average values of SP and VP across a duct section, measurements of each are made along traverses as was described for velocity measurement. SP and VP are measured by selectively adjusting the plumbing of a pitot tube and an inclined manometer (Figures 5.6 and 5.7). Each VP is converted to a velocity using Eq. (5.15). The velocities of a traverse are averaged arithmetically to calculate an average velocity. The average VP may then be calculated from Eq. (5.15).

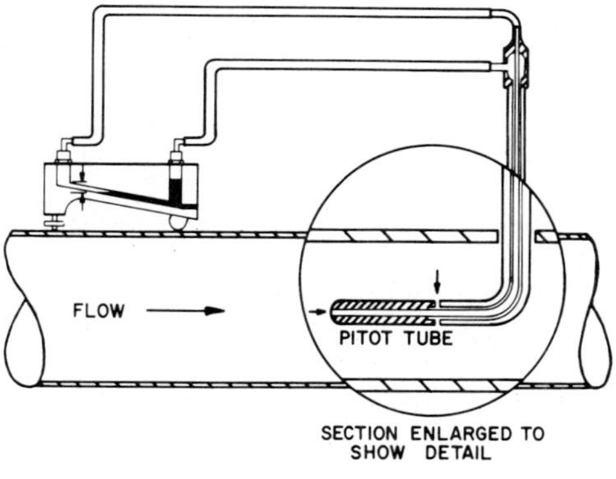

Figure 5.7 Pitot tube and inclined manometer (*NIOSH, 1973*).

III. Pressure Losses

Equation (5.14) indicates that air moving through a ventilation system loses pressure in moving from one point to another because of friction and dynamic losses. Increase in pressure (energy input) is provided by an air mover (fan or blower), usually located near the exhaust end so that most of the system operates at subatmospheric (negative) pressure. This prevents contaminants from leaking out of the ducting into the workplace. Since air entering most ventilation systems is at atmospheric pressure, the air mover needs only to be able to supply enough energy to overcome all losses.

The initial steps in designing a ventilation system are as follows:

1. Specify the quantity (rate) of air to enter each inlet and exit each outlet and the minimum duct velocity.
2. Specify duct components (fittings), including size, length, damper position, inlets and outlets, filters.
3. Calculate pressure losses in the system.
4. Calculate requirements of the air mover.
5. Refine design. (Larger fittings have less resistance to flow but cost more to build and require more space. Space may not be available. Smaller fittings cost less and require less space, but cause greater pressure drop. Greater losses require a larger air mover, which is more expensive initially as well as operationally. Final design is an economic compromise among all these factors.)

The key to the entire design procedure lies in being able to reasonably estimate losses in fittings. There are several different approaches to calculating pressure drop in a ventilation system. Two of the more commonly used approaches have been developed by the American Society of Heating, Refrigerating and Air-Conditioning Engineers (ASHRAE, 1985) and by the American Conference of Governmental Industrial Hygienists (ACGIH, 1984). Both methods are empirical. The ASHRAE method results in calculation of the loss of total pressure. The ACGIH method results in calculation of the change in static pressure. Both methods utilize velocity pressure in the calculation of pressure change. Thus, using the ASHRAE method, the total pressure loss is calculated and then the static pressure is calculated from Eq. (5.12). Using the ACGIH method, static pressure change is calculated and then Eq. (5.12) is used to arrive at total pressure. Which, if either, of these approaches gives a better representation of the pressure changes occurring in an actual system has not been systematically evaluated. It is not uncommon for the calculated pressure change using both methods to differ by no more than a few hundredths of an inch of water. There is, however, no guarantee that differences in the two calculational approaches will always be this small. The ASHRAE approach will be described and demonstrated because it is somewhat easier to follow. Two refinements developed by ACGIH will also be introduced. Thus the method used here is somewhat of a hybrid of the two methods.

TABLE 5.2 Duct Material Roughness

Duct material	Roughness category	Absolute roughness, ε (ft)	Roughness correction factor, K_m, at 2000 fpm and specified duct diameter	
			5 in.	100 in.
Uncoated carbon steel, clean	Smooth	0.00015	0.86	0.9
Galvanized steel, continuously rolled, spiral seams	Medium smooth	0.0003	0.93	0.96
Galvanized steel, sheets, hot dipped, longitudinal seams	Average	0.0005	1.0	1.0
Fibrous glass duct, rigid	Medium rough	0.003	1.52	1.31
Fibrous glass duct liner, air side spray coated	Rough	0.015	2.28	1.7

SOURCE: After ASHRAE, 1985.

Air moving through duct components will lose pressure due to friction between the moving air and the duct wall. The roughness of the duct surface is very important in determining the resulting loss. Table 5.2 lists various duct materials and categorizes the roughness of each. The roughness correction factor, K_m, is dimensionless. Friction loss in other than an average roughness duct is obtained by multiplying K_m by the loss in total pressure determined for average roughness duct. Figures 5.8*a* and 5.8*b* are graphical presentations of loss in total pressure as a function of any two of three duct variables (size, air velocity, and/or air flow rate) for average-roughness, clean, round, galvanized ducts. Finding the intersection of any two of duct size, air velocity, and/or air flow rate gives the loss of total pressure per unit length of straight duct.

Example 5.4 What is the loss in total pressure of 550 cfm of air moving through a 25-ft length of 8-in.-diameter, straight, average-roughness duct?

solution From Figure 5.8a, the friction loss is 0.5 in.H_2O per 100 ft of duct.

$$\text{Loss in total pressure} = \left(\frac{0.5 \text{ in.H}_2\text{O}}{100 \text{ ft duct}}\right)(25 \text{ ft duct})$$

$$= 0.125 \text{ in. H}_2\text{O} \approx 0.13 \text{ in. H}_2\text{O}$$

Figure 5.8 can also be used to calculate friction losses in rectangular ducts. The circular equivalent of the rectangular duct size can be obtained from Appendix A (from ASHRAE, 1985) at the end of this chapter and used as one of the two entry variables in Figures 5.8.

Example 5.5 What is the circular equivalent of a 10 in. by 24 in. rectangular duct?

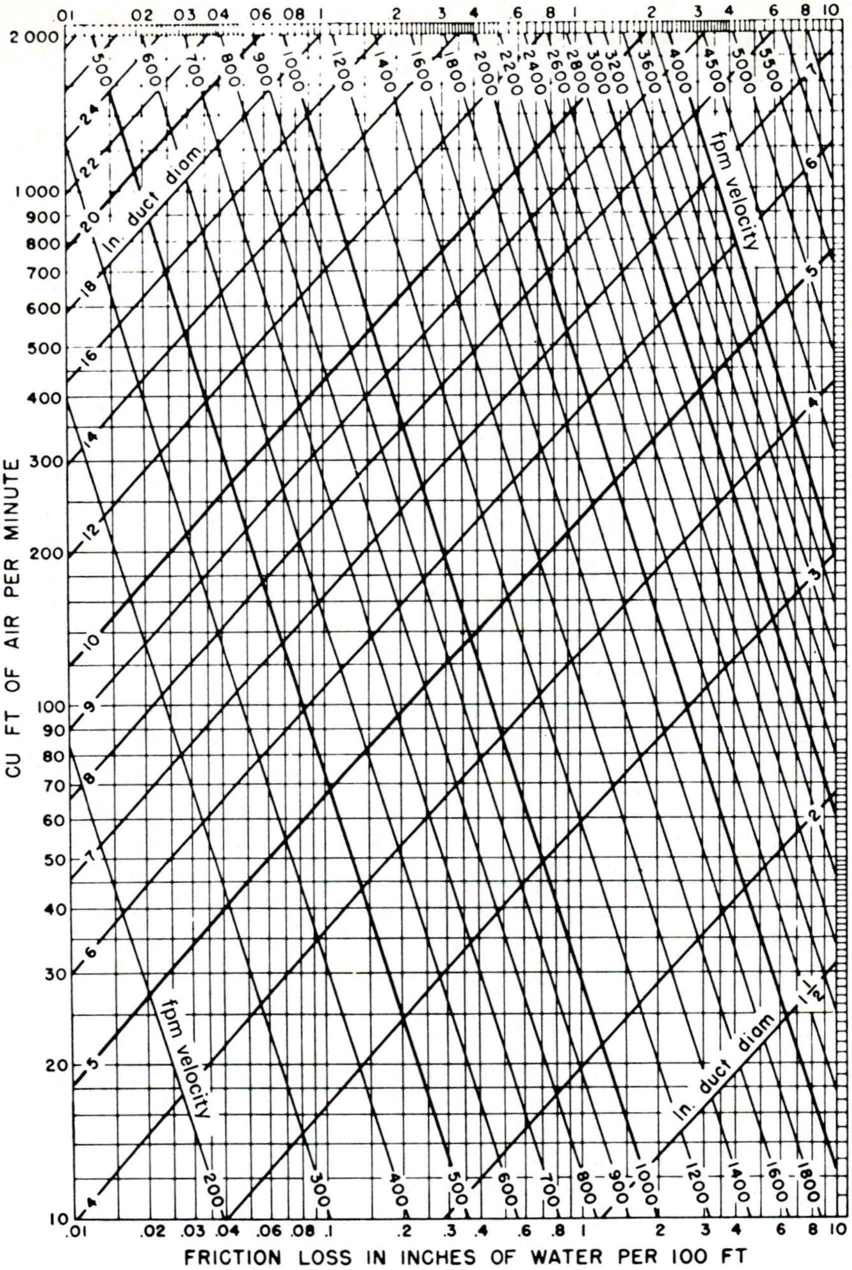

(Based on Standard Air of 0.075 lb per cu ft density flowing through average, clean, round, galvanized metal ducts having approximately 40 joints per 100 ft.)
Caution: Do not extrapolate below chart.

For proprietary duct, obtain data from manufacturer. Friction of Air in Straight Ducts for Volumes of 10 to 2000 cfm

$$\text{Friction Loss}/100' = \frac{2.74\left[\dfrac{V_{fpm}}{1000}\right]^{1.9}}{[D_{inches}]^{1.22}}$$

Note: Both "1.9" and "1.22" are exponents.

Figure 5.8a Friction loss (*ACGIH, 1984*).

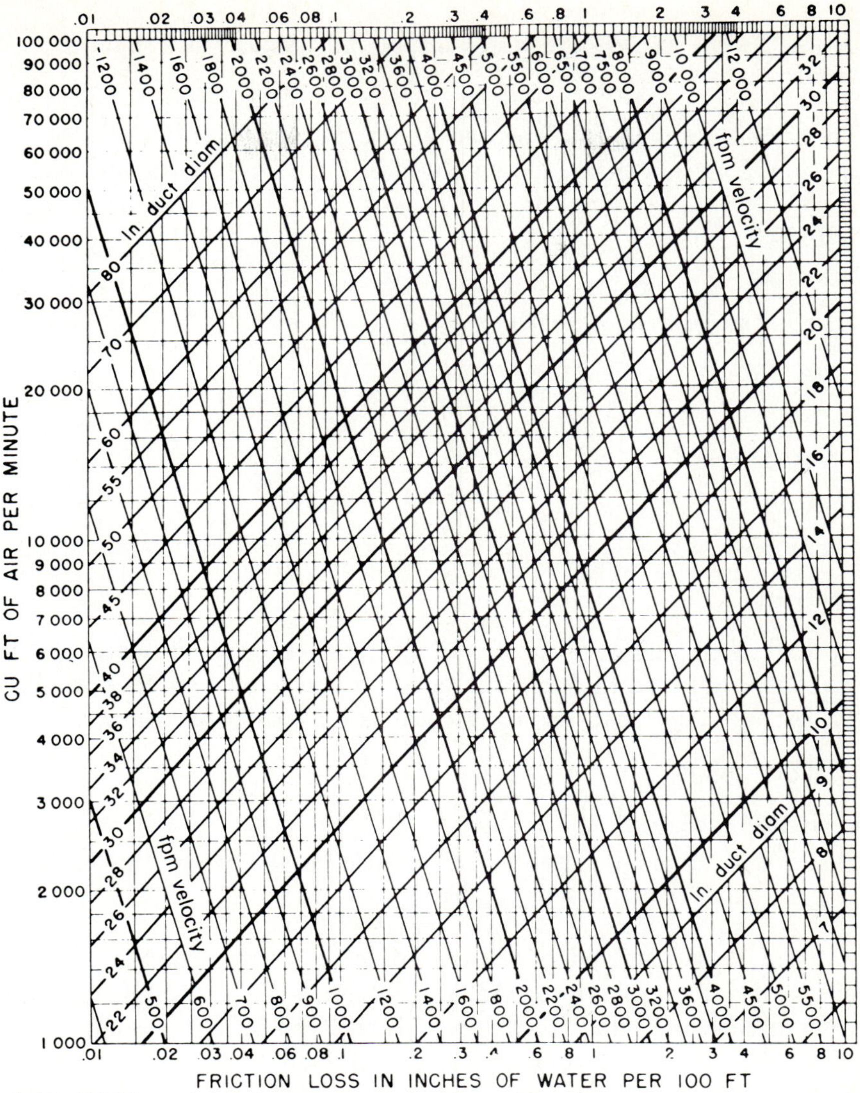

(Based on Standard Air of 0.075 lb per cu ft density flowing through average, clean, round, galvanized metal ducts having approximately 40 joints per 100 ft.)

For proprietary duct, obtain data from manufacturer. Friction of Air in Straight Ducts for Volumes of 1000 to 100,000 cfm

Reprinted from 37th Edition, Heating, Ventilating, Air Conditioning Guide. 1959, by permission of the American Society of Heating, Refrigerating and Air-Conditioning Engineers.

$$\text{Friction Loss}/100' = \frac{2.74\left[\dfrac{V_{fpm}}{1000}\right]^{1.9}}{[D_{inches}]^{1.22}}$$

Figure 5.8*b* Friction loss (*ACGIH, 1984*).

solution From Appendix A to this chapter, the circular duct would be 16.5 in. in diameter.

Dynamic losses in total pressure result from the air flow changing direction. Fittings that cause a change in flow direction include:

1. Inlets
2. Outlets
3. Diffusers (a type of outlet)
4. Nozzles
5. Orifices
6. Tees or wyes
7. Elbows
8. Dampers
9. Screens

Where dynamic losses occur, frictional losses also occur. The two kinds of loss are additive and are included in the local loss coefficient, C_0, for a fitting. The local loss coefficient is defined such that:

$$\Delta \mathrm{TP} = C_0 \, \mathrm{VP}_0 \tag{5.16}$$

where ΔTP = loss in total pressure

VP_0 = average velocity pressure at location 0 where the local loss coefficient applies

Appendix B at the end of this chapter is a tabulation of local loss coefficients taken from *ASHRAE Fundamentals* (ASHRAE, 1985). Diagrams are included that show which local loss coefficient and velocity pressure are appropriate for calculation of the loss in total pressure in a fitting or branch of a fitting.

Calculation of pressure loss through a ventilation system requires knowledge of the change in SP, VP, and TP from one location to another. Assumptions have been built into the procedure used to make the calculations. If the assumptions are violated, then the calculated results will likely be unreliable. One assumption is that the density of the gas in the ventilation system is 0.075 $\mathrm{lb_m/ft^3}$. If the gas density varies significantly, adjustments must be made in the calculations. The density correction to be made to VP is presented in Section VII of this chapter. Correction for density should not be overlooked in the calculations.

Example 5.6 What is the loss in total pressure from the branches to the common of the round, converging conical wye illustrated on Figure 5.9? The diameter of

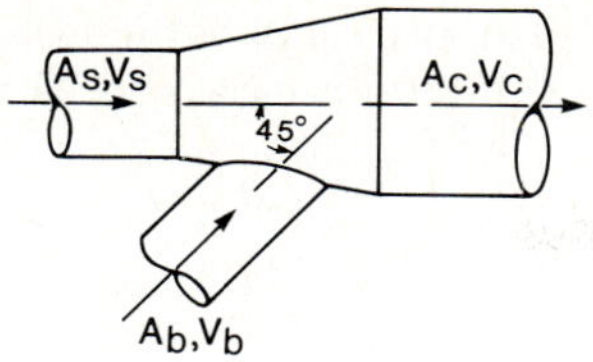

Figure 5.9 Converging conical wye for Example 5.6.

the duct at point c (the common) is 12 in., the diameter at point s (the straight route) is 8 in., and the diameter at point b (the branch route) is 6 in. The volumetric flow rate at c is 1800 cfm, and it is 1000 cfm at s.

solution From the continuity equation,

$$Q_b = Q_c - Q_s$$
$$= 1800 \text{ cfm} - 1000 \text{ cfm} = 800 \text{ cfm}$$

The velocity at point c is:

$$v_c = \frac{Q_c}{A_c}$$
$$= \frac{1800 \text{ ft}^3/\text{min}}{\left[\frac{(3.14)(12 \text{ in.})^2 (1 \text{ ft}/12 \text{ in.})^2}{4}\right]} = \frac{1800 \text{ ft}^3/\text{min}}{0.79 \text{ ft}^2}$$
$$= 2280 \text{ ft/min}$$

From Eq. (5.15),

$$\text{VP}_c = \left(\frac{v_s}{4005}\right)^2 = \left(\frac{2280}{4005}\right)^2 = 0.32 \text{ in.H}_2\text{O water}$$

Examination of Appendix B, Table 6-5, shows the loss in total pressure in either the straight (s) or main route as well as the branch (b) route as a function of the velocity pressure in the common (c) section of the fitting, hence the use of the subscripts of c, s and c, b on the local loss coefficient. The portion of the table for the local loss coefficient in the straight route ($C_{c,s}$) indicates that the ratios of A_s to A_c, A_b to A_c, and Q_b to Q_s are needed.

$$\frac{A_s}{A_c} = \frac{(3.14)(1/4)(8/12)^2 \text{ ft}^2}{(3.14)(1/4)(12/12)^2 \text{ ft}^2} = 0.44$$

$$\frac{A_b}{A_c} = \frac{(3.14)(1/4)(6/12)^2 \text{ ft}^2}{(3.14)(1/4)(12/12)^2 \text{ ft}^2} = 0.25$$

$$\frac{Q_b}{Q_s} = \frac{800 \text{ cfm}}{1000 \text{ cfm}} = 0.8$$

From Appendix B, Table 6-5, interpolation gives

$$C_{c,s} = 0.01$$

Similarly,

$$C_{c,b} = 3.3$$

The signs on the loss coefficients are important. There will be a decrease in total pressure in both routes. However, converging flow junctions may result in a decrease or an increase in total pressure. Note the differences in sign of the local loss coefficients in Table B-6, Appendix B. The changes in total pressure for the straight and branch routes are

$$\Delta TP_s = C_{c,s}\,VP_c$$

$$= (0.01)(0.32\ \text{in.}H_2O) = 0.00\ \text{in.}H_2O \qquad (5.16)$$

$$\Delta TP_b = C_{c,b}\,VP_c$$

$$= (3.3)(0.32\ \text{in.}H_2O) = 1.06\ \text{in.}H_2O \qquad (5.16)$$

Changes in duct velocity can occur downstream of converging flow when the common duct velocity exceeds the branch velocities. A difference of 0.10 in H_2O between the VP_c and resultant VP (VP_r) should be corrected (ACGIH, 1984). The VP_r is

$$VP_r = \left[\frac{Q_s + Q_b}{4005\,(A_s + A_b)}\right]^2 \qquad (5.17)$$

and the corrected static pressure at the wye common branch is

$$SP_c = SP_s - (VP_c - VP_r) \qquad (5.18)$$

where the algebraic sign of SP_c and SP_s is negative and $SP_s = SP_b$. Principles of good duct design require that the duct velocity remain relatively constant throughout the system. One suggested criteria for converging flow is that $A_c = (A_s + A_b)(\pm 20\%)$ (ACGIH, 1984).

Pressure loss estimates based on Appendix B do not include effects of upstream flow disturbance. In general, flow becomes fully developed if the distance between fittings is at least 7.5 duct diameters. Figure 5.10 presents the ACGIH method of calculating static regain for an expansion. Note that static regain (increase in static pressure) is only 70% of the amount expected if the flow entering the expansion is not fully developed.

STATIC PRESSURE REGAINS FOR EXPANSIONS

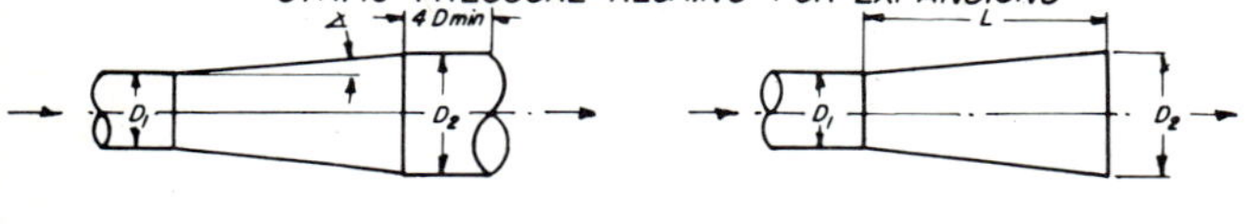

Within duct

Regain (R), fraction of VP difference					
Taper angle degrees	Diameter ratios D_2/D_1 1.25:1	1.5:1	1.75:1	2:1	2.5:1
3 1/2	0.92	0.88	0.84	0.81	0.75
5	0.88	0.84	0.80	0.76	0.68
10	0.85	0.76	0.70	0.63	0.53
15	0.83	0.70	0.62	0.55	0.43
20	0.81	0.67	0.57	0.48	0.43
25	0.80	0.65	0.53	0.44	0.28
30	0.79	0.63	0.51	0.41	0.25
Abrupt 90	0.77	0.62	0.50	0.40	0.25

Where: $SP_2 = SP_1 + R(VP_1 - VP_2)$

At end of duct

Regain (R), fraction of inlet VP						
Taper length to inlet diam L/D	Diameter ratios D_2/D_1 1.2:1	1.3:1	1.4:1	1.5:1	1.6:1	1.7:1
1.0:1	0.37	0.39	0.38	0.35	0.31	0.27
1.5:1	0.39	0.46	0.47	0.46	0.44	0.41
2.0:1	0.42	0.49	0.52	0.52	0.51	0.49
3.0:1	0.44	0.52	0.57	0.59	0.60	0.59
4.0:1	0.45	0.55	0.60	0.63	0.63	0.64
5.0:1	0.47	0.56	0.62	0.65	0.66	0.68
7.5:1	0.48	0.58	0.64	0.68	0.70	0.72

Where: $SP_1 = SP_2 - R(VP_1)$*

*When $SP_2 = 0$ (atmosphere) SP_1 will be (-)

The regain (R) will only be 70% of value shown above when expansion follows a disturbance or elbow (including a fan) by less than 5 duct diameters.

Figure 5.10 Static regain for expansion (*ACGIH, 1984*).

Appendix B also includes a number of different kinds of inlets, some of which are used in industrial ventilation systems. Loss in total pressure can be calculated using the local loss coefficients. Pressure loss in inlets is frequently expressed in terms of coefficient of entry, C_e, or hood entry loss, h_e. Both of these factors are dimensionless. The volumetric flow rate of air entering an inlet is [from Eq. (5.15)]

$$Q = vA = 4005A(\text{VP})^{0.5} \tag{5.15a}$$

The coefficient of entry is defined as

$$C_e = \left(\frac{\text{VP}}{|\text{SP}_{\text{hood}}|}\right)^{0.5} \tag{5.19}$$

where $|\text{SP}_{\text{hood}}|$ is the absolute value of SP in the inlet.

Substituting Eq. (5.19) into Eq. (5.15a) gives

$$Q = 4005\text{A}C_e(\text{SP}_{\text{hood}})^{0.5} \tag{5.20}$$

so the static pressure in the inlet (hood) is

$$\text{SP}_{\text{hood}} = \left(\frac{Q}{4005AC_e}\right)^2 \tag{5.21}$$

Given values of coefficient of entry as shown in Figure 5.11, SP_{hood} can be calculated. (Note that the actual SP in the inlet will be negative, since TP in front of the inlet is zero gauge pressure; TP must decrease for flow to occur, and VP is greater than zero.)

Entry loss, h_e, the loss in total pressure, is given in Figure 5.11. Entry loss is defined as

$$h_e = |\text{SP}_{\text{hood}}| - \text{VP} \tag{5.22}$$

Returning to Eq. (5.19),

$$C_e = \left(\frac{\text{VP}}{\text{SP}_{\text{hood}}}\right)^{0.5} = \left(\frac{\text{VP}}{h_e + \text{VP}}\right)^{0.5} \tag{5.19}$$

or

$$C_e^2 = \left(\frac{\text{VP}}{h_e + \text{VP}}\right)$$

Rearranging gives

$$h_e = \frac{(1 - C_e^2)}{C_e^2}(\text{VP}) \tag{5.23}$$

The local loss coefficient for inlets used by ASHRAE [see Eq. (5.16)] is therefore

$$C_0 = \frac{(1 - C_e^2)}{C_e^2} \tag{5.24}$$

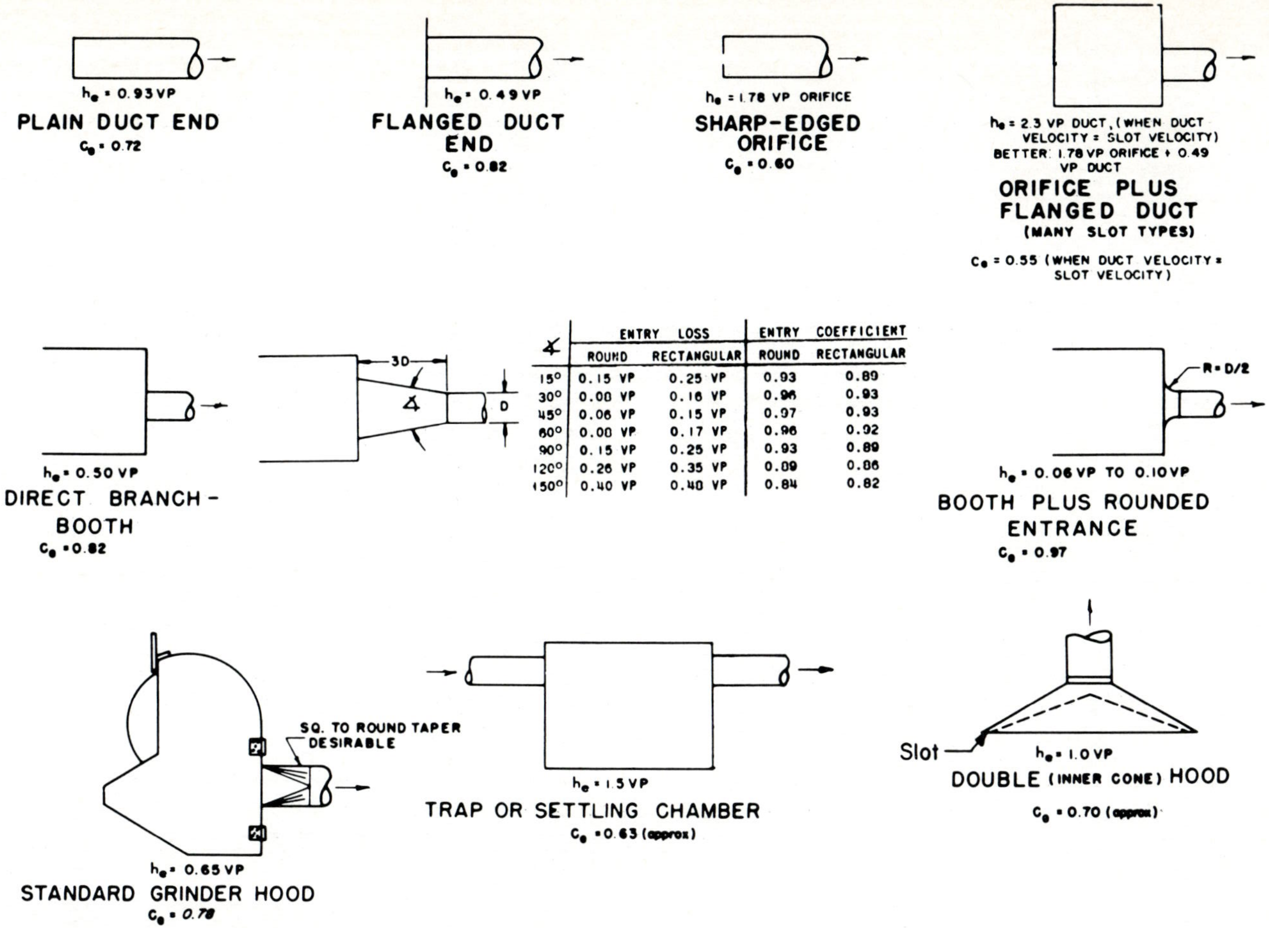

∡	Entry loss		Entry coefficient	
	Round	Rectangular	Round	Rectangular
15°	0.15 VP	0.25 VP	0.93	0.89
30°	0.08 VP	0.16 VP	0.96	0.93
45°	0.06 VP	0.15 VP	0.97	0.93
60°	0.08 VP	0.17 VP	0.96	0.92
90°	0.15 VP	0.25 VP	0.93	0.89
120°	0.26 VP	0.35 VP	0.89	0.86
150°	0.40 VP	0.40 VP	0.84	0.82

Figure 5.11a Hood coefficient of entry and entry loss (*ACGIH, 1984*).

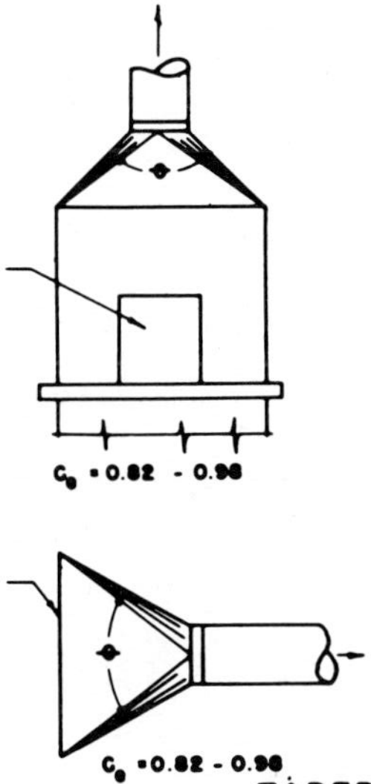

REPRODUCED BY PERMISSION FROM "INDUSTRIAL HEALTH ENGINEERING" BY A. D. BRANDT, PUBLISHED BY JOHN WILEY AND SONS, INC.

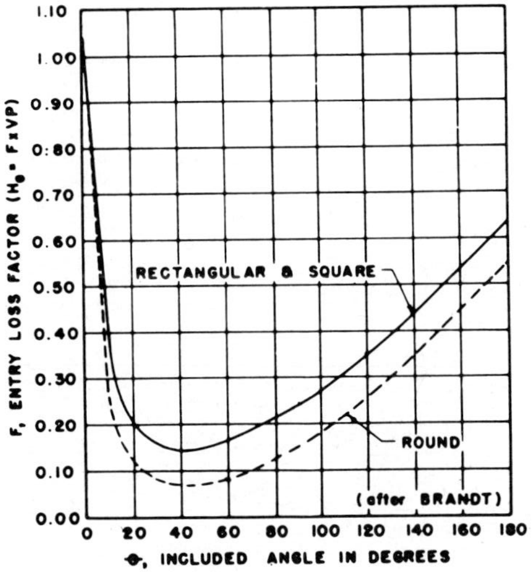

TAPERED HOODS

FLANGED OR UNFLANGED; ROUND, SQUARE OR RECTANGULAR.
Φ IS THE MAJOR ANGLE ON RECTANGULAR HOODS.

$$h_e = \frac{1-C_e^2}{C_e^2} \times VP$$

$$F = \frac{1-C_e^2}{C_e^2}$$

$$h_e = F \times VP$$

$$Q = 4005\, A\, C_e \sqrt{SP}$$

F = ENTRY LOSS FACTOR
VP = VELOCITY PRESSURE IN DUCT
SP = STATIC PRESSURE AT THROAT, INCHES WATER GAGE
h_e = ENTRY LOSS, INCHES WATER GAGE
Q = AIR VOLUME, CFM
A = CROSS SECTION AT THROAT, SQUARE FEET
C_e = COEFFICIENT OF ENTRY

ENTRY LOSS FOR COMPLICATED HOOD SHAPES:
1. BREAK HOOD INTO SIMPLE COMPONENTS
2. CALCULATE h_e FOR EACH COMPONENT
3. ADD VALUES OF h_e

MISCELLANEOUS VALUES

HOOD	ENTRY LOSS, F
ABRASIVE BLAST CHAMBER	1.0
ABRASIVE BLAST ELEVATOR	2.3
ABRASIVE SEPARATOR	2.3
ELEVATORS (ENCLOSURES)	0.69
FLANGED PIPE PLUS CLOSED ELBOW	0.8
PLAIN PIPE PLUS CLOSE ELBOW	1.60
TUMBLING MILLS (VARIES WITH DESIGN OF MILL)	AV. 2.0

Figure 5.11*a* *(Continued)*

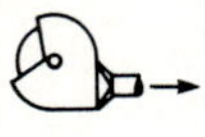	TYPICAL GRINDING HOOD	STRAIGHT TAKE-OFF	
		Ce=0.78	he=0.65VP
		TAPERED TAKE-OFF	
		Ce=0.85	he=0.40VP
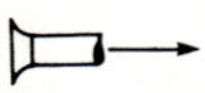	BELL MOUTH INLET	Ce=0.98	he=0.04VP

Figure 5.11*b* Hood coefficient of entry and entry loss (*ACGIH, 1984*).

Reduction of TP can be calculated given the local loss coefficient, coefficient of entry, or entry loss.

Up to this point the discussion has dealt with individual components of a ventilation system. A complete system is now examined.

Example 5.7 Figure 5.12 is a diagram with some specifications for a local exhaust ventilation system. Calculate TP, SP, and VP across the length of the system. Note that a new element has been introduced into the system: a fan (air mover). The fan can be ignored for purposes of calculation. Air movers will be discussed in a subsequent section of this chapter.

solution To set up calculations in an organized fashion, various locations in the ventilation system have been identified by circled numbers shown at the bottom of the diagram. Table 5.3 shows the calculation results. Note that before the air enters the system, and after the air exits the system, SP = 0. Also, before air enters the system, TP = VP = 0. For air to enter the system, TP must drop below atmospheric pressure at the inlet. For air to exit the system, TP must be above atmospheric pressure at the end of the system. In calculating TP, the approach is to work from the known pressure (TP = 0) at the start of the system up to the point where air enters the fan. The fan will increase TP, but at this point it is impossible to calculate the amount. Therefore, calculation of TP is also initiated at the end of the system, where TP is due entirely to the VP component and SP = 0. TP is then calculated backward to the point where air exits the fan.

Calculation of pressure loss is begun at the first fitting, the inlet (refer to Table 5.3). The first step is to calculate the velocity pressure at the inlet from Eq. (5.15). The pertinent velocity is equal to the volumetric flow rate divided by the duct cross sectional area. The next step is to find the appropriate local loss coefficient for the fitting. Table 1-1, Appendix B, requires ratios of L/D and t/D to arrive at a loss coefficient (L is the duct extension from the wall and t is the duct thickness). Since the inlet is mounted flush with the wall, $L = 0$. Hence $L/D = 0$. For $L/D = 0$, the local loss coefficient equals 0.5 regardless of the value of t/D. [Note that hood entry loss (h_e) could have been obtained from Figure 5.11*a*. For a flanged duct end, the entry loss is equal to 0.49VP, or the local loss coefficient is 0.49. The ASHRAE approach predicts a slightly larger loss for this inlet.] Multiplying the loss coefficient by the velocity pressure gives the loss in

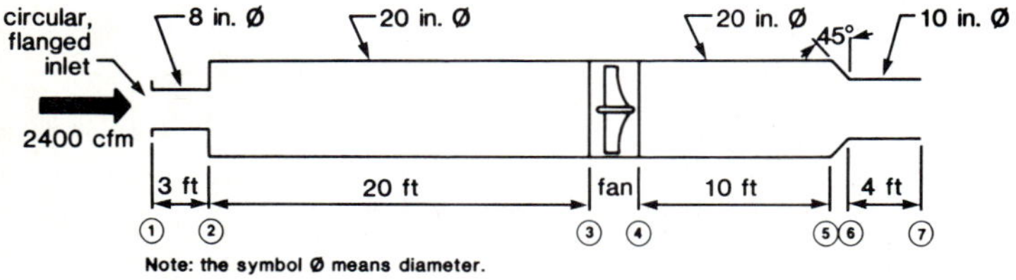

Figure 5.12 Ventilation system schematic (see Example 5.7 and Table 5.3 for results).

TABLE 5.3 Ventilation System Pressure Loss Calculations

System segment	Segment type	Air flow (cfm)	Duct size (ft)	Duct length (ft)	Velocity (fpm)	VP (in.H_2O)	Local loss coefficient (C_0)	Frictional loss (in.H_2O/100 ft)	ΔTP (in.H_2O)	TP (in.H_2O)	SP (in.H_2O)
1	Inlet	2400	0.67	—	6879	2.95	0.5[a]	—	1.48	−1.48	−4.43
1–2	Duct	2400	0.67	3	6879	2.95	—	8.5[b]	0.26	−1.74	−4.69
2	Abrupt expansion	2400	0.67–1.67	—	6879–1096	2.95–0.08	—	—	2.37	−4.11	−4.19[c]
2–3	Duct	2400	1.67	20	1096	0.08	—	0.08[b]	0.02	−4.13	−4.21
3–4	Fan	2400	—	—	—	—	—	—	—	—	—
4–5	Duct	2400	1.67	10	1096	0.08	—	0.08[b]	0.01	1.54	1.46
5–6	Contraction	2400	1.67–0.83	—	1096–4403	0.08–1.21	0.17[d]		0.21	1.53	1.45
6–7	Duct	2400	0.83	4	4403	1.21	—	2.8[b]	0.11	1.32	0.11
7	Exit	2400	0.83	—	4403	1.21	1.0[e]	—	1.21	1.21	0

[a] From Appendix B, Table 1-1; $L/D = 0$.
[b] From Figure 5.8*b*.
[c] From Figure 5.10. Abrupt: 90°, $D_2/D_1 = 2.5$, less than 5 duct diameters from last flow disturbance.
[d] From Appendix B, Table 5-1;

$$\frac{A_1}{A_0} = \frac{2.19}{0.55} = 4.0$$

[e] From Appendix B, Table 2-1.

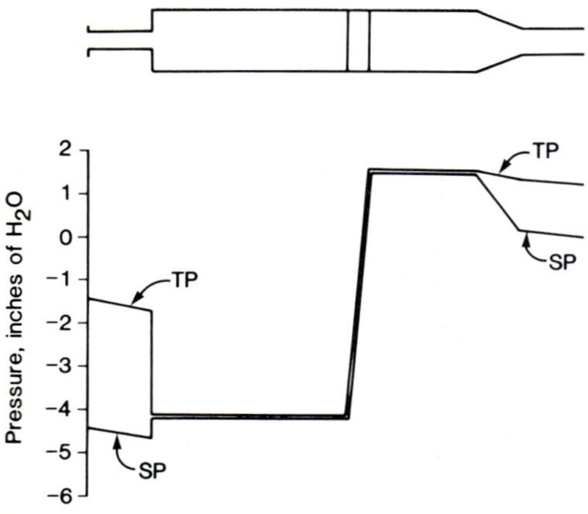

Figure 5.13 Gradelines for Example 5.7 (see Table 5.3).

total pressure for the fitting. Since the entry is open to the atmosphere, total pressure before the fitting is zero. Thus the total pressure in the fitting is zero minus the change in total pressure across the duct entry. Finally, static pressure can be obtained by subtracting velocity pressure from total pressure [Eq. (5.12)].

Calculation of pressure loss in subsequent fittings is approached in the same manner. Local loss coefficients, frictional loss, and static pressure regain used to initiate each calculation are identified in Table 5.3.

It should also be noted that the ventilation system depicted in this example is not a recommended design (although such an installation has been observed in practice). At least two things are undesirable. First, the large reduction in velocity in the middle of the system could permit particles to settle and accumulate. Second, the air mover is located midway through the ductwork. This means that much of the system is above atmospheric pressure, permitting contaminated air to leak from the duct (at joints) into the room.

After calculating and tabulating pressures through a ventilation system, it is common to display the results graphically. Plots of TP, SP, and VP vs. location in the ventilation system are known as gradelines. There is no need to plot all three pressures, because knowing two of the pressures defines the third. Commonly only TP and SP are plotted.

Example 5.8 Plot the gradelines for TP and SP from Example 5.7.

solution Plotted results are shown in Figure 5.13.

IV. Fan Specification

When calculations were begun in the previous example, it was noted that the fan put energy into the system, but the amount was unknown. The requirements of the fan can be specified after calculating the pressure losses through the ventilation system. Manufacturers typically rate fans and blowers on the basis of the relationship between flow

TABLE 5.4. Fan Rating Table for a Backward-Inclined Blade Fan from a Fan Catalog

CFM	4" SP		4½" SP		5" SP		5½" SP		6" SP		6½" SP		7" SP		7½" SP		8" SP		8½" SP	
	RPM	BHP	RPM	BHP	RPM	BHP	RPM	BHP	RPM	BHP	RPM	BHP	RPM	BHP	RPM	BHP	RPM	BHP	RPM	BHP
1289	2699	1.24	2841	1.42																
1375	2730	1.32	2868	1.49	3001	1.68														
1461	2762	1.39	2895	1.57	3024	1.76	3150	1.95	3277	2.16										
1547	2797	1.47	2927	1.66	3052	1.85	3180	2.05	3299	2.26	3417	2.47								
1633	2836	1.55	2963	1.75	3086	1.94	3206	2.15	3323	2.35	3438	2.57	3551	2.79	3660	3.01				
1719	2878	1.64	2999	1.84	3119	2.04	3238	2.25	3356	2.47	3465	2.68	3576	2.91	3684	3.14	3788	3.37	3892	3.61
1805	2922	1.73	3042	1.94	3160	2.15	3273	2.36	3386	2.58	3498	2.81	3603	3.03	3710	3.26	3812	3.51	3916	3.75
1891	2969	1.83	3087	2.04	3201	2.25	3312	2.47	3424	2.71	3530	2.93	3635	3.16	3740	3.41	3843	3.64	3940	3.89
1977	3023	1.94	3137	2.15	3245	2.37	3355	2.59	3462	2.82	3567	3.06	3671	3.31	3771	3.54	3868	3.78	3970	4.04
2063	3075	2.04	3185	2.26	3290	2.48	3398	2.72	3503	2.95	3604	3.19	3707	3.44	3807	3.69	3903	3.94	4000	4.21
2149	3131	2.16	3238	2.39	3340	2.61	3441	2.84	3543	3.08	3644	3.33	3746	3.58	3842	3.83	3938	4.09	4030	4.35
2235	3187	2.28	3290	2.51	3394	2.75	3493	2.98	3590	3.22	3693	3.49	3788	3.74	3881	3.99	3973	4.25	4065	4.51
2407	3306	2.54	3404	2.78	3501	3.02	3599	3.28	3690	3.53	3786	3.81	3879	4.07	3964	4.31	4059	4.61	4143	4.87
2579	3432	2.82	3524	3.07	3617	3.33	3708	3.69	3799	3.86	3888	4.13	3975	4.41	4059	4.68	4147	4.97	4232	5.26
2751	3559	3.13	3651	3.41	3738	3.67	3825	3.94	3910	4.21	3998	4.51	4079	4.78	4164	5.08	4245	5.37		
2923	3692	3.46	3779	3.74	3866	4.03	3947	4.31	4032	4.61	4110	4.89	4191	5.19						
3095	3830	3.83	3912	4.12	3994	4.42	4075	4.72	4154	5.02	4232	5.32								
3267	3969	4.23	4047	4.53	4130	4.85	4206	5.16												
3439	4110	4.66	4191	4.99	4267	5.31														

Selection within tinted area renders most efficient, quietest operation. BHP shown does not include belt drive losses.

SOURCE: McDermott, 1985.

rate of air and the static pressure against which the fan or blower must move the air (called fan static pressure or SP_{fan}). Fan static pressure is defined as:

$$SP_{fan} = TP_{fan} - VP_{out} \tag{5.25}$$

where TP_{fan} = total increase in TP due to the fan
VP_{fan} = VP of air stream leaving the fan

From Table 5.3, SP_{fan} is:

$$SP_{fan} = [(1.54 - (-4.13)) - 0.08] \text{ in.} H_2O$$
$$= 5.59 \text{ in.} H_2O$$

Therefore the fan in the system has to be able to move 2400 cfm against 5.59 in.H_2O static pressure.

Table 5.4 shows a typical fan manufacturer's data for centrifugal fans with backward-inclined blade design. By looking for the intersection of the required volumetric flow rate and fan static pressure, the fan speed and motor size can be obtained for a particular application.

Example 5.9 Using the SP_{fan} and flow rate from Example 5.7, choose an appropriate fan from Table 5.4.

solution The fan chosen would have to operate at 3690 rpm to deliver the required flow rate (2400 cfm) against the calculated static pressure (5.59 in.H_2O; use the 6-in. column). The motor would be rated at 3.53 BHP, which does not include the power requirement to overcome belt drive losses. (Actually, Example 5.7 shows an axial rather than a centrifugal design, but the procedure for selection is the same.)

Fan curves depend on the particular type (for example, blades that are forward curved or backward inclined) and the design characteristics of each manufacturer. Lightfoot et al. (1984) have suggested that because Reynolds numbers in fans and blowers ordinarily are quite high, it is possible to consolidate much of the performance data, such as appears in Table 5.4, using dimensionless measures of flow (Q^*), pressure drop (P^*) and efficiency (E^*). This procedure also allows a modest but systematic extrapolation of manufacturers' data to other operating conditions. The dimensionless variables are defined as

$$Q^* = \frac{Q}{ND^3} \tag{5.26}$$

$$P^* = \frac{SP_{fan}}{(\rho N^2 D^2)} \tag{5.27}$$

$$E^* = \frac{Q\ SP_{fan}}{HP} \tag{5.28}$$

where Q is volumetric throughput rate with the dimensions of length^3/time; N is rate of rotation of the rotor, time^{-1}; D is rotor diameter, length; SP_{fan} is increase in static pressure produced by the device, force/length^2; ρ is density of the gas, mass/length^3; and HP is the power supplied to the rotor, force · length/time. Each parameter must be expressed in consistent units. Figures 5.14, 5.15, and 5.16 show data from one fan manufacturer plotted using these dimensionless numbers for three design types. Performance data on four or five fan sizes form a single curve over a wide range of operating conditions. The procedure has also been reported as being successful for other manufacturers' data (Lightfoot et al., 1984).

Example 5.10 Using the data from Example 5.9 (SP_{fan} = 6 in.H_2O, Q = 2400 cfm, N = 3690 rpm), determine the wheel diameter and required horsepower from Figure 5.14 (for backward-inclined blades).

solution The determination of D is a trial-and-error procedure: (1) Assume D; (2) calculate Q^* from Eq. (5.26); (3) read P^* from Figure 5.14; (4) calculate D from Eq. (5.27). When the assumed and calculated values of D agree, look up E^* in Figure 5.14b and calculate HP from Eq. (5.28).

For ρ = 0.075 lb/ft^3, the calculational forms of Eqs. (5.25)–(5.27) will be:

$$Q^* = \frac{(1728)(Q,\ \text{ft}^3/\text{min})}{(N,\ \text{rpm})(D,\ \text{in.})^3} \tag{5.26a}$$

$$P^* = \frac{(1.16 \times 10^9)(SP_{fan},\ \text{in.}H_2O)}{(N,\ \text{rpm})^2(D,\ \text{in.})^2} \tag{5.27a}$$

$$E^* = \frac{(1.57 \times 10^{-4})(Q,\ \text{ft}^3/\text{min})(SP_{fan},\ \text{in.}H_2O)}{\text{HP}} \tag{5.28a}$$

Assumed D (in.)	Q^*	P^*	Calculated D (in.)	E^*	HP
11.5	0.739	3.11	12.8		
12	0.65	3.68	11.8 (close enough)	0.67	3.37

This HP value is in reasonable agreement with the 3.53 HP determined from Table 5.4, even though the table and figures are based on data from different manufacturers. If neither N nor D is known, Figures 5.14–5.16 will supply a series of acceptable design combinations of these two variables.

Figures 5.14–5.16 represent one way to consolidate fan information. However, the curves do not represent all possible fan designs. Particular aspects of fan specification, including blade design, speed, size of motor, and ability to vary speed, will ordinarily require more specialized evaluation that takes into account the specific characteristics of

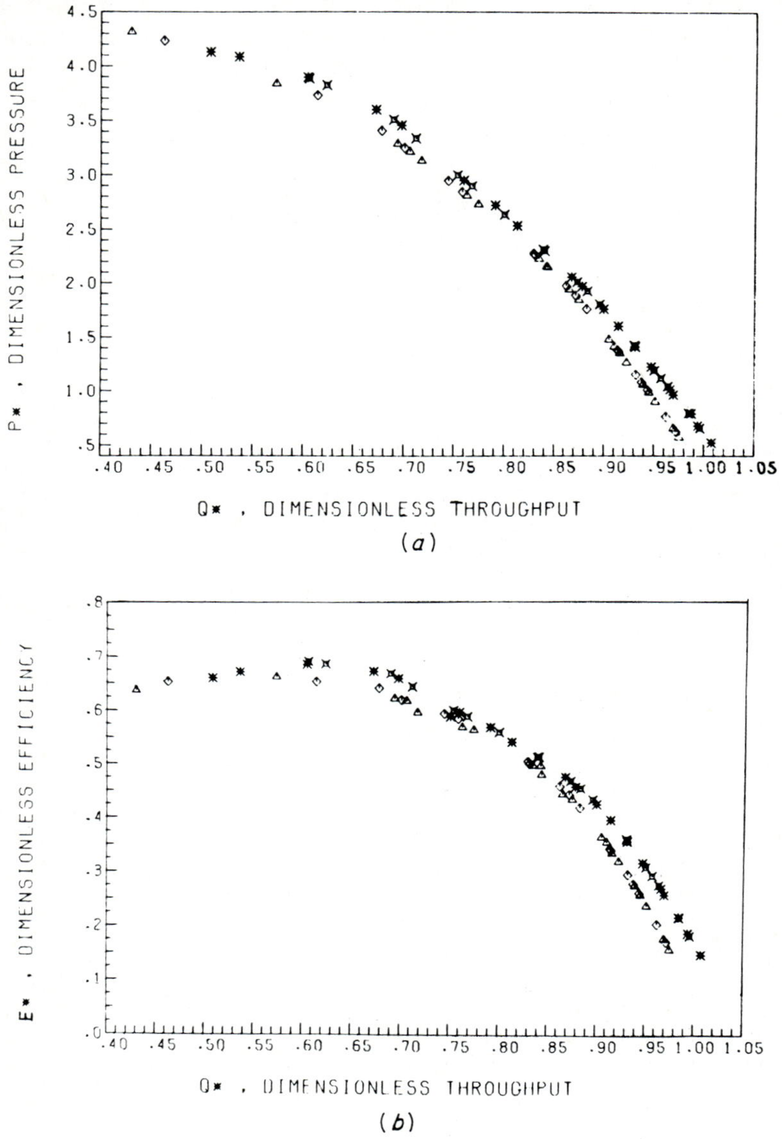

Figure 5.14 Dimensionless fan curves for belt drive, backward-inclined centrifugal fans (*Lightfoot et al., 1984*). Curves based on four fans sizes. Static pressures vary from 93.4 to 498 N/m^2; rotor speeds from 57.2 to 250 rad/sec; power from 59.7 to 3330 W; and delivery from 0.396 to 3.175 m^3/sec. Triangle = 0.343-m wheel diameter fan; diamond = 0.381-m; crossed square = 0.464-m; asterisk = 0.508-m. The curves for the 0.343- and 0.381-m fans are indistinguishable, as are the curves for the 0.464- and 0.508-m fans.

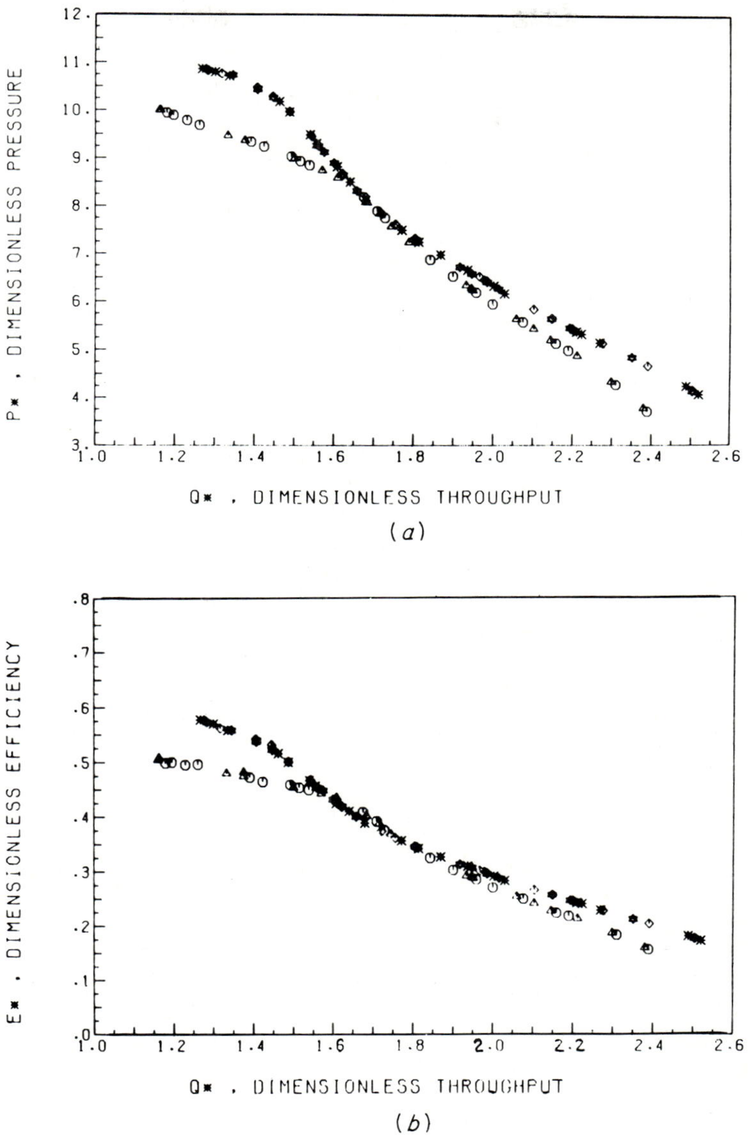

Figure 5.15 Dimensionless fan curves for belt drive, forward-curved centrifugal fans (*Lightfoot et al., 1984*). Curves based on five fan sizes. Operating conditions are static pressure, 93.4–498 N/m^2; rotor speed, 56.8–250 rad/sec; power, 112–3670 W; air delivery, 0.396–3.473 m^3/sec. Octagon = 0.343-m wheel diameter fan; triangle = 0.381-m; diamond = 0.464-m; asterisk = 0.508-m; star = 0.565-m.

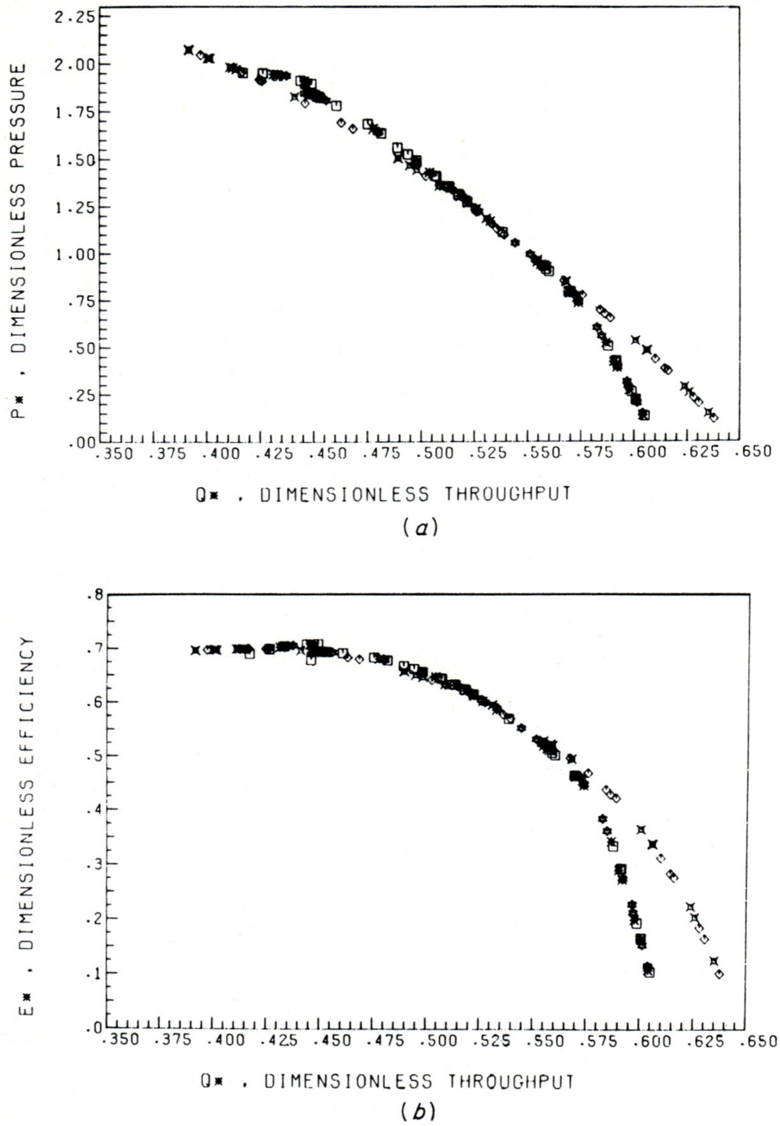

Figure 5.16 Dimensionless fan curves for axial flow fans (Trane Co., Model Q) (*Lightfoot et al., 1984*). Curves based on five fan sizes. Operating conditions are static pressure, 62.3–2864 N/m^2; rotor speed, 22.8–256 rad/sec; power, 544–257,590 W; air delivery, 3.68–74.1 m^3/sec. Diamond = 0.838-m wheel diameter fan; crossed square = 1.130-m; asterisk = 1.378-m; star = 1.854-m; square = 2.051-m. Performance curves for five fan sizes form a single curve over a very wide range of operating conditions.

the air and contaminant being transferred (AMCA, 1981). In general, the two most common types of fans in use for industrial ventilation are tube-axial propeller fans and straight or radial-blade centrifugal fans. Tube-axial fans are fabricated in a short section of round duct and are used for moving air against low static pressures. For such fans, the air volume exhausted is sensitive to added system resistance. However, this design is well suited for moving air that contains materials that can condense or collect on fan blades. Radial-blade centrifugal fans are used for applications where a heavy dust load passes through the fan. Centrifugal fans are most commonly used in industrial exhaust systems, whereas propeller fans are used for moving larger volumes of air for general or dilution ventilation applications (ACGIH, 1984).

V. Balancing

The ventilation system examined above is a very simple system in that it has only one inlet and one outlet. Industrial applications typically have more than one inlet. Consider a system that has multiple inlets (the same reasoning will apply to a system having multiple outlets). Air will move through the system by following the path of least resistance. If the design of a system calls for inlet branch A and inlet branch B to have equal flow rates, but branch B has less resistance to flow than branch A, then more air will flow through branch B and less through branch A. Such a situation is termed "unbalance." The concept of balancing is to provide sufficient resistance in branches so that the desired flow rates result. Adding unnecessary resistance will result in larger fan requirements. Sometimes the solution to an unbalanced design is to remove resistance (e.g., use larger ducting). Slight unbalance is usually not a problem.

The ASHRAE balancing method adjusts resistance such that loss in total pressure is equal in the branches. The ACGIH approach, used here, balances static and velocity pressure as well as total pressure. Criteria for evaluating balance at the junction of two ducts are:

1. If the ratio $SP_{higher}/SP_{lower} \leq 1.05$, consider the junction balanced. (Note that SP_{higher}/SP_{lower} are dealt with here as absolute values.)
2. If $1.05 < SP_{higher}/SP_{lower} \leq 1.2$, then the flow in the branch having SP_{lower} will be increased by the design flow rate times $(SP_{higher}/SP_{lower})^{0.5}$.
3. If $SP_{higher}/SP_{lower} > 1.2$, redesign the duct having SP_{lower} to have greater resistance; or redesign the duct having SP_{higher} to have lower resistance.

Example 5.11 A wye in a ventilation system joins flow from two inlets. The incoming main branch of the wye is 12 in. in diameter, has a design flow rate of 1500 cfm, and the predicted SP is -0.80 in.H_2O. In the branch the duct is 8 in. in diameter, is designed to carry 500 cfm, and has a predicted SP of -0.85 in.H_2O. Is the wye balanced?

solution The ratio $SP_{higher}/SP_{lower} = 0.85/0.80 = 1.06$. Using the criteria given above, the wye can be balanced by changing flow rates from the design values. The actual flow rates in each branch are:

$$Q_{s,actual} = Q_{s,design}\left(\frac{SP_{higher}}{SP_{lower}}\right)^{0.5}$$

$$= (1500\ \text{cfm})(1.06)^{0.5} = 1546\ \text{cfm}$$

$$Q_{b,actual} = 2000\ \text{cfm} - 1546\ \text{cfm} = 454\ \text{cfm}$$

There are two approaches to implementing system balance:

1. Design the system to be prebalanced without using adjustable balancing devices (dampers).
2. Design the system to be balanced with adjustable devices.

There are advantages and disadvantages to each approach. Use of a prebalanced design means that:

1. Air flow rate cannot be changed (or tampered with).
2. Design and design calculations must be accurate.
3. There is little flexibility to incorporate modifications into an already built system.
4. The only way to increase air flow in a branch is to increase total air flow in the entire system.

Use of balancing devices means that:

1. Air flow rate can be changed (and tampered with).
2. Design calculations need not be as accurate (but, if too inaccurate, may result in poorer operation).
3. The system can be adjusted to operate differently from design.
4. The system can be modified.

When all is said and done, most systems are designed with balancing devices. However, most systems that rely on these devices never attain the design balanced air flows.

Up to this point the velocity of air through a duct has been treated as a variable for which almost any value can be used. This is not the case. Large or heavy particles can settle out of a slow-moving air stream and may even clog the duct. Too rapid movement of abrasive materials

TABLE 5.5 Range of Duct Design Velocities

Nature of contaminant	Examples	Design velocity (fpm)
Vapors, gases, smoke		Any desired velocity (economic optimum velocity usually 1000–1200 rpm)
Fumes	Zinc and aluminum oxide fumes	1400–2000
Very fine light dust	Cotton lint, wood flour, litho powder	2000–2500
Dry dusts and powders	Fine rubber, Bakelite molding powder, jute lint, cotton dust, shavings (light), soap dust, leather shavings	2500–3500
Average industrial dust	Sawdust (heavy and wet), grinding dust, buffing lint (dry), wool jute dust (shaker waste), coffee beans, shoe dust, granite dust, silica flour, general material handling, brick cutting, clay dust, foundry (general), limestone dust, packaging and weighing asbestos dust in textile industries	3500–4000
Heavy dusts	Metal turnings, foundry tumbling barrels and shakeout, sand blast dust, wood blocks, hog waste, brass turnings, cast-iron boring dust, lead dust	4000–4500
Heavy or moist dusts	Lead dust with small chips, moist cement dust, asbestos chunks from transite pipe cutting machines, buffing lint (sticky), quick-lime dust	4500 and up

SOURCE: ACGIH, 1984.

can lead to erosion and perforation of the duct wall. Table 5.5 lists recommended duct design velocities for different kinds of contaminants. Corrosive and very abrasive materials are more efficiently removed from the air stream by an air cleaning device rather than permitting them to damage the ventilation system.

VI. Hood Design

Thus far the design of inlets has been dealt with in a very cursory fashion. In industrial ventilation systems most inlets are hoods. A hood, by definition, is a device that encloses a contaminant-releasing source and/or creates a zone of air flow to carry contaminated air to exhaust. It is the hood that captures contaminated air before the contaminants reach the worker's breathing zone. There are four kinds of hood arrangements (NIOSH, 1973):

1. An *enclosure* surrounds a contaminant-producing source. The contaminant is prevented from escaping by a continuous inflow of air.
2. An *exterior hood* is located some distance from the contaminant-generating source. The contaminants generated are drawn into the hood by air flow.
3. A *receiving hood* is an exterior hood except that the motion of contaminant ejection by the source is utilized for collection by the hood.
4. A *push-pull hood* uses a jet of air from a local supply to sweep contaminants into a receiving hood.

When a source is inside an enclosure, the velocity of air coming in through openings in the enclosure that is sufficient to prevent contaminants from escaping is called the control velocity (NIOSH, 1973). Control velocities for a number of different operations are shown in Table 5.6. When the source is not enclosed in a hood, the velocity of air at the contaminant source needed to draw off the contaminants is called the capture velocity. Capture velocities recommended for various situations are summarized in Table 5.7. Note the footnotes in Table 5.7 regarding use of the lower or upper end of capture velocity ranges. When dealing with explosive gases, it is also recommended that the flow rate of air be sufficient to maintain the concentration of the explosive gas at no greater than 25% of the LEL (see Chapter 4).

From the point of view of designing the ventilation system, the information needed regarding the hood is the volumetric flow rate of air to be removed and the consequent pressure drop. In all cases the flow rate is

$$Q = v_c A \tag{5.29}$$

where v_c is the control or capture velocity. Calculating a value for area, A, is more difficult. For an enclosure, A is equal to the sum of the area of all the openings in the enclosure. When dealing with exterior hoods, the size of the control surface, A_c (the area through which the air flows at the capture velocity), depends on the source-hood geometry. Furthermore, approximations differ depending on whether the source and hood are close together (near field) or far apart (far field) (ACGIH, 1984; Crawford, 1976). In the far field, air movement toward an exhaust hood is equal from all available directions. Since air flow in the far field is equally distributed, it is possible to predict the control surface area from simple geometric shapes such as cylinders and spheres. For instance, for a circular hood of diameter D, located in the center of a wall, the control surface will be that of a hemisphere:

$$A_c = 2\pi X^2 \tag{5.30}$$

where X is the distance from the center of the hood opening to the control surface. As a rule of thumb, $X > 3D$ constitutes a far-field condition.

In the near field, $X \leq 3D$, flow is not distributed equally with direction, and control surface area cannot be calculated from geometric shapes. Empirical relationships must be used. ACGIH has also developed designs for ventilation of specific processes. Velocity contours for flanged and unflanged hoods are shown on Figure 5.17. Hood velocity contours have also been predicted with a theoretical potential flow model that incorporates the time rate change of eccentricity (Flynn and Ellenbecker, 1985). This modification was reasonable given the elliptical shape of the experimental curves of Figure 5.17. The maximum difference between prediction and experiment was 23%, and the

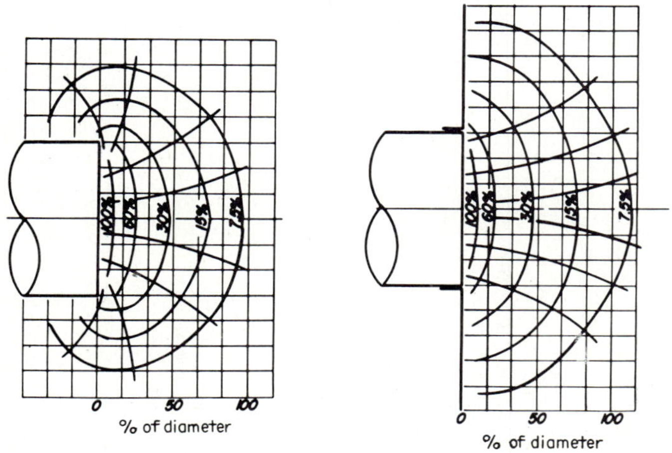

Figure 5.17 Velocity contours (expressed in percentage of opening velocity) and stream lines for round, flanged, and unflanged openings (*ACGIH, 1984*).

TABLE 5.6 Minimum Control Velocities and Exhaust Rates for Typical Specific Operations

Where both control velocity and exhaust rate are given, the air volume exhausted shall be based on the method that requires the larger volume.

Operating	Control velocity (fpm)	Control velocity basis	Exhaust rate (cfm)	Exhaust rate basis
Abrasive blasting				
Cabinets	500	Openings in enclosure	—	—
Rooms	60–100	Downdraft in room	—	—
Bag filling	500	Openings in enclosure	400–500	Nontoxic dust
			1000–1500	Toxic dust
Barrel filling	—	—	150	Per ft^2 open face area
Bin and hopper	200	Openings in enclosure	—	—
Belt conveyors				
Transfer point				
Belt speed <200 fpm	150	Openings in enclosure	350	Per ft belt width
>200 fpm	200	Openings in enclosure	500	Per ft belt width
Bucket elevators				
Casing	—	—	100–200	Per ft^2 casing cross section Tight casing required
Belt speed				
<200 fpm	—	—	350	Per ft belt width
>200 fpm	—	—	500	Per ft belt width
Foundry screens				
Cylindrical	400	Openings in enclosure	100	Per ft^2 circular cross section
Flat deck	200	Openings in enclosure	50	Per ft^2 screen area
Foundry shakeout				
Enclosure	200	Openings in enclosure	150–200	Cool or hot castings respectively per ft^2 grate area
Side hood	700–1000	Openings in enclosure	350–400	Cool castings per ft^2 grate area
			400–500	Hot castings per ft^2 grate area

Furnaces, melting	200–400	Openings in enclosure, flow rate corrected for temperature	2500	Per ton of charge, flow rate corrected for temperature
Granite cutting				
Surfacing machine	—		400–600	—
Abrasive blasting	500	Openings in enclosure	—	—
Grinding				
General	—	—	220–1570	Good enclosure, flow increases with wheel size and speed
Disk and portable	100	Face of booth	100	Per ft^2 of working area for floor downdraft grille
Swing frame	100–150	Face of booth	—	—
Laboratory hood	100–150	Face of hood, door open	—	—
Metal spraying				
Toxic material	200	Face of booth	NIOSH certified air-supplied respirator required	
Nontoxic	125	Face of booth	—	—
Mixer	150	Openings in enclosure	750–1575	Flow increases with mixer size
Packaging machines	100–400	Openings in complete enclosure	—	—
	50–100	Face of booth	—	—
	95–150	Downdraft	—	—
Paint spray	60–100	Openings in walk-in booth	—	—
Phamaceutical coating pans	100–150	Openings in enclosure	—	—
Quartz fusing	150–200	Face of booth on bench	—	—
Silver soldering	100	Point of origin	—	—
Steam kettles	150	Face of canopy	—	—
Welding	100	Face of booth	—	—

SOURCE: ACGIH, 1984.

TABLE 5.7 Range of Capture Velocities

Condition of dispersion of contaminant	Examples	Capture velocity (fpm)
Released with practically no velocity into quiet air	Evaporation from tanks, e.g., degreasing	50–100
Released at low velocity into moderately still air	Spray booths; intermittent container filling; low-speed conveyor transfers; welding; plating; pickling	100–200
Active generation into zone of rapid air motion	Spray painting in shallow booths; barrel filling; conveyor loading; crushers	200–500
Released at high initial velocity into zone of very rapid air motion	Grinding; abrasive blasting; tumbling	500–2000

In each category above, a range of capture velocity is shown. The proper choice of values depends on several factors:

Lower End of Range	*Upper End of Range*
1. Room air currents minimal or favorable to capture	1. Disturbing room air currents
2. Contaminants of low toxicity or of nuisance value only	2. Contaminants of high toxicity
3. Intermittent, low production	3. High production, heavy use
4. Large hood—large air mass in motion	4. Small hood—local control only

SOURCE: ACGIH, 1984.

average difference was 8.6%. This approach has also been used to model the effect of uniform cross drafts on velocity fields into a flanged circular hood (Flynn and Ellenbecker, 1986a). The model predictions have been empirically confirmed in wind tunnel experiments using hot-film anemometry.

Empirical relationships that have been developed to describe the centerline velocity in the near field of hoods are given in Figure 5.18. Note that the velocity is along the centerline at distance X from the hood. Strictly speaking, the value of X is less than $1.5D$. When $X > 1.5D$, the actual velocity decreases less rapidly than is indicated by these models.

Some concepts about hood design are:

1. The closer the source is to the hood, the lower the volumetric flow rate of air needed to collect the contaminant emission.

HOOD TYPE	DESCRIPTION	ASPECT RATIO, $\frac{W}{L}$	VELOCITY
	SLOT	0.2 or less	$\frac{Q}{3.7XL}$
	FLANGED SLOT	0.2 or less	$\frac{Q}{2.8XL}$
 A = WL (sq. ft.)	PLAIN OPENING	0.2 or greater and round	$\frac{Q}{10X^2+A}$
	FLANGED OPENING	0.2 or greater and round	$\frac{Q}{0.75(10X^2+A)}$
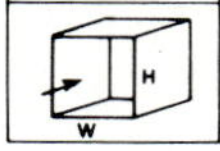	BOOTH	to suit work	$\frac{Q}{HW}$

Figure 5.18 Hood near-field velocity (*ACGIH, 1984*).

2. If a source is located in a hood's far field, the hood will be most useful in reducing the *area* concentration of the contaminant.
3. *Personal* contaminant exposure is most effectively controlled by locating the source in the hood's near field or in an enclosure.

Example 5.12 A round, flanged hood of diameter D is to be used to ventilate a source located as shown in Figure 5.19. Calculate the volumetric flow rate of air to be exhausted by the hood.

solution If far-field hemispherical symmetry is assumed, the area of the control surface is

$$A_c = 2\pi X^2 \tag{5.30}$$

and $$X = \left[D^2 + \left(\frac{D}{2}\right)^2\right]^{0.5} = 1.12D$$

Therefore

$$A_c = 2\pi(1.12D)^2 = 7.88D^2$$

and $$Q = 7.88D^2v$$

If near-field conditions are assumed and Figure 5.18 is used, the flow rate will be

$$Q = 0.75v(10X^2 + A) \tag{5.31}$$

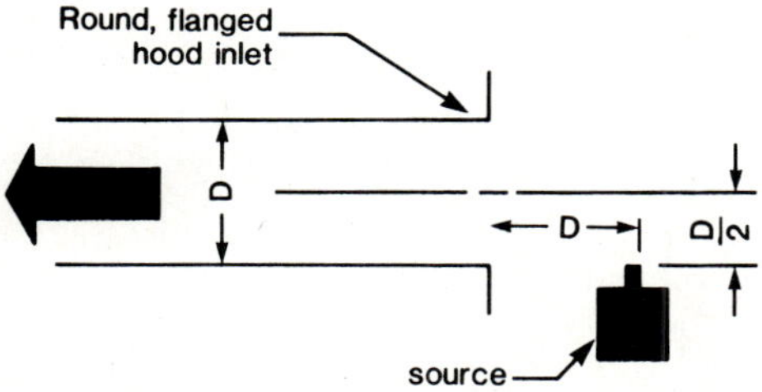

Figure 5.19 Arrangement for Example 5.12.

where X = centerline distance = D and $A = \pi D^2/4$. Therefore

$$Q = 0.75v\left(10D^2 + \frac{\pi D^2}{4}\right) \qquad (5.31)$$
$$= 8.09D^2v$$

Since the source is not located on the hood centerline, the velocity off the centerline will be somewhat less than the empirical model suggests, but will be greater than the far-field hemispherical model. The actual value may be estimated by adjusting the value of X. In the two models used, $X = D$ and $X = 1.12D$. The accepted approach is to use the average of these distance values in the near-field empirical model. Therefore the flow rate is

$$Q = 0.75v\left[10(1.06D)^2 + \frac{\pi D^2}{4}\right] \qquad (5.31)$$
$$= 9.02D^2v$$

Although Table 5.7 includes values of capture velocity that could be applied to operations involving evaporation of liquids or degreasing, the ACGIH has developed a more detailed procedure for calculating hood flow rate when a volatile liquid is involved. The procedure is outlined in Table 5.8. The first step is to develop a classification of the process based on the toxicity (hazard potential) and volatility (contaminant evolution rate) for the operation and chemicals being used. Table 5.9 contains the scheme to evaluate toxicity. It is based on both hygienic standards and the flash point of the material. Although three columns of possible classifications are given, the hazard potential to be selected is the highest one that any of the classification schemes might give. Table 5.10 shows the classification scheme for evolution rate. This classification is based on physical properties of the chemicals in use (liquid temperature, boiling point, and relative evaporation) as well

TABLE 5.8 Procedure for Calculation of Hood Flow Rate for Volatile Materials (no cross-drafts)

1. Determine hazard potential from Table 5.9 using information from threshold limit value (Table 4.2) or solvent flash point (Table 4.9); or determine from Table 5.11 or 5.12.

2. Determine contaminant evolution rate from Tables 5.10 and 4.9; or from Tables 5.11 and 5.12 employing number denoting highest range.

3. From Table 5.13, choose minimum control or capture velocity according to hazard potential, evolution rate, and hood design (see Tables 5.11 and 5.12 for typical processes).

4. From Table 5.14, select the cfm/ft^2 for tank dimensions and tank location.

5. Multiply tank area by value obtained from Table 5.14 to calculate required air volume.

SOURCE: ACGIH, 1984.

TABLE 5.9 Determination of Hazard Potential

Hazard potential	Hygienic standards (TLVs) Gas and vapor (ppm)	Mist (mg/m^3)	Flash point (°F)
A	0–10	0–0.1	—
B	11–100	0.11–1.0	Under 100
C	101–500	1.1–10	100–200
D	Over 500	Over 10	Over 200

SOURCE: ACGIH, 1984.

as the process (gassing). Additional information to identify relative evaporation is listed in Table 4.9. Again, all four possible classifications are examined and the highest ranking one is used. Tables 5.11 through 5.16 give hazard and evolution rate classifications for various processes, most involving metal treatment. Having classified the toxicity and contaminant evolution rate, Table 5.13 can be used to select the appropriate control or capture velocity. If lateral exhaust is to be used, the value from Table 5.13 is used to enter Table 5.14 as well as tank dimensions to arrive at an appropriate capture velocity.

Example 5.13 A chromium-based coating is to be stripped off steel base metal parts in a pickling tank. The tank contains a sodium hydroxide solution maintained at 90°F. The tank is 3 ft wide, 4 ft long, and 2.5 ft high. The emissions from the tank are to be collected by lateral exhaust on a free-standing tank following the ACGIH VS-504 design for a pickling tank shown in Figure 5.20. What flow rate of air has to be exhausted by the hood (in cfm)? What is the loss in total pressure as the air enters the hood?

TABLE 5.10 Determination of Rate of Gas, Vapor, or Mist Evolution

Rate	Liquid temperature (°F)	Degrees below boiling point (°F)	Relative evaporation[a] (time for 100% evaporation)	Gas release[b]
1	Over 200	0–20	Fast (0–3 hr)	High
2	150–200	21–50	Medium (3–12 hr)	Medium
3	94–149	51–100	Slow (12–50 hr)	Low
4	Under 94	Over 100	Nil (over 50 hr)	Nil

[a]Dry time relation can be used to estimate relative evaporation. The dry time relation categorization is: fast (0–5), medium (5–15), slow (15–75), and nil (over 75). See Table 4.9.

[b]Rate of gas release depends on rate of chemical or electrochemical action and therefore depends on the material treated and the solution used in the tank and tends to increase with: (1) amount of work in the tank at any one time; (2) strength of the solution in the tank; (3) temperature of the solution in the tank; and (4) current density applied to the work in electrochemical tanks.

SOURCE: ACGIH, 1984.

TABLE 5.11 Hazard and Evolution Rate Classification for Typical Processes

Operation	Contaminant	Hazard	Contaminant evolution
Anodizing aluminum	Chromic-sulfuric acids	A	1
Aluminum bright dip	Nitric + sulfuric acids	A	1
	Nitric + phosphoric acids	A	1
Plating			
Chromium	Chromic acid	A	1
Copper strike	Cyanide mist	C	2
Metal cleaning (boiling)	Alkaline mist	C	1
Hot water (if venting desired)			
Not boiling	Water vapor[a]	D	2
Boiling		D	1
Stripping			
Copper	Alkaline-cyanide mists	C	2
Nickel	Nitrogen oxide gases	A	1
Pickling—steel	Hydrochloric acid	A	2
	Sulfuric acid	B	1
Salt solution (bonderizing and parkerizing)			
Not boiling	Water vapor[a]	D	2
Boiling		D	1
Salt baths (molten)	Alkaline mists	C	1

[a]Where complete control of water vapor is desired, design as next highest class.
SOURCE: ACGIH, 1984.

solution The first step is to determine the hazard potential and contaminant evolution rate. From Table 5.16, assuming that chromium compounds are not emitted, the contaminants most likely to be released are sodium hydroxide (as an alkaline mist) and steam. The hazard class is C and the evolution rate 3. From Table 5.13, lateral exhaust requires a minimum capture velocity of 50

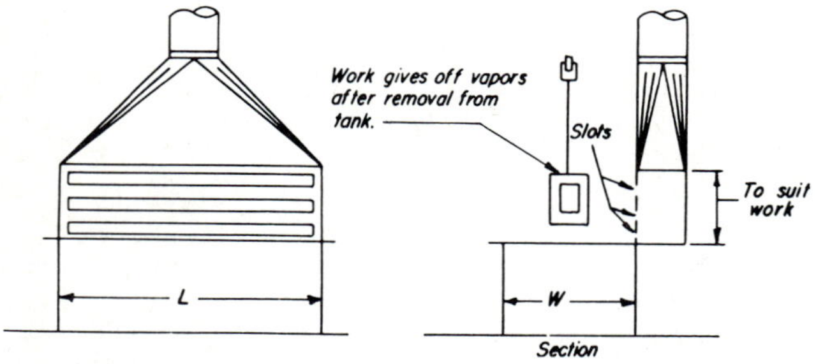

Figure 5.20 ACGIH design (VS-504) for a pickling tank (*ACGIH, 1984*).

$$h_e = 1.78\text{VP}_{\text{slot}} + 0.25\text{VP}_{\text{duct}}$$

$$v_{\text{slot}} = 2000 \text{ ft/min}$$

ft/min. Table 5.14 shows that a tank having a width-to-length ratio of 0.75 and a required capture velocity of 50 ft/min needs 110 cfm/ft^2 of tank surface area. Therefore the required flow rate for the hood is:

$$Q = (110 \text{ cfm/ft}^2)(3 \text{ ft})(4 \text{ ft})$$
$$= 1320 \text{ cfm}$$

The entry loss (loss in total pressure) for this hood can be estimated from (see Figure 5.20)

$$h_e = 1.78\text{VP}_{\text{slot}} + 0.25\text{VP}_{\text{duct}} \qquad (5.32)$$

It is recommended that the velocity of air entering the hood slots be 2000 ft/min. Therefore

$$\text{VP}_{\text{slot}} = \left(\frac{2000}{4005}\right)^2 = 0.25 \text{ in.H}_2\text{O}$$

From Table 5.5, the duct velocity can be 1000–1200 fpm. Assuming a velocity of 1200 fpm, the circular duct diameter would be:

$$D = \left(\frac{4Q}{3.14v}\right)^{0.5}$$
$$= \left[\frac{(4)(1320)}{(3.14)(1200)}\right]^{0.5}$$
$$= 1.40 \text{ ft} = 16.8 \text{ in.}$$

Ducts usually are not such an odd size, so a 16-in. diameter will be assumed. The actual duct velocity will then be

$$v = \frac{4Q}{\pi D^2}$$
$$= \frac{(4)(1320)}{(3.14)(16/12)^2}$$
$$= 1261 \text{ ft/min}$$

and the duct velocity pressure is

$$\text{VP}_{\text{duct}} = \left(\frac{1261}{4005}\right)^2$$
$$= 0.10 \text{ in.H}_2\text{O}$$

Finally, the entry loss is

$$h_e = (1.78)(0.25) + (0.25)(0.10)$$
$$= 0.47 \text{ in.H}_2\text{O} \qquad (5.32)$$

Since sodium hydroxide mist and vapor will condense in the hood, ductwork, and blower, provisions for easy cleaning and maintenance of the system should be included in the construction design.

Receiving hoods operate in a manner similar to exterior hoods except that the motion induced by the process is utilized. This is not as simple as it sounds because, as Figure 5.21 shows, eddies can occur and the

TABLE 5.12 Airborne Contaminants Released by Metallic Surface Treatment, Etching, Pickling, Acid Dipping, and Metal Cleaning Operations

Process	Type	Notes	Component of bath that may be released to atmosphere (13)	Physical and chemical nature of major atmospheric contaminant	Class (12)	Usual temperature range (°F)
Surface treatment	Anodizing aluminum		Chromic-sulfuric acids	Chromic acid mist	A-1	95
	Anodizing aluminum		Sulfuric acid	Sulfuric acid mist	B-1	60–80
	Black magic		Concentrated solution of alkaline oxidizing agents	Alkaline mist, steam	C-1	280–350
	Bonderizing	1	Boiling water	Steam	D-2,1 (14,15)	140–212
	Chemical coloring		None	None	D-4	70–90
	Descaling	2	Nitric-sulfuric, hydrofluoric acids	Acid mist, hydrogen fluoride gas, steam	B-2,1 (15)	70–150
	Ebonol		Concentrated solution of alkaline oxidizing agents	Alkaline mist, steam	C-1	260–350
	Galvanic-anodize	3	Ammonium hydroxide	Ammonia gas, steam	B-3	140
	Hard coating aluminum		Chromic-sulfuric acids	Chromic acid mist	A-1	120–180
	Hard coating aluminum		Sulfuric acid	Sulfuric acid mist	B-1	120–180
	Jetal		Concentrated solution of alkaline oxidizing agents	Alkaline mist, steam	C-1	260–350
	Magcote	4	Sodium hydroxide	Alkaline mist, steam	C-3,2 (15)	105–212
	Magnestium predye dip		Ammonium hydroxide-ammonium acetate	Ammonia gas, steam	B-3	90–180
	Parkerizing	1	Boiling water	Steam	D-2,1 (14,15)	140–212
	Zincete immersion	5	None	None	D-4	70–90
Etching	Aluminum		Sodium hydroxide-soda ash-trisodium phosphate	Alkaline mist, steam	C-1	160–180
	Copper	6	Hydrochloric acid	Hydrogen chloride gas	A-2	70–90
	Copper	7	None	None	D-4	70

Pickling	Aluminum		Nitric acid	Nitrogen oxide gases	A-2	70–90
	Aluminum		Chromic, sulfuric acids	Acid mists	A-3	140
	Aluminum		Sodium hydroxide	Alkaline mist	C-1	140
	Cast iron		Hydrofluoric-nitric acids	Hydrogen fluoride-nitrogen oxide gases	A-2,1 (15)	70–90
	Copper		Sulfuric acid	Acid mist, steam	B-3,2 (15)	125–175
	Copper	8	None	None	D-4	70–175
	Duralumin		Sodium fluoride, sulfuric acid	Hydrogen fluoride gas, acid mist	A-3	70
	Inconel		Nitric, hydrofluoric acids, steam	Nitrogen oxide, HF gases	A-1	150–165
	Inconel		Sulfuric acid	Sulfuric acid mist, steam	B-2	160–180
	Iron and steel		Hydrochloric acid	Hydrogen chloride gas	A-2	70
	Iron and steel		Sulfuric acid	Sulfuric acid mist, steam	B-1	70–175
	Magnesium		Chromic-sulfuric, nitric acids	Nitrogen oxide gases, acid mist, steam	A-2	70–160
	Monel and nickel		Hydrochloric acid	Hydrogen chloride gas, steam	A-2	180
	Monel and nickel		Sulfuric acid	Sulfuric acid mist, steam	B-1	160–190
	Nickel silver		Sulfuric acid	Acid mist, steam	B-3,2 (15)	70–140
	Silver		Sodium cyanide	Cyanide mist, steam	C-3	70–210
	Stainless steel	9	Nitric, hydrofluoric acids	Nitrogen oxide, hydrogen fluoride gases	A-2	125–180
	Stainless steel	9, 10	Hydrochloric acid	Hydrogen chloride gas	A-2	130–140
	Stainless steel	9, 10	Sulfuric acid	Sulfuric acid mist, steam	B-1	180
	Stainless steel immunization		Nitric acid	Nitrogen oxide gases	A-2	70–120
	Stainless steel passivation		Nitric acid	Nitrogen oxide gases	A-2	70–120

TABLE 5.12 Airborne Contaminants Released by Metallic Surface Treatment, Etching, Pickling, Acid Dipping, and Metal Cleaning Operations *(Continued)*

Process	Type	Notes	Component of bath that may be released to atmosphere (13)	Physical and chemical nature of major atmospheric contaminant	Class (12)	Usual temperature range (°F)
Acid dipping	Aluminum bright dip		Phosphoric, nitric acids	Nitrogen oxide gases	A-1	200
	Aluminum bright dip		Nitric, sulfuric acids	Nitrogen oxide gases, acid mist	A-2,1 (15)	70–90
	Cadmium bright dip		None	None	D-4	70
	Copper bright dip		Nitric, sulfuric acids mist	Nitrogen oxide gases, acid	A-2,1 (15)	70–90
	Copper semibright dip		Sulfuric acid	Acid mist	B-2	70
	Copper alloys bright dip		Nitric, sulfuric acids	Nitrogen oxide gases, acid mist	A-2,1 (15)	70–90
	Copper matte dip		Nitric, sulfuric acids mist	Nitrogen oxide gases, acid (15)	A-2,1	70–90
	Magnesium dip		Chromic acid	Acid mist, steam	A-2	190–212
	Magnesium dip		Nitric, sulfuric acids	Nitrogen oxide gases, acid mist	A-2,1 (15)	70–90
	Monel dip		Nitric, sulfuric acids	Nitrogen oxide gases, acid mist	A-2,1 (15)	70–90
	Nickel and nickel alloys dip		Nitric, sulfuric acids	Nitrogen oxide gases, acid mist	A-2,1 (15)	70–90
	Silver dip		Nitric acid	Nitrogen oxide gases	A-1	70–90
	Silver dip		Sulfuric acid	Sulfuric acid mist	B-2	70–90
	Zinc and zinc alloys dip		Chromic, hydrochloric acids	Hydrogen chloride gas (if HCl attacks Zn)	A-4,3 (15)	70–90

Metal cleaning	Alkaline cleaning	11	Alkaline sodium salts	Alkaline mist, steam	C-2,1 (15)	160–210
	Degreasing		Trichloroethylene-perchloroethylene	Trichloroethylene-perchloroethylene vapors	B (16)	180–250
	Emulsion cleaning		Petroleum-coal tar solvents	Petroleum-coal tar vapors	B-3,2 (15)	70–140
	Emulsion cleaning		Chlorinated hydrocarbons	Chlorinated hydrocarbon vapors	(17)	70–140

NOTES:
1. Also aluminum seal, magnesium seal, magnesium dye set, dyeing anodized magnesium, magnesium alkaline dichromate soak, coloring anodized aluminum.
2. Stainless steel before electropolishing.
3. On magnesium.
4. Also Manodyz, Dow-12.
5. On aluminum.
6. Dull finish.
7. Ferric chloride bath.
8. Sodium dichromate, sulfuric acid bath and ferrous sulfate, sulfuric acid bath.
9. Scale removal.
10. Scale loosening.
11. Soak and electrocleaning.
12. Class based on hazard potential and rate of evolution for usual operating conditions. Higher temperatures, agitation, or other conditions may result in a higher rate of evolution.
13. Hydrogen gas also released by many of these operations.
14. Rate where essentially complete control of steam is required. Otherwise, adequate dilution ventilation may be sufficient.
15. The higher rate is associated with the higher value in the temperature range.
16. For vapor degreasers, rate is determined by operating procedure.
17. Class of operation is determined by nature of the hydrocarbon.

SOURCE: ACGIH, 1984.

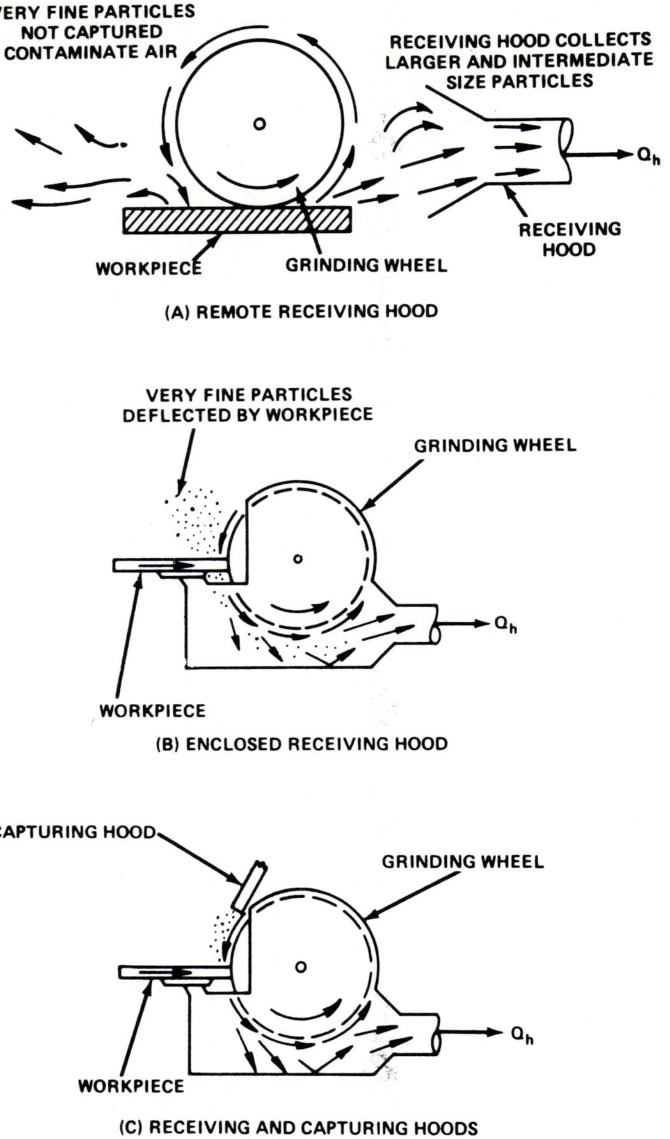

Figure 5.21 Receiving hoods (*ASHRAE, 1980*).

process-induced motion can be in more than one direction. Design practices for exterior hoods are largely experiential. In general it is advisable to follow ACGIH recommendations. An example is shown for a surface grinder receiving hood in Figure 5.22; information is given on flow rate, duct velocity, and hood entry loss, which are all necessary in the design of the remainder of the ventilation system.

Push-pull ventilation design includes an air jet on the opposite side

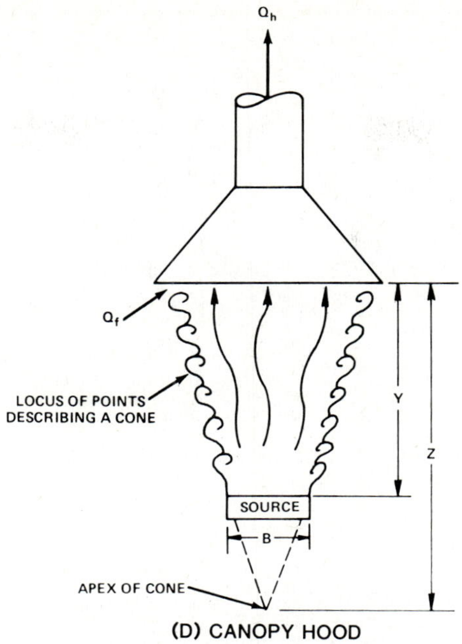

Figure 5.21 *(Continued)*

TABLE 5.13 Minimum Control of Capture Velocity (fpm) for Undisturbed Locations

Class (See Tables 5.8 and 5.9)	Enclosing hood		Lateral exhaust (note 1)	Canopy hoods	
	One open sides	Two open sides		Three open sides	Four open sides
A-1, and A-2 (note 2)	100	150	150	Do not use	Do not use
A-3 (note 2), B-1, B-2, and C-1	75	100	100	125	175
B-3, C-2, and D-1 (note 3)	65	90	75	100	150
A-4 (note 2), C-3, and D-2 (note 3)	50	75	50	75	125
For B-4, C-4, D-3 (note 3), and D-4, *adequate general room ventilation is required.*					

NOTES: 1. Use aspect ratio to determine air volume; see Table 5.14 for computation.
2. Do not use canopy hood for Hazard Potential A processes.
3. Where complete control of hot water is desired, design as next highest class.

SOURCE: ACGIH, 1984.

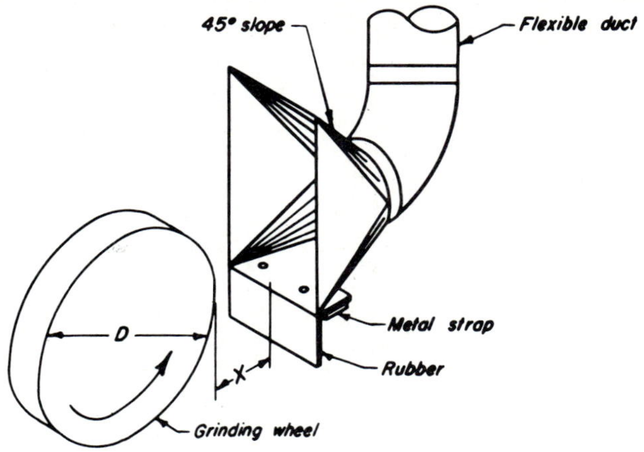

$Q = 0.0003\ Vs(10X^2 + A)$

Entry loss = 0.25 VP

Duct velocity = 3500 fpm minimum

X = Hood wheel distance in inches (measured from center of hood face to nearest point on wheel surface)

A = Hood face area (sq in)

Vs = Wheel surface speed (sfm)

$= 3.1416 \times rpm \times \frac{D}{12}$

D = Diam. in inches

Example for
X = 4"
A = 3" x 4 1/2"

Vs	Q, cfm
1 000	52
2 000	104
3 000	156
4 000	208
5 000	260
6 000	312
7 000	364
8 000	416
9 000	468
10 000	520

AMERICAN CONFERENCE OF GOVERNMENTAL INDUSTRIAL HYGIENISTS

SURFACE GRINDER

DATE 1-76 | VS-417

Figure 5.22 ACGIH recommended design for a receiving hood (*ACGIH, 1984*).

of the contaminant generation area from a receiving hood. As shown in Figure 5.23, the air to be exhausted by the "pull" end of the system is the sum of three components: the air supplied by the jet, the contaminant emitted by the process, and the induced ambient (room) air flow. Recent experiments have shown that such a design can reduce the pull flow rate (typically 100–150 cfm/ft^2 for plating tanks) by 25% to 50% (Huebner and Hughes, 1985; Hughes, 1985). Consideration of cross drafts up to 75 ft/min were also included in the test program. Based on these data, a preliminary method has been developed for

(*Continued on p. 242*)

TABLE 5.14 Minimum Flow Rate, cfm per square foot of Tank Area for Lateral Exhaust

Required minimum capture velocity (fpm) (From Table 5.13)	cfm per square foot to maintain required minimum capture velocities at following $\frac{\text{tank width}}{\text{tank length}}\left(\frac{W}{L}\right)$ ratios				
	0.0–0.09	0.1–0.24	0.25–0.49	0.5–0.99	1.0–2.0 (note 2)
Hood against wall or baffled (*see note 1*):					
50	50	60	75	90	100
75	75	90	110	130	150
100	100	125	150	175	200
150	150	190	225	[250] note 3	[250] note 3
Hood on free standing tank (*see note 1*):					
50	75	90	100	110	125
75	110	130	150	170	190
100	150	175	200	225	250
150	225	[250] note 3	[250] note 3	[250] note 3	[250] note 3

NOTES: 1. Use $W/2$ as tank width in computing W/L ratio for hood along centerline or two parallel sides of tank.

2. If L = 6 ft or greater, multiple takeoffs are desirable.
If L = 10 ft or greater, multiple takeoffs are necessary.
Tank width (W) means the effective width over which the hood must pull air to operate (i.e., where the hood face is set back from the edge of the tank, this set back must be added in measuring tank width).
If W = 20 in., slot on one side suitable.
If W = 20–36 in., slots on both sides are desirable.
If W = 36–48 in., slots on both sides are necessary unless all other conditions are optimum.
If W = 48 in. or greater, local exhaust is not usually practical. Enclosure is best.
It is not practicable to ventilate across the long dimension of a tank whose ratio W/L exceeds 2.0. It is undesirable to do so when W/L exceeds 1.0.

3. While bracketed values may not produce 150 fpm control velocity at all aspect ratios, the 250 cfm/ft^2 is considered adequate for control.

SOURCE: ACGIH, 1984.

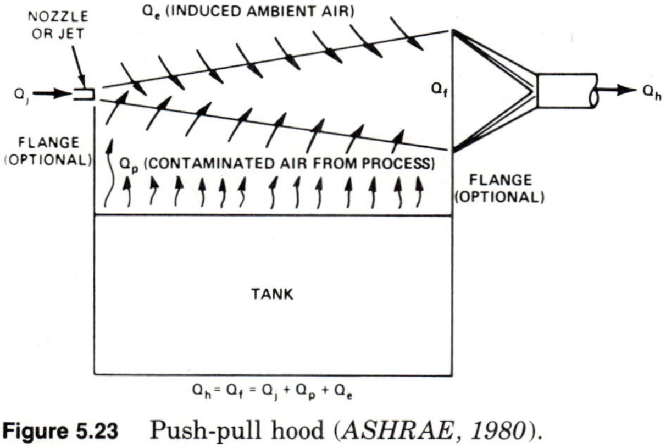

Figure 5.23 Push-pull hood (*ASHRAE, 1980*).

TABLE 5.15 Airborne Contaminants Released by Electropolishing, Electroplating, and Electroless Plating Operations

Process	Type	Notes	Component of bath that may be released to atmosphere (19)	Physical and chemical nature of major atmospheric contaminant	Class (18)	Usual temperature range (°F)
Electropolishing	Aluminum	1	Sulfuric, hydrofluoric acids	Acid mist, hydrogen fluoride gas, steam	A-2	140–200
	Brass, bronze	1	Phosphoric acid	Acid mist	B-3	68
	Copper	1	Phosphoric acid	Acid mist	B-3	68
	Iron	1	Sulfuric, hydrochloric, perchloric acids	Acid mist, hydrogen chloride gas, steam	A-2	68–175
	Monel	1	Sulfuric acid	Acid mist, steam	B-2	86–160
	Nickel	1	Sulfuric acid	Acid mist, steam	B-2	86–160
	Stainless steel	1	Sulfuric, hydrofluoric, chromic acids	Acid mist, hydrogen fluoride gas, steam	A-2,1 (20)	70–300
	Steel	1	Sulfuric, hydrochloric, perchloric acids	Acid mist, hydrogen chloride gas, steam	A-2	68–175
Strike solutions	Copper		Cyanide salts	Cyanide mist	C-2	70–90
	Silver		Cyanide salts	Cyanide mist	C-2	70–90
	Wood's nickel		Nickel chloride, hydrochloric acid	Hydrogen chloride gas, chloride mist	A-2	70–90
Electroless plating	Copper		Formaldehyde	Formaldehyde gas	A-1	75
	Nickel	2	Ammonium hydroxide	Ammonia gas	B-1	190
Electroplating, alkaline	Platinum		Ammonium phosphate, ammonia gas	Ammonia gas	B-2	158–203
	Tin		Sodium stannate	Tin salt mist, steam	C-3	140–170
	Zinc	3	None	None	D-4	170–180
Electroplating, fluoborate	Cadmium		Fluoborate salts	Fluoborate mist, steam	C-3,2 (20)	70–170
	Copper		Copper fluoborate	Fluoborate mist, steam	C-3,2 (20)	70–170
	Indium		Fluoborate salts	Fluoborate mist, steam	C-3,2 (20)	70–170
	Lead		Lead fluoborate-fluoboric acid	Fluoborate mist, hydrogen fluoride gas	A-3	70–90
	Lead-tin alloy		Lead fluoborate-fluoboric acid	Fluoborate mist	C-3,2 (20)	70–100

	Nickel		Nickel fluoborate	Fluoborate mist	C-3,2 (20)	100–170
	Tin		Stannous fluoborate, fluoboric acid	Fluoborate mist	C-3,2 (20)	70–100
	Zinc		Fluoborate salts	Fluoborate mist, steam	C-3,2 (20)	70–170
Electroplating, cyanide	Brass, bronze	4, 5	Cyanide salts, ammonium hydroxide	Cyanide mist, ammonia gas	B-4,3 (20)	60–100
	Bright zinc	5	Cyanide salts, sodium hydroxide	Cyanide, alkaline mists	C-3	70–120
	Cadmium	5	None	None	D-4	70–100
	Copper	5, 6	None	None	D-4	70–160
	Copper	5, 7	Cyanide salts, sodium hydroxide	Cyanide, alkaline mists, steam	C-2	110–160
	Indium	5	Cyanide salts, sodium hydroxide	Cyanide, alkaline mists	C-3	70–120
	Silver	5	None	None	D-4	72–120
	Tin-zinc alloy	5	Cyanide salts, potassium hydroxide	Cyanide, alkaline mists, steam	C-3,2 (20)	120–140
	White alloy	5, 8	Cyanide salts, sodium stannate	Cyanide, alkaline mists	C-3	120–150
	Zinc	5, 9	Cyanide salts, sodium hydroxide	Cyanide, alkaline mists	C-3,2 (7)	70–120
Electroplating, acid	Chromium		Chromic acid	Chromic acid mist	A-1	90–140
	Copper	10	Copper sulfate, sulfuric acid	Sulfuric acid mist	B-4,3 (20,21)	75–120
	Indium	12	None	None	D-4	70–120
	Indium	13, 14	Sulfamic acid, sulfamate salts	Sulfamate mist	C-3	70–90
	Iron		Chloride salts, hydrochloric acid	Hydrochloric acid mist, steam	A-2	190–210
	Iron	12	None	None	D-4	70–120
	Nickel	3	Ammonium fluoride, hydrofluoric acid	Hydrofluoric acid mist	A-3	102
	Nickel and black nickel	12, 15	None	None	C-4 (22)	70–150
	Nickel	9, 12	Nickel sulfate	Nickel sulfate mist	B-2	70–90
	Nickel	13, 14	Nickel sulfamate	Sulfamate mist	C-3	75–160

TABLE 5.15 Airborne Contaminants Released by Electropolishing, Electroplating, and Electroless Plating Operations *(Continued)*

Process	Type	Notes	Component of bath that may be released to atmosphere (19)	Physical and chemical nature of major atmospheric contaminant	Class (18)	Usual temperature range (°F)
	Palladium	15	None	None	D-4	70–120
	Rhodium	12, 17	None	None	D-4	70–120
	Tin		Tin halide	Halide mist	C-2	70–90
	Tin	12	None	None	D-4	70–120
	Zinc		Zinc chloride	Zinc chloride mist	B-3	75–120
	Zinc	12	None	None	D-4	70–120

NOTES:
1. Arsine may be produced due to the presence of arsenic in the metal or polishing bath.
2. Alkaline bath.
3. On magnesium.
4. Also copper-cadmium bronze.
5. HCN gas may be evolved due to the acidic action of CO_2 in the air at the surface of the bath.
6. Conventional cyanide bath.
7. Except conventional cyanide bath.
8. Albaloy, spekwhite, bonwhite (alloys of copper, tin, zinc).
9. Using insoluble anodes.
10. Over 90°F.
11. Mild organic acid bath.
12. Sulfate bath.
13. Sulfamate bath.
14. Air agitated.
15. Chloride bath.
16. Nitrite bath.
17. Phosphate bath.
18. Class on hazard potential and rate of evolution for usual operating conditions. Higher temperatures, agitation, high current density, or other conditions may result in a higher rate of evolution.
19. Hydrogen gas also released by many of these operations.
20. The higher rate is associated with the higher value in the temperature range.
21. Baths operated at a temperature of over 140°F with a current density of over 45 A per square foot and with air agitation will have a higher rate of evolution.
22. Local exhaust ventilation may be desired to control steam and water vapor.

SOURCE: ACGIH, 1984.

TABLE 5.16 Airborne Contaminants Released by Stripping Operations

Coating to be stripped	Base metal (notes)	Component of bath that may be released to atmosphere (*f*)	Physical and chemical nature of major atmospheric contaminant	Class (*e*)	Usual temperature range (°F)
Anodized coatings	1, 7	Chromic acid	Acid mist, steam	A-2	120–200
Black oxide coatings	14	Hydrochloric acid	Hydrogen chloride gas	A-3,2 (*g*)	70–125
Brass and bronze	8, 14 (*a*)	Sodium hydroxide, sodium cyanide	Alkaline, cyanide mists	C-3,2 (*g*)	70–90
Cadmium	8,14 (*a*)	Sodium hydroxide, sodium cyanide	Alkaline, cyanide mists	C-3,2 (*g*)	70–90
	2,4,14	Hydrochloric acid	Acid mist, hydrogen chloride gas	A-3,2 (*g*)	70–90
Chromium	7,8,14 (*a*)	Sodium hydroxide	Alkaline mist, steam	C-3	70–150
	2,4,8,14	Hydrochloric acid	Hydrogen chloride gas	A-2	70–125
	2,4,8,18 (*a*)	Sulfuric acid	Acid mist	B-2	70–90
Copper	8,14	Sodium hydroxide, sodium cyanide	Alkaline, cyanide mists	C-3,2 (*g*)	70–90
	7,12,14 (*b*)	None	None	D-4	70–90
	14 (*a*)	Alkaline cyanide	Cyanide mist	C-3,2 (*g*)	70–160
	1	Nitric acid	Nitrogen oxide gases	A-1	70–120
	18 (*a*)	Sodium hydroxide-sodium sulfide	Alkaline mist, steam	C-2	185–195
Gold	4,5,6,8,9,14 (*a*)	Sodium hydroxide, sodium cyanide	Alkaline, cyanide mists	C-3,2 (*g*)	70–90
	4,5,18 (*a*)	Sulfuric acid	Acid mist	B-3,2 (*g*)	70–100
Lead	13 (*c*)	Acetic acid, hydrogen peroxide	Oxygen mist	D-3	70–90
	14 (*a*),(*c*)	Sodium hydroxide	Alkaline mist, steam	C-3,2 (*g*)	70–140

TABLE 5.16 Airborne Contaminants Released by Stripping Operations (*Continued*)

Coating to be stripped	Base metal (notes)	Component of bath that may be released to atmosphere (*f*)	Physical and chemical nature of major atmospheric contaminant	Class (*e*)	Usual temperature range (°F)
Nickel	2,4	Sulfuric, nitric acids	Nitrogen oxide gases	A-2,1 (*g*)	70–90
	2,4 (*a*)	Hydrochloric acid	Hydrogen chloride gas	A-3	70–90
	2,4,14 (*a*)	Sulfuric acid	Acid mist	B-3	70–90
	7	Hydrofluoric acid	Hydrogen fluoride gas	A-3,2 (*g*)	70–90
	14	Fuming nitric acid	Nitrogen oxide gases	A-1	70–90
	(*a*),(*d*)	Hot water	Steam	D-2 (*h*)	200
	1,18,19 (*a*)	Sulfuric acid	Acid mist, steam	B-3,2 (*g*)	70–150
Phosphate coatings	15	Chromic acid	Acid mist, steam	A-3	165
	16	Ammonium hydroxide	Ammonia gas	B-3,2 (*g*)	70–90
Rhodium	10	Sulfuric, hydrochloric acids	Acid mist, hydrogen chloride gas	A-3,2 (*g*)	70–100
Silver	1	Nitric acid	Nitrogen oxide gases	A-1	70–90
	2,11	Sulfuric, nitric acids	Nitrogen oxide gases, steam	A-1	180
	8,14 (*a*)	Sodium hydroxide, sodium cyanide	Alkaline, cyanide mists	C-3	70–90
	17 (*a*)	Sodium cyanide	Cyanide mist	C-3	70–90
Tin	2,3,4	Ferric chloride, copper sulfate, acetic acid	Acid mist	B-4,3 (*g*)	70–90
	(*a*)	Sodium hydroxide	Alkaline mist	C-3	70–90
	2,4,14	Hydrochloric acid	Hydrogen chloride gas	A-3,2 (*g*)	70–90
	14 (*a*)	Sodium hydroxide	Alkaline mist, steam	C-2	70–200
Zinc	1	Nitric acid	Nitrogen oxide gases	A-1	70–90
	8,14	Sodium hydroxide, sodium	Alkaline, cyanide mists	C-3	70–90

Base metal:
1. Aluminum
2. Brass
3. Bronze
4. Copper

5. Copper alloys
6. Ferrous metals
7. Magnesium
8. Nickel
9. Nickel alloys
10. Nickel-plated brass
11. Nickel silver
12. Nonferrous metals
13. Silver
14. Steel
15. Steel (manganese-type coatings)
16. Steel (zinc-type coatings)
17. White metal
18. Zinc
19. Zinc-base die castings

NOTES: (*a*) Electrolytic process
(*b*) Refers only to steel (14) when chromic, sulfuric acids bath is used.
(*c*) Also lead alloys
(*d*) Sodium nitrate bath
(*e*) Class 3 based on hazard potential and rate of evolution classification 2 for usual operating conditions. Higher temperatures, agitation, or other conditions may result in a higher rate of evolution.
(*f*) Hydrogen gas also released by some of these operations.
(*g*) The higher rate is associated with the higher value in the temperature range.
(*h*) Rate where essentially complete control of steam is required. Otherwise, adequate dilution ventilation may be sufficient.

SOURCE: ACGIH, 1984.

determining appropriate push and pull flow rates for plating tanks. Given the area of the inlet (push) nozzle, A_D, in ft²/ft of nozzle length (which may be a slot or series of holes or nozzles that extend the length of the tank), the volumetric push flow rate, Q_D, in ft³/min · ft of nozzle length, can be estimated from

$$Q_D = 243 A_D^{0.5} \tag{5.33}$$

The corresponding required exhaust flow is specified at 75 ft³/min · ft² tank area (Hughes, 1985). The original articles should be consulted for geometric details of design, as these are important for proper functioning of the ventilation system.

Example 5.14 What are the push and exhaust volumes required for a chrome plating tank 5 ft × 5 ft given a push slot nozzle 0.125 in. wide?

solution

$$A_D = \frac{0.125 \text{ in.}}{12 \text{ in./ft}} = 0.0104 \text{ ft}^2\text{/ft nozzle length}$$

$$Q_D = (243)(0.0104)^{0.5} = 24.8 \text{ ft}^3\text{/min} \cdot \text{ft nozzle length} \tag{5.33}$$

or

$$Q_{D,\text{total}} = (24.8 \text{ ft}^3\text{/min} \cdot \text{ft})(5 \text{ ft}) = 124.0 \text{ cfm}$$

The exhaust volume, Q_E, will be:

$$Q_E = (75 \text{ cfm/ft}^2)(25 \text{ ft}^2) = 1875 \text{ cfm}$$

Note that the difference between Q_E and Q_D is attributable to flow emitted by the source and entrained air.

Capture velocities are usually restricted to centerline determinations. Consequently, they may not adequately reflect how a hood performs. Several investigators have suggested *capture efficiency*, η_e, as a more direct index of hood operation (Flynn and Ellenbecker, 1986b; Ellenbecker et al., 1983; Jansson, 1980). This parameter has the definition

$$\eta_e = \frac{G'}{G} \tag{5.34}$$

where G' is the pollutant mass capture rate of the hood and G is the pollutant mass generation rate of the source. Hood design based on η_e is conceptually appealing but, as yet, there have been only limited experimental verifications of its usefulness, usually in wind tunnels (Flynn and Ellenbecker, 1986b) or environmental chambers (Jansson, 1980).

In all of the discussion about hood design, nothing has been said about the need for make-up air. Make-up air must be provided:

1. To ensure that exhaust hoods operate as designed. If the room is

under negative pressure, the static pressure against which the exhaust fan operates will be increased and the volume flow decreased.

2. To ensure the proper operation of natural draft stacks, flues, and fuel-burning appliances.
3. To eliminate high-velocity cross-drafts through doors, windows, or cracks into the zone of influence of exhaust hoods.
4. To eliminate cold drafts on workers.
5. To eliminate additional dispersion of contaminants in the workroom by high-velocity drafts.
6. To prevent dust-laden air in adjacent areas from being directed into operating areas that must be kept clean for processing reasons.
7. To avoid difficulty in opening or closing doors to room or building.
8. To dilute low concentrations of airborne nuisance material that do not justify local exhaust systems.
9. To provide effective cooling for personnel, particularly in warm weather.

VII. Calculations vs. Reality

In going through the various ventilation design calculations, a fair question to ask is, "How good are they?" Perhaps equally important in these days of computers and calculators is the issue of accuracy and number of significant figures. Several factors need to be considered in addressing these issues. These factors will be combined into two broad categories: limitations on measurements, and design and operation problems.

Obviously the ability to verify a calculated result is limited by the ability to measure that which actually happens. This is true of ventilation work, too. The most useful measuring device, because it can be used to measure VP, SP, and TP directly, is the pitot tube and inclined manometer. A well-manufactured, properly leveled and calibrated pitot tube and inclined manometer can be read to approximately ±0.005 in. W.G. Therefore, the calculation of pressure losses beyond the nearest 0.01 in. is not justified.

A standard pitot tube and inclined manometer can be used with the accuracy as shown in Table 5.17. Because of the rapid increase in error at low velocity, use of the pitot tube is restricted to velocities above 600–800 ft/min.

If the temperature of the air stream differs more than 30°F from standard temperature (70°F), if altitude differs more than 1000 ft from sea level, or if the moisture content of the air is greater than 0.02 lb

TABLE 5.17 Pitot Tube and Inclined Manometer Accuracy

Air velocity (ft/min)	Error (±) (%)
4000	0.25
3000	0.3
2000	1.0
1000	4.0
800	6.0
600	15.0

SOURCE: ACGIH, 1984.

water/lb dry air, then correction for density change is necessary. The correction is:

$$v = 1096\left(\frac{VP}{0.075d}\right)^{0.5} \tag{5.35}$$

where v = gas velocity, fpm
d = density correction factor from Table 5.1
VP = velocity pressure, in.H_2O

Measurement of velocity should be corrected for gas density when conditions vary from ranges previously described by using

$$v = v_{meas}\left(\frac{1}{d}\right)^{0.5} \tag{5.36}$$

where v_{meas} = measured velocity and d is from Table 5.1. It should be kept in mind that density corrections do not improve on the limitations of the basic instrumentation (ACGIH, 1984).

A swinging-vane anemometer (velometer) or hot-wire anemometer can also be used to measure air stream velocity. Minimum velocity that can be measured with a velometer is 50 ft/min unless it is specially adapted for a lower range and 10 ft/min for a hot-wire anemometer. Velometers do tend to read low on the blowing side of a fan and high on the suction side. They are also reported to be inaccurate at the high end of the velocity scale. The sensing elements of a hot-wire anemometer can become coated with dirt, causing inaccurate measurements. Both devices require calibration (ACGIH, 1984).

Velometers and hot-wire anemometers are commonly used to measure the velocity of air movement outside the ventilation system, e.g., in the vicinity of the hood. (However, hot-wire anemometers do not indicate direction and must be used in conjunction with smoke tubes.) The purpose of such measurements is to determine the capture or control velocity for a functioning system. Large fluctuations can occur and, as a result, a time-weighted average should be used. Such an average can have an accuracy of ±10%, which ordinarily will not affect proper hood operation.

Under the category of design and operation problems are seven common problems:

1. Tampering
2. Dirt accumulation
3. Corrosion/erosion
4. Fan direction
5. Maintenance
6. System additions
7. System design

On systems where adjustable flow-controlling devices (dampers) are used, it is likely that the damper setting will be changed. No maliciousness may be intended in such acts; rather, someone is making an attempt to improve ventilation at one location and it is discovered that opening or closing a damper accomplishes the objective. This is done without knowledge of the impact on flow elsewhere. Clearly the system will operate differently than planned (calculated).

In making the loss calculations described earlier, it was assumed that the duct and fittings were clean. Significant build-up of dirt on the duct surface will change (usually increase) pressure loss, causing the system to operate differently from design. If dirt is deposited or settles into piles, the duct can become clogged. Dented ducts will behave similarly. Corrosion can also cause the roughness of the duct surface to increase. As corrosion progresses or if erosion occurs, the duct wall can be breached. This can also occur because of external forces, such as a hole punched out by a fork lift. Since this new opening will provide a route having less resistance to flow, air will flow through the hole rather than through upstream ducts.

Perhaps a more subtle, but not uncommon, problem is that of reversed fan direction. Fan blade design is such that air movement is not as good if the fan turns the wrong way. Typically air flow will be reduced on the order of 50%. When this problem occurs, it will occur immediately following the performance of maintenance on the fan or blower (Burton, 1984).

Maintenance, or the lack of maintenance, is a common problem with ventilation systems. Unfortunately, the attitude is typically taken that once the system is installed, the problem is solved. The system does require maintenance just like any other equipment. No matter what contaminants are being moved through the ventilation system, the air mover (blades, drive train, and motor) require periodic inspection and repair. Blades may become coated with dirt or eroded if abrasive contaminants are present. Belts may become loose, resulting in slipping. In addition to the motor, clean-out ports should be strategically included in the system to permit removal of dirt accumulating in ducting. Air cleaning devices need periodic cleaning; adsorbants, absorbants, and filters need to be replaced (see Chapter 6). This is not intended to be an exhaustive list of required maintenance, but rather to serve as a reminder that maintenance of the ventilation system is a necessary part of solving the personal or area exposure problem.

Another common problem with ventilation systems results from the relative physical ease with which additions can be made to the system, i.e., installation of an additional hood. Although the air mover may be able to move the increased air flow, chances are that an unbalance has been created. More likely, the air mover will not have the required capacity, with the result that some or all hoods no longer have an adequate flow rate. Ad hoc additions to a system should be avoided.

Finally, the quality of the original design of the system has much to do with whether or not the system will work as desired. For example, failure to provide make-up air can result in serious drafts that will interfere with hood performance. In other words, thought has to be given to design, the entire workspace has to be considered even in the design of a local system, and recommended practice should be followed whenever possible. Even then, situations will arise where a new ventilation solution will have to be developed.

REFERENCES

ACGIH (1984). *Industrial Ventilation,* 18th Ed. American Conference of Governmental Industrial Hygienists, Cincinnati, Ohio.

AMCA (1981). *Fan Application Manual Part I.* Air Movement and Control Association, Publication No. 201, Arlington Heights, Ill.

ASHRAE (1980). *ASHRAE Handbook, 1980 Systems.* American Society of Heating, Refrigerating and Air-Conditioning Engineers, Atlanta, Ga.

ASHRAE (1985). *ASHRAE Handbook, 1985 Fundamentals.* American Society of Heating Refrigerating and Air-Conditioning Engineers, Atlanta, Ga.

Burton, D. J. (1984). Investigating for clues to capture increased fugitive hood emissions. *Occup. Health Safety,* 76, June.

Crawford, M. (1976). *Air Pollution Control Theory.* McGraw-Hill, New York.

Ellenbecker, M. J., Gempel, R. J., and Burgess, W. A. (1983). Capture efficiency of local exhaust ventilation systems. *Amer. Ind. Hyg. Assoc. J., 44,* 752–755.

Flynn, M. R., and Ellenbecker, M. J. (1985). The potential flow solution for air flow into a flanged circular hood. *Amer. Ind. Hyg. Assoc. J., 46,* 318–322.

Flynn, M. R., and Ellenbecker, M. J. (1986a). Empirical validation of theoretical velocity fields into flanged circular hoods. *Amer. Ind. Hyg. Assoc. J.,* in press.

Flynn, M. R., and Ellenbecker, M. J. (1986b). Capture efficiency of local exhaust hoods. *Ann. Occup. Med.,* in press.

Huebner, D. J., and Hughes, R. T. (1985). Development of push-pull ventilation. *Amer. Ind. Hyg. Assoc. J., 46,* 262–267.

Hughes, R. T. (1985). Design criteria for plating tank push-pull ventilation. In: *Procedures of the 1st International Symposium on Ventilation for Contaminant Control,* Toronto. Elsevier, Amsterdam, The Netherlands.

Jansson, A. (1980). Capture efficiencies of local exhausts for hand grinding, drilling and welding. *Staub-Reinhalt. Luft., 40,* 111–113.

Lightfoot, E. N., Thorne, P. S., and Stoll, L. L. (1984). Dimensionless presentation of performance data for fans and blowers. *Amer. Inst. Chem. Eng. J., 30,* 341–346.

McDermott, H. J. (1985). *Handbook of Ventilation for Contaminant Control,* 2d ed. Ann Arbor Science, Ann Arbor, Mich.

NIOSH (1973). *The Industrial Environment—Its Evaluation and Control.* National Institute for Occupational Safety and Health, Cincinnati, Ohio.

Whitaker, S. (1968). *Introduction to Fluid Mechanics.* Prentice-Hall, Englewood Cliffs, N.J.

APPENDIX A

TABLE A.1 Circular Equivalents of Rectangular Ducts for Equal Friction and Capacity

Based on on $D_e = 1.3(ab)^{0.625}/(a + b)^{0.250}$

Lgth	Length of One Side of Rectangular Duct (a), in.																
Adj.[b]	4.0	4.5	5.0	5.5	6.0	6.5	7.0	7.5	8.0	9.0	10.0	11.0	12.0	13.0	14.0	15.0	16.0
3.0	3.8	4.0	4.2	4.4	4.6	4.7	4.9	5.1	5.2	5.5	5.7	6.0	6.2	6.4	6.6	6.8	7.0
3.5	4.1	4.3	4.6	4.8	5.0	5.2	5.3	5.5	5.7	6.0	6.3	6.5	6.8	7.0	7.2	7.5	7.7
4.0	4.4	4.6	4.9	5.1	5.3	5.5	5.7	5.9	6.1	6.4	6.7	7.0	7.3	7.6	7.8	8.0	8.3
4.5	4.6	4.9	5.2	5.4	5.7	5.9	6.1	6.3	6.5	6.9	7.2	7.5	7.8	8.1	8.4	8.6	8.8
5.0	4.9	5.2	5.5	5.7	6.0	6.2	6.4	6.7	6.9	7.3	7.6	8.0	8.3	8.6	8.9	9.1	9.4
5.5	5.1	5.4	5.7	6.0	6.3	6.5	6.8	7.0	7.2	7.6	8.0	8.4	8.7	9.0	9.3	9.6	9.9

Lgth	Length of One Side of Rectangular Duct (a), in.																				Lgth
Adj.[b]	6	7	8	9	10	11	12	13	14	15	16	17	18	19	20	22	24	26	28	30	Adj.[b]
6	6.6																				6
7	7.1	7.7																			7
8	7.6	8.2	8.7																		8
9	8.0	8.7	9.3	9.8																	9
10	8.4	9.1	9.8	10.4	10.9																10
11	8.8	9.5	10.2	10.9	11.5	12.0															11
12	9.1	9.9	10.7	11.3	12.0	12.6	13.1														12
13	9.5	10.3	11.1	11.8	12.4	13.1	13.7	14.2													13
14	9.8	10.7	11.5	12.2	12.9	13.5	14.2	14.7	15.3												14
15	10.1	11.0	11.8	12.6	13.3	14.0	14.6	15.3	15.8	16.4											15
16	10.4	11.3	12.2	13.0	13.7	14.4	15.1	15.7	16.4	16.9	17.5										16
17	10.7	11.6	12.5	13.4	14.1	14.9	15.6	16.2	16.8	17.4	18.0	18.6									17
18	11.0	11.9	12.9	13.7	14.5	15.3	16.0	16.7	17.3	17.9	18.5	19.1	19.7								18
19	11.2	12.2	13.2	14.1	14.9	15.7	16.4	17.1	17.8	18.4	19.0	19.6	20.2	20.8							19
20	11.5	12.5	13.5	14.4	15.2	16.0	16.8	17.5	18.2	18.9	19.5	20.1	20.7	21.3	21.9						20
22	12.0	13.0	14.1	15.0	15.9	16.8	17.6	18.3	19.1	19.8	20.4	21.1	21.7	22.3	22.9	24.0					22
24	12.4	13.5	14.6	15.6	16.5	17.4	18.3	19.1	19.9	20.6	21.3	22.0	22.7	23.3	23.9	25.1	26.2				24
26	12.8	14.0	15.1	16.2	17.1	18.1	19.0	19.8	20.6	21.4	22.1	22.9	23.5	24.2	24.9	26.1	27.3	28.4			26
28	13.2	14.5	15.6	16.7	17.7	18.7	19.6	20.5	21.3	22.1	22.9	23.7	24.4	25.1	25.8	27.1	28.3	29.5	30.6		28
30	13.6	14.9	16.1	17.2	18.3	19.3	20.2	21.1	22.0	22.9	23.7	24.4	25.2	25.9	26.6	28.0	29.3	30.5	31.7	32.8	30
32	14.0	15.3	16.5	17.7	18.8	19.8	20.8	21.8	22.7	23.5	24.4	25.2	26.0	26.7	27.5	28.9	30.2	31.5	32 .7	33.9	32
34	14.4	15.7	17.0	18.2	19.3	20.4	21.4	22.4	23.3	24.2	25.1	25.9	26.7	27.5	28.3	29.7	31.0	32.4	33.7	34.9	34
36	14.7	16.1	17.4	18.6	19.8	20.9	21.9	22.9	23.9	24.8	25.7	26.6	27.4	28.2	29.0	30.5	32.0	33.3	34.6	35.9	36
38	15.0	16.5	17.8	19.0	20.2	21.4	22.4	23.5	24.5	25.4	26.4	27.2	28.1	28.9	29.8	31.3	32.8	34.2	35.6	36.8	38
40	15.3	16.8	18.2	19.5	20.7	21.8	22.9	24.0	25.0	26.0	27.0	27.9	28.8	29.6	30.5	32.1	33.6	35.1	36.4	37.8	40
42	15.6	17.1	18.5	19.9	21.1	22.3	23.4	24.5	25.6	26.6	27.6	28.5	29.4	30.3	31.2	32.8	34.4	35.9	37.3	38.7	42
44	15.9	17.5	18.9	20.3	21.5	22.7	23.9	25.0	26.1	27.1	28.1	29.1	30.0	30.9	31.8	33.5	35.1	36.7	38.1	39.5	44
46	16.2	17.8	19.3	20.6	21.9	23.2	24.4	25.5	26.6	27.7	28.7	29.7	30.6	31.6	32.5	34.2	35.9	37.4	38.9	40.4	46
48	16.5	18.1	19.6	21.0	22.3	23.6	24.8	26.0	27.1	28.2	29.2	30.2	3!.2	32.2	33.1	34.9	36.6	38.2	39.7	41.2	48
50	16.8	18.4	19.9	21.4	22.7	24.0	25.2	26.4	27.6	28.7	29.8	30.8	31.8	32.8	33.7	35.5	37.2	38.9	40.5	42.0	50
52	17.1	18.7	20.2	21.7	23.1	24.4	25.7	26.9	28.0	29.2	30.3	31.3	32.3	33.3	34.3	36.2	37.9	39.6	41.2	42.8	52
54	17.3	19.0	20.6	22.0	23.5	24.8	26.1	27.3	28.5	29.7	30.8	31.8	32.9	33.9	34.9	36.8	38.6	40.3	41.9	43.5	54
56	17.6	19.3	20.9	22.4	23.8	25.2	26.5	27.7	28.9	30.1	31.2	32.3	33.4	34.4	35.4	37.4	39.2	41.0	42.7	44.3	56
58	17.8	19.5	21.2	22.7	24.2	25.5	26.9	28.2	29.4	30.6	31.7	32.8	33.9	35.0	36.0	38.0	39.8	41.6	43.3	45.0	58
60	18.1	19.8	21.5	23.0	24.5	25.9	27.3	28.6	29.8	31.0	32.2	33.3	34.4	35.5	36.5	38.5	40.4	42.3	44.0	45.7	60
62		20.1	21.7	23.3	24.8	26.3	27.6	28.9	30.2	31.5	32.6	33.8	34.9	36.0	37.1	39.1	41.0	42.9	44.7	46.4	62
64		20.3	22.0	23.6	25.1	26.6	28.0	29.3	30.6	31.9	33.1	34.3	35.4	36.5	37.6	39.6	41.6	43.5	45.3	47.1	64
66		20.6	22.3	23.9	25.5	26.9	28.4	29.7	31.0	32.3	33.5	34.7	35.9	37.0	38.1	40.2	42.2	44.1	46.0	47.7	66
68		20.8	22.6	24.2	25.8	27.3	28.7	30.1	31.4	32.7	33.9	35.2	36.3	37.5	38.6	40.7	42.8	44.7	46.6	48.4	68
70		21.1	22.8	24.5	26.1	27.6	29.1	30.4	31.8	33.1	34.4	35.6	36.8	37.9	39.1	41.2	43.3	45.3	47.2	49.0	70
72			23.1	24.8	26.4	27.9	29.4	30.8	32.2	33.5	34.8	36.0	37.2	38.4	39.5	41.7	43.8	45.8	47.8	49.6	72
74			23.3	25.1	26.7	28.2	29.7	31.2	32.5	33.9	35.2	36.4	37.7	38.8	40.0	42.2	44.4	46.4	48.4	50.3	74
76			23.6	25.3	27.0	28.5	30.0	31.5	32.9	34.3	35.6	36.8	38.1	39.3	40.5	42.7	44.9	47.0	48.9	50.9	76
78			23.8	25.6	27.3	28.8	30.4	31.8	33.3	34.6	36.0	37.2	38.5	39.7	40.9	43.2	45.4	47.5	49.5	51.4	78
80			24.1	25.8	27.5	29.1	30.7	32.2	33.6	35.0	36.3	37.6	38.9	40.2	41.4	43.7	45.9	48.0	50.1	52.0	80
82				26.1	27.8	29.4	31.0	32.5	34.0	35.4	36.7	38.0	39.3	40.6	41.8	44.1	46.4	48.5	50.6	52.6	82
84				26.4	28.1	29.7	31.3	32.8	34.3	35.7	37.1	38.4	39.7	41.0	42.2	44.6	46.9	49.0	51.1	53.2	84
86				26.6	28.3	30.0	31.6	33.1	34.6	36.1	37.4	38.8	40.1	41.4	42.6	45.0	47.3	49.6	51.7	53.7	86
88				26.9	28.6	30.3	31.9	33.4	34.9	36.4	37.8	39.2	40.5	41.8	43.1	45.5	47.8	50.0	52.2	54.3	88
90				27.1	28.9	30.6	32.2	33.8	35.3	36.7	38.2	39.5	40.9	42.2	43.5	45.9	48.3	50.5	52.7	54.8	90
92					29.1	30.8	32.5	34.1	35.6	37.1	38.5	39.9	41.3	42.6	43.9	46.4	48.7	51.0	53.2	55.3	92
96					29.6	31.4	33.0	34.7	36.2	37.7	39.2	40.6	42.0	43.3	44.7	47.2	49.6	52.0	54.2	56.4	96

SOURCE: ASHRAE, 1985.

Lgth Adj.[b]	32	34	36	38	40	42	44	46	48	50	52	56	60	64	68	72	76	80	84	88	Lgth Adj.[b]
	Length of One Side Rectangular Duct (a), in.																				
32	35.0																				32
34	36.1	37.2																			34
36	37.1	38.2	39.4																		36
38	38.1	39.3	40.4	41.5																	38
40	39.0	40.3	41.5	42.6	43.7																40
42	40.0	41.3	42.5	43.7	44.8	45.9															42
44	40.9	42.2	43.5	44.7	45.8	47.0	48.1														44
46	41.8	43.1	44.4	45.7	46.9	48.0	49.2	50.3													46
48	42.6	44.0	45.3	46.6	47.9	49.1	50.2	51.4	52.5												48
50	43.6	44.9	46.2	47.5	48.8	50.0	51.2	52.4	53.6	54.7											50
52	44.3	45.7	47.1	48.4	49.7	51.0	52.2	53.4	54.6	55.7	56.8										52
54	45.1	46.5	48.0	49.3	50.7	52.0	53.2	54.4	55.6	56.8	57.9										54
56	45.8	47.3	48.8	50.2	51.6	52.9	54.2	55.4	56.6	57.8	59.0	61.2									56
58	46.6	48.1	49.6	51.0	52.4	53.8	55.1	56.4	57.6	58.8	60.0	62.3									58
60	47.3	48.9	50.4	51.9	53.3	54.7	60.0	57.3	58.6	59.8	61.0	63.4	65.6								60
62	48.0	49.6	51.2	52.7	54.1	55.5	56.9	58.2	59.5	60.8	62.0	64.4	66.7								62
64	48.7	50.4	51.9	53.5	54.9	56.4	57.8	59.1	60.4	61.7	63.0	65.4	67.7	70.0							64
66	49.4	51.1	52.7	54.2	55.7	57.2	58.6	60.0	61.3	62.6	63.9	66.4	68.8	71.0							66
68	50.1	51.8	53.4	55.0	56.5	58.0	59.4	60.8	62.2	63.6	64.9	67.4	69.8	72.1	74.3						68
70	50.8	52.5	54.1	55.7	57.3	58.8	60.3	61.7	63.1	64.4	65.8	68.3	70.8	73.2	75.4						70
72	51.4	53.2	54.8	56.5	58.0	59.6	61.1	62.5	63.9	65.3	66.7	69.3	71.8	74.2	76.5	78.7					72
74	52.1	53.8	55.5	57.2	58.8	60.3	61.9	63.3	64.8	66.2	67.5	70.2	72.7	75.2	77.5	79.8					74
76	52.7	54.5	56.2	57.9	59.5	61.1	62.6	64.1	65.6	67.0	68.4	71.1	73.7	76.2	78.6	80.9	83.1				76
78	53.3	55.1	56.9	58.6	60.2	61.8	63.4	64.9	66.4	67.9	69.3	72.0	74.6	77.1	79.6	81.9	84.2				78
80	53.9	55.8	57.5	59.3	60.9	62.6	64.1	65.7	67.2	68.7	70.1	72.9	75.4	78.1	80.6	82.9	85.2	87.5			80
82	54.5	56.4	58.2	59.9	61.6	63.3	64.9	66.5	68.0	69.5	70.9	73.7	76.4	79.0	81.5	84.0	86.3	88.5			82
84	55.1	57.0	58.8	60.6	62.3	64.0	65.6	67.2	68.7	70.3	71.7	74.6	77.3	80.0	82.5	85.0	87.3	89.6	91.8		84
86	55.7	57.6	59.4	61.2	63.0	64.7	66.3	67.9	69.5	71.0	72.5	75.4	78.2	80.9	83.5	85.9	88.3	90.7	92.9		86
88	56.3	58.2	60.1	61.9	63.6	65.4	67.0	68.7	70.2	71.8	73.3	76.3	79.1	81.8	84.4	86.9	89.3	91.7	94.0	96.2	88
90	56.8	58.8	60.7	62.5	64.3	66.0	67.7	69.4	71.0	72.6	74.1	77.1	79.9	82.7	85.3	87.9	90.3	92.7	95.0	97.3	90
92	57.4	59.3	61.3	63.1	64.9	66.7	68.4	70.1	71.7	73.3	74.9	77.9	80.8	83.5	86.2	88.8	91.3	93.7	96.1	98.4	92
94	57.9	59.9	61.9	63.7	65.6	67.3	69.1	70.8	72.4	74.0	75.6	78.7	81.6	84.4	87.1	89.7	92.3	94.7	97.1	99.4	94
96	58.4	60.5	62.4	64.3	66.2	68.0	69.7	71.5	73.1	74.8	76.3	79.4	82.4	85.3	88.0	90.7	93.2	95.7	98.1	100.5	96

[a]Table based on Eq. (31)
[b]Length of adjacent side of rectangular duct (b), in.

APPENDIX B*

FITTING LOSS COEFFICIENTS

NOTES

1. To calculate the total pressure loss coefficient for this fitting ($A_1 \neq A_o$) with a screen, use Eq. (B.1).

$$C_o = C_o' + \frac{C_s}{(A_1/A_o)^2} \quad \text{(B.1)}$$

where

C_o = combination fitting and screen resistance coefficient referenced to section *o*, dimensionless
C_o' = fitting resistance coefficient referenced to section *o*, dimensionless
C_s = screen resistance coefficient, dimensionless (see Fitting 7-8)
A_1 = area at section 1, in.^{2}
A_o = area at section *o*, in.^{2}

2. To calculate the total pressure loss coefficient for this fitting (constant cross-sectional area), use Eq. (B.2).

$$C_o = C_o' + C_s \quad \text{(B.2)}$$

3. For elbows with angles other than 90°, multiply by the following factor (K_θ).

θ	0	20	30	45	60	75	90	110	130	150	180
K_θ	0	0.31	0.45	0.60	0.78	0.90	1.00	1.13	1.20	1.28	1.40

4. To calculate *Reynolds number* (Re) use Eq. (B.3).

$$Re = \frac{5 \varrho D V_o}{\mu} \quad \text{(B.3)}$$

where

ϱ = density, lb$_m$/ft^3
μ = viscosity, lb$_m$/(ft•h)
D = diameter, in.
V_o = velocity, ft/min

*Notes refer to footnotes referenced by number from the tables that follow. We have also indicated which duct designs are common for industrial applications (ASHRAE, 1986).

For *standard air,* Re can be calculated by:

$$\text{Re} = 8.56\,D\,V_o \qquad \text{(B.4)}$$

5. For this fitting, determine the Reynolds number correction factor (K_{Re}) from the following table.

$Re \cdot 10^{-4}$	1	2	3	4	6	8	10	≥14
K_{Re}	1.40	1.26	1.19	1.14	1.09	1.06	1.04	1.0

6. The cross section to which the coefficients are referenced is indicated in the following tabulation.

Table No.	Type of Fitting	Cross Section to Which Coefficient is Referenced
B-1	Entries	Downstream
B-2	Exits	Upstream
B-3	Elbows	Upstream or Downstream
B-4	Transitions (Diverging Flow)	Upstream
B-5	Transitions (Converging Flow)	Downstream
B-6	Junctions (Tees, Wyes)	Common
B-7	Obstructions	Upstream or Downstream

In the case of entries, exits, elbows, transitions and obstructions, the subscript *o* of A_o, Q_o and V_o shown on the sketches indicates the cross section to which the coefficients are referenced.

For converging and diverging junctions (tees, wyes) the subscript *c* of $C_{c,s}$ and $C_{c,b}$ signifies that the loss coefficient refers to the common section of the fitting. The subscripts *s* and *b* indicate that the loss coefficient is for the main and branch airflow paths.

The dynamic loss is based on the actual velocity in a rectangular duct, not the velocity in an equivalent round duct.

Table B-1 Local Loss Coefficients, ENTRIES

1-1 Duct Mounted in Wall (Hood, Non-Enclosing, Flanged and Unflanged)[1]

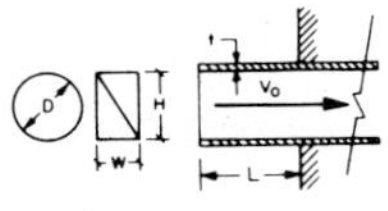

Rectangular: D = 2HW/(H + W)

t/D	L/D = 0	0.002	0.01	0.05	0.2	0.5	≥1.0
≈ 0	0.50	0.57	0.68	0.80	0.92	1.0	1.0
0.02	0.50	0.51	0.52	0.55	0.66	0.72	0.72
≥ 0.05	0.50	0.50	0.50	0.50	0.50	0.50	0.50

(C_o values)

Fitting coefficient (C_o) with screen at inlet.
Sharp Edge ($t/D \leq 0.05$):

$$C_o = 1 + C_s$$

Thick Edge ($t/D > 0.05$):

$$C_o = C_o' + C_s$$

where

C_o' = fitting coefficient (see Table above).
C_s = screen coefficient (see Fitting 7-8).

1-2 Smooth Converging Bellmouth without End Wall[1]

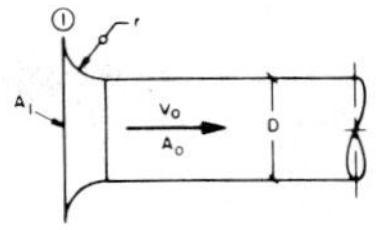

r/D	0	0.01	0.02	0.03	0.04	0.05
C_o	1.0	0.87	0.74	0.61	0.51	0.40
r/D	0.06	0.08	0.10	0.12	0.16	≥ 0.20
C_o	0.32	0.20	0.15	0.10	0.06	0.03

For loss coefficient with screen at section 1, see Note 1.

1-3 Smooth Converging Bellmouth with End Wall[1]

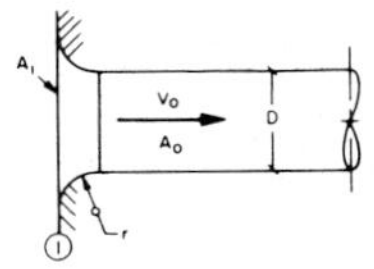

r/D	0	0.01	0.02	0.03	0.04	0.05
C_o	0.50	0.43	0.36	0.31	0.26	0.22
r/D	0.06	0.08	0.10	0.12	0.16	≥ 0.20
C_o	0.20	0.15	0.12	0.09	0.06	0.03

For loss coefficient with screen at section 1, see Note 1.

1-4 Conical Converging Bellmouth without End Wall, Round and Rectangular[1]

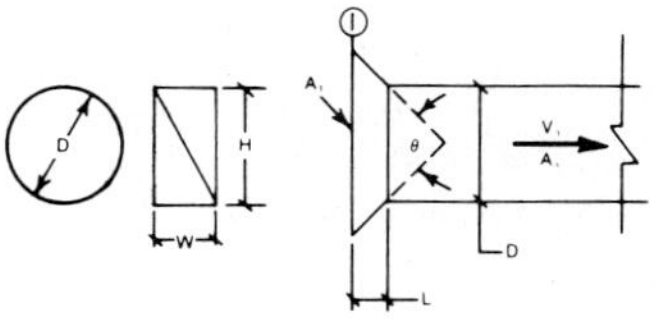

Rectangular: D = 2HW/(H + W)

L/D	θ, degrees = 0	10	20	30	40	60	100	140	180
0.025	1.0	0.96	0.93	0.90	0.86	0.80	0.69	0.59	0.50
0.05	1.0	0.93	0.86	0.80	0.75	0.67	0.58	0.53	0.50
0.10	1.0	0.80	0.67	0.55	0.48	0.41	0.41	0.44	0.50
0.25	1.0	0.68	0.45	0.30	0.22	0.17	0.22	0.34	0.50
0.60	1.0	0.46	0.27	0.18	0.14	0.13	0.21	0.33	0.50
1.0	1.0	0.32	0.20	0.14	0.11	0.10	0.18	0.30	0.50

(C_o values)

For loss coefficient with screen at section 1, see Note 1.

1-5 Conical Converging Bellmouth with End Wall, Round and Rectangular[1]

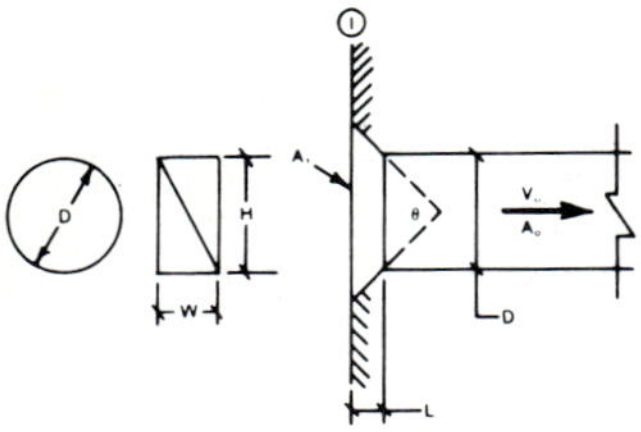

Rectangular: D = 2 HW/(H + W)

L/D	C_o θ, degrees 0	10	20	30	40	60	100	140	180
0.025	0.50	0.47	0.45	0.43	0.41	0.40	0.42	0.45	0.50
0.05	0.50	0.45	0.41	0.36	0.33	0.30	0.35	0.42	0.50
0.075	0.50	0.42	0.35	0.30	0.26	0.23	0.30	0.40	0.50
0.10	0.50	0.39	0.32	0.25	0.22	0.18	0.27	0.38	0.50
0.15	0.50	0.37	0.27	0.20	0.16	0.15	0.25	0.37	0.50
0.60	0.50	0.27	0.18	0.13	0.11	0.12	0.23	0.36	0.50

For loss coefficient with screen at section 1, see Note 1.

1-6 Intake Hood[1]

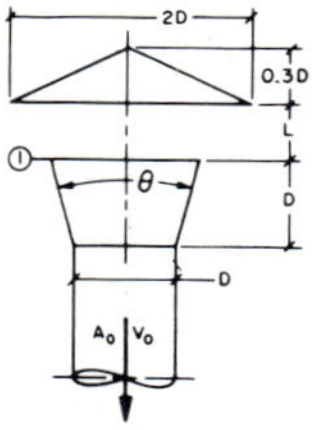

θ, deg	C_o L/D 0.1	0.2	0.3	0.4	0.5	0.6	0.7	0.8	⩾ 0.9
0	2.6	1.8	1.5	1.4	1.3	1.2	1.2	1.1	1.1
15	1.3	0.77	0.60	0.48	0.41	0.30	0.29	0.28	0.25

For hood loss coefficient with screen at section 1, see Note 1.

1-7 Hood, Tapered, Flanged or Unflanged[2]

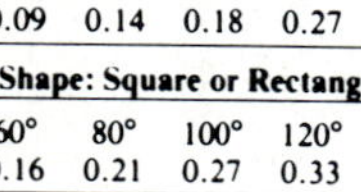

$A_1 \geqslant 2A_o$

θ is major angle for rectangular hoods

Hood Shape: Round

θ	0°	20°	40°	60°	80°	100°	120°	140°	160°	180°
C_o	1.0	0.11	0.06	0.09	0.14	0.18	0.27	0.32	0.43	0.50

Hood Shape: Square or Rectangular

θ	0°	20°	40°	60°	80°	100°	120°	140°	160°	180°
C_o	1.0	0.19	0.13	0.16	0.21	0.27	0.33	0.43	0.53	0.62

1-8 Air Inlet Openings on Side of Duct

a. First Opening[1]

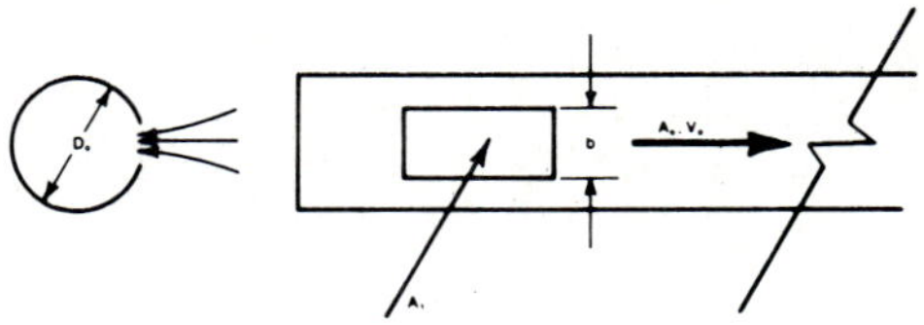

b/D_o	C_o A_1/A_o 0.1	0.2	0.3	0.4	0.5	0.6	0.7	0.8
0.13	335	85	—	—	—	—	—	—
0.26	305	85	42	23	16	12	—	—
0.38	280	79	38	23	16	12	9.3	6.4
0.48	260	75	36	22	15	11	8.8	6.9
0.62	235	61	33	20	14	10	8.0	6.5
0.70	230	63	30	18	13	9.4	7.4	6.0

b/D_o	C_o A_1/A_o 0.9	1.0	1.1	1.2	1.3	1.4	1.5	1.6
0.13	—	—	—	—	—	—	—	—
0.26	—	—	—	—	—	—	—	—
0.38	5.4	—	—	—	—	—	—	—
0.48	4.2	3.4	3.8	—	—	—	—	—
0.62	4.0	3.3	2.8	2.5	2.3	2.2	2.1	—
0.70	4.9	3.0	2.5	2.2	2.0	1.8	1.7	1.6

b. Succeeding Openings[3]

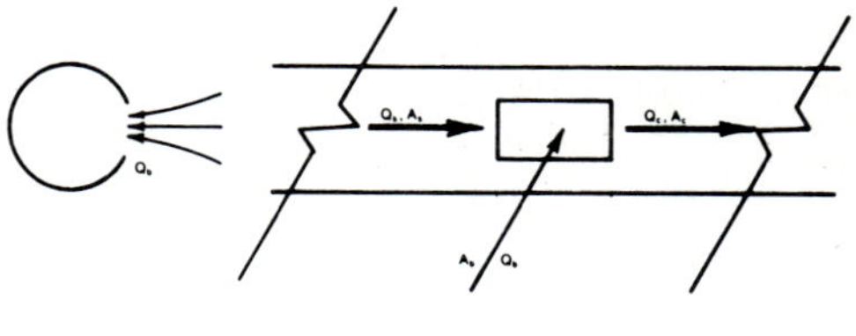

A_b/A_c	Branch (Opening), C_b Q_b/Q_c 0.1	0.2	0.3	0.4	0.5
0.1	0.8	1.3	1.4	1.4	1.4
0.2	−1.4	0.9	1.3	1.4	1.4
0.4	−10	0.2	0.9	1.2	1.3
0.6	−21	−2.5	0.3	1.0	1.2

A_b/A_c	Main, C_s Q_b/Q_c 0.1	0.2	0.3	0.4	0.5
0.1	0.1	−.1	−.8	−2.6	−6.6
0.2	0.1	0.2	0	−.6	−2.1
0.4	0.2	0.3	0.3	0.2	−.2
0.6	0.2	0.3	0.4	0.4	0.3

Table B-2 Local Loss Coefficients, EXITS

2-1 Exit, Abrupt, with Screen[1]

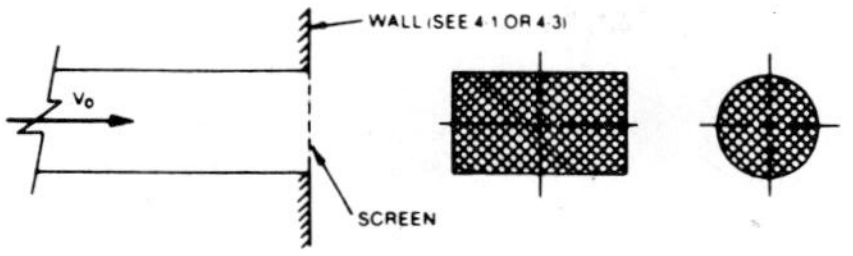

$C_o = 1 + C_s$, *where* C_s is obtained from 7-8.

If no screen is present, C_o is a function of velocity distribution and varies from 1 to 4.

2-2 Exit, Duct Flush with Wall, Flow along Wall[1]

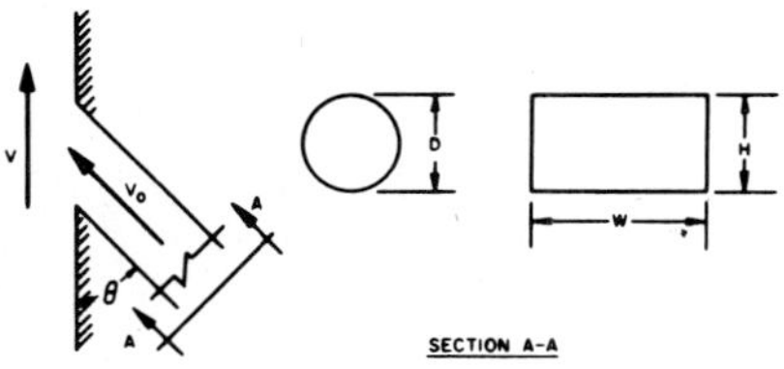

a. Round

	C_o				
	V/V_o				
θ, deg	0	0.5	1.0	1.5	2.0
30-45	1.0	1.0	1.1	1.3	1.6
60	1.0	0.90	1.1	1.4	1.6
90	1.0	0.80	0.95	1.4	1.7
120	1.0	0.80	0.95	1.3	1.7
150	1.0	0.82	0.83	1.0	1.3

b. Rectangular

		C_o				
		V/V_o				
Aspect Ratio (H/W)	θ, deg	0	0.5	1.0	1.5	2.0
0.1-0.2	30-90	1.0	0.95	1.2	1.5	1.8
	120	1.0	0.92	1.1	1.4	1.9
	150	1.0	0.75	0.95	1.4	1.8
0.5-2.0	30-45	1.0	1.0	1.1	1.3	1.6
	60	1.0	0.90	1.1	1.4	1.6
	90	1.0	0.80	0.95	1.4	1.7
	120	1.0	0.80	0.95	1.3	1.7
	150	1.0	0.82	0.83	1.0	1.3
5-10	45	1.0	0.92	0.93	1.1	1.3
	60	1.0	0.87	0.87	1.0	1.3
	90	1.0	0.82	0.80	0.97	1.2
	120	1.0	0.80	0.76	0.90	0.98

For loss coefficient with screen at outlet, see Note 2.

2-3 Exit, Conical, Round with Duct Upstream[1]

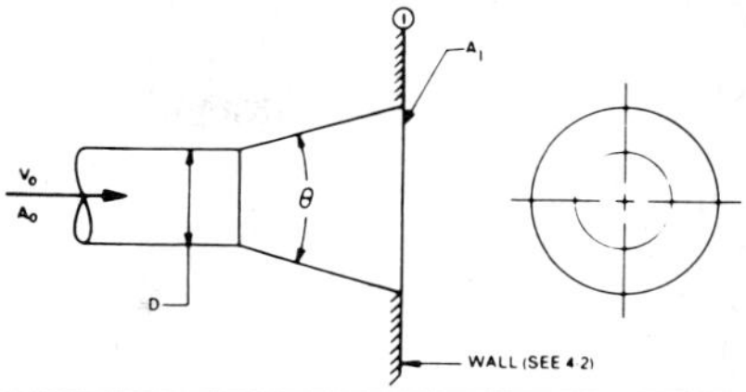

	C_o						
	θ, degrees						
A_1/A_o	8	10	14	20	30	45	≥60
2	0.36	0.33	0.37	0.51	0.90	1.0	1.0
4	0.24	0.21	0.28	0.40	0.70	0.99	1.0
6	0.20	0.19	0.26	0.37	0.67	0.99	1.0
10	0.18	0.16	0.24	0.36	0.68	0.99	1.0
16	0.16	0.16	0.20	0.36	0.66	0.99	1.0

For loss coefficient with screen at section 1, see Note 1

2-4 Exit, Plane Diffuser, Rectangular with Duct Upstream[1]

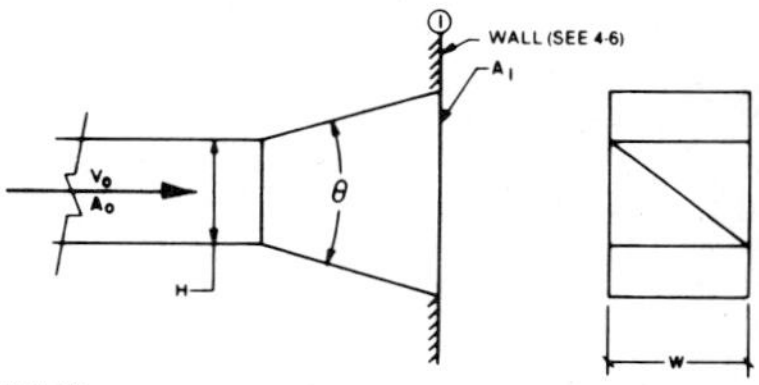

$0.5 \leq H/W \leq 2$

	C_o						
	θ, degrees						
A_1/A_o	8	10	14	20	30	34	≥60
2	0.50	0.51	0.56	0.63	0.80	0.96	1.0
4	0.34	0.38	0.48	0.63	0.76	0.91	1.0
6	0.32	0.34	0.41	0.56	0.70	0.84	0.96

For loss coefficient with screen at section 1, see Note 1.

2-5 Exit, Pyramidal Diffuser, with or without a Wall, Rectangular with Duct Upstream[1]

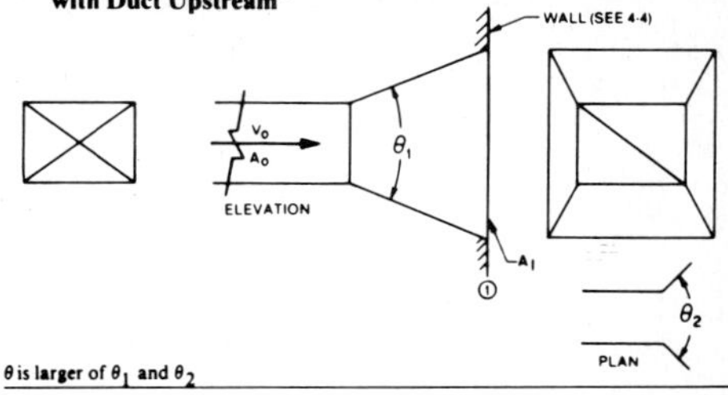

θ is larger of θ_1 and θ_2

	C_o						
	θ, degrees						
A_1/A_o	8	10	14	20	30	45	≥60
2	0.65	0.68	0.74	0.82	0.92	1.1	1.1
4	0.53	0.60	0.69	0.78	0.90	1.0	1.1
6	0.50	0.57	0.66	0.77	0.91	1.0	1.1
10	0.45	0.53	0.64	0.74	0.85	0.97	1.1

For loss coefficient with screen at section 1, see Note 1.

2-6 Air Discharge Openings on Side of Duct
a. First Opening[1]

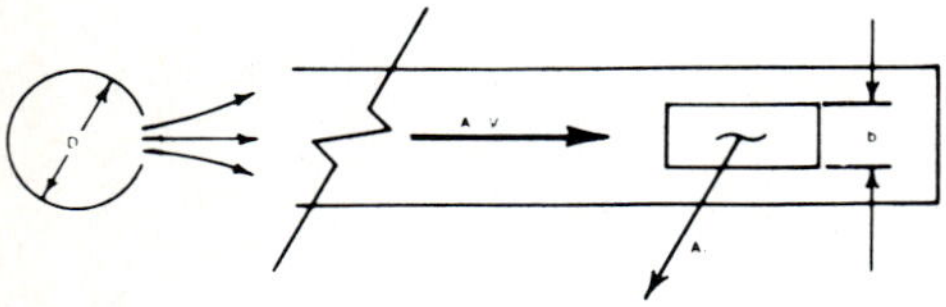

$\frac{b}{D_o}$	C_o A_1/A_o 0.1	0.2	0.3	0.4	0.5	0.6	0.7	0.8
0.13	253	63	28	16	—	—	—	—
0.26	248	62	28	16	10	7.2	5.4	—
0.38	244	61	27	16	10	7.1	5.4	4.3
0.48	240	60	27	15	9.9	7.0	5.3	4.2
0.62	228	57	26	15	9.6	6.8	5.2	4.1
0.70	220	55	25	14	9.3	6.6	5.1	4.0

$\frac{b}{D_o}$	C_o A_1/A_o 0.9	1.0	1.1	1.2	1.3	1.4	1.5	1.6
0.13	—	—	—	—	—	—	—	—
0.26	—	—	—	—	—	—	—	—
0.38	3.6	3.0	—	—	—	—	—	—
0.48	3.5	2.9	2.9	2.6	2.3	—	—	—
0.62	3.4	2.9	2.5	2.2	2.0	1.9	1.7	—
0.70	3.3	2.8	2.4	2.2	1.9	1.8	1.7	1.6

b. Succeeding Openings[3]

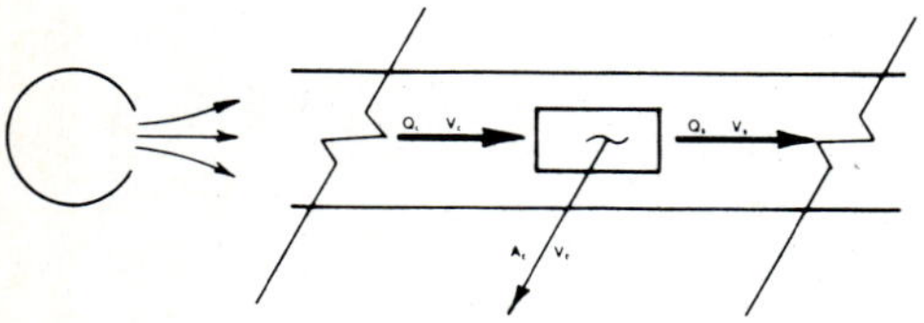

Branch (Opening), C_b									
V_b/V_c	0.4	0.6	0.8	1.0	1.2	1.4	1.6	1.8	2.0
C_b	1.8	1.7	1.7	1.8	1.9	2.1	2.3	2.6	3.0

Main, C_s					
V_s/V_c	0.4	0.5	0.6	0.8	1.0
C_s	0.06	0.01	−.03	−.06	−.03

2-7 Exit, Discharge to Atmosphere from a 90°Elbow, Rectangular and Round (Note: Elbow Loss Included)[1]

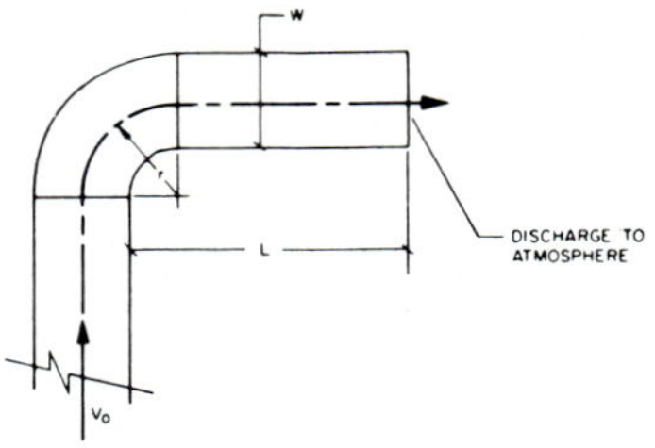

a. Rectangular

$\frac{r}{W}$	C_o L/W 0	0.5	1.0	1.5	2.0	3.0	4.0	6.0	8.0	12.0
0	3.0	3.1	3.2	3.0	2.7	2.4	2.2	2.1	2.1	2.0
0.75	2.2	2.2	2.1	1.8	1.7	1.6	1.6	1.5	1.5	1.5
1.0	1.8	1.5	1.4	1.4	1.3	1.3	1.2	1.2	1.2	1.2
1.5	1.5	1.2	1.1	1.1	1.1	1.1	1.1	1.1	1.1	1.1
2.5	1.2	1.1	1.1	1.0	1.0	1.0	1.0	1.0	1.0	1.0

b. Round (r/D = 1.0)

L/D	0.9	1.3
C_o	1.5	1.4

For loss coefficient with screen at outlet, see Note 2.

2-8 Plane Asymmetric Diffuser at Fan Outlet Without Ductwork[1]

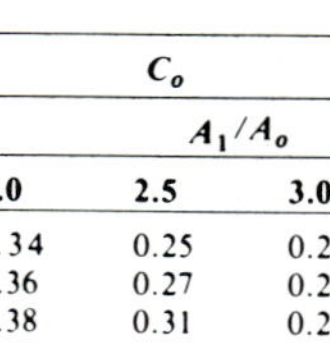

θ deg	C_o A_1/A_o 1.5	2.0	2.5	3.0	3.5	4.0
10	0.51	0.34	0.25	0.21	0.18	0.17
15	0.54	0.36	0.27	0.24	0.22	0.20
20	0.55	0.38	0.31	0.27	0.25	0.24
25	0.59	0.43	0.37	0.35	0.33	0.33
30	0.63	0.50	0.46	0.44	0.43	0.42
35	0.65	0.56	0.53	0.52	0.51	0.50

For loss coefficient with screen at section 1, see Note 1.

2-9 Pyramidal Diffuser at Fan Outlet Without Ductwork[1]

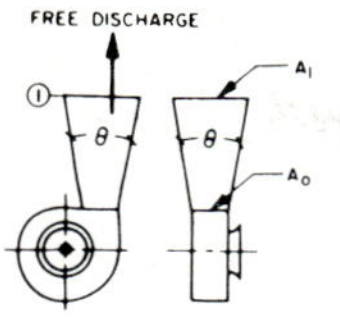

θ, deg	C_o A_1/A_o 1.5	2.0	2.5	3.0	3.5	4.0
10	0.54	0.42	0.37	0.34	0.32	0.31
15	0.67	0.58	0.53	0.51	0.50	0.51
20	0.75	0.67	0.65	0.64	0.64	0.65
25	0.80	0.74	0.72	0.70	0.70	0.72
30	0.85	0.78	0.76	0.75	0.75	0.76

For loss coefficient with screen at section 1, see Note 1.

2-10 Exhaust Hood[1]

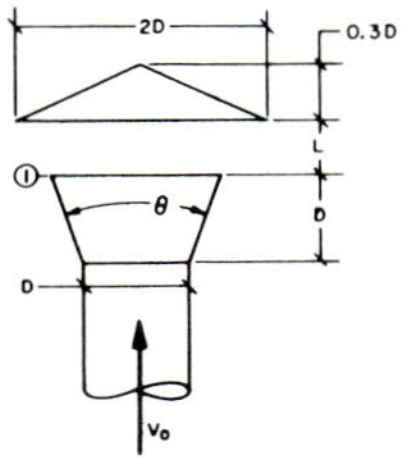

θ, deg	C_o L/D 0.1	0.2	0.25	0.3	0.35	0.4	0.5	0.6	0.8	1.0
0	4.0	2.3	1.9	1.6	1.4	1.3	1.2	1.1	1.0	1.0
15	2.6	1.2	1.0	0.80	0.70	0.65	0.60	0.60	0.60	0.60

For hood loss coefficient with screen at section 1, see Note 1.

2-11 Discharge through Nozzle (Stackheads)[1]

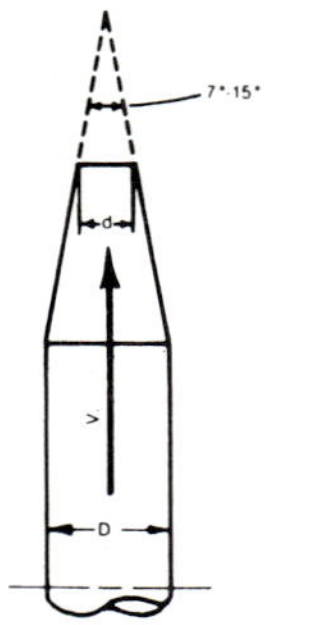

C_o	0.3	0.4	0.5	0.6	0.7	0.8	0.9	1.0
d/D_o	130	41	17	8.1	4.4	2.6	1.6	1.0

Table B-3 Local Loss Coefficients, ELBOWS

3-1 Elbow, Smooth Radius (Die Stamped), Round[1,4]

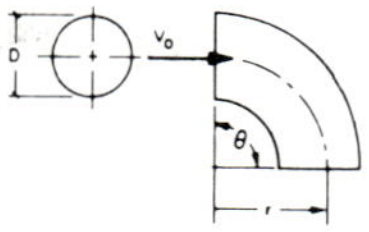

$C_o = K_\theta C_o'$

Coefficients for 90° Elbows:

r/D	0.5	0.75	1.0	1.5	2.0	2.5
C_o'	0.71	0.33	0.22	0.15	0.13	0.12

For angle correction factors K_θ, see Note 3.

3-2 Elbow, 90°; 3, 4 & 5-Pieces, Round[4]

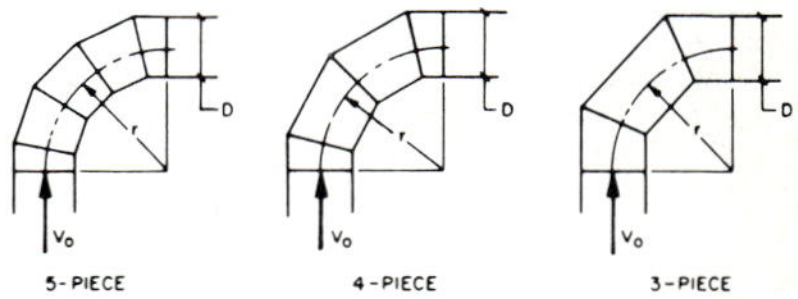

$C_o = K_\theta C_o'$

Coefficients for 90° Elbows:

No. of Pieces	C_o r/D 0.5	0.75	1.0	1.5	2.0
5	—	0.46	0.33	0.24	0.19
4	—	0.50	0.37	0.27	0.24
3	0.90	0.54	0.42	0.34	0.33

For angle correction factors K_θ, see Note 3.

3-3 Elbow, Mitered, Round[1]

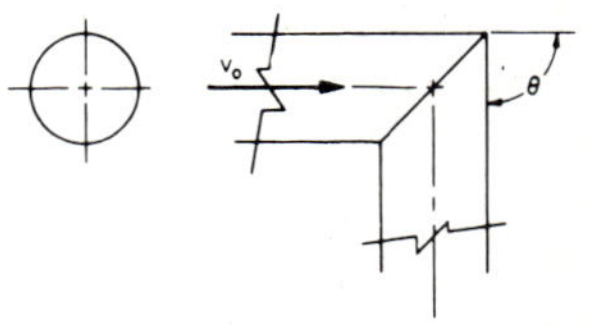

$C_o = K_{Re} C_o'$

θ, deg	20	30	45	60	75	90
C_o'	0.08	0.16	0.34	0.55	0.81	1.2

For Reynolds number correction factors K_{Re}, see Note 5.

3-4 Elbows, 30°, Z-Shaped, Round[1]

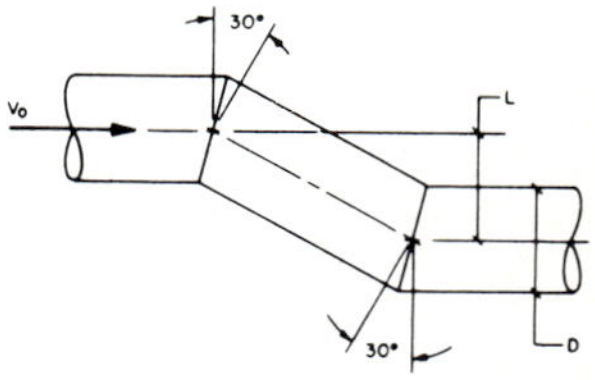

$C_o = K_{Re} C_o'$

L/D	0	0.5	1.0	1.5	2.0	2.5	3.0
C_o'	0	0.15	0.15	0.16	0.16	0.16	0.16

For Reynolds number correction factors K_{Re}, see Note 5.

3-5 Elbow, Smooth Radius without Vanes, Rectangular[1]

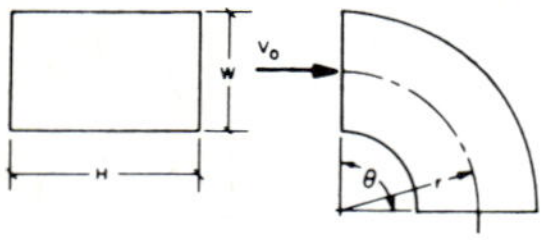

$C_o = K_\theta K_{Re} C_o'$

Coefficients for 90° Elbows:

	C_o'										
	H/W										
r/W	0.25	0.5	0.75	1.0	1.5	2.0	3.0	4.0	5.0	6.0	8.0
0.5	1.3	1.3	1.2	1.2	1.1	1.1	0.98	0.92	0.89	0.85	0.83
0.75	0.57	0.52	0.48	0.44	0.40	0.39	0.39	0.40	0.42	0.43	0.44
1.0	0.27	0.25	0.23	0.21	0.19	0.18	0.18	0.19	0.20	0.27	0.21
1.5	0.22	0.20	0.19	0.17	0.15	0.14	0.14	0.15	0.16	0.17	0.17
2.0	0.20	0.18	0.16	0.15	0.14	0.13	0.13	0.14	0.14	0.15	0.15

	K_{Re}								
	$Re \cdot 10^{-4}$								
r/W	1	2	3	4	6	8	10	14	≥20
0.5	1.40	1.26	1.19	1.14	1.09	1.06	1.04	1.0	1.0
≥0.75	2.0	1.77	1.64	1.56	1.46	1.38	1.30	1.15	1.0

For elbows with angles other than 90°, see Note 3. To calculate *Re*, see Note 4.

3-6 Elbow, Mitered, Rectangular[1]

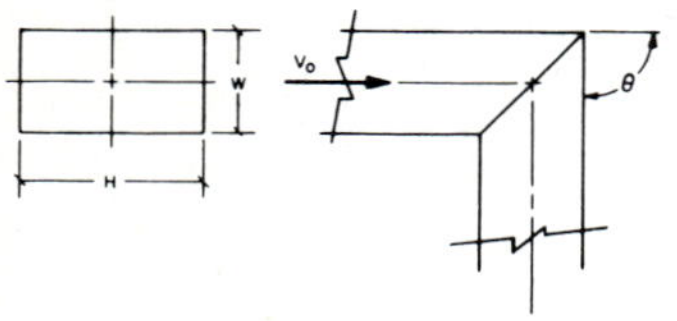

$C_o = K_{Re} C_o'$

	H/W										
θ, deg	0.25	0.5	0.75	1.0	1.5	2.0	3.0	4.0	5.0	6.0	8.0
20	0.08	0.08	0.08	0.07	0.07	0.07	0.06	0.06	0.05	0.05	0.05
30	0.18	0.17	0.17	0.16	0.15	0.15	0.13	0.13	0.12	0.12	0.11
45	0.38	0.37	0.36	0.34	0.33	0.31	0.28	0.27	0.26	0.25	0.24
60	0.60	0.59	0.57	0.55	0.52	0.49	0.46	0.43	0.41	0.39	0.38
75	0.89	0.87	0.84	0.81	0.77	0.73	0.67	0.63	0.61	0.58	0.57
90	1.3	1.3	1.2	1.2	1.1	1.1	0.98	0.92	0.89	0.85	0.83

For Reynolds number correction factors K_{Re}, see Note 5.

3-7 Elbow, Smooth Radius with Splitter Vanes, Rectangular[4,5]

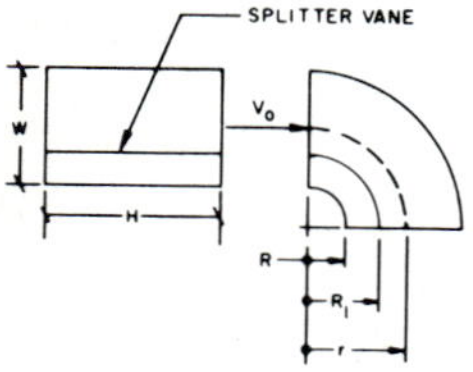

a. 1 Splitter Vane

$C_o = K_\theta C_o'$
$R_1 = R/CR$

where

R = throat radius
R_1 = splitter vane radius
CR = 'CURVE RATIO' (values from Table 3.7.a)
K_θ = angle factor (see Note 3 for values)

Table 3-7.a Coefficients for elbows with 1 splitter vane:

			C_o'										
			H/W										
R/W	r/W	CR	0.25	0.5	1.0	1.5	2.0	3.0	4.0	5.0	6.0	7.0	8.0
0.05	0.55	0.218	0.52	0.40	0.43	0.49	0.55	0.66	0.75	0.84	0.93	1.0	1.1
0.10	0.60	0.302	0.36	0.27	0.25	0.28	0.30	0.35	0.39	0.42	0.46	0.49	0.52
0.15	0.65	0.361	0.28	0.21	0.18	0.19	0.20	0.22	0.25	0.26	0.28	0.30	0.32
0.20	0.70	0.408	0.22	0.16	0.14	0.14	0.15	0.16	0.17	0.18	0.19	0.20	0.21
0.25	0.75	0.447	0.18	0.13	0.11	0.11	0.11	0.12	0.13	0.14	0.14	0.15	0.15
0.30	0.80	0.480	0.15	0.11	0.09	0.09	0.09	0.09	0.10	0.10	0.11	0.11	0.12
0.35	0.85	0.509	0.13	0.09	0.08	0.07	0.07	0.08	0.08	0.08	0.08	0.09	0.09
0.40	0.90	0.535	0.11	0.08	0.07	0.06	0.06	0.06	0.06	0.07	0.07	0.07	0.07
0.45	0.95	0.557	0.10	0.07	0.06	0.05	0.05	0.05	0.05	0.05	0.06	0.06	0.06
0.50	1.00	0.577	0.09	0.06	0.05	0.05	0.04	0.04	0.04	0.05	0.05	0.05	0.05

b. 2 Splitter Vanes

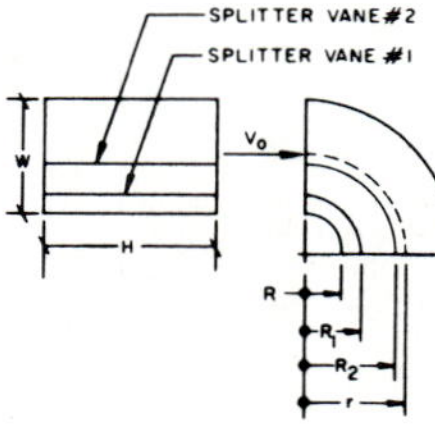

$C_o = K_\theta C_o'$
$R_1 = R/CR$
$R_2 = R_1/CR = R/CR^2$

where

R = throat radius
R_1 = splitter vane #1 radius
R_2 = splitter vane #2 radius
CR = 'CURVE RATIO' (values from Table 3.7.b)
K_θ = angle factor (see Note 3 for values)

Table 3-7.b Coefficients for elbows with 2 splitter vanes:

			C_o'										
			H/W										
R/W	r/W	CR	0.25	0.5	1.0	1.5	2.0	3.0	4.0	5.0	6.0	7.0	8.0
0.05	0.55	0.362	0.26	0.20	0.22	0.25	0.28	0.33	0.37	0.41	0.45	0.48	0.51
0.10	0.60	0.450	0.17	0.13	0.11	0.12	0.13	0.15	0.16	0.17	0.19	0.20	0.21
0.15	0.65	0.507	0.12	0.09	0.08	0.08	0.08	0.09	0.10	0.10	0.11	0.11	0.11
0.20	0.70	0.550	0.09	0.07	0.06	0.05	0.06	0.06	0.06	0.06	0.07	0.07	0.07
0.25	0.75	0.585	0.08	0.05	0.04	0.04	0.04	0.04	0.05	0.05	0.05	0.05	0.05
0.30	0.80	0.613	0.06	0.04	0.03	0.03	0.03	0.03	0.03	0.03	0.04	0.04	0.04

c. 3 Splitter Vanes

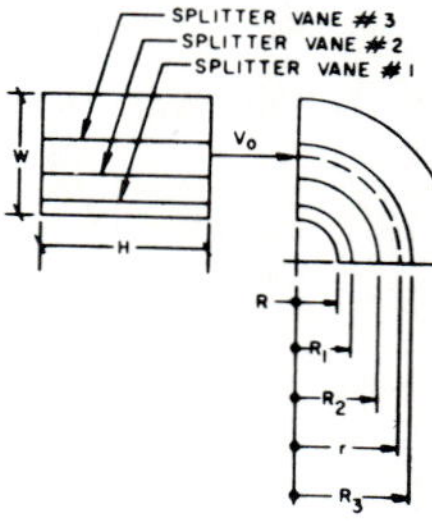

$C_o = K_\theta C_o'$
$R_1 = R/CR$
$R_2 = R_1/CR = R/CR^2$
$R_3 = R_2/CR = R/CR^3$

where

R = throat radius
R_1 = splitter vane #1 radius
R_2 = splitter vane #2 radius
R_3 = splitter vane #3 radius
CR = 'CURVE RATIO' (values from Table 3.7.c)
K_θ = angle factor (see Note 3 for values)

Table 3-7.c Coefficients for elbow with 3 splitter vanes:

			C_o'										
			H/W										
R/W	r/W	CR	0.25	0.5	1.0	1.5	2.0	3.0	4.0	5.0	6.0	7.0	8.0
0.05	0.55	0.467	0.11	0.10	0.12	0.13	0.14	0.16	0.18	0.19	0.21	0.22	0.23
0.10	0.60	0.549	0.07	0.05	0.06	0.06	0.06	0.07	0.07	0.08	0.08	0.08	0.09

3-8 Elbow, Mitered, with Single-Thickness Vanes, Rectangular[6]

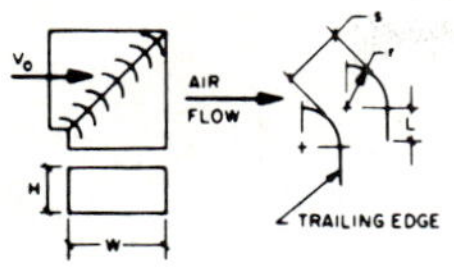

Design No.	Dimensions, in. r	s	L	C_o
1[a]	2.0	1.5	0.75	0.12
2	4.5	2.25	0	0.15
3	4.5	3.25	1.60	0.18

[a]When extension of trailing edge is not provided for this vane, losses are approximately unchanged for single elbows, but increase considerably for elbows in series.

3-9 Elbow, Mitered, with Double-Thickness Vanes, Rectangular[6]

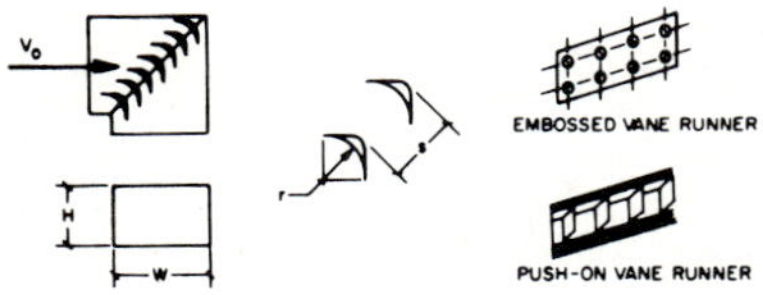

Design No.	Dimensions, in. r	s	C_o Velocity V_o, fpm 1000	2000	3000	4000	Remarks
1	2.0	1.5	0.27	0.22	0.19	0.17	Embossed Vane Runner
2	2.0	1.5	0.33	0.29	0.26	0.23	Push-On Vane Runner
3	2.0	2.13	0.38	0.31	0.27	0.24	Embossed Vane Runner
4	4.5	3.25	0.26	0.21	0.18	0.16	Embossed Vane Runner

3-10 Elbow, Converging or Diverging Flow, Rectangular[1]

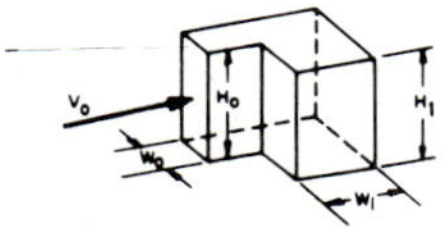

$C_o = K_{Re} C_o'$

H_o/W_o	C_o', W_1/W_o = 0.6	0.8	1.2	1.4	1.6	2.0
0.25	1.8	1.4	1.1	1.1	1.1	1.1
1.0	1.7	1.4	1.0	0.95	0.90	0.84
4.0	1.5	1.1	0.81	0.76	0.72	0.66
∞	1.5	1.0	0.69	0.63	0.60	0.55

For Reynolds number correction factors K_{Re}, See Note 5.

3-11 Elbows, 90°, Z-Shaped, Rectangular[1]

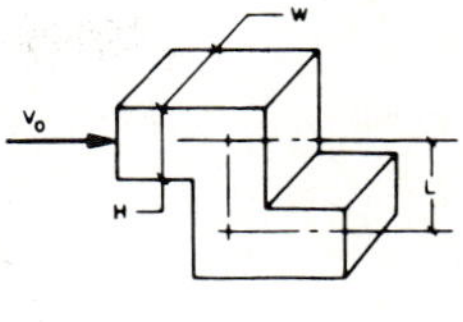

$C_o = K\,K_{Re} C_o'$

Coefficients for W/H = 1.0:

L/H	0	0.4	0.6	0.8	1.0	1.2	1.4	1.6	1.8	2.0
C_o'	0	0.62	0.90	1.6	2.6	3.6	4.0	4.2	4.2	4.2
L/H	2.4	2.8	3.2	4.0	5.0	6.0	7.0	9.0	10.0	∞
C_o'	3.7	3.3	3.2	3.1	2.9	2.8	2.7	2.6	2.5	2.3

For W/H values other than 1.0, apply the following factor:

W/H	0.25	0.50	0.75	1.0	1.5	2.0	3.0	4.0	6.0	8.0
K	1.10	1.07	1.04	1.0	0.95	0.90	0.83	0.78	0.72	0.70

For Reynolds number correction factors K_{Re}, see Note 5.

3-12 Combined 90° Elbows Lying in Different Planes, Rectangular[1]

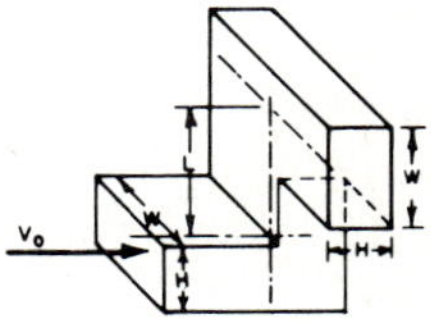

$C_o = K\,K_{Re} C_o'$

Coefficients for Square Ducts:

L/W	0	0.4	0.6	0.8	1.0	1.2	1.4	1.6	1.8	2.0
C_o'	1.2	2.4	2.9	3.3	3.4	3.4	3.4	3.3	3.2	3.1
L/W	2.4	2.8	3.2	4.0	5.0	6.0	7.0	9.0	10.0	∞
C_o'	3.2	3.2	3.2	3.0	2.9	2.8	2.7	2.5	2.4	2.3

Apply Following Factor for other than H/W=1.0:

H/W	0.25	0.50	0.75	1.0	1.5	2.0	3.0	4.0	6.0	8.0
K	1.10	1.07	1.04	1.0	0.95	0.90	0.83	0.78	0.72	0.70

For Reynolds number correction factors K_{Re}, see Note 5.

3-13 Offset, S-Shaped (Gooseneck), Rectangular and Round[1]

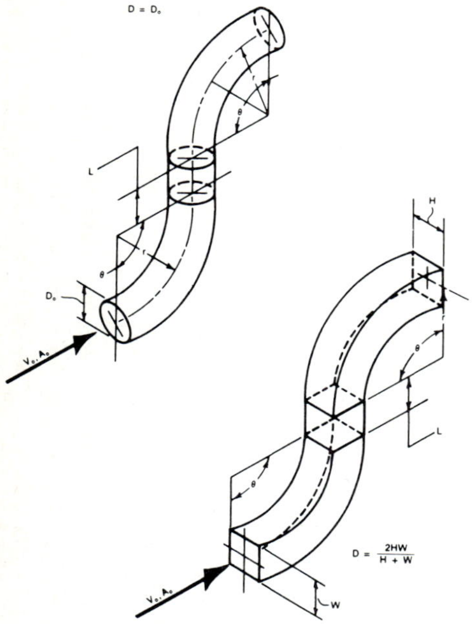

$$C_o = K\,C_o'$$

where

C_o = offset loss coefficient

C_o' = single elbow loss coefficient (see Fittings 3-1, 3-2, 3-5 and 3-7)

	K							
	L/D							
θ, deg	0	1	2	3	4	6	8	10
15	0.20	0.42	0.60	0.78	0.94	1.16	1.20	1.15
30	0.40	0.65	0.88	1.16	1.20	1.18	1.12	1.06
45	0.60	1.06	1.20	1.23	1.20	1.08	1.03	1.08
60	1.05	1.38	1.37	1.28	1.15	1.06	1.16	1.30
75	1.50	1.58	1.46	1.30	1.27	1.30	1.37	1.47
90	1.70	1.67	1.40	1.37	1.38	1.47	1.55	1.63

	K						
	L/D						
θ, deg	12	14	16	18	20	25	≥40
15	1.08	1.05	1.02	1.00	1.10	1.25	2.0
30	1.06	1.15	1.28	1.40	1.50	1.70	2.0
45	1.17	1.30	1.42	1.55	1.65	1.80	2.0
60	1.42	1.54	1.66	1.76	1.85	1.95	2.0
75	1.57	1.68	1.75	1.80	1.88	1.97	2.0
90	1.70	1.76	1.82	1.88	1.92	1.98	2.0

3-14 Offset, S-Shaped in Two Planes 90° Apart, Rectangular and Round[1]

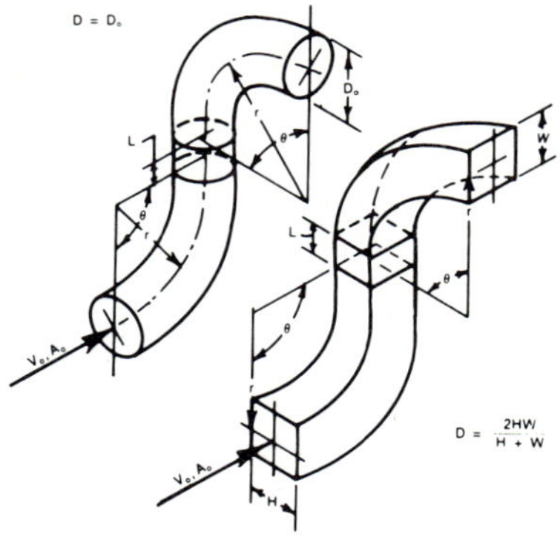

$$C_o = K\,C_o'$$

where

C_o = offset loss coefficient

C_o' = single elbow loss coefficient (see Fittings 3-1, 3-2, 3-5 and 3-7)

	K						
	L/D						
θ, deg	0	1	2	3	4	6	8
60	2.0	1.90	1.50	1.35	1.30	1.20	1.25
90	2.0	1.80	1.60	1.55	1.55	1.65	1.80

	K					
	L/D					
θ, deg	10	12	14	20	25	40
60	1.50	1.63	1.73	1.85	1.95	2.0
90	1.90	1.93	1.98	2.0	2.0	2.0

3-15 Elbows (4), 45°, Smooth Radius, Rectangular, Arranged to Go Around an Obstruction[7]

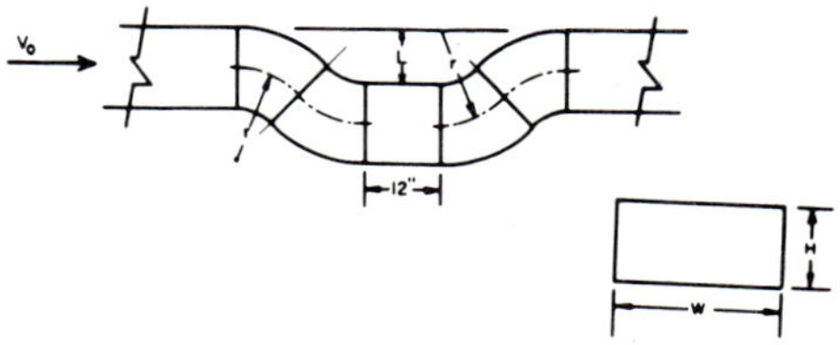

$W/H = 4, r/H = 1.5, L = 1.5H$

V_o, fpm	800	1200	1600	2000	2400
C_o	0.18	0.22	0.24	0.25	0.26

Table B-4 Local Loss Coefficients, TRANSITIONS (Diverging Flow)

4-1 Conical Diffuser, Round[1]

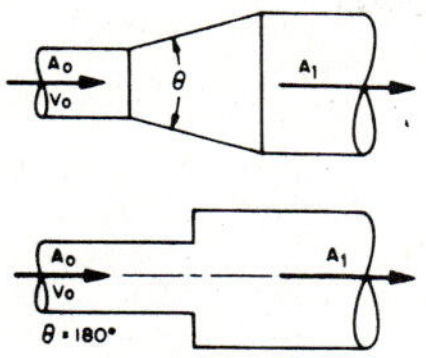

	C_o									
	θ, degrees									
A_1/A_o	8	12	16	20	30	45	60	90	120	180
2	0.11	0.11	0.14	0.19	0.32	0.33	0.33	0.32	0.31	0.30
4	0.15	0.17	0.23	0.30	0.46	0.61	0.68	0.64	0.63	0.62
6	0.17	0.20	0.27	0.33	0.48	0.66	0.77	0.74	0.73	0.72
10	0.19	0.23	0.29	0.38	0.59	0.76	0.80	0.83	0.84	0.83
≥16	0.19	0.22	0.31	0.38	0.60	0.84	0.88	0.88	0.88	0.88

4-2 Conical Diffuser, Stepped, Round[1]

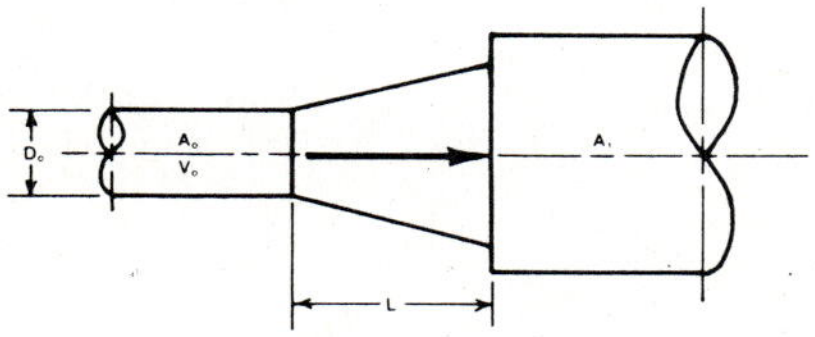

	C_o										
A_1/A_o	L/D_o										
	0.5	1.0	2.0	3.0	4.0	5.0	6.0	8.0	10	12	14
1.5	0.03	0.02	0.03	0.03	0.04	0.05	0.06	0.08	0.10	0.11	0.13
2.0	0.08	0.06	0.04	0.04	0.04	0.05	0.05	0.06	0.08	0.09	0.10
2.5	0.13	0.09	0.06	0.06	0.06	0.06	0.06	0.06	0.07	0.08	0.09
3.0	0.17	0.12	0.09	0.07	0.07	0.06	0.06	0.07	0.07	0.08	0.08
4.0	0.23	0.17	0.12	0.10	0.09	0.08	0.08	0.08	0.08	0.08	0.08
6.0	0.30	0.22	0.16	0.13	0.12	0.10	0.10	0.09	0.09	0.09	0.08
8.0	0.34	0.26	0.18	0.15	0.13	0.12	0.11	0.10	0.09	0.09	0.09
10	0.36	0.28	0.20	0.16	0.14	0.13	0.12	0.11	0.10	0.09	0.09
14	0.39	0.30	0.22	0.18	0.16	0.14	0.13	0.12	0.10	0.10	0.10
20	0.41	0.32	0.24	0.20	0.17	0.15	0.14	0.12	0.11	0.11	0.10

4-3 Pyramidal Diffuser, Rectangular[1]

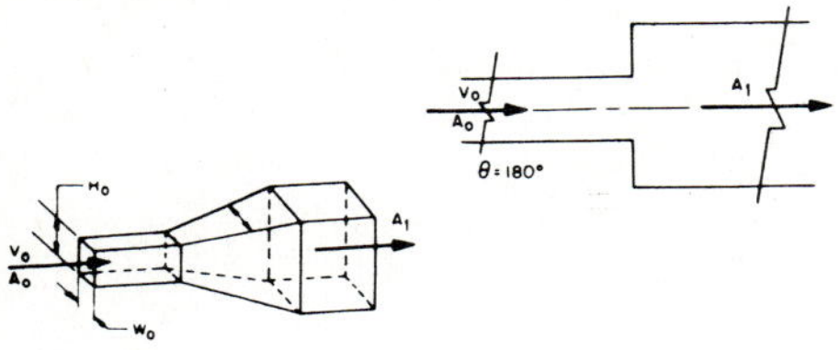

	C_o								
	θ, degrees								
A_1/A_o	8	10	14	20	30	45	60	90	180
2	0.14	0.15	0.20	0.25	0.30	0.33	0.33	0.33	0.30
4	0.20	0.25	0.34	0.45	0.52	0.58	0.62	0.64	0.64
6	0.21	0.30	0.42	0.53	0.63	0.72	0.78	0.79	0.79
≥10	0.24	0.30	0.43	0.53	0.64	0.75	0.84	0.89	0.88

4-4 Pyramidal Diffuser, Stepped, Rectangular[1]

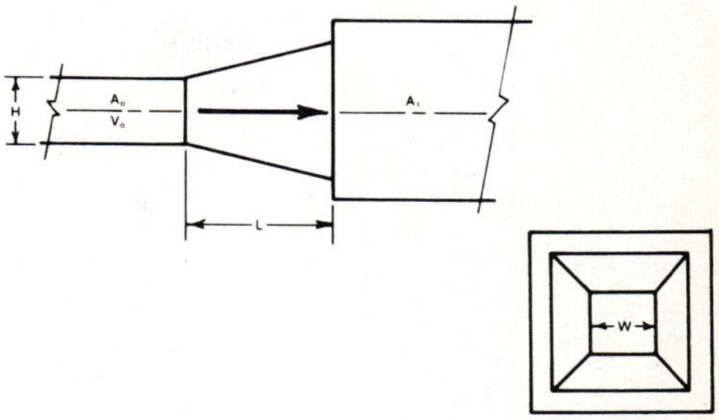

$D = 2HW/(H + W)$

	C_o										
A_1/A_o	L/D										
	0.5	1.0	2.0	3.0	4.0	5.0	6.0	8.0	10	12	14
1.5	0.04	0.03	0.03	0.04	0.05	0.05	0.06	0.08	0.10	0.11	0.13
2.0	0.11	0.08	0.06	0.06	0.06	0.06	0.07	0.07	0.08	0.09	0.10
2.5	0.16	0.13	0.09	0.08	0.08	0.07	0.08	0.07	0.08	0.08	0.09
3.0	0.21	0.17	0.12	0.10	0.09	0.09	0.09	0.09	0.09	0.09	0.09
4.0	0.27	0.22	0.17	0.14	0.12	0.11	0.11	0.11	0.11	0.10	0.10
6.0	0.36	0.28	0.21	0.18	0.16	0.15	0.14	0.13	0.12	0.12	0.11
8.0	0.41	0.32	0.24	0.21	0.18	0.17	0.16	0.14	0.13	0.12	0.12
10	0.44	0.35	0.26	0.22	0.20	0.18	0.17	0.15	0.14	0.13	0.13
14	0.47	0.37	0.28	0.24	0.21	0.20	0.18	0.16	0.15	0.14	0.14
20	0.49	0.40	0.30	0.26	0.23	0.21	0.19	0.17	0.16	0.15	0.14

4-5 Plane Diffuser, Rectangular[1]

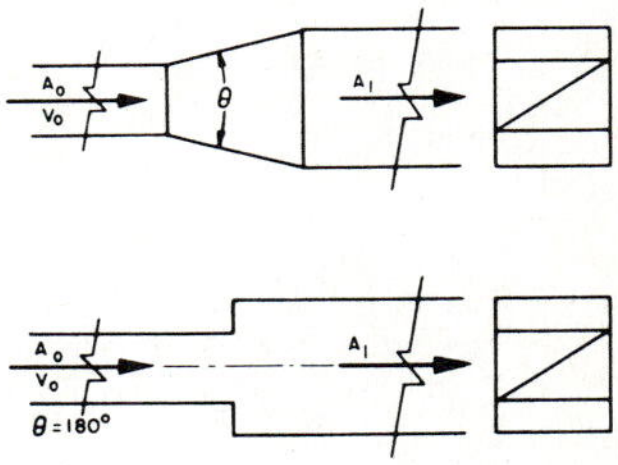

	C_o								
	θ, degrees								
A_1/A_o	8	10	14	20	30	45	60	90	180
2	0.14	0.14	0.13	0.15	0.24	0.35	0.37	0.38	0.35
4	0.19	0.17	0.18	0.25	0.42	0.60	0.68	0.70	0.66
6	0.22	0.21	0.21	0.30	0.48	0.65	0.76	0.83	0.80

4-6 Plane Diffuser, Stepped, Rectangular[1]

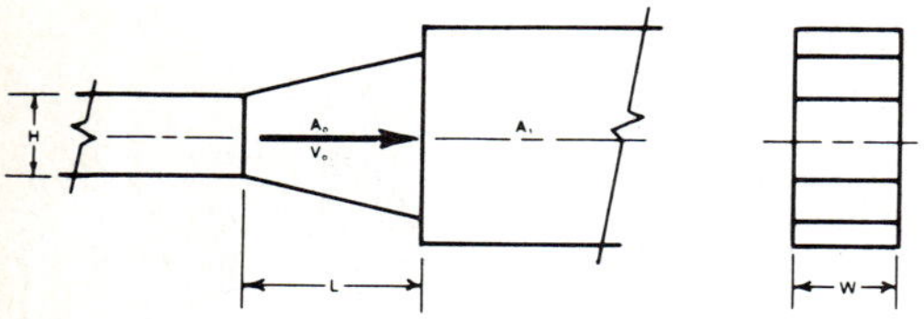

	C_o										
$\frac{A_1}{A_o}$	L/H										
	0.5	1.0	2.0	3.0	4.0	5.0	6.0	8.0	10	12	14
1.5	0.04	0.04	0.04	0.04	0.05	0.06	0.06	0.08	0.10	0.11	0.13
2.0	0.12	0.09	0.07	0.07	0.06	0.07	0.07	0.07	0.08	0.10	0.12
2.5	0.18	0.14	0.11	0.10	0.09	0.09	0.09	0.09	0.09	0.10	0.10
3.0	0.23	0.18	0.14	0.12	0.11	0.11	0.10	0.10	0.10	0.10	0.11
4.0	0.30	0.24	0.19	0.16	0.15	0.14	0.13	0.12	0.12	0.12	0.12
6.0	0.38	0.31	0.25	0.21	0.19	0.18	0.17	0.16	0.15	0.14	0.14
8.0	0.43	0.36	0.28	0.25	0.22	0.20	0.19	0.17	0.16	0.16	0.15
10	0.46	0.38	0.30	0.26	0.24	0.22	0.21	0.19	0.18	0.17	0.16
14	0.50	0.41	0.33	0.29	0.26	0.24	0.22	0.20	0.19	0.18	0.18
20	0.53	0.44	0.35	0.31	0.28	0.25	0.24	0.22	0.20	0.19	0.19

4-7 Transition Diffuser, Round to Rectangular or Rectangular to Round[1]

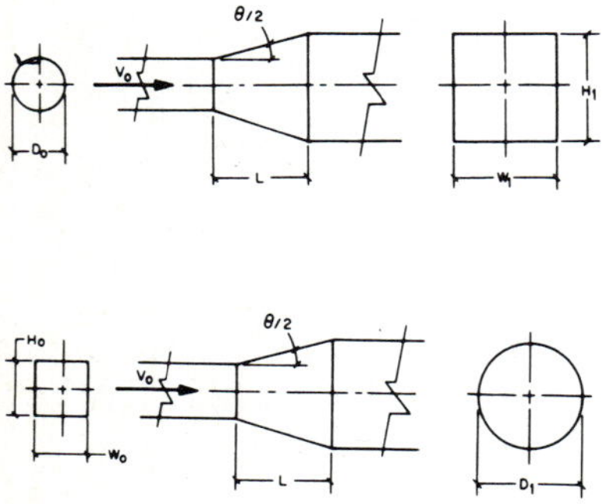

Use C_o from fitting 4-3 based on the following equivalent angles:

1. Transition from round to rectangular:

$$\tan(\theta/2) = (1.13\sqrt{H_1 W_1} - D_o)/2L$$

2. Transition from rectangular to round:

$$\tan(\theta/2) = (D_1 - 1.13\sqrt{H_o W_o})/2L$$

4-8 Plane Symmetric Diffuser at Fan Outlet with Ductwork[1]

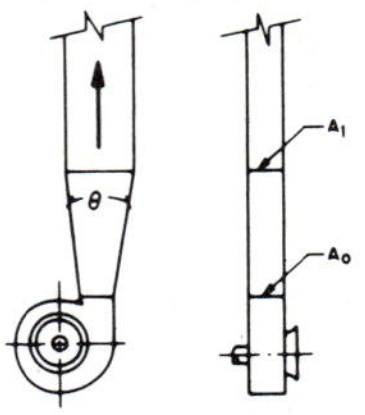

	C_o					
θ, deg	A_1/A_o					
	1.5	2.0	2.5	3.0	3.5	4.0
10	0.05	0.07	0.09	0.10	0.11	0.11
15	0.06	0.09	0.11	0.13	0.13	0.14
20	0.07	0.10	0.13	0.15	0.16	0.16
25	0.08	0.13	0.16	0.19	0.21	0.23
30	0.16	0.24	0.29	0.32	0.34	0.35
35	0.24	0.34	0.39	0.44	0.48	0.50

4-9 Plane Asymmetric Diffuser at Fan Outlet with Ductwork[1]

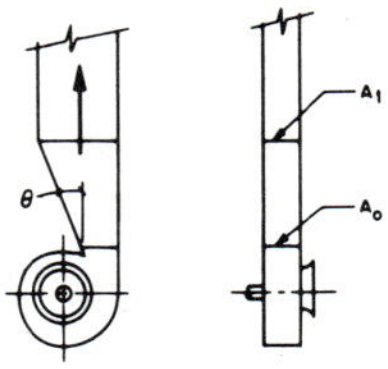

	C_o					
θ, deg	A_1/A_o					
	1.5	2.0	2.5	3.0	3.5	4.0
10	0.08	0.09	0.10	0.10	0.11	0.11
15	0.10	0.11	0.12	0.13	0.14	0.15
20	0.12	0.14	0.15	0.16	0.17	0.18
25	0.15	0.18	0.21	0.23	0.25	0.26
30	0.18	0.25	0.30	0.33	0.35	0.35
35	0.21	0.31	0.38	0.41	0.43	0.44

4-10 Plane Asymmetric Diffuser at Fan Outlet with Ductwork[1]

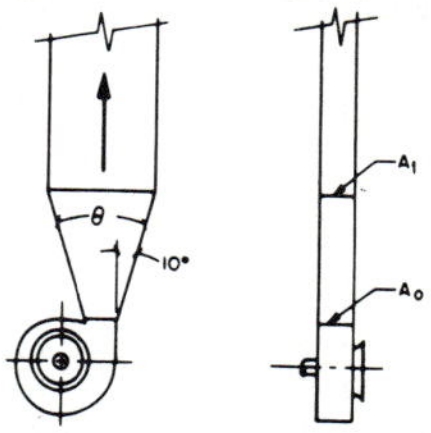

	C_o					
θ, deg	A_1/A_o					
	1.5	2.0	2.5	3.0	3.5	0.4
10°	0.05	0.08	0.11	0.13	0.13	0.14
15°	0.06	0.10	0.12	0.14	0.15	0.15
20°	0.07	0.11	0.14	0.15	0.16	0.16
25°	0.09	0.14	0.18	0.20	0.21	0.22
30°	0.13	0.18	0.23	0.26	0.28	0.29
35°	0.15	0.23	0.28	0.33	0.35	0.36

4-11 Plane Asymmetric Diffuser at Fan Outlet with Ductwork[1]

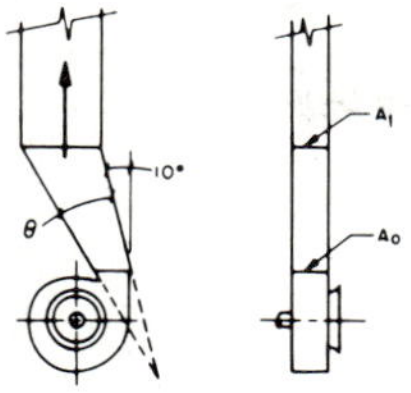

	C_o					
	A_1/A_o					
θ, deg	**1.5**	**2.0**	**2.5**	**3.0**	**3.5**	**4.0**
10	0.11	0.13	0.14	0.14	0.14	0.14
15	0.13	0.15	0.16	0.17	0.18	0.18
20	0.19	0.22	0.24	0.26	0.28	0.30
25	0.29	0.32	0.35	0.37	0.39	0.40
30	0.36	0.42	0.46	0.49	0.51	0.51
35	0.44	0.54	0.61	0.64	0.66	0.66

4-12 Pyramidal Diffuser at Fan Outlet with Ductwork[1]

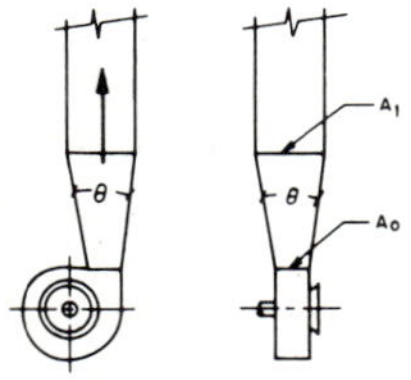

	C_o					
θ,	A_1/A_o					
deg	**1.5**	**2.0**	**2.5**	**3.0**	**3.5**	**4.0**
10	0.10	0.18	0.21	0.23	0.24	0.25
15	0.23	0.33	0.38	0.40	0.42	0.44
20	0.31	0.43	0.48	0.53	0.56	0.58
25	0.36	0.49	0.55	0.58	0.62	0.64
30	0.42	0.53	0.59	0.64	0.67	0.69

Table B-5 Local Loss Coefficients, TRANSITIONS (Converging Flow)

5-1 Contraction, Round & Rectangular[1]

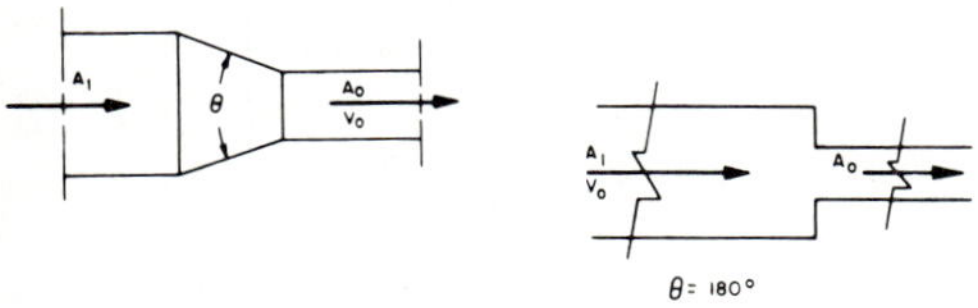

	C_o						
	θ, degrees						
A_1/A_o	**10**	**15-40**	**50-60**	**90**	**120**	**150**	**180**
2	0.05	0.05	0.06	0.12	0.18	0.24	0.26
4	0.05	0.04	0.07	0.17	0.27	0.35	0.41
6	0.05	0.04	0.07	0.18	0.28	0.36	0.42
10	0.05	0.05	0.08	0.19	0.29	0.37	0.43

5-2 Contraction, Conical, Round and Rectangular[1]

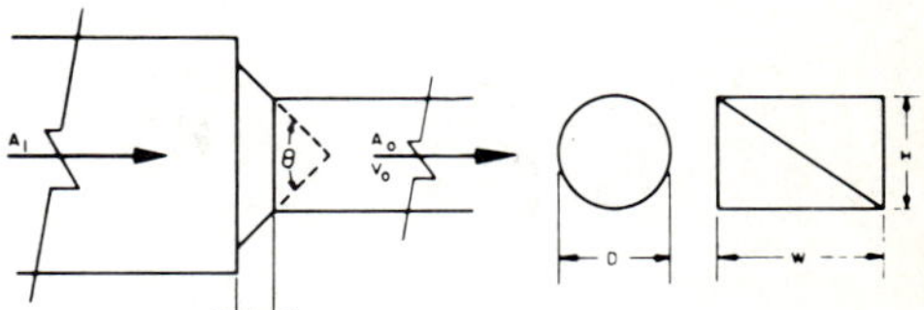

Rectangular: $D = 2HW/(H + W)$
θ = major angle

$C_o = K\, C_o'$

	C_o'								
	θ, degrees								
L/D	**0**	**10**	**20**	**30**	**40**	**60**	**100**	**140**	**180**
0.025	0.50	0.47	0.45	0.43	0.41	0.40	0.42	0.45	0.50
0.05	0.50	0.45	0.41	0.36	0.33	0.30	0.35	0.42	0.50
0.075	0.50	0.42	0.35	0.30	0.26	0.23	0.30	0.40	0.50
0.10	0.50	0.39	0.32	0.25	0.22	0.18	0.27	0.38	0.50
0.10	0.50	0.39	0.32	0.25	0.22	0.18	0.27	0.38	0.50
0.15	0.50	0.37	0.27	0.20	0.16	0.15	0.25	0.37	0.50
0.60	0.50	0.27	0.18	0.13	0.11	0.12	0.23	0.36	0.50

5-3 Contraction, Rectangular Slot to Round[1]

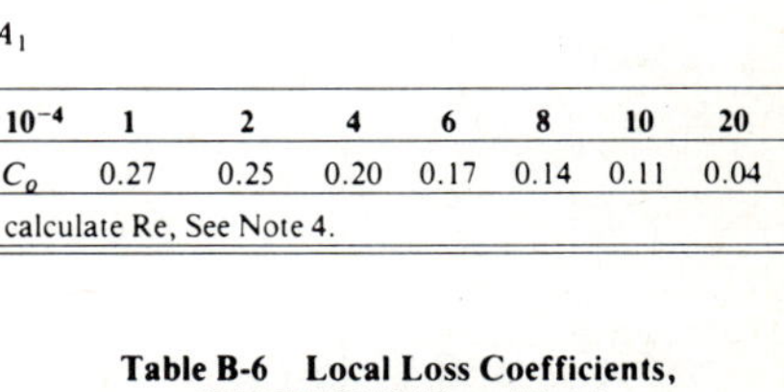

$A_o < A_1$

Re • 10^{-4}	1	2	4	6	8	10	20	≥40
C_o	0.27	0.25	0.20	0.17	0.14	0.11	0.04	0

To calculate Re, See Note 4.

Table B-6 Local Loss Coefficients, JUNCTIONS (Tees, Wyes)

6-1 Converging Wye (30°), Round[1]

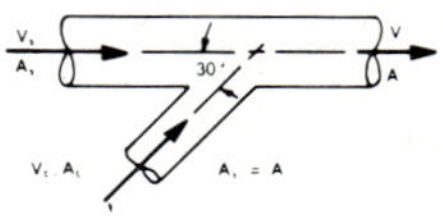

Branch, $C_{c,b}$							
Q_b/Q_c	A_b/A_c						
	0.1	0.2	0.3	0.4	0.6	0.8	1.0
0	-1.0	-1.0	-1.0	-1.0	-1.0	-1.0	-1.0
0.1	0.21	-.46	-.57	-.60	-.62	-.63	-.63
0.2	3.1	0.37	-.06	-.20	-.28	-.30	-.35
0.3	7.6	1.5	0.50	0.20	-.05	-.08	-.10
0.4	14	3.0	1.2	0.59	0.26	0.18	0.16
0.5	21	4.6	1.8	0.97	0.44	0.35	0.27
0.6	30	6.4	2.6	1.4	0.64	0.46	0.31
0.7	41	8.5	3.4	1.8	0.76	0.50	0.40
0.8	54	12	4.2	2.1	0.85	0.53	0.45
0.9	58	14	5.3	2.6	0.89	0.52	0.40
1.0	84	17	6.3	2.9	0.89	0.39	0.27

Main, $C_{c,s}$							
Q_b/Q_c	A_b/A_c						
	0.1	*0.2*	*0.3*	*0.4*	*0.6*	*0.8*	*1.0*
0	0	0	0	0	0	0	0
0.1	0.02	0.11	0.13	0.15	0.16	0.17	0.17
0.2	-.33	0.01	0.13	0.19	0.24	0.27	0.29
0.3	-1.1	-.25	-.01	0.10	0.22	0.30	0.35
0.4	-2.2	-.75	-.30	-.05	0.17	0.26	0.36
0.5	-3.6	-1.4	-.70	-.35	0	0.21	0.32
0.6	-5.4	-2.4	-1.3	-.70	-.20	0.06	0.25
0.7	-7.6	-3.4	-2.0	-1.2	-.50	-.15	0.10
0.8	-10	-4.6	-2.7	-1.8	-.90	-.43	-.15
0.9	-13	-6.2	-3.7	-2.6	-1.4	-.80	-.45
1.0	-16	-7.7	-4.8	-3.4	-1.9	-1.2	-.75

6-2 Converging Wye (45°), Round[8]

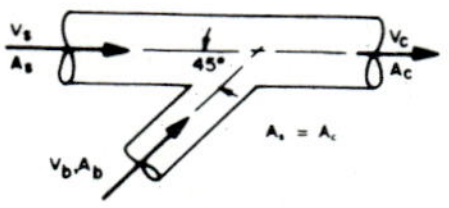

Branch, $C_{c,b}$							
V_b/V_c	A_b/A_c						
	0.1	0.2	0.3	0.4	0.6	0.8	1.0
0.4	-.56	-.44	-.35	-.28	-.15	-.04	0.05
0.5	-.48	-.37	-.28	-.21	-.09	0.02	0.11
0.6	-.38	-.27	-.19	-.12	0	0.10	0.18
0.7	-.26	-.16	-.08	-.01	0.10	0.20	0.28
0.8	-.21	-.02	0.05	0.12	0.23	0.32	0.40
0.9	0.04	0.13	0.21	0.27	0.37	0.46	0.53
1.0	0.22	0.31	0.38	0.44	0.53	0.62	0.69
1.5	1.4	1.5	1.5	1.6	1.7	1.7	1.8
2.0	3.1	3.2	3.2	3.2	3.3	3.3	3.3
2.5	5.3	5.3	5.3	5.4	5.4	5.4	5.4
3.0	8.0	8.0	8.0	8.0	8.0	8.0	8.0

Main, $C_{c,s}$							
V_s/V_c	A_b/A_c						
	0.1	0.2	0.3	0.4	0.6	0.8	1.0
	-8.6	-4.1	-2.5	-1.7	-.97	-.58	-.34
0.2	-6.7	-3.1	-1.9	-1.3	-.67	-.36	-.18
0.3	-5.0	-2.2	-1.3	-.88	-.42	-.19	-.05
0.4	-3.5	-1.5	-.88	-.55	-.21	-.05	0.05
0.5	-2.3	-.95	-.51	-.28	-.06	0.06	0.13
0.6	-1.3	-.50	-.22	-.09	0.05	0.12	0.17
0.7	-.63	-.18	-.03	0.04	0.12	0.16	0.18
0.8	-.18	0.01	0.07	0.10	0.13	0.15	0.17
0.9	0.03	0.07	0.08	0.09	0.10	0.11	0.13
1.0	0.01	0	0	0.10	0.02	0.04	0.05

6-3 Converging Tee, Round[1]

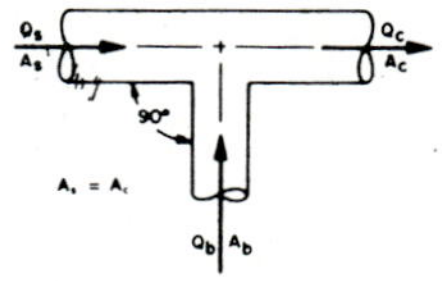

Branch, $C_{c,b}$							
	A_b/A_c						
Q_b/Q_c	0.1	0.2	0.3	0.4	0.6	0.8	1.0
0	-1.0	-1.0	-1.0	-.90	-.90	-.90	-.90
0.1	0.40	-.37	-.51	-.46	-.50	-.51	-.52
0.2	3.8	0.72	0.17	-.02	-.14	-.18	-.24
0.3	9.2	2.3	1.0	0.44	0.21	0.11	-.08
0.4	16	4.3	2.1	0.94	0.54	0.40	0.32
0.5	26	6.8	3.2	1.1	0.66	0.49	0.42
0.6	37	9.7	4.7	1.6	0.92	0.69	0.57
0.7	43	13	6.3	2.1	1.2	0.88	0.72
0.8	65	17	7.9	2.7	1.5	1.1	0.86
0.9	82	21	9.7	3.4	1.8	1.2	0.99
1.0	101	26	12	4.0	2.1	1.4	1.1

Main											
Q_b/Q_c	0	0.1	0.2	0.3	0.4	0.5	0.6	0.7	0.8	0.9	1.0
$C_{c,s}$	0	0.16	0.27	0.38	0.46	0.53	0.57	0.59	0.60	0.59	0.55

6-4 Converging Wye (30°), Conical, Round[9]

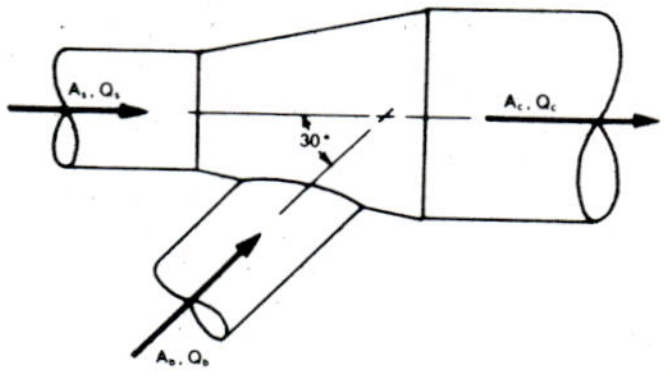

Branch, $C_{c,b}$											
A_s/A_c	A_b/A_c	Q_b/Q_s									
		0.2	0.4	0.6	0.8	1.0	1.2	1.4	1.6	1.8	2.0
0.3	0.2	-2.4	-.11	1.8	3.4	4.8	6.0	7.1	8.0	8.9	9.7
	0.3	-2.8	-1.3	0.14	0.72	1.4	2.0	2.4	2.8	3.2	3.5
0.4	0.2	-1.4	0.61	2.3	3.8	5.2	6.3	7.3	8.3	9.1	9.8
	0.3	-1.8	-.54	0.42	1.2	1.8	2.3	2.7	3.1	3.4	3.7
	0.4	-1.9	-.89	-.17	0.36	0.76	1.1	1.3	1.5	1.7	1.9
0.5	0.2	-.82	0.97	2.6	4.0	5.3	6.4	7.4	8.3	9.1	9.9
	0.3	-1.2	-.15	0.71	1.4	2.0	2.5	2.9	3.3	3.6	3.9
	0.4	-1.4	-.54	0.06	0.50	0.85	1.1	1.3	1.5	1.7	1.8
	0.5	-1.4	-.66	-.15	0.21	0.48	0.68	0.84	0.97	1.1	1.2
0.6	0.2	-.52	1.2	2.7	4.1	5.3	6.4	7.4	8.3	9.1	9.9
	0.3	-.93	0.06	0.85	1.5	2.1	2.6	3.0	3.4	3.7	4.0
	0.4	-1.1	-.37	0.16	0.55	0.86	1.1	1.3	1.5	1.6	1.8
	0.5	-1.1	-.49	-.06	0.25	0.48	0.66	0.79	0.90	1.0	1.1
	0.6	-1.2	-.55	-.15	0.12	0.31	0.45	0.56	0.65	0.71	0.77
0.8	0.2	-.27	1.3	2.7	4.0	5.2	6.3	7.3	8.2	9.0	9.7
	0.3	-.67	0.18	0.90	1.5	2.0	2.5	2.9	3.3	3.6	4.0
	0.4	-.85	-.27	0.16	0.49	0.75	0.97	1.2	1.3	1.4	1.6
	0.5	-.90	-.40	-.07	0.18	0.36	0.50	0.61	0.70	0.78	0.84
	0.6	-.92	-.46	-.16	0.04	0.18	0.29	0.37	0.44	0.49	0.53
	0.7	-.93	-.49	-.21	-.03	0.10	0.19	0.25	0.30	0.34	0.37
	0.8	-.93	-.50	-.24	-.07	0.05	0.13	0.19	0.23	0.27	0.29
1.0	0.2	-.26	1.2	2.6	3.9	5.1	6.1	7.1	8.0	8.8	9.5
	0.3	-.65	0.12	0.79	1.4	1.9	2.4	2.8	3.1	3.5	3.8
	0.4	-.83	-.34	0.04	0.33	0.58	0.78	0.95	1.1	1.2	1.3
	0.5	-.89	-.48	-.20	0	0.15	0.27	0.37	0.45	0.51	0.57
	0.6	-.91	-.54	-.31	-.14	-.03	0.06	0.12	0.18	0.22	0.25
	0.8	-.93	-.59	-.38	-.25	-.16	-.10	-.06	-.03	-.01	0.01
	1.0	-.93	-.60	-.40	-.28	-.20	-.14	-.11	-.08	-.07	-.06

Main, $C_{c,s}$

$\frac{A_s}{A_c}$	$\frac{A_b}{A_c}$	Q_b/Q_s 0.2	0.4	0.6	0.8	1.0	1.2	1.4	1.6	1.8	2.0
0.3	0.2	4.5	2.8	1.5	0.56	−.17	−.74	−1.2	−.16	−1.9	−2.1
	0.3	4.6	3.1	2.0	1.2	0.57	0.08	−.30	−.62	−.89	−1.1
0.4	0.2	1.6	0.85	0.16	−.43	−.92	−1.3	−1.7	−1.9	−2.2	−2.4
	0.3	1.7	1.1	0.58	0.13	−.24	−.56	−.82	−1.1	−1.3	−1.4
	0.4	1.8	1.3	0.80	0.42	0.11	−.15	−.37	−.55	−.72	−.86
0.5	0.2	0.67	0.18	−.33	−.79	−1.2	−1.5	−1.8	−2.1	−2.3	−2.5
	0.3	0.75	-0.42	0.07	−.25	−.54	−.80	−1.0	−1.2	−1.4	−1.5
	0.4	0.80	0.55	0.28	0.03	−.20	−.40	−.57	−.73	−.86	−.98
	0.5	0.82	0.62	0.41	0.20	0.02	−.15	−.29	−.42	−.53	−.63
0.6	0.2	0.26	−.11	−.54	−.95	−1.3	−1.6	−1.9	−2.1	−2.4	−2.5
	0.3	0.34	0.13	−.14	−.42	−.67	−.90	−1.1	−1.3	−1.4	−1.6
	0.4	0.39	0.25	0.06	−.14	−.33	−.51	−.66	−.80	−.93	−1.0
	0.5	0.41	0.32	0.18	0.03	−1.2	−.26	−.38	−.50	−.60	−.69
	0.6	0.43	0.37	0.26	0.14	0.02	−.09	−.19	−.29	−.37	−.45
0.8	0.2	−.01	−.30	−.67	−1.1	−1.4	−1.7	−2.0	−2.2	−2.4	−2.6
	0.3	0.07	−.07	−.29	−.58	−.76	−.97	−1.2	−1.3	−1.5	−1.6
	0.4	0.11	0.05	−.09	−.26	−.42	−.58	−.72	−.85	−.97	−1.1
	0.5	0.14	0.12	0.03	−.09	−.21	−.34	−.45	−.55	−.64	−.73
	0.6	0.15	0.17	0.11	0.02	−.07	−.17	−.26	−.34	−.42	−.49
	0.7	0.17	0.21	0.17	0.11	0.03	−.05	−.12	−.19	−.26	−.32
	0.8	0.17	0.23	0.22	0.17	0.11	0.05	−.02	−.07	−.13	−.18
1.0	0.2	−.05	−.33	−.70	−1.1	−1.4	−1.7	−2.0	−2.2	−2.4	−2.6
	0.3	0.03	−.10	−.31	−.55	−.78	−.98	−1.2	−1.3	−1.5	−1.6
	0.4	0.07	0.02	−.12	−.28	−.44	−.59	−.73	−.86	−.98	−1.1
	0.5	0.09	0.09	0.01	−.11	−.23	−.35	−.46	−.56	−.65	−.74
	0.6	0.11	0.14	0.09	0	−.09	−.18	−.27	−.35	−.43	−.50
	0.8	0.13	0.20	0.19	0.15	0.09	0.03	−.03	−.08	−.14	−.19
	1.0	0.14	0.24	0.25	0.24	0.20	0.16	0.12	0.08	0.04	0

6-5 Converging Wye (45°), Conical, Round[9]

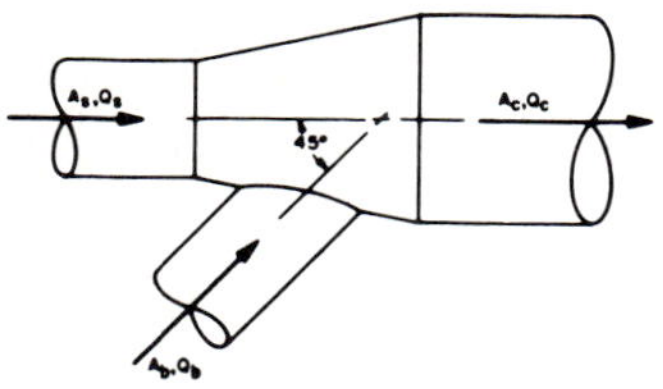

Branch, $C_{c,b}$

$\frac{A_s}{A_c}$	$\frac{A_b}{A_c}$	Q_b/Q_s 0.2	0.4	0.6	0.8	1.0	1.2	1.4	1.6	1.8	2.0
0.3	0.2	−2.4	−.01	2.0	3.8	5.3	6.6	7.8	8.9	9.8	11
	0.3	−2.8	−1.2	0.12	1.1	1.9	2.6	3.2	3.7	4.2	4.6
0.4	0.2	−1.2	0.93	2.8	4.5	5.9	7.2	8.4	9.5	10	11
	0.3	−1.6	−.27	0.81	1.7	2.4	3.0	3.6	4.1	4.5	4.9
	0.4	−1.8	−.72	0.07	0.66	1.1	1.5	1.8	2.1	2.3	2.5
0.5	0.2	−.46	1.5	3.3	4.9	6.4	7.7	8.8	9.9	11	12
	0.3	−.94	0.25	1.2	2.0	2.7	3.3	3.8	4.2	4.7	5.0
	0.4	−1.1	−.24	0.42	0.92	1.3	1.6	1.9	2.1	2.3	2.5
	0.5	−1.2	−.38	0.18	0.58	0.88	1.1	1.3	1.5	1.6	1.7
0.6	0.2	−.55	1.3	3.1	4.7	6.1	7.4	8.6	9.6	11	12
	0.3	−1.1	0	0.88	1.6	2.3	2.8	3.3	3.7	4.1	4.5
	0.4	−1.2	−.48	0.10	0.54	0.89	1.2	1.4	1.6	1.8	2.0
	0.5	−1.3	−.62	−.14	0.21	0.47	0.68	0.85	0.99	1.1	1.2
	0.6	−1.3	−.69	−.26	0.04	0.26	0.42	0.57	0.66	0.75	0.82
0.8	0.2	0.06	1.8	3.5	5.1	6.5	7.8	8.9	10	11	12
	0.3	−.52	0.35	1.1	1.7	2.3	2.8	3.2	3.6	3.9	4.2
	0.4	−.67	−.05	0.43	0.80	1.1	1.4	1.6	1.8	1.9	2.1
	0.6	−.75	−.27	0.05	0.28	0.45	0.58	0.68	0.76	0.83	0.88
	0.7	−.77	−.31	−.02	0.18	0.32	0.43	0.50	0.56	0.61	0.65
	0.8	−.78	−.34	−.07	0.12	0.24	0.33	0.39	0.44	0.47	0.50
1.0	0.2	0.40	2.1	3.7	5.2	6.6	7.8	9.0	11	11	12
	0.3	−.21	0.54	1.2	1.8	2.3	2.7	3.1	3.7	3.7	4.0
	0.4	−.33	0.21	0.62	0.96	1.2	1.5	1.7	2.0	2.0	2.1
	0.5	−.38	0.05	0.37	0.60	0.79	0.93	1.1	1.2	1.2	1.3
	0.6	−.41	−.02	0.23	0.42	0.55	0.66	0.73	0.80	0.85	0.89
	0.8	−.44	−.10	0.11	0.24	0.33	0.39	0.43	0.46	0.47	0.48
	1.0	−.46	−.14	0.05	0.16	0.23	0.27	0.29	0.30	0.30	0.29

Main, $C_{c,s}$

$\frac{A_s}{A_c}$	$\frac{A_b}{A_c}$	Q_b/Q_s 0.2	0.4	0.6	0.8	1.0	1.2	1.4	1.6	1.8	2.0
0.3	0.2	5.3	−.01	2.0	1.1	0.34	−.20	−.61	−.93	−1.2	−1.4
	0.3	5.4	3.7	2.5	1.6	1.0	0.53	0.16	−.14	−.38	−.58
0.4	0.2	1.9	1.1	0.46	−.07	−.49	−.83	−1.1	−1.3	−1.5	−1.7
	0.3	2.0	1.4	0.81	0.42	0.08	−.20	−.43	−.62	−.78	−.92
	0.4	2.0	1.5	1.0	0.68	0.39	0.16	−.04	−.21	−.35	−.47
0.5	0.2	0.77	0.34	−.09	−.48	−.81	−1.1	1.3	−1.5	−1.7	−1.8
	0.3	0.85	0.56	0.25	−.03	−.27	−.48	−.67	−.82	−.96	−1.1
	0.4	0.88	0.66	0.43	0.21	0.02	−.15	−.30	−.42	−.54	−.64
	0.5	0.91	0.73	0.54	0.36	0.21	0.06	−.06	−.17	−.26	−.35
0.6	0.2	0.30	0	−.34	−.67	−.96	−1.2	−1.4	−1.6	−1.8	−1.9
	0.3	0.37	0.21	−.02	−.24	−.44	−.63	−.79	−.93	−1.1	−1.2
	0.4	0.40	0.31	0.16	−0.1	−.16	−.30	−.43	−.54	−.64	−.73
	0.5	0.43	0.37	0.26	0.14	0.02	−.09	−.20	−.29	−.37	−.45
	0.6	0.44	0.41	0.33	0.24	0.14	0.05	−.03	−.11	−.18	−.25
0.8	0.2	−.06	−.27	−.57	−.86	−1.1	−1.4	−1.6	−1.7	−1.9	−2.0
	0.3	0	−.08	−.25	−.43	−.62	−.78	−.93	−1.1	−1.2	−1.3
	0.4	0.04	0.02	−.08	−.21	−.34	−.46	−.57	−.67	−.77	−.85
	0.5	0.06	0.08	0.02	−.06	−.16	−.25	−.34	−.42	−.50	−.57
	0.6	0.07	0.12	0.09	0.03	−.04	−.11	−.18	−.25	−.31	−.37
	0.7	0.08	0.15	0.14	0.10	0.05	−.01	−.07	−.12	−.17	−.22
	0.8	0.09	0.17	0.18	0.16	0.11	0.07	0.02	−.02	−.07	−.11
1.0	0.2	−.19	−.39	−.67	−.96	−1.2	−1.5	−1.6	−1.8	−2.0	−2.1
	0.3	−.12	−.19	−.35	−.54	−.71	−.87	−1.0	−1.2	−1.3	−1.4
	0.4	−.09	−.10	−.19	−.31	−.43	−.55	−.66	−.77	−.86	−.94
	0.5	−.07	−.04	−.09	−.17	−.26	−.35	−.44	−.52	−.59	−.66
	0.6	−.06	0	−.02	−.07	−.14	−.21	−.28	−.34	−.40	−.46
	0.8	−.04	0.06	0.07	0.05	0.02	−.03	−.07	−.12	−.16	−.20
	1.0	−.03	0.09	0.13	0.13	0.11	0.08	0.06	0.03	−.01	−.03

6-6 Converging Wye, Rectangular[1]

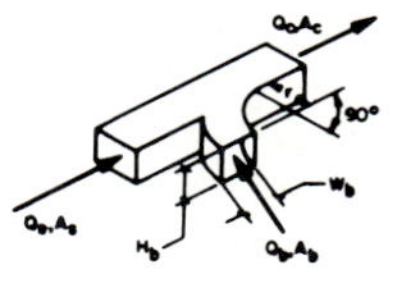

$r/w_b = 1$

Branch, $C_{c,b}$

A_b/A_s	A_b/A_c	Q_b/Q_c 0.1	0.2	0.3	0.4	0.5	0.6	0.7	0.8	0.9
0.25	0.25	−.50	0	0.50	1.2	2.2	3.7	5.8	8.4	11
0.33	0.25	−1.2	−.40	0.40	1.6	3.0	4.8	6.8	8.9	11
0.5	0.5	−.50	−.20	0	0.25	0.45	0.70	1.0	1.5	2.0
0.67	0.5	−1.0	−.60	−.20	0.10	0.30	0.60	1.0	1.5	2.0
1.0	0.5	−2.2	−1.5	−.95	−.50	0	0.40	0.80	1.3	1.9
1.0	1.0	−.60	−.30	−.10	−.04	0.13	0.21	0.29	0.36	0.42
1.33	1.0	−1.2	−.80	−.40	−.20	0	0.16	0.24	0.32	0.38
2.0	1.0	-2.1	-1.4	-.90	-.50	-.20	0	0.20	0.25	0.30

Main, $C_{c,s}$

A_s/A_c	A_b/A_c	Q_b/Q_c 0.1	0.2	0.3	0.4	0.5	0.6	0.7	0.8	0.9
0.75	0.25	0.30	0.30	0.20	.10	−.45	−.92	−1.5	−2.0	−2.6
1.0	0.5	0.17	0.16	0.10	0	−0.08	−.18	−.27	−.37	−.46
0.75	0.5	0.27	0.35	0.32	0.25	0.12	−.03	−.23	−.42	−.58
0.5	0.5	1.2	1.1	0.90	0.65	0.35	0	−.40	−.80	−1.3
1.0	1.0	0.18	0.24	0.27	0.26	0.23	0.18	0.10	0	−.12
0.75	1.0	0.75	0.36	0.38	0.35	0.27	0.18	0.05	−.08	−.22
0.5	1.0	0.80	0.87	0.80	0.66	0.55	0.40	0.25	0.08	−.10

6-7 Converging Tee, Round Tap to Rectangular Main[7]

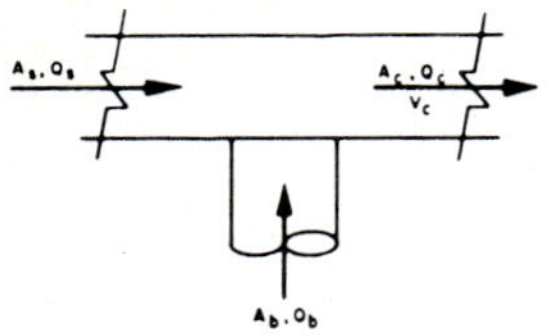

A_b/A_s	A_s/A_c	A_b/A_c
0.5	1.0	0.5

Branch, $C_{c,b}$

V_c	Q_b/Q_c									
(fpm)	**0.1**	**0.2**	**0.3**	**0.4**	**0.5**	**0.6**	**0.7**	**0.8**	**0.9**	**1.0**
<1200	-.63	-.55	0.13	0.23	0.78	1.30	1.93	3.10	4.88	5.60
>1200	-.49	-.21	0.23	0.60	1.27	2.06	2.75	3.70	4.93	5.95

For main coefficient ($C_{c,s}$), see Fitting 6-3.

6-8 Converging Tee, Rectangular, Main and Tap[7]

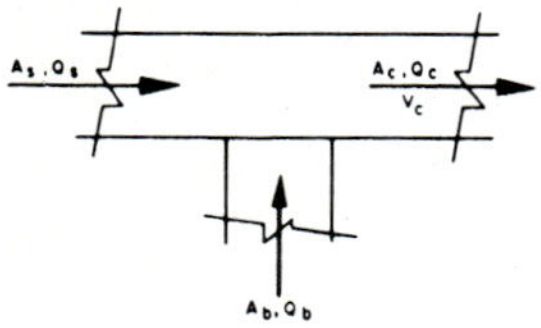

A_b/A_s	A_s/A_c	A_b/A_c
0.5	1.0	0.5

Branch, $C_{c,b}$

V_c	Q_b/Q_c									
(fpm)	**0.1**	**0.2**	**0.3**	**0.4**	**0.5**	**0.6**	**0.7**	**0.8**	**0.9**	**1.0**
<1200	-.75	-.53	-.03	0.33	1.03	1.10	2.15	2.93	4.18	4.78
>1200	-.69	-.21	0.23	0.67	1.17	1.66	2.67	3.36	3.93	5.13

For main coefficient ($C_{c,s}$), see Fitting 6-3.

6-9 Converging Tee, 45° Entry, Rectangular, Main and Tap[7]

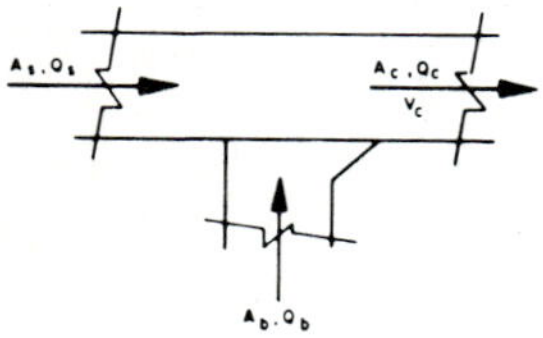

A_b/A_s	A_s/A_c	A_b/A_c
0.5	1.0	0.5

Branch, $C_{c,b}$

V_c	Q_b/Q_c									
(fpm)	**0.1**	**0.2**	**0.3**	**0.4**	**0.5**	**0.6**	**0.7**	**0.8**	**0.9**	**1.0**
<1200	-.83	-.68	-.30	0.28	0.55	1.03	1.50	1.93	2.50	3.03
>1200	-.72	-.52	-.23	0.34	0.76	1.14	1.83	2.01	2.90	3.63

For main coefficient ($C_{c,s}$), see Fitting 6-3.

6-10 Diverging 90° Conical Tee, Round[10]

$A_c = A_s$

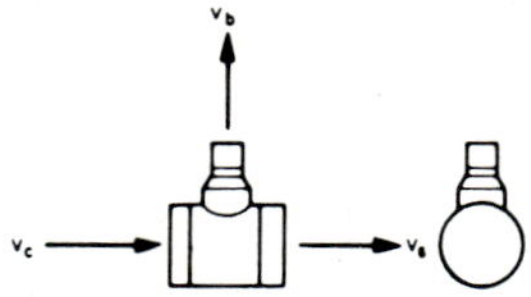

Branch

V_b/V_c	0	0.2	0.4	0.6	0.8	1.0	1.2	1.4	1.6	1.8	2.0
$C_{c,b}$	1.0	0.85	0.74	0.62	0.52	0.42	0.36	0.32	0.32	0.37	0.52

For main loss coefficient ($C_{c,s}$), see Fitting 6-23.

6-11 Diverging 45° Conical Wye, Round[10]

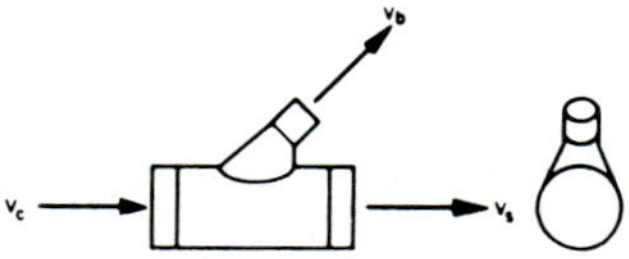

$A_c = A_s$

Branch

V_b/V_c	0	0.2	0.4	0.6	0.8	1.0	1.2	1.4	1.6	1.8	2.0
$C_{c,b}$	1.0	0.84	0.61	0.41	0.27	0.17	0.12	0.12	0.14	0.18	0.27

For main loss coefficient ($C_{c,s}$), see Fitting 6-23.

6-12 Diverging 90° Tee, Round, Rolled 90° with 90° Elbow, Branch 90° to Main[10]

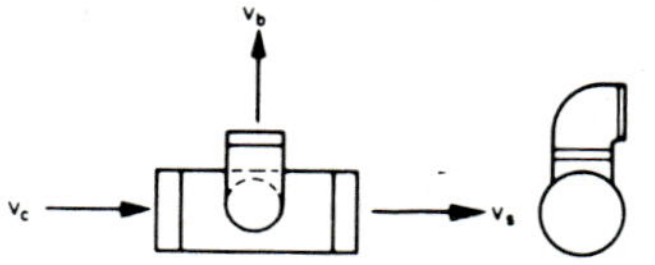

$A_c = A_s$

Branch

V_b/V_c	0	0.2	0.4	0.6	0.8	1.0	1.2	1.4	1.6	1.8	2.0
$C_{c,b}$	1.0	1.03	1.08	1.18	1.33	1.56	1.86	2.2	2.6	3.0	3.4

For main loss coefficient ($C_{c,s}$), see Fitting 6-23.

6-13 Diverging 90° Tee, Round, Rolled 45° with 45° Elbow, Branch 90° to Main[10]

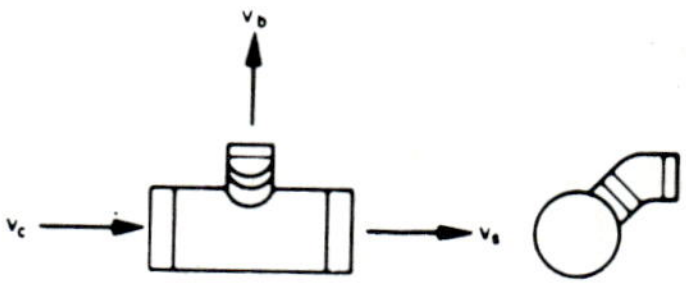

$A_c = A_s$

Branch

V_b/V_c	0	0.2	0.4	0.6	0.8	1.0	1.2	1.4	1.6	1.8	2.0
$C_{c,b}$	1.0	1.32	1.51	1.60	1.65	1.74	1.87	2.0	2.2	2.5	2.7

For main loss coefficient ($C_{c,s}$), see Fitting 6-23.

6-14 Diverging 90° Conical Tee, Round, Rolled 45° with 45° Elbow, Branch 90° to Main[10]

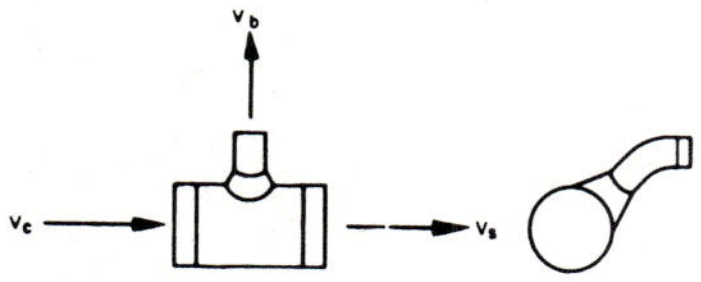

$A_c = A_s$

Branch											
V_b/V_c	0	0.2	0.4	0.6	0.8	1.0	1.2	1.4	1.6	1.8	2.0
$C_{c,b}$	1.0	0.94	0.88	0.84	0.80	0.82	0.84	0.87	0.90	0.95	1.02

For main loss coefficient ($C_{c,s}$), see Fitting 6-23.

6-15 Diverging 45° Wye, Round, Rolled 45° with 60° Elbow, Branch 90° to Main[11]

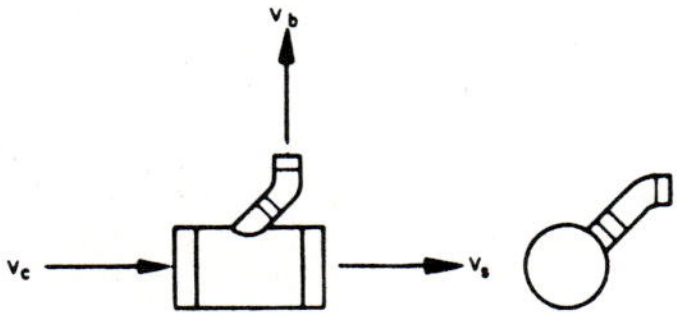

$A_c = A_s$

Branch											
V_b/V_c	0	0.2	0.4	0.6	0.8	1.0	1.2	1.4	1.6	1.8	2.0
$C_{c,b}$	1.0	0.88	0.77	0.68	0.65	0.69	0.73	0.88	1.14	1.54	2.2

For main loss coefficient ($C_{c,s}$), see Fitting 6-23.

6-16 Diverging 45° Conical Wye, Round, Rolled 45° with 60° Elbow, Branch 90° to Main[10]

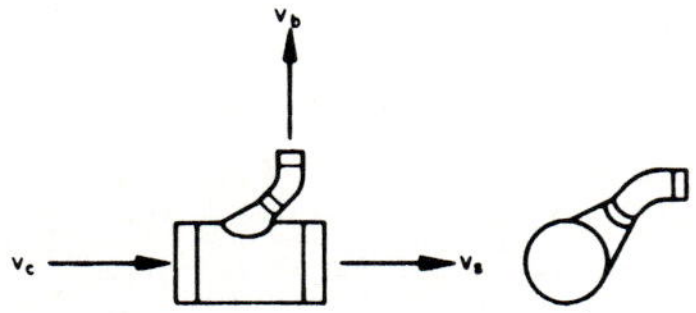

$A_c = A_s$

Branch											
V_b/V_c	0	0.2	0.4	0.6	0.8	1.0	1.2	1.4	1.6	1.8	2.0
$C_{c,b}$	1.0	0.82	0.63	0.52	0.45	0.42	0.41	0.40	0.41	0.45	0.56

For main loss coefficient ($C_{c,s}$), see Fitting 6-23.

6-17 Diverging 45° Wye, Conical Main and Branch, with 45° Elbow, Branch 90° to Main[1]

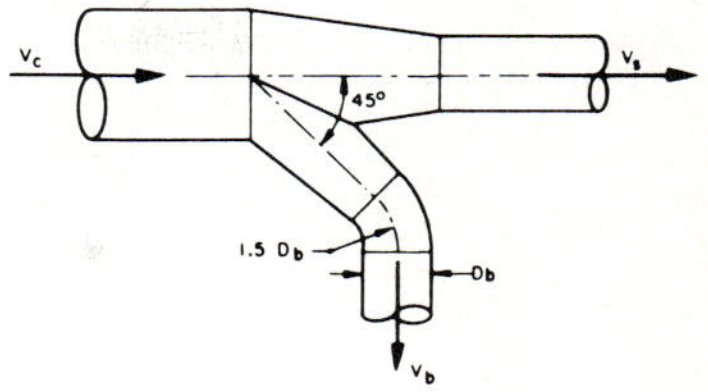

Branch									
V_b/V_c	0.2	0.4	0.6	0.7	0.8	0.9	1.0	1.1	1.2
$C_{c,b}$	0.76	0.60	0.52	0.50	0.51	0.52	0.56	0.61	0.68
V_b/V_c	1.4	1.6	1.8	2.0	2.2	2.4	2.6	2.8	3.0
$C_{c,b}$	0.86	1.1	1.4	1.8	2.2	2.6	3.1	3.7	4.2

Main										
V_s/V_c	0.2	0.4	0.6	0.8	1.0	1.2	1.4	1.6	1.8	2.0
$C_{c,s}$	0.14	0.06	0.05	0.09	0.18	0.30	0.46	0.64	0.84	1.0

6-18 Diverging 90° Tee, Round, Rolled 45° with 60° Elbow, Branch 45° to Main[10]

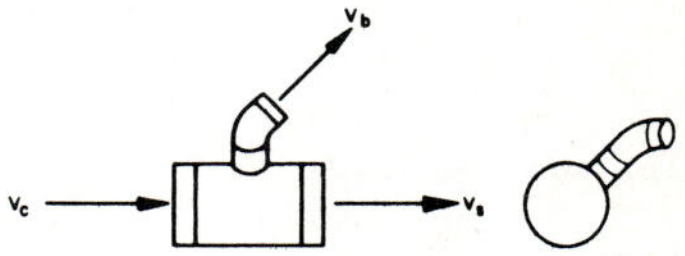

$A_c = A_s$

Branch											
V_b/V_c	0	0.2	0.4	0.6	0.8	1.0	1.2	1.4	1.6	1.8	2.0
$C_{c,b}$	1.0	1.06	1.15	1.29	1.45	1.65	1.89	2.2	2.5	2.9	3.3

For main loss coefficient ($C_{c,s}$), see Fitting 6-23.

6-19 Diverging 90° Conical Tee, Round, Rolled 45° with 60° Elbow, Branch 45° to Main[10]

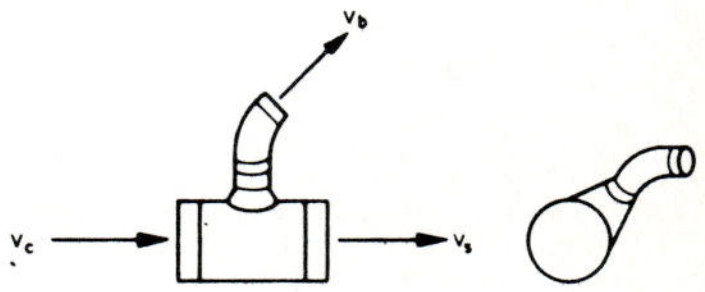

$A_c = A_s$

Branch											
V_b/V_c	0	0.2	0.4	0.6	0.8	1.0	1.2	1.4	1.6	1.8	2.0
$C_{c,b}$	1.0	0.95	0.90	0.86	0.81	0.79	0.79	0.81	0.86	0.96	1.10

For main loss coefficient ($C_{c,s}$), see Fitting 6-23.

6-20 Diverging 45° Wye, Round, Rolled 45° with 30° Elbow, Branch 45° to Main[11]

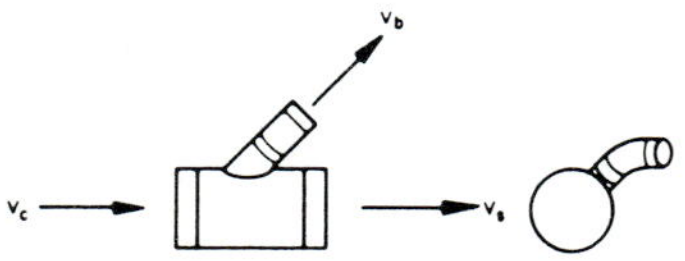

$A_c = A_s$

Branch										
V_b/V_c	0	0.2	0.4	0.6	0.8	1.0	1.2	1.4	1.6	1.8	2.0
$C_{c,b}$	1.0	0.84	0.72	0.62	0.54	0.50	0.56	0.71	0.92	1.22	1.66

For main loss coefficient ($C_{c,s}$), see Fitting 6-23.

6-21 Diverging 45° Conical Wye, Round, Rolled 45° with 30° Elbow, Branch 45° to Main[10]

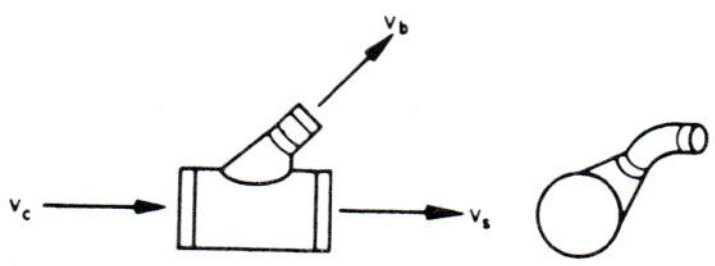

$A_c = A_s$

Branch											
V_b/V_c	0	0.2	0.4	0.6	0.8	1.0	1.2	1.4	1.6	1.8	2.0
$C_{c,b}$	1.0	0.93	0.71	0.55	0.44	0.42	0.42	0.44	0.47	0.54	0.62

For main loss coefficient ($C_{c,s}$), see Fitting 6-23.

6-22 Diverging Wye, Rectangular[1]

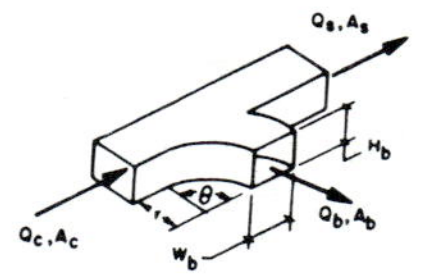

$\theta = 90°$

$r/W_b = 1.0$

Branch $C_{c,b}$

A_b/A_s	A_b/A_c	Q_b/Q_c 0.1	0.2	0.3	0.4	0.5	0.6	0.7	0.8	0.9
0.25	0.25	0.55	0.50	0.60	0.85	1.2	1.8	3.1	4.4	6.0
0.33	0.25	0.35	0.35	0.50	0.80	1.3	2.0	2.8	3.8	5.0
0.5	0.5	0.62	0.48	0.40	0.40	0.48	0.60	0.78	1.1	1.5
0.67	0.5	0.52	0.40	0.32	0.30	0.34	0.44	0.62	0.92	1.4
1.0	0.5	0.44	0.38	0.38	0.41	0.52	0.68	0.92	1.2	1.6
1.0	1.0	0.67	0.55	0.46	0.37	0.32	0.29	0.29	0.30	0.37
1.33	1.0	0.70	0.60	0.51	0.42	0.34	0.28	0.26	0.26	0.29
2.0	1.0	0.60	0.52	0.43	0.33	0.24	0.17	0.15	0.17	0.21

Main, $C_{c,s}$

A_b/A_s	A_b/A_c	Q_b/Q_c 0.1	0.2	0.3	0.4	0.5	0.6	0.7	0.8	0.9
0.25	0.25	-.01	-.03	-.01	0.05	0.13	0.21	0.29	0.38	0.46
0.33	0.25	0.08	0	-.02	-.01	0.02	0.08	0.16	0.24	0.34
0.5	0.5	-.03	-.06	-.05	0	0.06	0.12	0.19	0.27	0.35
0.67	0.5	0.04	-.02	-.04	-.03	-.01	0.04	0.12	0.23	0.37
1.0	0.5	0.72	0.48	0.28	0.13	0.05	0.04	0.09	0.18	0.30
1.0	1.0	-.02	-.04	-.04	-.01	0.06	0.13	0.22	0.30	0.38
1.33	1.0	0.10	0	0.01	-.03	-.01	0.03	0.10	0.20	0.30
2.0	1.0	0.62	0.38	0.23	0.13	0.08	0.05	0.06	0.10	0.20

6-23 Diverging Wye, Rectangular and Round[1]

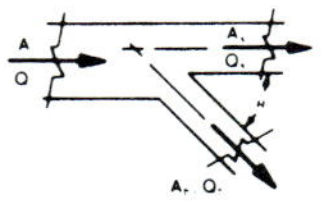

$A_c = A_s$; $H_b = H_c$, *where H* is height of rectangular duct

a. $\theta = 30°$

Branch, $C_{c,b}$

A_b/A_c	Q_b/Q_c 0.1	0.2	0.3	0.4	0.5	0.6	0.7	0.8	0.9
0.8	0.75	0.55	0.40	0.28	0.21	0.16	0.15	0.16	0.19
0.7	0.72	0.51	0.36	0.25	0.18	0.15	0.16	0.20	0.26
0.6	0.69	0.46	0.31	0.21	0.17	0.16	0.20	0.28	0.39
0.5	0.65	0.41	0.26	0.19	0.18	0.22	0.32	0.47	0.67
0.4	0.59	0.33	0.21	0.20	0.27	0.40	0.62	0.92	1.3
0.3	0.55	0.28	0.24	0.38	0.76	1.3	2.0	—	—
0.2	0.40	0.26	0.58	1.3	2.5	—	—	—	—
0.1	0.28	1.5	—	—	—	—	—	—	—

b. $\theta = 45°$

Branch, $C_{c,b}$

A_b/A_c	Q_b/Q_c 0.1	0.2	0.3	0.4	0.5	0.6	0.7	0.8	0.9
0.8	0.78	0.62	0.49	0.40	0.34	0.31	0.32	0.35	0.40
0.7	0.77	0.59	0.47	0.38	0.34	0.32	0.35	0.41	0.50
0.6	0.74	0.56	0.44	0.37	0.35	0.36	0.43	0.54	0.68
0.5	0.71	0.52	0.41	0.38	0.40	0.45	0.59	0.78	1.0
0.4	0.66	0.47	0.40	0.43	0.54	0.69	0.95	1.3	1.7
0.3	0.66	0.48	0.52	0.73	1.2	1.8	2.7	—	—
0.2	0.56	0.56	1.0	1.8	—	—	—	—	—
0.1	0.60	2.1	—	—	—	—	—	—	—

c. $\theta = 60°$

Branch, $C_{c,b}$

A_b/A_c	Q_b/Q_c 0.1	0.2	0.3	0.4	0.5	0.6	0.7	0.8	0.9
0.8	0.83	0.71	0.62	0.56	0.52	0.50	0.53	0.60	0.68
0.7	0.82	0.69	0.61	0.56	0.54	0.54	0.60	0.70	0.82
0.6	0.81	0.68	0.60	0.58	0.58	0.61	0.72	0.87	1.1
0.5	0.79	0.66	0.61	0.62	0.68	0.76	0.94	1.2	1.5
0.4	0.76	0.65	0.65	0.74	0.89	1.1	1.4	1.8	2.3
0.3	0.80	0.75	0.89	1.2	1.8	2.6	3.5	—	—
0.2	0.77	0.96	1.6	2.5	—	—	—	—	—
0.1	1.0	2.9	—	—	—	—	—	—	—

d. $\theta = 90°$

Branch, $C_{c,b}$

A_b/A_c	Q_b/Q_c 0.1	0.2	0.3	0.4	0.5	0.6	0.7	0.8	0.9
0.8	0.95	0.92	0.92	0.93	0.94	0.95	1.1	1.2	1.4
0.7	0.95	0.94	0.95	0.98	1.0	1.1	1.2	1.4	1.6
0.6	0.96	0.97	1.0	1.1	1.1	1.2	1.4	1.7	2.0
0.5	0.97	1.0	1.1	1.2	1.4	1.5	1.8	2.1	2.5
0.4	0.99	1.1	1.3	1.5	1.7	2.0	2.4	—	—
0.3	1.1	1.4	1.8	2.3	—	—	—	—	—
0.2	1.3	1.9	2.9	—	—	—	—	—	—
0.1	2.1	—	—	—	—	—	—	—	—

Main									
V_s/V_c	0	0.1	0.2	0.3	0.4	0.5	0.6	0.8	1.0
$C_{c,s}$	0.35	0.28	0.22	0.17	0.13	0.09	0.06	0.02	0

6-24 Diverging Wye, Rectangular[1]

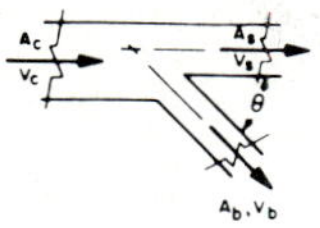

$\theta = 15° \text{ to } 90°$

$A_c = A_s + A_b$

Branch, $C_{c,b}$

θ, deg	V_b/V_c 0.1	0.2	0.3	0.4	0.5	0.6	0.8	1.0	1.2	1.4	1.6	1.8	2.0
15	0.81	0.65	0.51	0.38	0.28	0.20	0.11	0.06	0.14	0.30	0.51	0.76	1.0
30	0.84	0.69	0.56	0.44	0.34	0.26	0.19	0.15	0.15	0.30	0.51	0.76	1.0
45	0.87	0.74	0.63	0.54	0.45	0.38	0.29	0.24	0.23	0.30	0.51	0.76	1.0
60	0.90	0.82	0.79	0.66	0.59	0.53	0.43	0.36	0.33	0.39	0.51	0.76	1.0
90	1.0	1.0	1.0	1.0	1.0	1.0	1.0	1.0	1.0	1.0	1.0	1.0	1.0

Main, $C_{c,s}$

θ, deg	15—60	90				
V_s/V_c	A_s/A_c 0-1.0	0-0.4	0.5	0.6	0.7	≥0.8
0	1.0	1.0	1.0	1.0	1.0	1.0
0.1	0.81	0.81	0.81	0.81	0.81	0.81
0.2	0.64	0.64	0.64	0.64	0.64	0.64
0.3	0.50	0.50	0.52	0.52	0.50	0.50
0.4	0.36	0.36	0.40	0.38	0.37	0.36
0.5	0.25	0.25	0.30	0.28	0.27	0.25
0.6	0.16	0.16	0.23	0.20	0.18	0.16
0.8	0.04	0.04	0.17	0.10	0.07	0.04
1.0	0	0	0.20	0.10	0.05	0
1.2	0.07	0.07	0.36	0.21	0.14	0.07
1.4	0.39	0.39	0.79	0.59	0.39	—
1.6	0.90	0.90	1.4	1.2	—	—
1.8	1.8	1.8	2.4	—	—	—
2.0	3.2	3.2	4.0	—	—	—

6-25 Diverging Tee, Pyramidal Main and Take-off with 45° Leading Edge[12]

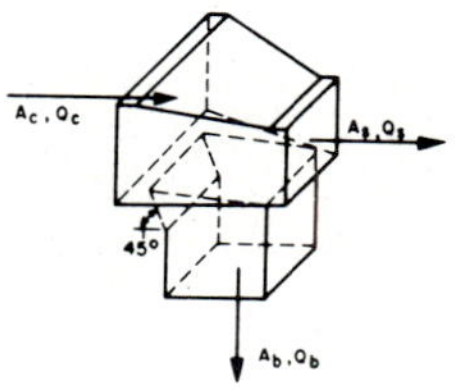

$0.5 \leqslant A_b/A_c \leqslant 1.0$
$0.5 \leqslant A_s/A_c \leqslant 1.0$

Branch

Q_b/Q_c	0	0.1	0.2	0.3	0.4	0.5	0.6	0.7	0.8	0.9	1.0
$C_{c,b}$	1.4	1.2	0.96	0.82	0.68	0.56	0.49	0.47	0.48	0.50	0.54

6-26 Diverging Tee, Rectangular Main to Round Tap[7]

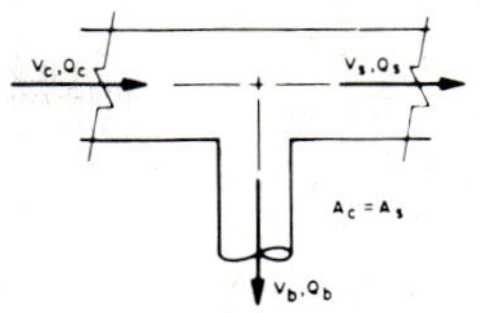

Branch, $C_{c,b}$

V_b/V_c	Q_b/Q_c 0.1	0.2	0.3	0.4	0.5	0.6	0.7	0.8	0.9
0.2	1.00								
0.4	1.01	1.07							
0.6	1.14	1.10	1.08						
0.8	1.18	1.31	1.12	1.13					
1.0	1.30	1.38	1.20	1.23	1.26				
1.2	1.46	1.58	1.45	1.31	1.39	1.48			
1.4	1.70	1.82	1.65	1.51	1.56	1.64	1.71		
1.6	1.93	2.06	2.00	1.85	1.70	1.76	1.80	1.88	
1.8	2.06	2.17	2.20	2.13	2.06	1.98	1.99	2.00	2.07

For main coefficient ($C_{c,s}$), see Fitting 6-23.

6-27 Diverging Tee, Rectangular Main to Round Tap[13]

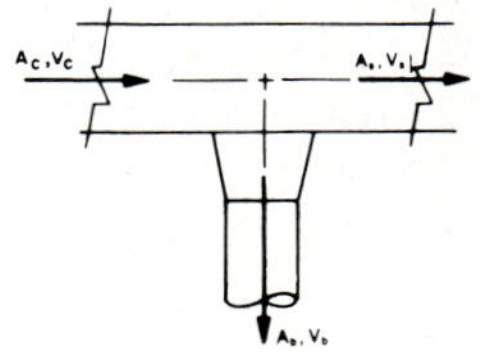

$A_c = A_s$

Branch

V_b/V_c	0.40	0.50	0.75	1.0	1.3	1.5
$C_{c,b}$	0.80	0.83	0.90	1.0	1.1	1.4

For main coefficient ($C_{c,s}$), see Fitting 6-23.

6-28 Diverging Tee, Rectangular Main and Tap[7]

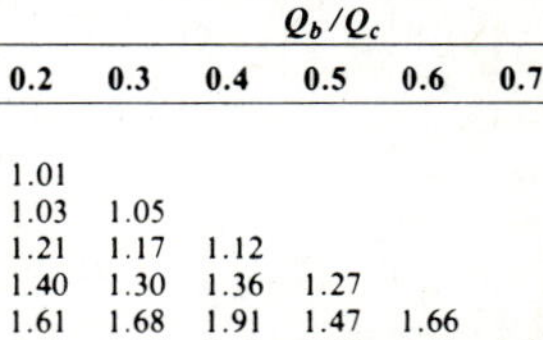

$A_c = A_s$

Branch, $C_{c,b}$

V_b/V_c	Q_b/Q_c 0.1	0.2	0.3	0.4	0.5	0.6	0.7	0.8	0.9
0.2	1.03								
0.4	1.04	1.01							
0.6	1.11	1.03	1.05						
0.8	1.16	1.21	1.17	1.12					
1.0	1.38	1.40	1.30	1.36	1.27				
1.2	1.52	1.61	1.68	1.91	1.47	1.66			
1.4	1.79	2.01	1.90	2.31	2.28	2.20	1.95		
1.6	2.07	2.28	2.13	2.71	2.99	2.81	2.09	2.20	
1.8	2.32	2.54	2.64	3.09	3.72	3.48	2.21	2.29	2.57

For main coefficient ($C_{c,s}$), see Fitting 6-23.

6-29 Diverging Tee, 45° Entry, Rectangular Main and Tap[7]

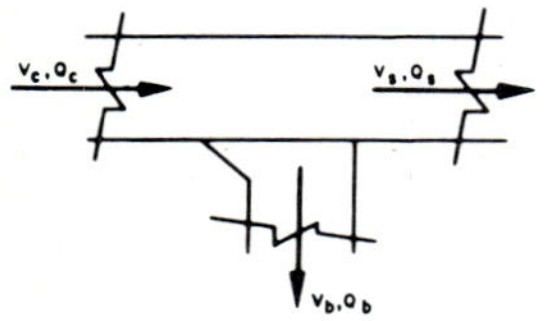

$A_c = A_s$

Branch, $C_{c,b}$									
	Q_b/Q_c								
V_b/V_c	0.1	0.2	0.3	0.4	0.5	0.6	0.7	0.8	0.9
0.2	0.91								
0.4	0.81	0.79							
0.6	0.77	0.72	0.70						
0.8	0.78	0.73	0.69	0.66					
1.0	0.78	0.98	0.85	0.79	0.74				
1.2	0.90	1.11	1.16	1.23	1.03	0.86			
1.4	1.19	1.22	1.26	1.29	1.54	1.25	0.92		
1.6	1.35	1.42	1.55	1.59	1.63	1.50	1.31	1.09	
1.8	1.44	1.50	1.75	1.74	1.72	2.24	1.63	1.40	1.17

For main coefficient ($C_{c,s}$), see Fitting 6-23.

6-30 Diverging Tee, 45° Entry, Rectangular Main and Tape, with Damper[7]

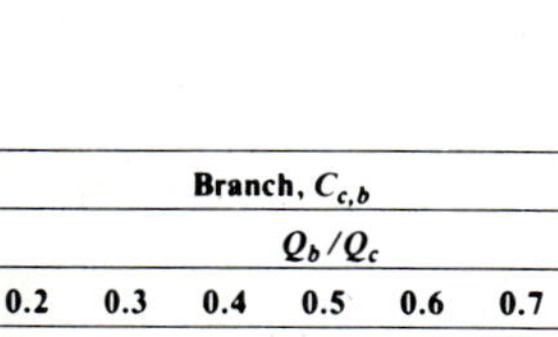

$A_c = A_s$

Branch, $C_{c,b}$									
	Q_b/Q_c								
V_b/V_c	0.1	0.2	0.3	0.4	0.5	0.6	0.7	0.8	0.9
0.2	0.61								
0.4	0.46	0.61							
0.6	0.43	0.50	0.54						
0.8	0.39	0.43	0.62	0.53					
1.0	0.34	0.57	0.77	0.73	0.68				
1.2	0.37	0.64	0.85	0.98	1.07	0.83			
1.4	0.57	0.71	1.04	1.16	1.54	1.36	1.18		
1.6	0.89	1.08	1.28	1.30	1.69	2.09	1.81	1.47	
1.8	1.33	1.34	2.04	1.78	1.90	2.40	2.77	2.23	1.92

For main coefficient ($C_{c,s}$), see Fitting 6-32.

6-31 Diverging Tee, Rectangular Main and Tap, with Damper[7]

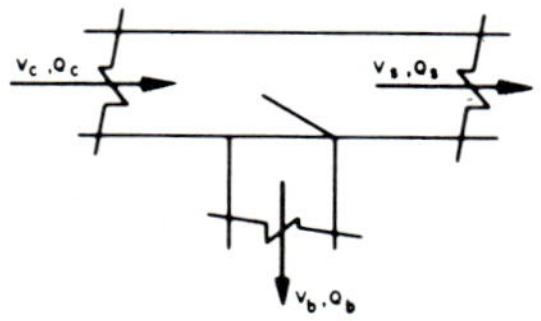

$A_c = A_s$

Branch, $C_{c,b}$									
	Q_b/Q_c								
V_b/V_c	0.1	0.2	0.3	0.4	0.5	0.6	0.7	0.8	0.9
0.2	0.58								
0.4	0.67	0.64							
0.6	0.78	0.76	0.75						
0.8	0.88	0.98	0.81	1.01					
1.0	1.12	1.05	1.08	1.18	1.29				
1.2	1.49	1.48	1.40	1.51	1.70	1.91			
1.4	2.10	2.21	2.25	2.29	2.32	2.48	2.53		
1.6	2.72	3.30	2.84	3.09	3.30	3.19	3.29	3.16	
1.8	3.42	4.58	3.65	3.92	4.20	4.15	4.14	4.10	4.05

For main coefficient ($C_{c,s}$), see Fitting 6-32.

6-32 Diverging Tee, Rectangular, with Extractor[7]

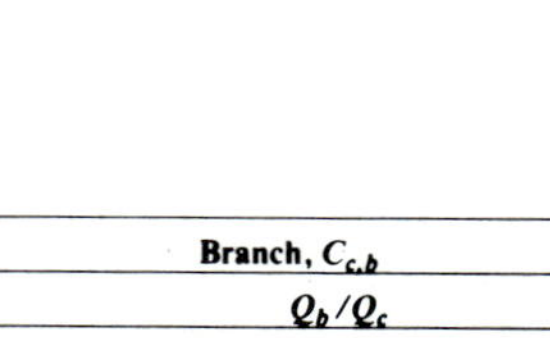

$A_c = A_s$

Branch, $C_{c,b}$									
	Q_b/Q_c								
V_b/V_c	0.1	0.2	0.3	0.4	0.5	0.6	0.7	0.8	0.9
0.2	0.60								
0.4	0.62	0.69							
0.6	0.74	0.80	0.82						
0.8	0.99	1.10	0.95	0.90					
1.0	1.48	1.12	1.41	1.24	1.21				
1.2	1.91	1.33	1.43	1.52	1.55	1.64			
1.4	2.47	1.67	1.70	2.04	1.86	1.98	2.47		
1.6	3.17	2.40	2.33	2.53	2.31	2.51	3.13	3.25	
1.8	3.85	3.37	2.89	3.23	3.09	3.03	3.30	3.74	4.11

Main									
V_b/V_c	0.2	0.4	0.6	0.8	1.0	1.2	1.4	1.6	1.8
$C_{c,s}$	0.03	0.04	0.07	0.12	0.13	0.14	0.27	0.30	0.25

6-33 Symmetrical Wye, Dovetail, Rectangular[1]

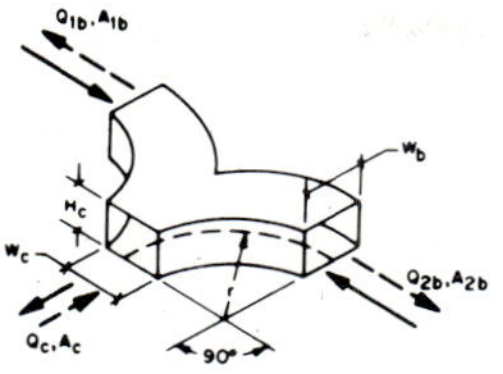

$r/W_c = 1.5$

$Q_{1b}/Q_c = Q_{2b}/Q_c = 0.5$

Converging		
A_{1b}/A_c or A_{2b}/A_c	0.50	1.0
$C_{c,1b}$ or $C_{c,2b}$	0.23	0.07
Diverging		
A_{1b}/A_c or A_{2b}/A_c	0.50	1.0
$C_{c,1b}$ or $C_{c,2b}$	0.30	0.25

6-34 Wye, Rectangular and Round[1]

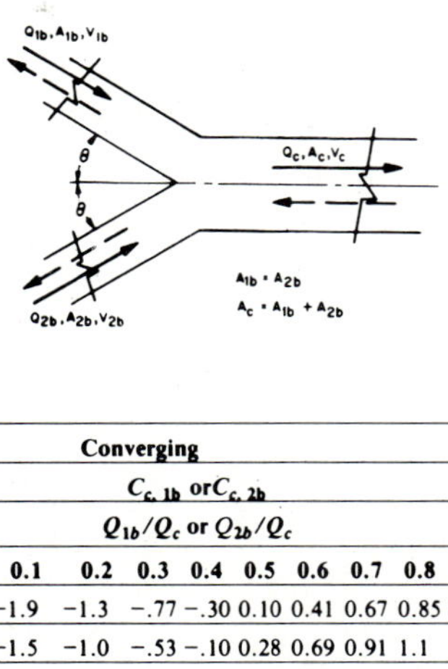

Converging

$C_{c,1b}$ or $C_{c,2b}$

θ, deg	Q_{1b}/Q_c or Q_{2b}/Q_c: 0	0.1	0.2	0.3	0.4	0.5	0.6	0.7	0.8	0.9	1.0
15	−2.6	−1.9	−1.3	−.77	−.30	0.10	0.41	0.67	0.85	0.97	1.0
30	−2.1	−1.5	−1.0	−.53	−.10	0.28	0.69	0.91	1.1	1.4	1.6
45	−1.3	−.93	−.55	−.16	0.20	0.56	0.92	1.3	1.6	2.0	2.3

Diverging

$C_{c,1b}$ or $C_{c,2b}$

θ, deg	V_{1b}/V_c or V_{2b}/V_c: 0.1	0.2	0.3	0.4	0.5	0.6	0.8	1.0	1.2	1.4	1.6	1.8	2.0
15	0.81	0.65	0.51	0.38	0.28	0.20	0.11	0.06	0.14	0.30	0.51	0.76	1.0
30	0.84	0.69	0.56	0.44	0.34	0.26	0.19	0.15	0.15	0.30	0.51	0.76	1.0
45	0.87	0.74	0.63	0.54	0.45	0.38	0.29	0.24	0.23	0.30	0.51	0.76	1.0
60	0.90	0.82	0.79	0.66	0.59	0.53	0.43	0.36	0.33	0.39	0.51	0.76	1.0
90	1.0	1.0	1.0	1.0	1.0	1.0	1.0	1.0	1.0	1.0	1.0	1.0	1.0

6-35 45° Double Wye, Rectangular and Round[1]

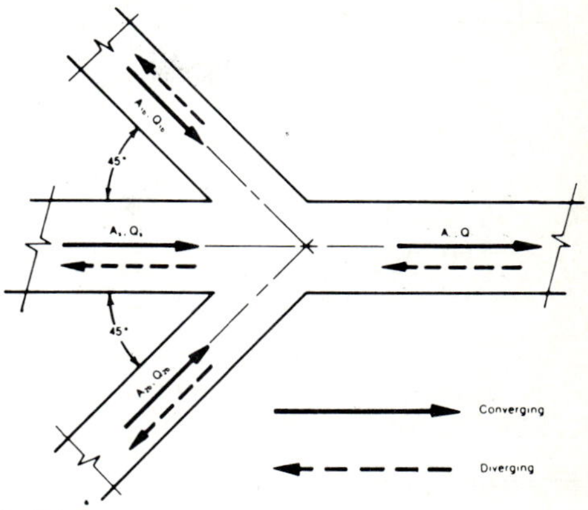

$A_{1b} = A_{2b}$

$A_s = A_c$

a. Converging Flow

Branch, $C_{c,b}$

Q_{2b}/Q_{1b}	Q_{1b}/Q_c: 0	0.1	0.2	0.3	0.4	0.5	0.6
			$A_{1b}/A_c = 0.2$				
0.5	−1.0	−.36	0.59	1.8	3.2	4.9	6.8
1.0	−1.0	−.24	0.63	1.7	2.6	3.7	—
2.0	−1.0	−.19	0.21	0.04	—	—	—
			$A_{1b}/A_c = 0.4$				
0.5	−1.0	−.48	−.02	0.58	0.92	1.3	16
1.0	−1.0	−.36	0.17	0.55	0.72	0.78	—
2.0	−1.0	−.18	0.16	−.06	—	—	—
			$A_{1b}/A_c = 0.6$				
0.5	−1.0	−.50	−.07	0.31	0.60	0.82	0.92
1.0	−1.0	−.37	0.12	0.55	0.60	0.52	—
2.0	−1.0	−.18	0.26	0.16	—	—	—
			$A_{1b}/A_c = 1.0$				
0.5	−1.0	−.51	−.09	0.25	0.50	0.65	0.64
1.0	−1.0	−.37	0.13	0.46	0.61	0.54	—
2.0	−1.0	−.15	0.38	0.42	—	—	—

Main, $C_{c,s}$

A_{2b}/A_{1b}	Q_s/Q_c: 0	0.1	0.2	0.3	0.4	0.5	0.6	0.7	0.8	0.9	1.0
					$A_{1b}/A_c = 0.2$						
0.5 & 2.0	−2.9	−1.9	−1.3	−.80	−.56	−.23	−.01	0.16	0.22	0.15	0
1.0	−2.5	−1.9	−1.3	−.80	−.42	−.12	0.08	0.20	0.22	0.15	0
					$A_{1b}/A_c = 0.4$						
0.5 & 2.0	−.98	−.61	−.30	−.05	0.14	0.26	0.33	0.34	0.28	0.17	0
1.0	−.77	−.44	−.16	0.05	0.21	0.31	0.36	0.35	0.29	0.17	0
					$A_{1b}/A_c = 0.6$						
0.5 & 2.0	−.32	0.08	0.11	0.27	0.37	0.43	0.44	0.40	0.31	0.18	0
1.0	−.18	−.04	0.21	0.34	0.42	0.46	0.46	0.41	0.31	0.18	0
					$A_{1b}/A_c = 1.0$						
0.5 & 2.0	0.11	0.36	0.46	0.53	0.57	0.56	0.52	0.44	0.33	0.18	0
1.0	0.29	0.42	0.51	0.57	0.58	0.58	0.54	0.45	0.33	0.18	0

b. Diverging Flow: Use Fitting 6-23.

6-36 90° Cross, Rectangular and Round[1]

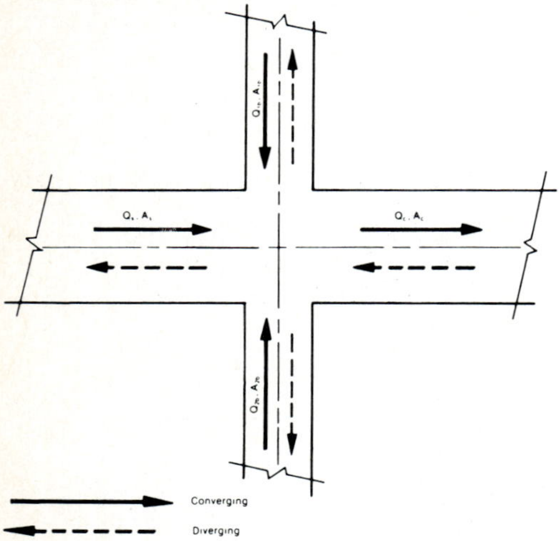

$A_{1b} = A_{2b}$

$A_s = A_c$

a. Converging Flow

	Branch, $C_{c,b}$						
Q_{2b}	Q_{1b}/Q_c or Q_{2b}/Q_c						
Q_{1b}	0	0.1	0.2	0.3	0.4	0.5	0.6
			$A_{1b}/A_c = 0.2$				
0.5	-.85	-.10	1.1	2.7	4.8	7.3	10
1.0	-.85	-.05	1.4	3.1	5.1	7.4	—
2.0	-.85	-.31	1.8	3.4	—	—	—
			$A_{1b}/A_c = 0.4$				
0.5	-.85	-.29	0.34	1.0	1.8	2.6	3.4
1.0	-.85	-.14	0.60	1.3	2.1	2.7	—
2.0	-.85	0.12	1.0	1.7	—	—	—
			$A_{1b}/A_c = 0.6$				
0.5	-.85	-.32	0.20	0.72	1.2	1.7	2.1
1.0	-.85	-.18	0.46	1.0	1.5	1.9	—
2.0	-.85	0.09	0.88	1.4	—	—	—
			$A_{1b}/A_c = 0.8$				
0.5	-.85	-.33	0.13	0.61	1.0	1.4	1.7
1.0	-.85	-.18	0.41	0.91	1.3	1.5	—
2.0	-.85	0.08	0.83	1.3	—	—	—
			$A_{1b}/A_c = 1.0$				
0.5	-.85	-.34	0.13	0.56	0.93	1.3	1.5
1.0	-.85	-.19	0.39	0.86	1.2	1.4	—
2.0	-.85	0.07	0.81	1.2	—	—	—

	Main					
Q_s/Q_c	0	0.1	0.2	0.3	0.4	0.5
$C_{c,s}$	1.2	1.2	1.2	1.1	1.1	0.96
Q_s/Q_c	0.6	0.7	0.8	0.9	1.0	
$C_{c,s}$	0.85	0.72	0.56	0.39	0.20	

b. Diverging Flow: Use Fitting 6-23.

Table B-7 Local Loss Coefficients, OBSTRUCTIONS

7-1 Damper, Butterfly, Round[1]

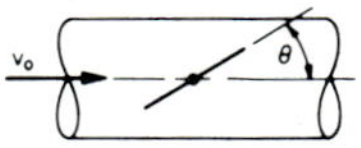

θ, deg	0	10	20	30	40	50	60
C_o	0.20	0.52	1.5	4.5	11	29	108

7-2 Damper, Butterfly, Rectangular[1]

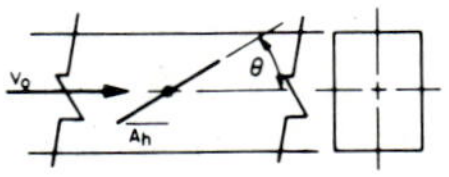

θ, deg	0	10	20	30	40	50	60
C_o	0.04	0.33	1.2	3.3	9.0	26	70

7-3 Damper, Gate, Round[1]

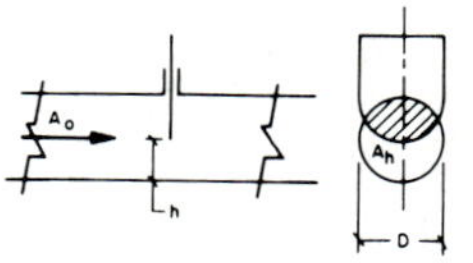

h/D	0.2	0.3	0.4	0.5	0.6	0.7	0.8	0.9
A_h/A_o	0.25	0.38	0.50	0.61	0.71	0.81	0.90	0.96
C_o	35	10	4.6	2.1	0.98	0.44	0.17	0.06

7-4 Damper, Gate, Rectangular[1]

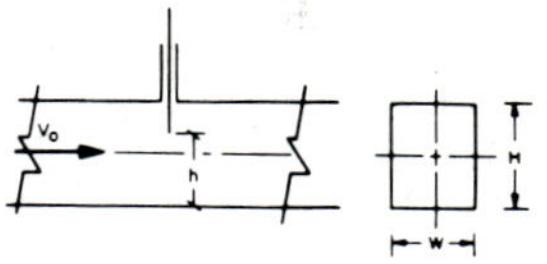

	h/H						
H/W	0.3	0.4	0.5	0.6	0.7	0.8	0.9
0.5	14	6.9	3.3	1.7	0.83	0.32	0.09
1.0	19	8.8	4.5	2.4	1.2	0.55	0.17
1.5	20	9.1	4.7	2.7	1.2	0.47	0.11
2.0	18	8.8	4.5	2.3	1.1	0.51	0.13

7-5 Damper, Butterfly, Airfoil Blade, Rectangular[1]

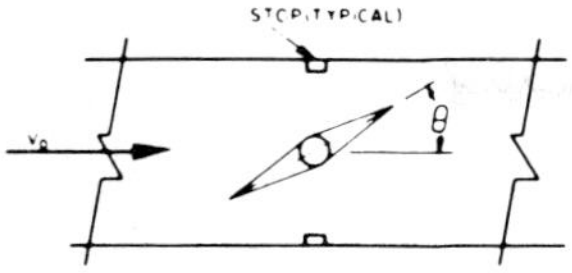

θ, deg	0	10	20	30	40	50	60
C_o	0.50	0.65	1.6	4.0	9.4	24	67

7-6 Damper, Rectangular, Parallel Blades[14]

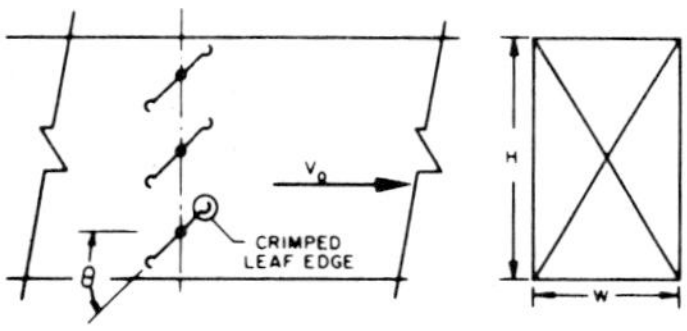

$$L/R = \frac{NW}{2\,(H + W)}$$

where

N = number of damper blades
W = duct dimension parallel to blade axis, in.
L = sum of damper blade lengths, in.
R = perimeter of duct, in.

$\frac{L}{R}$	C_o								
	θ, degrees								
	80	70	60	50	40	30	20	10	0
0.3	116	32	14	9.0	5.0	2.3	1.4	0.79	0.52
0.4	152	38	16	9.0	5.0	2.4	1.5	0.85	0.52
0.5	188	45	18	9.0	5.0	2.4	1.5	0.92	0.52
0.6	245	45	21	9.0	5.4	2.4	1.5	0.92	0.52
0.8	284	55	22	9.0	5.4	2.5	1.5	0.92	0.52
1.0	361	65	24	10	5.4	2.6	1.6	1.0	0.52
1.5	576	102	28	10	5.4	2.7	1.6	1.0	0.52

7-7 Damper, Rectangular, Opposed Blades[14]

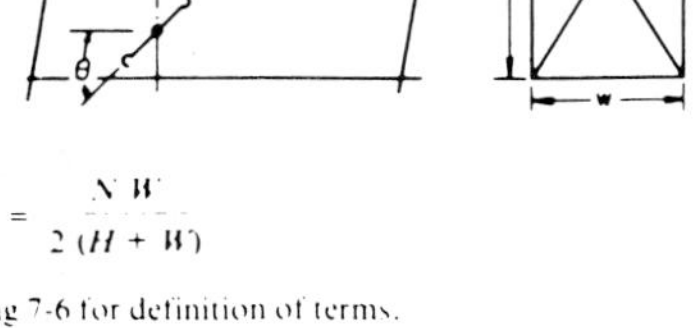

$$L/R = \frac{NW}{2\,(H + W)}$$

See Fitting 7-6 for definition of terms.

$\frac{L}{R}$	C_o								
	θ, degrees								
	80	70	60	50	40	30	20	10	0
0.3	807	284	73	21	9.0	4.1	2.1	0.85	0.52
0.4	915	332	100	28	11	5.0	2.2	0.92	0.52
0.5	1045	377	122	33	13	5.4	2.3	1.0	0.52
0.6	1121	411	148	38	14	6.0	2.3	1.0	0.52
0.8	1299	495	188	54	18	6.6	2.4	1.1	0.52
1.0	1521	547	245	65	21	7.3	2.7	1.2	0.52
1.5	1654	677	361	107	28	9.0	3.2	1.4	0.52

7-8 Obstruction, Screen in Duct, Round, and Rectangular[1]

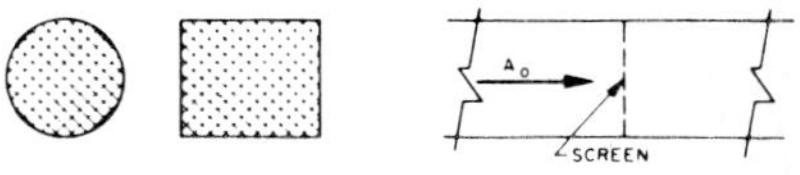

$$n = A_{or}/A_o$$

where

n = free area ratio of screen, dimensionless
A_{or} = total flow area of screen, in.^{2}
A_o = area of duct, in.^{2}

n	0.30	0.40	0.50	0.55	0.60	0.65	0.70	0.75	0.80	0.90	1.0
C_o	6.2	3.0	1.7	1.3	0.97	0.75	0.58	0.44	0.32	0.14	0

7-9 Obstruction, Perforated Plate, Thick, Round and Rectangular[1]

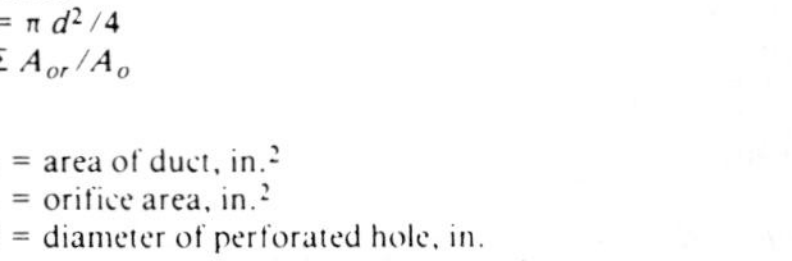

$t/d \geqslant 0.015$
$A_{or} = \pi d^2/4$
$n = \Sigma A_{or}/A_o$

where

A_o = area of duct, in.^{2}
A_{or} = orifice area, in.^{2}
d = diameter of perforated hole, in.
n = free area ratio of plate, dimensionless
t = plate thickness, in.

t/d	C_o							
	n							
	0.20	0.25	0.30	0.40	0.50	0.60	.70	0.80	0.90
0.015	52	30	18	8.2	4.0	2.0	0.97	0.42	0.13
0.2	48	28	17	7.7	3.8	1.9	0.91	0.40	0.13
0.4	46	27	17	7.4	3.6	1.8	0.88	0.39	0.13
0.6	42	24	15	6.6	3.2	1.6	0.80	0.36	0.13

7-10 Obstruction, Smooth Cylinder in Round and Rectangular Ducts[1]

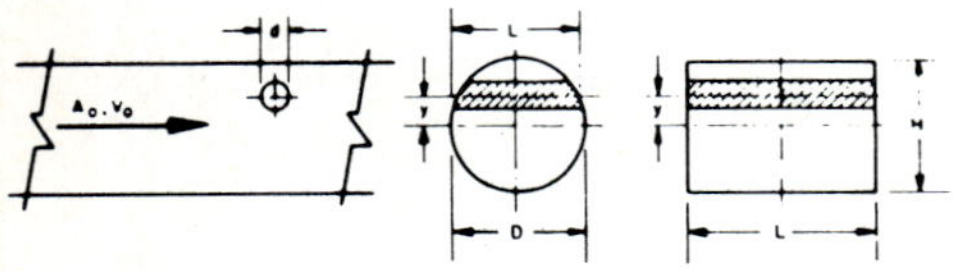

$S_m/A_o < 0.3$

$S_m = d\,L$

$R_e' = \rho d\,V_o/\mu$ (To calculate $R\acute{e}$, see Note 4.)

$C_o = K\,C_o'$

R_e'	C_o'			
	S_m/A_o			
	0.05	0.10	0.15	0.20
0.1	3.9	8.4	14	19
0.5	1.5	3.2	5.2	7.1
1	0.66	1.4	2.3	3.2
5	0.30	0.64	1.1	1.4
10	0.17	0.38	0.62	0.84
50	0.11	0.24	0.38	0.52
100	0.10	0.21	0.35	0.47
$(0.5\text{-}200)\times10^3$	0.07	0.15	0.24	0.33
3×10^5	0.07	0.16	0.26	0.35
4×10^5	0.05	0.11	0.19	0.25
5×10^5	0.04	0.09	0.14	0.19
$(6\text{-}10)\times10^5$	0.02	0.05	0.07	0.10

For obstruction offset from the centerline use the following factors:

y/D or y/H	0	0.05	0.10	0.15	0.20	0.25	0.30	0.35	0.40
K	1.0	0.97	0.93	0.89	0.84	0.79	0.74	0.67	0.58

7-11 Round Duct, Depressed to Avoid an Obstruction[7]

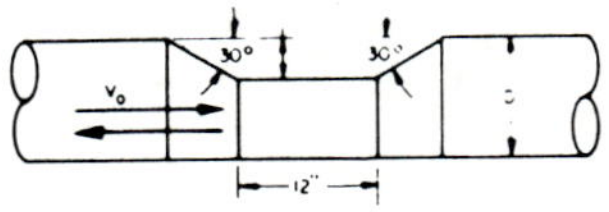

$L/D = 0.33$

$C_o = 0.24$

7-12 Rectangular Duct, Depressed to Avoid an Obstruction[7]

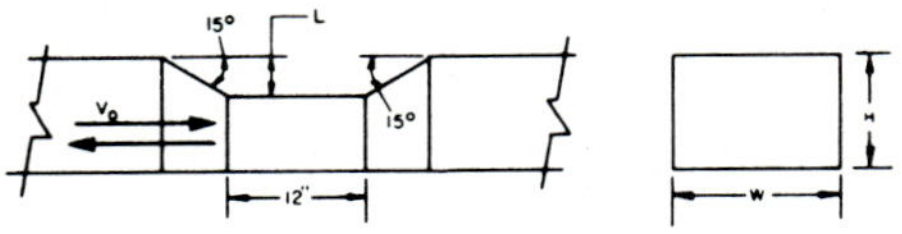

W/H	C_o			
	L/H			
	0.125	0.15	0.25	0.30
1.0	0.26	0.30	0.33	0.35
4.0	0.10	0.14	0.22	0.30

REFERENCES: APPENDIX B

[1] I.E. Idelchik: *Handbook of Hydraulic Resistances*, 2nd ed. (Mashinostroenia Publishing House, Moscow, USSR, 1975).

[2] A.D. Brandt and R.J. Steffy: Energy losses at suction hoods (ASHVE *Transactions*, Vol. 52, 1946, p. 205).

[3] *Designer's Handbook, Part II, Ventilation and Air Conditioning* (Stroyizdat, Moscow, 1977, pp. 256, 267).

[4] D.W. Locklin: Energy losses in *90-degree duct elbows* (ASHVE *Transactions*, Vol. 56, 1950, p. 479).

[5] R.D. Madison and J.R. Parker: Pressure losses in rectangular elbows (ASME *Transactions*, 1936, p. 167).

[6] J.M. Rozell: Duct turning vanes in 90-degree elbows (ASHRAE *Transactions*, Vol. 80, Part 2, 1974, p. 53).

[7] Unreported data based on tests sponsored by SMACNA in 1979 and 1980, and conducted at the ETL Testing Laboratories, Inc, Cortland, NY (Sheet Metal and Air Conditioning Contractors' National Association, Vienna, VA).

[8] H.F. Behls and W.K.Brown: Pressure loss coefficients for the 45-degree return air tee (*Building Science Series 39*, U.S. National Bureau of Standards, October 1971, p. 375).

[9] C.F. Sepsy and D.B. Pies: *An Experimental Study of the Pressure Losses in Converging Flow Fittings Used in Exhaust Systems* (prepared by Ohio State University for National Institute for Occupational Safety and Health, Document PB-221-130, April 1973).

[10] C.D. Jones and C.F. Sepsy: *Branch Loss Coefficients for 12 x 12 x 8 Plain and Combination Divided Flow Fittings* (United Sheet Metal, Div. of United McGill Corp., Columbus, OH, April 1969)

[11] C.D. Jones and C.F. Sepsy: *Friction Loss Coefficients for Four 12 x 10 x 8 Divided Flow Fittings and Various Combinations for Branch Crossover* (United Sheet Metal, Div. of United McGill Corp., Columbus, OH, June 1963).

[12] L.S. Kliatchko and V.F. Pustoshnaia: Research for the influence of some construction elements of rectangular tees on their aerodynamic characteristics (*Design and Installation of Sanitary Systems*, State Research Institute, Leningrad, 1970, p. 46).

[13] Korst, et al.: Fitting losses for extended-plenum forced air systems (ASHVE *Transactions*, Vol. 56, 1950, p. 259).

[14] E.J. Brown and J.R. Fellows: Pressure losses and flow characteristics of multiple-leaf dampers (*Heating, Piping and Air Conditioning*, August 1957, p. 119).

NOTE: Duct designs which are common for industrial applications (ASHRAE, 1980):

Frequently encountered: 1-1, 1-7, 3-1, 3-2, 6-4, 6-34

Occasionally encountered: 1-2, 1-3, 4-1, 4-8, 4-9, 4-10, 5-1, 5-2, 6-1, 6-2, 6-5, 6-6, 6-8, 6-17, 6-23, 7-1, 7-3, 7-10

Occasionally encountered but poor design: 2-7, 3-3, 4-7, 6-9

Chapter

6

Air Cleaning

R. A. Wadden

P. A. Scheff

R. J. Allen

I. Design Specifications

In Chapter 5 it was mentioned that the concentration of contaminants in air to be recirculated will ordinarily have to be reduced. Air exhausted outdoors may also need to be cleaned in order to avoid creating a community air pollution problem. The same types of devices are used in either case. Air cleaners used in the workplace for occupational safety and health reasons will be installed as a part of a ventilation system. Even if the air cleaner is free-standing, some ducting and an air mover will be needed. As with emission control on stack atmospheric releases, air cleaners used in ventilation systems need to be carefully designed. And, as with all such devices, field testing after installation must be carried out to determine whether design specifications have been met. Major considerations in the design of such equipment include (ACGIH, 1984; Spaite and Burckle, 1977):

1. The contaminant (particle or gas)
2. The required collection efficiency

3. The properties of the carrier gas and contaminant
4. The choice of an appropriate control device or collection process
5. The pressure drop across the air cleaner
6. The choice of recycle or outdoor exhaust

A. Collection efficiency

The collection efficiency of an air cleaner expresses the percentage or fraction of contaminant mass presented to an air cleaner that is removed from the air stream. Referring to Figure 6.1:

$$M_a = M_b + M_c \tag{6.1}$$

and

$$\eta = \left(\frac{M_c}{M_a}\right)(100) \tag{6.2}$$

where η = collection efficiency (%)
M_c = mass collected per unit time
M_a = mass available per unit time
M_b = mass escaping per unit time

Noting that the mass rate equals the product of flow rate, Q, and concentration, C, Eq. (6.2) can be written as

$$\eta = \left(\frac{M_c}{Q_aC_a}\right)(100) \tag{6.3}$$

or

$$\eta = \left(\frac{M_c}{Q_bC_b + Q_cC_c}\right)(100) \tag{6.4}$$

where the subscripts a, b, and c are shown in Figure 6.1. Penetration, Pt, is defined as the fraction or percentage of mass presented to the

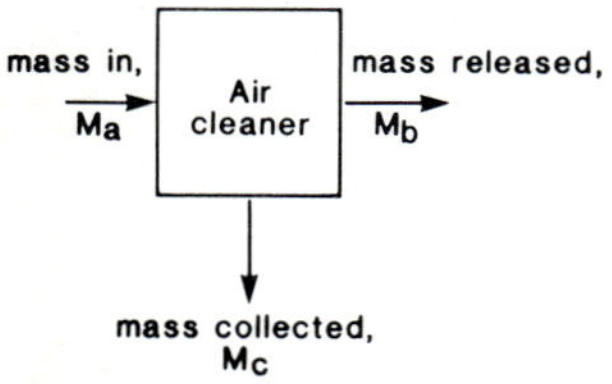

Figure 6.1 Air cleaner mass balance.

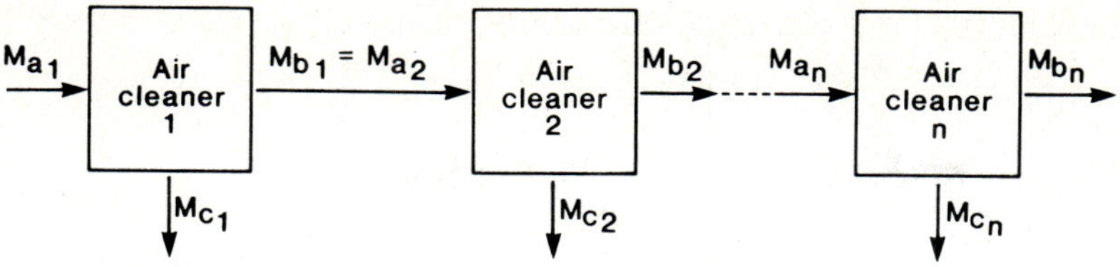

Figure 6.2 Air cleaners in series.

air cleaner that escapes collection. Referring to Figure 6.1:

$$\text{Pt} = \left(\frac{M_b}{M_a}\right)(100) \tag{6.5}$$

$$= \left(\frac{M_a - M_c}{M_a}\right)(100)$$

$$= 100 - \eta \text{ (as a percent)} \tag{6.6}$$

Now suppose that several air cleaners are operated in series as shown in Figure 6.2. The mass of contaminant escaping the first air cleaner is

$$M_{b_1} = \frac{\text{Pt}_1\, M_{a_1}}{100} = \frac{(100 - \eta_1)M_{a_1}}{100} \tag{6.7}$$

The mass escaping the second air cleaner is

$$M_{b_2} = \frac{(\text{Pt}_2)(M_{a_2})}{100}$$

$$= \frac{(\text{Pt}_2)(M_{b_1})}{100} \tag{6.8}$$

$$= \frac{(100 - \eta_1)(100 - \eta_2)(M_{a_1})}{(100)^2}$$

and the mass escaping the nth air cleaner can be shown to be

$$M_{b_n} = \frac{(100 - \eta_1)(100 - \eta_2)(100 - \eta_3)\cdots(100 - \eta_n)(M_{a_1})}{(100)^n} \tag{6.9}$$

giving

$$\text{Pt}_{\text{overall}} = (100)\left(\frac{M_{b_n}}{M_{a_1}}\right) = (100)\prod_{i=1}^{n}\left(\frac{100 - \eta_i}{100}\right) \tag{6.10}$$

The overall collection efficiency of the series of air cleaners is then

$$\eta_{\text{overall}} = 100 - \text{Pt}_{\text{overall}} \tag{6.11}$$

The efficiency of a particle removal device is a function of the size of the entering particles as well as the physical design of the collector. Most industrial particle emissions are composed of a wide range of sizes. The relationship between particle size and concentration is referred to as the particle size distribution. Because both particle size and collector efficiency vary with size, the overall collection efficiency for particle removal is a function of the product of the particle collection characteristics of the control device and the particle size distribution of the material being treated. Figure 6.3 shows typical performance characteristics of conventional air pollution control devices. The actual collection efficiency will vary with design conditions. For example, Figure 6.4 shows generalized efficiency curves for a large-diameter, low-efficiency cyclone and small-diameter, high-efficiency cyclones. These curves can be used to extend the data for cyclones on Figure 6.3 up to 50-μm-diameter particles. The effect of collector efficiency on overall efficiency is shown in Table 6.1. The first two columns in Table 6.1 show the uncontrolled particle size distribution of Portland cement kiln dust. Columns 3 and 4 show typical high- and low-efficiency cyclone collection efficiencies as shown in Figures 6.3 and 6.4 (use Figure 6.3 for particles <5 μm). Columns 5 and 6 show the weight percent of emission collected by the two devices (columns 2 times 3 and 2 times 4, respectively). As shown, the high-efficiency cyclone collected 85.84% of the total emission, whereas the low-efficiency device collected only 58.70%. Note that the same two devices on a different particle size distribution would have entirely different overall collection efficiencies.

B. Atmospheric release or recycle

What happens to the air stream after it passes through an air cleaner? Formerly air was simply exhausted outdoors and dispersed into the atmosphere. Unless there was poor ventilation design, the exhausted air did not reenter the workplace. However, a greater emphasis is now being placed on recycling air because of increased energy costs and the resulting increased costs for heating and cooling. This raises the question of how clean the recycled air should be. Clearly, recycled air should at least meet appropriate occupational health and safety guidelines (Chapter 4). If the contaminant concentration is being controlled through

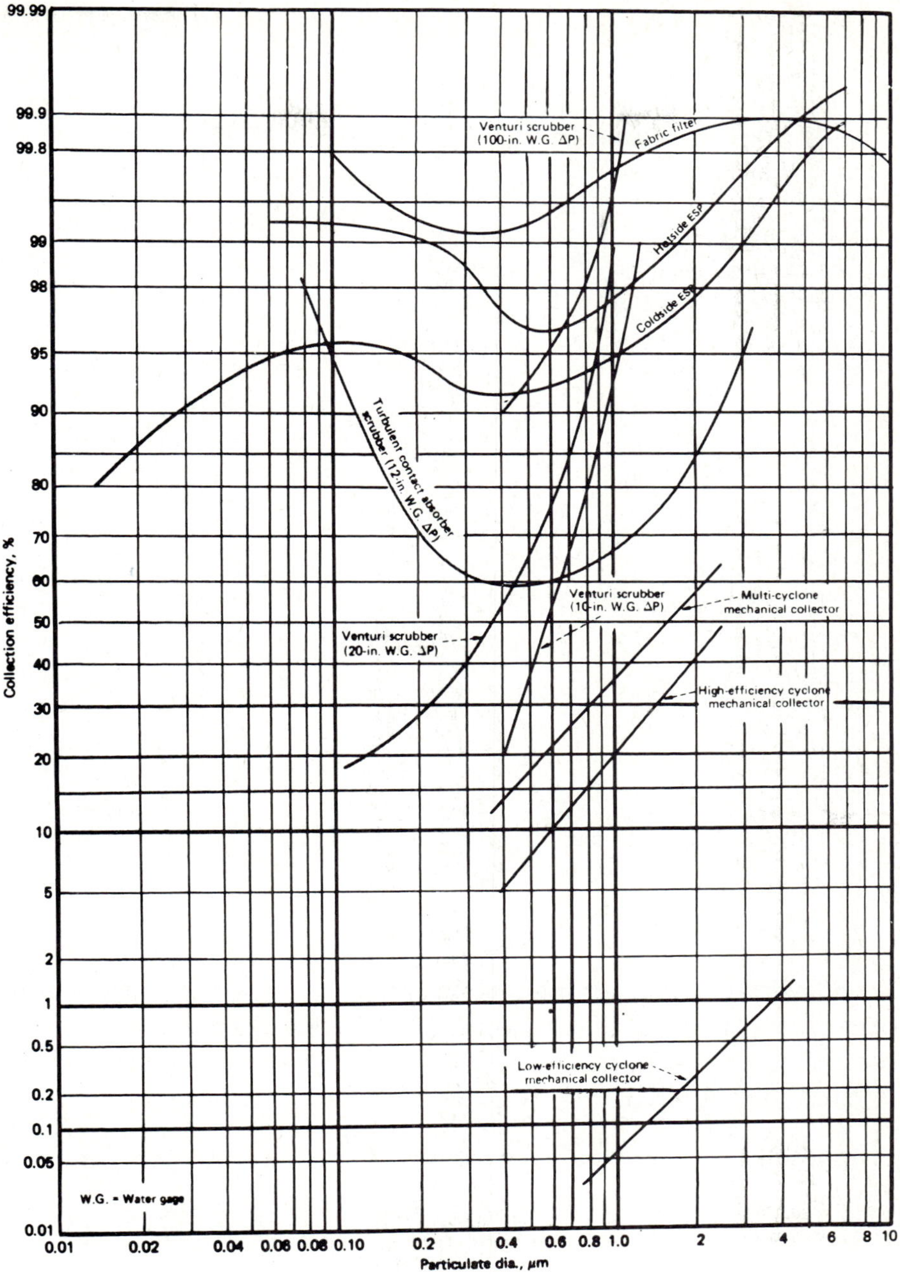

Figure 6.3 Typical removal efficiencies for conventional air pollution control devices (*Buonicore, 1980*).

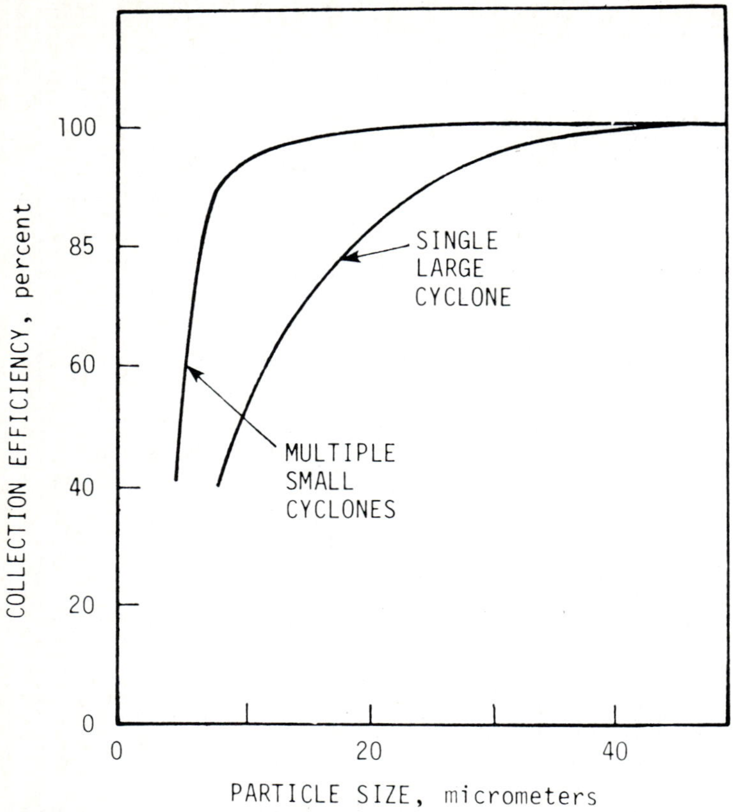

Figure 6.4 Typical relationship between cyclone collection efficiency and particle diameter for large and small tubes for the same inlet velocity (*EPA, 1982b*).

dilution ventilation, then reentering air most likely will need to be cleaner than the guidelines because of continued addition of pollutants. Even if local exhaust ventilation is used, it is still prudent to clean the air to a higher quality than the guidelines indicate because:

1. Guidelines for many substances have been lowered over time; and
2. There may be "more hazardous" contaminants present, such as suspected carcinogens, which have not yet been identified.

C. Air stream characteristics

Before proceeding to selection of an air cleaner, a number of questions must be answered. These answers will form the basis for the choice of a "generic" air cleaning device or process. Only then can a specific design be developed.

The first characteristic about the contaminant that needs to be de-

TABLE 6.1 Effect of Particle Size and Collector Efficiency on Overall Efficiency

Diameter (μm)	Portland cement kiln emissions (weight fraction)[a]	Cyclone collection efficiency (%)[c]		Percent collected	
		High efficiency	Low efficiency	High-efficiency cyclone	Low-efficiency cyclone
<1 (0.3)[b]	0.03	3	0.01	0.09	0.0003
1–5 (2.2)	0.20	60	0.5	12	0.1
5–10 (7.1)	0.15	85	34	12.75	5.1
10–20 (14.1)	0.20	95	69	19.0	13.8
20–30 (24.5)	0.16	100	88	16	14.1
30–40 (34.6)	0.10	100	96	10	9.6
40–50 (44.7)	0.06	100	100	6	6
50–60 (54.7)	0.03	100	100	3	3
>60 (77.5)	0.07	100	100	7	7
	1.00			85.84	58.70

[a]EPA, 1977.
[b]Median size of range.
[c]From Figures 6.3 and 6.4.

termined is whether it is present as a particle or a gas. The control technology that can be employed is different for particles and gases. Even if the contaminant consists of liquid, or solid and liquid droplets, it is still considered to be a particle.

The next question to be answered is that of the amount of contaminant and the volume of carrier gas to be cleaned. These quantities can be obtained in one of three ways:

1. For an existing ventilation system, the volumetric flow rate and contaminant concentration can be measured.
2. For a system yet to be built, flow rate and contaminant concentration from an existing system can be used if the two systems are very similar.
3. For a system yet to be built, rough estimates of flow rate and calculated source emissions (using emission factors) can be used.

In addition to the amount of contaminant and carrier gas, it is necessary to know other characteristics of both. These include temperature, water vapor concentration, combustion potential, and corrosion potential. Some control techniques are appropriate within certain temperature ranges (due to properties of materials); within ranges of water vapor concentration (due to properties of materials and clogging); when the contaminant and carrier gas are not combustible (using an electrostatic precipitator when the carrier gas is near an LEL can be devastating); and so that the control device will not be destroyed by corrosion.

Devices for particle control are usually referred to by a term indicating a generic type; for gas collection the process is specified. Generic particle control types include:

1. Cyclones
2. Electrostatic precipitators
3. Fabric filters
4. Wet scrubbers

Similarly, four general processes used for gas control are:

1. Adsorption
2. Absorption
3. Catalytic conversion
4. Thermal oxidation (combustion)

Examples of particle control devices used for selected stationary sources are listed in Table 6.2. Pressure drops, efficiencies, and operating data

TABLE 6.2 In-Place Particle Control Techniques for Selected Stationary Sources

Process	Control device	ΔP (kPa)[a]	Efficiency (%)	Operating data[b]
Stationary combustion				
Pulverized coal	Electrostatic precipitator	<0.25	≥99	$15 < \omega_e < 60$; $60 < A/Q < 130$
	Fabric filter	<1.5	≥99	$0.6 < A/C < 0.9$; shaker or reverse air cleaning
	Fabric filter[c]	0.6–2.5	>99	$0.3 < A/C < 0.9$; shake/deflate or reverse gas
	Venturi	4.0–7.0	95–99	$1.3 < L/G < 2.0$
Stoker boiler (<73 MW)	Electrostatic precipitator	< 0.25		$A/Q \simeq 100$
	Fabric filter	0.8–1.5	99	$1 < A/C < 1.5$; fiberglass bags with reverse air or pulse jet cleaning
Solid waste boiler	Electrostatic precipitator	<0.25		$11 < \omega_e < 15$; $A/Q \simeq 27$
	Fabric filter	0.5–0.75		$A/C \simeq 0.6$; coated fiberglass bags with shaker or reverse air cleaning
Municipal incinerator	Electrostatic precipitator	0.03–0.2	93–99	$6 < \omega_e < 12$; $24 < A/Q < 60$
	Fabric filter	0.75–1.75	>98	$0.6 < A/C < 1.2$; silicone-treated glass fiber bags; gas temperature must be between 120°C and 290°C
Chemical process industries				
Detergent manufacturing	Cyclone followed by fabric filter	~1.5	85.2 (cyclone) 95.6 (fabric filter)	$A/C < 1.8$
	Scrubber	2.5	98	Cyclone spray wet scrubber
		10.0	99.8	
Pesticide manufacturing				
Methyl parathion	Incinerator			Removes H_2S and mercaptans
	Scrubber			Removes phosphorus and HCl
Toxaphene	Fabric filter			
Pentachlorophenol	Venturi		90–100	Removes Cl_2, phenol, and acids
	Fabric filter		95–99	Collects pentachlorophenol
Potash dryers	Cyclone with fabric filters			
	Cyclone with scrubber			
Agricultural industry				
Grain handling	Cyclones	0.75–1.5		
	Fabric filters			Dacron felt bags; $A/C \simeq 3$
Cotton ginning	Cyclones	1.2	90	0.8 m diameter, high efficiency
	Fabric filters			Used less often
	Wet scrubbers			Used less often
Vegetable oil	Cyclones and fabric filters in combination		>99	$A/C \simeq 3$

TABLE 6.2 In-Place Particle Control Techniques for Selected Stationary Sources (*Continued*)

Process	Control device	ΔP (kPa)[a]	Efficiency (%)	Operating data[b]
Mineral products				
Crushing operations	Fabric filters			Cotton bags with shaker cleaner on 4-hr cleaning cycles; $A/C \simeq 0.6–0.9$
				Wool and synthetic felts on continuous clean; $1.8 < A/C < 3$
Clay manufacturing	Cyclones			Used as precleaners for the more efficient devices
	Wet scrubbers			Used when sulfur and floride concentration in clay are high
	Electrostatic precipitator			
	Fabric filters			
Lime kilns	Fabric filters	1.2		Glass fiber bags with graphite and silicone finishes; $0.5 < A/C < 1.0$
	Venturi	5.5		High-efficiency scrubber
	Electrostatic precipitator	<0.7	90–99	Horizontal flow construction
Cement kilns	Fabric filters			Glass and Nomex bags with reverse air cleaning with $0.37 < A/C < 0.46$; pulse air cleaning with $2.1 < A/C < 3.0$; shaker cleaning with $0.61 < A/C < 0.91$
	(combined cyclone and fabric filter common)			
Asphalt roofing products	Afterburner		86–97	Most popular method for controlling emissions from blowing stills (organic particulate matter and gases)
	Fabric filters		78–98	Controls emissions from saturators, coaters, mixers, and storage tanks.
	Electrostatic precipitator		92–94	Same as fabric filters
Glass manufacturing (welting furnace)	Fabric filters		72–95	$0.5 < A/C < 0.85$
	Venturi	8–10	80–85	Not used to control borosilicate emissions
	Electrostatic precipitator		91–97	$138 < A/Q < 290$; installed widely in industry
Metallurgical processes				
Sintering	Dry electrostatic precipitator			High resistivity causes collection problems
	Wet electrostatic precipitator	0.13	>99.5	Loading must be kept less than 0.57 g/Nm^3 with pretreatment cyclones
	Fabric filters	3.75	99.9	Oil content of exhaust must be kept low to minimize bag failure; $A/C \simeq 0.45$

	Venturi	>10		High collection efficiency for both particles and condensable hydrocarbons; $L/G \simeq 1.2$ with throat velocity $\simeq 140$ m/sec
Basic oxygen furnace	Venturi Electrostatic precipitator			Venturi preferred when high CO concentration poses an explosion hazard; ESP preferred when on-site water treatment is inadequate
Electric arc furnaces	Fabric filters			Gas temperature reduction by sprays, radiant coolers, or dilution with ambient air common
Primary aluminum	Venturi followed by electrostatic precipitator		99 for particulate matter 90 for fluorides	
	Counterflow packed scrubber		95 for particulate matter 90 for fluorides	
	Dry alumina scrubber		99 for particulate matter 99 for fluorides	
Lead smelting	Fabric filters	0.5–1.0	99	Wool and acrylic bags with shaker cleaning; $2 < A/C < 8.4$
	Electrostatic precipitator			Preferred when SO_3 concentration is high
Forest products				
Kraft recovery boiler	Electrostatic precipitator		99.5–99.8	$1.2 < A/Q < 1.5$
Lead-acid battery manufacturing	Fabric filters	1.2	99	Pulse-jet cleaning, $A/C \simeq 1.5$
	Fabric filters	1.5	98	Shaker cleaning, $1 < A/C < 2$

[a]0.249 kPa = 249 N/m^2 = 1 in.H_2O.

[b]ω_e = effective migration velocity, cm/sec; A/Q = collection plate area-to-volumetric flow ratio $m^2/(Am^3/min)$, with $Am^3 = m^3$ at actual operating temperature and pressure; A/C = air-to-cloth ratio, $(Am^3/min)/m^2$ net cloth area, where net cloth area is total filter area available for on-stream filtration exclusive of the filter area in compartments isolated for cleaning; L/G = liquid-to-gas ratio, liters/Am^3.

[c]Field data on 13 coal-fired electric generating plants, 25–354 mW (Carr and Smith, 1984).

SOURCE: EPA, 1982b.

shown are typical of each source-control device combination. Table 6.3 lists emission controls currently in use for gaseous pollutants in the air oxidation process industry. Techniques to estimate efficiency and pressure drop for control equipment and control processes are discussed in more detail in subsequent sections of this chapter. The emphasis of our approach will be on an estimation of collection efficiency and pressure loss based on the conditions of the application. This will permit the inclusion of an air cleaner in the design of a ventilation system as well as development of design specifications for the air cleaner manufacturer. This discussion is not intended to permit complete nuts-and-bolts design.

TABLE 6.3 Product Recovery and Emission Controls Currently Used to Collect Gaseous Organics in Synthetic Organic Manufacturing Plants Employing Major Air Oxidation Processes

Chemical	Process	Product or raw material recovery equipment	Emission control equipment
Acetaldehyde	Wacker	1, 2	
Acetic acid	Wacker	1, 2	
Acetic acid/formic acid/ methyl ethyl ketone	Butane	1, 2	
Acetone/phenol	Cumene peroxidation	1, 3	
Acrylonitrile	Propene ammoxidation	1, 2	5
Acrylic acid	Propene	1, 2	4B
1,3-Butadiene	Butene oxidative dehydrogenation	1, 2	4B
Cyclohexanol/cyclohexanone	Cyclohexane	1, 2	
Ethylene dichloride	Ethylene oxychlorination	1, 2	4B, 5
Ethylene oxide	Ethylene	1, 2	5
Formaldehyde	Silver catalyst	1, 2	
Formaldehyde	Mixed metal oxide catalyst	1, 2	
Hydrogen cyanide	Andrussow	2	6
Maleic anhydride	Benzene	1, 2, 3	4B, 5
Maleic anhydride	Butane	1, 2	4
Phthalic anhydride	Naphthalene	1, 2	4
Phthalic anhydride	Xylene	1, 2	4B
Propylene oxide/styrene	Ethylbenzene peroxidation	1, 2	
Terephthalic acid/dimethyl terephthalate	Xylene	1, 2, 3	

1 Condenser
2 Absorber
3 Carbon adsorber
4 Thermal incineration
4B Thermal incinerator–waste heat boiler
5 Catalytic incinerator
6 Process heater or boiler

SOURCE: EPA, 1984.

II. Particle Collection Equipment

Cyclones, fabric filters, electrostatic precipitators, and scrubbers are not interchangeable in a given application. Selection of a particle control device depends on several factors. Perhaps the most important is that of the size of the particles to be collected and the efficiency with which to collect those particles. Figure 6.3 is a generalized plot of particle size and collection efficiency for the four types of particle control. The actual collection efficiency will vary with design conditions. The point is that if large particles are to be captured, then a cyclone will probably be adequate. As the size of the particles to be collected gets smaller, more sophisticated control devices are necessary. The particle loading that a device can handle, gas flow rate, pressure loss, space requirements, and power and water requirements also affect the choice of a particle collector. These variables are summarized in Tables 6.4 and 6.5. The calculation of collection efficiency and pressure loss are discussed separately for each kind of control device.

A. Cyclone

In a cyclone, particulate matter is removed by centrifugal forces generated by providing a path for the carrier gas to spin. Separation is a function of the strength of these forces balanced against particle drag. Calculation of collection efficiency begins by determination of the cut size of the cyclone (Perry et al., 1984; Wark and Warner, 1981). The cut size is defined as the size of particle collected with 50% efficiency,

TABLE 6.4 Typical Characteristics of Particle Control Equipment

Particle collection device	Inlet concentration (mg/m^3)	Typical overall collection efficiency (weight %)	Pressure loss (in.H_2O)	Maximum gas flow rate (10^3 cfm)	Relative space required
Cyclone	2000	85	0.5–3	17.7	Medium
Multiple cyclone	2000	95	2–6	70.6	Small
Fabric filter	200	99	2–6	70.6	Large
Electrostatic precipitator	200	99	0.2–1	706.2	Large
Wet scrubbers					
Gravity spray	2000	70	1	35.3	Medium
Centrifugal	2000	90	2–6	35.3	Medium
Impingement	2000	95	2–8	35.3	Medium
Packed bed	600	90	1–10	17.7	Medium
Submerged orifice	200	90	2–6	17.7	Medium
Venturi	200	99	10–30	35.3	Small

SOURCE: ASHRAE, 1983.

d_c, and can be estimated from

$$d_c = \left[\frac{9\mu B_c}{2\pi N V_i(\rho_s - \rho)}\right]^{1/2} \tag{6.12}$$

where d_c = cut size
μ = gas viscosity
B_c = cyclone inlet width (see Figure 6.4)
N = effective number of outer turns in the cyclone (typically $\simeq 4$)
V_i = inlet gas velocity
ρ_s = particle density
ρ = gas density (usually considered $<< \rho_s$)

The effective number of outer turns made by the gas in the cyclone must ordinarily be determined from actual experiment. One method of estimation is (Wark and Warner, 1981):

$$N = \frac{1}{H_c}\left(L_c + \frac{Z_c}{2}\right) \tag{6.13}$$

TABLE 6.5 Particle Control Equipment Operating Requirements

Device	Typical capacity	Water requirement (gal/1000 ft^3 gas)	Pressure drop (in.H_2O)	Power requirement [W/(ft^3/min)]
Multiple cyclone (high efficiency)	2,500–3,500 ft^3/min per ft^2 inlet area	—	3–5	0.5–1.0
Fabric filter (automatic)	1–2 (ft^3/min)/ft^2 of fabric: reverse air cleaning 5–15 (ft^3/min)/ft^2 of fabric: pulse jet cleaning 2–5 (ft^3/min)/ft^2 of fabric: mechanical shaker cleaning	—	4–6	1.0–1.3
Electrostatic precipitator (dry single field)	2–8 ft^3/min per ft^2 of electrode collection area	—	0.2–0.5	0.3–1.0
Wet scrubbers				
Packed bed	500–700 ft^3/min per ft^2 of bed cross-sectional area	1–4	6–8	0.1–3.0
Venturi	6,000–30,000 ft^3/min per ft^2 throat area	5–10	10–50	4–12
Impingement baffle	400–600 ft^3/min per ft^2 of baffle area	1–5	2–5	0.2–1.0

SOURCE: ASHRAE, 1983; Wark and Warner, 1981; EPA 1982a.

where L_c = height of cyclone cylinder
Z_c = height of cyclone cone
H_c = height of inlet

Note that L_c, Z_c, and H_c must all have the same units. These dimensions are shown on Figure 6.5, together with standard proportions for conventional cyclone design.

Once d_c is determined for a particular cyclone, the ratio of actual particle size, d, to d_c, is calculated. This ratio is then used to obtain

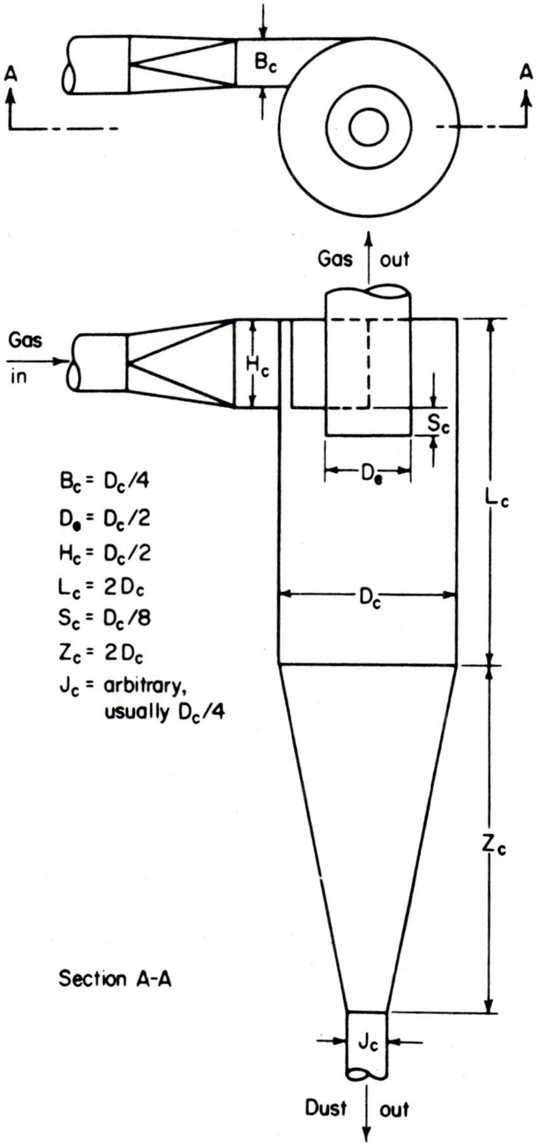

Figure 6.5 Conventional cyclone separator dimensions (*Perry et al., 1984*).

collection efficiency from Figure 6.6 (which was determined experimentally for cyclones with the proportions of Figure 6.5; Lapple, 1951).

The actual configuration of gas flow is more complex than that which would be inferred from Eqs. (6.12) and (6.13). However, more complicated design procedures that take flow patterns into account still result in about the same estimates of efficiency. Figure 6.7 shows reasonable agreement between the method presented here and a method that postulates three separate regions within the cyclone (Dietz, 1981). The cut size predicted from Eq. (6.12) was 3.09 μm compared with the three-region model prediction of 3.01 μm. A comparison of the three-region model with operating data also is reasonable, as shown in Figure 6.8. Ordinarily, Eqs. (6.12), (6.13), and Figure 6.6 will provide a satisfactory first estimate of efficiency for cyclones that have design proportions similar to those shown in Figure 6.5. The advantage of the more complex analysis is that it can also be used for cyclone designs not similar to Figure 6.5.

No general correlation of cyclone pressure drop data is presently available. One empirical relationship (Caplan, 1977) has the form

$$\Delta P = \frac{P\rho K Q^2}{T} \tag{6.14}$$

where ΔP = loss in total pressure, in.H_2O
Q = gas flow rate, actual ft^3/min
P = absolute pressure, atm
ρ = gas density, lb/ft^3
T = absolute temperature, °R
K = factor dependent on cyclone size (see Table 6.6)

Efficiency and pressure drop in cyclones are closely related (as is true with most other collectors). In general, increasing cyclone size (D_c), cylinder length (L_c), and exit diameter (D_e) decrease both pressure drop and efficiency while lengthening the cone (Z_c) and increasing velocity increases efficiency. Groups of small cyclones with higher ef-

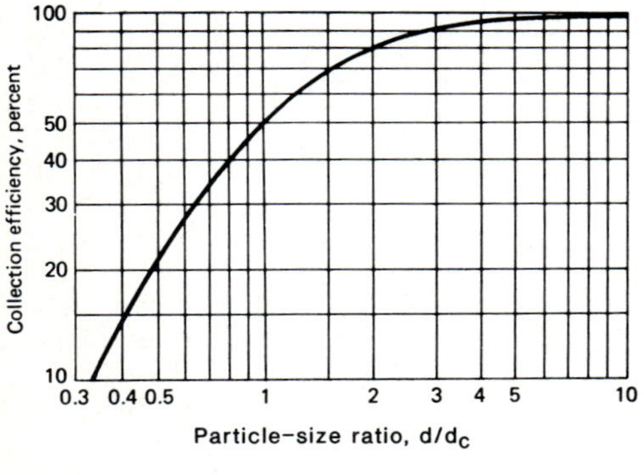

Figure 6.6 Cyclone efficiency (*Lapple, 1951*).

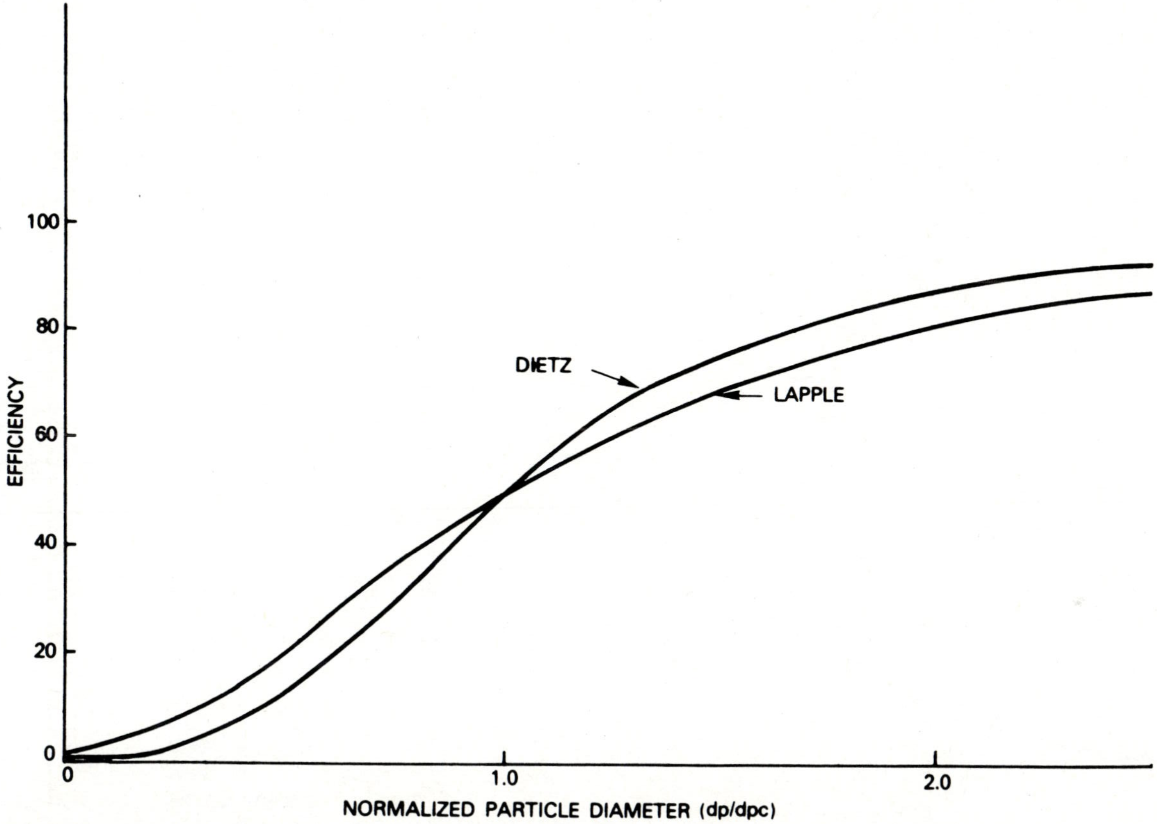

Figure 6.7 Comparison between three-region model and conventional design procedure (*Dietz, 1981*). (D_c = 0.2 m and assumed m = 0.7; *m*, the strength of free vortex in an inviscid fluid, varies only between 0.5 and 1.0.)

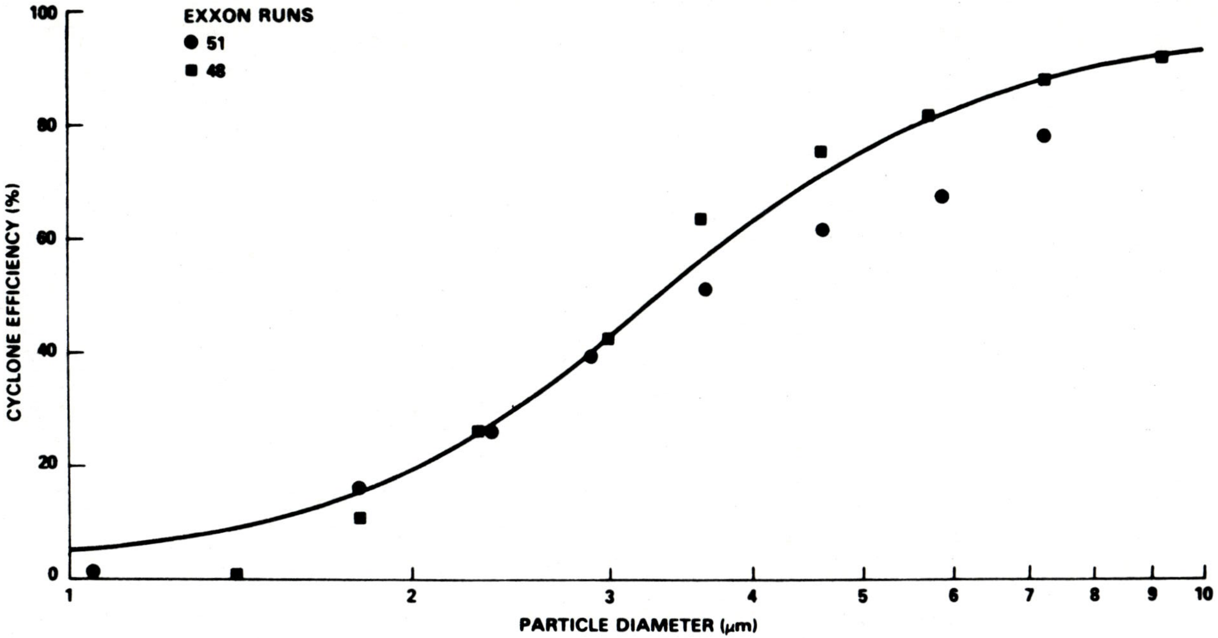

Figure 6.8 Comparison of theory and experiment for three-region cyclone model (*Dietz, 1981*). Points are from two separate experimental runs.

TABLE 6.6 Values of *K* for Calculating Cyclone Pressure Drop for Conventional Design

Cyclone diameter (in.)	K
29	10^{-4}
16	10^{-3}
8	10^{-2}
4	10^{-1}

SOURCE: Caplan, 1977.

ficiencies and pressure drops can be arranged in either parallel or series to both increase efficiency and decrease pressure drop.

B. Electrostatic precipitator

Electrostatic precipitators operate on the principle of imparting a charge to particles that are then collected at an oppositely charged electrode (Figure 6.9). The effective drift or migration velocity, ω_e, is the speed at which charged particles move to the collecting electrode and represents a balance between electrostatic and drag forces. The collection efficiency is a function of migration velocity and can be estimated from the Deutsch equation (White, 1977a):

$$\eta = 1 - e \exp[-A\omega_e/Q] \qquad (6.15)$$

where η = collection efficiency
A = area of collecting surfaces
ω_e = effective particle migration velocity
Q = volumetric flow rate

All variables in Eq. (6.15) must be in consistent units.

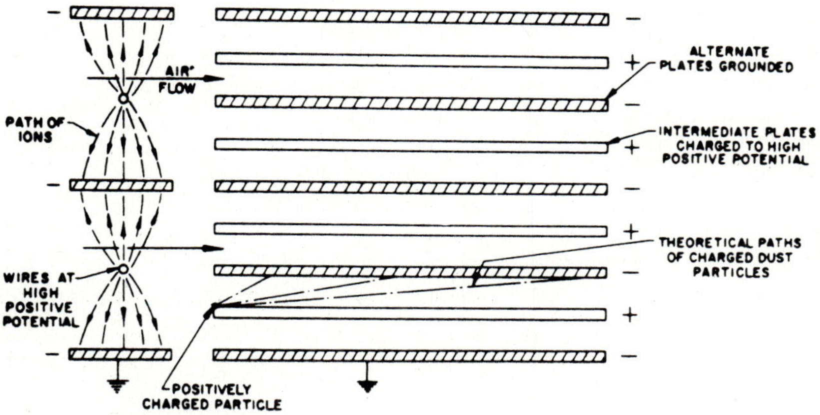

Figure 6.9 Diagrammatic cross section of ionizing-type electronic air cleaner (*ASHRAE, 1983*). Reprinted by permission of ASHRAE, New York.

For particles larger than 1 μm, the theoretical migration velocity will be directly proportional to the particle diameter. Smaller particles undergo a somewhat different charging mechanism and consequently will all have about the same mobility. Often, however, not enough information is available for a given installation to evaluate the migration velocity (or efficiency) by particle size. Under such circumstances an average ω_e, based on the measured overall efficiency, is used in Eq. (6.15). Table 6.7 gives typical ranges for various types of particle emissions.

Dust resistivity is also an important parameter in precipitator design. Particles of low resistivity (10^4–10^7 Ω·cm) are difficult to collect, as they easily lose their charge at the collection electrode. Unburned carbon in fly ash and carbon black are examples of particles with low resistivity. Particles with high resistivity are hard to collect because they are difficult to charge and, once collected, form a high-potential electric field between the dust layer and the collection electrode. The resistivity of a dust can be adjusted over a limited range by the addition of water, changing gas temperature, or by adding SO_3. The effective migration velocities in Table 6.7 incorporate the effects of dust resistivity as well as particle size on collection efficiency. For example, the wide range of ω_e for utility fly ash indicated in Table 6.7 results from the highly variable resistivity of fly ash and SO_2–SO_3 concentrations in emissions from such sources.

TABLE 6.7 Typical Effective Migration Velocities for Industrial Dusts

	ω_e (ft/sec)			
Application	Perry et al., 1984	ASHRAE, 1983	White, 1977b	EPA, 1982b
Utility fly ash	0.40	0.07–0.49	0.2–0.3[b]	
Sulfuric acid mist		0.20–0.26		
Dry cement	0.31	0.20–0.23		
Gypsum from kiln	0.64	0.43		
Catalyst dust		0.07–0.26		
Secondary blast furnace (80% foundry iron)		0.43		
Gray iron cupola (iron–coke ratio of 10)		0.10–0.13		
Smelter fume	0.13			0.19[a,c]
Coke oven gas				0.01–0.06[a]
Paper mills		0.26		
Glass melting furnace (lead glass)				0.03[a]

[a]Derived from test data.

[b]Representative of recent installations designed for 99+% efficiency and resistivity not much greater than 10^{10} Ω·cm; values as low as 0.05 ft/sec and as high as 0.8 ft/sec are sometimes encountered.

[c]Reverberatory copper furnace.

As indicated in Tables 6.4 and 6.5, the total pressure drop in an electrostatic precipitator is usually less than 1 in. H_2O (ACGIH, 1984; ASHRAE, 1983). No general equations are available for calculation of pressure loss, but 1 in.H_2O would constitute a conservative design.

C. Wet scrubbers

Wet collectors are devices that use a liquid to remove particles or gases from an air stream. Control device configurations include plate scrubbers, packed beds, orifice scrubbers, venturi scrubbers, spray towers, cyclone spray scrubbers, moving bed scrubbers, baffle spray scrubbers, and various combinations. General values for ΔP, η, and the water/gas ratio are given in Tables 6.4 and 6.5. Actual pressure drop and efficiency data for a variety of installations are given in Table 6.8. Predictions of pressure drop for some of the same installations are shown in Figure 6.10. Particle sizes are expressed as the aerodynamic cut diameter, d_{50A}. This is the aerodynamic diameter (the diameter of a sphere with density 1 g/cm^3 that has the same terminal setting velocity as the actual particle) that is collected with an efficiency of 50%.

Prediction of ΔP and η for specific equipment types depends on the geometry of the contacting device and often requires a somewhat com-

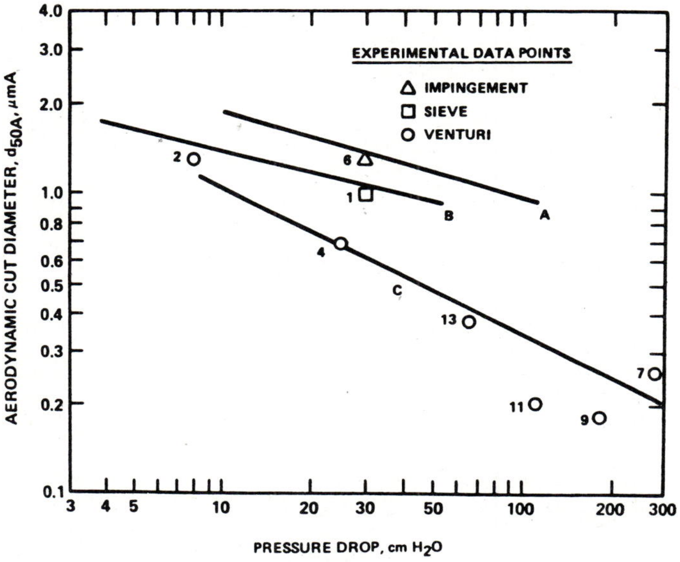

Figure 6.10 Theoretical and experimental cut diameters as a function of pressure drop for several scrubber types (*EPA, 1979*). (Numbers on data points from Table 6.8).

Prediction curve	*Device*
A	Impingement plate
B	Sieve
C	Venturi

TABLE 6.8 Performance Test Results for Various Wet Scrubbers

Number[a]	Source	Scrubber	Penetration (%)	ΔP(cmH_2O)	Mass median diameter (μm)
1	Urea prilling tower	Valve tray	54	30	1.1
2	KCl dryer	Vaned centrifugal	3	8	>100
3	Coal-fired utility boiler	Mobile bed (TCA)	6	30	3.0
4	Coal-fired utility boiler	Venturi	2.7	25	3.8
5	NaCl dryer	Wetted fiber	2.7	19	10
6	NaCl dryer	Impingement plate	10	30	>100
7	Foundry cupola	Venturi rod	0.4	273	1.0
8	Coal-fired utility boiler	Mobile bed (TCA)	8.1	22	9.0
9	Foundry cupola	Variable rod venturi	1.5	180	0.8
10	Foundry cupola	Gas-atomized spray	9.0	100	0.4
11	Borax fusing furnace	Venturi	2.5	110	1.0
12	Coal-fired utility boiler	Venturi	0.7	46	20
13	Asphalt batch plant	Venturi	0.3	66	10

[a]Number cited in Figure 6.10.
SOURCE: EPA, 1979.

plex calculational procedure. Probably the best-developed estimation method on theoretical grounds is that for venturi scrubbers (Boll, 1973; Goel and Hollands, 1977; Crowder et al., 1982; Yung et al., 1977; Yung et al., 1978). Although detailing the calculational approach for all types of contractors is beyond the scope of this chapter, it is perhaps useful to describe that for venturis.

A typical venturi design is shown in Figure 6.11. Liquid, usually water, is injected at the venturi entrance, accelerated with the gas and particles through the throat, removed, and usually recycled after cleaning. Many installations include a device to adjust the throat size.

One procedure for determining pressure drop in a venturi has been given by Yung et al. (1977):

$$\Delta P = \frac{2\rho_L U_G^2}{g_c}\left(\frac{Q_L}{Q_G}\right)[1 - x^2 + (x^4 - x^2)^{0.5}] \tag{6.16}$$

$$x = \frac{3LC_{Do}\rho_G}{16 d_D \rho_L} + 1 \tag{6.17}$$

$$C_{Do} = 0.22 + \frac{24}{N_{\mathrm{Re}_1}}(1 + 0.15N_{\mathrm{Re}_1}^{0.6}) \tag{6.18}$$

$$N_{\mathrm{Re}_1} = \frac{\rho_G U_G d_D}{\mu_G} \tag{6.19}$$

$$d_{D\ (\text{for water})} = \frac{49.7}{U_G} + 91.1\left(\frac{Q_L}{Q_G}\right)^{1.5} \tag{6.20}$$

where ΔP = pressure drop, dynes/cm²
ρ_L, ρ_G = liquid and gas densities, respectively, g/cm³
U_G = gas velocity, cm/sec
g_c = 1 g·cm·sec⁻²/dyne
Q_L, Q_G = volumetric flow rate for liquid and gas, respectively, cm³/sec
L = throat length, cm
d_D = liquid drop diameter, cm
N_{Re_1} = Reynolds number based on drop diameter at throat entrance
C_{Do} = dimensionless drag coefficient for the liquid droplet
μ_G = gas viscosity, poise (g/cm·sec)

Equation (6.18) is for flow around spheres and is valid for Reynolds numbers from 1 to 500. If the water drops accelerate to the gas velocity in the throat, $[1 - x^2 + (x^4 - x^2)^{0.5}] = 0.5$, and Eq. (6.16) reduces to

$$\Delta P = \frac{\rho_L U_G^2}{g_c}\frac{Q_L}{Q_G} \tag{6.21}$$

or, for water droplets and pressure drop in inches of water,

$$\Delta P = (4.02 \times 10^{-4})U_G^2 \frac{Q_L}{Q_G} \tag{6.22}$$

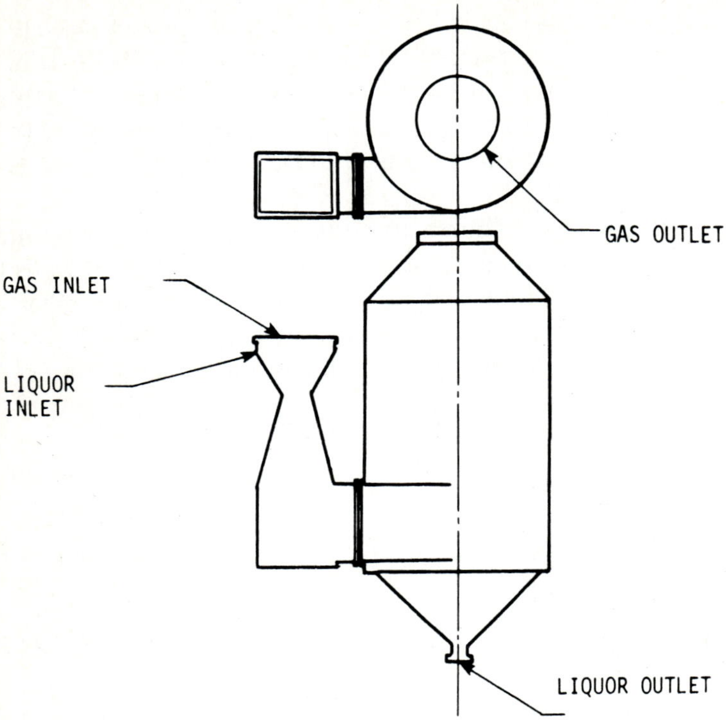

Figure 6.11 Venturi scrubber (*EPA, 1982a*).

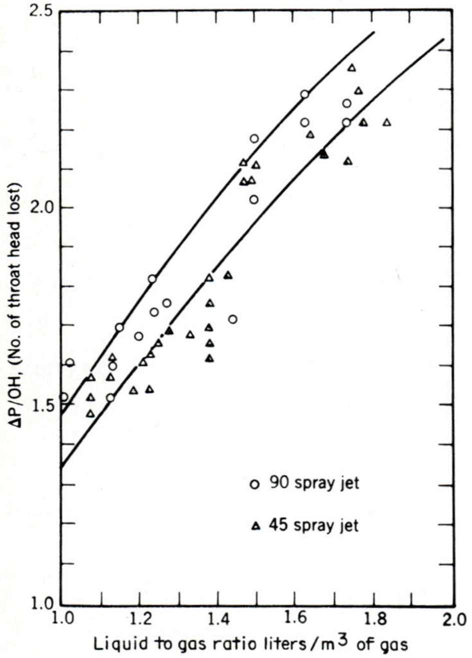

Figure 6.12 Comparison of theoretical estimates of ΔP for a venturi scrubber with operating data (*Yung et al., 1977*). Lower line predicted from Eq. (6.16), upper line from model of Boll (1973); data from Brink and Coutant (1958). $\Delta P = (\rho_G U_g^2/2g_c) \cdot$ (y-coordinate value).

Equation (6.22) will usually give a reasonable first estimate. However, it is well to recognize that for industrial venturis the residence time in the throat section is rarely long enough to allow acceleration to the gas velocity. Consequently, Eq. (6.22) will predict a higher pressure drop than is experimentally measured, except at low liquid rates where friction losses from gas flow predominate (Yung et al., 1977).

Equations (6.16) and (6.21) are derived from mass and momentum balances across the venturi. A comparison between Eq. (6.16) and experimental data is shown in Figure 6.12. The industrial scrubber tested had a rectangular cross section of 15.2 × 86.4 cm and a throat length of 30.5 cm. Also shown are predicted values for another theoretical model (Boll, 1973). ΔP may be calculated from values on the y coordinate by multiplying by $\rho_G U_G^2/2g_c$.

The corresponding efficiency relationship (Yung et al., 1978) is shown graphically in Figure 6.13. The procedure for using the plot is as follows:

1. Calculate the Cunningham slip correction, K_c, for the particle size of interest. K_c (dimensionless) for particles in air at 1 atm and 25°C is given by

$$K_c = 1 + \frac{0.1340}{d_p}[1.257 + 0.4\exp(-8.21 d_p)] \tag{6.23}$$

   where the particle diameter, d_p, is in micrometers.
2. Calculate water drop diameter, d_D, from Eq. (6.20).
3. Calculate N_{Re_1} from Eq. (6.19) using the gas velocity in the venturi and the water drop diameter, and consistent units.
4. Calculate C_{Do} from Eq. (6.18).
5. Calculate the dimensionless throat length, L^*, from

$$L^* = \frac{3C_{Do}\rho_G L}{2d_D\rho_L} \tag{6.24}$$

   using consistent units.
6. Calculate N_1 (dimensionless), the inertial impaction parameter, from

$$N_1 = \frac{K_c\rho_p d_P^2 U_G}{18\mu_G d_D} \tag{6.25}$$

   with the parameters in consistent units.
7. Read $-(\ln \mathrm{Pt})/B$ from Figure 6.13.

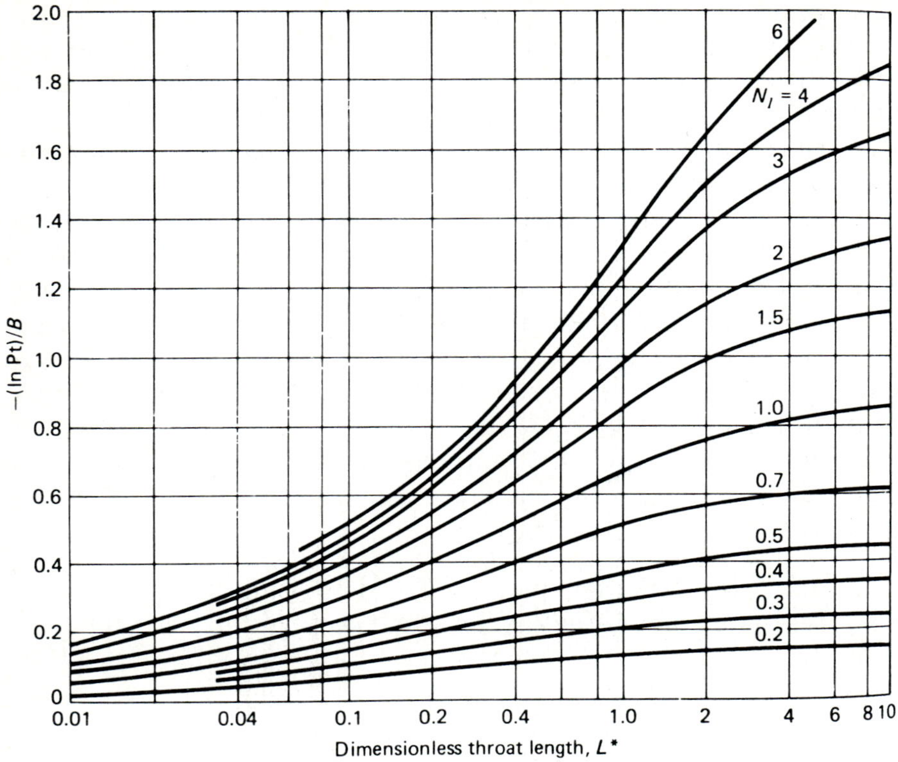

Figure 6.13 Pentration in venturi scrubber (*Yung et al., 1978; redrawn by Wark and Warner, 1981;* $B = Q_L\rho_L/Q_G\rho_G C_{Do}$, $L^* = 3C_{Do}\rho_g L/2d_D\rho_L$, $N_1 = K_c\rho_p d_p^2 U_G/18\mu_G d_D$). This plot is developed from the following equations (*Rudnick et al., 1986*):

$$U^* = 2[1 - x^2 + (x^4 - x^2)^{0.5}]$$

$$-\frac{(\ln \text{Pt})}{B} = \left[\frac{2.1 + 4N_1 - 3.55N_1^{0.5}(1 + 0.35/N_1)\tan^{-1}(N_1/0.35)^{0.5}}{0.35 + N_1}\right]$$

$$- \left\{\frac{4N_1(1 - U^*)^{1.5} + 2.1(1 - U^*)^{0.5} - 3.55N_1^{0.5} \times (1 - U^* + 0.35/N_1)\tan^{-1}[N_1(1 - U^*)/0.35]^{0.5}}{0.35 + N_1(1 - U^*)}\right\}$$

where U^* is the ratio of droplet velocity at exit of throat to gas velocity in throat, dimensionless; the argument of the arctan is expressed in radians.

8. Calculate B (dimensionless) from

$$B = \frac{Q_L\rho_L}{Q_G\rho_G C_{Do}} \tag{6.26}$$

using consistent units.

9. Calculate the penetration, Pt (which will be a fraction).

At values of $L^* > 10$, $-\ln \text{Pt}/B$ becomes constant for given values of N_1 (which is a reflection of particle size). Increasing throat length will increase efficiency, at the cost of increased pressure drop, only up to this asymptotic value, Beyond this point pressure drop will increase but not efficiency. Consequently, design of the venturi will involve a compromise between the two. For most particle sizes, $L^* = 2–3$ (Figure 6.13) appears to be sufficient. A reasonable comparison between measured and predicted penetration for an industrial venturi is shown in Figure 6.14 (Yung et al., 1978). A recent study on three laboratory venturis (Rudnick et al., 1986) indicated that this model also gave a better fit between observed and predicted penetration than did those of Calvert et al. (1972) or Boll (1973). The performance of each model improved with increasing venturi diameter. Three diameters, 32 mm, 54 mm, and 76 mm, were tested, as well as several methods of water injection, and penetration was determined for particle sizes from 0.32 to 2.8 μm.

D. Fabric filtration

Industrial fabric filter control devices are typically designed with nondisposable filters that are cleaned in place. Cleaning mechanisms in-

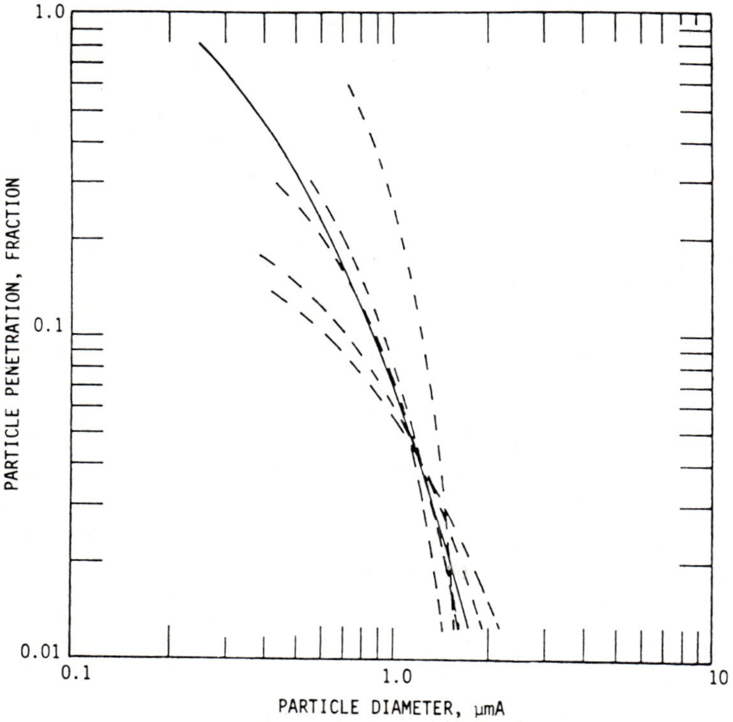

Figure 6.14 Comparison between experimental data for venturi scrubbers and prediction procedure of Yung et al., 1978 (*EPA, 1982a*). (--------experimental; ———model; μmA is the aerodynamic diameter in micrometers.)

clude mechanical shaking, reverse air flow, and pulse jets. Fabric cleaning may require that the control device be taken off-line. Many installations, therefore, utilize a multicompartment arrangement so that active filtration is available at all times. The overall size of the filter area is determined by the particle loading, filter and dust cake resistance to flow, an acceptable maximum pressure drop before cleaning, and the air volume-to-cloth ratio. Typical air-to-cloth ratios are given in Tables 6.2 and 6.5. High air-to-cloth ratios can result in excessive pressure drop, reduced collection efficiency from bag failure, and rapid wear. Table 6.9 shows temperature characteristics of fibers used for industrial fabric filtration. Typical small baghouse designs are shown in Figures 6.15 and 6.16. Overall collection efficiencies are ordinarily quite high, greater than 99% for well-maintained installations (Tables 6.2, 6.4; Figure 6.3). Reliable prediction of efficiency using only particle and fabric characteristics has not yet been developed. Ordinarily, test data on penetration are required for the same particle-fiber system to be used in the field installation. The other major design factor is the pressure loss across the bag and the collected dust layer. The following procedure provides an initial estimate of pressure drop for a given particle-fabric combination.

The filter drag, S, in $\text{N}\cdot\text{min}\cdot\text{m}^{-3}$, can be represented by (Dennis and Klemm, 1979) as

$$S = \frac{\Delta P}{v} = S_E + K_2 W \tag{6.27}$$

where ΔP is the pressure drop across the filter and collected dust layer (N/m^2); v is the superficial gas velocity (m/min) and is equal to Q/A_{filter}, the ratio of the volumetric gas flow rate to the filter area; S_E is the effective residual drag of the filter (i.e., after cleaning) ($\text{N}\cdot\text{min/m}^3$); W is the filter loading (g/m^2); and K_2 is the dust specific resistance

TABLE 6.9 Nominal Operating Temperatures for Fabric Filters

Material	Temperature (°F)
Cotton	180
Polypropylene	180
Nylon	200
Acrylic	275
Polyester	275
Teflon	400
Nomex	425
Fiber glass	550

SOURCE: Neveril et al., 1978a.

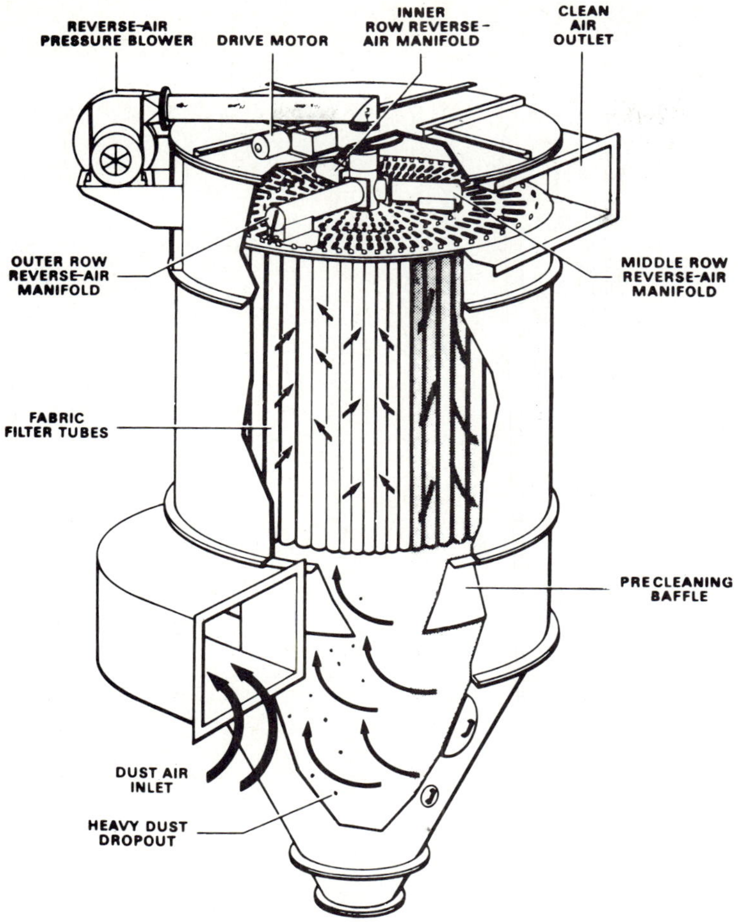

Figure 6.15 Reverse-air baghouse (*courtesy of Carter-Day Company; EPA, 1982a*).

coefficient (N·min/g·m). Equation (6.27) assumes a linear relationship between filter drag and loading. Some particle-fabric combinations display less collection and ΔP early in the filtration cycle and may justify a nonlinear approach.

The most accurate way to determine values for S_E and K_2 is experimentally. A schematic is shown in Figure 6.17. Curve 1 in Figure 6.17 illustrates the maximum cleaning possible, although some residual dust will remain. The other curves indicate the effect of cyclic operation and that an increasing residual quantity of dust will remain on the filter as cleaning efficiency decreases. Dust cake is actually dislodged from the fabric as slabs or flakes, leaving "clean" areas (i.e., with residual dust in the pores but no surface accumulation) and "uncleaned" portions, which have the same dust loading as occurred on the filter at the end of the previous collection cycle (Dennis et al., 1978).

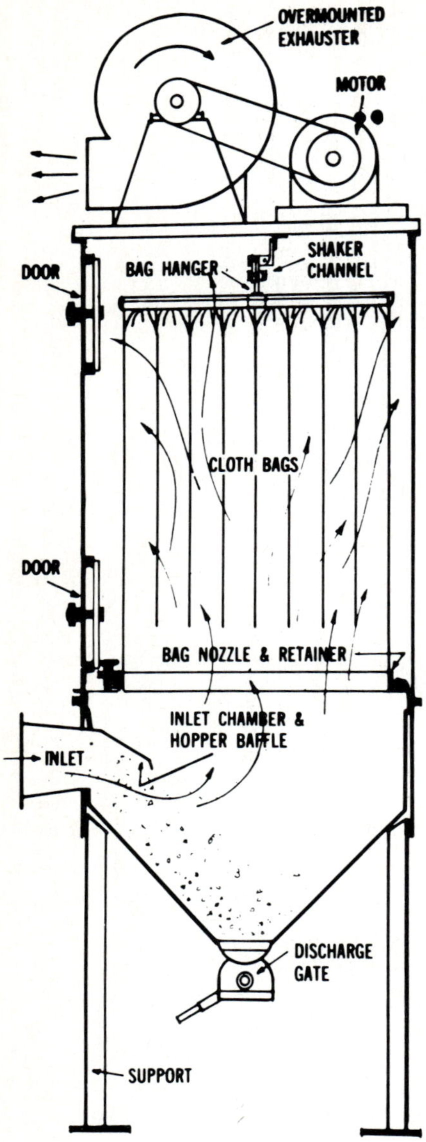

Figure 6.16 Small shaker-type baghouse (*courtesy of Carborundum Division of Flakt, Inc.; EPA, 1982a*).

If experimental data are not available, S_E may be chosen from Table 6.10. Data are given for cycling and one-time cleaning test systems. Some values are also reported for nonsoiled filters. These may be used for order of magnitude estimates but with recognition that residual dust would additionally contribute to pressure drop.

Values of K_2 depend on dust characteristics and probably on the type of filter as well (Dennis et al., 1978). Some typical values are listed in Table 6.11. An estimate of K_2 may also be made from the following equation, but with the proviso that the predicted value should not be expected to be better than ±50% (Dennis and Klemm, 1979):

$$K_2 = \frac{\mu_G S_0^2}{2\rho_p K_c}\left(\frac{3 + 2\beta^{5/3}}{3 - 4.5\beta^{1/3} + 4.5\beta^{5/3} - 3\beta^2}\right) \tag{6.28}$$

where $$\beta = \frac{\bar{\rho}}{\rho_p} \tag{6.29}$$

($1 - \beta$ = void fraction or porosity)

μ_G = gas viscosity, poise = 1.8×10^{-4} poise for air at 70°F, 1 atm

S_0 = specific surface dust parameter, $\text{cm}^{-1} = 6/d_p$ for monodisperse spherical particles with diameter d_p in cm; or for a log-normal distribution of spherical particles with mass median diameter, d_{MMD} in cm, and geometric standard deviation, σ_G, (6.30a)

$$S_0 = \frac{6 \times 10^{1.151(\log \sigma_G)^2}}{d_{\text{MMD}}} \tag{6.30b}$$

ρ_p = particle density, g/cm³

$\bar{\rho}$ = cake bulk density, g/cm³

K_c = Cunningham correction

In order to use Eq. (6.28), it is necessary to have values for ρ_p, $\bar{\rho}$, and the inlet particle diameter or particle size distribution. K_c can be calculated from Eq. (6.23). For coal fly ash porosity typically varies from 0.6 to 0.9. Porosities for lignite fly ash have been reported as 0.4–0.5; for granite dust, 0.6–0.7; and 0.7–0.85 for talc (Dennis et al., 1977).

With known values of S_E and K_2, ΔP can be determined from Eq. (6.27) for any dust loading. Alternatively, if we assume that essentially

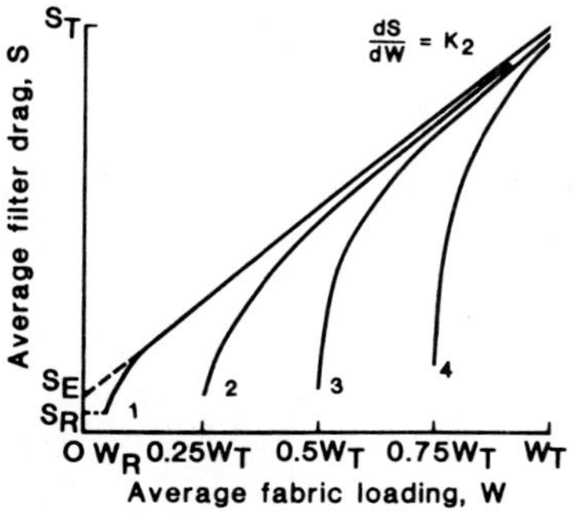

Figure 6.17 Typical drag vs. fabric loading relationships for various levels of cleaning (*Dennis et al., 1978*). (W_T = maximum loading before cleaning; S_T = terminal filter drag.) Curve 1 indicates maximum possible cleaning; 2, highly efficient cleaning; 3, average cleaning range (mechanical-shaking); and 4, average cleaning range (collapse with reverse flow).

Curve	*Description*
1	Maximum possible cleaning
2	Highly efficient cleaning
3	Average cleaning range (mechanical shaking)
4	Average cleaning range (collapse with reverse flow)

TABLE 6.10 Some Values of Effective Residual Drag, S_E

Fabric		S_E (N·min/m³)[a]			
		Fabric resistance with residual dust			Fabric resistance without residual dust[e]
Type	Weight (oz/yd²)	Fly ash	Coal	Liquids	
Plain-weave dacron	10	245–286[b]			
Crawfoot dacron	10	351–384[b] 10[c]			
Napped cotton sateen	10	188–547[b]			
Unnapped cotton sateen	10	547–596[b] 92[c]			17
Woven glass	9.2	12–110[c]	28–51[c]	12[c]	
Woven glass	10.5	12–30[c]	134[c]	41[c]	
Woven glass		350[d]			
Wool	11.5				5
Teflon	8.7				21
Wool felt	21.2–23.1				14–15
Orlon felt	10.9				16–21
Orlon felt	24				21–41
Acrilan felt	17.9				21–27
Dynel felt	24				10–14
Dacron felt	9.9				2
Dacron felt	17.9				21–27
Nylon felt	24				10–14
Arnel felt	24				10–14
Teflon felt	15.6				5
Teflon felt	43.5				19

[a]1 N·min/m³ = 1.22 × 10⁻³ in.H_2O·min/ft.

[b]Dennis, 1974; 30-min filtration cycle, cleaning with 360 shakes, 1-in. amplitude, 8 Hz.

[c]Dennis et al., 1978; single fabric cleaning by flexure.

[d]Dennis and Klemm, 1979. General recommended value for cyclically cleaned fly ash-glass fiber systems.

[e]Perry et al., 1984, pp. 20–103. Values reported at 70°F, 0.5 in.H_2O ΔP, and an air viscosity of 0.0181 cP.

all of the particles are collected,

$$W = C_i v \, \Delta t \tag{6.31}$$

where C_i is the inlet particle concentration (g/m³); and Δt is the interval between fabric cleanings (min). Equation (6.27) then becomes

$$\Delta P = vS_E + K_2 v^2 C_i \, \Delta t \tag{6.32}$$

Figure 6.18 shows experimental and predicted drag values using a slightly more complex version of the above procedure.

The limiting pressure loss at cleaning, ΔP_L, can be represented by

$$\Delta P_L = S_E v + K_2 v (W_p - W_R) \tag{6.33}$$

TABLE 6.11 Specific Resistance Coefficient, K_2, for Various Dusts

	K_2 (N·min/g·m)[a]							
	Cumulative particle distributions[b]				Mass mean diameter			
Dust[h]	<2 μm	<20 μm	<45 μm	<90 μm	3.5 μm[c]	9 μm[d]	17 μm[c]	18.5 μm[c]
Granite		3.3						
Gypsum		3.2						
Feldspar		4.6						
Stone				1.0				
Limestone								0.8–2.6[e]
Amorphous silica							3.1–7.3[e]	
Fly ash					0.2–1.8[e]	$1.8v^{0.5}$[f]		
Lampblack	7.9					$1.48v^{0.5}$[g]		
Wood				1.0				
Resin (cold)		4.2						
Oats			1.8	1.6				
Corn			1.5	0.6				

[a] 1 N·min/g·m = 5.99 in.H_2O·min·ft/lb.
[b] Perry et al., 1984.
[c] Dennis, 1974.
[d] Dennis et al., 1978.
[e] Tests on 4–5 oz/yd^2 twill-weave Dacron of various thread counts and conformance.
[f] v, superficial velocity, in m/min; typical values were 0.31–1.5 m/min. Tests on 9.2–10.5 oz/yd^2 woven glass fiber of various thread counts and conformance.
[g] v, superficial velocity, in m/min; typical values were 0.31–1.5 m/min. Tests on 10-oz/yd^2 sateen-weave cotton, unnapped.
[h] Range of K_2 values for other materials and processes, N·min/g·m (Turner and McKenna, 1984): <0.1, abrasives, alumina, aluminum, calcium; 0.2–0.5, asphalt, asbestos, bismuth, cadmium, pigments, soap; 0.5–1.0, flour, milk powder, sand; 1.0–2.0, sinter dust, lime kiln, stucco; 2.0–5.0, chrome salts, feldspar, hypochlorite manufacture, PVA, lead chloride; 5.0–10.0, carbon black, cement, copper, lamp black, lead blast furnace, lead oxide, tobacco, zinc, zinc oxide; basic oxygen furnace (20–39), dolomite (112), electric furnace (7.5–120), oil fly ash (0.79–21), incinerator fly ash (30), foundry (0.1–20), iron oxide (11–120), kish (38), titanium dioxide (15–35).

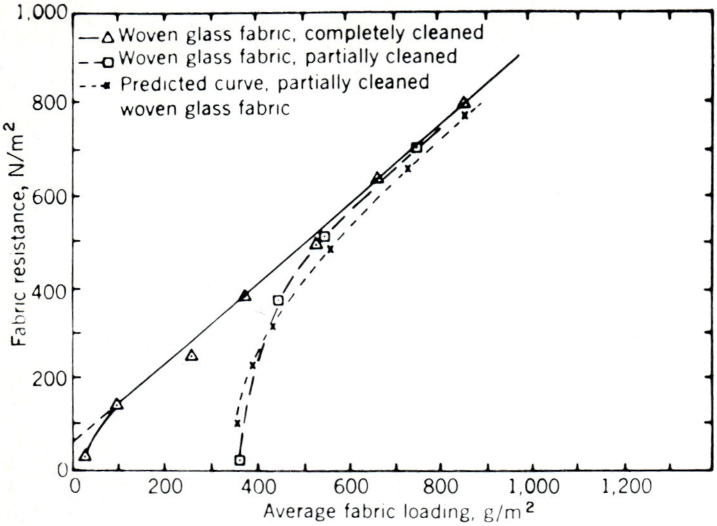

Figure 6.18 Measured and predicted drag values for fly ash on glass fiber fabric (*Dennis et al., 1978*); $v = 0.61$ m/min, $C_i = 6.9$ g/m^3, clean fabric fraction $= 0.485$, $K_2(N{\cdot}\text{min/g}{\cdot}\text{m}) = 1.81v^{0.5}$.

where W_p is the fabric loading at ΔP_L and W_R is the fabric loading when clean, since not all collected dust is removed. A typical value for the fly ash-glass fiber combination is $W_R = 50$ g/m^2, which is less than 10% of normal loading at initiation of cleaning (Dennis and Klemm, 1979). Carr and Smith (1984) also found 8 g/m$^2 \leq W_R \leq 46$ g/m^2 for field data from coal-burning power plants. Figure 6.19 shows the typical relationship among ΔP, ΔP_L, and v for given inlet gas loadings and other system parameters. Forty percent ($a_c = 0.40$) of the bag surface area is effectively cleaned at each cycle. The rest of the surface

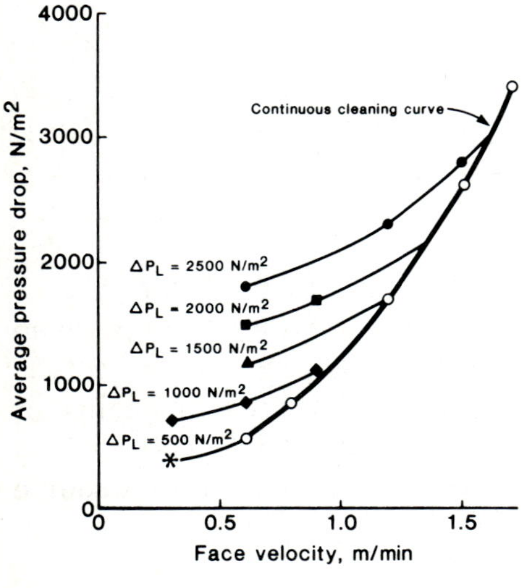

Figure 6.19 Effect of face velocity (v) and limiting pressure drop (ΔP_L) on average pressure loss for a 10-compartment system, with a cleaning time of 30 min (*Dennis and Klemm, 1979*); $C_i = 6.87$ g/m^2, $a_c = 0.4$, $K_2 = 1$ $N{\cdot}$min/g$\cdot$m.

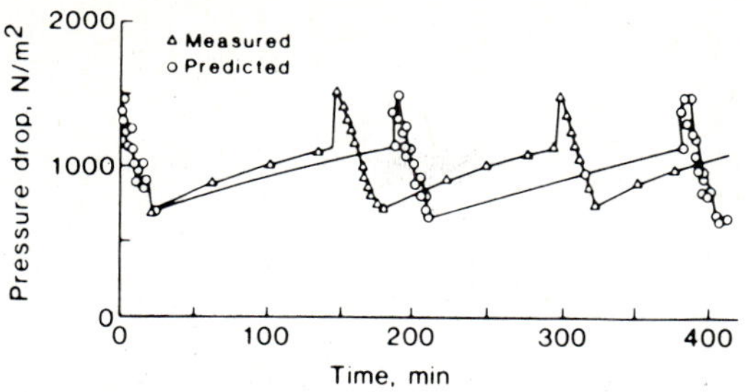

Figure 6.20 Power plant baghouse pressure drop (*Dennis and Klemm, 1979*); v = 0.824 m/min, C_i = 2.6 g/m^3, S_E (clean) = 434 N·min/m^3, K_2 = 0.76 N·min/g·m, W_R = 50 g/m^2, a_c = 0.375.

retains the dust loading that exists just before cleaning is instituted. The lowest curve (ΔP_L = 500 N/m^2) represents a continuous cleaning requirement. Points that fall to the right of this curve will result in pressure increase over time even with continuous cleaning. A more complex procedure, based on the above equations, has also been developed (Dennis and Klemm, 1979). A reasonable comparison between measured and predicted pressure drop for a power plant baghouse using woven glass bags is shown in Figure 6.20.

E. Particulate control for an asphalt concrete plant

Asphalt concrete is used to pave highways, streets, and parking surfaces. Asphalt hot-mix production heats and combines aggregate and liquid asphalt in quantities for the required paving mix. The rotary dryer is the principal source of particulate emission. Figure 6.21 shows typical flows for a 3.5 × 10^5 lb/hr asphalt concrete plant. Flow streams 1 and 9 are the raw materials input to the process, 3 is an intermediate product stream, and 10 is the final asphalt product. Rotary dryer and asphalt mixer emissions (streams 2 and 7A, respectively) are cleaned sequentially by primary and secondary collectors (either the fabric filter or venturi scrubber). Dust collected by the primary collector is recycled back into the process (stream 4). Emissions collected by the secondary collector (fabric filter or venturi) are discarded or recycled (streams 11A or 11B). Final atmospheric dust emissions are shown as stream 12A or 12B. Exhaust gas characteristics for the plant are given in Table 6.12. As shown on Figure 6.21, control devices used in the hot-mix industry typically include a cyclone as a primary collector followed by a high-efficiency secondary collector (EPA, 1982b). The secondary collector is necessary if the process is to meet the new source performance standard of 0.04 gr/dscf (Pahl, 1983).

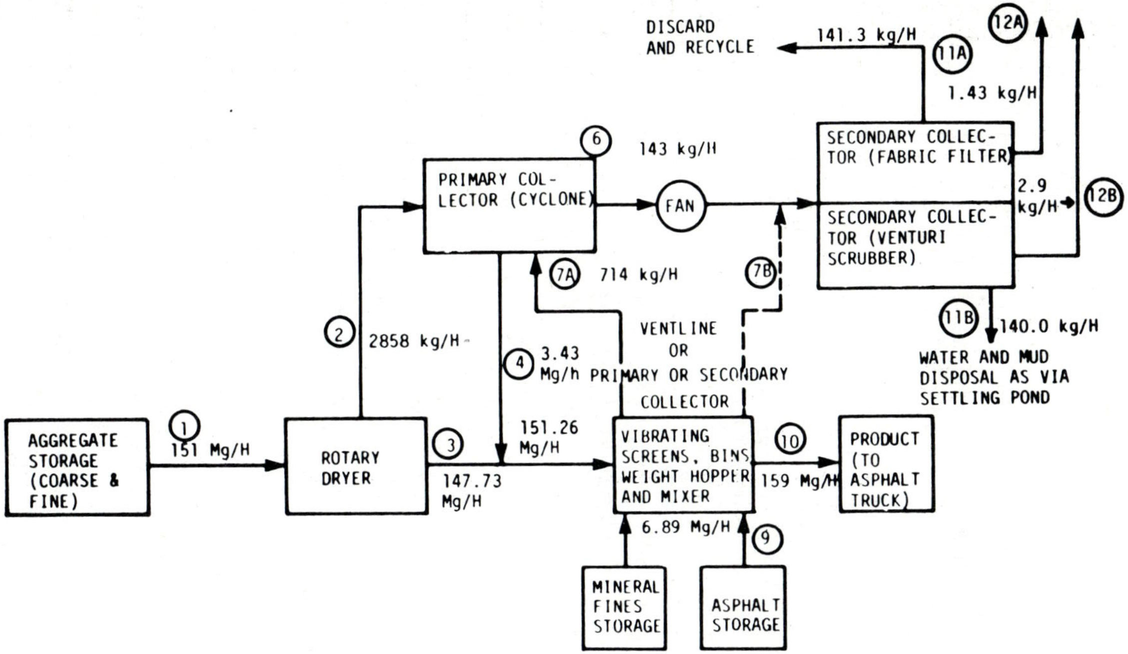

Figure 6.21 Material flow for representative 159-Mg/hr asphalt concrete plant (*EPA, 1982b*).

TABLE 6.12 Characteristics of Rotary Dryer Emissions from an Asphalt Concrete Plant

Stack gas temperature	93°C (199°F)
Pressure, atm	1.0
Gas flow rate, dNm^3/sec[a]	16.7
Acfm[b]	43,455
Gas density, lb/ft^3	0.0739 at 25°C
Particle density, g/cm^3	2.0
Rotary dryer emissions, g/dNm^3:	
Total uncontrolled	55.5
>15.3 μm	49.28
15.3–12.9 μm	1.66
12.9–10.1 μm	1.17
10.1–7.28 μm	1.22
7.28–5.00 μm	0.87
5.00–2.5 μm	0.90
2.5–1.01 μm	0.29
<1.01 μm	0.11

[a]Dry normal cubic meters per second (25°C).
[b]Actual cubic feet per minute.
SOURCE: EPA, 1982b.

Problem. Design a primary collector and secondary collectors using a fabric filter, venturi scrubber, and electrostatic precipitator for the asphalt plant.

Solution. The design process is not a once-through calculational procedure. Ranges of acceptable pressure drops, collector sizes, operating parameters, and emission rates must be initially specified. A review of the literature, (e.g., EPA, 1982b) will often be helpful in selecting these design parameters. However, the final specifications of a control system will often weigh heavily on the economics of the operation. The following ground rules and initial design specifications were used in developing the solution to this design example. The pressure drops for the primary (cyclone) and secondary collectors (fabric filter and venturi) are 1 and 10 in.H_2O, respectively. The fabric filter system will use pulse-jet cleaning with a filtering velocity of 2 $m^3/min \cdot m^2$. The venturi will be operated with a liquid-to-gas ratio of 7.5 gal/1000 ft^3 and will be designed with a throat length of 1 ft. Finally, the electrostatic precipitator will be sized to meet the new source performance standard. Note that many of these initial design parameters would have to be adjusted to fine-tune the design for an actual installation, and postinstallation in-place testing would be needed to evaluate the efficacy of the control system.

1. Design of the cyclone primary collector. The cyclone collector will be sized for a maximum pressure drop of 1.0 in.W.G. From Eq. (6.14) and

Table 6.12:

$$
\begin{aligned}
K &= \frac{\Delta PT}{P\rho Q^2} \\
&= \frac{(1\text{ in.})(460 + 199, °\text{R})}{(1\text{ atm})(0.0739\text{ lb/ft}^3) \times [(273 + 25)/(273 + 93)](43{,}455\text{ ft}^3/\text{min})^2} \\
&= 5.8 \times 10^{-6}
\end{aligned}
\tag{6.14}
$$

The data from Table 6.6 plotted on log paper forms a straight line. At $K = 5.8 \times 10^{-6}$, the required cyclone diameter is 70 in. Therefore, from Figure 6.5, the dimensions of the cyclone are as follows: $B_c =$ 17.5 in.; $D_e = 35$ in.; $H_c = 35$ in.; $L_c = 140$ in.; $S_c = 8.75$ in.; $Z_c = 140$ in.; $J_c = 17.5$ in. The cut size of the cyclone (particle size removed with 50% efficiency) is determined from Eq. (6.12):

$$
d_c = \left[\frac{9\mu B_c}{2\pi \text{NV}_i(\rho_s - \rho)}\right]^{1/2} \tag{6.12}
$$

$$
\begin{aligned}
\text{where } N &= \frac{1}{H_c}\left(L_c + \frac{Z_c}{2}\right) \\
&= \frac{1}{35}\left(140 + \frac{140}{2}\right) = 6
\end{aligned}
\tag{6.13}
$$

$$
\begin{aligned}
V_i &= \frac{\text{volumetric flow rate}}{\text{inlet cross-sectional area}} \\
&= \frac{(43{,}455\text{ ft}^3/\text{min})(144\text{ in}^2/\text{ft}^2)(60\text{ min/hr})}{(35\text{ in.})(17.5\text{ in.})} \\
&= 6.13 \times 10^5\text{ ft/hr} \\
\mu &= 0.052\text{ lb/ft}\cdot\text{hr at } 200°\text{F} \\
B_c &= 17.5\text{ in.} = 1.46\text{ ft} \\
\rho_s &= 2\text{ g/ml} = 124.6\text{ lb/ft}^3 \\
\rho &= (0.0739)\left(\frac{273 + 25}{273 + 93}\right) = 0.0602\text{ lb/ft}^3\text{ at } 93°\text{C}
\end{aligned}
$$

and substituting into Eq. (6.12):

$$
\begin{aligned}
d_c &= \left[\frac{(9)(0.052)(1.46)}{(2\pi)(6)(6.13 \times 10^5)(124.54)}\right]^{1/2} \\
&= 1.54 \times 10^{-5}\text{ ft} = 4.7\ \mu\text{m}
\end{aligned}
$$

Table 6.13 summarizes the cyclone design. Columns 1, 2, and 3 show the particle size distribution data from Table 6.12. Column 4 shows the ratio of the median size of each interval to the cut size of the collector. Column 5, taken from Figure 6.6, shows the collection effi-

TABLE 6.13 Cyclone Design for Asphalt Concrete Plant

Particle size range (μm)	Median of size range (μm)	Mass in size range (g/dNm^3)	Ratio of median size to cut size	Fractional collection efficiency	Mass removed (g/dNm^3)	Mass penetrating collector (g/dNm^3)	Cumulative mass penetrating collector (g/dNm^3)	Cumulative percent less than upper bound of size interval
<1.01	0.32	0.11	0.07	0	0	0.11	0.11	3.3
1.01–2.5	1.59	0.29	0.34	0.10	0.029	0.261	0.371	11.1
2.5–5.00	3.53	0.90	0.75	0.35	0.315	0.585	0.956	28.5
5.00–7.28	6.03	0.87	1.28	0.62	0.539	0.331	1.287	38.4
7.28–10.1	8.56	1.22	1.82	0.76	0.927	0.293	1.580	47.1
10.1–12.9	11.4	1.17	2.42	0.86	1.006	0.164	1.744	52.0
12.9–15.3	14.0	1.66	2.98	0.92	1.527	0.133	1.877	55.9
>15.3	39.1	49.28	8.32	0.97	47.802	1.478	3.355	100
Total		55.5			52.145	3.355		

ciency as a function of the previously calculated ratio. The mass removed is the collection efficiency times the mass concentration in each size range, and the mass penetrating is the mass in minus the mass removed. The overall collection efficiency of this cyclone on the asphalt concrete plant emissions is:

$$\eta = \left(\frac{\text{mass}_{\text{in}} - \text{mass}_{\text{out}}}{\text{mass}_{\text{in}}}\right)(100)$$

$$= \left(\frac{55.5 - 3.355}{55.5}\right)(100) = 94\% \qquad (6.2)$$

Note that this cyclone at the same flow rate cleaning a dust with a larger fraction of particles <10 μm would have a much lower overall collection efficiency. As shown in Table 6.13, the mass concentration penetrating the cyclone is 3.355 g/dNm^3 or 1.47 gr/dscf. This emission is 37 times the new source performance standard and, therefore, will require further treatment.

2. Design of a fabric filter as a secondary collector. About 40% of secondary control devices used in the asphalt hot-mix industry are fabric filters. Key design criteria are collector size (from the ratio of volumetric flow to surface area for a given fabric-particle-cleaning method combination), selection of fabric and cleaning system, and maximum pressure drop before cleaning is required. The most commonly installed systems for asphaltic concrete plant control use pulse-jet cleaning with filtering velocities of about 2 m^3/min·m^2 of cloth (EPA, 1982b). Typical fibers include glass yarns, polyesters, and Nomex. Exhaust gases being treated contain large quantities of water vapor, and the filter units need to be insulated to keep the gas above the dewpoint. With a design-limiting pressure drop of 10 in.W.G. (2259 N/m^2), the maximum filter drag before cleaning (Eq. 6.27) is:

$$S = \frac{\Delta P}{v} = \frac{2259\ \text{N/m}^2}{2\ \text{m/min}} = 1130\ \frac{\text{N·min}}{\text{m}^3} = S_E + K_2W \qquad (6.27)$$

As shown by Eq. (6.27), the filter drag is composed of the linear combination of S_E (residual drag of filter after cleaning) and K_2W, the drag due to the collected dusts. It is best to determine S_E from experimental data. Without such data, Table 6.10 can be consulted as a first estimate. Using woven glass bags and assuming that the residual dust is similar to fly ash, a value of 350 N·min/m^3 will represent a conservative first estimate of S_E. An estimate of K_2 can be made from Eq. (6.28):

$$K_2 = \frac{\mu_G S_0^{\,2}}{2\rho_p K_c}\left(\frac{3 + 2\beta^{5/3}}{3 - 4.5\beta^{1/3} + 4.5\beta^{5/3} - 3\beta^2}\right) \qquad (6.28)$$

where

$$\beta = \frac{\bar{\rho}}{\rho_p} \tag{6.29}$$

Given a porosity $(1 - \beta)$ of 0.6, $\beta = 0.4$ and

$$\bar{\rho} = (\beta)(2) = 0.8 \text{ g/ml}$$

$$\mu_G = 0.052 \text{ lb/ft·hr} = 2.15 \times 10^{-4} \text{ poise}$$

$$S_0 = \frac{6 \times 10^{1.151(\log^2 \sigma_G)}}{d_{\text{MMD}}} \tag{6.30b}$$

where d_{MMD} and σ_G are the geometric mean and geometric standard deviation of the particle size distribution entering the collector. These parameters can be determined by plotting the cumulative size distribution penetrating the cyclone on log probability paper. The 50% percentile diameter is the d_{MMD}, and the 50%/16% diameter is σ_G. This is shown in Figure 6.22 and

$$\sigma_G = 3.8$$

$$d_{\text{MMD}} = 11.8 \ \mu\text{m} = 0.00118 \text{ cm}$$

Therefore,

$$S_0 = \frac{6 \times 10^{1.151(\log^2 3.8)}}{0.00118} = 12{,}393$$

$$K_c \simeq 1 \qquad \text{(as a first approximation)}$$

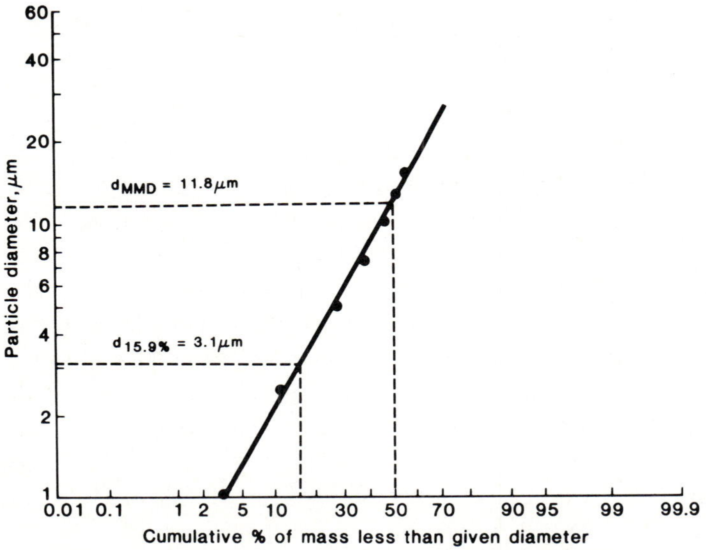

Figure 6.22 Size distribution of asphalt concrete rotary dryer dust after the primary collector ($\sigma_G = 11.8/3.1 = 3.8$).

Substituting into Eq. (6.28),

$$K_2 = \frac{(2.15 \times 10^{-4})(12{,}393)^2}{(2)(2)(1)} \left(\frac{3 + 2\beta^{5/3}}{3 - 4.5\beta^{1/3} + 4.5\beta^{5/3} - 3\beta^2} \right) \tag{6.28}$$

and for $\beta = 0.4$,

$$K_2 = 1.56 \times 10^5 \text{ sec}^{-1} = 2.60 \frac{\text{N·min}}{\text{g·m}}$$

This value is consistent with the value for limestone with mass median diameter of 18.5 μm and for cumulative distributions less than 20 μm for various stone dusts as shown in Table 6.11.

Equation (6.32) can now be solved for the time between cleaning cycles:

$$\begin{aligned} \Delta t &= \frac{\Delta P - vS_E}{K_2 v^2 C_i} \\ &= \frac{(2259 \text{ N/m}^2) - (2 \text{ m/min})(350 \text{ N·min/m}^3)}{(K_2 \text{ N·min/m·g})(2 \text{ m/min})^2 (3.355 \text{ g/m}^3)} \\ &= \frac{116}{K_2} \text{ min} \end{aligned} \tag{6.32}$$

For $K_2 = 2.60$ N·min/g·m, $\Delta t = 45$ min.

Note that changing the porosity $(1 - \beta)$ assumption of 0.6 would have a significant effect on the solution. For example, an assumed porosity of 0.8 ($\beta = 0.2$) results in $K_2 = 0.78$ N·min/g·m and $\Delta t = 151$ min.

3. Design of a venturi scrubber as a secondary collector. Wet scrubbers are the second most popular method of secondary control at asphaltic concrete plants (EPA, 1982b). From a survey of in-place control devices, venturi scrubbers represent 28% of the wet scrubbers installed. For venturi scrubbers, the collection efficiency is closely related to pressure drop. One therefore needs to consider the cost of operating a high-pressure-loss device against the collection efficiency obtained. Typical operating parameters for venturi collectors are given in Tables 6.4 and 6.5. For this calculation, we will consider an average water requirement of 7.5 gal/1000 ft^3 of gas and moderate pressure drop of 10 in.W.G. as design criteria for evaluating collection efficiency.

Pressure drop across a venturi is described by Eq. (6.16) or, in simplified form, Eq. (6.21) or (6.22). With the design specification of $\Delta P =$ 10 in.W.G. and $Q_L/Q_G = 7.5$ gal/1000 ft^3 $= 1.01 \times 10^{-3}$ (cm^3/sec)/

(cm^3/sec) and solving Eq. (6.22) for velocity:

$$U_G = \left[\frac{\Delta P}{4.02 \times 10^{-4}}\left(\frac{Q_G}{Q_L}\right)\right]^{0.5}$$
$$= \left[\frac{10}{4.02 \times 10^{-4}}\left(\frac{1}{1.01 \times 10^{-3}}\right)\right]^{0.5}$$
$$= 4963 \text{ cm/sec} \tag{6.22}$$

The throat cross-sectional area can then be obtained by dividing the total volumetric flow by U_G, and

$$\text{Throat area} = \frac{(43{,}455 \text{ ft}^3/\text{min})(30.48 \text{ cm/ft})}{(4963 \text{ cm/sec})(60 \text{ sec/min})}$$
$$= 4.45 \text{ ft}^2$$

Note that for the case when the water drops do not accelerate to U_G, the pressure drop-to-gas velocity relationship is also affected by the drag coefficient for the liquid drop, the diameter of the liquid drop, and the throat length. Given a throat length, the actual pressure drop can be determined from Eqs. (6.16) through (6.20). For example, with U_G and Q_L/Q_G as before, and throat length $L = 1 \text{ ft} = 30.48 \text{ cm}$,

$$d_D = \frac{49.7}{4963} + 91.1(1.01 \times 10^{-3})^{1.5}$$
$$= 0.0129 \text{ cm}$$

$$N_{\mathrm{Re}_1} = \frac{(1.183 \times 10^{-3} \text{ g/cm}^3)[273 + 25/273 + 93](4963 \text{ cm/sec})(0.0129 \text{ cm})}{(2.15 \times 10^{-4} \text{ g/cm·sec})}$$
$$= 287$$

$$C_{Do} = 0.22 + \frac{24}{287}[1 + 0.15(287)^{0.6}]$$
$$= 0.678$$

$$x = \frac{(3)(30.48 \text{ cm})(0.678)(1.183 \times 10^{-3} \text{ g/cm}^3)}{(16)(0.0129 \text{ cm})(1 \text{ g/cm}^3)} + 1$$
$$= 1.355$$

$$\Delta P = \frac{(2)(1 \text{ g/ml})(4963 \text{ cm/sec})^2}{1 \text{ g·cm/sec}^2\text{·dyne}}(1.01 \times 10^{-3})[1 - x^2 + (x^4 - x^2)^{0.5}]$$
$$= 20{,}050 \text{ dynes/cm}^2$$

and with 4.016×10^{-4} in.W.G./(dyne/cm^2), the actual pressure drop will be

$$\Delta P = 8.05 \text{ in.W.G.}$$

The fractional efficiency is a function of particle size and is determined using the procedure outlined in Section II.C. The calculations are summarized in Table 6.14. Column 3 shows the mass penetrating the primary collector (Table 6.13). This is the particle size distribution and concentration entering the venturi. Column 4 shows the Cunningham slip correction factors (Eq. 6.23) for the median size of each interval. The dimensionless throat length L^* is calculated for $L = 1$ ft from Eq. (6.24),

$$L^* = \frac{(3)(0.678)(1.183 \times 10^{-3}\ \text{g/cm}^3)(30.48\ \text{cm})}{(2)(0.0129\ \text{cm})(1\ \text{g/cm}^3)}$$

$$= 2.84 \tag{6.24}$$

and the inertial impaction parameter N_1 is calculated from Eq. (6.25),

$$N_1 = \frac{(K_c)(d_p,\ \text{cm})^2(2\ \text{g/cm}^3)(4963\ \text{cm/sec})}{(18)(2.15 \times 10^{-4}\ \text{g/cm·sec})(0.0129\ \text{cm})} \tag{6.25}$$

and is listed in column 6. With N_1 and L^*, $-\ln \text{Pt}/B$ is read from Figure 6.13. B is calculated from Eq. (6.26):

$$B = \frac{Q_L}{Q_G}\frac{\rho_L}{\rho_g}\frac{1}{C_{Do}}$$

$$= (1.01 \times 10^{-3})\left(\frac{1}{1.183 \times 10^{-3}}\right)\left(\frac{1}{0.678}\right)$$

$$= 1.26 \tag{6.26}$$

The last two columns of Table 6.14 show the fraction of each size range (Pt) and the mass concentration of aerosol penetrating the venturi. The overall efficiency of the venturi is:

$$\frac{3.355 - 0.1492}{3.355}(100) = 95.5\%$$

The overall efficiency of the cyclone and venturi in series is:

$$\frac{55.5 - 0.1492}{55.5}(100) = 99.7\%$$

The mass concentration penetrating the venturi is 0.1492 g/dNm^3 = 0.065 gr/dscf, just over the new source performance standard of 0.04 gr/dscf. Operating the venturi at a higher pressure drop would lower the output concentration to below the standard.

4. Design of a electrostatic precipitator as a secondary collector. Although the use of electrostatic precipitators is not widespread in the industry, they can be used as secondary collectors. The required collection efficiency

TABLE 6.14 Venturi Design for Asphaltic Concrete Plant

Particle size range (μm)	Median size (μm)	Mass in size range (g/dNm^3)	K_c	L^*	N_1	$-\dfrac{\ln Pt}{B}$[a]	Pt (fraction)	Mass concentration penetrating collector (g/dNm^3)
<1.01	0.32	0.11	1.54	2.84	0.314	0.24	0.74	0.0814
1.01–2.5	1.59	0.261	1.11	2.84	5.58	1.68	0.12	0.0313
2.5–5.00	3.53	0.585	1.05	2.84	26.0	2.56[b]	0.04	0.0234
5.00–7.28	6.03	0.331	1.03	2.84	74.5	3.11[b]	0.02	0.0066
7.28–10.1	8.56	0.293	1.02	2.84	149	3.48[b]	0.01	0.0029
10.1–12.9	11.4	0.164	1.01	2.84	261	3.77[b]	0.008	0.0013
12.9–15.3	14.0	0.133	1.01	2.84	394	4.00[b]	0.006	0.0008
>15.3	39.1	1.478	1.00	2.84	3040	5.06[b]	0.001	0.0015
Total		3.355						0.1492

[a]From Figure 6.13.

[b]From linear extrapolation of $-\ln Pt/B = a + b(\ln N_1)$ at $L^* = 2.84$, where $a = 0.851$, $b = 0.525$, and $r = 0.995$ from least-squares statistics.

for a precipitator that meets the new source emission standard is

$$\eta = \left(\frac{1.47 - 0.04}{1.47}\right) 100 = 97.3\%$$

where 1.47 are the emissions penetrating the primary collector (cyclone) in gr/dscf. Given this required removal efficiency and an estimate of the effective migration velocity, ω_e, the Deutsch equation (6.15) can be used to size the collector surface area. Table 6.7 lists typical effective migration velocities for a number of applications. Assuming that the asphalt concrete dust reacts similarly to dry cement process dust, ω_e will be approximately 0.2 ft/sec. From Eq. (6.15):

$$A = -\frac{Q \ln(1 - \eta)}{\omega_e}$$

$$= [-\ln(1 - 0.973)] \frac{(43{,}455 \text{ ft}^3/\text{min})}{(60 \text{ sec/min})(0.2 \text{ ft/sec})}$$

$$= 13{,}080 \text{ ft}^2 \text{ of collecting surface required} \qquad (6.15)$$

Actual measurements of ω_e at realistic operating conditions are needed for a more accurate estimate of A. However, for a first-cut estimate, 13,000 ft^2 can be used to evaluate the overall cost effectiveness of the precipitator alternative compared to the venturi and fabric collector.

5. Discussion. Field testing at asphalt concrete plants has shown that a two-collector system (shown in Figure 6.21) with a fabric filter or venturi secondary collector can be designed and operated to meet the new source performance standard (EPA, 1982b). From tests at operating plants, primary collector efficiency (usually a cyclone) is typically 96% with emissions of 143 kg/hr. This agrees well with our estimated cyclone efficiency of 94% and emissions of:

$$\left(\frac{3.355 \text{ g}}{\text{m}^3}\right)\left(\frac{16.7 \text{ m}^3}{\text{sec}}\right)\left(\frac{3600 \text{ sec}}{\text{hr}}\right)\left(\frac{10^{-3} \text{ kg}}{\text{g}}\right) = 202 \text{ kg/hr}$$

Tests at operating plants have also shown typical secondary emissions of 1.43 kg/hr and 2.9 kg/hr from fabric filter and venturi secondary collectors (99% and 98% efficiency), respectively. Our venturi design example has an overall efficiency of 95.5% and emissions of

$$\left(\frac{0.1492 \text{ g}}{\text{m}^3}\right)\left(\frac{16.7 \text{ m}^3}{\text{sec}}\right)\left(\frac{3600 \text{ sec}}{\text{hr}}\right)\left(\frac{10^{-3} \text{ kg}}{\text{g}}\right) = 8.9 \text{ kg/hr}$$

While this collection efficiency is somewhat lower than desired (higher emissions), we started with a relatively low venturi pressure drop (8.05 in.H_2O). As shown in Figure 6.10, increasing pressure drop would significantly increase overall collection efficiency.

III. Pollutant Gas Removal

In removal of pollutant gases from air, collection is discussed in terms of the process employed rather than in terms of equipment. Adsorption is the removal of a contaminant from the gas or liquid phase by contact with a solid. Absorption is the collection of pollutant gas by a liquid. Catalytic conversion is a process where a catalyst (a material that increases the rate of reaction, but is not consumed in the reaction) converts a pollutant to a chemical form not considered a pollutant. Thermal oxidation (combustion) is a process where the contaminant is oxidized by burning, carbonaceous material being converted to CO_2 and H_2O.

Because removal of gaseous pollutants is more dependent on the specific chemistry involved than was the case for particulate matter, it is more difficult to generalize about ranges of efficiency and pressure drop. Usually the final choice of process will be dictated by economics, i.e., the alternative that results in the lowest operating cost and/or investment to accomplish a given removal or separation. However, for certain industries, release of product or raw materials into ambient air will also constitute pollution. Since such discharges represent economic loss as well as pollution, control may be variously interpreted as efficient process operation or pollution control. For instance, in the manufacture of synthetic organic chemicals, air oxidation is a commonly used process (EPA, 1984). Almost all of these air oxidation processes require a combination of absorption devices, condensers, and/or carbon adsorption units for product or unreacted raw material recovery. In the past, design of such devices would have been based on recovering only as much of the volatile organic material as was economically feasible, and thus they would not be considered to be control devices. However, present environmental and occupational regulations may encourage design based on greater recovery, previously not thought to be economically justified, but now the device fulfills both process and pollution control requirements. Table 6.3 gives some examples of such applications. We describe several similar examples in Chapter 7.

A. Adsorption

Adsorption is often the method of choice for removal of gaseous pollutants. Since the adsorptive capacity for a given size bed is finite, adsorption systems for continuous use are usually designed with two or more beds in parallel. As a given bed becomes saturated, flow can be switched to a clean bed. The saturated bed is taken out of service and the pollutant removed by stripping with steam or low-pressure gas.

Although many different types of adsorbents are available, activated carbon is the most widely used for pollutant control because of its general effectiveness for many organic materials. Activated carbon is also less sensitive to reduction in capacity due to the presence of water vapor than other adsorbents. For instance, no effect was seen in bed capacity for chloroform for relative humidities $<40\%$; and for a dry bed no effect up to 95% RH (Jonas et al., 1985). (Some reduction in capacity does occur even for activated carbon when the adsorbent has already been in contact with water vapor, i.e., a "wet" bed). In a second study, the effect of relative humidity on the adsorption of trichloroethylene (TCE) was greatest at low influent concentration (300 mg/m^3) and high humidity (85%). Under these conditions, the bed capacity was reduced to 9% of the amount absorbed with the same TCE concentration and 5% humidity. The effect of humidity was less pronounced at higher TCE concentrations (e.g., no effect seen at 1000 mg/m^3 and 25% relative humidity)(Werner, 1985).

Bed breakthrough occurs when the exit gas concentration exceeds some preset level. Ordinarily pollutant penetration will be only a small fraction of the incoming adsorber load. Typically, design efficiencies will be greater than 90%, and for more hazardous materials $>99\%$. Given an inlet concentration, flow rate, and a control efficiency, it is then necessary to estimate the amount of activated carbon required to remove the pollutant mass for a given time, or, alternatively, the time interval during which a given quantity of adsorbent can be effective before breakthrough.

The adsorption of gas molecules depends on the number of active sites available in the adsorbent. Initially, when many sites are available, this removal can be described by pseudo-first-order kinetics with respect to pollutant concentration in the gas phase. Wheeler and Robell (1969) have derived a useful equation, based on a mass balance on the contaminant, which can be used to predict the time during which a mass of adsorbent will effect a specified removal. The equation applies while the bed is adsorbing essentially all of the incoming material and has the form

$$t_s = \frac{W_e}{Cq}\left[W + \frac{\rho_B q}{k_v}\ln(1 - \eta)\right] \tag{6.34}$$

where t_s is the time of adsorbent service to breakthrough (hr); W is the weight of adsorbent (kg); C is entering pollutant concentration (kg/m^3); q is air flow rate (m^3/hr); W_e is the kinetic adsorption capacity (kg_{vapor}/kg_{carbon}) at the arbitrarily chosen *fractional* efficiency at breakthrough, η; ρ_B is the bulk density of the packed bed (kg_{carbon}/m^3); and k_v is the pseudo-first-order adsorption rate constant (hr^{-1}). The actual efficiency before breakthrough will ordinarily be essentially 100%.

Equation (6.34) requires two parameters: k_v and W_e. The values of W_e and k_v for a particular carbon and a specific vapor can be determined experimentally from plots of t_s vs. W at a selected η (Sansone and Jonas, 1981; Sansone et al., 1979; Golovoy and Braslaw, 1982). However, it would be desirable to be able to predict these parameters from more basic considerations.

It has been observed that a well-packed bed of activated carbon granules displays an adsorption capacity very close to the adsorption capacity at equilibrium (Jonas and Rehrmann, 1973). This finding provides a means for estimating W_e for a variety of compounds on a particular carbon, given kinetic capacity measurements with only one reference vapor. The Polanyi-Dubinin-Radushkevich equilibrium equation (Dubinin, 1975) for fine-grained carbon has then been used to extend the information:

$$\ln W_e = \ln(W_0 \rho_L) - \kappa \frac{[\mathrm{RT} \ln(P_0/P)]^2}{\beta^2} \tag{6.35}$$

where W_0 is the maximum adsorption space for condensed adsorbate (cm^3/g); κ is a structural constant of the adsorbent [(mol/cal)^2]; P_0 is the saturated vapor pressure of the condensed adsorbate at gas temperature T (K); P is the equilibrium pressure of adsorbate vapor at T (i.e., the partial pressure of material in the gas phase entering the bed); ρ_L is the liquid adsorbate density (g/cm^3); R is the ideal gas constant (1.987 cal/g-mole·K); and β is the dimensionless affinity coefficient. The values of κ and W_0 are parameters of a particular carbon and independent of the vapor adsorbed (although there may be some effect of water vapor). The affinity coefficient can be estimated from

$$\beta = \frac{[(n_i^2 - 1)/(n_i^2 + 2)]M_i/\rho_{L_i}}{[(n_r^2 - 1)/(n_r^2 + 2)]M_r/\rho_{L_r}} \tag{6.36}$$

where n is the refractive index determined at the sodium D wavelength; M is the molecular weight; ρ_L is the liquid density; and the subscripts i and r refer to a particular compound and the reference vapor, respectively (Reucroft et al., 1971).

If $\ln W_e$ is plotted for one reference vapor vs. $[RT \ln(P_0/P)]^2$, W_0 may be determined from the $\ln W_e$ intercept and the slope of the line will be $-\kappa$ (since $\beta = 1$ for the reference material). Some typical values of W_0 and κ are given in Table 6.15. W_e for another compound is calculated using Eqs. (6.36) and (6.35), recognizing that W_0 and κ will remain constant. k_v for the new compound can be estimated from (Sansone et al., 1979) as

$$k_{v_i} = k_{v_r}\left(\frac{M_r}{M_i}\right)^{0.5} \tag{6.37}$$

TABLE 6.15 Adsorption Parameters for Various Commercial Activated Carbons

Description	Surface area (m^2/g)	Packed density (g/cm^3)	W_0 (cm^3/g)[a]	κ $(mol/cal)^2$ $\times 10^8$
Urano et al. (1982); 25°C				
A. Tsurumi 4GS-S, from coal and coconut shell	1170	0.43	0.485	5.1
B. Tsurumi HC-8, from coconut shell	1270	0.44	0.525	4.7
C. Takeda Sx, from coconut shell	1090	0.41	0.455	4.2
D. Hokuetsu Y-20 from coconut shell	1098	0.45	0.440	5.1
E. Fujisawa D-CG, from coconut shell	1240	0.43	0.495	5.2
F. Fujisawa A, from coal	840	0.42	0.375	4.6
G. Kureha G-BAC, from oil pitch	1000	0.51	0.430	4.6
Golovoy and Braslaw (1981, 1982)				
H. Columbia JXC 4/6 pellets; 25°C, 737 mm	1194	0.461	0.404[b]	1.44[b]
			0.413[c]	4.7[c]
			0.409[d]	1.78[d]
Sansone and Jonas (1981)				
I. Barnebey–Cheney 6–10 mesh BC-AC from lot 0993; 23°C	—	—	0.481	1.5
Jonas and Sansone (1986); Jonas et al. (1983)				
J. North American Carbon, NACAR G-352, lot 2905, 6–10 mesh granular	—	0.347	0.700	14.9

[a]Reference vapors: benzene for A–G; *n*-heptane, *n*-butanol, and MIBK for H; CCl_4 for I–J.
[b]*n*-Heptane.
[c]*n*-Butanol.
[d]MIBK.

where k_v and M have been defined for Eqs. (6.34) and (6.36). Equation (6.34) is then used to calculate time to saturation, or alternatively, required quantity of adsorbent, for known values of ρ_B (ordinarily approximately 500 kg/m³), C, and q, and the same value of η specified when testing the reference vapor.

Golovoy and Braslaw (1982) also reported a method for estimating k_v from transport properties of the pollutant, the gas flow characteristics, and carbon particle dimensions. When the main resistance to mass transfer is in the gas phase, k_v is related to the mass transfer coefficient, k_G, by

$$k_v = \frac{k_G a P_T v}{G_m} \tag{6.38}$$

where a is the external adsorbent surface per unit volume of adsorber; P_T is the total pressure; v is the superficial linear gas velocity; and G_m is the molar flux of pollutant entering the adsorber in moles per time per unit of adsorber cross section. For spherical particles, and for cylindrical particles with length equal to diameter,

$$a = \frac{6(1 - \varepsilon)}{d_p} \tag{6.39}$$

where ε is the void volume fraction in the adsorber and d_p is the average carbon particle diameter in the bed. Using Eq. (6.38), the Colburn–Chilton relationship for mass transfer, and an empirical correlation between mass transfer and flow properties (Hsiung and Thodos, 1977), and substituting into Eq. (6.38) gave:

$$k_v = 2.88\left(\frac{1-\varepsilon}{\varepsilon}\right)\left(\frac{v}{d_p}\right)\left(\frac{\mu}{d_p v \rho}\right)^{0.39}\left(\frac{\rho D_{AB}}{\mu}\right)^{2/3} \tag{6.40}$$

where μ is the gas viscosity; ρ is the gas density; and D_{AB} is the diffusivity of the pollutant in air. All parameters should be in consistent units. Excellent agreement was found for four paint solvents between measured values of k_v and those predicted using Eq. (6.40). These are shown in Table 6.16.

Equation (6.34) is properly applicable only at high values of η. Extensive testing has been carried out at $\eta = 0.99$ on a Barnebey–Cheney activated carbon (carbon I in Table 6.15) for carbon tetrachloride (CCl_4), chloroform, benzene, *p*-dioxane, *sec*-butylamine, 1,2-dichloroethane, chlorobenzene, and acrylonitrile (Sansone and Jonas, 1981; Sansone et al., 1979). Carbon tetrachloride was used as a reference vapor and the estimation procedure described above was compared to actual measurements. Calculated values agreed with experimental observations within 8% for W_e and within 15% for k_v. Based on Eqs. (6.35) and (6.37), W_e and k_v were estimated for $\eta = 0.99$ for a variety of carcinogenic materials and the specific carbon (Table 6.17; Sansone and Jonas, 1981). Testing of paint solvents (*n*-butanol, *m*-xylene, cellosolve acetate, and 2-heptanone) on Columbia JXC activated carbon (carbon H, Table 6.15) also gave good agreement (within 14%) with prediction using Eq. (6.34) at all $\eta \geq 0.90$; and matched reasonably well for $\eta > 0.5$ (Golovoy and Braslaw, 1982).

Equations (6.34)–(6.37) have also been applied to binary vapor mixtures. Various combinations of CCl_4, chloroform, and benzene were

TABLE 6.16 Predicted and Experimental Adsorption Rate Constants for Paint Solvents on JXC Carbon

Solvent	Diffusivity (in air at 25°C, 737 mmHg) (cm^2/sec)	k_v (sec^{-1})	
		Experimental	Calculated[a] (Eq. 6.40)
Butanol	0.089	48.0 ± 2.2	54.1
m-Xylene	0.071	55.4 ± 2.7	46.6
Cellosolve acetate	0.063	45.5 ± 0.9	43.0
2-Heptanone	0.071	49.4 ± 1.9	46.7

[a] v = 58 cm/sec; d_p = 0.37 cm; ε = 0.457; μ/ρ (air, 25°C) = 0.155 cm^2/sec.

SOURCE: Golovoy and Braslaw, 1982.

TABLE 6.17 Predicted Values of W_e and k_v for Carcinogenic Vapors on 6–10 Mesh BC-AC Activated Carbon[a]

Vapor	Liquid density,[b] ρ_L (g/cm^3)	Refractive index,[c] n	k_v (min^{-1})	W_e (kg/kg)
Acetamide	1.159	1.4274[78]	1186	0.494
Acrylonitrile	0.8060	1.3911	1251	0.357
Benzene	0.8761[23]	1.5011	1031	0.409
Carbon tetrachloride	1.5881[23]	1.4607	735	0.741
Chloroform	1.4832	1.4459	834	0.688
bis(Chloromethyl) ether	1.315	1.4346	850	0.608
Chloromethyl methyl ether	1.0605	1.3974	1016	0.480
1,2-Dibromo-3-chloropropane	2.093[11]	1.553[11]	593	0.992
1,1-Dibromoethane	2.0555	1.5128	665	0.962
1,2-Dibromoethane	2.1792	1.5383	665	1.020
1,2-Dichloroethane	1.2492[33]	1.4448	916	0.575
Diepoxy butane (meso)	1.1157	1.4330	982	0.510
1,1-Dimethyl hydrazine	0.791[22]	1.4075[22.3]	1176	0.359
1,2-Dimethyl hydrazine	0.8274	1.4204	1176	0.375
Dimethyl sulfate	1.332	1.3874	812	0.615
p-Dioxane	1.0333[23]	1.4220	971	0.475
Ethylenimine	0.8321	1.412[25]	1388	0.354
Hydrazine	1.0083	1.4698[22.3]	1610	0.380
Methyl methane sulfonate	1.2943	1.4140	869	0.595
1-Naphthylamine	1.229[25]	1.6703[51]	762	0.585
2-Naphthylamine	1.0614[98]	1.6493[98]	762	0.506
N-Nitrosodiethylamine	0.9422	1.4386	902	0.442
N-Nitrosodimethylamine	1.0059	1.4368	1059	0.458
N-Nitroso-N-methylurethane	1.133	1.4363	793	0.534
N-Nitrosopiperidine	1.0631[18.5]	1.4933[18.5]	853	0.501
N-Nitrosodipropylamine	0.9163	1.4437	799	0.434
1,3-Propane sultone	1.393[10]	1.450[10]	825	0.646
β-Propiolactone	1.1460	1.4118	1074	0.508
Propylenimine	0.802[25]	1.409[25]	1206	0.361
Safrole	1.096	1.5383	716	0.522
Urethane	0.9862[21]	1.4144[52]	966	0.456
Vinyl chloride	0.9114	1.3700	1153	0.404

[a]Barnebey–Cheney Co., lot 0993, carbon I, Table 6.15; T = 23°C, P/P_0 = 0.0936, W_0 = 0.481 cm^3/g, κ = 1.5 × 10^{-8} (mole/cal)^2.

[b]Liquid densities at 20°C relative to water unless otherwise stated.

[c]Refractive indices are for D line of the sodium spectrum and at 20°C unless otherwise stated (superscript numbers).

SOURCE: Sansone and Jonas, 1981.

adsorbed at η = 0.99 on carbon J in Table 6.15 (Jonas et al., 1983). For given estimates or measurements of W_0 and κ, the kinetic adsorption capacity for each component, W_{e_i}, is calculated from

$$\ln W_{e_i} = \ln(W_0 \rho_{L_i} y_i) - \kappa \frac{[RT \ln(P_0/P)_i]^2}{\beta^2} \tag{6.41}$$

where y_i is the relative mole fraction of component i (i.e., the sum of all the relative mole fractions = 1.0 but does not include the mole fraction of air). For example, for a mixture of CCl_4 ($P_{CCl_4}/P_0 = 0.0471$, P_0 = 104 mmHg; $y_{CCl_4} = 0.380$) and chloroform ($P_{CHCl_3}/P_0 = 0.0454$, P_0 = 178 mmHg; $y_{CHCl_3} = 0.620$) at 23°C and carbon J from Table 6.15:

$$\beta_{CCl_4} = 1.0 \qquad \text{(reference vapor)}$$

$$\beta_{CHCl_3} = 0.808 \qquad \text{(Eq. 6.36 and Table 6.17)}$$

$$\ln W_{e,CHCl_3} = \ln[(0.7)(1.4832)(0.620)] - 14.9 \times 10^{-8} \times \left[\frac{(1.987)(296)\ln(1/0.0454)}{(0.808)^2}\right]^2$$

$$W_{e,CHCl_3} = 0.302 \text{ kg/kg} \qquad \text{(found experimentally to be 0.331 kg/kg)}$$

$$\ln W_{e,CCl_4} = \ln[(0.7)(1.5881)(0.380)] - 14.9 \times 10^{-8} \times \left[\frac{(1.987)(296)\ln(1/0.0471)}{1.0}\right]^2$$

$$W_{e,CCl_4} = 0.261 \text{ kg/kg} \qquad \text{(found experimentally to be 0.290 kg/kg)}$$

The experimental values of W_{e_i} deviated from those calculated from Eq. (6.41) over a range from −1.0% to +20% with an absolute mean deviation of 5.9%. Values of k_v, calculated from Eq. (6.37), ranged from −30.1% to +26.9% with an absolute mean deviation of 14.8%. The greater deviation probably occurs because Eq. (6.37) is based only on molecular diffusion and does not take into account the effect of mass flow. In general it appeared that each vapor's adsorption space was reduced from the space it would have occupied at the same relative pressure if the bed had been exposed to each vapor separately (Jonas and Sansone, 1986; Jonas et al., 1983).

Pressure drop through adsorption beds depends considerably on bed configuration. Cylindrical canisters, pleated cells, and beds with aggregates of flat cells are all typical designs. Superficial velocities for flat bed designs vary between 0.12 and 0.25 m/s. In general for turbulent flow of air at 25°C, 1 atm, through beds packed with spherical or near-spherical particles the pressure drop can be estimated from (Perry et al., 1984)

$$\Delta P \simeq 0.24\left(\frac{L}{d_p}\right)(v^2) \tag{6.42}$$

where bed length, L, and particle diameter, d_p, have the same units; v is the superficial velocity in m/sec; and ΔP is in in.H_2O.

Example A booth for repainting cars measures 14 ft wide, 9 ft high, and 26 ft long. Between six and eight cars are painted per shift. The booth is ventilated at a rate of 100 cfm/square foot cross-sectional area (pass-through area), the recommended rate from the ACGIH ventilation manual (ACGIH, 1984). A toluene-based paint is used, and the concentration of toluene in the booth is 39 mg/m^3 at 25°C (O'Brien and Hurley, 1981). It has been decided to control and recover these emissions with an activated carbon air filter.

Given an air cleaner with two adsorption beds, each with 500 kg of Columbia JXC carbon (carbon H in Table 6.15), estimate the length of service of a bed (time to saturation when the bed must be taken out of service and cleaned), where breakthrough is defined as an outlet concentration of 1.0 mg/m^3. The company is also considering switching to a paint with toluene and *m*-xylene solvents. If the concentration of toluene remains at 39 mg/m^3 and the concentration of *m*-xylene is 10 mg/m^3, estimate the length of service of a bed where breakthrough is defined as 1.0 mg/m^3 each of toluene and *m*-xylene.

solution The characteristics of Columbia JXC carbon are as follows (Table 6.15): $\rho_B = 0.461$ g/cm^3; $W_0 = 0.404$ cm^3/g; $\kappa = 1.44 \times 10^{-8}$ (mole/cal)^2; and k_v for *n*-heptane reference vapor is 2730 min^{-1}. Note that W_0 and κ are also based on *n*-heptane as a reference vapor. The removal efficiency at breakthrough for toluene is $(39 - 1)/39 = 0.975$ and for *m*-xylene is $(10 - 1)/10 = 0.9$. The room conditions are 25°C and 750 mmHg. The following data are available from Perry et al., 1984:

	Toluene	*m*-Xylene	*n*-Heptane
ρ_L, g/cm^3	0.8669	0.8684	0.6838
MW	92.13	106.16	100.21
Vapor pressure P_0, mmHg @ 25°C	28.0	6.1	48.3
Refractive index, *n*	1.4961	1.4973	1.3876
Diffusivity in air, cm^2/sec	0.086	0.0709	0.07

Volumetric flow through the adsorber is:

$$\left(100\ \frac{\text{ft}^3/\text{min}}{\text{ft}^2}\right)(14\ \text{ft})(9\ \text{ft})\left(0.02832\ \frac{\text{m}^3}{\text{ft}^3}\right) = 357\ \text{m}^3/\text{min}$$

$$= 5.95\ \text{m}^3/\text{sec}$$

The vapor pressure equivalent of 39 mg toluene/m^3 at 25°C is:

$$\left(\frac{39\ \text{mg}}{\text{m}^3}\right)\left(\frac{\text{mole}}{92{,}130\ \text{mg}}\right)\left(\frac{0.0245\ \text{m}^3}{\text{mole}}\right)\left(\frac{\text{mm/mm}}{\text{m}^3/\text{m}^3}\right)(750\ \text{mm})$$

$$= 0.0078\ \text{mmHg}$$

The vapor pressure equivalent of 10 mg *m*-xylene/m^3 is:

$$\left(\frac{10\ \text{mg}}{\text{m}^3}\right)\left(\frac{\text{mole}}{106{,}160\ \text{mg}}\right)\left(\frac{0.0245\ \text{m}^3}{\text{mole}}\right)\left(\frac{\text{mm/mm}}{\text{m}^3/\text{m}^3}\right)(750\ \text{mm})$$

$$= 0.0017\ \text{mmHg}$$

In order to calculate the time of service of the carbon bed for toluene, we need to determine the kinetic adsorption capacity, W_e, and the kinetic adsorption rate constant, k_v. W_e is determined from Eq. (6.35), where β, the affinity coefficient (Eq. 6.36) is:

$$\beta = \frac{[(1.4961^2 - 1)/(1.4961^2 + 2)]92.13/0.8669}{[(1.3876^2 - 1)/(1.3876^2 + 2)]100.21/0.6838}$$
$$= 0.8987 \qquad (6.36)$$

k_v can be estimated using the n-heptane reference vapor and Eq. (6.37):

$$k_{v_{\text{toluene}}} = 2730 \text{ min}^{-1}\left(\frac{100.21}{92.13}\right)^{1/2}$$
$$= 2847 \text{ min}^{-1} = 47.4 \text{ sec}^{-1} \qquad (6.37)$$

k_v can also be estimated using Eq. (6.40). Given a superficial linear velocity v of 58 cm/sec (adsorption bed cross-sectional area of 10.3 m^2), average carbon particle diameter d_p = 0.37 cm, void fraction = 0.457, and μ/ρ for air = 0.155 cm^2/sec at 25°C (Golovay and Braslaw, 1982):

$$k_{v_{\text{toluene}}} = (2.88)\left(\frac{1 - 0.457}{0.457}\right)\left(\frac{58}{0.37}\right)\left(\frac{0.155}{(0.37)(58)}\right)^{0.39}\left(\frac{0.086}{0.155}\right)^{2/3}$$
$$= 52.9 \text{ sec}^{-1} \qquad (6.40)$$

This is in close agreement with the value from Eq. (6.37). The kinetic adsorption capacity of Columbia JXC carbon for toluene is calculated using Eq. (6.35):

$$\ln W_e = \ln[(0.404)(0.8669)] - 1.44 \times 10^{-8}\,\frac{[(1.987)(298)\ln(28.0/0.0078)]^2}{(0.8987)^2} \qquad (6.35)$$
$$W_e = 0.2304 \text{ kg/kg}$$

The time to breakthrough of each adsorption bed is estimated using Eq. (6.34):

$$t_s = \frac{0.2304}{(39 \times 10^{-6})(5.95)}\left[500 + \frac{(461)(5.95)}{(52.9)}\ln(1 - 0.975)\right]$$
$$\frac{\text{kg/kg}}{(\text{kg/m}^3)(\text{m}^3/\text{sec})}\left[\text{kg} + \frac{(\text{kg/m}^3)\ \text{m}^3/\text{sec}}{(\text{sec}^{-1})}\right]$$
$$= 3.06 \times 10^5 \text{ sec}$$
$$= 85 \text{ hr} \qquad (6.34)$$

For the case with the binary vapor mixture, the kinetic adsorption capacity for each component W_{e_i} is calculated from Eq. (6.41), where

$$y_{C_7H_8} = \frac{0.0078 \text{ mmHg}}{(0.0078 + 0.0017) \text{ mmHg}} = 0.821$$

$$y_{C_8H_{10}} = \frac{0.0017 \text{ mmHg}}{(0.0078 + 0.0017) \text{ mmHg}} = 0.179$$

The affinity coefficient for toluene is as before ($\beta_{C_7H_8} = 0.8987$), and

$$\beta_{C_8H_{10}} = \frac{[(1.4973^2 - 1)/(1.4973^2 + 2)]106.16/0.8684}{[(1.3876^2 - 1)/(1.3876^2 + 2)]100.21/0.6838} = 1.0356 \qquad (6.36)$$

Substituting into Eq. (6.41):

$$\ln W_{e,C_7H_8} = \ln[(0.404)(0.8669)(0.821)] - 1.44 \times 10^{-8}\left[\frac{(1.987)(298)\ln(28/0.0078)}{(0.8987)}\right]^2$$

$$W_{e,C_7H_8} = 0.1891 \text{ kg/kg}$$

$$\ln W_{e,C_8H_{10}} = \ln[(0.404)(0.8684)(0.179)] - 1.44 \times 10^{-8}\left[\frac{(1.987)(298)\ln(6.1/0.0017)}{(1.0356)}\right]^2$$

$$W_{e,C_8H_{10}} = 0.0458 \text{ kg/kg}$$

The kinetic adsorption rate for toluene is as before ($k_{v_{C_7H_8}} = 52.9 \text{ sec}^{-1}$) and:

$$k_{v,C_8H_{10}} = (2.88)\left(\frac{1 - 0.457}{0.457}\right)\left(\frac{58}{0.37}\right)\left[\frac{0.155}{(0.37)(58)}\right]^{0.39}\left(\frac{0.0709}{0.155}\right)^{2/3} = 46.4 \text{ sec}^{-1} \qquad (6.40)$$

The time to breakthrough for each component is then:

$$t_{s,C_7H_8} = \frac{0.1891}{(39 \times 10^{-6})(5.95)}\left[500 + \frac{(461)(5.95)}{(52.9)}\ln(1 - 0.975)\right] = 2.52 \times 10^5 \text{ sec} = 70 \text{ hr} \qquad (6.34)$$

$$t_{s,C_8H_{10}} = \frac{0.0458}{(10 \times 10^{-6})(5.95)}\left[500 + \frac{(461)(5.95)}{(46.5)}\ln(1 - 0.9)\right] = 2.80 \times 10^5 \text{ sec} = 78 \text{ hr} \qquad (6.33)$$

For the binary vapor case, the bed would have to be regenerated every 70 hr. The pressure drop across the bed can be estimated from Eq. (6.42), where the bed length L is

$$L = (500\ \text{kg})\left(\frac{m^3}{461\ \text{kg}}\right)\left(\frac{\text{sec}}{5.95\ \text{m}^3}\right)\left(\frac{0.58\ \text{m}}{\text{sec}}\right)$$

$$= 0.11\ \text{m} = 11\ \text{cm}$$

and the pressure drop is

$$\Delta P \simeq 0.24\left(\frac{11}{0.37}\right)(0.58)^2$$

$$\simeq 2.4\ \text{in.H}_2\text{O} \qquad (6.42)$$

B. Absorption

Soluble pollutant gases are sometimes removed by dissolution in a liquid. As is indicated in Table 6.3, this unit operation is used more frequently for product recovery. However, absorption is also applied for air pollution control, particularly when used in conjunction with another control device such as an incinerator. The inverse operation, called stripping, transfers volatile components from a liquid into the gas phase by contact with an inert gas such as nitrogen or steam. Often absorption and stripping are used together to separate a gaseous product or pollutant from its original stream and then to recover or dispose of the material in concentrated form in the gas phase. Figure 6.23 shows such a combination. The “treated gas” would have a very low concentration of pollutant, whereas the “solute gas” or “recovered solute” would be much more concentrated. This process of concentration is implemented by operating the stripper and adsorber at different temperatures or at different pressures. The absorption process may depend only on physical solubility or may include chemical reaction(s) in the liquid phase as well. Typical absorbing liquids are water, mineral oils, nonvolatile hydrocarbon oils, ethanol amines, and aqueous solutions of sodium carbonate or sodium hydroxide.

The ultimate capacity of an absorption system is limited by the vapor-liquid equilibrium relationship between gas pollutant and absorbing material at the operating pressure and temperature. The rate at which the pollutant is transferred into the liquid is proportional to the pollutant concentration difference between the gas phase and the equilibrium concentration at the same temperature (see Eq. 7.6). The transfer rate is also a function of the area for mass transfer at the gas-liquid interface. This interfacial area in turn depends on the design of the gas-liquid contactor. Towers filled with various types of packing

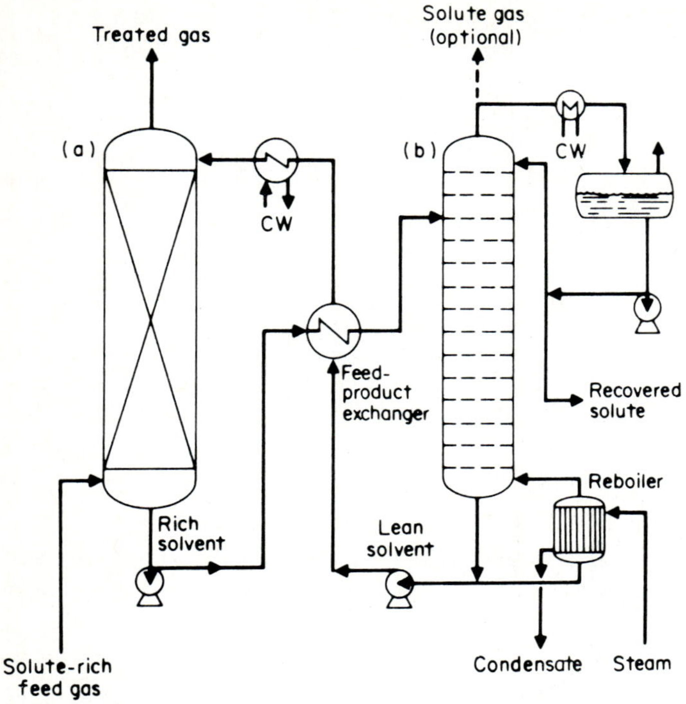

Figure 6.23 Gas absorber using a solvent regenerated by stripping: (*a*) absorber; (*b*) stripper (*Perry et al., 1984*).

materials designed to enhance gas-liquid contact (packed columns); or stacked with perforated trays separated by 1–2 foot spacings (plate columns) are the usual choice for absorption operations. As indicated in Figure 6.23, liquid flow is usually countercurrent to the direction of gas flow. Ordinarily, plate columns are preferred for processing greater gas flow rates, i.e., those that require column diameters >0.6 m (2 ft) (Perry et al., 1984). Efficiency of pollutant gas removal is quite dependent on the gas-liquid system used. Pressure drop is also a complex function of the relative gas and liquid flow rates, the types of packing or trays, and the dimensions of the column. Description of the design procedures for overall efficiency of the collection and pressure drop for absorption columns is somewhat too specialized for the scope of this chapter. However, detailed analyses are given in Chapters 14 and 18 of Perry et al. (1984). Generally, absorption becomes less feasible as pollutant gas concentration becomes smaller, although this observation must be understood in light of the specific chemistry involved. For air oxidation processes, absorption has been reported as not being applicable at pollutant concentrations in the gas phase ≤200–300 ppm (EPA, 1984).

C. Thermal oxidation (combustion)

The purpose of combustion is to oxidize pollutants such as hydrocarbons to carbon dioxide and water, and carbon monoxide to CO_2:

$$C_mH_n + \left(m + \frac{n}{4}\right)O_2 \longrightarrow mCO_2 + \frac{n}{2}H_2O \tag{6.43}$$

Some oxygen may come from the gas to be oxidized, but additional air will usually need to be supplied.

Most thermal oxidation applications are directed toward control of relatively low concentrations of organics or other materials. Sulfur and halogen-containing organic compounds may also be treated in this manner. However, in these cases, an auxiliary scrubber (absorber) must be added to remove the products of combustion; SO_2 from sulfur and HCl (with some Cl_2) from chlorinated compounds. Since the exhaust stream from a thermal oxidation unit will still contain some unoxidized material, as well as the products of combustion, it is not appropriate to recycle this gas into the workplace.

The heat content of the waste gas to be processed and the combustion temperature are important design variables. Low heat contents (<20 Btu/scf, standard cubic feet of gas at 25°C and 1 atm) require addition of auxiliary fuel such as natural gas or fuel oil to maintain combustion temperatures (650–900°C). Heat contents between 13 and 20 Btu/scf correspond approximately to 25–40% of the LEL (see Chapter 4). Moderate heat content gases (20–100 Btu/scf) correspond to ~40–200% of the LEL and must be diluted with inert gases or enriched with auxiliary fuel before oxidation to meet flammability safety limits imposed by insurance companies. Combustion temperatures are typically 750–1200°C. When the heat content is greater than ~100 Btu/scf, the gas stream has enough energy to maintain the flame by itself and may be considered a potential fuel gas (combustion at >1200°C).

Residence time and the mixing characteristics of the combustion chamber are other significant design parameters. Typical values of residence time are on the order of a second. The combined effect of residence time, combustion temperature, excess air, heating value, and the completeness of flame and fuel mixing determine the efficiency of combustion. Theoretical prediction of efficiency for a multicomponent mixture of volatile organic compounds (VOC) has not yet been thoroughly validated. However, some actual test data for various commercial units are shown in Table 6.18. The destruction efficiencies are typically in excess of 95%.

Extensive testing on steam- and air-assisted flares has indicated that

TABLE 6.18 Results from Actual Thermal Oxidizer Tests

Company	Residence time/inlet flow (scfm)	Temperature (°F)	Number of test runs	Inlet VOC (ppmv)[a]	Outlet VOC (ppmv)	VOC[b] destruction efficiency (%)
Union Carbide	2–3 sec/	1,160	6[c]	11,900	243	96.1
	20,600	1,475	3[c]	11,900	10	99.9
Rohm & Haas	1 sec/tank farm vent, (TFV)—12,500 plus	1,425	3[c]	TFV 2,580 OXV 11,600	1,330	82.6
	Oxidizer vent, (OXV)—40,000	1,510	4[c]	TFV 2,600 OXV 12,800	150	98.3
		1,545	1[c]	TFV 2,410 OXV 12,200	25	99.7
Denka	0.6 sec/33,000	1,400	3[c]	950	13	98.5
Koppers[d]	0.6 sec/30,000	1,800	Inlet—4[c]	(Set 1) 850	7	99.0
			Outlet—6[c]	(Set 2)	11	97.2
Petro-tex[e]	0.6 sec/14,400	1,400	N/A	(Set 1) 10,300	1,000	70.3
				(Set 2) 10,650	215	94.1
				(Set 3) 10,300	10	99.6

[a] VOC does not include methane or ethane.
[b] VOC destruction efficiency is weight percent.
[c] Sampling conducted with integrated bags.
[d] The data in Set 1 and Set 2 for Koppers were taken during different time periods.
[e] Inlet and outlet VOC for Petro-tex reported as ppmv methane. This case involves the use of an existing boiler to control a process fume. The data in Set 1 were taken prior to adjustment of the boiler; the data in Sets 2 and 3, after adjustment. The specific alterations made by Petro-tex involve changes in mixing induced by retrofit baffles.
SOURCE: EPA, 1980.

combustion efficiencies >98% are typical under conditions of representative industrial practice (McDaniel, 1983). Varying rates of exhaust gas (0.5–640 scfm, standard cubic feet per minute) had no effect on steam-assisted flare combustion efficiencies as long as the exit velocity was <62.5 ft/sec (19 m/sec). Variation in heat content of 200–2200 Btu/scf in the steam-assisted system resulted in a slight decline in efficiency below 300 Btu/scf (but only to 98%), and essentially no observed efficiency effects at higher energy inputs.

As has been noted, VOC streams containing sulfur or chloride compounds require an auxiliary scrubber. With this device, SO_2 removal efficiency is typically a maximum of 90%, whereas ~99.9% of HCl can be removed (EPA, 1980). Because HCl is much more easily absorbed than Cl_2, the combustion product-gas equilibrium mixture containing the two is forced toward a higher relative concentration of HCl. This is ordinarily accomplished by maintaining combustion temperatures above 980°C or adding steam to the combustion mixture.

D. Catalytic conversion (oxidation)

Catalytic oxidation accomplishes the same conversion as does thermal oxidation (Eq. 6.43). However, in this case, the destruction of VOC and CO is brought about by contact with air (oxygen) in the presence of a catalyst. Noble metals (such as platinum) deposited on a ceramic base, shaped to enhance good surface contact, or on a metal mesh-pad structure, are typical catalyst configurations. Bed depths are normally 12 in., with variation between 8 and 24 in. (EPA, 1980).

In contrast to thermal oxidation, lower temperatures (200–550°C) are an important characteristic of such systems. This reduces the amount of NO formed, but also is necessary to maintain the integrity of the catalyst. Because of safety considerations, VOC concentrations will ordinarily be kept at <25–30% of the LEL. The incoming gas must be cleaned of materials such as lead, antimony, iron oxide, liquid and solid particles as these may poison the catalyst. Even with such cleaning and temperature control, catalysts will need to be regenerated and ultimately replaced as their efficiency declines.

Specific applications usually require pilot experimentation or considerable experience with the gas stream to be controlled. One cubic foot of catalyst/1000 scfm flue gas to achieve 90% VOC destruction and 1.5 ft^3 catalyst/1000 scfm to achieve 99% destruction have been suggested for one type of catalyst system design (EPA, 1980).

Several applications for a proprietary precious metal catalyst on ceramic substrate are described in Table 6.19 (Martin, 1981). One installation was on the exhaust stream of a plastic film printing press. The film is used for paper tissue packaging. The central impression cylinder flexographic press was 60 in. wide with six stations. The exhaust stream (4.72 m^3sec^{-1}, ≤85°C) contained ethanol (the major component), *n*-propanol, *n*-propyl acetate, heptane, and several other organics that were volatilized during drying of the polyamid printing inks. Two cylindrical catalyst elements, each 1.5 in. in diameter by 3 in. deep, were used with a total catalyst volume of 0.006 ft^3. Only a portion of the total exhaust stream was treated. Flow rates are expressed as space velocities, the actual volumetric flow rate in the bed divided by bed volume. Efficiencies were usually >90% except at catalyst temperatures of 260°C. At 370°C it was estimated that ≥95% efficiency would be realized for ethanol and *n*-propanol for all space velocities even after 3 years of catalyst use. This was also indicated for space velocities <50 hr^{-1} for 316°C.

The second application was on an off-gas (32–43°C) from a process to make formaldehyde (Formax). The proprietary precious metal formula was deposited on a honeycomb ceramic substrate in block form, layered between stainless steel wire and supported by an open grating

TABLE 6.19 Efficiency of a Proprietary Catalytic Oxidation System on Two Industrial Processes

A. Plastic Film Printing Press Ink Drying Exhaust[a]

Catalyst volume = 0.006 ft^2

Component	Inlet concentration (σ) in ppmv
Ethanol	445 (42)
n-Propanol	328 (33)
n-Propyl acetate	24 (8)
Heptane	20 (12)
Total organics	609 (26)

	Catalyst exit temperature (°C)	Efficiency, η^a				
		Ethanol	*n*-Propanol	*n*-Propyl acetate	Heptane	Total organics
Space velocity[b] = 70 hr^{-1}; ΔP = 630 Pa	371	98.3	99.9	96.2	—	97.7
	316	97.2	96.3	86.5	83.3	94.3
	266	95.7	95.0	52.9	57.7	82.6
Space velocity[b] = 50 hr^{-1}; ΔP = 456 Pa	371	99.1	99.9	99.6	—	98.0
	316	98.0	97.3	93.0	63.8	95.0
	260	97.6	97.2	51.0	62.0	85.8
Space velocity[b] = 30 hr^{-1}; ΔP = 266 Pa	362	99.8	99.9	99.7	—	98.0
	316	99.7	99.8	98.5	97.7	95.2
	260	99.7	99.7	61.5	78.0	87.5

B. Formaldehyde Plant (Formax Process) Off-Gas

Components	Inlet concentration (σ) ppmv
Carbon monoxide	5173 (1039)
Methyl alcohol	556 (250)
Dimethyl ether	3042 (364)
Formaldehyde	490 (600)

Flow rate m^3/hr (25°C)	Catalyst exit temperature (°C)	Efficiency, η^c (%)			
		CO	CH_3OH	CH_3OCH_3	HCHO
5200	540	98.9	97.2	96.8	95.0
5149	532	99.1	98.6	97.2	98.6
2957	540	99.3	99.2	96.6	97.0
4124	505	99.0	98.9	96.7	99.1
4590	510	99.0	97.1	96.5	98.7

[a] Efficiencies after 3 months of bed use.
[b] Space velocity = actual volumetric flow rate in catalyst bed/bed volume.
[c] Testing was carried out over a period of 4 months.

SOURCE: Martin, 1981.

of stainless steel. Catalyst dimensions were not given. The overall testing program indicated that the catalytic system operated with efficiencies similar to those shown in Table 6.19 for a year (Martin, 1981). Both of the applications above require preheating to achieve appropriate catalyst temperatures. A model for predicting the behavior of platinum-alumina catalyst pellets during warm-up has been developed and successfully compared with test data (Oh and Cavendish, 1985a, 1985b).

IV. Collection Device Auxiliaries

It has not been the intent of the discussion on air cleaning to create the impression that once a gas stream passes through a cleaning device, all problems are solved. With the exception of combustion and catalysis, all of the equipment and processes reviewed transfer the contaminant(s) from one phase to another, i.e., gas to liquid or solid. Particles

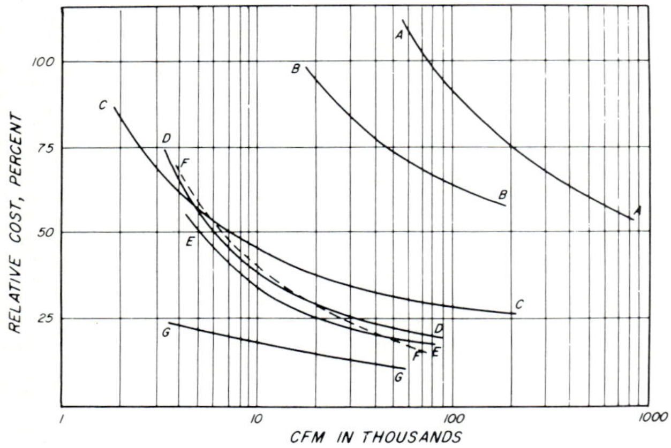

Figure 6.24 Relative cost comparison of particle control equipment (*ACGIH, 1984*).

A. High-voltage precipitator (minimum cost range)
B. Continuous-duty high-temperature fabric collector (2:1)[a]
C. Continuous-duty reverse-pulse fabric collector (8:1)[a]
D. Wet collector
E. Intermittent-duty fabric collector (2:1)[a]
F. Low-voltage precipitator
G. Cyclone

---

[a] Air-to-cloth ratio, cfm/ft^2.

NOTE 1: Cost based on collector section only. Does not include ducts, dust disposal devices, pumps, exhausters, or other accessories not an integral part of the collector.

NOTE 2: Price of high-voltage precipitator will vary substantially with applications and efficiency requirements. Costs shown are for fly ash applications where velocities of 200 to 300 fpm are normal.

TABLE 6.20 Average Cost Factors for Estimating Capital Costs

	Precipitators	Scrubber	Fabric filters	Incinerators (combustion)	Adsorber
		Direct Costs			
1. Purchased equipment costs					
a. Control device	As req'd.	As req'd.	As req'd.	As req'd.	As req'd.
b. Auxiliary equipment	As req'd.	As req'd.	As req'd.	As req'd.	As req'd.
c. Instruments and controls	0.10	0.10	0.10	0.10	0.10
d. Taxes	0.03	0.03	0.03	0.03	0.03
e. Freight	0.05	0.05	0.05	0.05	0.05
Total	1.00	1.00	1.00	1.00	1.00
2. Installation direct costs					
a. Foundations and supports	0.04	0.06	0.04	0.08	0.08
b. Erection and handling	0.50	0.40	0.50	0.14	0.20
c. Electrical	0.08	0.01	0.08	0.04	0.08
d. Piping	0.01	0.05	0.01	0.02	0.05
e. Insulation	0.02	0.03	0.07	0.01	0.02
f. Painting	0.02	0.01	0.02	0.01	0.01
g. Site preparation	As req'd.	As req'd.	As req'd.	As req'd.	As req'd.
h. Facilities and buildings	As req'd.	As req'd.	As req'd.	As req'd.	As req'd.
Total	1.67	1.56	1.72	1.30	1.44
		Indirect Costs			
3. Installation indirect costs					
a. Engineering and supervision	0.20	0.10	0.10	0.10	0.10
b. Construction and field expense	0.20	0.10	0.20	0.05	0.05
c. Construction fee	0.10	0.10	0.10	0.10	0.10
d. Start-up	0.01	0.01	0.01	0.02	0.02
e. Performance test	0.01	0.01	0.01	0.01	0.01
f. Model study	0.02	—	—	—	—
g. Contingencies	0.03	0.03	0.03	0.03	0.03
Total	2.24	1.91	2.17	1.61	1.75

SOURCE: Neveril et al., 1978e.

collected dry have to be removed from the collection device and disposed of. In the case of a cyclone, particles normally fall out. In an electrostatic precipitator or fabric filter, the particles collected have to be dislodged, which requires additional equipment. Depending on where the air cleaner is located, material handling equipment may be necessary to transport collected particles to the next step in disposal. Wet collectors result in a liquid (usually water) that needs to be cleaned. It is frequently desirable to recycle the liquid, so the collected contaminants are ultimately removed as solids. These solids must be disposed of. Adsorbents have a finite life and must be reprocessed or replaced. Combustion and catalysis may create new contaminants, which also have to be treated. Contaminated air can be cleaned effectively, but the air cleaner is just the first step in the overall treatment process. It is always desirable to consider process modifications that would

TABLE 6.21 Basis for Estimating Annual Air Cleaner Operating Costs (1977 dollars)

	Cost factor
Direct Operating Costs	
1. Operating labor	
a. Operator	\$7.87/labor-hour
b. Supervisor	15% of 1*a*
2. Maintenance	
a. Labor	\$8.66/labor-hour
b. Material	100% of 2*a*
3. Replacement parts	As required
4. Utilities	
a. Electricity	\$0.0432/kW·hr
b. Fuel oil	\$0.47/gal
c. Natural gas	\$1.98/1000 ft^3
d. Water	\$0.25/1000 gal
e. Steam	\$5.04/1000 lb
f. Compressed air	\$0.02/1000 ft^3
5. Waste disposal	\$5–10/ton
Indirect Operating Costs	
6. Overhead	80% of 1*a* + 1*b* + 2*a*
7. Property tax	1% of capital costs
8. Insurance	1% of capital costs
9. Administration	2% of capital costs
10. Capital recovery costs	0.16275
a. Interest of 10%	
b. Equipment life of 10 years	
Credits	
11. Recovered product	As required

SOURCE: Neveril et al., 1978e.

reduce the amount of contaminants generated rather than deciding to treat the existing waste products (Chapter 7 gives some examples).

V. Relative Cost Comparisons

The cost of a specific control option is difficult to estimate without a reasonable degree of design specification. Rough estimates can be made based on the type of air cleaner to be used and the volumetric flow rate of gas to be cleaned. A comparison of costs of particle collection equipment is shown in Figure 6.24. Note that the relative values do not include the cost of auxiliaries. In general, the more efficient the collection device, the more expensive it is.

There are a number of other costs involved in the design and installation of control equipment. Table 6.20 presents a listing of these and provides information on the relative cost of different control devices. Note that both particle and gas collection equipment are included. Table 6.21 gives some information (in 1977 dollars) for estimating air cleaner operation and maintenance. For a more detailed approach to estimating the cost of specific air pollution control systems, the reader is referred to the articles by Neveril et al. (1978a–1978e).

REFERENCES

ACGIH (1984). *Industrial Ventilation,* 18th ed. American Conference of Governmental Industrial Hygienists, Lansing, Mich.

ASHRAE (1983). *ASHRAE Handbook and Product Directory, 1983 Equipment.* American Society of Heating, Refrigerating and Air-Conditioning Engineers, Atlanta, GA.

Boll, R. H. (1973). Particle collection and pressure drop in venturi scrubbers. *Ind. Eng. Chem. Fund., 12,* 40–50.

Brink, J. A., and Coutant, C. G. (1958). Experiments on an industrial venturi scrubber. *Ind. Eng. Chem., 50,* 1157–1160.

Buonicore, A. J. (1980). Air pollution control. *Chem. Eng., 87,* 81–101, June 30.

Calvert, S., Lundgren, D., and Mehta, D. S. (1972). Venturi scrubber performance. *J. Air Pollut. Control Assoc., 22,* 529–532.

Caplan, K. J. (1977). Source control by centrifugal force and gravity. In: Stern, A. C., ed., *Air Pollution,* vol. IV, 3d ed. Academic Press, New York, pp. 98–148.

Carr, R. C., and Smith, W. B. (1984). Fabric filter technology for utility coal-fired power plants. *J. Air Pollut. Control Assoc., 31,* 281–293.

Crowder, J. W., Noll, K. E., and Davis, W. T. (1982). Modeling of venturi scrubber efficiency. *Atmos. Environ., 16,* 2009–2013.

Dennis, R. (1974). Collection efficiency as a function of particle size, shape and density: Theory and experience. *J. Air Pollut. Control Assoc., 24,* 1156–1163.

Dennis, R., Cass, R. W., Cooper, D. W., Hall, R. R., Hampl, V., Klemm, H. A., Langley, J. E., and Stern, R. W. (1977). *Filtration Model for Coal Fly Ash with Glass Fabrics.* U.S. EPA Report No. EPA-600/7-77-084, August.

Dennis, R., Cass, R. W., and Hall, R. R. (1978). Dust dislodgement from woven fabrics versus filter performance. *J. Air Pollut. Control Assoc., 28,* 47–52.

Dennis, R., and Klemm, H. A. (1979). Verification of projected filter system design and operation. In: Venditti, F. P., Armstrong, J. A., and Durham, M., eds. *Symposium on the Transfer and Utilization of Particulate Control Technology: Volume II.* U.S. EPA Report No. EPA-600/7-79-044b, February, pp. 143–160.

Dietz, P. W. (1981). Collection efficiency of cyclone separators. *Amer. Inst. Chem. Eng. J., 27,* 888–892.

Dubinin, M. M. (1975). Physical adsorption of gases and vapors in micropores. *Prog. Surf. Membr. Sci., 9,* 1–70.

EPA (1977). *Compilation of Air Pollution Emission Factors,* 2d ed. U.S. EPA AP-42, April.

EPA (1979). *Symposium on the Transfer and Utilization of Particulate Control Technology,* vol. 3, Venditti, F. P., Armstrong, J. A., and Durham, M., eds. U.S. EPA Report No. EPA-600/7-79-044c, February.

EPA (1980). *Organic Chemical Manufacturing. Volume 4: Combustion Control Devices.* U.S. EPA Report No. EPA-450/3-80-026, December.

EPA (1982a). *Control Techniques for Particulate Emissions from Stationary Sources,* vol. 1. U.S. EPA, Emissions Standards and Engineering Division, Report No. EPA-450/3-81-005a, September.

EPA (1982b). *Control Techniques for Particulate Emissions for Stationary Sources,* vol. 2. U.S. EPA, Emissions Standards and Engineering Division, Report No. EPA-450/3-81-005b, September.

EPA (1984). *Control of Volatile Organic Compound Emissions from Air Oxidation Processes in Synthetic Organic Chemical Manufacturing Industry.* U.S. EPA, Emissions Standards and Engineering Division, Report No. EPA-450/3-84-015, December.

Goel, K. C., and Hollands, K. G. T. (1977). A general method for predicting particulate collection efficiency of venturi scrubbers. *Ind. Eng. Chem. Fund., 16,* 186–193.

Golovoy, A., and Braslaw, J. (1981). Adsorption of automotive paint solvents on activated carbon: I. Equilibrium adsorption of single vapors. *J. Air Pollut. Control Assoc., 31,* 861–865.

Golovoy, A., and Braslaw, J. (1982). Adsorption of automotive paint solvents on activated carbon: II. Adsorption kinetics of single vapors. *Environ. Prog., 1,* 89–95.

Hsiung, T. H., and Thodos, G. (1977). Mass-transfer factors from actual driving forces for the flow of gases through packed beds. *Int. J. Heat Mass Transfer, 20,* 331–340.

Jonas, L. A., and Rehrmann, J. A. (1973). Predictive equations in gas adsorption kinetics. *Carbon, 11,* 59–64.

Jonas, L. A., and Sansone, E. R. (1986). Prediction of activated carbon performance for sequential adsorbates. *Amer. Ind. Hyg. Assoc. J., 47,* 509–511.

Jonas, L. A., Sansone, E. B., and Farris, T. S. (1983). Prediction of activated carbon performance for binary vapor mixtures. *Amer. Ind. Hyg. Assoc. J., 44,* 716–719.

Jonas, L. A., Sansone, E. B., and Farris, T. S. (1985). The effect of moisture on the adsorption of chloroform by activated carbon. *Amer. Ind. Hyg. Assoc. J., 46,* 20–23.

Lapple, C. E. (1951). Processes use many collection types. *Chem. Eng., 58,* 144–151, May.

Martin, N. A. (1981). *Catalytic Incineration of Low Concentration Organic Vapors.* U.S. EPA Report No. EPA-600/2-81-017, February.

McDaniel, M. (1983). *Flare Efficiency Study.* U.S. EPA Report No. EPA-600/2-83-052, July.

Neveril, R. B., Price, J. U., and Engdahl, K. L. (1978a). Capital and operating costs of selected air pollution control systems—I. *J. Air Pollut. Control Assoc., 28,* 829–836.

Neveril, R. B., Price, J. U., and Engdahl, K. L. (1978b). Capital and operating costs of selected air pollution control systems—II. *J. Air Pollut. Control Assoc., 28,* 963–968.

Neveril, R. B., Price, J. U., and Engdahl, K. L. (1978c). Capital and operating costs of selected air pollution control systems—III. *J. Air Pollut. Control Assoc., 28,* 1069–1072.

Neveril, R. B., Price, J. U., and Engdahl, K. L. (1978d). Capital and operating costs of selected air pollution control systems—IV. *J. Air Pollut. Control Assoc., 28,* 1171–1174.

Neveril, R. B., Price, J. U., and Engdahl, K. L. (1978e). Capital and operating costs of selected air pollution control systems—V. *J. Air Pollut. Control Assoc., 28,* 1253–1256.

O'Brien, D. M., and Hurley, D. E. (1981). *An Evaluation of Engineering Control Technology for Spray Painting.* NIOSH Report No. 81-121, June.

Oh, S. H., and Cavendish, J. C. (1985a). Mathematical modeling of catalytic converter light off. Part II: Model verification by engine-dynamometer experiments. *Amer. Inst. Chem. Eng. J., 31,* 935–942.

Oh, S. H., and Cavendish, J. C. (1985b). Mathematical modeling of catalytic converter light off. Part III. Prediction of vehicle exhaust emissions and parametric analysis. *Amer. Inst. Chem. Eng. J., 31,* 943–949.

Pahl, D. (1983). EPA's program for establishing standards of performance for new stationary sources of air pollution. *J. Air Pollut. Control Assoc., 33,* 468–482.

Perry, R. H., Green, D. W., and Maloney, J. O. (1984). *Perry's Chemical Engineers Handbook,* 6th ed. McGraw-Hill, New York.

Reucroft, P. J., Simpson, W. J., and Jonas, L. A. (1971). Sorption properties of activated carbon. *J. Phys. Chem., 75,* 3526–3531.

Rudnick, S. M., Koehler, J. L. M., Martin, K. P., Leith, D., and Cooper, D. W. (1986). Particle collection efficiency in a venturi scrubber: Comparison of experiments with theory. *Environ. Sci. Technol., 20,* 237–242.

Sansone, E. B., and Jonas, L. A. (1981). Prediction of activated carbon performance for carcinogenic vapors. *Amer. Ind. Hyg. Assoc. J., 42,* 688–691.

Sansone, E. B., Tewari, Y. B., and Jonas, L. A. (1979). Prediction of removal of vapors from air by adsorption on activated carbon. *Environ. Sci. Technol., 13,* 1511–1513.

Spaite, P. W., and Burckle, J. O. (1977). Selection, evaluation, and application of control devices. In: A. C. Stern, ed., *Air Pollution,* vol. IV, 3d ed. Academic Press, New York, pp. 44–47.

Turner, J. H., and McKenna, J. D. (1984). Control of particles by filters. In: Calvert, S. and England, H. M., eds., *Handbook of Air Pollution Technology*, John Wiley and Sons, New York, pp. 253–255.

Urano, K., Omori, S., and Yamamoto, E. (1982). Prediction method for adsorption capacities of commercial activated carbons in removal of organic vapors. *Environ. Sci. Technol., 16,* 10–14.

Wark, K., and Warner, C. F. (1981). *Air Pollution, Its Origin and Control.* Harper & Row, New York.

Werner, M. D. (1985). The effects of relative humidity on the vapor phase adsorption of trichloroethylene by activated carbon. *Amer. Ind. Hyg. Assoc. J., 46,* 585–590.

Wheeler, A., and Robell, A. J. (1969). Performance of fixed-bed catalytic reactors with poison in the feed. *J. Catal., 13,* 299–305.

White, H. J. (1977a). Electrostatic precipitation of fly ash. *J. Air Pollut. Control Assoc., 27,* 15–21.

White, H. J. (1977b). Electrostatic precipitation of fly ash. Precipitator design. *J. Air Pollut. Control Assoc., 27,* 206–217.

Yung, S-C., Barbarika, H. F., and Calvert, S. (1977). Pressure loss in venturi scrubbers. *J. Air Pollut. Control Assoc., 27,* 348–351.

Yung, S-C., Calvert, S., Barbarika, H. F., and Sparks, L. E. (1978). Venturi scrubber performance model. *Environ. Sci. Technol., 12,* 456–459.

Chapter

7

Process Alteration

R. A. Wadden

I. Design Fundamentals

The intent of this chapter is not to convert all practicing engineers to process designers. Process changes to effect workplace control depend on the particular chemistry of each process. Consequently, the solution of such problems is quite specific for a given process and sometimes even for an individual plant. However, it is useful to recognize the underlying principles that are used to develop chemical processes. Understanding these principles will allow engineers with responsibility for hazard control at least to encourage designers to minimize or eliminate toxic material problems in the conceptual stage of plant design; and to seek innovative solutions for existing installations. At the end of this chapter we have included example design solutions to process problems in a variety of chemical settings.

Underlying any process design are the chemical and physical changes required to form products from raw materials. For instance, nitric acid (HNO_3) is usually produced by high-pressure oxidation of ammonia to nitric oxide (NO) over a heated platinum-rhodium catalyst (400–800 kPa, 750–800°C). The NO is subsequently converted to NO_2, which is absorbed in water to form 55–65% HNO_3. A flow sheet of the process is shown in Figure 7.1 (Drabkin, 1979). The principal stoichiometric reactions are:

$$4NH_3 + 5O_2 \xrightarrow[\text{catalyst}]{\text{Pt-Rh}} 4NO + 6H_2O \quad \text{(gas phase)} \tag{7.1}$$

$$2NO + O_2 \longrightarrow 2NO_2 \longrightarrow N_2O_4 \quad \text{(gas phase)} \tag{7.2}$$

$$3NO_2 + H_2O \longrightarrow 2HNO_3 + NO \quad \text{(liquid phase)} \tag{7.3}$$

The nitric oxide formed in Eq. (7.3) is further reacted with secondary air to NO_2. Overall the conversion is:

$$NH_3 + 2O_2 \longrightarrow HNO_3 + H_2O \tag{7.4}$$

The process requires three chemical transformations and a change of phase.

Chemical process designs are based on energy and mass balances, and the speed at which chemical and physical transformations take place. Energy balances must meet the restrictions of the laws of thermodynamics. For instance, heat energy added to a system must be accounted for in the temperature and composition of the exiting streams, or may be converted to useful work, subject to the second law. The mechanical energy balance (Bernoulli's equation), which was introduced in Chapter 5, is a form of the first law. The mass that goes into a chemical process as raw material must exit as product, by-product, or unreacted raw material.

Mass and energy balances can be evaluated on either steady-state (no accumulation or depletion within the process) or non-steady-state bases. The choice depends on the type of process and the conditions of

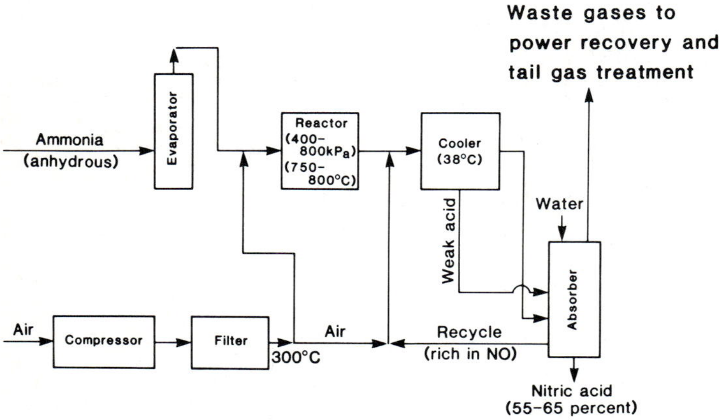

Figure 7.1 Single-pressure nitric acid manufacturing process (*revised from Drabkin, 1979*).

operation. In general, continuous processes are conveniently analyzed under steady-state assumptions, and batch operations are considered as being at non-steady-state. The balance of mass or energy determines how much comes into and leaves a process, but does not supply any information on how quickly transformations occur.

The amounts of materials and energy contained in an industrial process—or, alternatively, the size of the equipment, piping, and vessels—presume a knowledge of the speed of conversion or transfer. [For example, the catalyst contact time for Eq. (7.1) is about 3×10^{-4} sec (Drabkin, 1979).] Such an analysis requires rate information: heat transfer rates for heat exchangers, kinetic reaction rate constants for chemical reactions, or mass transfer rates for movement of material between phases. Each rate also requires a driving force: temperature differences for heat transfer, concentration differences for mass transfer, or concentrations of reacting materials for reaction rates. These relationships can be expressed symbolically as

$$\dot{Q} = UA_h \,\Delta T \qquad \text{(heat transfer)} \tag{7.5}$$

$$\dot{M} = KA_m \,\Delta C \qquad \text{(mass transfer)} \tag{7.6}$$

$$\dot{N}_i = kN_i \qquad \text{(first-order reaction, } i \longrightarrow j\text{)} \tag{7.7}$$

where $\dot{Q}$ is a change in energy per time; U is the heat transfer coefficient, energy/area·temperature difference·time; A_h is the heat transfer area; ΔT is the temperature difference in the direction of heat flow; $\dot{M}$ is the change in mass/time; K is the mass transfer coefficient, length/time; A_m is the mass transfer area; ΔC is the concentration difference in the direction of mass transfer, mass/volume; $\dot{N}_i$ is the decrease (for reactants) in molar concentration of component i/time; k is the chemical reaction rate constant (first-order), time^{-1}; and N_i is the molar concentration of component i. In addition, chemical reaction rates, at least for simple reactions, often vary with temperature, such that

$$\frac{k_{T_2}}{k_{T_1}} = \exp\left[-\frac{E}{R}\left(\frac{T_1 - T_2}{T_1 T_2}\right)\right] \tag{7.8}$$

where k_{T_1} and k_{T_2} are reaction rates at absolute temperatures (K or °R) T_1 and T_2, respectively; E is the activation energy; and R is the ideal gas constant.

Some *physical* systems, where the rates of conversion or transfer are quite rapid, relative to residence time of the material within the process, can be characterized as being at equilibrium. The dissolution of salt and water is an example. However, most *chemical* systems of industrial interest will not be at equilibrium. Indeed, the equilibrium condition is often not desirable, since there is ordinarily an objective to maximize the formation of product in the shortest period of time. Within practical limits, reactions are forced to completion by addition of cheaper reactants in excess amounts relative to the limiting (most expensive) reactant.

Even with favorable conditions, reactions may not go to completion. If we consider the HNO_3 process described by Eqs. (7.1)–(7.4), we would expect each mole of NH_3 to produce a mole of HNO_3. In reality, only about 95% of the NH_3 is actually converted, even with 50% excess air over and above what is stoichiometrically required. The limitation occurs because of various inefficiencies in the process, mostly due to less than ideal mixing during the available contact time, and side reactions to undesirable by-products.

II. Process Description

The design procedure has not been successful if we wait until workers or the community are exposed to hazardous materials. An examination of the process before it is built or put into operation will often reveal agents, points of release, and quantities of emission that may pose an environmental problem. And these potential releases can be guarded against by appropriate controls. So, we have to know where to look for problems. The general considerations to be aware of were listed in Chapter 3.

A. Design conditions

One of the first things to recognize is whether the process is batch or continuous. As our previous discussion in Chapter 3 indicated, batch processes are intrinsically more difficult to control in terms of workplace emissions. The process of Figure 7.1 is continuous, and so does not pose special problems of raw material addition or product removal. On the other hand, all of the processing steps in a gray iron foundry, as illustrated by Figure 7.2, are batch operations (Burgess, 1981).

Analysis of any chemical process ordinarily will begin with consideration of the flow train. This consists of a diagram, such as Figure 7.1, which blocks out the major process steps. Ordinarily, flow rates and operating conditions will be available at this stage. At the next level of design a more detailed description will be available, which includes schematics of major pieces of equipment and control devices. Final design drawings will usually be quite specific in terms of piping, foundation placement, and vessel and equipment detail.

Information on design streams indicates how the plant is supposed to operate. Recognized hazardous substance release points are identi-

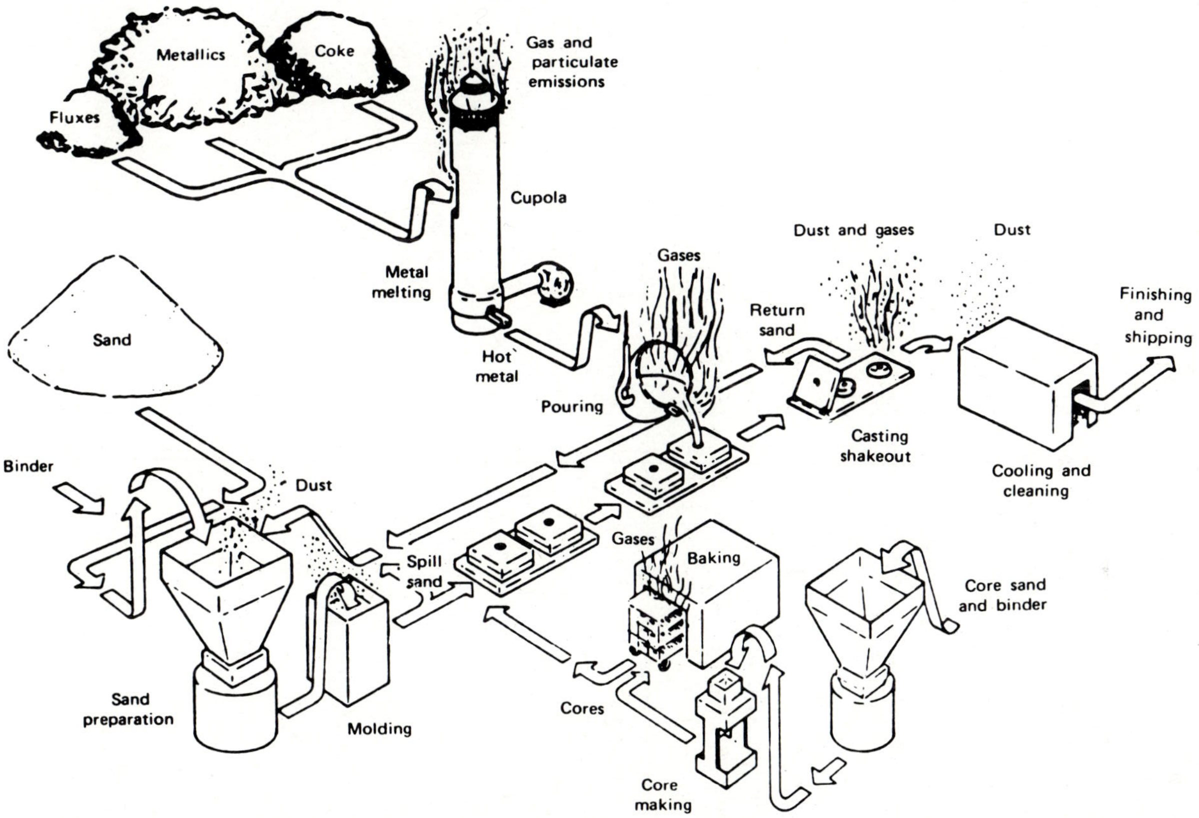

Figure 7.2 Gray iron cupola foundry (*Burgess, 1981*).

fied. For instance the absorber waste gas stream in Figure 7.1 will require further treatment for removal of nitric acid mist and unreacted nitrogen oxides (about 0.25% by volume) (Drabkin, 1979). The quantities and types of raw material (anhydrous ammonia, ≃0.3 tons NH_3/ton HNO_3), and products (≃800 tons/day of 60% HNO_3) are also specified. The health hazards of nitric acid and ammonia are well documented (ACGIH, 1984a), as are appropriate handling procedures (NIOSH/OSHA, 1981). Ammonia also presents a potential fire and explosion hazard, and the feed stream to the conversion reactor must be maintained at <15.5% NH_3, the lower flammability limit in air (see Chapter 4). Storage and transfer of ammonia must conform to specified safe practice procedures (NFPA, 1984).

Often design conditions will change over time due to efforts to maximize product yield and throughput. If these process changes are not integrated with appropriate changes in process and pollution control, the capability of dealing with emergencies may no longer exist. A useful example has been presented by Russell (1975) involving a catalyzed exothermic batch polymerization in a jacketed reactor.

Example The polymerization reaction rate is proportional to the square root of catalyst concentration and doubles for every 10°C rise in temperature. During a critical part of the reaction, the temperature is maintained at 110°C by circulating water through the jacket at an average temperature of 60°C. The *minimum* available water temperature for emergency cooling is 35°C. Careful testing has indicated that it will be possible to reduce reaction time and significantly increase production by increasing the reaction temperature to 112°C. Our problem is to examine the capability of the cooling system to remove heat from the process under emergency conditions when the catalyst charge is accidently doubled.

solution At this point is is useful to consider Eqs. (7.5) and (7.8). Substituting what we know in Eq. (7.8) gives:

$$\frac{k_{T_2}}{k_{T_1}} = \frac{k_{(T_1+10)}}{k_{T_1}} = 2 = \exp\left\{-\frac{E}{R}\left[\frac{-10}{(383)(393)}\right]\right\}$$

$$\frac{E}{R} \simeq 10{,}400 \text{ K} \tag{7.8}$$

In order to maintain isothermal conditions, the amount of heat generated by the reaction must just equal that removed by heat transfer; and the quantity of heat generated will be approximately proportional to the reaction rate. Consequently, from Eq. (7.5),

$$\frac{k_{112^\circ C}}{k_{110^\circ C}} \simeq \frac{Q_{112^\circ C}}{Q_{110^\circ C}} \simeq \frac{UA(112 - T_{2C})}{UA(110 - 60)}$$

$$= \frac{\text{cooling under new design conditions}}{\text{cooling under original design conditions}} \tag{7.9}$$

where T_{2C} is the required cooling water temperature at the new design conditions.

Then, for the new design conditions, we can also estimate that

$$\frac{k_{112°C}}{k_{110°C}} = \exp\left\{-10{,}400\left[\frac{-2}{(383)(385)}\right]\right\} = 1.15 \tag{7.8}$$

and substituting in Eq. (7.9) gives

$$\frac{UA(112 - T_{2C})}{UA(110 - 60)} = 1.15 \tag{7.9}$$

giving

$$T_{2C} = 112 - (1.15)(50) \simeq 55°C$$

At a reaction temperature of 112°C, the cooling water must be at 55°C to maintain control; and the difference between 60°C and 55°C does not seem to be a serious loss of emergency cooling capacity. However, before the design change, an accidental double charge of catalyst would require a greater cooling water temperature drop than for design conditions in order to maintain control. That is,

$$\frac{k_{110°C,DC}}{k_{110°C}} \simeq (2)^{0.5} \simeq \frac{Q_{110°C,DC}}{Q_{110°C}} \simeq \frac{UA(110 - T_{1C,DC})}{UA(110 - 60)} \tag{7.9}$$

where $k_{110°C,DC}$ and $T_{1C,DC}$ are the reaction rate and required cooling water temperature, respectively, under double charge conditions, and the $2^{0.5}$ is the change in reaction rate under double charge conditions. Solving, we find that:

$$T_{1C,DC} = 39°C$$

At the new design operating temperature, an accidental double catalyst charge will require a cooling water temperature, $T_{2C,DC}$, of

$$\frac{k_{112°C,DC}}{k_{110°C}} = (1.15)[(2)^{0.5}] = \frac{UA(112 - T_{2C,DC})}{UA(110 - 60)}$$

$$T_{2C,DC} = 31°C \tag{7.9}$$

Since this cooling water temperature is not available, the process is potentially always out of control. But this may only be discovered, possibly disastrously, when an operating error is actually made. Hence the need for an integrated approach to design changes.

B. Nonroutine operation

Design values of temperature, pressure, and composition show how the process *should* operate, and probably *will* operate, under steady-state or regular batch conditions. However, there will be identifiable periods when design values will not be attained. These will occur during start-up and shut-down, and during upset and breakdown. Depending on the plant size and complexity, start-up may take anywhere from a few hours up to several months. So appropriate control is a necessary part of the overall process design. Blanketing vessels and reactors with inert gas, and recycling unreacted, or partially reacted, raw materials until design operating conditions are reached are desirable methods of control. The production of 1,2-dichloroethane, which is described in detail later in this chapter, provides an example of how such control is implemented.

1. Leaks. Leaks are unintentional release points and ordinarily are not evident from process description and flow sheets. However, leaks may constitute major sources for personnel exposure as well as for environmental contamination. Some equipment is designed so that a limited amount of leakage occurs. A case in point is "self-sealing" doors on coke batteries, which require some of the coal charged to the oven to become plastic and flow into any openings between door and oven. Another example is leakage for purposes of lubrication, illustrated by the pump packings shown in Figures 7.3–7.5. A packed seal (Figure 7.3) requires a lubrication rate of about 1 drop/sec between the packing and the rotating shaft. Absence of lubrication will cause packing to lose resiliency, which in turn increases friction, shortens packing life, and increases operating costs. A single mechanical seal also requires lubrication but only at a rate of 1 drop/min (Figure 7.4). And double mechanical seals are designed to incorporate a sealing fluid such as water or steam, glycols, or methanol, usually at a higher pressure than that of the process material (Figure 7.5). Consequently, no process liquid can leak out. Sealless pumps, in which the cavity housing the motor rotor and the pump casing are interconnected and the process fluid becomes the lubricant, and diaphragm pumps, which completely contain the process material, are both devices that offer no pathway for process liquid release. In the synthetic organic chemical industry it has been estimated that about 10% of pumps are packed seals, 72% are single mechanical, 17% are double mechanical, and 1% are sealless (Van Wagenen, 1981).

Table 7.1 summarizes nonmethane hydrocarbon leak rates from a

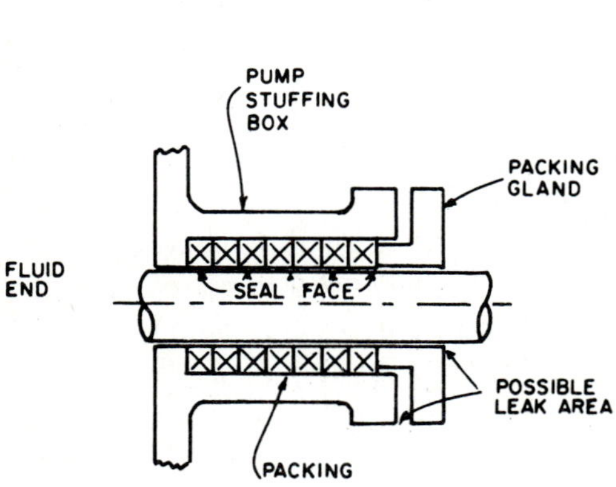

Figure 7.3 Simple packed seal (*EPA, 1982b*).

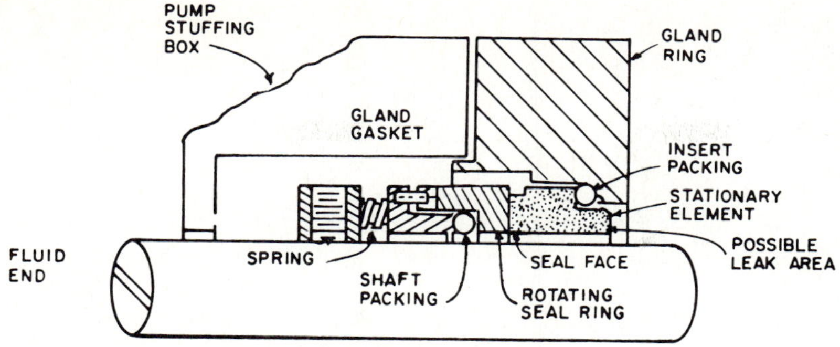

Figure 7.4 Single mechanical seal (*EPA, 1982b*).

variety of services in petroleum refineries (Wetherold and Provost, 1979). The estimates are based on an extensive testing survey of 13 operating plants. A survey of six California refineries found 4% of all flanges and 9% of all valves leaking. Valve leakage was significantly higher in certain areas, such as LPG storage (20% leaking) and alkylation units (18% leaking). Of the valves, 1.6% were > 2.86 kg/day and an additional 7.2% were between 0.068 and 2.86 kg/day (Morgester et al., 1979). The results of a survey of four different petrochemical processes are shown in Table 7.2 (Hughes et al., 1979). The relatively high emission rates from agitators emphasizes the difficulty of effectively sealing equipment that is supported at only one end.

All seals will eventually fail. This will usually be a catastrophic, all-or-nothing type of failure as opposed to a gradual increase in leakage over time. Consequently, the solution to control of leakage is a regular inspection and maintenance program (EPA, 1983b; Freeberg and Aarni, 1982). Table 7.3 shows the decrease in leak rates upon implementation of an inspection program and replacement of leaking seals. It has been demonstrated that a two-person monitoring team can inspect between 400 and 500 components a day (EPA, 1983b).

Some design changes may also be effective. These include controlled

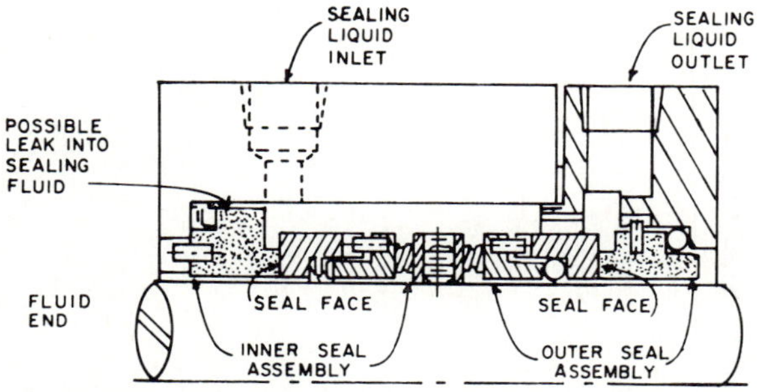

Figure 7.5 Double mechanical seal (back to back) (*Van Wagenen, 1981*).

TABLE 7.1 Estimated Vapor Emission Factors for Nonmethane Hydrocarbons from Selected Sources in Petroleum Refineries[a]

Source type	Estimated percent leaking	95% confidence interval for percent leaking	Average emission factor estimate,[b] lb/hr/source (kg/day)	95% confidence interval for emission factor, lb/hr/source	Relationship between leak rate, L (lb/hr) and source screening value, S (ppmv as hexane)[c]
Valves					
Gas/vapor streams	29.3	25.9–32.7	0.047 (0.51)	0.027–0.084	$L = 8.59 \times 10^{-7} S^{1.16}$
Light liquid/two-phase streams[d]	36.5	33.6–39.5	0.023 (0.25)	0.016–0.034	$L = 4.75 \times 10^{-5} S^{0.76}$
Heavy liquid streams[e]	6.7	4.6–8.9	0.0007 (0.0076)	0.0002–0.002	
Flanges	3.1	2.2–3.9	0.00058 (0.0063)	0.0002–0.001[f]	
Pump seals					$L = 4.9 \times 10^{-4} S^{0.73}$
Light liquid streams[d]	63.8	58.9–68.8	0.26 (2.84)	0.17–0.39	
Heavy liquid streams[e]	22.6	17.1–28.1	0.045 (0.49)	0.02–0.11	
Compressor seals					$L = 4.9 \times 10^{-4} S^{0.73}$
Hydrocarbon service	70.3	62.9–77.7	0.98 (10.7)	0.46–2.0	
Hydrogen service	81.2	71.2–88.8	0.10 (1.09)	0.04–0.24	
Drains	19.2	14.4–24.0	0.070 (0.76)	0.02–0.20	$L = 4.9 \times 10^{-4} S^{0.73}$
Relief valves	39.2	25.1–40.6	0.19 (2.07)	0.07–0.52	$L = 4.9 \times 10^{-4} S^{0.73}$

[a]Leaking sources were defined as sources with screening values ≥200 ppmv or sources with measured leak rates greater than 0.00001 lb/hr (Wetherold and Provost, 1979).

[b]Average emission factor based on all source screened. Includes leakers and nonleakers.

[c]Based on using J. W. Bachrach TLV sniffer at the source calbrated to hexane (by volume). Concentration measurements were made at each source for screening purposes. Later these sources were enveloped with plastic bags, and total nonmethane hydrocarbon emissions (mass/time) were determined. The equations shown above represent the least-squares fit between screening concentrations and emission rates. The correlation coefficient varied between 0.6 and 0.8.

[d]Light liquid defined as a fluid with a vapor pressure >0.3 kPa at 20°C (e.g., naphtha).

[e]Heavy liquid defined as a fluid with a vapor pressure ≤0.3 kPa at 20°C (e.g., kerosene).

[f]Same equation as for valves in similar service.

degassing vents for reciprocating compressors; caps on the ends of open-ended valves; closed purge sampling for system sample collection; sealed bellows valves; and double mechanical seals for pumps (EPA, 1983b). Recently, the EPA has proposed 10,000 ppm of nonmethane hydrocarbons at the source as defining a leak action level (EPA, 1983b). As indicated by footnote *c* in Table 7.1, the screening concentration is reasonably correlated with the mass emission rate. Whether this is an appropriate value for workplace exposures depends on the toxicity of the material and the proximity of workers.

2. Safety valves and flares. In order to prevent explosions, relief devices are required by construction codes whenever maximum working design pressures of vessels or piping may be exceeded by system pressure. Typically, spring-loaded pressure relief valves are used for this purpose (Figure 7.6). When excess process pressures occur, the valve opens by unseating a disk and closes on return to design pressures. Because the disk may not reseat to form an effective seal, relief valves may be a source of leaks. Rupture disks are sometimes used to accomplish the same end. The material of the disk maintains a tight seal until the set pressure is exceeded. It then ruptures. The disadvantage of the rupture disk is that once it is broken, system pressure will drop to atmospheric and large volumes of hazardous gas or liquid may be released.

Releases from safety valves pose some unique problems for worker exposures. Since emission temperatures are often close to ambient and release velocities are very high (from 20 m/sec to sonic), discharges are characterized as momentum sources (as opposed to buoyant emissions from combustion sources). Consequently, safety valve emissions, which may contain high concentrations of hazardous materials, tend

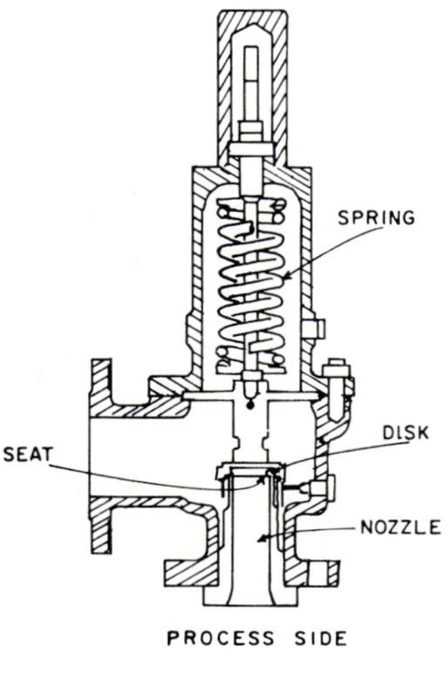

Figure 7.6 Spring-loaded relief valve (*Van Wagenen, 1981*).

TABLE 7.2 Nonmethane Hydrocarbon Emission Rates for Fugitive Sources by Chemical Process

Source type	Monochlorobenzene	Butadiene	Dimethylterephthalate	Ethylene oxide/glycol	
	Average Emission Rate for Significant Fugitive Sources,[a,b] (kg/day)				
Pump seals	0.55	3.84	0.48	1.97	
Compressor seals	—	1.42	—	0.26	
Valves	0.036	2.88	0.77	0.038	
Flanges	1.97	0	2.64	0.024	
Relief devices	—	0.34	0	0	
Process drains	—	—	—	1.63	
Agitator	4.80	—	5.23	—	
Sample valves	—	—	2.18	—	
	Average Emission Rate for All Potential Sources[a,c]				
					USEPA[d]
Pump seals	0.18	1.51	0.079	0.31	
Light liquid	—	—	—	—	1.19
Heavy liquid	—	—	—	—	0.51

Compressor seals	—	1.3	—	0.14	5.47
Valves	0.0012	0.41	0.036	0.0017	
Gas	—	—	—	—	0.73
Light liquid	—	—	—	—	0.17
Heavy liquid	—	—	—	—	0.0055
Flanges	0.053	0	0.082	0.0007	0.020
Relief devices	—	0.12	0	0	2.50
Process drains	—	—	—	0.96	
Agitator	4.8	—	3.48	—	
Sample valves	—	—	0.96	—	0.36
Open-ended lines	—	—	—	—	0.046

[a]Hughes et al., 1979 (dashes indicate that the source type is not in the process).

[b]Significant fugitive sources are those having an emission rate greater than or equal to 0.012 kg/day (0.001 lb/hr) as determined by sampling and analysis.

[c]Emission rates were determined by calculating the mass of fugitive emissions from the total emission rates for all significant sources in each process. This total was divided by the total number of sources screened to arrive at an average fugitive emission rate for all sources.

[d]EPA, 1982a; based on leak frequency in 24 processes in synthetic organic chemical and polymer manufacturing industry and emission rates from petroleum refining (Wetherold and Provost, 1979).

TABLE 7.3 Valve and Flange Inspection Results

Inspection period	Number of valves inspected	Percent valves leaking	Number of flanges inspected	Percent flanges leaking
Jan.–June 1979	24,482	4.3	—	—
July–Dec. 1979	35,394	2.2	—	—
1980	31,487	3.1	26,330	0.56
1981	31,108	2.8	31,943	0.41

SOURCE: Freeberg and Aarni, 1982.

to be dispersed differently than steady-state discharges from recognized design release points. The following procedure has been suggested to predict the 1-hr critical ground level concentration, X_c. X_c is the highest 1-hr ambient concentration, resulting from a horizontal release, at any downwind location or meteorological condition (ASME, 1979; Hanzevack, 1982). It is characterized by:

$$X_c = C_2 \frac{Q}{h^{1.29} D^{0.71}\ v_s} \left(\frac{\rho_s}{\rho_a}\right) \tag{7.10}$$

where Q is the steady-state pollutant emission rate (μg·sec^{-1}); D is stack or relief valve exit diameter (m); v_s is stack or relief valve exit velocity (m·sec^{-1}); ρ_s and ρ_a are density of emissions and ambient air, respectively; h is actual stack height or relief valve elevation above ground level (m); and C_2 is an empirical constant, such that for $\rho_s \leq \rho_a$, $C_2 = 0.012$, and for $\rho_s > \rho_a$, $C_2 = 0.027$. D and v_s are characteristics of the relief system design, and their values are usually determined so that a sufficient volume of gas is vented in a time interval short enough to prevent overpressurization of process vessels, piping, and/or equipment (Leung, 1986).

For steady-state emissions, critical ground-level concentrations can be estimated for other averaging times, t, from

$$X_t = X_c t^{-0.5} \tag{7.11}$$

where t is in hours and recognizing that X_c is a 1-hr average. These equations were validated for dispersion of SF_6 and CO_2, although the density correction was considered tentative because of limited data at elevated densities (Hanzevack, 1982). X_c represents the worst-case concentration and, although downwind distance does not appear in the prediction, it is likely that the maximum concentration will occur in the immediate vicinity of the emission.

It has also been suggested that the maximum concentration can be calculated for non-steady-state *emission* times, T, from 0.08 to 0.25 hr, and any *averaging* time in the range $T \le t \le 5T$ (i.e., t up to 1.25 hr) by using

$$X_t^T = 0.5\left(\frac{T}{t}\right)^{0.8} T^{-0.5}\left[C_2 \frac{Q}{h^{1.29}D^{0.71}v_s}\left(\frac{\rho_s}{\rho_a}\right)\right] \tag{7.12}$$

where T is in hours (DeFaveri et al., 1982) and X_t^T symbolizes the maximum ground-level concentration over averaging time t for an emission period T. For $Q = 0.22 \times 10^6$ μg·sec^{-1}, h = 19 m, D = 0.1 m, ρ_s/ρ_a = 1.5, v_s = 9 m/s, T = 0.25 hr (15 min), and t = 1 hr, Eq. (7.12) gives $X_c^{0.25}$ = 37.5 μg/m^3. This is the critical ground-level concentration (i.e., 1-hr averaging time) for a 15-min emission period.

One solution to episodic releases, such as from relief valves, is a closed-vent collection system with appropriate control devices. Such a system gathers gaseous emissions from pump and compressor barrier fluids, and from relief valve leakage and overpressurization, into a common piping manifold. These gases are then conducted to a device to remove or destroy the hazardous components. The control element may be an adsorber or absorber, a gas recovery compressor that will recycle the vapors to the process, or an incinerator, flare or boiler. Adsorption, absorption, and incineration are discussed in Chapter 6. If a combustion device is used, flame arrestors will be required to prevent flame flashback into the process.

The most desirable state-of-the-art flare design is the smokeless flare. This type inducts combustion air into the flare by steam injection. The injection of steam increases the mixing of materials within the flame zone, which promotes more efficient combustion. For instance, such devices can oxidize 98% of the volatile organic compounds (VOC) to CO_2 and H_2O, provided that the heat content of the flared gas is maintained above 200 Btu/scf (for smokeless steam-assisted elevated design) or 300 Btu/scf (for smokeless elevated flares with no assist); and the exit velocity of the flare is 22 m/sec or less (EPA, 1983a).

Because safety-valve discharges are intermittent in nature, a number of engineering practices have been used to maintain flare flows at reasonable velocities. One of these designs, illustrated in Figure 7.7, is a staged flare system. The small-diameter, elevated flare receives continuous low-flow releases such as seal fluid degassing from compressor shaft seal enclosures or double mechanical seal purges. Operating in tandem with the elevated device is a large ground flare that accepts high, intermittent flows (Figure 7.7). Another approach is to install a small, separate conveyance line to the flare tip in order to maintain high exit velocities for the continuous, low-flow, low-pressure gas flow. Depending on the quantity of low-pressure gas, a compressor is sometimes used to recycle the continuous emissions to the process;

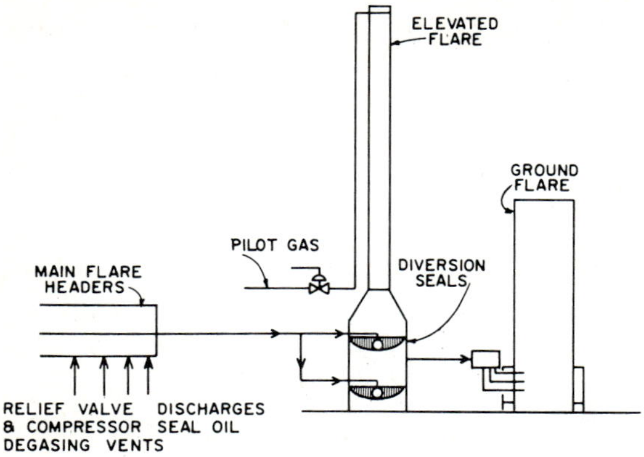

Figure 7.7 Simplified closed-vent system with dual flares (*EPA, 1982b*). The diversion seals, usually water, maintain a slight back pressure (<6 in. H_2O) on the flare header to prevent flashbacks.

whereas intermittent flows above the base load are flared (EPA, 1982b, 1983a). More detailed information on vent and flare system sizing are given in Leung (1986) and Banerjee et al. (1985).

3. Recycle. Recycle streams are a common feature of chemical processing. They are usually used to achieve more complete conversion of reactants or more precise separation of a multicomponent mixture. In Figure 7.1, a recycle stream rich in NO is brought from the bottom tray of the absorber, further contacted with air, and returned to the process. This step effectively utilizes the NO formed in the production of HNO_3 (Eqs. 7.2 and 7.3). Eventually this nitric oxide is mostly converted to product and, as a result, is not combined with other waste gases that will need further treatment before release. Consequently, recycle streams are often desirable, both to maximize process yield and to minimize environmental discharges.

However, recycle streams may be the cause of hazardous material production under several conditions. One of these is when the stream to be processed contains materials that are not affected by the reaction or separation. The nitrogen associated with oxidation air is an example. The inert components will continue to increase in the recycle streams relative to the amounts of other materials, and may eventually interfere with the process. Before this occurs, it is common practice to purge such streams of inerts. But while a component such as nitrogen is present in relatively high concentration and is easily recognized, other materials, in trace amounts but with more toxic properties, may also be present, both in the recycle and in the purge stream released to the environment. A similar situation exists for absorption oils in petroleum and natural gas processing, which gradually change characteristics (e.g., boiling point and molecular weight) because of inadequate stripping of low- or high-boiling hydrocarbons. Another example was a recycle stream in an early coal gasification plant, which was found, over time, to have an increase in high-boiling fractions. Samples of this material exhibited very high toxicity for test animals (Ketcham and Norton, 1960); and workers at the plant exhibited elevated skin cancer prevalence (Palmer, 1979).

Another potential problem with some recycle streams occurs when side reactions producing hazardous materials take place at very slow rates. If the reactions are not recognized and the hazardous products are not appropriately removed in subsequent separation steps, the levels will eventually build up. An example of this involving the formation of tetrachlorodibenzodioxin during the production of chlorophenol is given later in this chapter.

III. Control by Process Modification

As was introduced in Chapter 3, control of hazardous material releases can be accomplished by substitution of a less toxic or nontoxic material; process (re-)design; equipment (re-)design; or changes in equipment or process operation. The following examples, drawn from actual experience, illustrate one or more of these approaches.

A. Lube oil upgrading

Several methods are used by oil refineries to upgrade lube oil stocks by reducing viscosity, color, and carbon residue content. Comparison of two of these, hydrofinishing and furfural extraction followed by bauxite filtration, illustrate the reduction in hazardous material exposure possible through change of process.

Traditional refinery lube oil upgrading requires extraction of solvent dewaxed bright and neutral lube oil stocks with furfural in a packed column, steam stripping for removal of excess furfural, and adsorption on bauxite for removal of impurities and color. The process is a continuous one save for the batch adsorption step. After the bauxite capacity is exhausted, it is washed with naphtha, steam-stripped to remove the naphtha wash, dumped, and regenerated by burning off adsorbed hydrocarbons in an open hearth furnace (Emmel et al., 1983).

In the hydrofinishing process, sulfur, nitrogen, and oxygen in light and heavy waxy distillates are converted to H_2S, NH_3, and CO by reaction with hydrogen over a catalyst (Emmel et al., 1983). In addition, the heavy distillate is cracked to reduce viscosity, light hydrocarbons and other gases are flashed and removed, and the distillates are treated through a methyl ethyl ketone dewaxing unit. All processes are continuous and completely contained.

Measurements made at both installations are given in Table 7.4.

TABLE 7.4 Employee Exposure at Two Lube Oil Upgrading Units

Process	Substance or agent	TLV[a]	Exposure range
Hydrofinishing	H_2S	10 ppm	None detected
	NH_3	25 ppm	None detected
	PAH (phenanthrene)	None	0.25–0.54 μg/m^3
	Noise	85 dBA	52% of TLV
Furfural extraction followed by clay filtering	Furfural	2 ppm	1.2–1.8 ppm
	PAH (phenanthrene)	None	0.26–0.30 μg/m^3
	Noise	85 dBA	16–59% of TLV
	PAH (phenanthrene)	None	0.71 μg/m^3
	Respirable dust	5 mg/m^3	0–0.28 mg/m^3

[a]ACGIH, 1984b.
SOURCE: Emmel et al., 1983.

Polycyclic aromatic hydrocarbon (PAH) concentrations are reported because some of these are known to be carcinogenic. The major PAH component was phenanthrene. The values for nuisance dusts at the adsorption unit were thought to be unrepresentatively low because operation was substantially below capacity during sampling. In addition bauxite regeneration did not take place. Furfural levels are fairly high, given that samples were collected over only 2 days. The hydrofinishing process is intrinsically more amenable to control, since it does not include the bauxite adsorption or regeneration step.

B. Production of 1,2-dichloroethane

The oxychlorination of HCl to 1,2-dichloroethane is an example of the effect that a change in raw material can have on process operation (Markeloff, 1984). Vinyl chloride (an intermediate for plastics) is produced by the pyrolysis of 1,2-dichloroethane (EDC):

$$C_2H_4Cl_2 \longrightarrow C_2H_3Cl + HCl \tag{7.13}$$

Efficient use of the feed materials requires that the HCl produced in Eq. (7.13) be reacted with oxygen and ethylene back to $C_2H_4Cl_2$, which is carried out through the oxychlorination process:

$$C_2H_4 + 2HCl + \tfrac{1}{2}O_2 \longrightarrow C_2H_4Cl_2 + H_2O \tag{7.14}$$

The oxygen can be supplied in the pure state or as air. Both ethylene and oxygen are added in excess of the stoichiometric amounts shown in Eq. (7.14). Other, nonuseful by-products formed at this stage are CO, CO_2, and C_2H_5Cl:

$$C_2H_4 + 2O_2 \longrightarrow 2CO + 2H_2O \tag{7.15}$$

$$C_2H_4 + 3O_2 \longrightarrow 2CO_2 + 2H_2O \tag{7.16}$$

$$C_2H_4 + HCl \longrightarrow C_2H_5Cl \tag{7.17}$$

When O_2 was supplied as air, it was staged to the primary reactors to prevent formation of flammable mixtures (Figure 7.8). It was also necessary to have an additional chlorination step to convert the nonreacted ethylene to EDC, which was then recovered (Figure 7.8):

$$C_2H_4 + Cl_2 \longrightarrow C_2H_4Cl_2 \tag{7.18}$$

This step also caused two important side reactions. One converted some of the desired EDC product to nonuseful 1,1,2-trichloroethane:

$$C_2H_4Cl_2 + Cl_2 \longrightarrow C_2H_3Cl_3 + HCl \tag{7.19}$$

The other reaction, which, since vinyl chloride is a hazardous substance, was desirable from an occupational and environmental standpoint, acted to convert any vinyl chloride in the oxychlorination product stream to 1,1,2-trichloroethane:

$$C_2H_3Cl + Cl_2 \longrightarrow C_2H_3Cl_3 \tag{7.20}$$

The vent gas to the atmosphere typically contained mostly nitrogen, EDC ($\approx$0.21 mole %), vinyl chloride ($\approx$100 ppmv), and <5 ppm ethylene. The oxychlorination vent was a major source of organic chloride emissions for the vinyl chloride plant. Although incineration would provide the lowest emissions, the large volume of vent gas and the presence of over 95% N_2, which would require high fuel gas usage, made this control alternative economically unattractive. For the air-

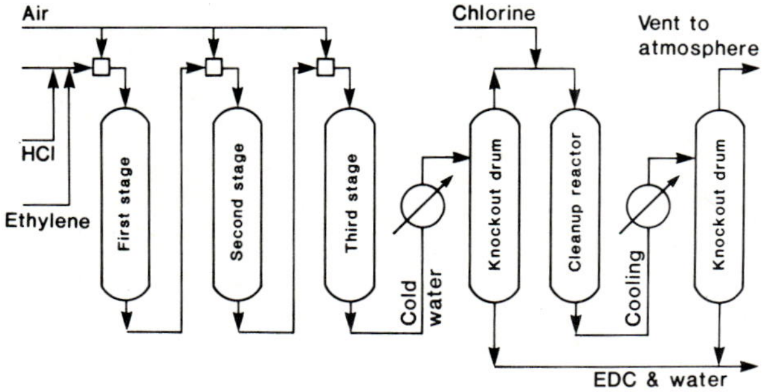

Figure 7.8 Air oxychlorination process (*Markeloff, 1984*).

based process, emission control of vinyl chloride remaining after the conversion described by Eq. (7.20) was essentially by dilution with N_2. Another undesirable aspect was that, upon start-up, additional ethylene and some vinyl chloride were released in the vent gas before the process reached steady state.

Solution: The use of oxygen instead of air in Eq. (7.14) was investigated in order to increase the conversion of ethylene to EDC. As a result of pilot plant work, the process was also revised as shown in Figure 7.9. The major process changes were the removal of the chlorination step (Eq. 7.19) and addition of an extra ethylene recycle stream. As indicated in Table 7.5, the conversion of ethylene to EDC was increased from 95.11% for the air-based process to 97.28% for the oxygen-based plant; and the production of nonuseful by-products was correspondingly reduced. More significantly from a health hazard perspective, the volume of the vent stream, because N_2 was no longer present, was reduced to 5% of the air-based process value. This reduction made installation and operation of a final incinerator a much more feasible control option.

In addition, because of the recycle ethylene stream, emissions during start-up were much reduced. Process vessels are initially filled with N_2. Reactants are added and all reaction products are recycled until the process reaches design conditions. In the air-based process, ethylene flow was established through the reactors and exhausted to the atmosphere. Then HCl flow was begun, followed by air to the first reactor. When reaction began in the first reactor, air was added to the second, and so on. The process had to steady out before chlorination of the atmospheric vent gases could begin, and led to significant ethylene and HCl emissions, and sometimes vinyl chloride as well.

One complication of the replacement of air with oxygen is that nitrogen is no longer available to prevent flammable mixtures. Both

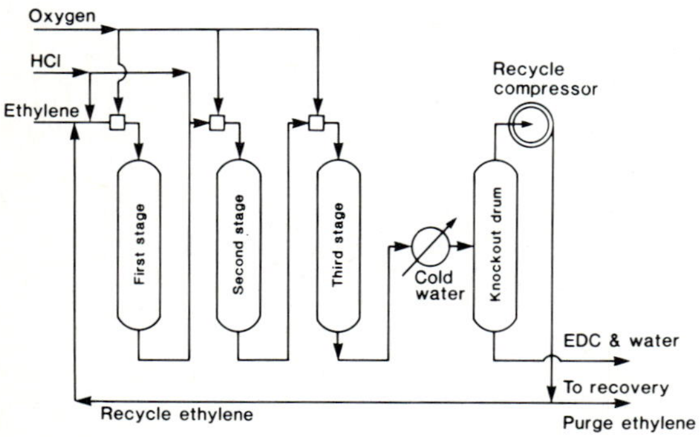

Figure 7.9 Oxygen oxychlorination process (*Markeloff, 1984*).

TABLE 7.5 Oxychlorination Process Yields

Ethylene to (%)	Air based	Oxygen based
EDC	95.11	97.28
Ethylchloride, (C_2H_5Cl)	1.73	1.50
Oxides (CO, CO_2)	1.78	0.68
1,1,2-Trichloroethane	0.88	0.08
Other components	0.50	0.46
Percent HCl converted	99.13	99.83

SOURCE: Markeloff, 1984.

processes are designed to operate in the nonflammable regime. In order to do this, HCl and ethylene are mixed first, and then air or oxygen are added. In the oxygen case, ethylene was mixed in greater excess over stoichiometric than in the air-based process; and special controls, oxygen mixing technology, and operating procedures were required to minimize the amount of flammable gas inventory. However, with these modifications came better control of the process, since the reactors operate at cooler temperatures, and a consequent increase in product.

C. VOC fugitive emissions

An ethylene oxide/glycol manufacturing plant contains the following sources of fugitive volatile organic compounds:

	Number	Average VOC emission rate (from Table 7.2) (kg/day·source)
Pump seals	31 single mechanical 4 packed	0.31
Valves	1050	0.0017
Flanges	3000	0.0007
Compressor seals	2	0.14
Process drains	1	0.96

About 10% of the emissions are ethylene oxide.

a. If the unit is to be enclosed in a 10-m-high shed covering 0.5 acre, what ventilation rate is required to maintain an air concentration ≤TLV for ethylene oxide (EtO)? (A reasonable mixing factor from Chapter 3 is $k_{mix} = 0.3$.)

b. Is it likely that this ventilation rate will be achieved with natural ventilation?

c. If the pump seals are all converted to double mechanical design, what will be the resulting concentration of ethylene oxide with natural ventilation?

Solution

a. The shed volume will be 2.02×10^4 m^3. From Chapter 4, Table 4.2, the TLV for ethylene oxide is 2 mg/m^3. Total emissions are:

$$\begin{aligned} S &= [(35)(0.31) + (1050)(0.0017) + (3000)(0.0007) \\ &\quad + (2)(0.14) + (1)(0.96)] \frac{\text{kg}}{\text{day}} \cdot \left(10^6 \frac{\text{mg}}{\text{kg}}\right)\left(\frac{\text{day}}{24\text{ hr}}\right)\left(0.1 \frac{\text{g EtO}}{\text{g VOC}}\right) \\ &= 6.7 \times 10^4 \text{ mg/hr} \end{aligned}$$

Using Eq. (3.6) at steady state (with C_o, q_1, q_0, and E set to zero) gives:

$$C_{i,ss} = \frac{S}{kq_2} = \frac{S}{kV(q_2/V)} \tag{3.6}$$

$$\frac{q_2}{V} = \frac{S}{C_{i,ss}kV} = \frac{6.7 \times 10^4 \text{ mg/hr}}{(2 \text{ mg/m}^3)(0.3)(2.02 \times 10^4 \text{ m}^3)} = 5.5 \text{ ach}$$

b. Since natural ventilation rates are typically $\simeq$ 1 ach (Chapter 3), it is unlikely control will be achieved by this means.

c. Well-maintained double mechanical seals should not release VOC. The resulting EtO emissions without pump contributions will be

$$S = 2.1 \times 10^4 \text{ mg EtO/hr}$$

and, with 1 ach,

$$C_{i,ss} = \frac{2.1 \times 10^4 \text{ mg/hr}}{(0.3)(2.02 \times 10^4 \text{ m}^3)(1 \text{ ach})} = 3.5 \text{ mg/m}^3$$

which is still in excess of the TLV. Ultimate solution of the problem requires a regular inspection, replacement, and maintenance program.

D. Chlorophenol recycle

A semi-batch process for chlorinating phenol is shown in simplified form in Figure 7.10. Initially virgin phenol (ph) is used at a rate of 9400 lb/day. Chlorine, the limiting reactant, is added such that C = 5000 lb/day. The main reaction, which consumes 96% of the Cl_2, is

$$C_6H_5OH + Cl_2 \rightarrow \underset{\text{chlorophenol (CP)}}{C_6H_4OHCl} + HCl \tag{7.21}$$

and the major side reaction, which uses up about 4% of the Cl_2, is

$$C_6H_5OH + 2Cl_2 \rightarrow \underset{\text{dichlorophenol (DCP)}}{C_6H_3OHCl_2} + 2HCl \qquad (7.22)$$

The chlorine and phenol are fed to the reactor over 24 hr. All of product stream D (Figure 7.10) is collected in a storage tank during this period and later is sent on for further processing. The system is reasonably characterized as a continuous-flow, completely mixed reactor. While mass balances for the major components are described in a form similar to Eq. (3.5), the reactions can be considered to be at steady state. Flow rates and steady-state compositions are given in Table 7.6.

After a time it is decided to recycle some of the product stream in order to increase the chlorophenol (CP) yield. Once again product is collected in a holding tank. Now, however, some of this material is returned to the chlorinator the next day in place of some of the phenol in stream V. The recycle ratio is $R/(R + D) = 0.8$ and is based on supplying 25% of the required phenol from the recycle stream. The total phenol rate to the chlorinator remains the same. After a few runs

TABLE 7.6 Parameters for TCDD Estimation

ρ = 10 lb/gal
υ = 2500 gal
ν = 3×10^{-8} lb TCDD formed/day·lb CP in system
α = 0.8 (recycle ratio)

Material balance (lb/day) (see Figure 7.10)

Nonrecycle						
	V	C	G	D	x	
CP				8,687	0.735	
ph	9,400			2,913	0.246	
DCP				230	0.019	
Cl_2		5,000				
HCl			2,571			
	9,400	5,000	2,571	11,830	1.000	
Recycle						
	V	C	G	D	x	R
CP				8,687	0.915	34,748
ph	7,069			583	0.061	2,331
DCP				229	0.024	917
Cl_2		5,000				
HCl			2,570			
	7,069	5,000	2,570	9,499	1.000	37,996

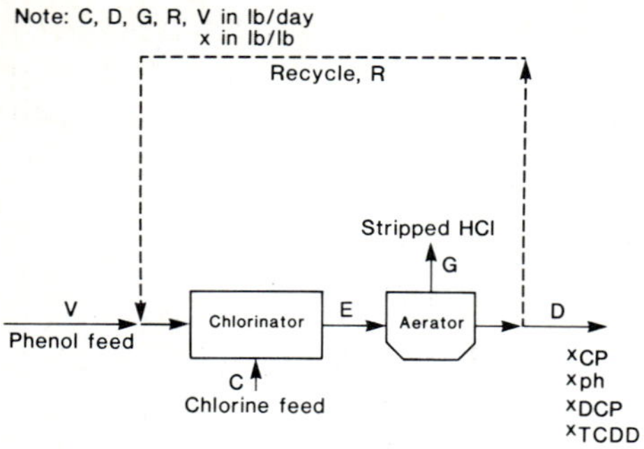

Figure 7.10 Phenol chlorination recycle process.

a new steady-state condition is achieved, which is shown in Table 7.6. The CP composition of the product stream is now higher than for the nonrecycle configuration.

What is not recognized is that a side reaction is also occurring, which converts some of the CP to tetrachlorodibenzodioxin (TCDD). Certain isomers of TCDD are known to be quite toxic (e.g., see Esposito et al., 1980). The kinetics of the TCDD reaction are not thoroughly understood, but the conversion may be approximated by

$$\dot{M}_{\mathrm{TCDD}} = \nu \rho x_{\mathrm{CP}} v \tag{7.23}$$

where $\dot{M}_{\mathrm{TCDD}}$ is the mass of TCDD produced per unit time; ν is the mass of TCDD produced per time/mass of CP; ρ is the density of the reacting material in the chlorinator and aerator; x_{CP} is the mass of CP/mass of stream D; and v is the volume of the chlorinator and aerator.

After a period of time, measurable concentrations of TCDD were found in stream D although none had been detected during nonrecycle operation. (The detection limit for TCDD at the time of the measurements was 20 ppb). Our problem is to suggest a reason for these differences.

Solution: The key to evaluating the TCDD build-up is recognizing that its composition will reach steady state much more slowly than the major components. (We suspect this is the case because no TCDD was found for the nonrecycle case.) A first step is to write a TCDD material balance across the chlorinator and aerator based on one run (i.e., time, t, is ≤ 1 day). No TCDD is ever present in stream V (virgin phenol) or G (HCl vapor) (Figure 7.10). Accordingly, for the chlorinator-aerator combination and the stream designations in Figure 7.10,

$$R x'_{\mathrm{TCDD}} + \nu \rho x_{\mathrm{CP}} v - (D + R) x_{\mathrm{TCDD}} = v \rho \frac{dx_{\mathrm{TCDD}}}{dt} \tag{7.24}$$

TCDD in feed to chlorinator	TCDD produced in chlorinator and aerator	TCDD leaving the system to storage	Net TCDD accumulated in the system

where x_{TCDD} is the weight fraction of TCDD in the product stream D and x'_{TCDD} is the TCDD composition in the product stream recycled from the previous day's run. (At steady state, $x'_{TCDD} = x_{TCDD}$). The general solution of Eq. (7.24) is

$$x_{TCDD} = x_{TCDD_0} \exp\left[-\left(\frac{R+D}{v\rho}\right)t\right] + \left(\frac{R}{R+D}\right)x'_{TCDD} \times \left\{1 - \exp\left[-\left(\frac{R+D}{v\rho}\right)t\right]\right\} + \frac{v x_{CP} v\rho}{R+D} \times \left\{1 - \exp\left[-\left(\frac{R+D}{v\rho}\right)t\right]\right\} \tag{7.25}$$

where $t \leq 1$ day and x_{TCDD_0} is the initial concentration of TCDD in the system at $t = 0$ and will usually be

$$x_{TCDD_0} = \left(\frac{R}{R+V+C}\right)x'_{TCDD} \simeq \left(\frac{R}{R+D}\right)x'_{TCDD} \tag{7.26}$$

For convenience,

$$\delta = 1 - \exp\left[-\left(\frac{R+D}{v\rho}\right)t\right] \tag{7.27}$$

$$\beta = \frac{v x_{CP} v\rho\delta}{R+D} \tag{7.28}$$

$$\alpha = \left(\frac{R}{R+D}\right) \tag{7.29}$$

After substituting Eqs. (7.26)–(7.29) in Eq. (7.25), and writing the solution for each run (day), n, after recycling has begun, the TCDD composition in the product, x_{TCDD_n}, becomes:

$$x_{TCDD_1} = \alpha x'_{TCDD_0} + \beta \qquad \text{(run 1)}$$

$$x_{TCDD_2} = \alpha x'_{TCDD_1} + \beta = \alpha(\alpha x'_{TCDD_0} + \beta) + \beta \qquad \text{(run 2)}$$

$$x_{TCDD_3} = \alpha x'_{TCDD_2} + \beta = \alpha^3 x'_{TCDD_0} + \alpha^2\beta + \alpha\beta + \beta \qquad \text{(run 3)}$$

$$\vdots$$

$$x_{TCDD_n} = \alpha^n x'_{TCDD_0} + \beta \cdot \sum_{i=1}^{i=n} \alpha^{i-1} \qquad \text{(run } n\text{)}$$

or, in generalized form,

$$x_{TCDD_n} = \alpha^n x'_{TCDD_0} + \beta\left(\frac{1-\alpha^n}{1-\alpha}\right) \qquad \text{(for } n \geq 1\text{)} \tag{7.30}$$

For our example, $x'_{\mathrm{TCDD}_0} = 0$ because for the first run the phenol charged to the system comes only from stream V (no recycle is available). Referring to Table 7.6 for values of the various parameters,

$$\alpha = \frac{R}{R + D} = \frac{37{,}996}{9{,}499 + 37{,}996} = 0.80 \tag{7.29}$$

$$\begin{aligned}\delta &= 1 - \exp\left[-\left(\frac{R + D}{v\rho}\right)t\right] \\ &= 1 - \exp\left\{-\left[\frac{9{,}499 + 37{,}996,\ \text{lb/day}}{(2{,}500\ \text{gal})(10\ \text{lb/gal})}\right](1\ \text{day})\right\} \\ &= 0.8504\end{aligned} \tag{7.27}$$

$$\begin{aligned}\beta &= \frac{\nu x_{\mathrm{CP}} v\rho\delta}{R + D} = \frac{(3 \times 10^{-8}\ \text{day}^{-1})(0.915)(2{,}500\ \text{gal})(10\ \text{lb/gal})(0.8504)}{(9{,}499 + 37{,}996\ \text{lb/day})} \\ &= 12.3 \times 10^{-9}\end{aligned} \tag{7.28}$$

Substituting these values in Eq. (7.30) results in

$$x_{\mathrm{TCDD}_n} = \alpha^n x'_{\mathrm{TCDD}_0} + \beta\left(\frac{1 - \alpha^n}{1 - \alpha}\right) \qquad (\text{for } n \geq 1) \tag{7.30}$$

For the nonrecycle case, Eq. (7.25) becomes

$$x_{\mathrm{TCDD,NR}} = \frac{\nu x_{\mathrm{CP}} v\rho}{D}\left[1 - \exp\left(\frac{Dt}{v\rho}\right)\right] \tag{7.25a}$$

Substituting appropriate values from Table 7.6 gives

$$\begin{aligned}x_{\mathrm{TCDD,NR}} &= \left[\frac{(3 \times 10^{-8}\ \text{day}^{-1})(0.735)(2{,}500\ \text{gal})(10\ \text{lb/gal})}{11{,}830\ \text{lb/day}}\right] \\ &\quad \times \left(1 - \exp\left\{-\left[\frac{(11{,}830\ \text{lb/day})(1\ \text{day})}{(2{,}500\ \text{gal})(10\ \text{lb/gal})}\right]\right\}\right) \\ &= 17.4 \times 10^{-9}\ \text{lb TCDD/lb stream } D \text{ or } 17.4\ \text{ppb}\end{aligned} \tag{7.25a}$$

which is less than the detection limit and may be the reason why TCDD was not found during straight-through operation.

For the recycle case, x_{TCDD} will eventually increase to a steady-state value that is greater than that for nonrecycle operation. For instance, if recycle is continued for 20 runs, the resulting TCDD concentration will be

$$\begin{aligned}x_{\mathrm{TCDD}_{20}} &= (12.3 \times 10^{-9})\left[\frac{1 - (0.8)^{20}}{1 - 0.8}\right] \\ &= 6.06 \times 10^{-8}\ \text{lb TCDD/lb stream } D \text{ or } 60.6\ \text{ppb}\end{aligned} \tag{7.30}$$

This level is considerably above the detection limit and, if released, would be an undesirable material in the workplace. It is evident that for this process, subject to equipment limitations, greater recycle ratios (higher values of α) will result in even higher TCDD concentrations. The process has been simplified from the actual installation for purposes of demonstration.* However, the essential elements for analysis have been retained and are applicable for other settings. This type of problem illustrates how certain recycle conditions may lead to hazardous concentrations.

E. Pesticide dusts

A granular formulation process requires that a very toxic pesticide solution be coated onto corn grit granules (14–40 mesh). Static charges build up on the granules and cause the fines to agglomerate. The agglomeration interferes with removal of fines during screening of the final product, resulting in dust hazard to workers during bagging operations.

Solution: The plant now adds powdered charcoal during the mixing-coating operation, which dispels the static change because the charcoal is more conductive than the granules alone. Consequently, the fines do not agglomerate and are removed during screening. The charcoal also makes the final product much darker, which aids in housekeeping by indicating contaminated areas (SRI, 1980).

F. Gravity transfer of PVC

Mass polymerization of polyvinylchloride (PVC) commonly takes place in several steps. Prepolymerization is the first of these. The prepolymizer reactor is an agitated, vertical cylindrical tank where PVC nuclei are produced from vinyl chloride monomer (VCM). When a conversion level of ≃8–10% is reached, the batch is fed to a horizontal cylindrical autoclave where the reaction goes to completion (Burgess, 1982).

Although the slurry from the prepolymizer could be pumped to the autoclave, it is much more desirable to design the reactor building and vessel placement in such a way that gravity flow is possible. This type of design removes the problems of VCM, a potential carcinogen, leaking from the pump seals and packings.

What height and pipe size would be required to effect transfer of

---

* For instance, the daily average value of x'_{TCDD} may be a more appropriate value for the recycle stream rather than the end-of-day value we have shown. However, the form of the solution is similar to our example, although somewhat more complex, and gives a concentration of about 44 ppb after 20 runs.

6600 gal of a PVC slurry in 60 sec to an autoclave located 20 ft away in the horizontal direction? The slurry has a density $\rho = 62.36\ \text{lb/ft}^3$, and a viscosity $\mu = 1\ \text{cp} = 6.72 \times 10^{-4}\ \text{lb/ft·sec}$. Because of the fouling characteristics of the partially reacted monomer, the surface roughness, ε, is $\simeq 0.1$ in. A butterfly valve is also included in the system, which has a head loss equivalent to 20 velocity heads.

Solution: The required height will be in the $\simeq$10–20 ft range, since the available head will be limited by the normal space between floors. Within this range, flow will be limited by the friction of the pipe and valve. An appropriate form of the mechanical energy balance, introduced in Chapter 5, is:

$$x = \left[F + K_H\left(\frac{v^2}{2g}\right)\right] = \left[4f\frac{(L^2 + x^2)^{0.5}}{D}\left(\frac{v^2}{2g}\right)\right] + 20\left[\left(\frac{v^2}{2g}\right)\right] \tag{5.10a}$$

where x is the pipe height (ft); L is the horizontal run (ft); F is the friction head loss (ft of fluid flowing); f is the Fanning friction factor (dimensionless); D is the pipe diameter (ft); g is the acceleration due to gravity (32.2 ft/sec^2); K_H is the number of valve velocity heads; and v is the pipe velocity (ft/sec). A sketch of the system is shown in Figure 7.11. The friction factor is a function of the Reynolds number, $N_{\text{Re}} = v\rho D$, and the roughness factor, ε/D, where ρ is the fluid density (g/cm^3); μ is the fluid viscosity (g/m·sec); and ε is the surface roughness. Rearranging Eq. (5.10a) and substituting $L = 20$ ft gives:

$$x = 4f\left[\frac{(400 + x^2)^{0.5}}{D}\,\frac{v^2}{64.4}\right] + 20\,\frac{v^2}{64.4} \tag{7.31}$$

The calculational procedure is: (1) Assume D; (2) calculate v and N_{Re}; (3) look up f for N_{Re} and ε/D on appropriate plots (e.g., Perry et al., 1984); (4) calculate x from Eq. (7.31). If the height is in the appropriate

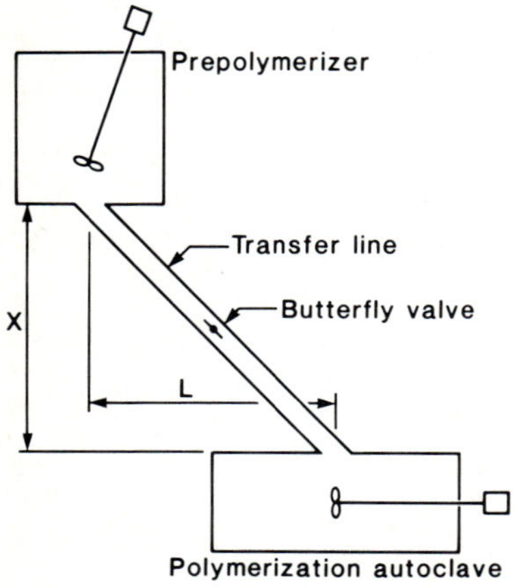

Figure 7.11 Transfer line for polyvinylchloride polymerization.

range, then the chosen D is an acceptable solution. If not, repeat from (1).

Inner diameter D, ft(in.)	v, ft/sec	N_{Re}	ε/D	f	x, ft
1 (12)	18.7	1.7×10^6	8.3×10^{-3}	0.0089	135.0 (Too great)
1.5 (18)	8.3	1.2×10^6	5.6×10^{-3}	0.0079	22.1 All acceptable,
1.75 (21)	6.1	9.9×10^5	4.8×10^{-3}	0.0075	11.8 depending on
2.0 (24)	4.7	8.7×10^5	4.2×10^{-3}	0.0072	7.0 distance between floors

G. Amine vapor collection

A 10% (weight) solution of methylamine (CH_3NH_2) is used in a reaction to produce a pesticide. The solution is added to a weighing container in ≈50 gal lots, 50 times/24-hr day, before being charged to the amination reactors. Since gravity flow is used to the reactors, the weighing container is open to the atmosphere and some amine escapes during each weighing. The vapor space, V, in the filled weighing container and distribution system is ≈13 gal. Each weighing requires 2 min. The density of the solution is 1 g/ml, and the molecular weight of methylamine is $MW_A = 31.06$.

a. Estimate the maximum amount of methylamine released per day.

b. An activated carbon adsorbent can be used to collect the gaseous amine, provided the volume/adsorbent cross-sectional area is ≤231 cfm/ft^2. Under these conditions, the adsorbent can retain amine equivalent to 30% of the weight of the carbon. Will this flow limitation be met for a barrel with $A_m = 314$ in.^{2} of cross-sectional area? If so, how long will 50 lb of adsorbent last?

c. If the cross-sectional area of the weigh container is also 314 in.^{2}, and the mass transfer coefficient is 0.1 cm/min, how reasonable is the assumption of equilibrium?

Methylamine in aqueous solution,[a] weight % (25°C)	Vapor pressure, psia
2.5	0.54
10	1.1
20	1.95
30	3.5
40	6.7
50	11.1

[a] From Gallant and Railey (1984).

Solution

a. There are a number of ways to approach this problem. Perhaps the most straightforward is to assume that vapor-liquid equilibrium is established instantaneously between the amine solution and the vapor space. We can then estimate the maximum CH_3NH_2 that might be released during each charging operation. From Dalton's law of partial pressures, the equilibrium volume fraction of amine in the gas phase, $y_{A,e}$, assuming atmospheric pressure of 14.65 psia, will be

$$y_{A,e} = \frac{p_A}{14.65 + p_A} \tag{7.29}$$

where p_A is the vapor pressure of methylamine. Substituting (for a 10% solution):

$$y_{A,e} = \frac{1.1}{14.65 + 1.1} = \frac{0.070 \text{ ft}^3 \text{ amine}}{\text{ft}^3 \text{ mixture}}$$

For each charge, the total volume of amine, v_A, will be:

$$v_A = y_{A,e}V = \left(0.070 \frac{\text{ft}^3}{\text{ft}^3}\right)(13 \text{ gal})\left(\frac{\text{ft}^3}{7.48 \text{ gal}}\right)$$

$$= 0.12 \frac{\text{ft}^3}{\text{charge}} \quad \text{or} \quad 6.1 \text{ ft}^3 \text{ amine/day}$$

b. The total gas flow rate is;

$$\frac{V}{t} = \frac{(13 \text{ gal})(\text{ft}^3/7.48 \text{ gal})}{2 \text{ min}} = 0.87 \text{ ft}^3/\text{min}$$

Checking the adsorber flow restrictions

$$\frac{V/t}{A_m} = \frac{0.87 \text{ ft}^3/\text{min}}{(314 \text{ in.}^2)(\text{ft}^2/144 \text{ in.}^2)} = 0.4 \text{ cfm/ft}^2$$

which is much less than the guideline of 231 cfm/ft^2. Consequently, 50 lb of adsorbent will last for t_s days:

$$t_s = \frac{(50 \text{ lb adsorbent})(0.3 \text{ lb amine/lb adsorbent})(379 \text{ ft}^3 \text{ at } 60°\text{F/lb mole})}{(50 \text{ charges/day})(0.12 \text{ ft}^3 \text{ amine/charge})(31.06 \text{ lb/lb mole})}$$

$$= 30 \text{ days}$$

c. We can check our assumption about instantaneous equilibrium by using Eq. (7.6) and a reasonable estimate of the mass transfer coefficient, $K = 0.1$ cm/min (e.g., Liss and Slater, 1974) based on liquid phase resistance predominating:

$$\dot{M} = KA_m\Delta C \tag{7.6}$$

ΔC is the difference between the liquid concentration, C_A (10% of 1 g/ml or 0.1 g/cm^3), and the liquid concentration, $C_{A,e}$, in equilibrium with the amine vapor composition, y_A. For a closed system, the total pressure, P, is

$$P = p_A + 14.65 = y_A P + 14.65 \tag{7.32}$$

or

$$p_A = \left(\frac{y_A}{1 - y_A}\right)(14.65) \tag{7.33}$$

A reasonable fit of the given equilibrium data (e.g., by least squares) is

$$C_{A,e}(\text{g/cm}^3) = 0.1593 \ln p_A \text{ (psia)} + 0.1026 \tag{7.34}$$

or

$$C_{A,e} = 0.1593 \ln\left(\frac{y_A}{1 - y_A}\right) + 0.5302 \tag{7.35}$$

Recognizing that for a closed system

$$\dot{M} = \frac{V\,\text{MW}_A}{24{,}400}\frac{dy_A}{dt} \qquad \frac{(\text{cm}^3)(\text{g A/g-mole A})}{(\text{cm}^3/\text{g-mole})}\;\frac{\text{g-moles A}}{(\text{g-moles})(\text{sec})} \tag{7.36}$$

and combining Eqs. (7.6), (7.35), and (7.36) gives

$$\int_0^{Y_A} \frac{dy_A}{C_A - 0.1593 \ln\left(\dfrac{y_A}{1 - y_A}\right) - 0.5302} = \frac{KA_m(2.44 \times 10^4)}{V\,\text{MW}_A}\int_0^T dt \tag{7.37}$$

Equation (7.37) can be integrated numerically. Since $y_A \rightarrow y_{A,e}$ asymptotically, it is appropriate to set the upper limit of integration $Y_A = 0.9(y_{A,e}) = (0.9)(0.070) = 0.063$. Substituting $C_A = 0.1$ g/cm³ (reasonably assumed constant during charging), $K = 0.1$ cm/min, $A_m = 2026$ cm², $V = 49{,}210$ cm³, $\text{MW}_A = 31.06$; and reorganizing Eq. (7.37) gives

$$\int_0^T dt = (0.3092)\int_0^{0.0629} \frac{dy_A}{0.1593 \ln[(1 - y_A)/y_A] - 0.4302} \tag{7.38}$$

The solution becomes:

Y_A *at beginning and end of interval*	Δt, *min (Eq. 7.38)*
0.00	
	0.0072
0.01	
	0.0129
0.02	
	0.0200
0.03	
	0.0314

Y_A at beginning and end of interval	Δt, min (Eq. 7.38)
0.04	
	0.0556
0.05	
	0.1583
0.06	
	0.4820
0.0629	
	T = 0.7674 min

(The integration was carried out using Simpson's rule with each interval in Y_A divided into subintervals small enough so that subinterval size did not affect the answer.) Since 90% of equilibrium is achieved in less than half of the 2-min charging time, our assumption of a close approach to equilibrium is reasonable.

REFERENCES

ACGIH (1984a). *Documentation of the Threshold Limit Values*. American Conference of Governmental Industrial Hygienists, Cincinnati, Ohio.

ACGIH (1984b). *Threshold Limit Values for Chemical Substances and Physical Agents in the Work Environment Adapted by ACGIH for 1984–1985*. American Conference of Governmental Industrial Hygienists, Cincinnati, Ohio.

ASME (1979). *Recommended Guide for the Prediction of the Dispersion of Airborne Effluents,* 3d ed. American Society of Mechanical Engineers.

Banerjee, K., Cheremisinoff, N. P., and Cheremisinoff, P. N. (1985). *Flare Gas Systems Handbook*. Gulf Publishing Co., Houston, Tex.

Burgess, R. H. (1982). *Manufacture and Processing of PVC*. Macmillan, New York.

Burgess, W. A. (1981). *Recognition of Health Hazards in Industry*. Wiley, New York.

DeFaveri, D. M., Hanzevack, E. L., and Delaney, B. T. (1982). Dispersion from safety valves and other momentum emission sources: Intermittent. *Atmos. Environ., 16,* 1239–1242.

Drabkin, M. (1979). *A Review of Standards of Performance for New Stationary Sources—Nitric Acid Plants*. U.S. EPA Report No. EPA-450/13-79-013, March.

Emmel, E., Lee, B. B., and Simonson, A. V. (1983). *Control Technology Assessment of Selected Petroleum Refinery Operations*. NIOSH Contract No. 210-81-7102, June.

EPA (1981). *Control of Volatile Organic Fugitive Emissions from Synthetic Organic Chemical, Polymer, and Resin Manufacturing Equipment*. EPA Report Office of Air Quality Planning and Standards, January.

EPA (1982a). *Fugitive Emission Sources of Organic Compounds—Additional Information on Emissions, Emission Reductions, and Costs*. EPA Report No. EPA 450/3-82-010, April.

EPA (1982b). *VOC Fugitive Emissions in Petroleum Refining Industry—Background Information for Proposed Standards*. EPA Report No. EPA-450/3-81-015a, November.

EPA (1983a). Standards of performance for new stationary sources; VOC fugitive emission sources; petroleum refineries. *Fed. Reg., 48,* 279–299, January 4.

EPA (1983b). *Petroleum Fugitive Emissions—Background Information for Promulgated Standards*. EPA Report No. EPA-450/3-81-015b, October.

EPA (1984). *Control of Volatile Organic Compound Leaks from Synthetic Organic Chemical and Polymer Manufacturing Equipment*. EPA Report No. EPA-450/3-83-006, March.

Esposito, M. P., Tierman, T. O., and Dryden, F. E. (1980). *Dioxins*. EPA Report No. EPA-600/2-80-197, November.

Freeberg, C. R., and Aarni, C. W. (1982). Hydrocarbon emission control in petroleum refineries. *Chem. Eng. Prog.*, 35–39, June.

Gallant, R. W., and Railey, J. M. (1984). *Physical Properties of Hydrocarbons*, vol. 2, 2d ed. Gulf Publishing Co., Houston, Tex.

Hanzevack, E. L. (1982). Dispersion from safety valves and other momentum emission sources: Continuous. *Atmos. Environ., 16,* 1231–1237.

Hughes, T. W., Tierney, D. R., and Khan, Z. S. (1979). Measuring fugitive emissions from petrochemical plants. *Chem. Eng. Prog.*, 35–39, August. 1979.

Ketcham, N. H., and Norton, R. W. (1960). The hazards to health in the hydro-generation of coal. III. The industrial hygiene studies. *Arch. Environ. Health, 1,* 194–207.

Leung, J. C. (1986). Simplified vent sizing equations for emergency relief requirements in reactors and storage vessels. *Amer. Inst. Chem. Engin. J., 32,* 1622–1634.

Liss, P. S., and Slater, P. G. (1974). Flux of gases across the air-sea interface. *Nature, 247,* 181–184.

Markeloff, R. G. (1984). Convert oxychlorination from air to oxygen. *Hydrocarbon Proc.*, 91–94, November.

Morgester, J. J., Frisk, D. L., Zimmerman, G. L., Vincent, R. C., and Jordon, G. H. (1979). Control of emissions from refinery valves and flanges. *Chem. Eng. Prog.*, 40–45, August.

NFPA (1984). *Fire Protection Guide on Hazardous Materials,* 8th ed. National Fire Protection Association, Quincy, Mass.

NIOSH/OSHA (1981). *Occupational Health Guidelines for Chemical Hazards.* NIOSH Publication 81-123.

Palmer, A. (1979). Mortality experience of 50 workers with occupational exposure to the products of coal hydrogenation processes. *J. Occup. Med., 21,* 41–44.

Perry, R. H., Green, D. W., and Maloney, J. O. (1984). *Perry's Chemical Engineering Handbook,* 6th ed. McGraw-Hill, New York, pp. 5–24.

Russell, W. M. (1975). Hazard control of plant process changes. Presented at the 81st national meeting of the AIChE, Kansas City, April 11–14.

SRI (Stanford Research Institute) (1980). *Control Technology of the Pesticides Manufacturing and Formulating Industry.* NIOSH Contract No. 210-77-0093, May.

Van Wagenen, H. (1981). *Control of Emissions from Seals and Fittings in Chemical Process Industries.* NIOSH Publication 81-118.

Wetherold, R., and Provost, L. (1979). *Emission Factors and Frequency of Leak Occurrence for Fittings in Refinery Process Units.* EPA Report No. EPA-600/2-79-044, February.

Chapter

8

Worker-Machine Interactions

P. A. Scheff

I. Human Factors Engineering

The study of human-machine interactions falls under the disciplines of human factors engineering and ergonomics. These are multidisciplinary activities describing people's capacities, limitations, and capabilities for performing work. Proper engineering and design of the workplace consequently requires contributions from many fields, including industrial hygiene, industrial safety engineering, psychology, physiology, anatomy, and systems engineering.

The objectives of the human factors approach include the maintenance and enhancement of health, safety, satisfaction, efficiency, and effectiveness of work. The approach to the problem involves the systematic application of relevant information about human abilities, characteristics, and behavior as well as information describing the task to be accomplished. This chapter will examine the human-machine interface. This interface is influenced by a variety of factors, including workplace design, equipment design, information transfer, and environmental conditions (e.g., heat, light, noise, humidity, vibration). Specific components of the human-machine system include:

1. Anthropometry, the measurement of people
2. Biomechanics, the study of human anatomy and how it relates to work
3. Workplace design, the specification of workplace dimensions
4. Equipment design, the specification of equipment dimensions and operation to ensure safety, including the proper location and specification of displays and controls
5. Information transfer, the principles of interactions between workers and machines
6. Human factors engineering, the systematic analysis and design of work

The specification of controls of the physical environment including light, heat, and noise are covered in Chapters 11, 12, and 13, respectively.

Modern industrial systems are complex. The costs of retrofitting a system to correct a deficiency in design can be prohibitively expensive. It is therefore essential that the initial design of a workplace be satisfactory in terms of human-machine interactions. The improper selection of a warning signal, overloading of an operator with extraneous information, improper selection of a critical control, or incorrect specification of the level of illumination can all lead to process upsets with subsequent risk of injury as well as inefficient operation.

Industrial systems traditionally have been designed more to accommodate the machines and materials than the workers involved. The relationship between worker safety and productivity has often not been considered in workplace design. The goal of the human factors approach is to shift the emphasis away from the machine to the human-machine system. The roles of the worker, machine, environment, and safety must all be considered in system design. The human-machine interface plays a key role with respect to:

1. Workplace design, including specification of traffic flow, work position, layout of materials and tools, communications, illumination, environment (e.g., temperature, ventilation, noise), and hazards (e.g., materials handling, electrical shock, toxic substances)
2. Design of machines, tools, and equipment, including visual displays, controls and handles, machine guarding, ease of maintenance, and supplementary illumination

3. Design of supporting devices, including vehicles, scaffolds, ladders, stairs, lifts, cranes, and process monitoring consoles.

A human-machine system can be defined as a combination of one or more people and one or more physical components interacting to bring about a desired output. With this general definition, a machine is virtually any type of physical object, device, equipment, or facility that people use in carrying out an activity. A simple example is a worker with a hand tool. A complex example is an automated oil refinery with operating control engineers. Functions in this system include the sensing or receiving of information, information storage, information processing and decision making, and the operation that occurs from the decision-making process. Human beings sense with vision, touch, and audition, store information with memory or aids such as data tables, and decide on a course of action such as the activation of a control or modification of a process. Machines sense by electrical, photographic, or mechanical devices, can electronically store and retrieve data quickly, and make preprogrammed responses to indicate a course of action.

As part of the process of system development, the design engineer must decide what tasks are appropriate for human operators; where, when, and how the operators will interact with components of the system; and what human constraints impact on system design and performance. Human characteristics to consider include age, size, experience, mobility, dexterity, reaction time, sensory response, cognitive response, motor response, handicaps, physiological tolerance, fatigue, strength, metabolic requirements, and adaptive limits. System features to consider include operating procedures, instructions, training, information display, body control and support, reach and clearance, demands on speed, accuracy, duration, and strength, task complexity, and inconvenience.

The allocation of tasks between humans and machines is then a critical step in the design. Table 8.1 lists human capabilities with machine alternatives. Functional advantages of people over machines include ability to sense unusual, unexpected, or low-probability events; to detect signals masked or partially obscured by noise; to recognize and interpret complex patterns; to recognize objects from different distances and angles (perceptual consistency); to adapt and make new, unprogrammed decisions; and to be able to adapt to overload conditions, selectively focusing attention to critical events. Table 8.2 lists human limitations with machine alternatives. In general, machines are excellent for routine, repetitive tasks; are highly reliable; respond at great speed; apply massive force smoothly and precisely; compute, differentiate, and integrate; operate complex simultaneous functions; sense stimuli outside the normal range of human sensitivity (e.g., x-rays,

TABLE 8.1 Comparison of Human Capabilities with Machine Alternatives

Human	Machine
Can recognize and use information redundance (pattern) in the real world to simplify complex situations.	Has limited perceptual constancy and is very expensive.
Has high tolerance for ambiguity, uncertainty, and vagueness.	Is highly limited by ambiguity and uncertainty in input
Can interpret an input signal even when subject to distraction, high noise, or message gap.	Performs well only in a generally clean, noise-free environment.
Is a selecting mechanism and can adjust to sense specific inputs.	Is a fixed sensing mechanism, operating only on that which has been programmed for it.
Has very low absolute thresholds for sensing (e.g., vision, audition, and touch).	To have the same capability, becomes extremely expensive.
Has excellent long-term memory for related events.	To have the same capability, becomes extremely expensive.
Can become highly flexible in terms of task performance.	Is relatively inflexible.
Can improvise and exercise judgment on the basis of long-term memory and recall.	Cannot do this; is best at routine, repetitive functions.
Can perform under transient overload; performance degrades gradually.	Stops under overload; generally fails all at once.
Can make inductive decisions in novel situations; can generalize.	Has little or no capability for induction or generalization.
Can modify performance as a function of experience; can "learn to learn."	Is not characterized by trial-and-error behavior.
Can override own actions, should the need arise.	Can do only what it is built to do.
Is reasonably reliable; can add reliability to system performance by selection of alternatives.	Is reliable only at the expense of increased complexity and cost, and then only for routine functions.
Complements the machine, i.e., can use it in spite of design failures, can use it for a different task, or can use it more efficiently than it was designed to be used.	Has no such capability.
Complements the machine by aiding in sensing, extrapolating, decision making, goal setting, monitoring, and evaluating.	Has no capacity for performance different from what was originally designed.
Can acquire and report information incidental to the primary mission.	Cannot do this.
Can perform time-contingency analyses and predict events in unusual situations.	Does very poorly at this.

TABLE 8.1 Comparison of Human Capabilities with Machine Alternatives (*Continued*)

Human	Machine
Is relatively inexpensive for corresponding complexity and is generally in good supply, but must be trained.	Is more limited in terms of complexity and supply by cost and time.
Is light in weight and small in size for function achieved for most situations.	To have functional equivalence of the human, requires more weight, power, and cooling facilities.
Is relatively easy to maintain; demands a minimum of "in-task" extras.	Maintenance problems become disproportionately serious as complexity increases.

SOURCE: Woodson, 1981.

TABLE 8.2 Comparison of Human Limitations with Machine Alternatives

Human	Machine
Is a poor monitor of infrequent events or of events that occur frequently over a long period of time.	Can be constructed to reliably detect infrequent events or events that occur frequently over a long period of time.
Has a limited channel capacity.	May have as much channel capacity as can be afforded.
Is subject to coriolis effects, motion sickness, disorientation, etc.	Is not subject to these effects.
Has extremely limited short-term memory for factual material.	May have as much short-term (buffer) memory as can be afforded.
Is not well suited to data coding, amplification, or transformation tasks.	Is well suited to these tasks.
Performance is degraded by fatigue and boredom.	Performance is degraded only by wearing out or by lack of calibration.
Performance is degraded by long duty periods, repetitive tasks, and cramped or unchanged positions.	Is less affected by long duty periods and performs repetitive tasks well; some may be restricted by position.
Becomes saturated quickly in terms of the number of things that can be done and the duration of the effort.	Can do one thing at a time so fast that it seems to do many things at once, for a long period of time.
May introduce errors by misidentification, reintegration, or closure.	Utilizes these processes.
Expectation or cognitive set may lead an operator to see what he or she expects or wants to see.	Does not exercise these processes.
Much human mobility is predicated and based on gravity relationships.	May be built to perform independently of gravity.
Is adversely affected by high *g* forces.	Is unaffected by *g* forces.
Can generate only relatively small forces and cannot exert large forces for very long or very smoothly.	Can generate and exert forces as needed.

TABLE 8.2 Comparison of Human Limitations with Machine Alternatives (*Continued*)

Human	Machine
Generally requires a review or rehearsal period before making decisions based on items in memory.	Goes directly to stored information for a decision.
When performing a tracking task, requires frequent reprogramming; does best when changes are under 3 radians.	Has no such limitations.
Has a built-in response latency of about 200 μsec in a go/no-go situation.	Has no response latency.
Is not well adapted to high-speed, accurate search of large volumes of information.	Computers are designed to do just this.
Does not always follow an optimum strategy.	Will always follow the strategy designed into it.
Has physiological, psychological, and ecological needs.	Has only ecological needs.
Is subject to anxiety, which may affect performance efficiency.	Is not subject to this factor.
Is dependent on the social environment, both present and remembered.	Has no social environment.
Diurnal cycle imposes cyclic degradation of behavior.	The machine cycle may be whatever is desired.
Is subject to stress as a result of interpersonal problems.	Is not affected by such problems.
Great differences exist among unselected individuals.	There are no unselected machines.

SOURCE: Woodson, 1981.

microwaves, etc.); have excellent short-term memory and short access times. Machine limitations include poor signal detection; are subject to interference from noise; are not well suited to handle unexpected events; and can be very complex, with large maintenance requirements. Table 8.3 compares some human and machine functions in terms of memory and sensing, identifying, filtering and interpreting displays.

Once the basic descriptions of operator and machine functions are complete, the design of an industrial system should be based on the simple fact that people will operate and maintain the system. The following sections, describing human characteristics, and workplace and equipment design specifications, present information to be used to properly fit the job to the operator.

TABLE 8.3 Some Comparisons of the Limitations of Machine and Human Functions, in Relative Terms

Conditions for function	Limitations	
	Human	Machine
Sensing		
Display	Limited to certain ranges of energy change affecting human senses Sensitivity: very good	Range extends far beyond human senses (x-rays, infrared, etc.) Sensitivity: excellent
Filtering	Easy to reprogram	Difficult to reprogram
Identifying		
Display	Can be varied over relatively wide range of physical dimensions Channel capacity: small	Can be varied only in very narrow range of physical dimensions Channel capacity: large
Filtering	Easy to reprogram	Difficult to reprogram
Memory	Limits to complexity of models probably fairly high, but not precisely known Limits to length of sequential routines fairly high, but time-consuming to train	Potential limits of complexity very high Potential limits of length of routines very high
Interpreting		
Display	Same as Identifying	Same as Identifying
Filtering	Easy to reprogram Highly flexible, i.e., adaptable May be reprogrammed by self-instruction following input changes contingent on previous response (dynamic decision making)	Difficult to reprogram Relatively inflexible
Shunting	Can be readily reprogrammed to lower levels of functioning	Difficult to reprogram
Memory	Limitation to rule storage not known Speed of reinstatement of rule sequences relatively low (as in computing) Use of novel rules possible (inventing)	Limits of rule storage ("logic") quite high Speed of using rule sequences high (e.g., computing) Limited in use of novel rules

SOURCE: Woodson, 1981.

II. Human Characteristics

A. Statistical distributions

The structure and mechanical function of the human body are obvious considerations in human-machine design and operation. Failure to provide a few critical inches in a component's layout can jeopardize performance and safety. With forethought, these critical inches can be provided without compromising design. At the same time, however, a workplace cannot be designed to fit only a specific size operator. People vary in height, weight, body type, sensory ability, accuracy, ability to adapt, and capacity to withstand fatigue and stress. As a design benchmark, therefore, the workplace should be designed to accommodate at least 90% and preferably 98% of the population of operators (Damon et al., 1966; Van Cott and Kinkade, 1972). Designing for the average worker is never a satisfactory solution.

Frequency plots of the variations in human structure and performance are approximately normal; that is, the variation follows the gaussian or bell-shaped curve. The normal distribution is defined by the population (or sample) mean (a measure of central tendency) and standard deviation (a measure of variability). The standard deviation, or root-mean-square deviation from the mean, is defined as:

$$\sigma = \left[\frac{\sum_{i=1}^{n} (X_i - \overline{X})^2}{n - 1} \right]^{1/2} \tag{8.1}$$

where X_i is the value of the ith individual, n is the total number of people considered, and $\overline{X}$ is the mean, where

$$\overline{X} = \sum_{i=1}^{n} \frac{X_i}{n} \tag{8.2}$$

Figure 8.1 shows the normal curve along with percentages of the distribution covered by multiples of the standard deviation. Ninety percent of a normal population (5 to 95 percentiles) represents the mean $\pm 1.645\sigma$ (standard deviations), 98% (1 to 99 percentiles) represents the mean $\pm 2.326\sigma$, and 99% (0.5 to 99.5 percentiles) represents the mean $\pm 2.576\sigma$. It should be apparent that it may be necessary to make the dimensions of the workplace adjustable to accommodate the variability among people. The use of normal statistics can therefore be useful in specifying the range of adjustment necessary.

Machines should be designed to suit the operator as he/she will perform the job under actual operating conditions. These can include extreme environments with respect to heat, cold, pressure, and noise, all of which may involve the use of special protective devices. It is therefore necessary to consider the size of the operator with clothes and personal protective equipment when dimensioning the workplace.

B. Anthropometric data

Table 8.4 lists anthropometric data for males and females; and Figure 8.2 illustrates some of the characteristic scales. These data are based on static measurements of minimally clothed subjects standing or sitting erect and are useful for specifying minimum, maximum, and adjustable dimensions of workplaces and equipment. Columns 1 and 2 of the table are derived primarily from military studies (Kodak, 1983). These data, therefore, will tend to exclude the smallest (shortest) and largest (tallest) people. The third and forth columns are derived from a national survey of the civilian population (Damon et al., 1966) that included 3091 men and 3581 women, ages 18–79. The data show directional dimensions; i.e., the measurements reflect hand and body postures typical of work situations. Note that the 5th to 95th percentile range can be calculated as the mean $\pm 1.64\sigma$, the 2.5 to 97.5 percentile range as the mean $\pm 1.96\sigma$, the 1 to 99 percentile range as the mean $\pm 2.33\sigma$, and the 0.5 to 99.5 percentile range as the mean $\pm 2.58\sigma$. These ranges correspond to 90, 95, 98, and 99% of the population, respectively.

In general, measurements are made on nude or lightly clothed subjects standing or sitting erect. When applying these data to industrial workers, it is necessary to adjust for clothing and posture. Increments

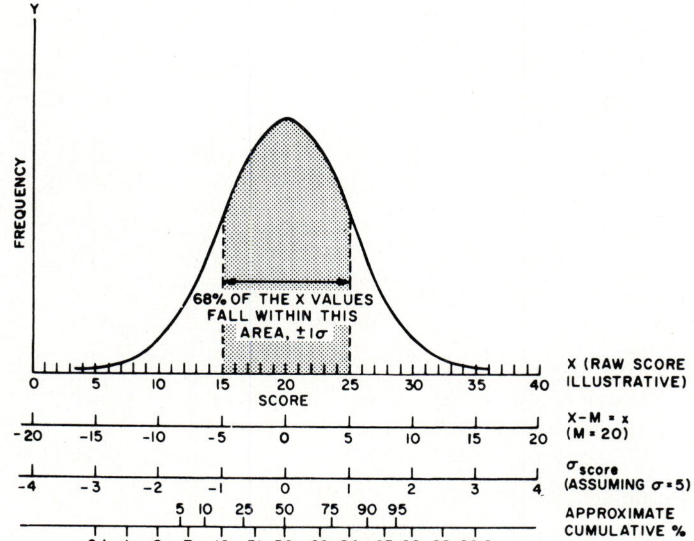

Figure 8.1 Normal curve showing percentages of the distribution covered by multiples of the standard deviation (*Van Cott and Warrick, 1972*).

TABLE 8.4 Anthropometric Data

Measurement[c]	Military population[a] (cm) Males median (σ)	Military population[a] (cm) Females median (σ)	Civilian population[b] (cm) Males median (σ)	Civilian population[b] (cm) Females median (σ)
Standing				
1. Forward functional reach				
a. Includes shoulder depth	82.6 (4.8)	74.1 (3.9)	79.3 (5.6)	71.3 (4.4)
b. Acromial process to functional pinch	63.8 (4.3)	62.5 (3.4)		
c. Abdominal extension to functional pinch[d]			62.1 (8.9)	60.4 (6.7)
2. Abdominal extension depth	23.1 (2.0)	20.9 (2.1)		
3. Waist height	106.3 (5.4)	101.7 (5.0)	104 (6.3)	98.5 (5.5)
4. Tibial height	45.6 (2.8)	42.0 (2.4)		
5. Knuckle height	75.5 (4.1)	71.0 (4.0)		
6. Elbow height	110.5 (4.5)	102.6 (4.8)	114.6 (6.3)	107.1 (6.8)
7. Shoulder height	143.7 (6.2)	132.9 (5.5)	146.4 (7.8)	135.3 (6.6)
8. Eye height	164.4 (6.1)	151.4 (5.6)		
9. Stature	174.5 (6.6)	162.1 (6.0)	173.5 (7.36)	159.8 (6.3)
10. Functional overhead reach	209.6 (8.5)	199.2 (8.6)		
Seated				
11. Thigh clearance height	14.7 (1.4)	12.8 (1.2)	14.5 (2.0)	13.7 (2.2)
12. Elbow rest height	24.1 (3.2)	23.1 (3.0)	24.1 (3.2)	23.4 (3.0)
13. Midshoulder height	62.4 (3.2)	58.0 (2.7)	63.1 (3.2)	58.0 (2.7)
14. Eye height	78.7 (3.6)	73.7 (3.1)	79.2 (3.02)	74.1 (3.02)
15. Sitting height, normal slump	86.6 (3.8)	81.8 (4.0)	86.6 (3.9)	82.0 (4.0)
16. Functional overhead reach	128.4 (8.5)	119.8 (6.6)		
17. Knee height	54.0 (2.7)	51.0 (2.6)	54.4 (3.2)	49.8 (2.8)
18. Popliteal height	44.6 (2.5)	41.0 (1.9)	43.9 (2.9)	39.9 (2.7)
19. Leg length	105.1 (4.8)	100.7 (4.3)		
20. Upper leg length	59.4 (2.8)	57.4 (2.6)	59.2 (3.0)	56.9 (3.3)
21. Buttocks to popliteal length	49.8 (2.5)	48.0 (3.2)	49.5 (3.3)	48.0 (3.1)
22. Elbow to fist length	38.5 (2.1)	34.8 (2.3)		
23. Upper-arm length	36.9 (1.9)	34.1 (2.5)	37.8 (1.9)	35.2 (2.5)
24. Shoulder breadth	45.4 (1.9)	39.0 (2.1)		
25. Hip breadth	35.6 (2.3)	38.0 (2.6)	35.6 (2.9)	36.3 (3.7)
Elbow to elbow breadth			41.9 (4.8)	38.4 (5.4)

Measurement[c]	Military population[a] (cm)		Civilian population[b] (cm)	
	Males median (σ)	Females median (σ)	Males median (σ)	Females median (σ)
		Foot		
26. Foot length	26.8 (1.3)	24.1 (1.1)	26.8 (1.3)	24.1 (1.1)
27. Foot breadth	10.0 (0.6)	8.9 (0.5)	10.0 (0.6)	8.9 (0.5)
		Hand		
28. Hand thickness	3.3 (0.2)	2.8 (0.2)		
29. Hand length	19.0 (1.0)	18.4 (1.0)	19.2 (1.0)	17.4 (1.0)
30. Digit two length	7.5 (0.7)	6.9 (0.8)		
31. Hand breadth	8.7 (0.5)	7.7 (0.5)	9.1 (0.5)	8.1 (0.5)
32. Digit one length	12.7 (1.1)	11.0 (1.0)		
33. Breadth of digit one, interphalangeal joint	2.3 (0.1)	1.9 (0.1)		
34. Breadth of digit three, interphalangeal joint	1.8 (0.1)	1.5 (0.1)		
35. Grip breadth	4.9 (0.6)	4.3 (0.3)		
36. Hand spread, digit one to two, first phalangeal joint	12.4 (2.4)	9.9 (1.7)		
37. Hand spread, digit one to two, second phalangeal joint	10.5 (1.7)	8.1 (1.7)		
		Head		
38. Head breadth	15.3 (0.6)	14.5 (0.6)	15.8 (0.6)	15.0 (0.6)
39. Interpupillary breadth	6.1 (0.4)	5.8 (0.4)		
40. Biocular breadth	9.2 (0.5)	9.0 (0.5)		
		Other		
41. Flexion–extension—range of motion, wrist (degrees)	134 (19)	141 (15)		
42. Ulnar–radial—range of motion, wrist (degrees)	60 (13)	67 (14)		
43. Weight (kg)	83.2 (15.1)	6.4 (13.9)	74.5 (12.6)	61.4 (13.2)

[a] Kodak, 1983.
[b] Damon et al., 1966; Kodak, 1983.
[c] Numbers and letters refer to Figure 8.2.
[d] Add the following number of cm (σ) for bending forward. Male: waist 25 (7), hips 42 (8); female: waist 20 (5), hips 36 (9).

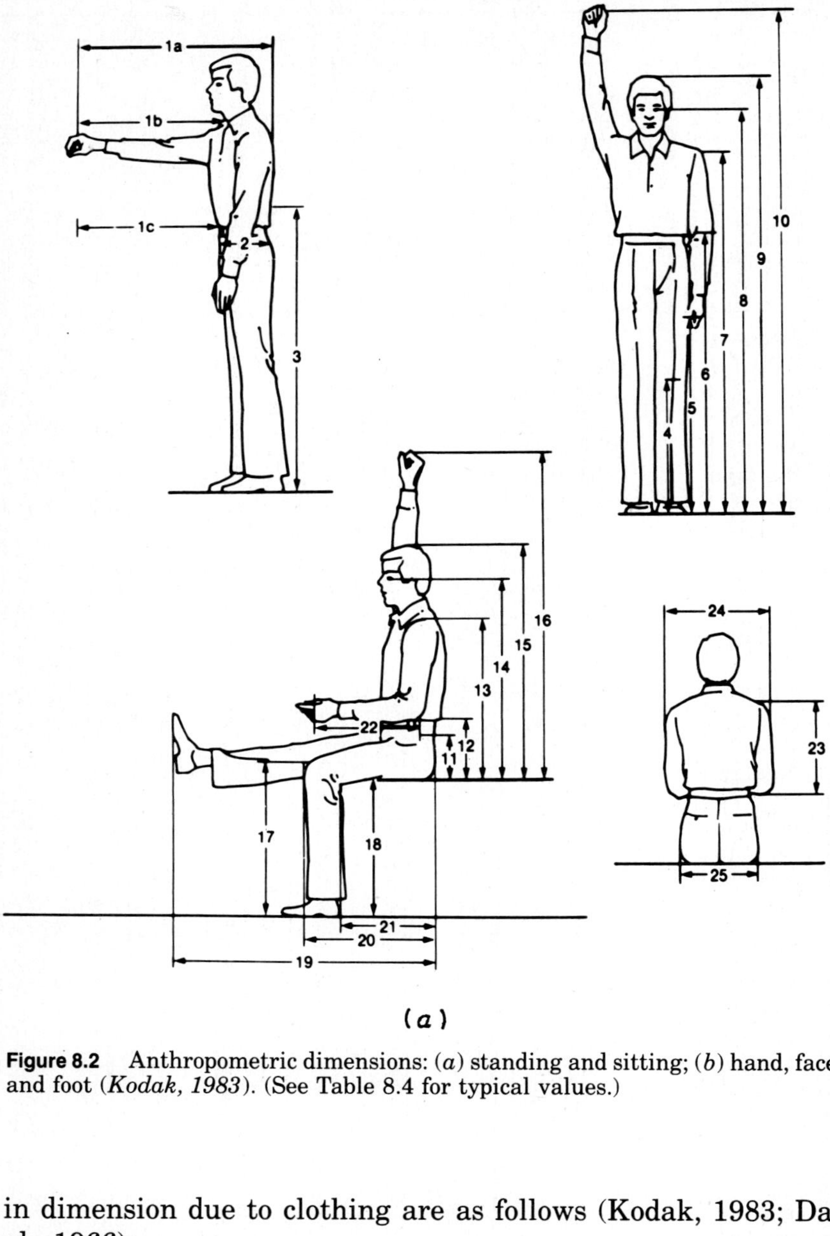

Figure 8.2 Anthropometric dimensions: (*a*) standing and sitting; (*b*) hand, face, and foot (*Kodak, 1983*). (See Table 8.4 for typical values.)

in dimension due to clothing are as follows (Kodak, 1983; Damon et al., 1966):

Weight	(men) 1.8–2.7 kg	(women) 1.4–1.8 kg
Standing heights	(men) 2.5 cm	(women) 1.0–7.6 cm
Sitting heights	0.5 cm	
Knee height	(men) 2.5 cm	(women) 1.0–7.6 cm
Elbow breadth	2.5 cm	
Hip breadth	2.0 cm	
Foot breadth	0.5–0.75 cm	
Foot length	3.0–4.0 cm	

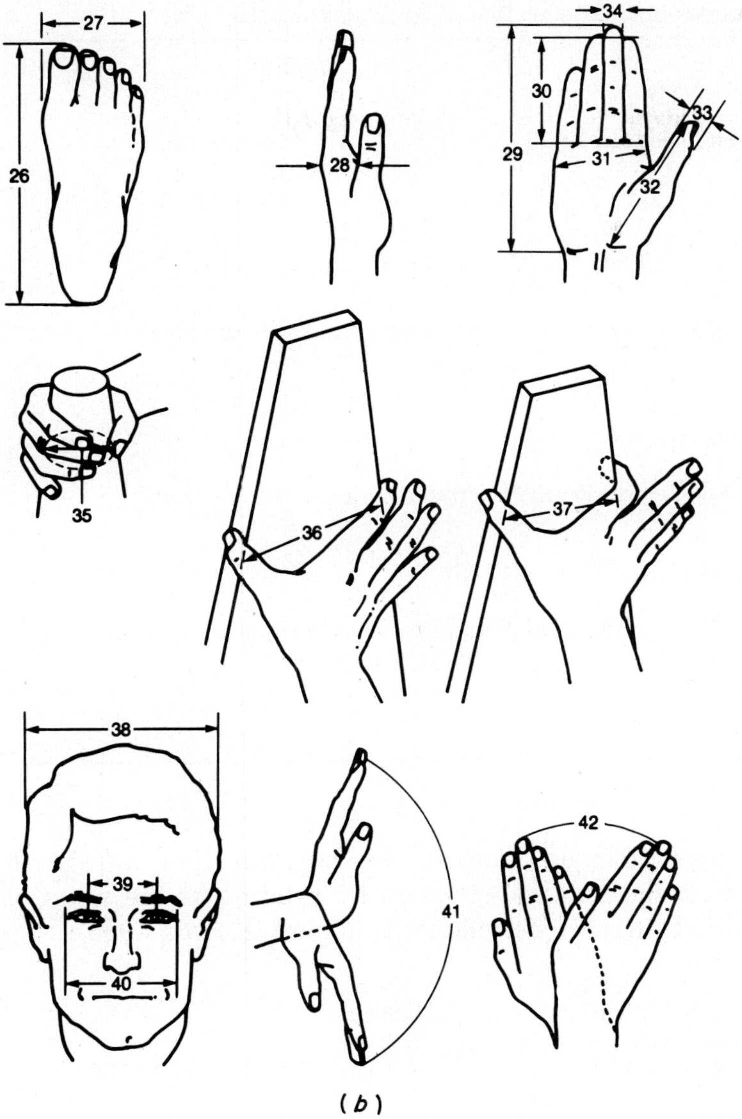

Figure 8.2 *(Continued)*

Postural slump should be subtracted from the data in Table 8.4. Slump decrements have the following average values:

Standing height	2.0 cm
Sitting height	4.5 cm

Additional allowance may be needed for specialized personal protection devices.

The range of acceptable heights for displays and shelves can be broad-

TABLE 8.5 Anthropometric Data for Typical Working Positions

Measurement	Median (cm)	σ (cm)
Kneeling height (on hands and knee)	81.3	4.0
Kneeling length	109.2	8.3
Crawling height	72.1	3.3
Crawling length	135.1	6.6
Prone height	36.8	3.3
Prone length	228.8	8.7
Standing, Forward Reach, Both Arms (with hands grasping)		
Depth of reach	56.5	3.8
Breadth of aperture	45.1	3.0
Floor to bottom of aperture	142.2	5.2
Vertical dimension of aperture	42.5	—
Standing, Lateral Reach Preferred Arm (with hand grasping)		
Depth of reach	62.9	3.8
Breadth of aperture	25.4	—
Floor to bottom of aperture	142.2	5.2
Seated, Forward Reach, Both Arms (with hands grasping)		
Depth of reach	45.1	5.5
Breadth of aperture	40.6	3.2
Floor to bottom of aperture	90.2	4.0
Vertical dimension of aperture	31.1	—

SOURCE: Woodson, 1981.

ened by going up on toes or squatting for short-duration, infrequent tasks. Forward reaches can be extended by bending at the waist or hips. These adjustments are estimated as follows (Kodak, 1983):

Additional height on tiptoe	7.5 cm
Reduced height by squatting	15.0 cm
Extended reach, bending at waist	20.0 cm
Extended reach, bending from hips	36.0 cm

For industrial equipment design, data for civilians from the National Health Examination Survey are most appropriate (Damon et al., 1966). Data from military studies, however, are more complete. Table 8.5 lists additional anthropometric data for typical working positions (Woodson, 1981). These data are derived from a survey of 40 industrial workers.

C. Human capacities and limitations

1. Biomechanics. Biomechanical work tolerance is a state in which a worker can effectively perform a task at levels of physiological and emotional well-being acceptable to the individual (ANSI, 1983). Work tolerance is affected by environmental inputs, workplace layout, and state of health. Work stress can be defined as any action on the body by any environmental or physical process intrinsic to the industrial environment. The physiological response to work stress is work strain. For example, a hot industrial environment produces heat stress, resulting in increased sweating or heat strain. Force on the musculoskeletal system produces work stress. This results in increased metabolic activity or work strain. The human-machine interface should be designed to minimize work stress, thus increasing work tolerance.

Table 8.6 shows the prerequisites for biomechanical work tolerance (Tichauer, 1978). All of these prerequisites directly affect the human interface in equipment design. For example, providing a place to rest elbows by the proper selection of workbench height can reduce fatigue and increase efficiency by as much as 50%. Faulty workplace design can also impose great stress on joints. For example, the rotation of the forearm such as is required for the operation of a screwdriver is far more efficient with the elbow bent from 85° to 100°. An outstretched arm (angle around 130°) greatly increases the sore elbow rate. In addition, the work station should be designed to avoid the need for excessive head movements during visual scanning.

Engineering also plays a central role. Ischemia is the obstruction of blood flow to tissues. Tool design must be compatible with the location of blood vessels vulnerable to compression. For example, poorly designed hand tools can squeeze the ulnary artery, leading to numbness and tingling of the fingers. It is generally better to bend the tool than bend the wrist. This is demonstrated in Figure 8.3, which compares two groups of trainees in electronics assembly. With bent pliers and wrists straight, workers become stabilized in the second week of training. With wrists in ulnar deviation, a gradual increase in disease is observed. Note that only 15 of 40 workers remained unimpaired after 12 weeks (Tichauer, 1978). Chairs should be designed to match the

TABLE 8.6 Prerequisites for Biomechanical Work Tolerance

Postural	Engineering	Kinesiological
1. Keep elbows down	1. Avoid compression ischemia	1. Keep forward reach short
2. Minimize moments on spine	2. Avoid critical vibrations	2. Avoid muscular insufficiency
3. Consider sex differences	3. Individualize chair design	3. Avoid straight-line motions
4. Optimize skeletal configuration	4. Avoid stress concentration	4. Consider working gloves
5. Avoid head movement	5. Keep wrist straight	5. Avoid antagonist fatigue

SOURCE: Tichauer, 1978.

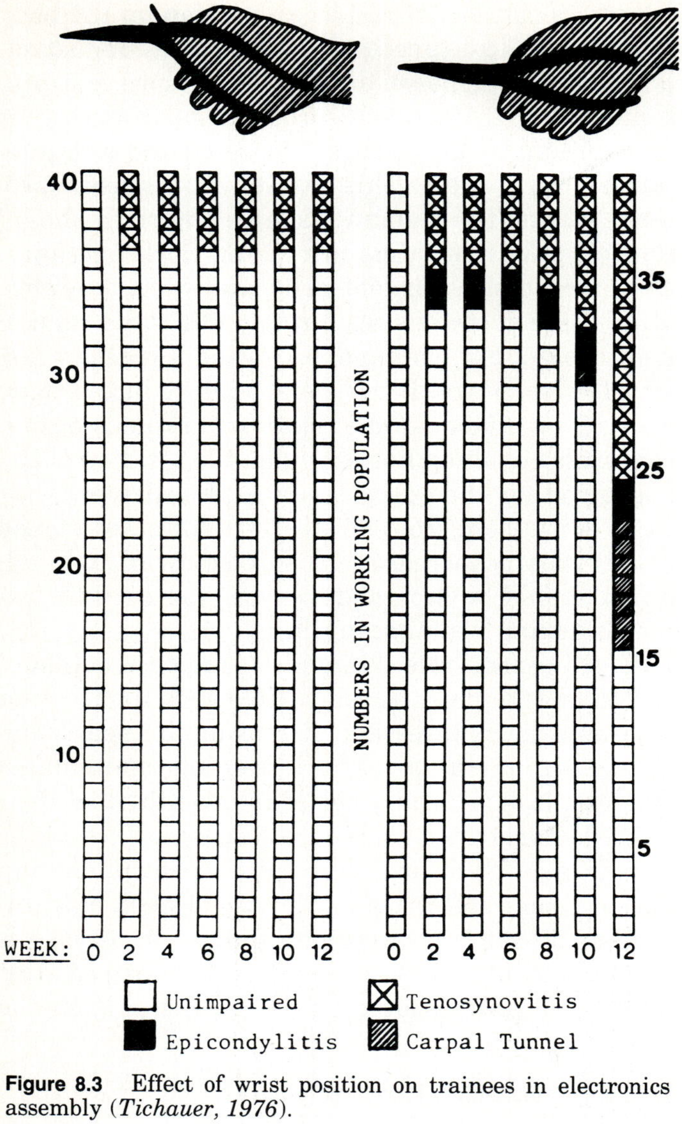

Figure 8.3 Effect of wrist position on trainees in electronics assembly (*Tichauer, 1976*).

individual work situation. A properly designed work chair should: (1) have an adjustable backrest that swivels about a horizontal axis; (2) be small so as not to interfere with the free movement of elbows; (3) have a seat adjustable in height; and (4) be complemented by an adjustable footrest to relieve pressure by edge of seat bottom on thigh.

Kinesiology is the study of human movements as a function of the musculoskeletal system. As indicated on Table 8.6, frequent and rapid forward reaches (beyond 16 in.) should be avoided, as they put stress on lastissimus dorsi muscle and vertebral column. Straight-line movements are also undesirable, since they involve the coordination of several muscles and therefore require a higher skill level. Finally, physiological work stress should not exceed the capacity of the smallest muscle involved.

Data on the range of joint movement help in the specification of control placement as well as the operator's movements in performing tasks. Extensive summaries of these data have been published (e.g., Woodson, 1981; Damon et al., 1966; Hertzberg, 1972; Huchingson, 1981). For example, Table 8.7 shows the range of movement at the joints of the arm and hand. These movements are illustrated on Figure 8.4. The data are based on measurements of 39 male workers performing forced movements and therefore represent maximum values. Normal movements would have less excursion. The data in Table 8.8 show the useful range of wrist flexion and extension while grasping a control.

Human strength is defined as the maximum static force that muscles can exert isometrically in a single voluntary effort. Table 8.9 shows

TABLE 8.7 Range of Movement at the Joints of the Arm and Hand[a]

Joint	Type of movement[b]	Range of movement (degrees) Mean	σ
Wrist	Flexion	90	12
	Extension	99	13
	Total	189	21
	Adduction	27	9
	Abduction	47	7
	Total	74	13
Forearm	Supination[c]	113	22
	Pronation[c]	27	24
	Total[c]	190	30
Elbow	Flexion	142	10
Shoulder	Flexion	188	12
	Extension	61	14
	Total	249	19
	Adduction	48	9
	Abduction	134	17
	Total	182	20
	Rotation: medial	97	22
	Rotation: lateral	34	13
	Total	131	24

[a] Thirty-nine male subjects representing varied types of body build.
[b] Refer to Figure 8.4.
[c] Elbow at 90° angle.
SOURCE: Damon et al., 1966.

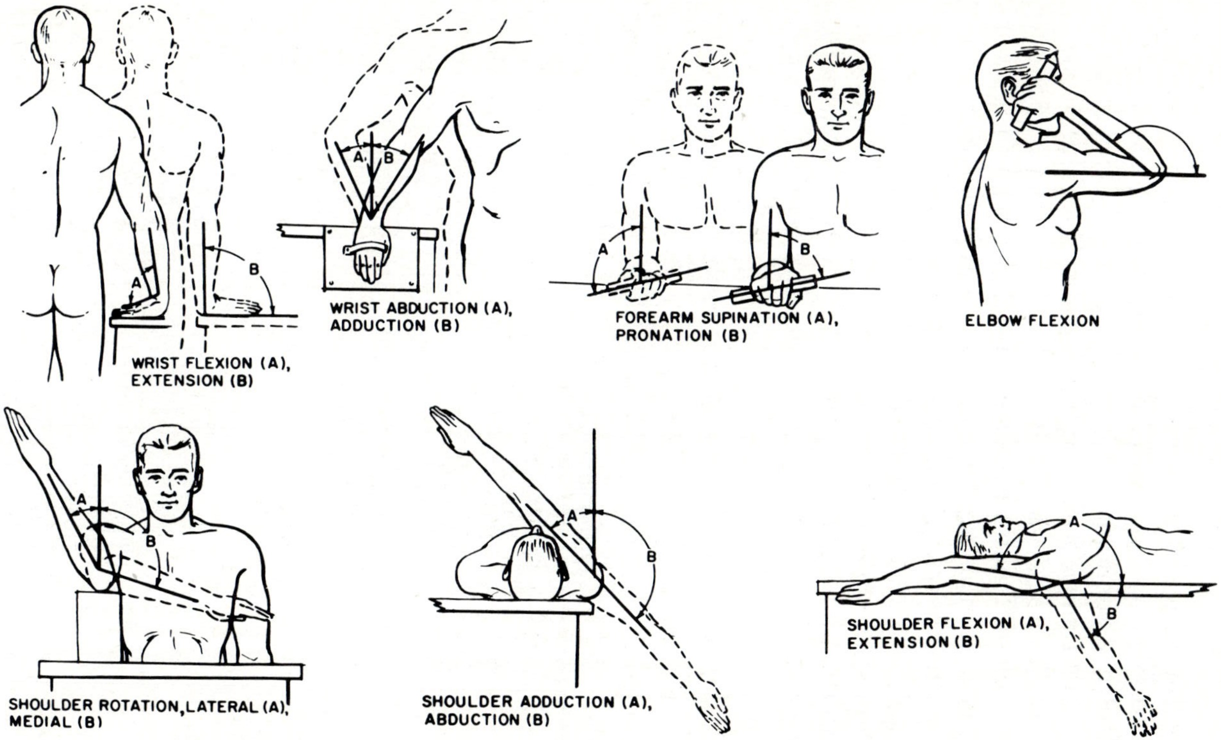

Figure 8.4 Methods of measuring range of motion at the joints of the arm and hand (*Damon et al., 1966*). (See Table 8.7 for typical values.)

TABLE 8.8 Range of Wrist Flexion and Extension While Grasping a Control[a]

Type of movement	Range of movement (degrees) Mean	σ
Flexion—to left of neutral position		
Comfortable, useful limit	46.0	15.7
Extreme possible limit	91.0	16.6
Extension—to right of neutral position		
Comfortable, useful limit	33.6	13.7
Extreme possible limit	71.8	16.0
Total movement, extension-flexion		
Comfortable, useful limit	76.6	23.3
Extreme possible limit	164.2	22.6

[a] Seventy-nine male subjects, average age 28 years, grasping a vertical handgrip located 19 in. forward and 13.5 in. above seat level. Flexion and extension of right wrist measured from "neutral" or resting position of handgrip selected as most comfortable by each subject.
SOURCE: Damon et al., 1966.

maximal right-hand static forces exerted on a vertical hand grip by standing subjects (Hertzberg, 1972). These values represent maximum values that must not be exceeded in design of hand controls. The measurements in Table 8.9 were made by free-standing subjects. Much higher push forces may be exerted if there are rigid surfaces available to brace the body against. The data suggest that push-pull operations are best done with the handle away from the body. However, left-right, up-down, and rotational strength are greater with the handle closer to the body. Summaries of maximum strength data for various body positions are available (e.g., Hertzberg, 1972; Damon et al., 1966; Woodson, 1981).

People do not always use the best method for applying force. Consequently, design should not be based on maximum capability. Table 8.10 lists nominal values for various common force-application activities. These values represent the 5th percentile values and are absolute maximum limits for equipment/operation design.

Maximum strength is a short-term measurement. Endurance is the ability to continue work or exert force. Typical endurance time vs. strength curves suggest that maximum strength is reduced by 50% after 1 min, and by 80% after 10 min exertion (Woodson, 1981). Normal people will increase their strength by 30–50% after 12 weeks of training.

2. Sensory. We have the capacity to sense low levels of energy. For example, we can detect electromagnetic radiation between 380 and

TABLE 8.9 Maximal Right-Handed Static Forces and Torques Exerted on a Vertical Handgrip[a]

Movement	Angle[b] (degrees)	Location of handgrip: Force (lb) 50%[c]	75%[c]	100%[c]
Push horizontal	30	16	24	32
	0	30	35	40
	−30	28	30	32
	−60	28	32	36
Pull horizontal	30	19	22	26
	0	23	26	29
	−30	28	30	31
	−60	23	28	34
Left horizontal	30	35	30	24
	0	42	33	24
	−30	42	34	26
	−60	33	30	26
Right horizontal	30	24	22	21
	0	30	25	20
	−30	33	27	22
	−60	25	23	21
Up vertical	30	28	24	19
	0	34	26	18
	−30	50	40	28
	−60	63	51	41
Down vertical	30	76	58	41
	0	56	40	33
	−30	35	33	30
	−60	39	36	32
Rotation		**Torque (ft-lb)[d]**		
Clockwise	30	12.9	12.2	5.8
	0	13.5	11.7	4.3
	−30	11.5	9.8	3.6
	−60	8.5	6.7	4.9
Counterclockwise	30	12.4	11.0	8.0
	0	13.2	11.4	8.0
	−30	15.3	13.4	8.7
	−60	16.8	14.8	10.1

[a] Standing subjects, feet parallel, 12 in. apart, without any structure against which to brace themselves.
[b] Degrees from horizontal (up or down).
[c] Percentage of maximal grip distance away from body.
[d] About a horizontal axis at center of the handgrip.
SOURCE: Hertzberg, 1972.

TABLE 8.10 General Strength-Limit Expectations

Operation	Force (lb)[a]
Lifting (back muscles only)	350
Forearm lift (palm up, elbow 90° angle)	60
Forearm lift (palm down, elbow 90° angle)	48
Finger pull	8
Finger pinch	21
Finger press	8
Finger squeeze	23
Palm squeeze (on hand tool)	95
Two-hand scissor-action force, hands 6–8 in. apart, at elbow level: Pull apart	25
Squeeze together	30
Pedal force, seated position	60
Lever force, seated position: Fore-aft	45
Lateral	18
Standing, optimal position:	
Lever control push	130
Lever control pull	120
Cranking; wheel diameter 4.5–8.0 in.:	
Wheel parallel to operator	30[b]
Wheel perpendicular to operator	40[b]

[a] Divide values by 2 for females, except for lifting, where value is 160 lb.
[b] Torque, in inch-pounds.
SOURCE: Developed from Woodson, 1981.

700 nm at levels as low as 10^{-6} millilamberts, and audio energy in the 20- to 20,000-Hz range as low as 0.0002 dyne/cm^2. Table 8.11 summarizes the stimulation intensity ranges of human senses (Van Cott and Warrick, 1972). The lower sensitivity threshold is not a specific value, but rather the median of a statistical distribution of values.

In many operator and maintenance tasks it is necessary to determine whether the relative intensity, frequency, etc., of two stimuli are different or changing. Such tasks are relative discrimination tasks, and involve a person's ability to detect a small difference between two stimuli. This difference is also referred to as the just noticeable difference (JND). Table 8.12 summarizes JND for physical intensities and frequency. Although Table 8.12 can be used as a guide in evaluating the characteristics of a proposed display, the values are based on ideal and static conditions and should be considered too optimistic for most practical purposes.

Identification of a stimulus must often be made on an absolute rather than a relative basis. This is usually true when it is impractical to make a reference stimulus available. Table 8.13 summarizes the absolute discrimination of stimuli. Note that the number of absolute judgments is much less than the number of relative discriminations.

TABLE 8.11 Stimulation-Intensity Ranges of Human Senses

Sensation	Smallest detectable (threshold)	Largest tolerable or practical
Sight	10^{-6} millilamberts[a]	10^{4} millilambert[a]
Hearing	2×10^{-4} dyne/cm^2	$< 10^{3}$ dynes/cm^2
Mechanical vibration	25×10^{-5} mm average amplitude at 200 Hz[b]	40 dB above threshold[b]
Touch	0.04–1.1 erg	—
Smell	Very sensitive for some substances (0.0047–0.00047 ppm for H_2S)[c]	—
Taste	Very sensitive for some substances (4×10^{-7} molar concentration of quinine sulfate)	
Temperature	15×10^{-5} (cal/cm^2)/sec for 3-sec exposure of 200 cm^2 skin	22×10^{-2} (cal/cm^2)/sec 3-sec exposure, 200 cm^2 skin
Position and movement	0.2–0.7 deg at 10 deg/min for joint movement	—
Acceleration	0.02*g* linear acceleration[d] 0.08*g* linear deceleration[d] 0.12 deg/sec^2 rotational	5–8*g* positive[d] 3–4*g* negative[d] Disorientation vertigo or blackout
Color (hue)	300 nm	800 nm
Interrupted white light	Unlimited	50 interruptions/sec
Pure tones	20 Hz	20,000 Hz
Vibration	Unlimited	10,000 Hz at high intensities[b]

[a] See Chapter 11 for definition of lambert.
[b] See Chapter 13 for definition of dB and Hz.
[c] Leonardos et al., 1969.
[d] In terms of multiples of the acceration of gravity.

SOURCE: Van Cott and Warrick, 1972.

TABLE 8.12 Relative Discrimination of Physical Intensity and Frequency

Sensation	Number discriminable
Brightness	570 intensities, white light
Hues	128 hues at medium intensity
White light	375 interruption rates between 1 and 45 per sec
Loudness	325 intensities at 2,000 Hz
Pure tones	1,800 tones between 20 and 20,000 Hz at 60 dB
White noise	460 interruption rates between 1 and 45 per sec
Vibration	75 amplitudes in chest region, 0.05–0.5 mm amplitude
Vibration	180 frequencies between 1 and 320 Hz

SOURCE: Van Cott and Warrick, 1972.

TABLE 8.13 Absolute Identification of Intensity and Frequency

Sensation	Number identifiable
Brightness	3–5 intensities with white light, 0.1–50 millilamberts
Hues	12–13 hues
White light	5 or 6 interruption rates
Loudness	3–5 intensities with pure tones
Pure tones	4 or 5 discriminable tones
Vibration	3–5 amplitudes

SOURCE: Van Cott and Warrick, 1972.

The ability to discriminate information from the sensory dimensions so that each level conveys a distinct meaning is referred to as channel capacity. Table 8.14 summarizes the channel capacity of normal people. This table, which is based on absolute judgments, shows that absolute discrimination is greatly increased for multidimensional stimuli. For example, size, wavelength, and luminance alone can be discriminated into five to nine categories. Taken together, however, they can be discriminated into 18 categories.

TABLE 8.14 Channel Capacity of Senses for Unidimensional and Multidimensional Stimuli

	Stimulus dimension	Channel capacity (bits)	Discriminable categories
	Unidimensional Stimuli		
Vision	Dot position in space	3.2	10
	Size of squares	2.2	5
	Dominant wavelength	3.1	9
	Luminance	2.3	5
	Area	2.6	6
	Line length	2.6–3.0	7–8
	Direction of line inclination	2.8–3.3	7–11
	Line curvature	1.6–2.2	4–5
Taste	Salt concentration	1.9	4
Audition	Intensity	2.3	5
	Pitch	2.5	7
Vibration (on chest)	Intensity	2.0	4
	Duration	2.3	5
	Location	2.8	7
Electrical shock	Intensity	1.7	3
	Duration	1.8	3
	Multidimensional Stimuli		
Size, brightness, and hue varied together		4.1	18
Frequency, intensity, rate of interruption, on-time fraction, duration, location		7.2	150
Colors of equal luminance		3.6	13
Loudness and pitch		3.1	9
Position of points in a square		4.6	24

SOURCE: Van Cott and Warrick, 1972.

3. Performance. Human performance is affected by many variables. For example, exceeding the sensory limitations given in Table 8.14 will result in an increase in judgment errors. For many industrial situations, errors in judgment can lead to unsafe conditions. It is known that speed and stress have significant effects on human performance and accuracy. This is demonstrated by the relationship between the frequency with which signals are presented and the percentage of signals detected. This relationship is shown in Figure 8.5. Optimum performance is obtained in the range of 100–300 signals/hr. The effect of stress, as determined by the complexity of the stimulus (load), is shown in Figure 8.6. These data come from a study of comparing numbers in columns of numbers displayed on a panel. The plot shows that errors increase with load even though the number of decisions per minute remains constant. Increased speed, as measured by rate of decision, also had an adverse effect (McCormick and Sanders, 1982).

Human performance is also adversely affected by the addition of irrelevant signals or distractions (Grandjean, 1980). Results from vigilance research suggests that:

1. Sustained alertness decreases after 30 min.
2. Performance is improved if signals are more frequent, stronger, operator gets immediate feedback about critical system conditions, and signals are distinct in shape and contrast.
3. Performance is degraded if intervals between signals are great, the operator is under work or load stress, and the operator is in unfavorable environmental conditions, such as excessive noise or heat.

Operators are also frequently required to perform time-sharing tasks, for example, to monitor more than one indicator or panel simultane-

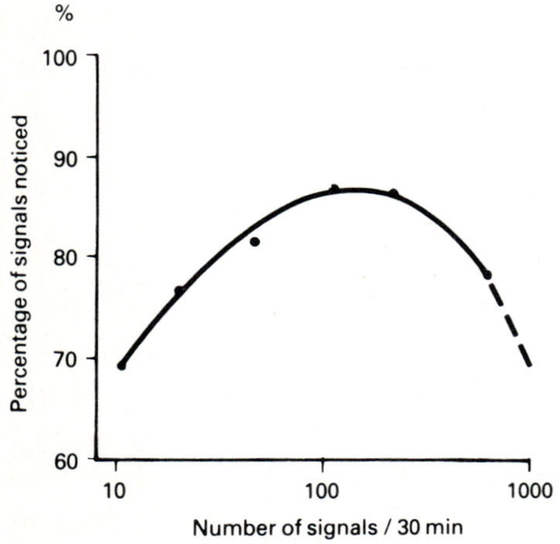

Figure 8.5 Relation between frequency of signals requiring reaction and observed performance (*Grandjean, 1980*).

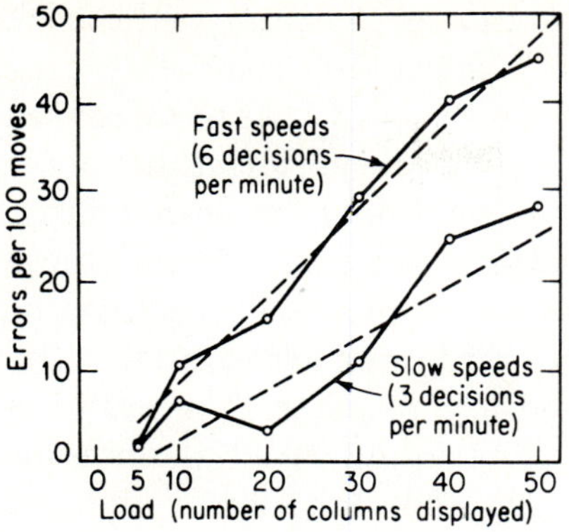

Figure 8.6 Effects of increasing load (number of channels used) on errors (*Mackworth and Mackworth, 1958*).

ously. Some techniques to improve human performance in time-sharing tasks include (Huchingson, 1981):

1. Minimize the number of potential competing information sources.
2. Provide priorities for attention. Weight signals by importance and relative probability of occurrence.
3. Minimize the need for short-term memory or responding to low-probability events.
4. Avoid signals that occur at intervals of less than 0.25 sec.
5. Allow operator to control the pace of input.
6. Display only signals that are to be responded to immediately or in the near future.
7. Use auditory signals where considerable visual time sharing would be required.
8. Direct attention to important information with advanced cues.
9. When coding information, it is better to increase the number of input dimensions than to add new steps along a single dimension.

For example, when displaying several audio messages simultaneously, prioritize the most relevant message by increasing its intensity, giving it special spectral characteristics, or selectively filtering less important messages.

In addition to the complexity of sound, the total amount of sound can affect performance. Examples from case studies demonstrate that a reduction in sound level of 25 dB resulted in a 50% reduction in rejected pieces in a machine shop; a reduction of 20 dB raised production by 30% in an assembly shop; and a reduction of 25 dB resulted in a 30% reduction of typing errors in a typing pool (Grandjean, 1980). Studies of the combined effects of heat and noise on human performance suggest that while the two stressors do not interact within the ranges commonly experienced in the occupational setting, each stressor individually will significantly impair human performance (Hancock and Pierce, 1985). For example, Figure 8.7 shows that the mean error ratio (a measure of error rate) increases under conditions of increased heat or noise (Viteles and Smith, 1946). The six performance tasks summarized in Figure 8.7 are mental multiplication, number checking, lathe operation, type coding, location testing, and pursuit (tracking).

Reaction time, as a measure of human performance, should also be considered in the human-machine design. Simple reaction time is the shortest time between the moment a sensory receptor is stimulated and the time some body element reacts. The simple reaction time for hearing or touch is around 0.16 sec, for sight and sensation of cold and warmth about 0.20 sec, for smell 0.3 sec, and for pain 0.7 sec. Some general characteristics of human performance with respect to reaction time include (Hutchingson, 1981; Woodson, 1981):

1. Greater signal intensities result in shorter reaction times.
2. Reaction time is shorter for anticipated signals.
3. Practice reduces reaction time.
4. Self-pacing an operation can reduce reaction time.
5. Signal quality affects reaction time. Fast reaction is associated with high-pitched sound, bright lights, large visual targets, long-duration signals, and centralized visual signals.
6. Least likely signals have longest reaction times.
7. Signals grouped in a logical way decrease reaction times.
8. Two signals within 0.15 sec are treated as one.
9. Reaction time is 20% faster with the hand than with the foot.

These principles should be followed in designing operations that require rapid, critical response from the operator.

Complex reaction time includes both the recognition of a stimulus

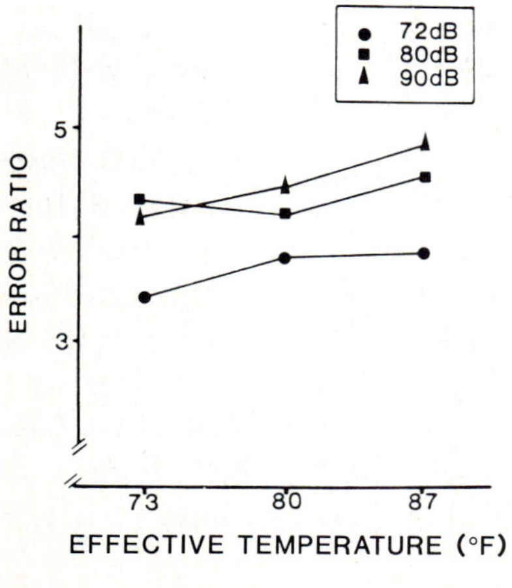

Figure 8.7 Mean error ratio for six performance tasks against effective temperature and noise condition (*Viteles and Smith, 1946; drawn by Hancock and Pierce, 1985*).

TABLE 8.15 Common Reaction Times

I. Effect of number of response choices

Number of choices	Mean reaction time (sec)
1	0.20
2	0.35
4	0.45
10	0.65

II. Effect of reach distance

Reach distance (in.)	Reaction time (sec)
10	0.03
20	0.045
30	0.062

III. Movement response to an end point

	Response time (sec)	
Distance (in.)	Mechanical stop	Operator-controlled stop
6	0.28	0.47
12	0.36	0.57
20	0.46	0.67
30	0.58	0.79

IV. Effect of end-point precision

Precision (in.)	Response time (sec)
0.50	0.11
0.25	0.27
0.125	0.48
0.0625	0.67

V. Effect of angular variation (from operator's normal line of sight)

Angle (deg)	Response time (sec)
20	0.40
40	0.40
80	0.80
100	1.30

SOURCE: Adapted from Woodson, 1981.

and a response by selecting one of several alternatives. Such reaction times are more akin to real-world situations. They depend on the sensory channel through which the stimulus is initiated, the signal characteristics (including complexity and rate), whether anticipatory provisions are present, and the response required of the operator. For example, a complex reaction time may include 0.1 sec for stimulus detection, 0.4 sec for brain recognition, 4.0 sec for decision-making time, and 6.0 sec for motor response time. Some common reaction times for different situations are given in Table 8.15. As shown in the table, reaction and response times are affected by the number of choices, reach distance, precision at the end point of a movement, and the angle of a warning display from the operator's normal line of sight.

System response time can also affect human performance. The following considerations are important for the human-machine interface design (Woodson, 1981):

1. Increased delay increases frustration and degrades quality of performance. (There is a linear decrease in efficiency with increase in delay.)
2. Most individuals become aware of waiting 20 sec.
3. Equipment response time of >2.0 sec should only follow completed tasks.
4. Response times >5.0 sec should be accompanied by a signal to operator that operation is in progress.
5. Response time variation contributes to frustration.
6. Delays >15 sec rule out captive interaction between operator and machine.
7. Delays of 4–15 sec inhibit problem-solving activity.
8. Delays of 2–4 sec inhibit tasks requiring increased concentration.
9. Response time ≤2.0 sec are good for short-term memory of operator.
10. Response feedback should be ≤100 msec.

Human performance is also affected by temperature, light levels, and noise. These considerations are addressed in Chapters 11, 12, and 13.

III. Design Specifications

A. Workplace design

General principles of workplace design include (Kodak, 1983):

1. Services needed by several people should be in a central location.
2. Lines of sight and other visual requirements for an operation should be clear.
3. Heat, noise, and odor production should be modified to minimize effects on other operations.
4. Work areas should be arranged so that the product can flow in one direction with a minimum of rehandling.
5. Offices should provide people minimum separations of 122 cm, with 244 cm being more desirable.

6. Postural flexibility and change should be provided. A person should be able to change posture during a shift.

Choice of the type of workplace depends on the task to be performed. Sitting workspaces are best if all items needed can be supplied and handled within the seated area; items being handled do not require work greater than 15 cm above the work surface; no large forces are required ($\leq$4.5 kg weight); the job requires fine assembly or writing for the majority of the shift; a high degree of body stability is required; use of both hands and feet are required for an operation; or precise foot actions with large force or range of movement are required (Kodak, 1983; Van Cott and Kinkade, 1972). Standing workspaces are more desirable if no knee clearance can be provided; objects handled weigh more than 4.5 kg; high and low extended reaches are required; operations are physically separated and require frequent movement between work stations; or downward forces must be exerted. Table 8.16 summarizes the choice of workplace by task variables. Note that for standing operations, provisions for sitting down during machine time are desirable. Also, designing the operation to allow the operator to move outside the immediate work area several times per hour is desirable.

1. Seated reaches and clearances. Objects to be grasped frequently should be located in the range of 15–36 cm from the front of the work surface, with larger objects located closer. Leaning forward to reach a component of a task should be restricted to a few times per hour. Most manual tasks (such as light assembly) are best performed at elbow height. Fine detail work may be raised to bring it closer to the eyes. The forward reach capability of a small operator (5th percentile female operator, right hand) 5 to 10 cm above the work surface is 45 cm within 22 cm of centerline, 35 cm within 38 cm of centerline, and 20 cm within 52 cm of centerline. Reaching beyond 10 cm above the work surface will also decrease forward reach capabilities. Table 8.17 summarizes recommended dimensions for a seated workplace with and without a footrest (Kodak, 1983). These data are consistent with the anthropometric data in Table 8.4. For example, the minimum leg clearance of 20 cm (Table 8.17) represents adequate clearance for the 99th percentile person (i.e., from Table 8.4, the thigh clearance height + 2.326σ for males is 18 cm and 19 cm for females). Sitting workspaces may be raised to provide extra arm support. However, the maximum surface height should be 91 cm, and a footrest should be provided. Table 8.17 also shows recommended work surface heights for various tasks broken down for males and females (Huchingson, 1981). The critical dimensioning factors for seated operations are proper eye position relative

TABLE 8.16 Choice of Workplace by Task Variables[a]

Parameters	Heavy load and/or forces	Inter-mittent work	Extended work envelope	Variable tasks	Variable surface height	Repetitive movements	Visual attention	Fine manipu-lation	Duration >4 hr
Heavy load and/or forces	—	ST	ST	ST	ST	S/ST	S/ST	S/ST	ST/C
Intermittent work		—	ST	ST	ST	S or S/ST	S or S/ST	S or S/ST	S or S/ST
Extended work envelope			—	ST	ST	S/ST	S/ST	S/ST	ST/C
Variable tasks				—	ST	S/ST	S/ST	S/ST	ST/C
Variable surface height					—	S	S	S	S
Repetitive movements						—	S	S	S
Visual attention							—	S	S
Fine manipulation								—	S
Duration >4 hr									—

[a]S = sitting; ST = standing; S/ST = sit/stand (an alternative to standing all day; a standing workplace could be used but would not be the preferred choice); ST/C = standing, with chair available.

SOURCE: Kodak, 1983.

TABLE 8.17 Recommended Dimensions for a Seated Workplace

Dimension	Length (cm) With footrest	Length (cm) No footrest
Work surface height	81	66
Work surface thickness (maximum)	5	5
Knee clearance[a] (minimum)	—	46
Leg clearance[a]	66	100
Thigh clearance[b] (minimum)	20	—
Adjustable seat height	51–66	39–51
Footrest depth	30	—
Adjustable footrest height	2–23	
Sitting well width (minimum)	81	81
Work surface height (cm)	**Male**	**Female**
Fine assembly	99–105	89–95
Precision work (mechanical assembly)	89–94	82–88
Light assembly or writing	74–78	70–75
Coarse or medium manual work	68–72	66–70

[a]Horizontal distance from front of table to back of opening for legs.
[b]Seat to bottom of table.
SOURCES: Kodak, 1983; Hutchingson, 1981.

to viewing tasks; seat height, back angle, and depth; hand and foot reach requirements; and adjustable seat height for common eye position for large and small operators (Van Cott and Kinkade, 1972).

2. Standing reaches and clearances. Standing workspaces should be designed to eliminate excessive reaches, stooping, bending, twisting, and unnatural head positions. Table 8.18 summarizes the standing reach area for a 5th-percentile person. This table is based on data from subjects where no bending or leaning forward was permitted. Most people can reach 46 cm in front of their body if the object is 110–165 cm above floor. Acceptable forward reaches for two-handed operations are less. Bending at hips and waist can add 36 and 20 cm to the forward reach. Table 8.19 lists recommended dimensions for standing workspaces. For standing manual work, the surface height should be positioned so that the operator's hands are 5–10 cm below elbow height.

3. Environmental factors. Performance in the workplace can be affected by excessive heat or humidity, noise, vibration, and lighting levels. Design criteria based on ergonomic considerations for efficient operation are presented in this section. Table 8.20 gives guidelines for intensity of lighting. The level of illumination should be set to meet the specifications of task complexity and contrast. Excessive light lev-

TABLE 8.18 The Standing Forward Reach Area[a]

Hand height above floor (cm)	Distance from centerline (cm)	Forward reach (cm)	
		One arm[b]	Two arms
86	15–30[c]	35	25
	46	25	5
	61–76	12–0	—[d]
112	15–30	50	46
	46	43	28
	61–76	33–18	—[d]
137	15–30	56	50
	46	46	33
	61–76	38–20	—[d]
162	15–30	50	46
	46	41	25
	61–76	30–10	—[d]

[a]Subject's eye height 162 cm above floor (5th percentile).
[b]Right hand reach to the right.
[c]For one-arm reach only.
[d]Cannot reach two-handed.
SOURCE: Adapted from Kodak, 1983.

els have been associated with frequency of eye complaints (Grandjean, 1980). Table 8.21 lists recommended sound pressure levels for several categories of activity. Note that excessive background noise can interfere with complex mental activities. Thermal comfort is also necessary to maintaining a comfortable workplace. Table 8.22 defines the comfort zone as a function of humidity, air velocity, workload, clothing insu-

TABLE 8.19 Recommended Dimensions for a Standing Workplace

Dimension	Length (cm)	
	Male	Female
Work surface height:		
Precision work	100–120[a]	95–113[a]
Light assembly/writing	90–110	85–98
Heavier work	75[b]–100	70–94
Knee clearance	10	10
Foot depth clearance	13	13
Foot height	10	10
Overhead clearance (floor to ceiling)	203	203
Height of objects above work surface	0–20[c]	0–20[c]
Workbench depth	50–75	50–75

[a]With supported elbows.
[b]Best for applying downward force.
[c]Function of object being handled.
SOURCES: Kodak, 1983; Hutchingson, 1981; Grandjean, 1980; Van Cott and Warrick, 1972.

TABLE 8.20 German and U.S. Norms for Intensity of Lighting[a]

Visual task	Light intensity (lux) DIN[b]	IES[c]
Rough assembly work	250	320
Precise assembly work	1,000	5,400
Delicate assembly work	1,500	10,800
Rough work on tool-making machine	250	540
Fine work on tool-making machine	500	5,400
Precise work on tool-making machine	1,000	10,800
Technical drawing	1,000	2,200
Bookkeeping; office work	500	1,600

[a]See also Chapter 11 (Table 11.12) for lighting requirements for specific industrial settings.
[b]European guidelines.
[c]Illuminating Engineering Society; U.S. guidelines (ANSI, 1973).
SOURCE: Grandjean, 1980.

lation, and radiant heat. The comfort zone is defined as conditions judged acceptable by 95% of the population (Kodak, 1983).

B. Equipment design

Appropriate human-machine interface design can be used to improve system performance by assigning people to tasks they perform best as well as designing equipment within the capabilities of people. The result will be to optimize their performance and simultaneously reduce the chance of error. The examples given in Chapter 1 for stereo cabinet spray painting and a conveyor for corrugated box trim illustrate this philosophy. Tables 8.1 and 8.2 summarized the relative advantages of people and machines. Once a task has been allocated to either an

TABLE 8.21 Recommended Sound Pressure Levels for Several Categories of Activity[a]

Acoustical requirements	Sound pressure level (dBA)
Excellent listening conditions	≤30
Good listening conditions	≤42
Normal conversation	38–47
Moderately good listening conditions	42–52
Fair listening conditions	47–56
Just acceptable speech communication	56–66
Speech not required—possible hearing damage at upper end of range	66–80

[a]See also Chapters 2, 4, and 13.
SOURCE: Kodak, 1983.

TABLE 8.22 The Thermal Comfort Zone[a]

Factor	Level	Ambient dry-bulb temperature[b] (°C) Lower limit	Upper limit
Relative humidity (percent)	20	20	26
	50	19	25.5
	80	18.5	24
Air velocity (m/sec)	0.1	18	24
	0.36	21	27
	0.71	23	29
Workload, 8-hr average, multiples of resting values[c]	×2.0	19	25.5
	×3.5	17	23
	×5.0	15.5	20
Clothing insulation[d] (clo)	0.25	27	28
	1.25	19	22
	2.50	11	16
Radiant heat; Δ°C that globe temperature exceeds dry-bulb temperature	0	19	25.5
	2.8	16	23
	5.6	13	20

[a]See also Chapters 4 and 12.
[b]Unless otherwise noted, the following values are used to calculate thermal comfort: air velocity = 0.25 m/sec; workload = light assembly, ≤2 times resting metabolic level; clothing insulation = 0.6 clo in heat and 1.25 clo in cold; no radiant heat load; 50% humidity.
[c]Is a function of work/rest pattern. Typical units are kcal/min.
[d]Light cotton = 0.3 clo; light clothes with long sleeves = 0.6 clo; light business suit = 1.0 clo; suit with lab coat = 1.5 clo; heavy wool suit = 1.5–2.0 clo.
SOURCE: Kodak, 1983.

TABLE 8.23 Anthropometric Data for the Design of Instrument Consols (in.)

Type of console	Maximum console height from standing surface A	Console depth at base B	Vertical dimension of panel, including sills C	Console panel angle, from vertical D	Minimum pencil shelf depth E	Minimum writing surface depth, including pencil shelf F
1. Sit-stand	62.0	Optional	26	15°	4	16
2. Sit (with vision over top)	47.5[b]–58.0	Optional	22	15°	4	16
3. Sit (without vision over top)	51.5[c]–62.0	Optional	26	15°	4	16
4. Stand (with vision over top)	62.0	Optional	26	15°	4	16
5. Stand (without vision over top)	72.0	Optional	36	15°	4	16

[a]When seat-to-standing surface exceeds 18 in., a heel catch should be provided.
[b]A must never be more than 29.5 in. greater than L.
[c]A must never be more than 33.5 in. greater than L.

operator or machine, the designer should consider the interface, or points of interaction, between the machine and the operator. These interactions include being able to reach and operate all parts of the equipment, planning for maintenance, and providing safe and comfortable conditions for the operators in the workplace.

The design of individual workplaces should incorporate human factors data on both human characteristics and workplace design. For example, Table 8.23 and Figure 8.8 show anthropometric data used to design instrument consoles. Principles of control panel design include (Grandjean, 1980; Van Cott and Kinkade, 1972):

1. Locate displays close to the controls that affect them.
2. If controls and displays must be separate, arrange each in exactly the same order and pattern.
3. Locate identification labels above the control or display.
4. Sequential controls should be arranged from left to right.
5. If controls are not operated sequentially, arrange them in a fractional order.
6. Primary displays and warning lights (e.g., hazard indicators or critical monitor displays) should be in the operator's normal line of sight so as not to require excessive head or eye movement.

Minimum knee clearance G	Foot support to seat[a] H	Seat adjustability I	Minimum thigh clearance at midpoint of I J	Writing surface height from standing surface K	Seat height at midpoint of I L	Maximum console panel breadth H
18	18	4	6.5	36.0	28.5	36
18	18	4	6.5	25.5–36.0	18.0–28.5	36
18	18	4	6.5	25.5–36.0	18.0–28.5	36
—	—	—	—	36.0	—	36
—	—	—	—	36.0	—	36

NOTE: Standard values for critical dimensions used in the design of instrument consoles for the seated and/or standing operator, with and without a requirement on the operator to maintain horizontal visual contact with other displays or test apparatus beyond the console. Design values for each console established to accommodate 95 plus percent of U.S. Air Force population.

SOURCE: Van Cott and Warrick, 1972; see also Figure 8.8.

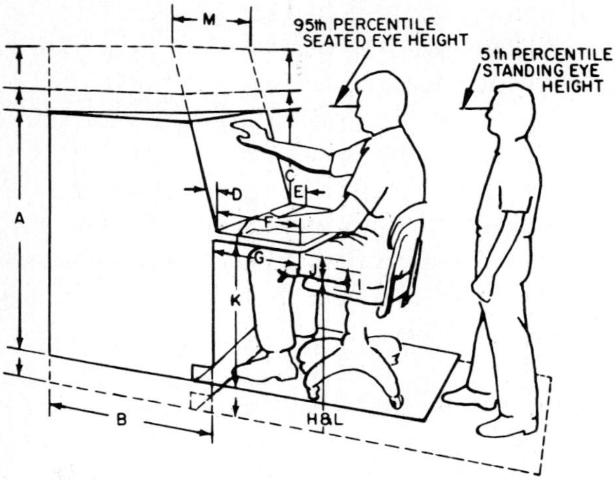

Figure 8.8 General workplace dimensions adapted to large and small men (*Van Cott and Warrick, 1972*). (See Table 8.23 for typical values.)

7. Secondary displays (frequently used operational displays) may require eye movement, but should minimize head movement for viewing.
8. Auxiliary displays (e.g., power indicators) may require head movement for viewing.

Workplace and equipment clearances should be dimensioned to accommodate the largest worker. Critical clearances (e.g., head clearances) should accommodate the 99th percentile worker. Table 8.24 summarizes selected clearances for hands, arms, and the full body.

1. Displays. Information displays can be designed to convey: quantitative information about a dynamic parameter such as temperature or speed; qualitative information such as an approximate value, direction of a trend, or rate of change; status information showing the condition of a system such as on/off or go/no go; warnings of unsafe or emergency conditions, or the presence or absence of a condition or object; representational information including pictorial or graphical representations of objects or configurations; identification information concerning the state of a process coded as alphanumeric, symbolic, or auditory; and time-phase information as pulsed signals coded by duration, intensity, or intersignal interval. The proper selection of sensory modality depends on the intrinsic advantage of each input. Table 8.25 summarizes the rationale for selecting auditory or visual presentation. Tactual devices attached to the surface of the skin can be useful as warning devices when visual and auditory modalities are not practical or are overloaded.

TABLE 8.24 Selected Clearances for Arms, Hands, and the Full Body

	Dimension (cm)
Full Body	
A. Entering open-top vessels	
1. Overhead clearance	91
2. Horizontal leg clearance	76
3. Vessel height	51–61
B. Full-body access ports	
1. Circular hatch diameter	76
2. Horizontal hatch (width × height)	61 × 51
Arm	
1. Both arms, one opening	75% of depth of reach plus 15 cm
2. Arm to elbow (diameter)	11[a]
3. Arm to shoulder (diameter)	13[a]
Hand	
1. Empty—flat	5 × 10
2. Empty—fist clenched	10 × 13
3. Holding small object	10 × 10
4. Turning screwdriver	10 × 11
5. Turning socket wrench	14 × 16
6. Pushbutton access (diameter)	4
7. Two-finger twist (diameter)	9

[a]Add 8 cm for winter clothing.
SOURCE: Kodak, 1983.

TABLE 8.25 When to Use the Auditory or Visual Form of Presentation

Use auditory presentation if:	Use visual presentation if:
1. The message is simple.	1. The message is complex.
2. The message is short.	2. The message is long.
3. The message will not be referred to later.	3. The message will be referred to later.
4. The message deals with events in time.	4. The message deals with locations in space.
5. The message calls for immediate action.	5. The message does not call for immediate action.
6. The visual system is overburdened.	6. The auditory system is overburdened.
7. The receiving location is too bright, or dark-adaptation integrity is necessary.	7. The receiving location is too noisy.
8. The person's job requires him or her to move about continually.	8. The person's job allows him or her to remain in one position.

SOURCE: Deatherage, 1972.

TABLE 8.26 Types and Recommended Information Displays[a]

Information type	Preferred display	Comments	Examples in industry
Quantitative reading	Digital readout or counter	Minimum reading time; minimum error potential	Number of units produced by a production machine
Qualitative reading	Moving pointer or graph	Position easy to detect; trends apparent	Temperature changes in a work area
Check reading	Moving pointer	Deviation from normal easily detected	Pressure gauges on a utilities' console
Adjustment	Moving pointer or digital readout	Direct relation between pointer movement and motion of control or accuracy	Calibration charts on test equipment
Status reading	Lights	Color-coded; indication of status (e.g., "on")	Consoles in production lines
Operating instructions	Annunciator lights	Engraved with action required; blinking for warnings	Manufacturing lines in major production systems

[a]Six types of information to be gathered from displays are indicated in column 1. For each type the preferred display is indicated in column 2. Additional information is provided in column 3 about the reasons for choosing the preferred display or about characteristics of that display that make information transfer from display to operator more effective (e.g., color-coded lights). Some examples of industrial uses of these displays are given in column 4.

SOURCE: Kodak, 1983.

Visual displays should be designed to conform with the following principles:

1. Use the simplest display commensurate with the information transfer needs of the human-machine system. Complexity requires increased reading time and increases the chance of misinterpreting the information.
2. Use the least precise display format commensurate with the readout accuracy required. Requiring extra precision increases response time, stress, and fatigue and will contribute to reading errors.
3. Use the most natural or expected display format.
4. Match the display technique to the anticipated environment. Consider such factors as light levels, vibration, position, and viewing angle.
5. Optimize the visibility (distance to scale size ratios, angles, contrast, and glare), conspicuousness (contrast with background), legibility (style, pattern, and color), and interpretability (minimize the need for extrapolation, training, and memory).

Table 8.26 summarizes types of information displays. The table gives recommended display formats along with typical industrial examples. Figure 8.9 shows various types of displays for quantitative information, including fixed scale with moving pointers, moving scales with fixed pointers, and digital formats (McCormick and Sanders, 1982). The digital format is superior to analog if precise numerical values are required. Note, however, that values on digital displays must remain long enough to be read. Mean response times and error rates to digital information are lower compared to circular displays. Fixed-scale moving-pointer displays are useful when values frequently or continually change, when it is important to observe rate of change of values presented, and in general are superior to displays with fixed pointers and moving scales. Slight variations in output are more apparent with moving-pointer displays. A moving scale read with a fixed pointer displayed through a window can, however, be read more quickly than an open-faced, fixed-scale dial. Ergonomic considerations in the design of dials and gauges include (Woodson, 1981; Grandjean, 1980; Kodak, 1983; McCormick and Sanders, 1982; Grether and Baker, 1972):

1. Avoid double pointers or double scales.
2. The direction of pointer movement is more clear in a rectangular (linear) format.
3. Gauges side by side are effective for indicating relative information.

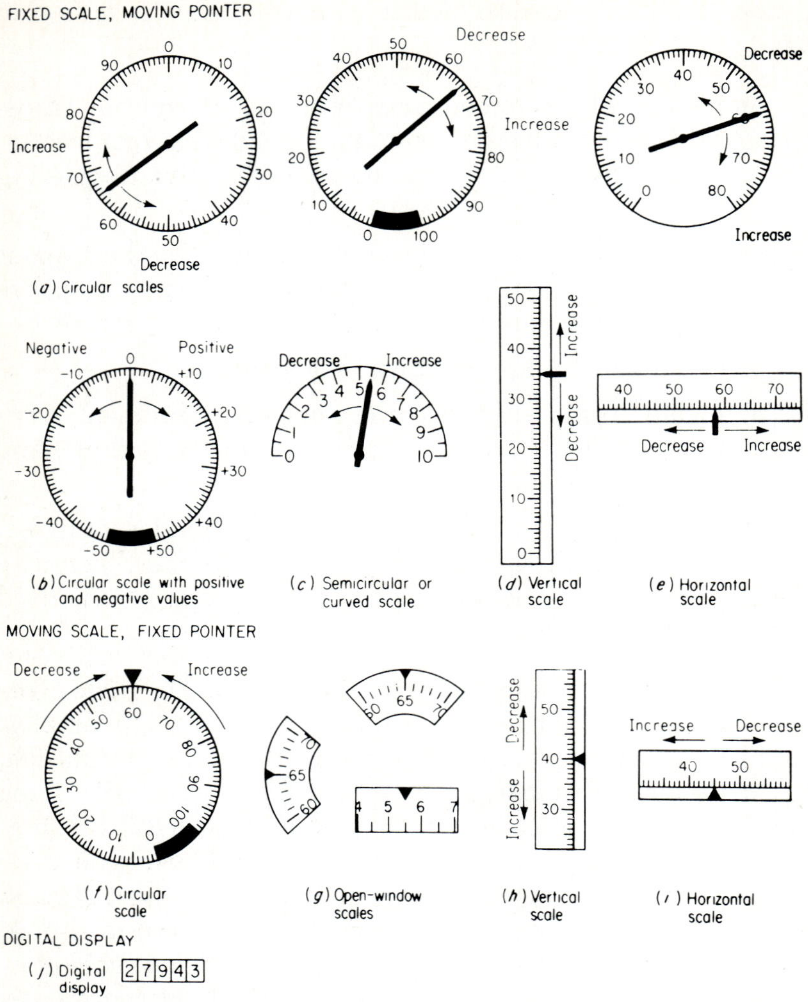

Figure 8.9 Examples of certain types of displays used in presenting quantitative information (*McCormick and Sanders, 1982*).

4. When clustering dials together, orient the display so that all pointers are aligned when the instrument is in normal condition. Adjacent indicators on a control panel should have the same layout of marks and numbers.

5. Select dial size in relationship to the number of scale marks and viewing distance. Table 8.27 shows recommended dial diameters for various scales and viewing distances.

6. Direction of scale numbering and indicator movement should reflect expected direction. For example, increase should be represented as movement right, up, or clockwise.

TABLE 8.27 Minimum Diameter of Inner Annulus for Various Viewing Distances[a]

Number of scale marks	Viewing distance				
	20 in.	3 ft	6 ft	12 ft	20 ft
50	—	1.3	2.6	5.0	9.0
100	1.4	2.6	5.0	10.0	17.0
150	2.0	3.9	8.0	15.0	26.0
200	2.9	5.0	10.0	21.0	34.0
250	3.5	6.4	13.0	26.0	43.0
300	4.0	7.7	15.0	31.0	51.0
350	5.0	9.0	18.0	36.0	60.0

[a]Diameter given in inches.
SOURCE: Woodson, 1981.

7. Scale intervals should be kept as simple as possible. The length of a scale unit should never exceed the smallest unit to which the scale is to be read. Avoid requiring an operator to interpolate between scale marks.

8. Mark scales at intervals of powers of 10 or 2. Avoid all other formats. Mark scales with a maximum of 10 subdivisions (9 marks) between numbered graduations.

9. Place the scale's zero at 9 to 12 o'clock on round dials with conditions scales. If scale does not fill the dial perimeter, locate unused space in lower part of dial.

10. Recommended numeral and letter heights for 28-in. viewing distance are as follows (heights in inches) (Grether and Baker, 1972):

	Low luminance	*High luminance*
Critical markings with movable positions	0.20–0.30	0.12–0.20
Critical markings with fixed position	0.15–0.3	0.10–0.20
Noncritical markings	0.05–0.20	0.05–0.20

Low luminance represents light levels between 0.03 and 1.0 ft-lamberts. High luminance represents light levels above 1.0 ft-lamberts. For other viewing distances, increase or decrease values proportionately. Stroke width should equal one-sixth to one-eighth of numeral height. Numeral width should be three-fifths of height, except that the numbers 4 and 1 should be one stroke wider and thinner, respectively. Single-word labels should be printed in capital letters only. Recommended letter height can also be estimated as viewing distance (in millimeters) divided by 200.

11. Recommended dimensions for scale graduations can be calculated as follows (given a = greatest viewing distance in millimeters):

    Height of largest major graduation = $a/90$

    Height of middle major graduation = $a/125$

    Height of smallest major graduation = $a/200$

    Thickness of graduations = $a/500$

    Distance between smallest graduations = $a/600$

    Distance between largest graduations = $a/50$

12. Orient all numbers upright, not radially.
13. Using a coded zone on a scale greatly decreases reading time and reading errors. For example, indicate warning areas of a dial with a red color. Do not use color coding if ambient lighting is poor.
14. Use white markings, pointers, and numbers on black background for display in reduced illumination.
15. Avoid glare and shadows from light sources and optical distortion from glass covers.
16. Remove or cover unused displays.

Indicator lights can be used to warn of a hazardous condition or display status information. Color should be used as an indicator of operational conditions as follows (Woodson, 1981; Kodak, 1983; Grether and Baker, 1972):

1. Red: system inoperative, error, failure or hazard.
2. Yellow: caution, recheck display, dangerous situation possible.
3. Green: satisfactory operation, successful operation.
4. White: status indicator, selection of function, test in progress. Does not imply success or failure.
5. Blue: Does not have standard meaning.
6. Flashing: Indicates more important condition than continuous light. Flashing rate should be between 3 and 10 per second with on duration $\geq$0.05 sec.
7. Brightness: Function of environment.

Electronic displays include light-emitting diodes (LEDs), liquid crystals, characters generated in dot-matrix form (cathode-ray tubes, CRTs), plasma displays, electroluminescent displays, and printed characters. In general, segmented displays (LED and liquid crystal formats) are less legible than dot matrix. LED and liquid crystal displays can be used only in situations with well-controlled general lighting. Dot matrix 7 × 7 and 7 × 9 dots/character formats are preferred to the 5 × 7 format. Ergonomic considerations in the design of CRT displays include (Kodak, 1983; Woodson, 1981):

1. Display diameter should be ≥7 in., with 12–19 in. for typical console.
2. For dot-matrix formats, optimum dot dize = 0.8–1.1 mm with the ratio of dot spacing to dot size <0.05.
3. Visual angle subtended by characters should be 15–25 min of arc.
4. Display resolution should be at least 10 scan lines/character height with 125 lines/inch; bandwidth 4.0–10.0 MHz; and scan lines should not be visible.
5. Brightness should be adjustable with luminances up to 350 candelas/m^2 (100 foot-lamberts) for areas with high ambient lighting.
6. Contrast ratio should be at least 0.88.
7. Phosphor persistence of 10 μsec to 1.0 msec (P4 and P7 phosphors, respectively) are best for alphanumeric displays.
8. Viewing distance should be limited to 20 ft and viewing angle ≤30° from the perpendicular axis.
9. Locate critical information near center of screen, static information at top, right of screen, and error messages near bottom of screen.

Symbols can be coded in a variety of ways on CRT displays. Table 8.28 summarizes coding methods for symbols. Typical symbol spacing is 50% of character width between characters and 75–100% of character width between groups of characters. Table 8.29 summarizes important parameters from visual display terminal standards.

Auditory displays are used to warn of dangerous or important status conditions. Such displays do not require the intended listener to face the source of signal and can work in all but impossibly loud environments. Signals should be at least 10 dBA above ambient noise, should be discriminable (>1 octave separation), and should be attention-getting (Kodak, 1983). Table 8.30 lists types and characteristics of audio alarms. Design recommendations for auditory alarm and warning devices include (Deatherage, 1972; McCormick and Sanders, 1982):

1. If distance to listener is great:	Use high intensities and avoid high frequencies (≤1000 Hz).
2. If sound must bend around obstacles or pass through partitions:	Use low frequencies (≤500 Hz).
3. If background noise is present:	Select alarm frequency in a region where noise is minimal.

4. To demand attention:	Modulate signal to given intermittent beeps or modulate frequency at the rate of 1–3 Hz.
5. To acknowledge warning:	Provide signal with manual shut-off, so that it sounds until action is taken.

2. Controls. The operator's response to information from audio and visual displays is communicated through controls and data entry devices. A control can be a switch, lever, pedal, button, knob, or keyboard. Performance can be enhanced if the controls operate as one expects them to; they are dimensioned to fit the human body; and they can be operated within the strength and precision capabilities of most people. Ergonomic considerations in the selection, design, and use of controls include (Woodson, 1981; Chapanis and Kinkade, 1972):

1. Distribute controls so that no one limb is overburdened. Rapid or precise settings should be assigned to hands; large force applications to feet. Avoid two-hand operations unless great force is required.
2. Select, locate, and orient controls so their motion is compatible with the natural motion of the operator's arm, wrist, finger, leg, ankle, or foot, and with the movement of the associated display.
3. Select multirotation controls when precise settings are required over a wide range of adjustment.
4. Select discrete-adjustment (detent) controls or pushbutton arrays when control object is to be set in discrete positions.
5. Provide operator with feedback showing machine response to input.
6. Provide sufficient resistance to dampen spurious inputs but not too much resistance in order to avoid muscle fatigue.
7. All critical and emergency controls should be positioned to reflect their importance and be identifiable both visually and by touch with shape coding.
8. Combine functionally related controls to reduce movements, aid in sequential operations, and economize panel space.

People expect things to behave in certain ways when they operate controls. Although individuals can be taught to operate systems that do not conform to population stereotypes, their performance may deteriorate when placed in emergency situations. Examples of stereotypes include: Very loud sounds, or sounds repeated in rapid succession, and visual displays that blink or are very bright, imply urgency and excitement; red signifies stop or danger, yellow indicates caution, green indicates go or on, and flashing blue indicates an emergency vehicle.

TABLE 8.28 Summary of Coding Method for Symbols

Code	Number of steps in code	Evaluation	Comments
Alphanumerics	Unlimited	Excellent	High information-handling rate; unlimited number of coding steps
Geometrics	20 or more	Excellent	Certain shapes easily recognized; many coding steps
Color	4	Excellent	Difficulty in techniques of reproducing for CRT; objects easily and quickly identified
Blink	2	Poor	Distracting and fatiguing; interacts poorly with other codes; best for attention getting; few steps in code
Brightness	2	Poor	Limited number of steps; fatiguing; detrimental to decoding performance
Line lengths	4	Fair	Limited number of steps; clutters displays
Angular orientation	12	Fair	95% of estimates correct within 15%
Inclination	24 or more	Fair	Many coding steps, especially with combinations
Visual number (dots)	6	Fair	Few steps
Combinations	Unlimited	Good	Avoid overloading symbols with too much information; complex combination can degrade decoding speed and accuracy

SOURCE: Woodson, 1981.

TABLE 8.29 Visual Display Terminal Standards

Parameter	Standard[a]	Value
Symbol contrast	DIN	3:1 minimum; 6:1–10:1 preferred; 15:1 maximum
	VDT	8:1–10:1 with background luminance between 15 and 20 candellas/m^2
Symbol luminance	VDT	45 candellas/m^2 minimum; 80–160 candellas/m^2 preferred
Color	DIN	Green through orange
Refresh rate	DIN	50 Hz minimum
	VDT	50–60 Hz
Glare control	DIN	Use of (1) diffusing surfaces, (2) micromesh filters, (3) optical coatings, (4) sprays, (5) hoods, or (6) combination filters recommended
Symbol size (capital letter)	DIN	18 min of arc for viewing distances ≥50 cm; 2.6 mm high for viewing distances <50 cm
Character format	DIN	5 × 7 dot matrix minimum plus one dot upward for ascenders and two dots downward for descenders; letter width 50–70% of height
Line/column spacing	DIN	minimum one dot or 10% of width between characters; minimum one dot or 10% of height between lines (including ascenders and descenders)

Distortions	VDT	character size variation $\leq 10\%$ in either horizontal or vertical direction; jitter $\leq 0.05\%$ of screen diagonal
Ambient light levels	DIN	300–500 lux with negative image; 500 lux minimum with positive image
Reflectivities	DIN	Keys: 0.2–0.7; work surface: 0.3–0.5.
Keyboard factors:	VDT	
Actuation force		0.25–1.5 N
Feedback		Tactile, acoustic, or snap-action
Key travel (displacement)		0.8–4.8 mm
Key spacing		18–20 mm (between centers)
Keytop dimensions		12–15 mm
Keyboard slope		5–15 deg
Numeric block		Telephone format preferred

[a]DIN: Draft standard 66234, "Characteristic values for the adaptation of work stations with fluorescent screens to humans," DIN, Deutsches Institut fur Normung (German Standards Organization). (Also under consideration by the International Standards Organization, ISO.)

VDT: Cakir, A., Hart, D. J., and Stewart, T. S. M., *The VDT Manual.* International Research Association for Newspaper Technology, Darmstadt, FRG, 1979.

SOURCE: Banks et al., 1982.

TABLE 8.30 Alarm Types, Characteristics, and Special Features

Alarm	Intensity	Frequency	Attention-getting ability	Noise-penetration ability	Special features
Diaphone (foghorn)	Very high	Very low	Good	Poor in low-frequency noise Good in high-frequency noise	
Horn	High	Low to high	Good	Good	Can be designed to beam sound directionally Can be rotated to get wide coverage
Whistle	High	Low to high	Good if intermittent	Good if frequency is properly chosen	Can be made directional by reflectors
Siren	High	Low to high	Very good if pitch rises and falls	Very good with rising and falling frequency	Can be coupled to horn for directional transmission
Bell	Medium	Medium to high	Good	Good in low-frequency noise	Can be provided with manual shut-off to ensure alarm until action is taken
Buzzer	Low to medium	Low to medium	Good	Fair if spectrum is suited to background noise	Can be provided with manual shut-off to ensure alarm until action is taken
Chimes and gong	Low to medium	Low to medium	Fair	Fair if spectrum is suited to background noise	
Oscillator	Low to high	Medium to high	Good if intermittent	Good if frequency is properly chosen	Can be presented over intercom system

SOURCE: Deatherage, 1972.

Table 8.31 summarizes stereotypes for controls. Note that the stereotypes for rotary controls and valves are reversed.

Table 8.32 summarizes many characteristics of common controls. Important control characteristics include displacement (linear or angular), operating force, friction, inertia or drag forces, number of positions, direction of movement, detents or stops, identification, and compatibility with associated displays and size. Controls that require little manual effort include pushbuttons, tumbler switches, small hand levers, and rotating and bar knobs. Controls that require muscular effort include handwheels, cranks, and heavy levers and pedals. Some ergonomic principles for controls are (Grandjean, 1980):

1. Use fingers and hands for quick, precise movements; arms and feet for operations requiring force.
2. Hand-operated controls are easily grasped when located between elbow and shoulder height in full view.
3. Knobs and switches operated with fingers should be at least 15 mm apart and, when operated with a full hand, at least 50 mm apart.
4. Pushbuttons, tumbler switches, and rotary knobs are well suited for jobs requiring little movement or travel, and high precision.
5. Long-arm levers, cranks, handwheels, and pedals are well suited for applications requiring muscular effort but little precision.

TABLE 8.31 Population Stereotypes for Controls

Switches	
Up	On, Start, High, In, Fast, Raise, Increase, Open, Engage, Forward, Alternating, Positive
Down	Off, Stop, Low, Out, Slow, Lower, Decrease, Close, Disengage, Reverse, Direct, Negative
Rotary controls:	
Clockwise	On, Increase, Right, Up
Counterclockwise	Off, Decrease, Left, Down
Vertical levers:	
Movement away from body	Decreasing action
Movement toward body	Increasing action
Push/pull control	
Outward from panel	On
Pushed in	Off
Overhead controls	
Pushing forward	Increasing activity
Pulling back	Decreasing activity
Valves controlling liquids	
Counterclockwise	Open
Clockwise	Close

SOURCES: Adapted from Kodak, 1983; Woodson, 1981; Chapanis and Kinkade, 1972.

TABLE 8.32 Comparison of the Characteristics of Common Controls

	Type of control										
	Discrete adjustment					Continuous adjustment					
Characteristic	Rotary selector switch	Thumb-wheel	Hand push-button	Foot push-button	Toggle switch	Knob	Thumb-wheel	Hand-wheel	Crank	Pedal	Lever
Large forces can be developed	—	—	—	—	—	No	No	Yes	Yes	Yes	Yes
Time required to make control setting	Medium to quick	—	Very quick	Quick	Very quick	—	—	—	—	—	—
Recommended number of control positions (settings)	3 to 24	3 to 24	2	2	2 to 3	—	—	—	—	—	—
Space requirements for location and operation of control	Medium	Small	Small	Large	Small	Small to medium	Small	Large	Medium to large	Large	Medium to large
Likelihood of accidental activation	Low	Low	Medium	High	Medium	Medium	High	High	Medium	Medium	High
Desirable limits to control movement	270°	—	$\frac{1}{8} \times 1\frac{1}{2}$ in.	$\frac{1}{2} \times 4$ in.	120°	Unlimited	180°	±60°	Unlimited	Small[a]	±45°
Effectiveness of coding	Good	Poor	Fair to good	Poor	Fair	Good	Poor	Fair	Fair	Poor	Good

Effectiveness of visually identifying control position	Fair to good	Good	Poor[b]	Poor	Fair to good	Fair[c] to good	Poor	Poor to fair	Poor[d]	Poor	Fair to good
Effectiveness of nonvisually identifying control position	Fair to good	Poor	Fair	Poor	Good	Poor to good	Poor	Poor to fair	Poor[d]	Poor to fair	Poor to fair
Effectiveness of check-reading to determine control position when part of a group of like controls	Good	Good	Poor[b]	Poor	Good	Good[c]	Poor	Poor	Poor[d]	Poor	Good
Effectiveness of operating control simultaneously with like controls in an array	Poor	Good	Good	Poor	Good	Poor	Good	Poor	Poor	Poor	Good
Effectiveness as part of a combined control	Fair	Fair	Good	Poor	Good	Good[e]	Good	Good	Poor	Poor	Good

[a]Except for rotary pedals, which have unlimited range.
[b]Exception: when control is back-lighted and light comes on when control is activated.
[c]Applicable only when control makes less than one rotation. Round knobs must also have a pointer attached.
[d]Assumes control makes more than one rotation.
[e]Effective primarily when mounted concentrically on one axis with other knobs.

SOURCE: Chapanis and Kinkade, 1972.

TABLE 8.33 Summary of Selected Data Regarding Design Recommendations for Control Devices[a]

Device	Size (in.)		Recommended separation (in.)		Displacement		Resistance	
	Minimum	Maximum	Minimum	Desirable	Minimum	Maximum	Minimum	Maximum
Hand pushbutton								
Fingertip operation	$\frac{1}{2}$	None	$\frac{1}{4}$	1	$\frac{1}{8}$ in.	15 in.	10 oz	40 oz
Foot pushbutton	$\frac{1}{2}$	None						
Normal operation					$\frac{1}{2}$ in.			
Wearing boots					1 in.			
Ankle flexion only						$2\frac{1}{2}$ in.		
Leg movement						4 in.		
Will not rest on control							4 lb	20 lb
May rest on control							10 lb	20 lb
Toggle switch			$\frac{1}{2}$	1	30°	120°	10 oz	40 oz
Control tip diameter	$\frac{1}{8}$	1						
Lever arm length	$\frac{1}{2}$	2						
Rotary selector switch							10 oz	40 oz
Length	1	3						
Width	$\frac{1}{2}$	1						
Depth	$\frac{1}{2}$							
Visual positioning					15°	40°[a]		
Nonvisual positioning					30°	40°[a]		
Knob, continuous adjustment[b]								
Finger-thumb in/oz			1	2				$4\frac{1}{2}$–6
Depth	$\frac{1}{2}$	1						
Diameter	$\frac{3}{8}$	4						
Hand/palm diameter	$1\frac{1}{2}$	3	3	5				
Crank[b]			2	4				
For light loads, radius	$\frac{1}{2}$	$4\frac{1}{2}$						

For heavy loads, radius	$\frac{1}{2}$	20						
Rapid, steady turning								
<3–5 in. radius							2 lb	5 lb
5–8 in. radius							5 lb	10 lb
>8–in. radius							7.5 lb	—
For precise settings							$2\frac{1}{2}$ lb	8 lb
Lever[c]			2	4				
Fore-aft (one hand)						14 in.		
Lateral (one hand)						38 in.		
Finger grasp, diam.	$\frac{1}{2}$	3					12 oz	32 oz
Hand grasp, diam.	$1\frac{1}{2}$	3					2 lb	20–100 lb
Handwheel[b]						90°–120°	5 lb	30 lb[d]
Diameter	7	21						
Rim thickness	$\frac{3}{4}$	2						
Pedal			6	10				
Length	$2\frac{1}{2}$							
Width	1							
Normal use					$\frac{1}{2}$ in.			
Heavy boots					1 in.			
Ankle flexion						$2\frac{1}{2}$ in.		10 lb
Leg movement						7 in.		180 lb
Will not rest on control							4 lb	
May rest on control							10 lb	

[a]When special requirements demand large separations, maximum should be 90°.

[b]Displacement of knobs, cranks, and handwheels should be determined by desired control-display ratio.

[c]Length depends on situation, including mechanical advantage required. For long movements, longer levers are desirable (so movement is more linear).

[d]For two-handed operation, maximum resistance of handwheel can be up to 50 lb.

SOURCES: Kodak, 1983; McCormick and Sanders, 1982.

Table 8.33 summarizes data regarding design recommendations for control devices. Varying the size and shape of controls can help the operator identify specific controls in areas of low illumination and can help reduce the potential for error. Toggle switches are good with limited panel space and for two-position operations only. Pushbuttons are useful for information entry. However, they should be accompanied by a display to inform the operator of their status. Pushbuttons can be recessed to avoid inadvertent activation. Rocker switches are preferred for horizontal layout. Rotary switches are useful for applications where from 3 to 24 values must be selected. Avoid large angular separations; use only as much control rotation as needed. Knobs should be used for

TABLE 8.34 Maximum Torque and Force That Can Be Applied to a Round Knob, Crank, and Lever

	Maximum torque (N · m)			
	Finger-thumb grip		Hand-palm grip	
	Knob depth (mm)		Knob depth (mm)	
Knob diameter (mm)	12	25	12	25
12	0.43	0.51	0.90	1.42
19	0.51	0.68	1.70	2.49
25	0.68	0.85	5.08	6.10
38	1.11	1.27	7.12	10.17
51	1.70	2.03	11.19	13.22
76	0.51	0.51	14.24	16.27
	Horizontal		Vertical	
Distances above floor (cm):	91		91	120–140
Crank radius (cm)				
3.8	2.3		2.3	—
6.4	2.3		4.5	4.5
11.4	4.5		10.2	10.2
19.0	10.2		10.2	>10.2
	Force (N)			
Lever movement	Finger grasp			Palm grasp
Front-to-back	9			130
Side-to-side	3			90

SOURCES: Adapted from Woodson, 1981; Kodak, 1983.

operations where continuous smooth rotation is required. Table 8.34 summarizes maximum force that can be applied to a round knob. Thumbwheels should be used for nonprecise operations only. Expose at least 1 in. of a thumbwheel knob edge. Cranks require a large space. However, they can provide coarse and/or fine control over a wide adjustment. Maximum force that can be applied to a crank is also shown in Table 8.34. Levers are useful for accurate adjustments over a small range. Installation of a knob on the end of a lever allows for simultaneous operation of two controls with one hand. Foot pedals leave hands free for other work activities but are not recommended for standing workspaces.

3. Hand tools. Hand tools are used throughout industry to operate, assemble, or repair equipment. Hand tool design can affect the productivity and health of an operator. Factors that affect performance include static loading of arm and shoulder muscles, awkward hand position, pressure on palm and fingers, vibration and noise exposure with power tools, and pinch points with double-handed tools (Kodak, 1983). Static muscle loading and resulting fatigue can be minimized by angling tools to reduce the need to raise the arm and arranging the workplace so that the elbows can be kept at 90° during the work cycle. Handles should be designed with flanges or knurled surfaces to minimize the need to grip.

Awkward hand positions may result in wrist soreness and difficulty in sustaining a grip. This problem can be reduced by redesigning the tool or repositioning the work piece to minimize extreme wrist deviations. Figure 8.3 showed the advantage of using properly bent pliers in an electronic assembly shop. Tool handles should be long enough so they do not put pressure at the base of the thumb. Pressure here can affect the blood vessels and nerves that pass through the hand and can cause pain and swelling. Pressure on the skin or joints of fingers should be kept below 150 kPa. Table 8.35 summarizes hand-tool design criteria based on biomechanical analysis.

The primary concern in hand-tool design is to make the tool suitable for the operations to be performed. Table 8.36 summarizes design recommendations for hand tools. Handles should not have protruding sharp edges; should be hard enough to resist embedding of work particles; should be impervious to absorbing oil, solvents, and other chemicals; and should have a slight ripple texture to aid in avoiding slippage. Design handles to provide as great a force-bearing area as possible. Power-tool weight and trigger activation forces are given in Table 8.37. The weight of a hand tool will determine how long it can be held or used and how precisely it can be manipulated. Tools weighing more than 2.3 kg can fatigue arm muscles when held away from the body.

TABLE 8.35 Hand-Tool Design Criteria Based on Biomechanical Analysis

Design criteria	Effects of incorrect design
Use offset tool handles for pliers; keep wrists straight while rotating forearm and hand.	Radial deflection while rotating wrists leads to tennis elbow; deflection of wrist outward while rotating outward may inflame wrist tendons (tenosynovitis).
Size the handle to the user; avoid form-fitted grips that are incompatible with hand sizes.	If too large or too small, the metal ridges cause discomfort to joints of fingers and reduce the force that can be applied.
Use plastic grips on metal handles of pliers.	Gripping tightly with bare hand can cause stress concentration.
Redesign handles of tools such as paint scrapers so they do not press against the palms.	Obstruction of the blood flow to tissue (compression ischemia) may be due to the handle shape or how it is held in the palm.
Use handles that absorb vibration on pneumatic drills and other vibrating hand tools.	Vibrating power tools may cause "white finger syndrome" (Chapter 2). Parts of the body will react in resonance with certain critical frequencies, causing physical ailments (numbness, swelling of hand tissues).
Avoid use of rachet screwdrivers over long periods.	Outward rotation of the forearm uses strong biceps, but inward rotation uses weaker muscles that quickly fatigue.
Apply forces and torques with the elbow at 85°. Move close to the work object.	Torquing a screwdriver with extended arm resulted in sore arms for 23 of 35 female workers.
Provide for secure grasp of tools and objects.	If glove fingers are too thick, pressures between the fingers may result in an insecure grasp. Also, flexing the wrist (forward) reduces ability to hold a rod securely. Keep wrist straight.

SOURCE: Huchingson, 1981.

Tools heavier than 0.4 kg are not recommended for precision operations without counterbalancing systems. For some jobs, heavy tools may be necessary to help absorb impact vibrations. These tools should be mounted on a truck to reduce the strain on the operator.

C. Information transfer

Information can be transferred from person to person, from equipment to person (displays), from product to person (inspection), or from person to equipment (controls). This section will focus on design guidelines

TABLE 8.36 Hand-Tool Design Guidelines

Characteristic	Recommendation	Comment
Handle length	13 cm (10 cm minimum)	Add 13 mm when tools are used with gloves
Handle diameter		For circular handles
Power grips	4 cm (3–5 cm)	
Precision grips	12 mm (8–16 mm)	
Handle span	6.5–9.0 cm (range of maximum grip strength)	For tools with two handles (e.g., pliers, scissors)
Insulation	Good heat and electrical insulation	Compressible rubber and plastic

SOURCE: Kodak, 1983.

for improving information transfer between people and in inspection tasks.

The transfer of information via written instructions, forms, questionnaires, codes, labels, and signs can be subject to error. In general, misinterpretation of information from signs or instructions will be less likely to occur for an experienced compared to an unexperienced person. However, visual information displays should be designed for the novice or casual visitor to a work area. Since emergency situations often require rapid, complex reactions rather than analytical troubleshooting, well-designed and well-written information is needed for even the most experienced worker. Some general human-factors guidelines for the design of instructions are as follows (Kodak, 1983):

1. The sequence of the instructions should follow the sequence of actions required.
2. Information should be located in specific places in order to avoid misuse or unsafe operations.
3. Instructions should be integrated into the equipment or the production work rather than set aside on a separate sheet.
4. Short sentences, flow diagrams, algorithms, lists, and tables are superior to prose.
5. The main topic of the instruction should appear at the beginning of the sentence.
6. The choice of format is situation-dependent.
7. All instructions should be tested on naive users before being finalized.

Table 8.38 summarizes factors that affect the written communication process. Key design criteria for word displays are legibility, size,

TABLE 8.37 Power-Tool Weights and Trigger Activation Forces

Tool type	Weight (kg)[a]	Trigger type	Grip	Average force to activate trigger (N)
$\frac{1}{4}$-in. electric drill	2.3	Index finger trigger	Pistol	17–22
$\frac{3}{8}$-in. electric drill	4.3	Index finger trigger	Pistol	30
$\frac{1}{2}$-in. electric drill	4.5	Index finger trigger	Pistol	52
7-in. sand grinder	7	Thumb bar	Straight	33–36
Air hammer	7	Thumb bar	Straight	10
Chopper hammer	8	Thumb bar	Straight	32
Air saw	3	Thumb bar	Straight	16
Air saw	3	Index finger trigger	Pistol	10
Air angle drill	3	Thumb bar	Straight	9

[a]Tools weighing more than 3 kg and requiring activation forces in excess of 10 N are difficult for many people to use.

SOURCE: Kodak, 1983.

TABLE 8.38 Factors Affecting the Written Communication Process

Design of message by sender	Factors affecting message transmission	Elements influencing receipt of the message
Comprehensibility	Environment	Discrimination
Purpose	Viewing distance	Visual abilities
User knowledge	Viewing angle	Interpretation
Brevity	Illumination	Language skills
Accuracy	Deterioration	Situation knowledge
Clarity	Competing displays	Recall
Legibility	Timing pressure	Time delay
Font style		Interference
Font size		
Color		
Readability		
Borders		
Layout		
Abbreviations		
Spacing		
Case		

SOURCE: Kodak, 1983.

location, and clarity of meaning (Woodson, 1981). Warning signs must be conspicuous, legible, simple, intelligible to clarify the hazard, visible, constructed of materials that resist wear, and standardized (e.g., OSHA regulations).

Message comprehension can be improved by following these guidelines (Woodson, 1981):

1. Use common words (e.g., brightness or adjust instead of intensity and calibrate).
2. Use whole words if space permits (e.g., range instead of RNG).
3. Use terms to indicate what is measured (e.g., RPM instead of speedometer).
4. Avoid words that can be interpreted as nouns or verbs.
5. Use capital letters for short instructions.
6. Keep message brief.
7. Avoid the number 1 in combination with letters.
8. Select words that imply the level of hazard. Do not leave it up to the observer to decide if a hazard exists.

Message legibility can be improved by the following recommendations:

1. Use size and color to help discriminate and differentiate. Tables 8.39 and 8.40 summarize size and color recommendations for labels.

TABLE 8.39 Letter or Number Height vs. Viewing Distance for Labels[a]

Viewing distance	Critical labels	Routine labels
0.7 m (28 in.)	2–5 mm (0.1–0.2 in.)	1–4 mm (0.05–0.2 in.)
0.9 m (3 ft)	3–7 mm (0.1–0.3 in.)	2–5 mm (0.06–0.2 in.)
1.8 m (6 ft)	7–13 mm (0.3–0.5 in.)	3–10 mm (0.1–0.4 in.)
6.1 m (20 ft)	22–43 mm (0.9–1.7 in.)	11–33 mm (0.4–1.3 in.)

[a]The distance from the operator to the display when it is read (column 1) will determine how high the letters or numbers should be for legibility (columns 2 and 3). Critical labels refer to key control or component identifiers and to position markers on such controls (column 2). Routine labels refer to overall instrument identifiers or any markings required only for initial familiarization (column 3).

SOURCE: Kodak, 1983.

2. Use specifications for display labels (see "Displays") for letter and number size with respect to width-to-height ratio and stroke width-to-height ratio.
3. Abstract symbols can be useful. Figure 8.10 shows examples of symbols developed by the engineering professions.
4. Keep fonts simple; avoid curlicues and flourishes.
5. Use borders to separate blocks of numbers or specific messages that are clustered with several labels.

TABLE 8.40 Legibility of Color Combinations in White Light[a]

Legibility	Color combination
Very good	Black characters on a white background Black on yellow
Good	Yellow on black White on black Dark blue on white Green on white
Fair	Red on white Red on yellow
Poor	Green on red Red on green Orange on black Orange on white
Very poor	Black on blue Yellow on white

[a]The ease with which written information can be distinguished from its background is indicated in the first column for different color combinations (column 2).

SOURCE: Kodak, 1983.

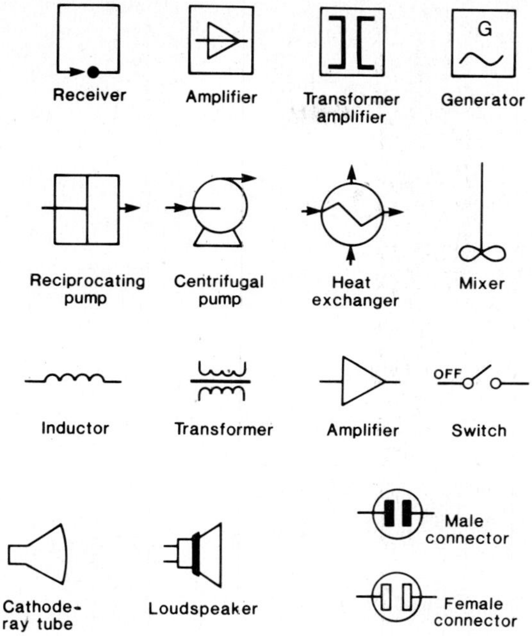

Figure 8.10 Standardized symbols developed by various engineering professions (*Woodson, 1981*).

6. Characters should be at least two times as bright or dark as background. Avoid glossy or reflective materials (Woodson, 1981).

In industry, the transfer of information from product to person usually involves inspection. The methods of inspection include direct visual scanning, manual measurement, or automatic measurement. For a large number of industrial manufacturing systems, visual scanning for defects is the primary inspection method. Individual, environmental, task, and organizational factors may influence the performance of an inspector. Table 8.41 summarizes many of the factors that can affect inspection performance. Inspection operations where defects are rare are difficult to perform (Kodak, 1983). This is consistent with human performance data that showed decreased performance for very frequent or infrequent stimuli (Figure 8.5). Defect rates below 1.5% can result in reduced detection performance. Table 8.42 summarizes the results of studies that illustrate the effect of task factors on inspection. In general, the more time a person has to make an inspection, the better is his or her detection performance. The following guidelines have been developed for the design of inspection tasks (Kodak, 1983):

1. Make the angle of viewing downward instead of straight ahead or upward.
2. Allow a moving product to move toward rather than away from the inspector. It does not make a significant difference if the product

TABLE 8.41 Factors That May Influence Inspection Performance[a]

Individual factors	Physical and environmental factors	Task factors	Organizational factors
Visual acuity	Lighting	Inspection time	Number of inspectors*
Static*	General*	Stationary*	Briefing/instructions
Dynamic	Surround luminance	Conveyor-paced*	Feedback
Peripheral	Lighting for color	Paced vs. unpaced	Feed forward*
Color vision*	Specialized*	Direction of movement	Training*
Eye movement scanning strategies*	Aids	Viewing area	Selection*
	Magnification*	Shape of viewing area	Standards*
Age*	Overlays*	Density of items*	Time on task*
Experience*	Viewing screen*	Spatial distribution of items	Rest pauses
Personality	Closed-circuit TV	Defect probability*	Shift*
Sex	Partitioning of display	Defect mix	Sleep deprivation
Intelligence	Automatic scanner	Defect conspicuity*	Social factors
Subjective probability of defect occurrence	Background noise	Product physical factors*	General*
	Music-while-you-work*	Complexity	Isolation of inspectors*
	Workplace design	Two- or three-dimensional	Working in pairs
		Specularity	Effects on sampling scheme*
		Hue	Motivation*
		Size	Incentives*
		Defect physical factors	Product price information
		Shape	Job rotation*
		Size	
		Specularity	
		Contrast	

[a]The ability of a person to perform an inspection task may be influenced by individual (column 1), environmental (column 2), task (column 3), or organizational (column 4) factors. Those factors that have been identified in industrial experiments are indicated by an asterisk (*); the other factors have been shown to influence inspection performance in laboratory tasks or military studies.

SOURCE: Kodak, 1983.

TABLE 8.42 Effects of Task Variables on Inspection Performance[a]

Task variable	Explanation of inspection task	Percent errors	Conditions	Percent detection
1. Complexity	*a.* Students identified defects in a series of geometric shapes on a 10 × 13 cm (4 × 5 in.) target. The shapes were 10 mm ($\frac{3}{8}$ in.) high on the 5 × 5 grid and 6 mm ($\frac{1}{4}$ in.) high on the 7 × 7 grid.	4.7	5 × 5 grid	—
		11.2	7 × 7 grid	—
	b. Process control inspectors on two inspection tasks searched for 50 or 200 defects on a product.	—	50 defects	81
		—	200 defects	42
2. Velocity	Students searched for defects on targets (geometric shapes) moving at 13 or 18 cm/sec (5 or 7 in./sec).	5.1	13 cm/sec (5 in./sec)	—
		10.8	18 cm/sec (7 in./sec)	—
3. Spacing	Students searched for defective shapes at 13 cm/sec (5 in./sec) with different spacings between the targets (25, 38, and 50 cm, or 10, 15, and 20 in.).	10.1	25 cm (10 in.) spacing	—
		4.5	38 cm (15 in.)	—
		2.4	50 cm (20 in.)	—
4. Direction	Inspectors looked for defects on a moving web. Their position relative to the web's direction of movement was varied.	No significant difference	Left to right vs. right to left	—
			Movement toward vs. away from inspector	Toward better than away from

[a]Four task variables (column 1) that affect inspection performance are presented in terms of errors (column 3) and detection rates (column 5) for some specific inspection tasks (columns 2 and 4). Error rates increased with increasing complexity (1*a*), or information density, and with higher speeds of product being inspected (2). The more defects being searched for, the lower was the percentage detected (1*b*). In a dynamic task, increased spacing (up to 50 cm, or 20 in. between defects) resulted in fewer errors (3). Detection of defects was better when the product moved toward the inspector (4).

SOURCE: Kodak, 1983.

moves from left to right or from right to left. At least 30 cm of viewing area should be available for every 18 m/min of product movement.

3. Evaluate the vision of inspectors who have to detect fine color differences or low-contrast defects.
4. Whenever possible, remove the process control inspector from direct machine pacing.
5. If the defect rate is low, provide environmental aids (e.g., special lighting) to assist the inspector in detecting the defect.
6. Permit the inspector to adjust the lighting to fit his or her size and visual requirements.
7. Develop a standardized training program for inspectors. This program should include procedures for continued training and quality performance checks.
8. Provide the inspector with an opportunity to break up the inspection periods with auxiliary events (e.g., paperwork). Continuous monitoring should not be required for more than 30 min.
9. Minimize environmental distractions at inspection stations.

IV. Design Examples

A. Bottle packing workplace (Kodak, 1983)

Figure 8.11 shows the existing workplace. Operators are lined up on each side of a conveyor belt from which they remove product, inspect it, put it in trays, and remove the defective items. The conveyor is 100 cm above the floor, 30 cm wide, with a 5-cm guardrail. The height of the conveyor cannot be adjusted. The bench top is 85 cm high, 5 cm thick, and 30 cm deep. Forward leg clearance is 35 cm. Forward reach to the bottles on the conveyor is 50 cm. The operators currently have the option of sitting or standing at the work station. The chair height is adjustable up to 63 cm. How can this workplace be modified to reduce complaints of sore arms and shoulders?

Solution: The workplace as shown on Figure 8.11 is not well suited for standing or seated operation. Recommended work-surface heights (Tables 8.17 and 8.19) for seated and standing workplaces are 81 cm and about 100 to 110 cm, respectively. The 85-cm workbench height is, therefore, too high for seated operations and too low for standing operations. As discussed under "Seated Reaches and Clearances," objects to be frequently grasped should be located between 15 and 36 cm from the front of the work surface. Leaning forward, which is required to reach the conveyor, should be restricted to no more than a few times

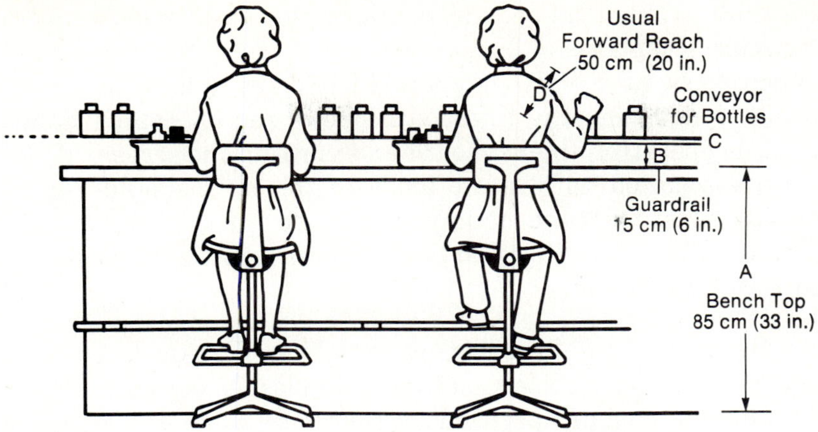

Figure 8.11 Layout of a bottle-packing workplace (*Kodak, 1983*).

per hour. Seated workspaces should also be restricted to situations where all items handled do not require work greater than 15 cm above the work surface. With the conveyor 15 cm above the benchtop and the guardrail 5 cm above that, bottles must routinely be manipulated above the 15-cm limit. Also note that minimum recommended thigh clearance (distance from seat height to bottom of workbench) is 20 cm, making the highest recommended seat adjustment 20 cm below the bottom of the workbench (60 cm above the floor). It is therefore not possible to raise the chairs to partially compensate for the extreme forward reaching required. Even as a standing workplace, 50 cm is the maximum forward reach with hand height 112 cm above the floor. Although this workplace is not well suited for a seated operation, Table 8.16 does suggest that seated workspaces are preferred for repetitive operations of extended durations.

The best way to improve this workplace is to dimension it to a seated operation and bring the conveyor and work surface to the same height. Because the conveyor cannot be lowered, the best solution is to raise the operator. Since the limited thigh clearance makes raising the chair impractical, a full-floor platform should be used to raise the workbench and chair. A 20-cm platform with an 80-cm table on the platform would work well. Moving the operator close to the conveyor would also help reduce muscle fatigue. This can be accomplished by using a narrower table or cutting out a semicircular opening in the workbench at each station to reduce the reach requirements. A 15-cm cutout would bring the operator within the 36-cm forward reach limit for repetitive tasks. A footrest on the chair or workbench is also highly recommended to reduce discomfort of the legs.

B. Video display units

The use of video display units (VDUs) has increased dramatically in the past decade. There is, therefore, a need to optimize the design of these workplaces. Considerations in the design of VDU displays were discussed under "Displays."

Continuous work at a VDU is characterized by constrained body postures that result from the fixed positions of the operator's hands and head while operating the keyboard and viewing the screen. These postures are associated with static loads in the neck-shoulder-arm-hand area. Long-lasting static loadings can cause muscle fatigue, stiffness, and pain in this area. How can the VDU work station be designed to minimize these problems?

Solution: Because of the long-lasting muscle loads associated with VDU operation, the workplace design should closely match operator size. An adjustable design approach is, therefore, very desirable for fitting the workplace to the operator. Adjustable components should include chair height, keyboard home-row height, and visual display height and angle. Chair, keyboard, and display height adjustments are needed to fit the workplace to the operator. Display angle adjustments are needed to both fit the operation to the worker by maintaining the screen perpendicular to the line of sight and controlling reflections and glare on the screen. The keyboard should also be movable to allow for each operator's preferred keying position. Numerous recommendations for VDU workplace dimensions are available. Operator-preferred workplace dimensions from this study are summarized in Table 8.43. Table 8.44 shows additional design information. Note that these recommendations and standards are consistent with information in Tables 8.4 and 8.17. A field study of VDT operators at adjustable work stations showed that preferred body postures (and therefore work station dimensions) can differ from recommended values (Grandjean et al., 1983).

TABLE 8.43 Operator-Preferred Workplace Dimensions

Dimension	Mean[a] (cm)	Range (cm)	Comment
Keyboard height	80	71–87	Height of low-row keys
Screen height	103	92–116	Center of screen above floor
Seat height	48	43–57	63% between 46 and 48 cm
Screen height above desk	32	21–40	
Eye level above floor	115	107–127	
Viewing angle	−9°	+2° to −26°	Eye to screen center (below horizontal)
Viewing distance	76	61–93	
Screen inclination	94°	88–103°	Angle from horizontal

[a]All data in centimeters except viewing angle and screen inclination, which are in degrees from horizontal.

SOURCE: Grandjean et al., 1983.

TABLE 8.44 Seated Visual Display Unit Workplace Dimensions

Dimension	(cm)[a]	Source[b]	Comment
Table height	72 fixed	TUB, DIN	
	65–72 adj	TUB, DIN	
	74–79 adj	MIL-STD	
width	120 min	DIN	
	160	DIN	Recommended if other tasks are performed at work station
	76 min	MIL-STD	
depth	40 min	MIL-STD	
Keyboard base height	64	IBM	Workplace clearance for legs
	69	TUB, DIN	
	65 min	DIN	
	64 min	MIL-STD	
width	51 min	MIL-STD	
depth	46 min	MIL-STD	
Keyboard height	66–71 adj	IBM	Height of home-row keys
	75 max	TUB	
	3 max	TUB	Height above table top
Screen height	94–110 adj	IBM	Height to center of screen
	37–52 adj	DIN	Height of upper edge of screen above work surface
	—[c]	TUB	Top edge of screen below eye level
Viewing distance	50	TUB	
	50–70 adj	DIN	
	25 min	MIL-STD	
	41	MIL-STD	Preferred viewing distance

[a]Fixed = fixed workplace dimension; adj = adjustible range; min = minimum requirement; max = maximum requirement.

[b]DIN standard 66234, "Characteristic values for the adaptation of work stations with fluorescent screens to humans."

TUB: Cakir, A., Reuter, H. J., Von Schmude, L., and Armbruster, A., Technical University of Berlin, "Standard for visual display units," April 1978.

MIL-STD-1472 B and C, U.S. Government Printing Office, Washington D.C., September 1981.

IBM 3270 Information Display Systems: A Human Factors Study of Work Station Design, IBM Report No. GA27-2759-0.

[c]Line of sight to center of screen at 20° below horizontal.

SOURCE: Kodak, 1983; Banks et al., 1982.

C. Coin inspection work station (Fox, 1977)

The quality control section of a mint is responsible for identifying and removing defective coins before distribution. An evaluation was carried out to determine the efficiency of an existing off-line batch inspection process. At the mint, single bags of coins were emptied into a vibrating hopper, which poured the coins onto a moving belt across an inspection table. One inspector examined the top face of the coin as it moved along the conveyor. The coins were then flipped mechanically, and a second inspector examined the back side of the coin. Coins were placed randomly on the conveyor. Each inspection cycle lasted 24 min, of which about 14 min were spent examining the coins. Although the system was considered to be 100% effective, there was concern that the effectiveness of the system could not be properly maintained. In fact, the evaluation found that the average inspection rate was 815 coins/min; one in 600 coins were defective; only 54.5% of the defective coins were being detected; and that less than 1% of the coins classed as defective were in fact not defective. Because of the high cost of equipment to automate the process, the mint decided to redesign the manual inspection work station to improve efficiency and throughput.

Solution: An examination of the human-machine interface identified the following factors for evaluation: display, layout, and design of the job with respect to such factors as visual acuity, perceptual organization, scanning behavior, pacing behavior, defect frequency, and arousal and vigilance; illumination; and workplace layout and anthropometric considerations. The performance evaluation had revealed that visual acuity has an important effect on inspection performance. Studies of perceptual organization suggest that good displays should have simple figure-to-ground relationships. This translates, for example, into the design recommendation that clusters of dials be oriented so that all pointers are aligned during normal conditions (see "Displays"). The random placing of coins on the conveyor resulted in a complex figure-to-ground relationship, making the inspection task more difficult. Studies of scanning behavior suggest that open-ended displays result in uneven viewing of the whole display. There was no defined viewing area at the inspection stations, as the operators were able to see along the entire conveyor. The guidelines for inspection tasks (under "Information Transfer") also suggest that operator-paced tasks are preferred to machine-paced tasks. While the inspectors were controlling the rate at which bags were inspected, the conveyor paced the actual inspection task. Information on human performance (see "Performance") also suggests that performance is degraded for infrequent events. This leads to the recommendation that special aids are needed for inspection tasks for low defect rates. The low defect rate (1 in 600) provided a difficult task for the inspectors. Finally, arousal and vigilance research suggests that continuous monitoring of displays should be limited to 30 min (Kodak, 1983). The actual inspection of each batch of coins was found to last only 14 min.

Given these considerations, the job was redefined so that the coins were presented for inspection in a regular-patterned, bounded display

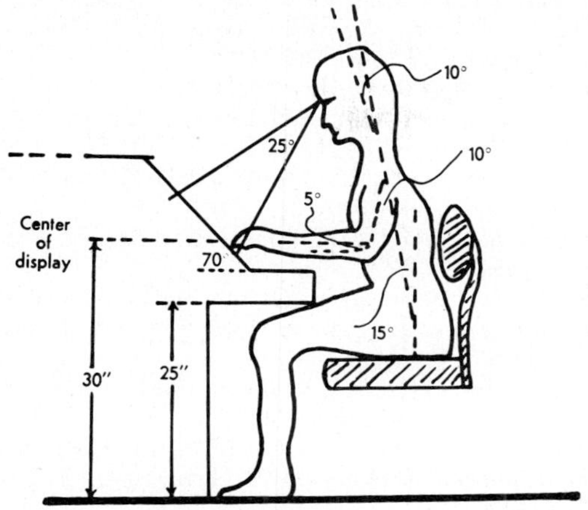

Figure 8.12 Layout of ergonomically designed coin inspection work station (*Fox, 1977*).

12 in. long by 8 in. high. This size of display allowed for easy access to remove defective coins. The display was paced to move coins past the inspector at 8 ft/min, and the maximum inspection period between breaks was fixed at 30 min. This is well within design recommendations ("Information Transfer"). The intensity of illumination was set at 600 lux, a level consistent with standard design recommendations (e.g., ANSI, 1973; Table 11.12). Figure 8.12 shows the redesigned work station. This work station was fitted to the 50th percentile adult male such that in the seated position the thigh was parallel to horizontal, the angle from the vertical at the hip was less than 15°, the maximum angle of the head and neck was 10°, and the maximum downward angle of the eyes was 25°. This resulted in a work station in good agreement with the design recommendations of "Seated Reaches and Clearances" and Table 8.23.

A validation study was conducted to evaluate the redesigned coin inspection task. The redesigned operation was found to have an improved defect recovery rate of 74%, up from the original 54%, with increased throughput of 972 coins/min, up from 815 coins/min.

REFERENCES

ANSI (1973). *American National Standard Practice for Industrial Lighting, A11.1–1973.* Illuminating Engineering Society, New York.

ANSI (1983). *ANSI Z94.1.* American National Standards Institute, New York.

Banks, W. W., Gertman, D. I., and Petersen, R. J. (1982). *Human Engineering Design Considerations for Cathode Ray Tube-Generated Displays.* U.S. Nuclear Regulatory Commission, NUREG/CR-2496, EGG-2161, Washington, D.C.

Chapanis, A., and Kinkade, R. G. (1972). Design of controls. In: Van Cott, H. P., and Kinkade, R. G., eds., *Human Engineering Guide to Equipment Design.* U.S. Government Printing Office, Washington, D.C., pp. 345–380.

Damon, A., Stoudt, H. W., and McFarland, R. A. (1966). *The Human Body in Equipment Design.* Harvard University Press, Cambridge, Mass.

Deatherage, B. H. (1972). Auditory and other forms of information presentation. In: Van Cott, H. P., and Kinkade, R. G., eds., *Human Engineering Guide to Equipment Design.* U.S. Government Printing Office, Washington, D.C., pp. 123–160.

Fox, J. G. (1977). Quality control of coins. In: Maule, H. G., and Weiner, J. S., eds., *Human Factors in Work, Design and Production.* Taylor and Francis, London, pp. 101–133.

Grandjean, E. (1980). *Fitting the Task to the Man: An Ergonomic Approach.* Taylor and Francis, London.

Grandjean, E., Hunting, W., and Pidermann, M. (1983). VDT workstation design: Preferred settings and their effects. *Human Factors, 25*(2), 161–175.

Grether, W. F., and Baker, C. A. (1972). Visual presentation of information. In: Van Cott, H. P., and Kinkade, R. G., eds., *Human Engineering Guide to Equipment Design.* U.S. Government Printing Office, Washington, D.C., pp. 41–122.

Hancock, P. A., and Pierce, J. D. (1985). Combined effects of heat and noise on human performance: A review. *Amer. Ind. Hyg. Assoc. J., 46,* 555–566.

Hertzberg, H. T. E. (1972). Engineering anthropology. In: Van Cott, H. P., and Kinkade, R. G., eds., *Human Engineering Guide to Equipment Design.* U.S. Government Printing Office, Washington, D.C., pp. 467–584.

Huchingson, R. D. (1981). *New Horizons for Human Factors in Design.* McGraw-Hill, New York.

Kodak (1983). The Human Factors Section, Eastman Kodak Company, *Ergonomic Design for People at Work: Volume 1.* Lifetime Learning Publications, Belmont, Calif.

Leonardos, G., Kendall, D., and Barnad, N. (1969). Odor threshold determination of 53 odorant chemicals. *J. Air Pollut. Control Assoc., 19,* 91–95.

McCormick, E. J., and Sanders, M. S. (1982). *Human Factors in Engineering and Design,* 5th ed. McGraw-Hill, New York.

Mackworth, N. H., and Mackworth, J. F. (1958). The visual search for successive decisions. *Brit. J. Psych., 49,* 210–221.

Tichauer, E. R. (1978). *The Biomechanical Basis of Ergonomics.* Wiley, New York.

Tichauer, E. R. (1976). Biomechanics sustains occupational safety and health. *Industrial Engin., 8,* 46–56, February.

Van Cott, H. P., and Kinkade, R. G., eds. (1972). *Human Engineering Guide to Equipment Design.* U.S. Government Printing Office, Washington, D.C.

Van Cott, H. P., and Warrick, M. J. (1972). Man as a system component. In: Van Cott, H. P., and Kinkade, R. G., eds., *Human Engineering Guide to Equipment Design.* U.S. Government Printing Office, Washington, D.C., pp. 17–40.

Viteles, M. S., and Smith, K. R. (1946). An experimental investigation of the effect of change in atmospheric conditions and noise on human performance. *Heating, Piping and Air Cond., 18,* 107–112.

Woodson, W. E. (1981). *Human Factors Design Handbook.* McGraw-Hill, New York.

Chapter

9

Guarding

F. G. Miller

R. A. Wadden

In Chapter 2 there is a summary of the types and frequency of occupational injuries. It is evident that many of these were caused by inadvertent or unwise contact with moving machine parts. Because of the great diversity of machine designs and functions, appropriate guarding to protect workers from such hazards may also have numerous forms. However, certain principles are basic to any effective safeguarding design. The object of this chapter is to discuss these principles, and to illustrate their generality in a variety of example applications. More detailed design requirements for specific machine types may be found in OSHA (1983), NSC (1980), and the publications of the American National Standards Institute.

Mechanical hazards usually occur in three general areas: the point of operation; the power transmission train; and at other parts that move when the machine is operating (OSHA, 1980). Chapter 4, Section IV.A, points out the OSHA requirements limiting worker presence in these areas. The point of operation is that area where material is positioned for processing or change. It is an obvious hazard location, since the operator often needs to be close to the point of tool-workpiece engagement in order to set up, observe and direct the process. The power transmission train supplies energy to the machine to carry out work. Components to be safeguarded include gears, cams, shafts, pulleys, belts, connecting rods, flywheels, couplings, chains, and cranks. Examples of other moving parts are feed and ejection devices.

Machine motions may be rotating, transverse, or reciprocating and may result in cutting, punching, shearing, or bending motions. Rotating shafts or flywheels, even when smooth, can grip clothing or skin surfaces and cause injury. The danger is increased when bolts, nicks, or projecting set screws or keys are exposed on rotating parts. Nip points are common rotating hazards; examples are shown in Figure 9.1. Reciprocating motion may cause a problem if a worker is caught between moving and stationary parts during back-and-forth or up-and-down movement. Similarly, transverse motion (in a straight, continuous line, as with a conveyor belt) may catch the worker by the moving part in a punch or shear point.

Cutting of wood, metal, and other materials can be hazardous to limbs (particularly fingers) and, because of flying chips or sparks, to

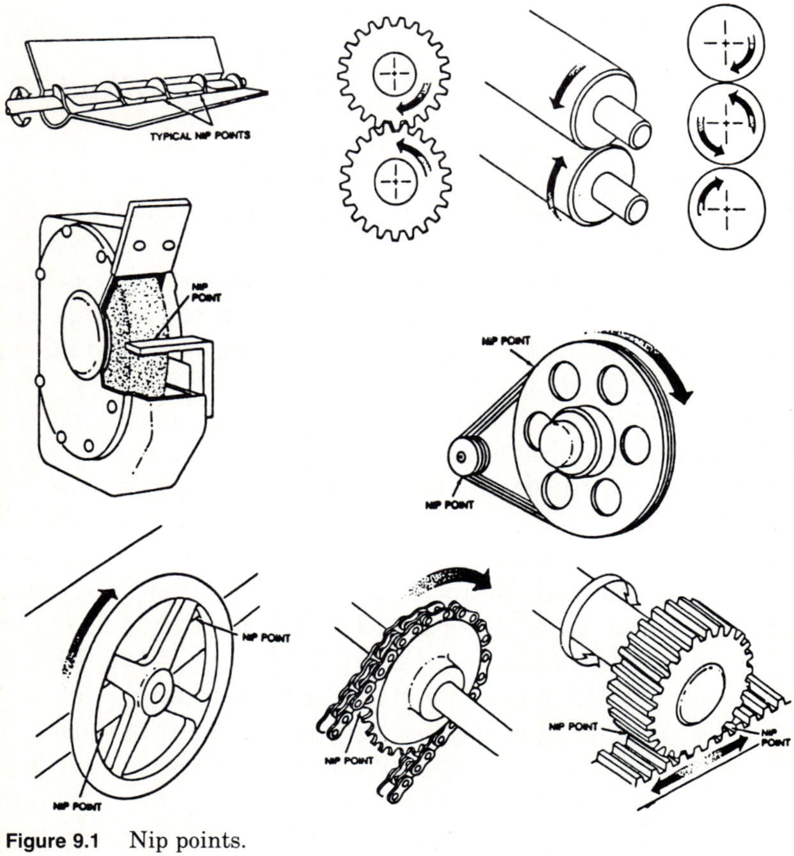

Figure 9.1 Nip points.

other parts of the body as well. Various types of saws, boring and drilling machines, and lathes and milling machines require cutting actions. Punching metal or other materials to blank, draw, or stamp feed stock, such as with power presses, is typically most hazardous at the point of operation, where material is inserted, held, and withdrawn. The point of material entry and removal is also a problem zone for mechanical or hydraulic shears, and for the bending actions of press brakes and tube benders.

Safeguarding workers from mechanical hazards requires, at a minimum, the following design responses:

1. The possibility of contact between the worker's body and dangerous moving parts and/or flying objects must be eliminated.
2. The safeguard should not be easily removed or tampered with, should be firmly secured to the machine, and should be made of a durable material that will withstand the conditions of normal use.
3. No new hazard, such as a shear point or jagged edge, should be created by the safeguard.
4. No interference should be produced that will impede the worker from performing the job quickly and comfortably. In particular, ease of feeding stock to the machine and good visibility at the point of operation must be maintained.
5. If possible, lubrication, and other regular maintenance that does not require shutting down the machine, should be possible without removing the safeguard.

In addition, the effectiveness of safeguards must always be evaluated in the field. Worker input and acceptance of retroactive and new-machine design solutions may well be the most important aspect of protection from mechanical hazards.

Since many machines used in manufacturing processes can suit several purposes and/or operate on different feed stocks, their safeguarding requires consideration of the whole production system. Each component of the system, from tooling to feeding to removal of the finished part and scrap, needs to be evaluated as well as the power component (machine). A safe combination of components for one production system may not constitute a safe combination for another (NSC, 1980). In addition, a hazard point may appear to be guarded when the machine is at rest; but when it is in operation, moving parts may be exposed or interact with other components to produce an unsafe situation. Besides mechanical motion, consideration should also be given to proper electrical grounding; appropriate maintenance and management of high-pressure systems; minimization of noise production; and the anchoring of machinery designed for a fixed location to prevent walking or other movement. Operation under nonproduction conditions, such as during maintenance or set-up of press dies, should also be a design concern. While there is no substitute for clear and careful backout procedures to prevent inadvertent operation while workers are in close proximity to hazard points, cumbersome safeguard design will encourage circumvention and exacerbate the danger. Appropriate safe operating procedures are described in detail in OSHA (1983), OSHA (1980), and NSC (1980). In general, we will classify safeguards as guard barriers that prevent access to dangerous areas; safety devices to remove or prevent the operator from being in a hazardous zone; and feed and ejection methods that make it unnecessary for the operator to be in a dangerous setting.

I. Guard Barriers

Various types of guard barriers and some of their advantages and disadvantages are given in Table 9.1. A *fixed guard* is a permanent part of the machine. It does not depend on moving parts to fulfill its intended function. Figure 9.2 shows a fixed guard on a power press that completely encloses the point of operation. Stock is fed through one side into the die area, and scrap exits out the other side. A fixed guard on an egg carton folding machine is shown in Figure 9.3. The guard protects the operator from the folding mechanism and is not removed except to do maintenance. Blade and wheel guards on a bandsaw are illustrated in Figure 9.4. The wheel guards protect from broken blades and teeth, while the blade covering shields the point of operation (ANSI, 1974; OSHA, 1983). The guards need to be securely fastened while the saw is in use and normally will be opened or removed only when the blade is changed or for maintenance. Fixed guards on a veneer clipper and a power press brake are shown in Figures 9.5 and 9.6.

The openings for fixed guards must often accommodate some stock manipulation at the point of operation. Appropriate dimensions can be determined based on typical men's and women's hand sizes. An appropriate guide diagram, based on anthropometric measurements, is shown in Figure 9.7. As indicated, no openings over $\frac{1}{4}$ in. (6 mm) are considered safe in the first $1\frac{1}{2}$ in. (38 mm) from the danger line. Most men and women have fingertips that will not travel any considerable distance through a $\frac{3}{8}$-in. opening (NSC, 1980).

Example 9.1 Stock is fed along a table to the mill rolls of Figure 9.8. The distance of a $\frac{3}{8}$-in. nip zone from the near contact point between the rolls is 4 in., and the guard must allow entrance of $1\frac{3}{4}$-in. stock. Specify the distance, X, from the near contact point to the guard location.

(*Solution continues on p. 454.*)

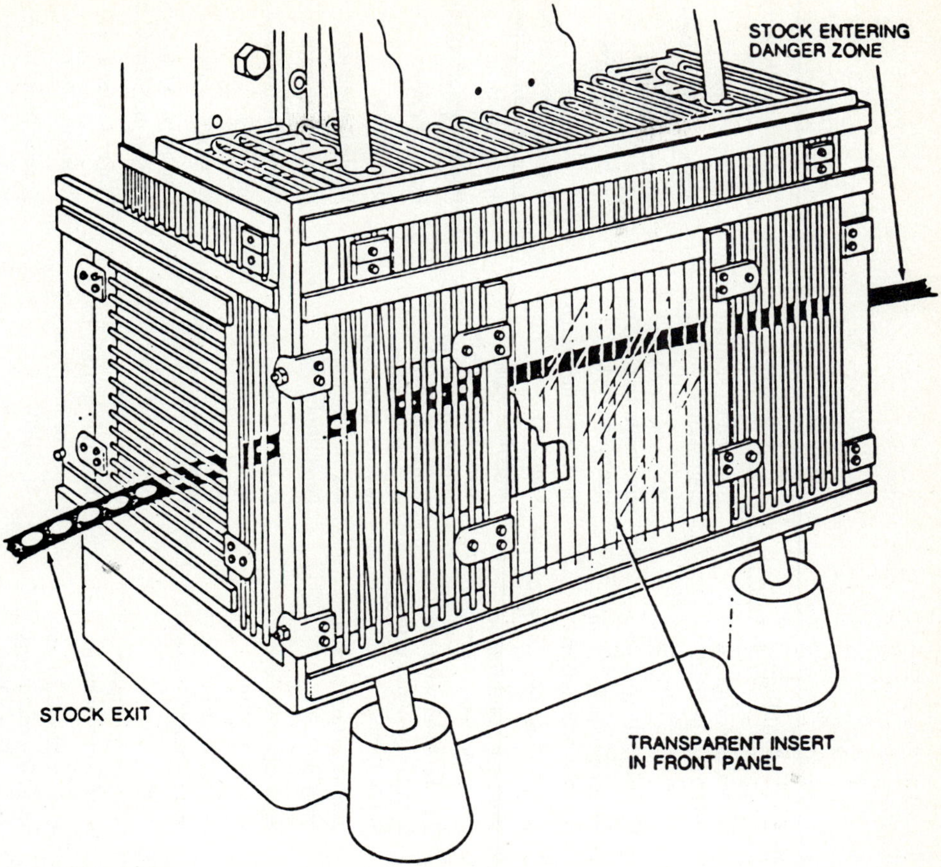

Figure 9.2 Fixed guard on power press (*OSHA, 1980*).

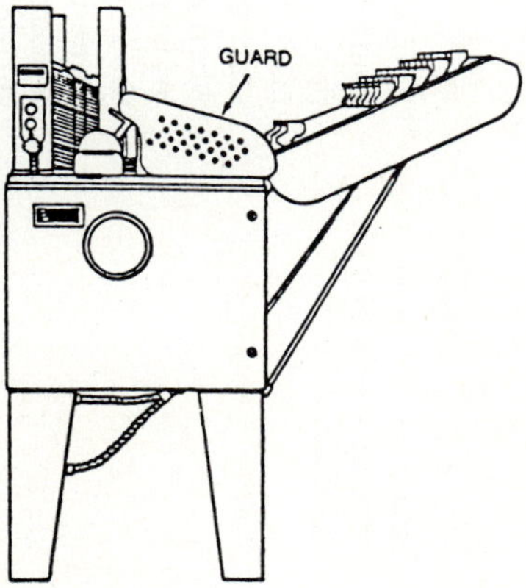

Figure 9.3 Fixed guard on egg carton folding machine (*OSHA, 1980*).

TABLE 9.1 Types of Guard Barriers

Type	Action	Advantages	Limitations	Typical applications
Complete, simple fixed enclosure	Barrier or enclosure that admits the stock but that will not admit hands into danger zone because of feed opening size, remote location, or unusual shape	Provides complete enclosure if kept in place Both hands free Generally permits increased production Easy to install Ideal for blanking on power presses Can be used with automatic or semiautomatic feeds In-plant construction often possible	Limited to specific operations May require special tools to remove jammed stock May interfere with visibility Machine adjustment and repair often require its removal, thereby requiring other means of protecting maintenance personnel	Bread slicers Embossing presses Meat grinders Metal squaring sheers Nip points of in-running rubber, paper, or textile rolls Paper corner cutters Power presses Bandsaws Veneer clippers Power transmission
Enclosure with electrical or mechanical interlock	Enclosure or barrier shuts off or disengages power (and may brake machine) and prevents starting of machine when guard is open; prevents opening of the guard while machine is under power or coasting; (interlocks should not prevent	Does not interfere with production Hands are free; operation of guard is automatic Provides complete and positive enclosure Allows access to machine for removing jams without time-	Requires careful adjustment and maintenance Operator may be able to make guard inoperative Does not protect in event of mechanical repeat	Dough brakes and mixers Foundry tumblers Laundry extractors, driers, and tumblers Power presses Tanning drums Textile pickers, cards

	manual operation or "inching" by remote control)	consuming removal of fixed guards		Corn cutters
Adjustable enclosures	Provides a barrier that may be adjusted to facilitate a variety of production operations	Makes "hard-to-guard" machines safer Generally does not interfere with production Easy to install Admits varying sizes of stock	Hands may enter danger zone; enclosure not complete at all times Danger of operator not using guard Often requires frequent adjustment and careful maintenance May interfere with visibility	Bandsaws Circular saws Cloth cutters Dough brakes Ice crushers Jointers Leather strippers Rock crushers Wood shapers Power presses
Self-adjusting	Provides a barrier that moves according to the size of the stock entering the danger area	Off-the-shelf guards are often available commercially	Does not provide maximum protection May interfere with visibility May require frequent maintenance and adjustment	Circular saws Jointers Table saws

SOURCES: NSC, 1980; OSHA, 1980.

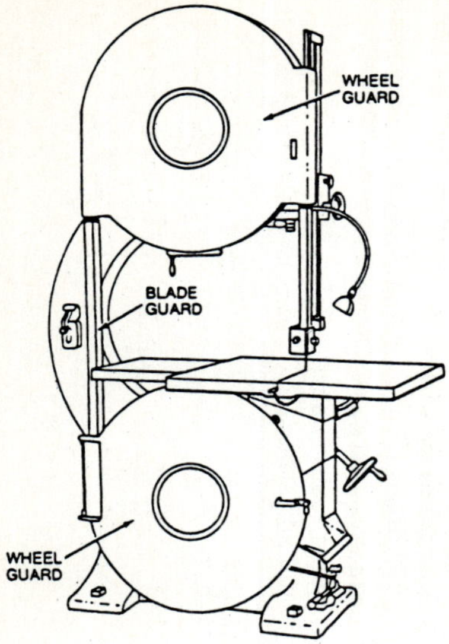

Figure 9.4 Fixed guards on a bandsaw (*OSHA, 1980*).

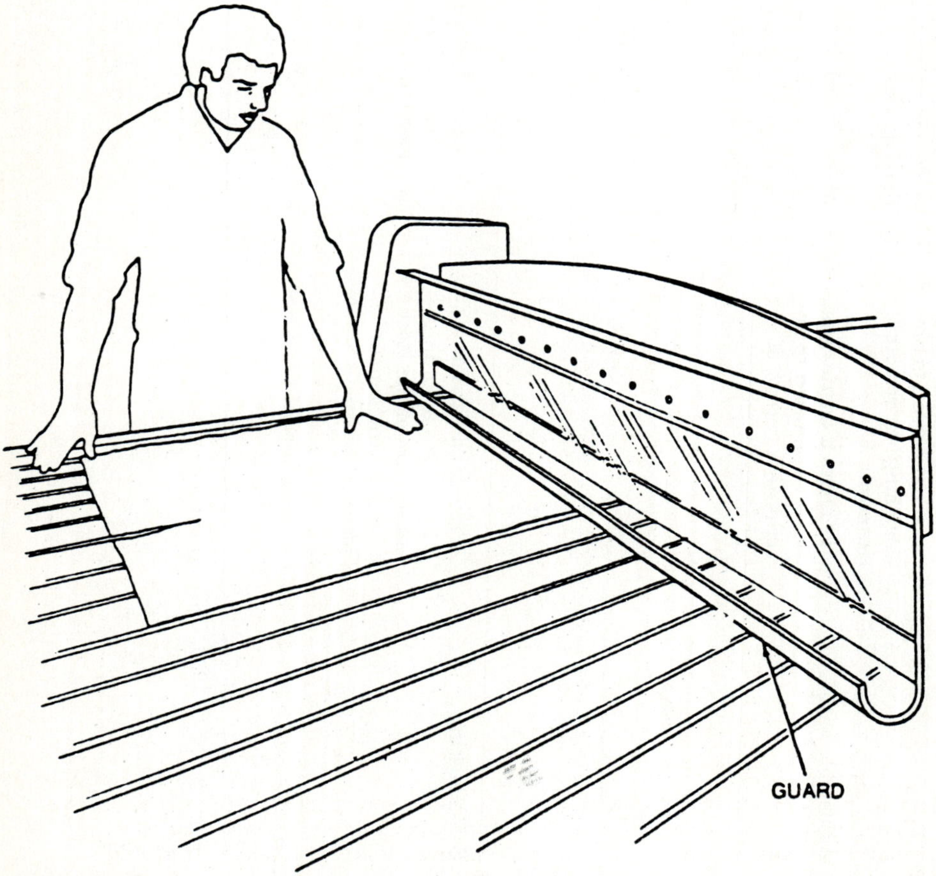

Figure 9.5 Fixed guard on veneer clipper (*OSHA, 1980*).

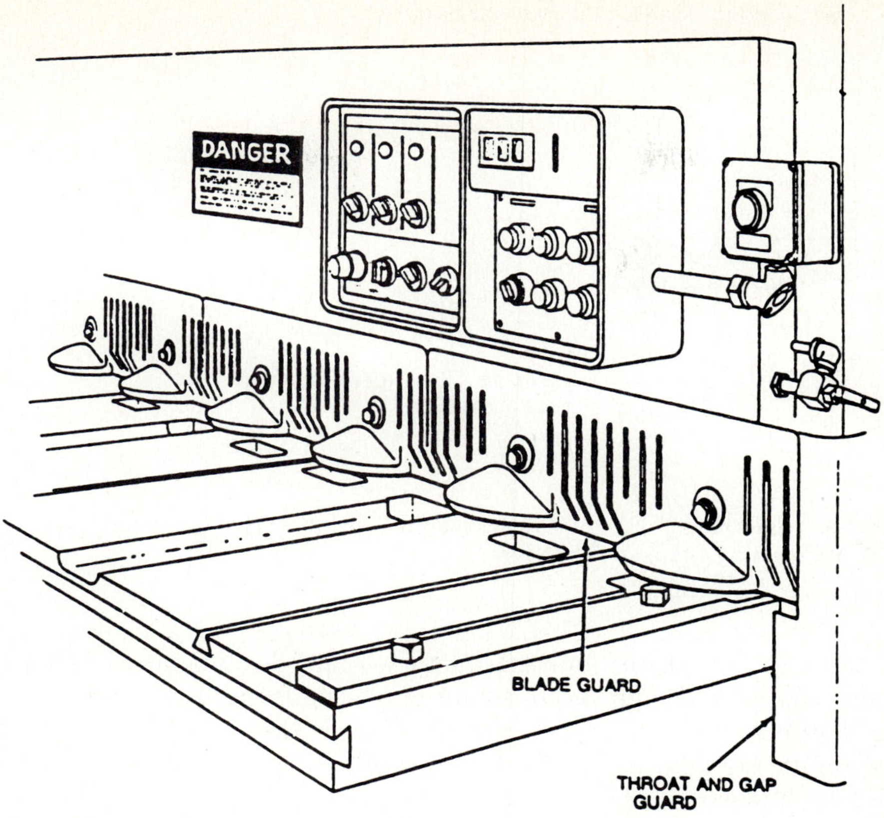

Figure 9.6 Fixed guards on a power press brake (*OSHA, 1980*).

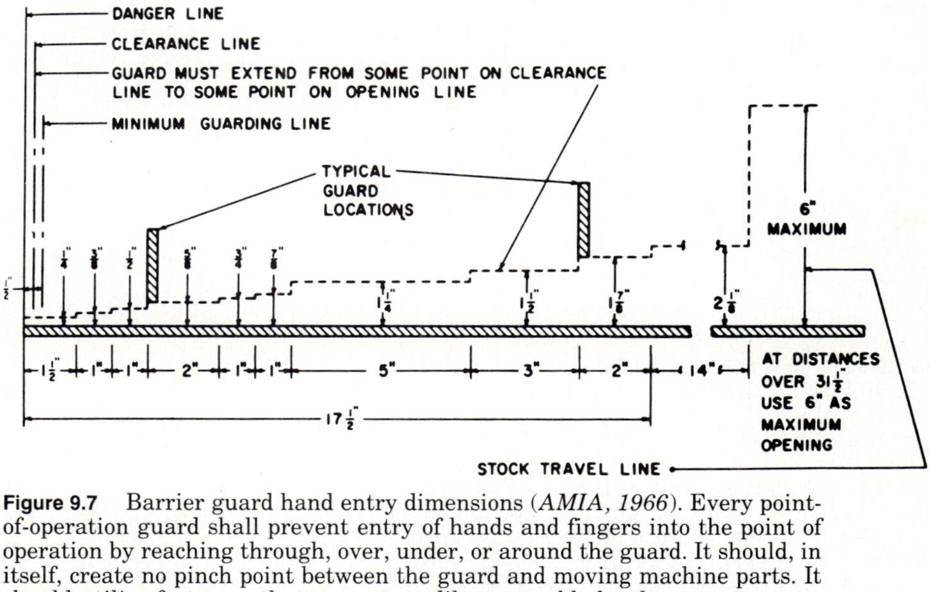

Figure 9.7 Barrier guard hand entry dimensions (*AMIA, 1966*). Every point-of-operation guard shall prevent entry of hands and fingers into the point of operation by reaching through, over, under, or around the guard. It should, in itself, create no pinch point between the guard and moving machine parts. It should utilize fasteners that are not readily removable by the operator so as to minimize the possibility of misuse or removal of essential parts. It should facilitate its inspection. And it should offer maximum visibility of the point of operation consistent with other requirements.

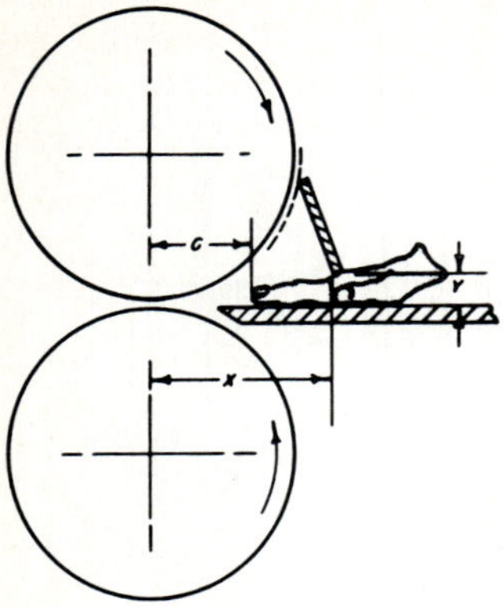

Figure 9.8 Guard design for in-running rolls.

solution Referring to Figure 9.7, the required distance from the danger line for a $1\frac{7}{8}$-in. opening is $15\frac{1}{2}$ in. Therefore:

$$X = 15\tfrac{1}{2} + 4 = 19\tfrac{1}{2} \text{ in.}$$

Figure 9.9 illustrates four satisfactory guard designs, any of which would fulfill the design requirement of a $\frac{5}{8}$-in. opening for feed stock. The distance from the near contact point of the rolls to the guard opening at *a* is determined from Figure 9.7 to be $3\frac{1}{2}$ in. This distance should be measured along the centerline and is equal to $X - C$. Note that these four guard designs also may act as nip points at the feed stock openings. An additional consideration, integral to all guard design, is that installations be checked in the field to assure that safeguarding is actually effective.

A major problem with the design of enclosure guards is how to conveniently incorporate access for normal maintenance. If lubrication is

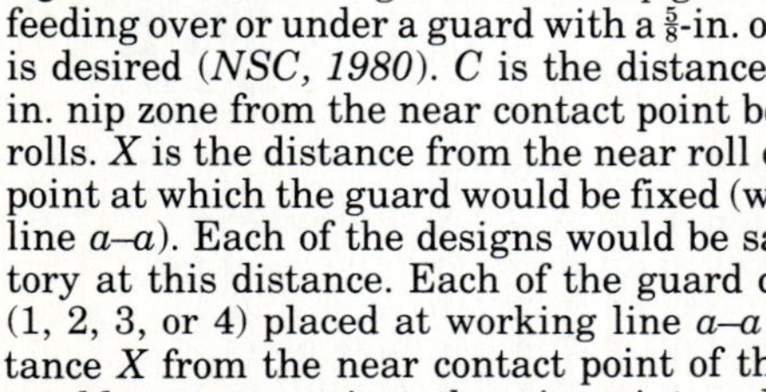

Figure 9.9 Four designs for a roll nip guard where feeding over or under a guard with a $\frac{5}{8}$-in. opening is desired (*NSC, 1980*). *C* is the distance of a $\frac{3}{8}$-in. nip zone from the near contact point between rolls. *X* is the distance from the near roll contact point at which the guard would be fixed (working line *a–a*). Each of the designs would be satisfactory at this distance. Each of the guard designs (1, 2, 3, or 4) placed at working line *a–a* at distance *X* from the near contact point of the rolls would protect against the nip point and allow $\frac{5}{8}$-in. stock to be fed to the rolls.

a regular requirement and this cannot be done without removing the guard, it is likely at some point that the barrier will not be replaced. One solution is to automate lubrication or bring appropriate fittings outside the guard so that maintenance can be carried out while the machine is operating.

Another response is to use *interlocked guards* that, when removed, automatically shut off or disengage power and/or tripping mechanisms. The machine cannot cycle or be restarted until the guard is back in position. Electrical, mechanical, hydraulic, or pneumatic power, or combinations of these, may be used to impose the interlock. Such guards should not automatically restart the machine when they are put back in place.

All machine functions are not necessarily overridden by interlocking guard action. For instance, "inching" is an intermittent motion of the slide or ram of machines such as hydraulic or mechanical power presses or power press brakes (ANSI, 1973a, 1982a, 1982b). Its purpose is to aid in proper positioning of dies and other components. Operation of the inching control mechanism allows the ram to move only a small portion of one stroke, and the procedure is not intended for production purposes. Ordinarily interlocked guards do not prevent inching by remote control.

Figure 9.10 illustrates an interlocked guard on a corn cutter, a device that strips corn from the cob with fast-turning cutter blades. If the guard panel protecting the operator from the cutting zone is opened or removed while the machine is running, the power disengages and a brake stops the blades before hands can reach into the hazard area. The automatic bread bagging machine of Figure 9.11 will not operate

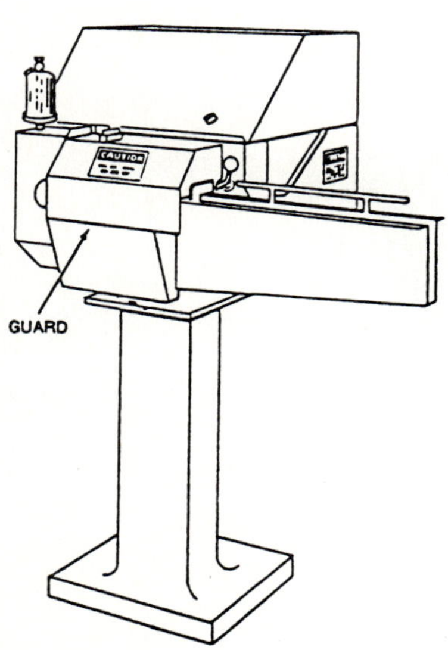

Figure 9.10 Interlocked guard on corn cutter (*OSHA, 1980*).

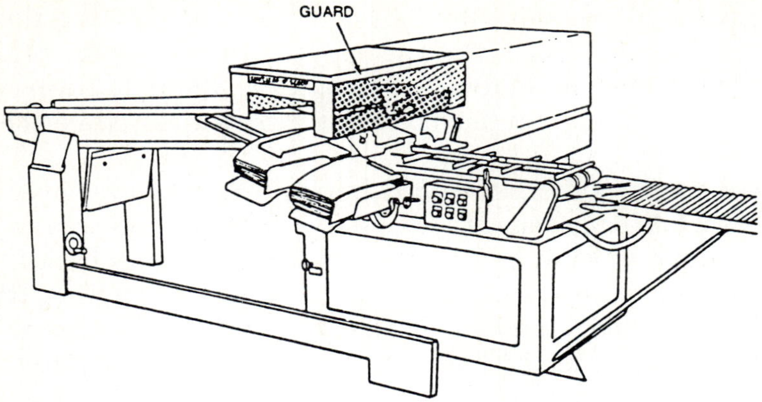

Figure 9.11 Interlocked guard on automatic bread bagging machine (*OSHA, 1980*).

when the guard is removed. Otherwise the operator would be exposed to pinch points in the bagging mechanism. The beater mechanism of a textile picker machine is protected by a guard that cannot be raised while the machine is running (Figure 9.12); nor can the machine be restarted while the guard is in the raised position. In Figure 9.13 an interlocked guard covers the rotating cylinder of the dividing head of a hamburger and hotdog roll make-up machine.

Adjustable guards allow flexibility in accommodating different sets of tools or various stock sizes. Their major disadvantage is that their

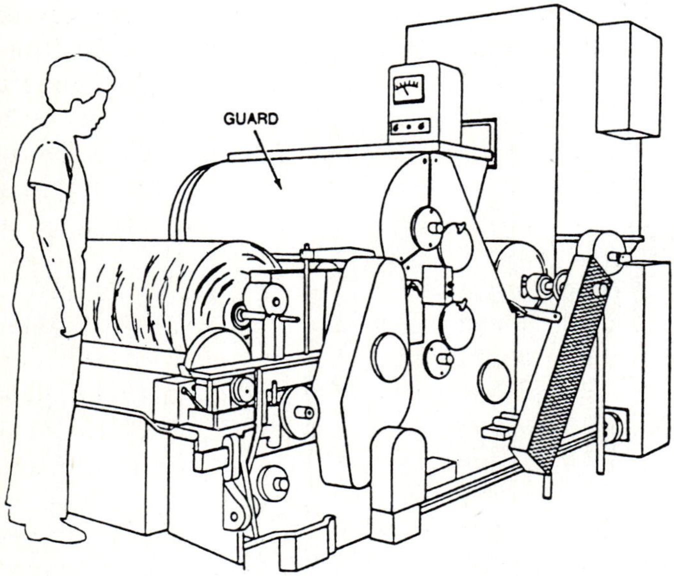

Figure 9.12 Interlocked guard on picker machine (*OSHA, 1980*).

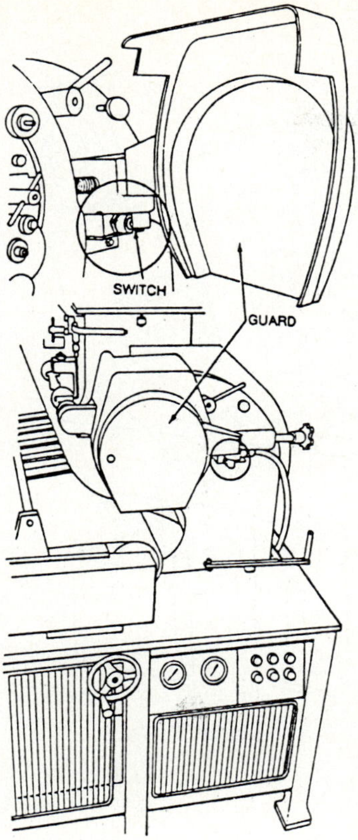

Figure 9.13 Interlocked guard on roll make-up machine (*OSHA, 1980*).

protective purpose may be more easily circumvented, purposely or through neglect, than is possible for fixed guards. Figure 9.14 illustrates a bandsaw guard that protects the operator from the unused portion of the blade. The guard can be adjusted to the size of the stock. The bars on the power press guard of Figure 9.15 are adjustable to accommodate the size and shape of stock. Guarding of the shaper blade (Figure 9.16) and vertical bandsaw blade (Figure 9.17) also may be adjusted to accept various workpiece sizes.

The openings of *self-adjusting guards* are determined by the movement of stock. As the stock is moved into the danger zone, the barrier is pushed away only enough to allow admission of the material. Upon removal of the stock the barrier returns to its rest position. The table saw in Figure 9.18 has two transparent guards that ride up over the workpiece as it encounters different positions of contact with the blade.

While all of the guards that have been described are appropriate for point-of-operation guarding, only fixed guards should be used for power transmission components. Figure 9.19 shows a typical pulley guard with an inspection panel. Power transmission guards must cover all

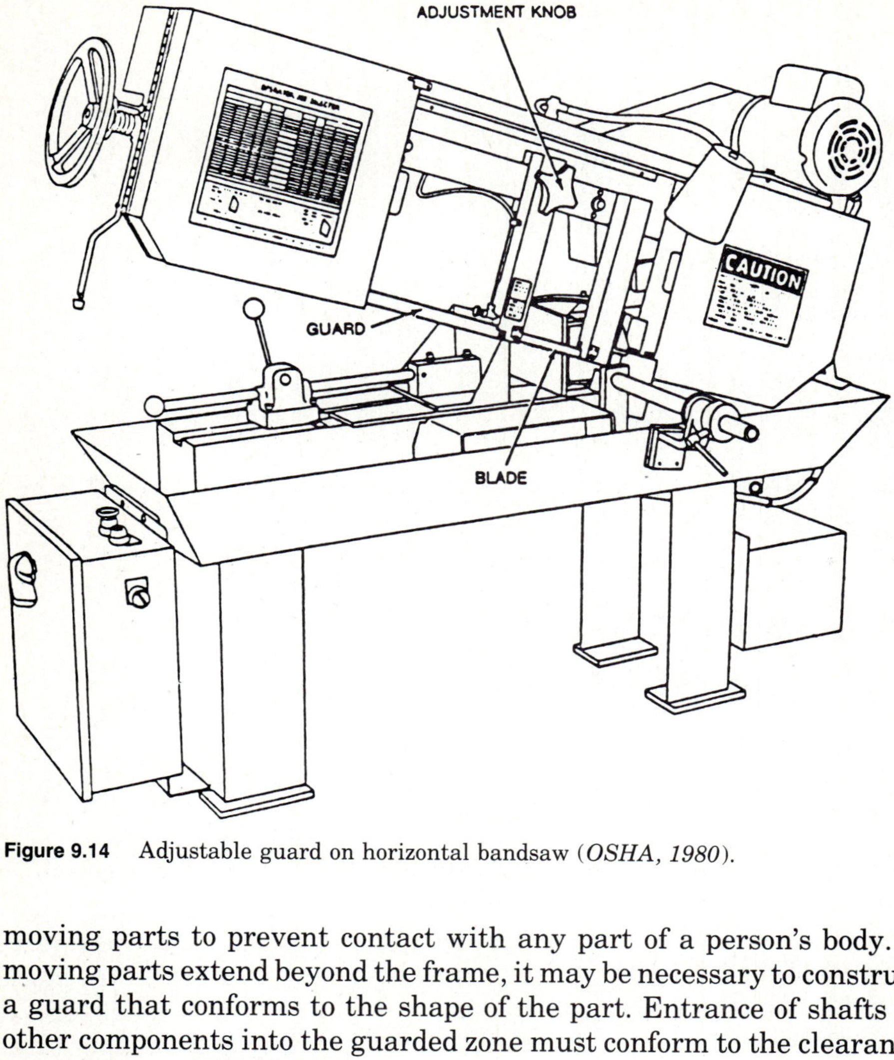

Figure 9.14 Adjustable guard on horizontal bandsaw (*OSHA, 1980*).

moving parts to prevent contact with any part of a person's body. If moving parts extend beyond the frame, it may be necessary to construct a guard that conforms to the shape of the part. Entrance of shafts or other components into the guarded zone must conform to the clearance requirements of Table 9.2.

Table 9.2 also indicates typical filler materials. Frameworks are usually of metal (e.g., structural shapes, pipe, bar, or rod stock) and filler materials may be expanded metal or, when good visibility is required, plastic or safety glass. All construction materials should be strong enough to protect the operator and others in the vicinity from flying chips, broken grinding wheels, saw blades, and teeth, or other materials that may be released with energies great enough to cause injury (ANSI 1973a, 1973b, 1974, 1982a, 1982b; OSHA, 1983; ANSI/ASME, 1986). Because of the possibility of machine-induced vibration in the guard structure and subsequent elevated noise levels, framework and filler materials should be securely attached, constructed of ma-

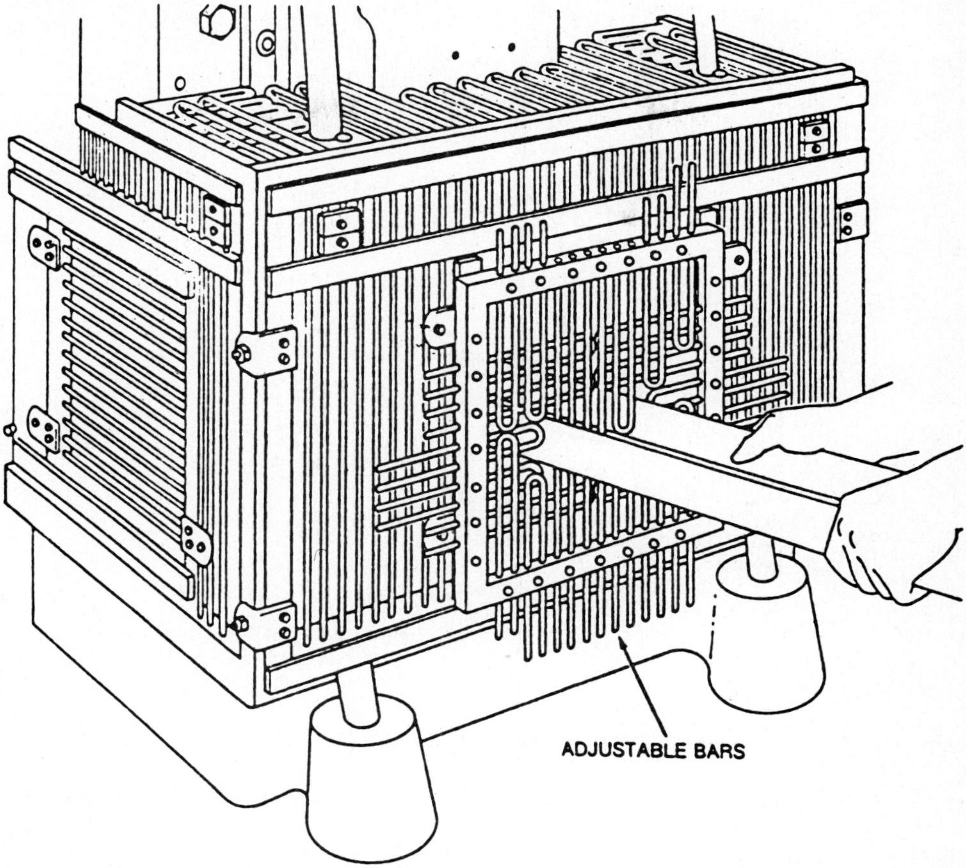

Figure 9.15 Adjustable guard on power press (*OSHA, 1980*).

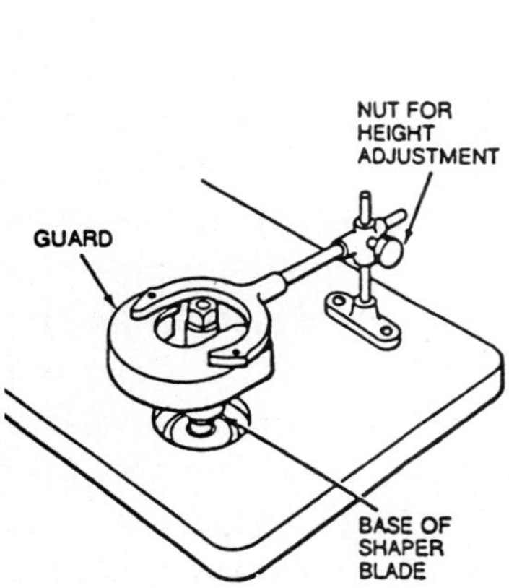

Figure 9.16 Adjustable guard on shaper (*OSHA, 1980*).

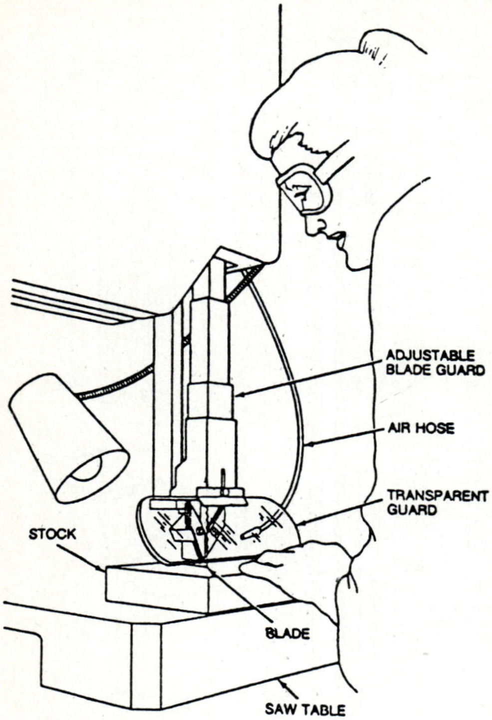

Figure 9.17 Adjustable guard on a vertical bandsaw (*OSHA, 1980*).

terials substantial enough to absorb such energies, and be appropriately isolated or cushioned (see also Chapter 13). Adequate ventilation is also required to prevent overheating of the enclosed machinery.

II. Safety Devices

Safety devices for preventing contact with moving machine parts, stock, scrap, or product pieces either prevent the operator from being in the danger zone or prevent the machine from operating when hands or other body parts enter the danger zone. *Presence-sensing* devices fall into the latter category. Photoelectric (optical) presence sensing utilizes a light field that, if broken, prevents the machine from completing its cycle. Figure 9.20 shows such a light source-and-detecting device on a part-revolution power press. When the light beam is broken, either the ram will not start to cycle or, if the press is already functioning, the stopping mechanism will be activated. Figure 9.21 shows a similar arrangement on a press brake. The device may be swung up or down to accommodate different production requirements. Note that this machine is actuated by a foot pedal, to free the operator's hands for stock feeding. The presence sensor may also use radiofrequency (capacitance) as a means of interrupting machine operation. When the capacitance

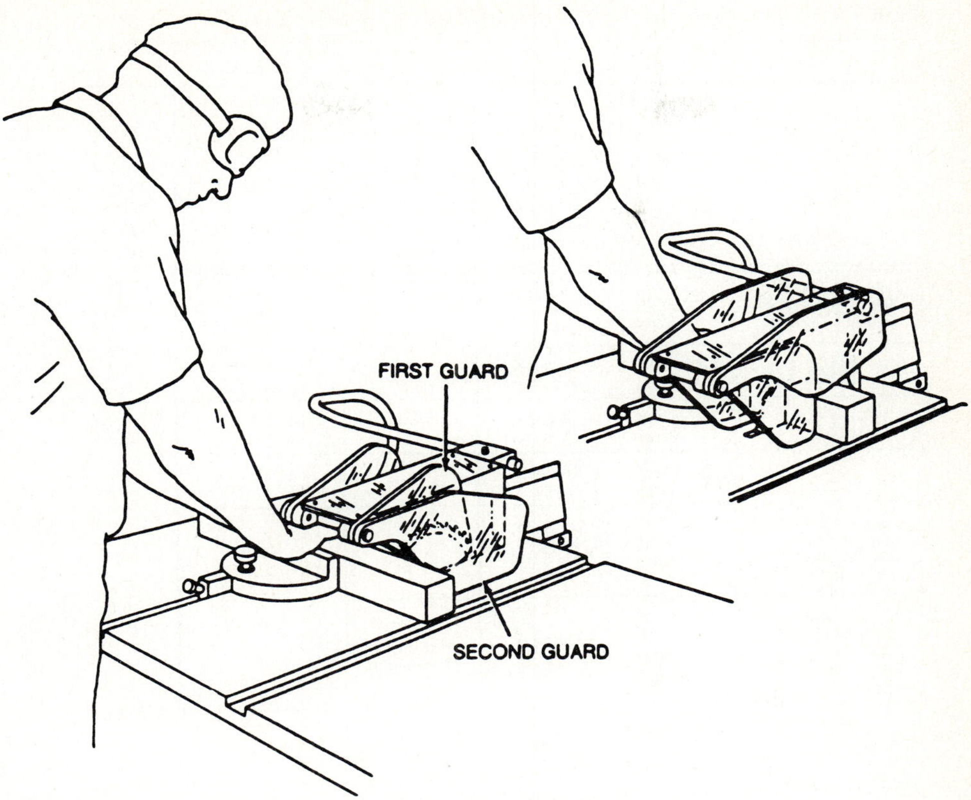

Figure 9.18 Self-adjusting guard on table saw (*OSHA, 1980*).

field is broken, the machine will stop or not activate. As with optical sensing, these devices should be used only on machines that can be stopped (e.g., having a friction clutch) before the worker reaches the danger area. Electromechanical sensing utilizes a probe or contact bar that descends a predetermined distance when the operator initiates the cycle. If there is an obstruction to this descent, the control circuit will not activate the machine cycle. Figure 9.22 illustrates such a system for an eyeletter.

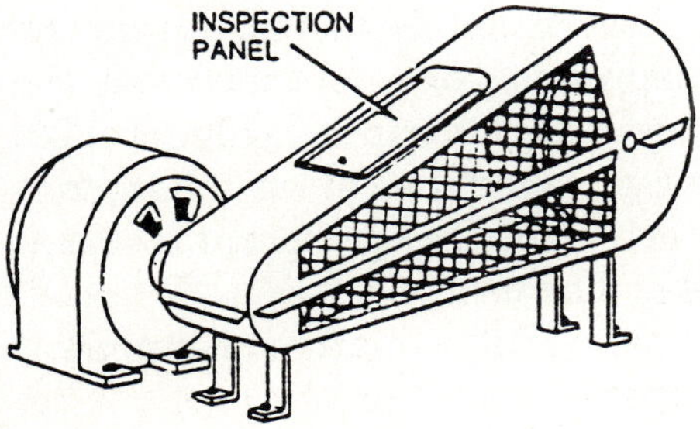

Figure 9.19 Fixed guard enclosing belt and pulleys (*OSHA, 1980*).

TABLE 9.2 Standard Materials and Dimensions for Machinery Guards

	SIZE OF FILLER MATERIALS			
Material 1 in. = 2.5 cm; 1 ft = 0.305 m	*Clearance from Moving Part at All Points (inches)*	*Largest Mesh or Opening Allowable B (inches)*	*Minimum Gage (U.S. Standard) or Thickness*	*Min. Height of Guard from Floor or Platform Level (ft, in.)*
Woven Wire	Under 2 2-4 4-15	3/8 1/2 2	No. 16-3/8 in. No. 16-1/2 No. 12-2	8-0* 8-0 8-0
Expanded Metal	Under 4 4-15	1/2 2	No. 18-1/2 in. No. 13-2	8-0 8-0
Perforated Metal	Under 4 4-15	1/2 2	No. 20-1/2 in. No. 14-2	8-0 8-0
Sheet Metal	Under 4 4-15	. . .	No. 22 No. 22	8-0 8-0
Wood or Metal Strips Crossed	Under 4 4-15	3/8 2	3/4 in. wood or No. 16 metal	8-0
Wood or Metal Strips Not Crossed	Under 4 4-15	1/2 the width One width		
Plywood, Plastic or Equivalent	Under 4 4-15	. . .	1/4 in. 1/4 in.	8-0
Standard Railing	Min. 15 Max. 20	. . .	. . .	3-6

*Guards for rotating protruding objects should extend to a minimum height of 9 ft from the floor or platform.

SOURCE: NSC, 1980.

Pullback involves fastening cables to the operator's hands, wrists, and/or arms. Ordinarily used on machines with stroking action, the operator is allowed access to the danger zone when the slide (ram) is up. When the slide begins to descend, a mechanical linkage automatically withdraws the operator's hands from the point of operation. Figure 9.23 shows such a device on a small power press. *Restraint* (holdout) requires that the operator's hands be limited by straps or cables to a safe area. There is no extending or retracting action. Consequently, hand-feed tools are usually required if stock is to be placed in a danger zone. *Safety trip controls,* through depression of a pressure-sensitive tripwire (Figure 9.24), body bar (Figure 9.25), or trip rod (Figure 9.26), provide a quick means of deactivating a machine in an emergency. The trip rod must be intentionally activated by the operator and so must be conveniently located for his or her use.

Two-hand control, shown for a part-revolution clutch power press in

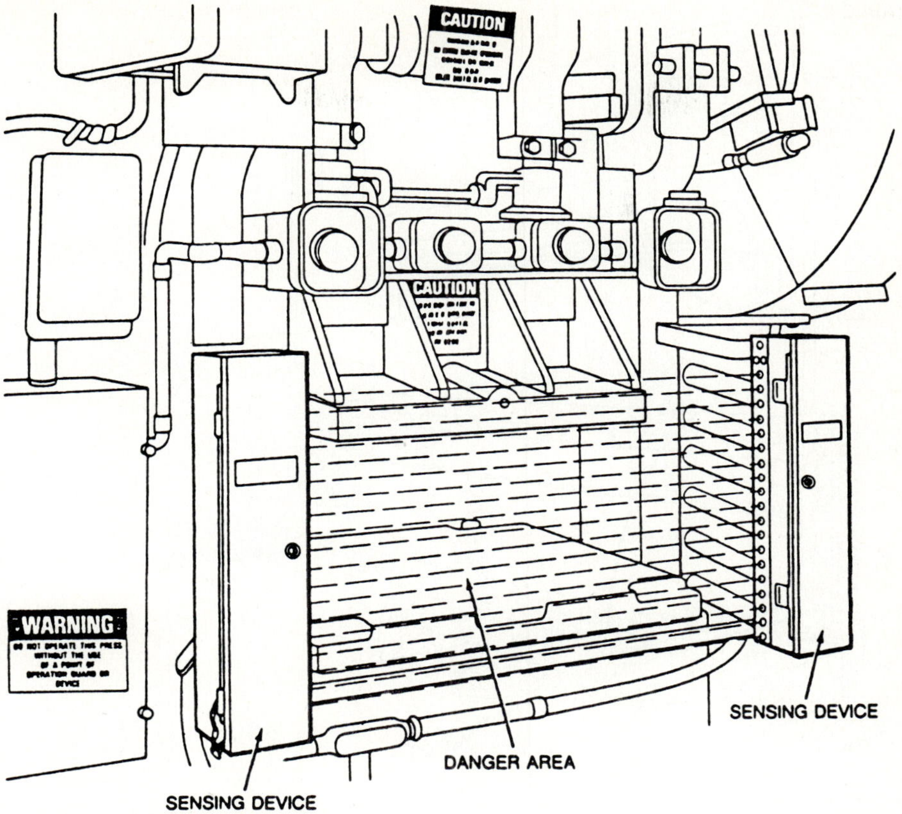

Figure 9.20 Photoelectric presence-sensing device on power press (*OSHA, 1980*).

Figure 9.27, requires constant concurrent pressure by the operator to activate the machine. Rims are included around the buttons to prevent inadvertent operation. Location of such control buttons is based on a typical maximum hand speed of 63 in./sec (OSHA, 1983). Consequently, the minimum safety distance, D_s (in inches), from the point of operation to the buttons can be estimated from

$$D_s = 63T_s \tag{9.1}$$

where T_s is in seconds and is the stopping time of a power press measured at approximately the 90° position of crank shaft rotation (ANSI, 1982a); or is the stopping time of a hydraulic press measured on the rapid-approach portion of the die-closing cycle (ANSI, 1982b); or is an appropriate stopping time of other machines.

Two-hand trip requires that both hands be used to activate the machine cycle and consequently be away from the point of operation. After

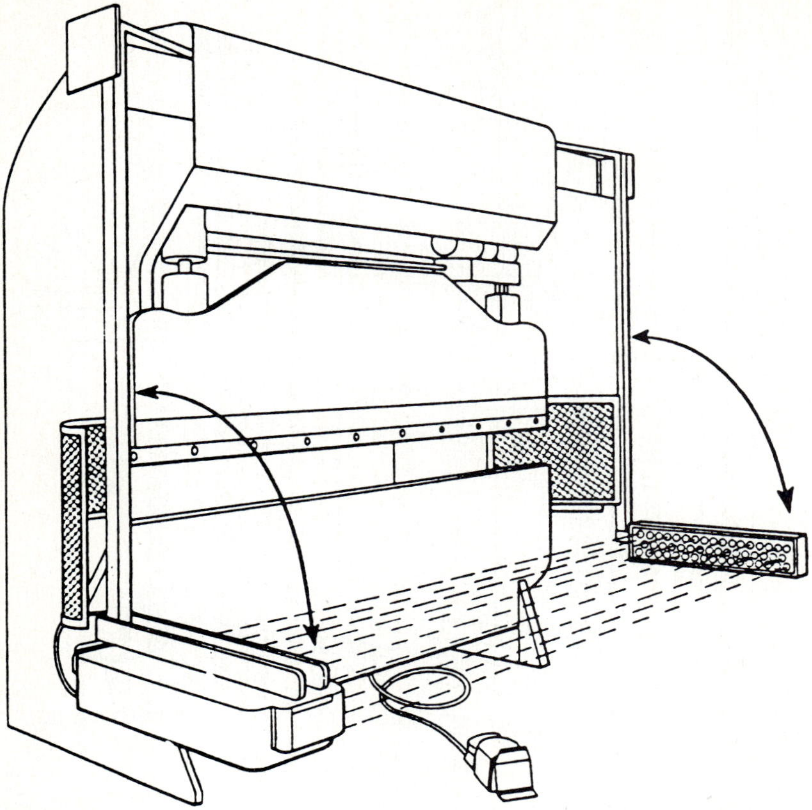

Figure 9.21 Photoelectric presence-sensing device on foot pedal-actuated press brake (*OSHA, 1980*).

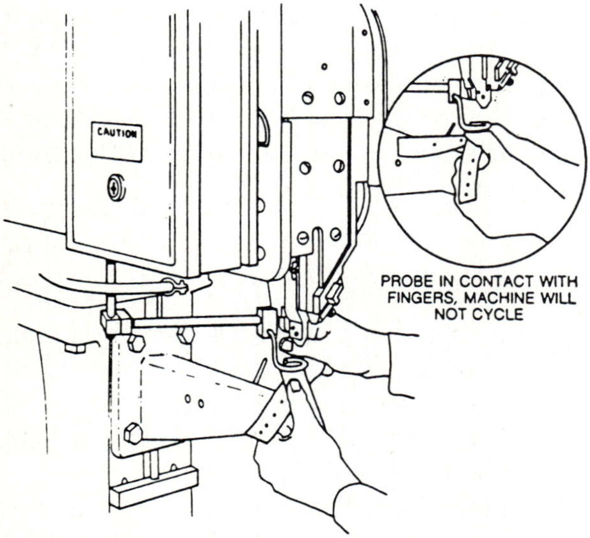

Figure 9.22 Electromechanical sensing device on an eyeletter (*OSHA, 1980*).

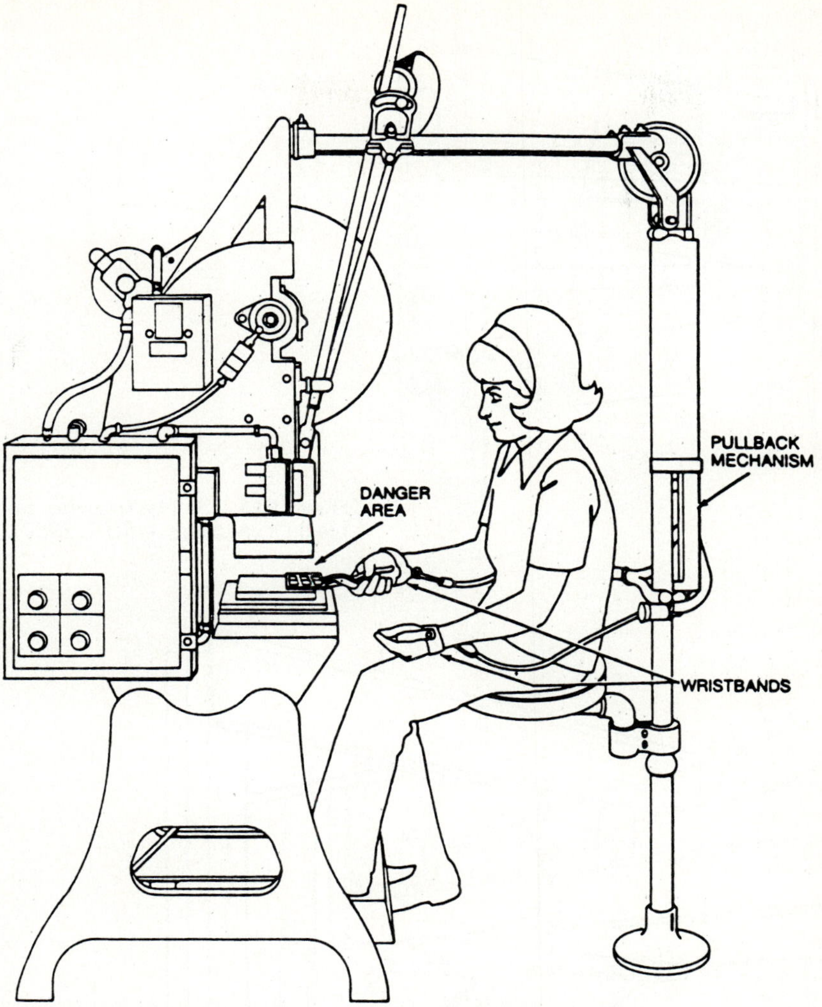

Figure 9.23 Pullback device on a power press (*OSHA, 1980*).

activation the hands are free. Such devices are ordinarily used on machines with full-revolution clutches. The trip buttons must be far enough away from the point of operation so that the operator cannot move his or her hands into this area before the first half of the cycle is completed (the ram has reached the full down position). Appropriate distances, D_m (inches), can be estimated from (ANSI, 1982b):

$$D_m = 63T_m \qquad (9.2)$$

where, for full-revolution clutch presses and one engaging point,

$$T_m = \text{time in seconds necessary for } 1\tfrac{1}{2} \text{ revolutions of the crankshaft} \qquad (9.3a)$$

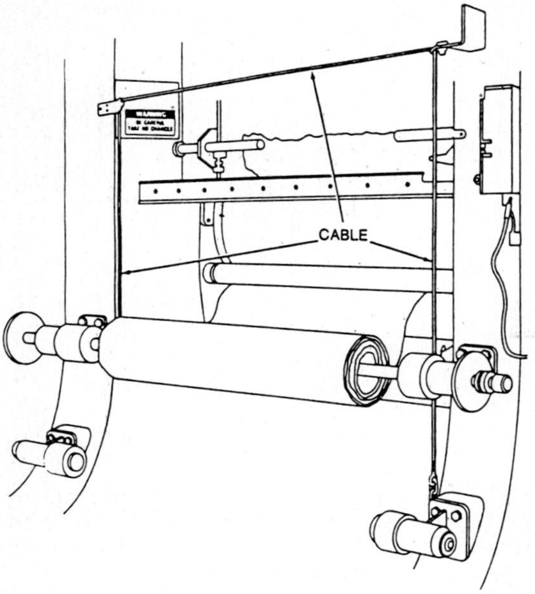

Figure 9.24 Safety tripwire cable on a calender (*OSHA, 1980*).

Figure 9.25 Pressure-sensitive body bar on a rubber mill (*OSHA, 1980*).

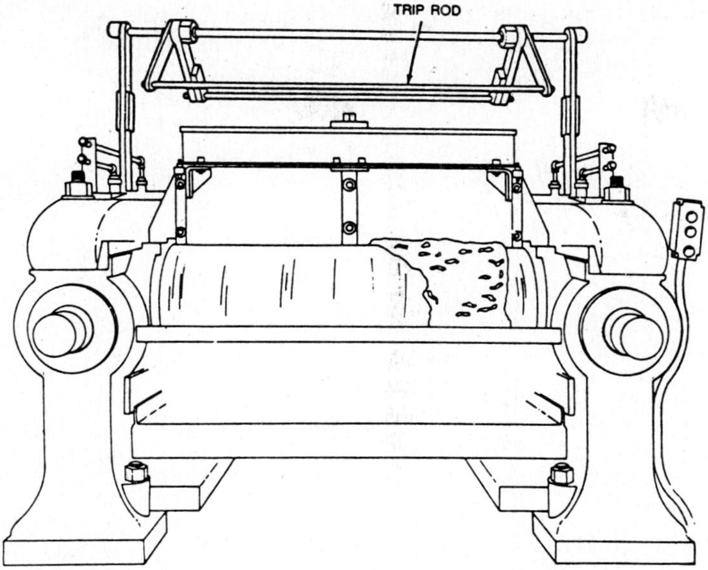

Figure 9.26 Safety trip rod on a rubber mill (*OSHA, 1980*).

and for full-revolution clutch presses and more than one engaging point,

$$T_m = \left(\frac{1}{2} + \frac{1}{\text{Number of engaging points per revolution}}\right)\left(\begin{array}{c}\text{time in seconds necessary}\\ \text{for one complete revolu-}\\ \text{tion of crankshaft}\end{array}\right) \qquad (9.3b)$$

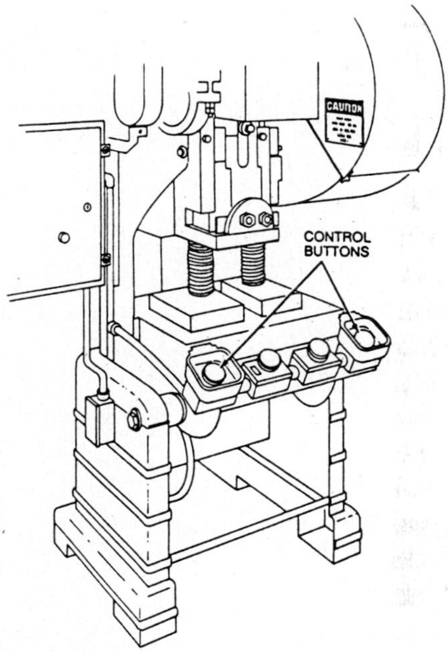

Figure 9.27 Two-hand control buttons on part-revolution clutch power press (*OSHA, 1980*).

Example 9.2 Determine the distance at which two-hand trip buttons must be placed from the die of a power press that operates at 90 strokes/min with a full-revolution clutch and three engaging points per revolution.

solution

$$T_m = \left(\frac{1}{2} + \frac{1}{3}\right)\left(\frac{60 \text{ sec/min}}{90 \text{ strokes/min}}\right) \tag{9.3b}$$

$$= 0.56 \text{ sec}$$

$$D_m = (63 \text{ in./sec})(0.56 \text{ sec}) \tag{9.2}$$

$$= 35 \text{ in.}$$

Two-hand trips and controls should be installed in a fixed position (ANSI, 1982a,1982b). An antirepeat feature must be incorporated for full-revolution clutches; i.e., the press is limited to a single stroke even if the tripping control is hand-operated.

Gates are movable barriers at the point of operation that must be closed before a machine action can be initiated. They are often designed to be operated at each machine cycle. Figure 9.28 shows such a device on a power press; and Figure 9.29 illustrates a gate on a molding machine. Both of these gates need to be closed before the machines will function. Type A gates protect operators by enclosing the point of operation before a machine cycle (press stroke) can be initiated. Type B gates, which should not be used on full-revolution presses, are devices designed to make impossible an act of reaching into the point of operation prior to die-closing on the downward stroke.

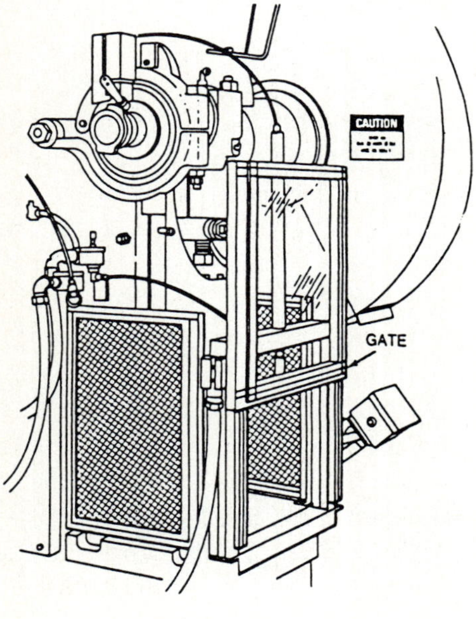

Figure 9.28 Power press with gate (*OSHA, 1980*).

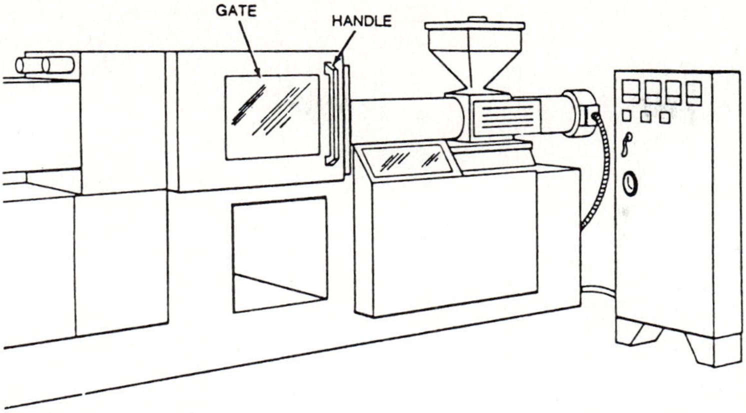

Figure 9.29 Horizontal injection molding machine with gate (*OSHA, 1980*).

A summary of safeguarding devices and typical applications are given in Table 9.3. As with barrier guards, the efficacy of design must be carefully tested with the device in place. Appropriate training of operators and others involved in machine operation and maintenance to recognize the nature of the hazard and the reasons supporting the design is a necessary concomittant of the safeguarding system.

III. Feeding and Ejection

Table 9.4 lists stock feeding and ejection methods that make it unnecessary for the operator's hands to be placed in the danger zone are a useful adjunct to barrier guards and/or safeguarding devices. Stock feeders are also likely to be economical for larger jobs, e.g., greater than a few thousand pieces for power presses (Doyle et al., 1985). However, some feeding and ejection methods create hazards in themselves. While the use of a robot for feeding material to a press removes the operator from the point of operation, this approach produces a new problem due both to the robot arm, which requires safeguarding (Figure 9.30, p. 475), and to other aspects of robot operation (e.g., the example given in Chapter 1).

Automatic feeding is illustrated in Figure 9.31 (p. 476). The saw has an automatic indexing mechanism that moves the stock a preset distance for each cut. The traveling head also automatically recycles for each cut. Figure 9.32 (p. 477) shows stock being fed as a strip to a power press. The roller feed unit is actuated through a linkage to an eccentric on the press crankshaft. The stock is released when the ram descends but is gripped and drawn as the ram rises. Note the transparent fixed enclosure guard on the press. The pinch points on the scrap winder wheel and stock feed roll must also be adequately guarded.

TABLE 9.3 Devices for Safeguarding Mechanical Hazards

Type	Action	Advantages	Limitations	Typical applications
Photoelectric (optical)	Machine will not start cycling when the light field is interrupted When the light field is broken by any part of the operator's body during the cycling process, immediate machine braking is activated	Can allow freer movement for operator	Does not protect against mechanical failure May require frequent alignment and calibration Excessive vibration may cause lamp filament damage and premature burnout Limited to machines that can be stopped Does not protect against mechanical repeat	Embossing presses Power presses Press brakes
Radiofrequency (capacitance)	Machine cycling will not start when the capacitance field is interrupted When the capacitance field is disturbed by any part of the operator's body during the cycling process, immediate machine braking is activated	Can allow freer movement for operator	Does not protect against mechanical failure Antennae sensitivity must be properly adjusted Limited to machines that can be stopped Does not protect against mechanical repeat	Embossing presses Power presses Press brakes
Electromechanical	Contact bar or probe travels a predetermined distance between the operator and the danger area	Can allow access at the point of operation	Contact bar or probe must be properly adjusted for each application; this adjustment must be maintained properly	Embossing presses Power presses Press brakes

	Interruption of this movement prevents the starting of machine cycle			
Pullback	As the machine begins to cycle, the operator's hands are pulled out of the danger area	Eliminates the need for auxiliary barriers or other interference at the danger area	Limits movement of operator May obstruct workspace around operator Adjustments must be made for specific operations and for each individual Requires frequent inspection and regular maintenance Requires close supervision of the operator's use of the equipment	Embossing presses Power presses Press brakes
Restraint (holdback)	Prevents the operator from reaching into the danger area	Little risk of mechanical failure	Limits movements of operator May obstruct workspace Adjustments must be made for specific operations and each individual Requires close supervision of the operator's use of the equipment	Embossing presses Power presses Press brakes
Safety trip controls: Pressure-sensitive body bar Safety trip rod Safety tripwire	Stops machine when tripped	Simplicity of use	All controls must be manually activated May be difficult to activate controls because of their location	Embossing presses Power presses Press brakes

TABLE 9.3 Devices for Safeguarding Mechanical Hazards *(Continued)*

Type	Action	Advantages	Limitations	Typical applications
			Protects only the operator May require special fixtures to hold work May require a machine brake	
Two-hand control	Concurrent use of both hands is required, preventing the operator from entering the danger area	Operator's hands are at a predetermined location Operator's hands are free to pick up a new part after first half of cycle is completed	Requires a partial cycle machine with a brake Some two-hand controls can be rendered unsafe by holding with arm or blocking, thereby permitting one-hand operation Protects only the operator Does not protect against mechanical repeat unless blocks or stops are used	Dough mixers Embossing presses Power cutters Pressing machines Power presses Power press brakes
Two-hand trip	Concurrent use of two hands on separate controls prevents hands from being in danger area when machine cycle starts	Operator's hands are away from danger area Can be adapted to multiple operations No obstruction to hand feeding Does not require adjustment for each operation	Operator may try to reach into danger area after tripping machine Some trips can be rendered unsafe by holding with arm or blocking, thereby permitting one-hand operation Protects only the operator	Dough mixers Embossing presses Power cutters Pressing machines Power presses Power press brakes

			May require special fixtures Does not protect against mechanical repeat unless blocks or stops are used	
Gate	Provides a barrier between danger area and operator or other personnel	Can prevent reaching into or walking into the danger area	May require frequent inspection and regular maintenance May interfere with operator's ability to see the work Usually limited to machines on which the part or material being processed are fully within the point-of-operation area	Power presses Plastic injection molding machines Compression molding machines Die casting machines

SOURCES: OSHA, 1980; NSC, 1980.

TABLE 9.4 Feeding and Ejection Methods

Method	Safeguarding action	Advantages	Limitations
Automatic feed	Stock is fed from rolls, indexed by machine mechanism, etc.	Eliminates the need for operator involvement in the danger area	Other guards are also required for operator protection, usually fixed barrier guards Requires frequent maintenance May not be adaptable to stock variation
Semiautomatic feed	Stock is fed by chutes, movable dies, dial feed, plungers, or sliding bolster		
Automatic ejection	Workpieces are ejected by air or mechanical means		May create a hazard of blowing chips or debris Size of stock limits the use of this method Air ejection may present a noise hazard
Semiautomatic ejection	Workpieces are ejected by mechanical means that are initiated by the operator	Operator does not have to enter danger area to remove finished work	Other guards are required for operator protection May not be adaptable to stock variation
Robots	Perform work usually done by operator	Operator does not have to enter danger area Suitable for operations where high stress factors are present, such as heat and noise	Can themselves create hazards Require maximum maintenance Are suitable only to specific operations

SOURCE: OSHA, 1980.

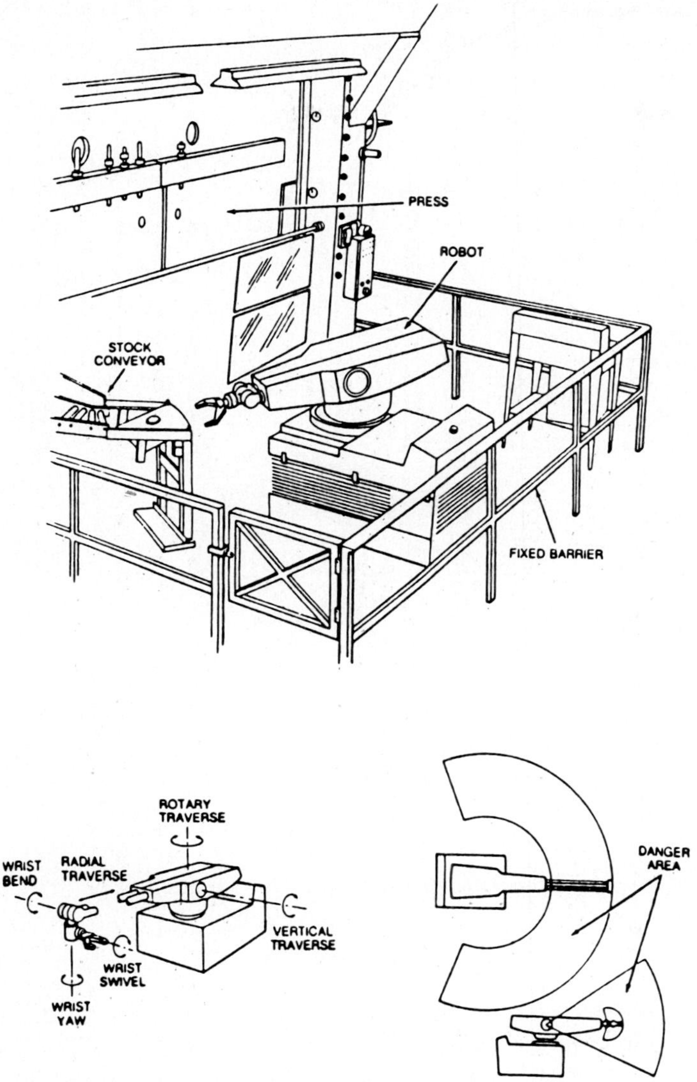

Figure 9.30 Robot feed of a power press (*OSHA, 1980*).

Semiautomatic feeding usually involves the operator actuating a mechanism to place the stock. Dial station feed (Figure 9.33, p. 477) requires placement of stock in a nest in a round table that rotates into the die with each press stroke. Stations are loaded and unloaded on the side away from the punch. Figure 9.34 (p. 478) shows a mechanically controlled arrangement where blanks are placed in the nest one at a time by a plunger, which then pushes each under the ram. When the plunger is pushed in, pin B is allowed to rise up through hole A,

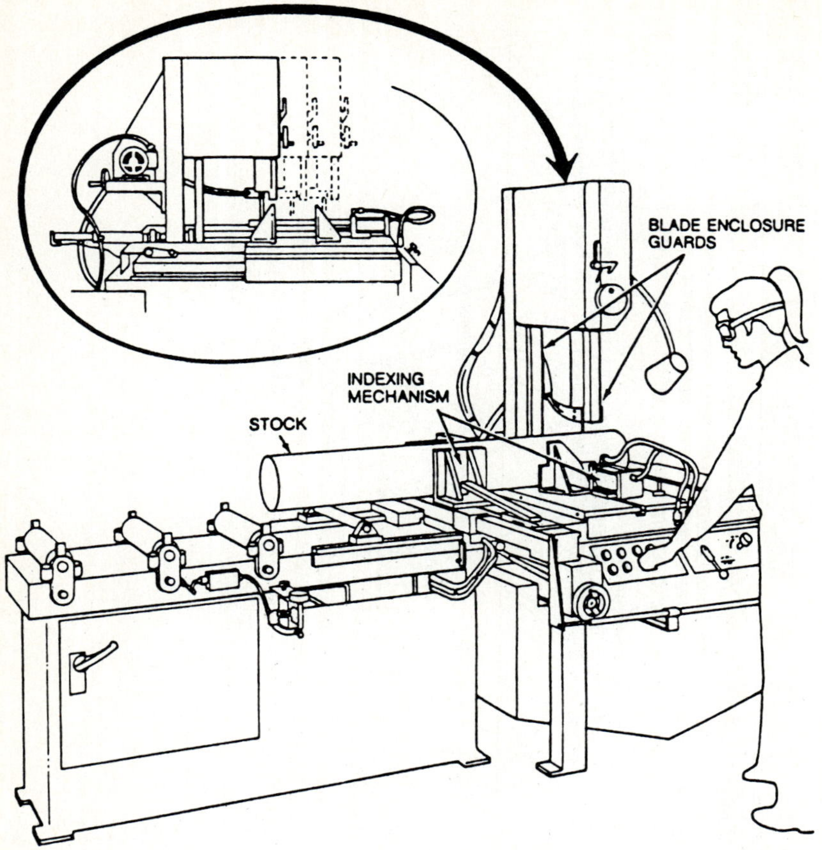

Figure 9.31 Saw with automatic indexing mechanism and traveling head (*OSHA, 1980*).

allowing yoke C to release so that the press can be tripped. The sliding die of Figure 9.35 moves in and out by a foot lever or by hand. The die should be interlocked with the press to prevent tripping when the die is not aligned with the ram. Stops prevent the die from being inadvertently pulled out of the slides (OSHA, 1980). Other one-piece-at-a-time feeding devices include inclined chutes and magazine feeds (stacks of blanks or pieces in a holder that feeds into a slide arrangement similar to that in Figure 9.34).

Accumulation of stock and scrap at a guarded point of operation encourages removal of guards or use of inadequate barriers. Consequently, efficient, reliable part and scrap ejection is an important consideration in guard design. Two general problems must be solved for adequate ejection: (1) how to strip the piece from the machine; and (2) how to eject the piece away from the machine to a container or conveyor. For instance, stripping methods for presses include positive stripper

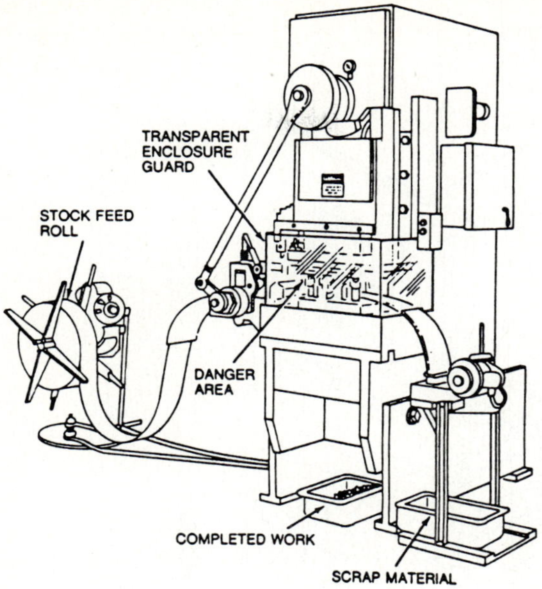

Figure 9.32 Power press with automatic feed (*OSHA, 1980*).

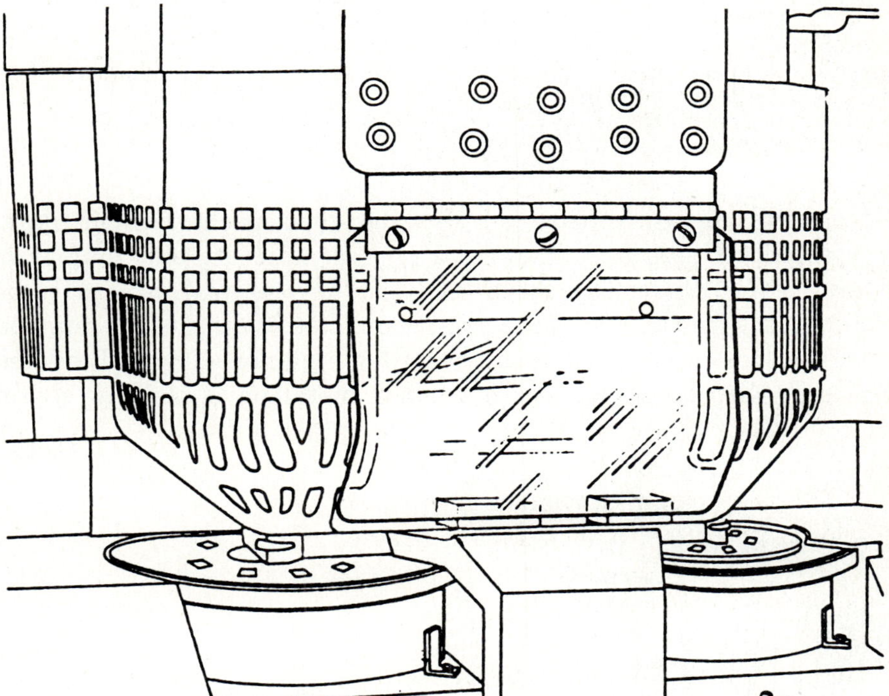

Figure 9.33 Power press with double-dial feed (*OSHA, 1980*).

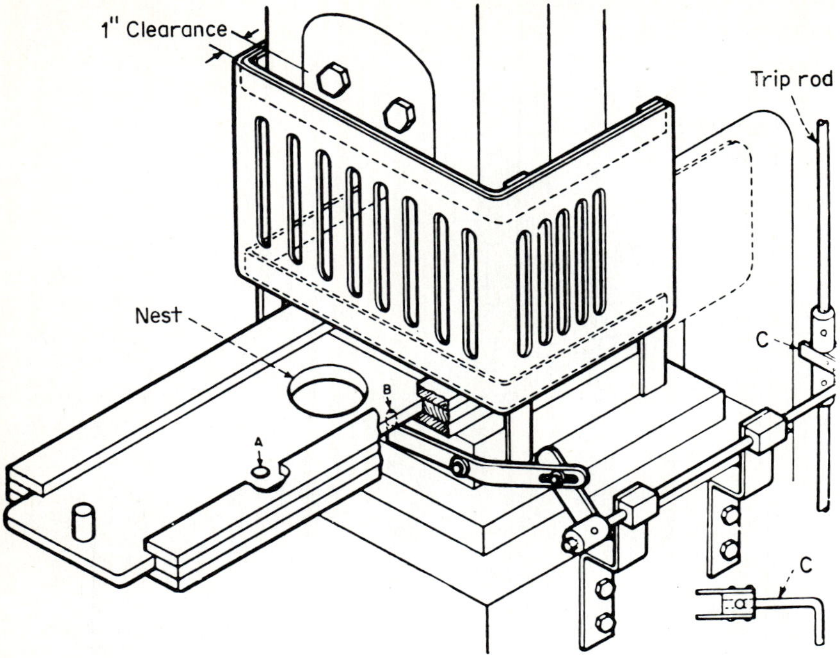

Figure 9.34 Manually operated plunger feed (*NSC, 1980*). Note that the press cannot be tripped until the part is pushed to the next location. At this point, hole A is directly over tapered pin B; it can rise and release yoke C so that the press can be tripped.

plates, spring-pressure pads or pins, latch-type mechanical lift dogs, compressed air jets, and pneumatic or hydraulic lift pins or pickup fingers (NSC, 1980). The action of these devices will ordinarily be coordinated with the use of the press ram. Use of air jets for removal of small pieces carries with it the hazard of high noise levels. Elevated noise may also occur because of ejected metal parts striking metal

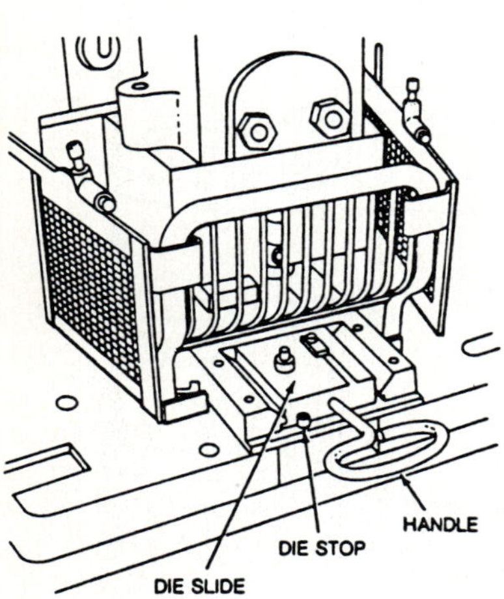

Figure 9.35 Power press with sliding die (*OSHA, 1980*).

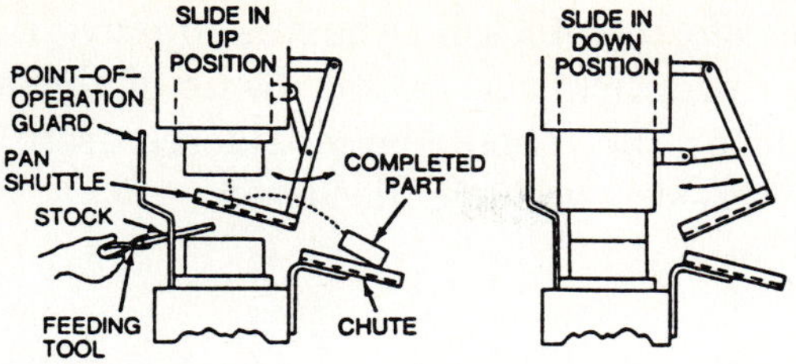

Figure 9.36 Shuttle ejection mechanism (*OSHA, 1980*).

product chutes and bins. Noise-deadening materials may be used to cover the metal surfaces; or, often more effectively, a layer of damping material may be sandwiched between the metal surface of the chute and another metal surface for support. Heavy pieces should be removed from presses using positive mechanical methods, such as knockout pins pneumatically actuated by the press upstroke. Figure 9.36 illustrates a mechanical ejection method from a power press. The pan shuttle mechanism moves under the finished part on the upstroke of the press ram. The piece is stripped by knockout pins (not shown) and deflected into a product chute. As the ram moves down toward the next blank, the shuttle moves out of the way (NSC, 1980).

With any machine, some jamming of feed stock or scrap will eventually occur. Holding tools of various types are available for helping to clear obstructions or place or remove stock while keeping the operator's hands away from the point of operation. Figures 9.23 and 9.36 illustrate the use of such devices for power presses; and Figure 9.37 shows a push stick and push block being used while feeding stock to a saw blade. Kickbacks are another saw hazard. These occur when a saw blade binds in the stock being fed and hurls it back toward the operator. Use of a spreader, a post mounted behind the saw blade that prevents the stock from gripping the blade after the cut, and anti-

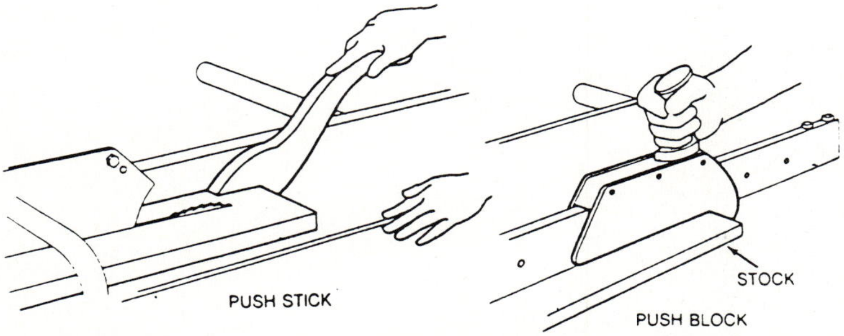

Figure 9.37 Push stick and block for saw operation (*OSHA, 1980*). (Note that the push block fits over the nip fence, a structure that also provides protection from the blade and helps to align the workpiece.)

kickback dogs or fingers, which hold the stock if it begins to move in the opposite direction, will minimize kickback action. None of these tools are machine safeguards. Rather, they should be considered useful supplements to the barriers and devices we have discussed.

REFERENCES

AMIA (1966). *Safe Openings for Some Point of Operation Guards.* American Mutual Insurance Alliance Technical Guide 02-678, Chicago.

ANSI (1973a). *American National Standard Safety Requirements for the Construction, Care, and Use of Power Press Brakes.* American National Standards Institute B11.3-1973, New York.

ANSI (1973b). *American National Safety Code for the Use, Care, and Protection of Abrasive Wheels.* American National Standards Institute B7.1-1970 including Supplement B7.1a-1973, New York.

ANSI (1974). *American National Standard Safety Requirements for the Construction, Care, and Use of Metal Sawing Machines.* American National Standards Institute B11.10-1974, New York.

ANSI (1982a). *American National Standards for Machine Tools—Mechanical Power Presses—Safety Requirements for Construction, Care, and Use.* American National Standards Institute B11.1-1982, New York.

ANSI (1982b). *American National Standards for Machine Tools—Hydraulic Presses—Safety Requirements for Construction, Care, and Use.* American National Standards Institute B11.2-1982, New York.

ANSI/ASME (1986). *Safety Standard for Mechanical Power Transmission Apparatus.* American Society of Mechanical Engineers, ANSI/ASME B15.1a-1986, New York.

Doyle, L. E., Keyser, C. A., Leach, J. L., Schrader, G. F., and Singer, M. B. (1985). *Manufacturing Processes and Materials for Engineers,* 3d ed. Prentice-Hall, Englewood Cliffs, N.J.

NSC (1980). *Accident Prevention Manual for Industrial Operations,* 8th ed. (McElroy, F. E., ed.). National Safety Council, Chicago.

OSHA (1980). *Concepts and Techniques of Machine Guarding.* Occupational Safety and Health Administration Report No. OSHA 3067, Washington, D.C.

OSHA (1983). *General Industry Standards Part 1910.* Occupational Safety and Health Administration 2206, Washington, D.C., rev. March 11.

Chapter

10

Ionizing Radiation

P. A. Scheff

I. Characteristics of Ionizing Radiation

Ionizing radiation is functionally defined as radiation with energy sufficient to ionize matter. Although there is no sharp dividing line, the generally accepted value of the minimum ionizing energy is 10 electron volts (eV). Ionizing radiation can consist of particles or electromagnetic radiation. The frequency of electromagnetic radiation associated with 10 eV can be calculated from

$$E = hf \tag{10.1}$$

where E is energy (joules), h is Plank's constant (6.6262×10^{-34} J·sec), and f is frequency (cycles per second or hertz). Substituting 10 eV [(10 eV)(1.6×10^{-19} J/eV) = 1.6×10^{-18} J] into Eq. (10.1) and solving for frequency gives $f = 2.4 \times 10^{15}$ Hz. Since the speed of all electromagnetic waves in free space is constant, the relationship between frequency and wavelength is

$$c = f\lambda \tag{10.2}$$

where c is the speed of light (3×10^{10} cm/sec) and λ is the electromagnetic wavelength (cm). Therefore the maximum wavelength for ionizing electromagnetic radiation is

$$\lambda = \frac{c}{f} = \frac{3 \times 10^{10}\ \text{cm/sec}}{2.4 \times 10^{15}/\text{sec}}$$
$$= 1.24 \times 10^{-5}\ \text{cm}$$

Particle forms of ionizing radiation include alpha-particles (α, helium nuclei), beta-particles (β, electrons), positrons (β^+, positively charged electrons, ${}^{\ 0}_{+1}e$), and neutrons. Ionizing particles can be charged (α, β, and β^+) or neutral in charge (neutrons).

When sufficient energy is imparted to an electron to raise it to an infinitely great orbit (to remove the electron from the electrical field of the nucleus), the atom is said to be ionized. The ionized atom and the ejected electron are called an ion pair. Ionization can occur when either a photon or charged particle of sufficient energy collides with an orbital electron. The ionization potential of hydrogen is 13.6 eV. Therefore, if a collision with a rapidly moving particle or photon transfers more than 13.6 eV to the electron of hydrogen, the result is the ionization of the atom. The excess energy above 13.6 eV that is transferred appears as kinetic energy of the electron and positive ion formed.

Energy is also released when outer-shell electrons move closer to the nucleus to fill a vacant electron orbit. Such a release is called a characteristic x-ray. A second mechanism of x-ray production, bremsstrahlung, involves the inelastic interaction of a charged particle and a nucleus. Unlike characteristic x-rays, bremsstrahlung x-rays are emitted with a continuous energy distribution, with E_{max} = incident energy of charged particle. This type of x-radiation is an important consideration in the design of shields for β particles.

A. Radioactivity

Radioactivity can be described as spontaneous nuclear transformation that results in the formation of new elements. This transformation is accomplished by one of four mechanisms: α-particle emission, β-emission, β^+-emission, or orbital electron capture. Each pathway can be accompanied by the emission of a gamma photon. With this definition, α-, β-, and gamma (γ)-radiation are defined as intranuclear phenomena, and x-radiation as extranuclear. One important exception is the production of two 0.51-MeV gamma photons from the annihilation of a β, β^+ pair. Radioactive transformations change the neutron-to-proton ratio of the atom toward a more stable ratio.

Isotopes are defined as nuclear species of the same element with different atomic mass numbers. Therefore, isotopes contain the same number of protons and have the same extranuclear electron structure. They differ only in the number of neutrons. The common nomenclature is:

$$ {}^{A}_{z}N \tag{10.3} $$

where N is the chemical symbol, z is the atomic number (number of protons), and A is the atomic mass (sum of number of protons and neutrons). For example, carbon-12, an element with 6 protons and 6 neutrons, is written ${}^{12}_{6}C$. The atomic number (z) is frequently not included because it is redundant with the chemical symbol, e.g., ${}^{12}C$ (or C-12). Atomic weight is the weighted average of the natural isotopes of a given element. For example, natural chlorine is made up of 75.4% ${}^{35}Cl$ and 24.6% ${}^{37}Cl$. Therefore, the atomic weight of chlorine is $(0.754)(35) + (0.246)(37) = 35.5$.

Alpha-particle emissions from unstable isotopes are essentially monoenergetic; that is, most particles are emitted with their maximum theoretical energy. Table 10.1 lists characteristics of commonly available α-emitters. Note that the daughter isotope will have a higher neutron-to-proton ratio than the parent isotope.

In contrast to α-, β-particles are emitted with a continuous energy distribution from zero to the theoretically maximum value. The difference between the actual β-energy and the theoretical maximum energy is carried off by a neutrino. In general, the stated energy of a β is the theoretical maximum. For the purposes of shielding design, this maximum energy is of primary concern. Table 10.2 lists characteristics of some commonly available β-emitters. Note that the β-emission lowers the neutron-to-proton ratio of the atom.

TABLE 10.1 Commonly Available α-Emitters

Isotope	Energy (MeV)	Half-life	Percent yield
${}^{238}U$	4.15	4.51×10^9 years	23
${}^{238}U$	4.20	4.51×10^9 years	77
${}^{210}Bi$	4.57	3.0×10^6 years	6
${}^{210}Bi$	4.92	3.0×10^6 years	36
${}^{210}Bi$	4.96	3.0×10^6 years	58
${}^{226}Ra$	4.599	1602 years	5
${}^{226}Ra$	4.782	1602 years	95
${}^{210}Po$	5.305	138.4 days	100
${}^{241}Am$	5.443	456 years	13
${}^{241}Am$	5.486	456 years	86
${}^{222}Rn$	5.490	3.823 days	100
${}^{218}Po$	6.002	3.05 min	100
${}^{214}Po$	7.687	164 μsec	100

SOURCE: PHS, 1970.

TABLE 10.2 Commonly Available β-Emitters

	Energy (MeV)			
Isotope	Ave.	Max.	Half-life	Gamma (MeV)[a]
^{14}C	0.049	0.158	5700 years	None
^{24}Na	0.553	4.170	15 hours	1.369
^{32}P	0.694	1.709	14.4 days	None
^{35}S	0.048	0.167	88 days	None
^{42}K	1.446	3.559	12.5 hours	1.524
^{45}Ca	0.076	0.254	165 days	None
^{47}Ca	0.341	2.000	4.6 days	1.308
^{60}Co	0.094	1.478	5.2 years	1.332
^{85}Kr	0.249	0.672	4.5 hours	0.305[b]
^{90}Sr	0.200	0.544	28 years	None
^{90}Y	0.931	2.245	64 hours	None
^{115}Cd	0.318	1.110	53.5 hours	0.53[c]
^{124}Sb	0.385	2.313	60 days	0.603[d]
^{131}I	0.180	0.810	8.1 days	0.364[e]
^{134}Cs	0.152	1.453	2.05 years	0.605[f]
^{137}Cs	0.195	1.167	30 years	0.662[g]
^{140}Ba	0.282	1.010	12.8 days	0.537[h]

[a]Gamma-energy, if present.
[b]Kr x-ray.
[c]In x-rays.
[d]97% at 0.603 MeV; maximum at 2.088 MeV.
[e]Xe x-rays; maximum at 0.723 MeV.
[f]98% at 0.605 MeV, maximum at 1.365 MeV.
[g]Ba x-ray.
[h]Maximum La x-ray.
SOURCE: PHS, 1970.

Positrons (β^+) are positively charged electrons and are emitted from the nucleus of an atom by radioactive transformation when the neutron to proton ratio is too low and an α-emission is not energetically possible. This process results in the transformation of a proton to a neutron within the atom. The β^+-emission decays on the order of microseconds after formation by combination with an electron. The two particles are annihilated resulting in the formation of two gamma photons, each with energy equal to the rest mass of an electron (0.51 MeV). Annihilation radiation represents a significant radiation hazard and must be considered along with the β^+ energy.

Orbital electron capture is the process to overcome a neutron deficiency when the parent isotope does not exceed the daughter by two electron masses (the minimum energy needed for β^+-emission). In this process, an extranuclear electron is captured and unites with an intranuclear proton to form a neutron. An atom transformed by orbital capture will emit a characteristic x-ray from the replacement of the captured K-shell electron. This x-ray can be a significant factor in radiation dose calculations.

Gamma-photons are monochromatic electromagnetic radiation emitted from the nuclei of atoms following radioactive transformation. Gamma-emission is an important mechanism for removing excess energy from excited nuclei, which then return to a ground state. Table 10.3 lists some commonly available γ-emitters. Note that γ-photons are also associated with positron emission.

Internal conversion is an alternative mechanism to rid the excess energy from excited nuclei. With this mechanism, a tightly bound electron (K or L shell) absorbs the excitation energy of a nucleus and is ejected from the atom. Internal conversion electrons are monoenergetic. After conversion, characteristic x-rays are emitted, as outer-shell electrons fill the vacancies left by the photoelectrons.

B. Units of radioactivity

Quantities of radioactive materials are given in units of activity or transformations per second (tps). The standard international (SI) unit of activity is the Becquerel (Bq), defined as 1 tps. Becquerels are, therefore, a measure of the quantity of a radioactive material. Commonly used multiples are the kilo- (10^3), mega- (10^6), giga- (10^9) and tera- (10^{12}) Bq.

The Curie (Ci) is the unit of activity that preceded the Bq and is defined as 3.7×10^{10} tps. Therefore 1 Ci = 3.7×10^{10} Bq. Commonly

TABLE 10.3 Commonly Available γ-Emitters

Isotope	Dominant energy (MeV)	Half-life	Additional radiation (maximum energy, MeV)
^{7}Be	0.477	53 days	None
^{24}Na	1.369 and 2.754	15.0 hours	β (1.389)
^{42}K	1.524	12.4 hours	β (3.52)
^{47}Ca	1.308	4.53 days	β (1.98)
^{55}Fe	≤0.23	2.6 years	None
^{60}Co	1.173 and 1.332	5.26 years	β (1.48)
^{85}Kr	0.514	10.76 days	β (0.67)
^{89}Sr	0.91	52 days	β (1.463)
^{109}Cd	0.088	453 days	e (0.062)[a]
^{124}Sb	0.603	60 days	β (2.31)
^{130}I	0.538, 0.699, 0.743	12.4 hours	β (1.7)
^{131}I	0.364	8.05 days	β (0.806), e[a]
^{133}Ba	0.356	7.2 years	e (0.319)[a]
^{134}Cs	0.605 and 0.796	2.05 years	β (0.662)
^{137}Cs	0.662	30.0 years	β (0.67)
^{192}Ir	0.317–0.589	74.2 days	β (0.67)
^{210}Pb	0.047	21 years	β (0.061), e (0.043)
^{210}Bi	0.262 and 0.30	5.01 days	α (4.96)
^{210}Po	0.803	138.4 days	α (5.305)

[a]Photoelectron from internal conversion.

SOURCE: PHS, 1970.

used multiples of the curie are the milli- (10^{-3}), micro- (10^{-6}), nano- (10^{-9}), pico- (10^{-12}), and femto- (10^{-15}) Ci.

Specific activity (SA) is defined as the number of transformations per second per unit mass or volume. Examples of SA include Bq/g, Bq/ml, and mCi/ml. If N is defined as the number of atoms/unit mass or volume and λ is the decay constant (time^{-1}), then SA = λN.

C. Interaction of radiation with matter

The mechanisms for the transfer of energy from radiation to matter provide the basis for radiation dosimetry and the calculation of shielding requirements. These processes involve the interaction of radiation with atomic nuclei and extranuclear electrons and result in the excitation and ionization of absorber atoms.

1. Particles. Beta-particles have a finite range in matter. The property of matter to absorb energy from a β is a function of the number of absorbing electrons in the path of the particle. The number of electrons or areal density (electrons/cm^2) is a function of density and linear thickness or density-thickness (cm·g/cm^3). Figure 10.1 shows the maximum range of β particles (in mg/cm^2) as a function of their maximum energy. This curve applies to all materials with known density. Table 10.4 lists densities of common materials.

Example 10.1 As an example of β-attenuation, what is the minimum thickness of Plexiglas needed to stop 100% of β-particles from ^{90}Sr?

solution From Table 10.2, the maximum energy of ^{90}Sr is 0.54 MeV. However, the daughter of ^{90}Sr is ^{90}Y, which emits a β with maximum energy of 2.24 MeV. It is therefore necessary to design a shield to stop 2.24 MeV β. From Figure 10.1, the maximum range of 2.24-MeV β particles is 1.1 g/cm^2. The required minimum Plexiglas thickness is:

$$(1.1\ \text{g/cm}^2)/(1.18\ \text{g/cm}^3) = 0.93\ \text{cm}$$

Therefore, a Plexiglas shield 1 cm thick will stop 100% of the β-particles from ^{90}Sr.

Bremsstrahlung radiation is emitted from β-interactions with a continuous energy distribution. The maximum bremsstrahlung photon energy is equal to the maximum kinetic energy of the particle. The fraction of β-particles converted to x-ray photons (f) can be estimated from (Cember, 1983):

$$f = (3.5 \times 10^{-4})ZE \tag{10.4}$$

where Z = atomic number of absorber and E = maximum energy of the β (MeV). Equation (10.4) shows that the occurrence of bremsstrah-

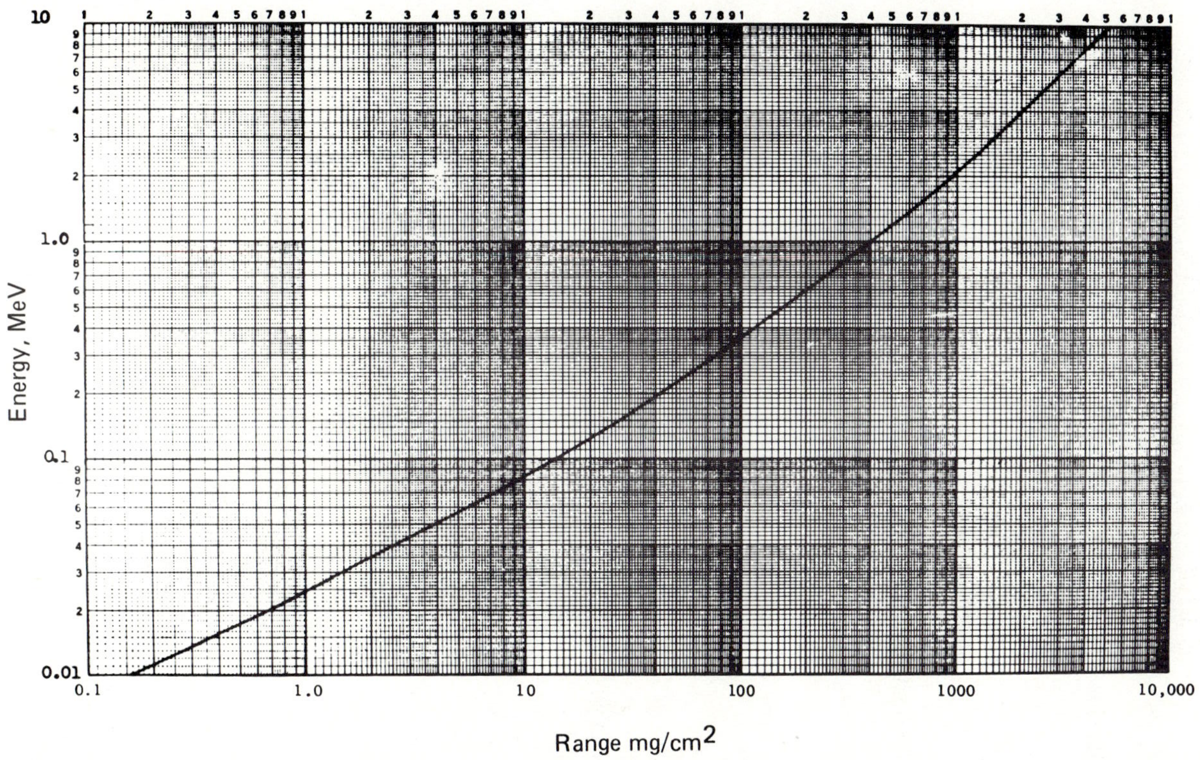

Figure 10.1 Range of β-particles as a function of absorber density and maximum energy (*PHS, 1970*).

TABLE 10.4 Density of Common Materials

Material	Range of density (g/cm^3)	Average density (g/cm^3)
Air	—	0.001293
Aluminum	—	1.18
Asbestos	2.0–2.8	—
Asphalt	1.1–1.5	—
Barium sulfate	—	4.5
Barytes, concrete	3.6–4.1	3.6
Bone	1.7–2.0	—
Brick, soft	1.4–1.9	1.65
hard	1.8–2.3	2.05
Cement	2.7–3.0	—
Concrete	2.25–2.4	2.35
Earth (dry packed)	—	1.5
Ebonite	—	1.15
Ferrophosphorus, aggregate concrete	—	4.8
Gelatin	—	1.27
Glass (common)	2.4–2.8	—
Glass (flint)	2.9–5.9	—
Granite	2.6–2.7	2.65
Gypsum	2.3–2.33	—
Lead	—	11.36
Lead glass	—	3.27
Lead glass (high-density)	—	6.22
Limestone	2.1–2.8	2.46
Linoleum	—	1.18
Marble	2.6–2.86	2.7
Paraffin	0.87–0.91	—
Plaster, sand	—	1.54
Plexiglas	—	1.18
Polyethylene	—	0.95
Pressed wood, pulpboard	—	0.19
Sand	1.6–1.9	—
Sandstone	—	1.90
Slate	2.6–3.3	—
Steel	—	7.8
Tile	1.6–2.5	1.9
Water	—	1.000
Water (heavy)	—	1.105
Wood: oak	0.60–0.90	—
pine	0.35–0.60	—

SOURCES: PHS 1970; NCRP 49, 1976.

lung photons is directly proportional to the atomic number of the absorber. Because of this relationship, β-shields are made with materials of minimum atomic number, seldom greater than 13 (aluminum). The flux of bremsstrahlung photons at distance r from a shielded β-source is estimated by:

$$\Phi = \frac{f\overline{E}_\beta}{4\pi r^2 E} \tag{10.5}$$

where Φ is in photons/cm^2·sec; $\overline{E}_\beta$ is the average β-energy emission rate [average β-energy (MeV/transformation) times activity (tps)] in MeV/sec; r is distance (radius of a sphere) in cm; and E is the energy of the bremsstrahlung photons (assumed equal to the maximum β-energy for exposure or dose calculations) in MeV/photon.

Alpha-particles are the least penetrating of the common forms of ionizing radiation. This radiation is essentially monoenergetic and has a finite range in matter. The range of α-particles in air (R, cm) at standard temperature and pressure is given by:

$$R = 0.056E\ (\text{MeV}) \qquad \text{for } E < 4 \text{ MeV} \tag{10.6a}$$

$$= 1.24E - 2.62 \qquad \text{for } 4 \leq E \leq 8 \text{ MeV} \tag{10.6b}$$

The range of α-particles in another medium R_i (mg/cm^2) is given by:

$$R_i = 0.56A^{1/3}R \tag{10.6c}$$

where A is the atomic number of the absorber (material i) and R is the range in air (cm).

Neutrons are classified by their kinetic energy. Those with energy >0.1 MeV are fast neutrons. Those with kinetic energy equal to that of gas molecules are thermal neutrons. The most probable energy, E_{mp}, of a thermal neutron is:

$$(E_{mp}) = kT \tag{10.7}$$

where k is Boltzmann's constant (1.38×10^{-23} J/K) and T is temperature (K). Neutrons between thermal and fast are classified as intermediate or slow neutrons. Fast neutrons are produced from neutron sources. These lose energy through elastic collisions with nuclei until they reach near-thermal energies. They are then captured by the nuclei of absorber atoms, which in turn may become unstable and undergo radioactive decay.

The removal of neutrons is exponential and follows the relationship

$$I = I_0 e^{-\sigma N t} \tag{10.8}$$

where I and I_0 are the final and initial intensity of neutrons, respec-

TABLE 10.5 Characteristics of (α, n) and (γ, n) Sources

Important (α, n) Sources					
Sources	Half-life	Maximum neutron energy (MeV)	Average neutron energy (MeV)	Yield $\left(\left[\frac{n}{sec \cdot Ci}\right] \times 10^{-6}\right)$	Remarks
$^{210}Po - Li$	138.40 days	1.32	0.48	0.05	
$^{210}Po - Be$	138.40 days	10.87	4.2	2.5	
RaD E F − Be	19.4 years	10.87	4.5	2.5	Po − Be with a long half-life
Ra − Be	1,622 years	13.08	3.9	15	
$^{222}Em - Be$	3.825 days	13.08		15	
$^{239}Pu - Be$	24,400 years	10.74	4.5	0.064 (per g)	
$^{227}Ac - Be$	21.8 years	12.79	4.6		Made by irradiating radium in reactor
$^{208}Po - Be$	2.93 years	10.71			
$RaBeF_4$	1,622 years	13.08		2.53	Proposed std. source
$^{210}Po - B$	138.40 days	^{10}B 6.29, ^{11}B 4.48		0.6	Relatively monoenergetic
Ra − B	1,622 years	^{10}B 8.58, ^{11}B 7.25		7	
$^{210}Po - F$	138.40 days	2.8	1.4	0.2	
$^{210}Po - Na$	138.40 days	4.45		0.04	
$^{241}Am - Be$	462 years				Suggested for stoichiometric std. source
$^{242}Cm - Be$	162.5 days				
Mock fission	138.40 days	10.87	1.6	0.4	

Important (γ, n) Sources

Sources	Half-life	E_γ	E_n	Standard yield[a]	Actual source yield[b]
$^{24}Na + Be$	14.8 hr	2.76	0.83	13	
$^{24}Na + D_2O$	14.8 hr	2.76	0.22	27	
$^{72}Ga + Be$	14.1 hr	1.87, 2.21, 2.51	(0.78)	5	
$^{88}Y + Be$	87 days	1.9, 2.8	0.158 ± 0.005	10	
$^{116}In + Be$	54 min	1.8, 2.1	0.30	0.82	
$^{124}Sb + Be$	60 days	1.7	0.024 ± 0.003	19	1.6
$^{140}La + Be$	40 days	2.50	0.62	0.3	
RdTh + D_2O	1.90 years	2.62 (ThC″)	0.197 ± 0.010	9.5	1.2[c]
MsTh + Be	6.7 years	1.80, 2.62	0.827 ± 0.030	3.5	
MsTh + D_2O	6.7 years	2.62 (ThC″)	0.197 ± 0.010	9.5	1.2[c]
Ra + Be	1622 years	1.69, 1.75, 1.82, 2.09, 2.20, 2.42	0.7 max		1.3

[a]This is the neutron yield × 10^{-4} for a 1-Ci γ-source with 1 g of target material placed 1 cm away from the γ-source.

[b]10^6 n/sec-Ci.

[c]Ms-Th and Rd-Th sources emit some neutrons through (α, n) reactions with light elements in the carrier and container walls.

NOTE: All photoneutron sources possess intense γ-ray backgrounds of at least 10^3 γ-rays per neutron.

SOURCE: NBS, 1960.

tively; σ is the microscopic cross section or probability of removal in barns (1 barn = 10^{-24} cm^2); N is the number of absorber atoms/cm^3; and t is the thickness of the absorber, cm. The macroscopic cross section, Σ, is defined as σN.

Absorbers with small atomic mass numbers are best at slowing fast neutrons. Once slowed to thermal energies, the capture of neutrons is a function of σ. The absorption cross section of many nuclei for thermal or near-thermal neutrons varies inversely with the square root of neutron energy (or inversely with velocity). Thermal neutron cross sections are usually given for a most probable energy of 0.025 eV. The cross section at other energies can be calculated from:

$$\frac{\sigma}{\sigma_0} = \left(\frac{E_0}{E}\right)^{0.5} \tag{10.9}$$

where σ_0 is the cross section of the absorber at energy E_0 (0.025 eV) and σ is the cross section of the absorber at energy E. Important thermal neutron capture reactions include:

^{1}H	(n, γ)	^{2}H	σ = 0.33 barns
^{14}N	(n, p)	^{14}C	σ = 1.70 barns
^{10}B	(n, α)	^{7}Li	σ = 4,010 barns
^{113}Cd	(n, γ)	^{114}Cd	σ = 21,000 barns

The reactions with H and N are important for radiation dose calculations, and the reactions with B and Cd are important for shielding applications. Note that the capture of a neutron by ^{1}H or ^{113}Cd produces a high-energy γ-photon, whereas the capture of a neutron by ^{14}N results in the release of a 0.6-MeV proton.

Radioactive neutron sources are of two types, alpha-neutron (α, n) and gamma-neutron (γ, n). In the (α, n) source, the α-emitter is mixed or alloyed with target material. In (γ, n) sources, the γ-emitter usually surrounds the neutron-producing material (e.g., Be or D_2O). The neutron spectra from most (α, n) sources extend up to the neighborhood of 10 MeV, whereas neutrons from (γ, n) sources are below 1 MeV (NCRP 38, 1971). Table 10.5 gives characteristics of (α, n) and (γ, n) sources. Special precautions must be taken when using sources based on reactions with ^{210}Po, ^{239}Pu, and ^{241}Am, as these elements are among the most dangerous when ingested or inhaled.

Constant-voltage accelerators can be used for the production of neutrons. For example, 100 to 200-KeV deuterons focused on a target of tritium (^{3}H) will produce 14-MeV neutrons. Large numbers of neutrons ($>10^{12}$/sec) can be produced by bombarding a thick beryllium target with charged particles. Neutron energies up to 20 MeV are obtainable from Van de Graaf accelerators for protons and deuterons, with higher energies for α-particles and heavy ions (NCRP 38, 1971). Figures 10.2*a*, 10.2*b*, and 10.2*c* show the most prolific neutron-producing reactions from accelerated protons, deuterons, and electrons, respectively, as a function of incident ion beam energy and target material. Figure 10.2*d* shows the trend of neutron energies as a function of incident particle energy for selected proton and deuteron reactions These figures indicate, for example, that the ^3H (d, n) ^4He reaction provides a high fluence rate of 14-MeV neutrons at deuteron energies above 0.1 MeV. Note that at low bombarding energies, the (p, n) and (γ, n) reactions are not as prolific in the production of neutrons as the (d, n) reactions.

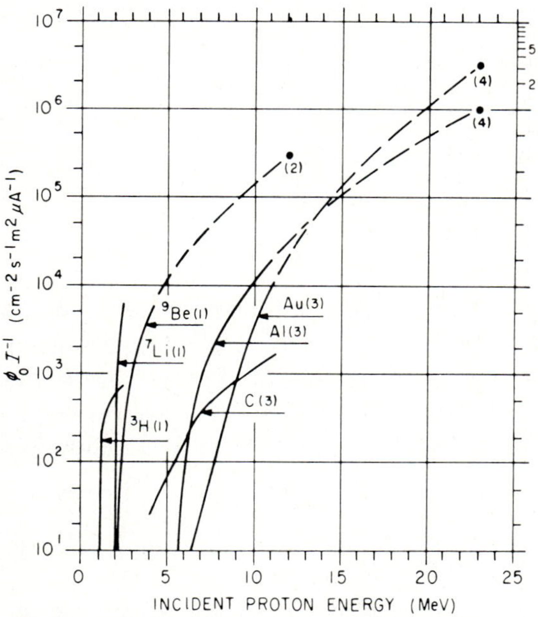

Figure 10.2*a* Thick-target neutron fluence rates for (p, n) reactions (*NCRP 51, 1977*). Thick-target neutron fluence rates from several (p, n) reactions, as a function of incident proton energy; measured in the 0° direction at 1 m from the target and divided by the target current ($cm^{-2} \cdot sec^{-1} \cdot m^2 \cdot \mu A^{-1}$). Neutron fluence rates at higher incident proton energies are estimated as follows:

Proton energy (MeV)	$\phi_0 I^{-1}$ (0°) ($cm^{-2} \cdot sec^{-1} \cdot m^2 \cdot \mu A^{-1}$)		
	C	Al	Ta or Pb
30	1.3×10^6	6×10^6	6×10^6
40	6×10^6	1.3×10^7	2.2×10^7

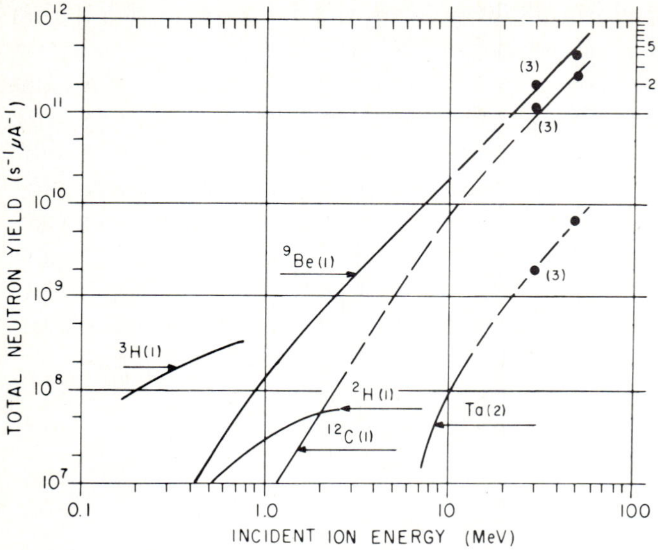

Figure 10.2*b* Thick-target neutron yields for (d, n) reactions (*NCRP 51, 1977*). Thick-target neutron yields (in 4π solid angle) per microampere of target current from several (d, n) reactions, as a function of incident deuteron energy. The yield from the Al (d, n) reaction is slightly less than that from the C (d, n) reaction.

2. Photons. The absorption of γ- and x-ray photons is a statistical phenomenon; that is, photon flux can be reduced by increasing absorber thickness. Unlike charged particles, all incident photons cannot be stopped. Good geometry is defined as a well-collimated, narrow beam of monoenergetic photons. Under these conditions, the relationship between absorber thickness and intensity is given by:

$$I = I_0 e^{-\mu_t t} \tag{10.10}$$

where I_0 = intensity at absorber thickness $t = 0$, I = intensity at thickness t, and μ_t = total attenuation coefficient (slope of absorption curve). The attenuation coefficient may be reported as a linear coefficient, $\mu_{t,l}$ with units cm^{-1}; or a mass coefficient, $\mu_{t,m}$ with units $cm^2 \cdot g^{-1}$. [Note that $\mu_{t,l} = \mu_{t,m}\rho$ when density, ρ, is in $g \cdot cm^{-3}$.] Table 10.6 gives mass attenuation coefficients for a variety of materials and energies. This is the coefficient used in shield design. Note that the mass attenuation coefficients for different materials are approximately equal for energies in the 1- to 5-MeV range.

The fraction of energy actually removed by the absorber is necessary for the estimation of radiation dose. This is characterized by the energy absorption coefficient, μ_e, and values are given in Table 10.7. This coefficient may also be expressed in mass or linear terms.

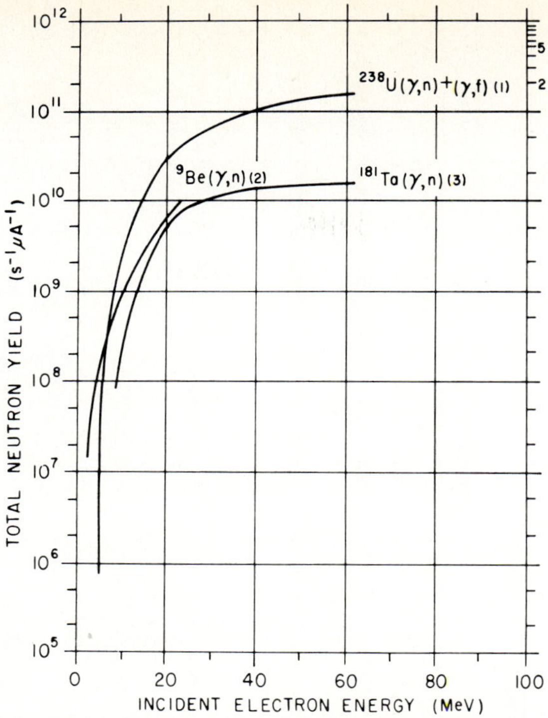

Figure 10.2c Thick-target neutron yields for (γ, n) reactions (*NCRP 51, 1977*). Thick-target neutron yields (in 4π solid angle) per microampere of target current from several photonuclear reactions, as a function of incident electron energy on the thick-target x-ray source. In the case of the ^{9}Be (γ, n) yields, a mass of Be was placed around the x-ray source. The ^{181}Ta target thickness was 6.3 g·cm^{-2}, serving as x-ray source as well as neutron source. The ^{238}U target was 5.9 g·cm^{-2} thick; the yields include the neutrons from the photofission of ^{238}U.

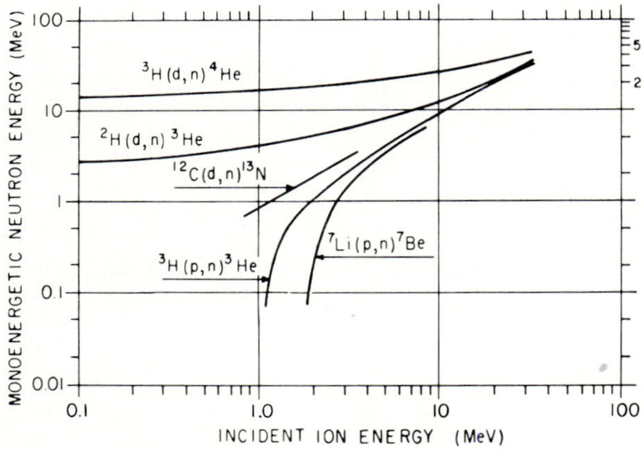

Figure 10.2*d* Maximum energies of neutrons produced from proton and deuteron reactions (*NCRP 51, 1977*). Energies of monoenergetic neutrons (in the laboratory system) from several nuclear reactions, as a function of incident proton or deuteron energy, emitted from the target in the 0° direction. Neutrons emitted in other directions are less energetic.

TABLE 10.6 Mass Attenuation Coefficients

Photon energy (MeV)	Attenuation coefficient, $\mu_{t,m}$, cm^2/g (for use in Eq. 10.10)								
	C	Al	Fe	Pb	H_2O	Air	Concrete	Bone	Muscle
0.010	2.28	26.3	173.	133.	5.18	4.99	26.9	20.3	5.27
0.015	0.787	7.93	56.4	115.	1.58	1.55	8.24	6.32	1.63
0.020	0.429	3.41	25.5	85.7	0.775	0.752	3.59	2.79	0.793
0.030	0.251	1.12	8.13	29.7	0.370	0.349	1.19	0.962	0.373
0.050	0.187	0.369	1.94	7.81	0.227	0.208	0.392	0.349	0.227
0.080	0.161	0.203	0.595	2.33	0.184	0.167	0.213	0.209	0.183
0.100	0.152	0.171	0.370	5.40	0.171	0.154	0.179	0.180	0.170
0.150	0.135	0.138	0.196	1.97	0.151	0.136	0.144	0.149	0.149
0.200	0.123	0.122	0.146	0.991	0.137	0.123	0.127	0.133	0.136
0.300	0.107	0.104	0.110	0.404	0.119	0.107	0.108	0.114	0.118
0.500	0.0872	0.0844	0.0840	0.161	0.0968	0.0870	0.0877	0.0927	0.0960
0.800	0.0709	0.0684	0.0669	0.0885	0.0786	0.0707	0.0709	0.0752	0.0779
1.0	0.0637	0.0613	0.0599	0.0708	0.0707	0.0636	0.0637	0.0676	0.0700
1.5	0.0519	0.0500	0.0488	0.0517	0.0575	0.0518	0.0519	0.0550	0.0570

2.0	0.0445	0.0432	0.0425	0.0455	0.0494	0.0445	0.0448	0.0473	0.0489
3.0	0.0357	0.0354	0.0362	0.0418	0.0397	0.0358	0.0365	0.0383	0.0393
5	0.0271	0.0284	0.0314	0.0424	0.0303	0.0275	0.0290	0.0297	0.0300
8	0.0216	0.0244	0.0298	0.0459	0.0243	0.0223	0.0245	0.0244	0.0240
10	0.0196	0.0231	0.0298	0.0484	0.0222	0.0204	0.0231	0.0226	0.0219
15	0.0170	0.0219	0.0307	0.0548	0.0194	0.0181	0.0215	0.0204	0.0192
20	0.0158	0.0216	0.0321	0.0606	0.0181	0.0170	0.0210	0.0194	0.0179
30	0.0147	0.0219	0.0345	0.0696	0.0171	0.0162	0.0210	0.0189	0.0168
50	0.0142	0.0230	0.0382	0.0804	0.0167	0.0161	0.0218	0.0190	0.0164
80	0.0144	0.0244	0.0416	0.0896	0.0170	0.0165	0.0229	0.0197	0.0167
100	0.0146	0.0251	0.0432	0.0934	0.0172	0.0168	0.0235	0.0201	0.0170
150	0.0150	0.0263	0.0458	0.0996	0.0178	0.0174	0.0247	0.0210	0.0175
200	0.0153	0.0271	0.0475	0.103	0.0182	0.0179	0.0254	0.0215	0.0179
300	0.0159	0.0282	0.0494	0.108	0.0188	0.0185	0.0264	0.0223	0.0185
500	0.0164	0.0292	0.0514	0.112	0.0195	0.0192	0.0273	0.0231	0.0192
800	0.0169	0.0300	0.0527	0.115	0.0200	0.0197	0.0281	0.0237	0.0197
1000	0.0171	0.0302	0.0532	0.116	0.0202	0.0199	0.0283	0.0239	0.0199

SOURCE: PHS, 1970.

TABLE 10.7 Mass Energy-Absorption Coefficients

Photon energy (MeV)	Mass energy-absorption coefficient, μ_e, cm^2/g (for use in Eqs. 10.13–10.15)			
	Water	Air	Compact bone	Muscle
0.010	4.89	4.66	19.0	4.96
0.015	1.32	1.29	5.89	1.36
0.020	0.523	0.516	2.51	0.544
0.030	0.147	0.147	0.743	0.154
0.040	0.0647	0.0640	0.305	0.0677
0.050	0.0394	0.0384	0.158	0.0409
0.060	0.0304	0.0292	0.0979	0.0312
0.080	0.0253	0.0236	0.0520	0.0255
0.10	0.0252	0.0231	0.0386	0.0252
0.15	0.0278	0.0251	0.0304	0.0276
0.20	0.0300	0.0268	0.0302	0.0297
0.30	0.0320	0.0288	0.0311	0.0317
0.40	0.0329	0.0296	0.0316	0.0325
0.50	0.0330	0.0297	0.0316	0.0327
0.60	0.0329	0.0296	0.0315	0.0326
0.80	0.0321	0.0289	0.0306	0.0318
1.0	0.0311	0.0280	0.0297	0.0308
1.5	0.0283	0.0255	0.0270	0.0281
2.0	0.0260	0.0234	0.0248	0.0257
3.0	0.0227	0.0205	0.0219	0.0225
4.0	0.0205	0.0186	0.0199	0.0203
5.0	0.0190	0.0173	0.0186	0.0188
6.0	0.0180	0.0163	0.0178	0.0178
8.0	0.0165	0.0150	0.0165	0.0163
10.0	0.0155	0.0144	0.0159	0.0154

SOURCE: PHS, 1970.

II. Sources and Uses of Ionizing Radiation

The principal sources of exposure to ionizing radiation include cosmic radiation, medical, industrial, and research use of x-rays, naturally occurring radionuclides, artificially produced nuclides, nuclear reactor operations, and weapons testing. Medical uses include diagnostic x-rays, isotopes for medical diagnosis, use of x-rays and radionuclides for therapeutic treatment, and the use of radionuclides in medical research. Industrial applications involve the use of x-rays for radiographic inspection and quantitative analysis (x-ray diffraction and fluorescence) and the use of isotopes for radiography (^{60}Co and ^{137}Cs), static elimination, and gauging (liquid levels in tanks, thickness of sheet materials and film, leveling of charge in foundry cupolas). Additional practical applications include tracing flow of materials in pipes, studying wear rates on metal parts, tracers in chemical and biological research, sterilization of food and medical supplies, energy sources for navigational beacons, nuclear power reactors, and chemical analysis (e.g., neutron activation). Some of the characteristics of neutron production devices have already been described.

A significant industrial hazard from the use of ionizing radiation involves radiography. Radiography is used for the examination of metal fabrications, including welds, castings, and forgings. It may be performed in shielded rooms (for components that can be transported) or in open shop areas, construction sites, on board ships, and along pipelines. Both x-ray and γ-ray sources are used. Gamma-ray sources can be small and can be used to inspect from the inside of a pipe. Short-wavelength γ-rays are effective for thick material sections. Neutron radiography can also be used for nondestructive testing of materials. In this process, neutrons strike a transfer screen to produce a radiation image after they have passed through the material. Because neutron absorption is qualitatively different from photon absorption (e.g., neutrons are more readily absorbed by hydrogen than by lead), it can show defects not detected by conventional radiography (Doyle et al., 1985).

Radiological hazards have also been reported in the liquefied petroleum gas (LPG) industry. The decay products of Ra-222, a contaminant in natural gas, are concentrated on the interior surfaces of plant machinery. For example, a pump impeller from an LPG refinery was found to emit 30 mrad per hour of combined β- and γ-radiation at a distance of 1 ft. When the surface of the impeller was cleaned, the airborne activity reached the level of 30 pCi/liter of ^{210}Po as measured in the breathing zone (Summerlin and Prichard, 1985). The maximum permissible exposure for a 40-hr-week air concentration of ^{210}Po is 2×10^{-10} μCi/cm^3 (Table 10.12). Cleaning the entire impeller (30 min) would be equivalent to breathing air at the MPC for 60 hr.

The design and manufacture of industrial x-ray generators are regulated by the Food and Drug Administration. The design standard for a nonmedical x-ray unit specifies a maximum leakage rate (radiation outside the useful beam) of $\leq$1 roentgen/hr at 1 m from the x-ray tube. (The design of shields and barriers for leakage and primary beam radiation is covered later in this chapter.) Radiographic operators are also required to wear personal monitoring devices (e.g., film badges, thermoluminescent dosimeters, or direct-reading pocket dosimeters), and must be trained in the use of radiation survey instruments.

The principal γ-sources used for industrial radiography are iridium-192 and cobalt-60. Typical sources contain up to 200 Ci of ^{192}Ir and 1000 Ci of ^{60}Co. The sources are sealed inside stainless steel capsules and stored in shielded containers. Radioisotope sources are favored for use in field construction sites or cramped environments, as they are easier to move and position than conventional x-ray generators. Additional safety precautions are required, as isotope sources continuously emit radiation and cannot be shut off like x-ray sources. Areas in which radiography is being performed must be secured and posted with warning signs (OSHA, 1983). In addition, operators must wear a direct-reading pocket dosimeter and a film badge or thermoluminescent dosimeter, and use calibrated radiaton survey equipment during all radiographic operations to assure that the source is properly shielded (Burgess, 1981).

III. Exposure Evaluation

Evaluation of an exposure to ionizing radiation includes an understanding of dosimetry and protection guidelines. A comparison of the estimated dose and appropriate protection guideline will provide the basis for engineering control calculations.

A. External radiation dosimetry

Exposure is a measure of photon flux and is defined as the energy transferred to a unit mass of air. The quantity of x- or γ-radiation that produces 1 coulomb of charge (C) per kilogram of air is defined as one X-unit. The X-unit is the SI unit of exposure. The more familiar unit of exposure is the roentgen (R), which is defined as the quantity of x- or γ-radiation that produces ions of one stat-coulomb charge (sc) per cubic centimeter of air at 0°C, 760 mmHg:

$$1 \text{ X-unit} = \frac{1 \text{ C}}{\text{kg air}} \qquad (10.11a)$$

$$1 \text{ R} = \frac{1 \text{ sc}}{\text{cm}^3 \text{ air}} \qquad (10.11b)$$

Exposure units are limited to photon energies below 3 MeV. Above this energy, exposure is expressed as $\text{W}\cdot\text{sec/m}^2$ and exposure rate as W/m^2.

A radiation dose is defined as the fraction of exposure energy absorbed (usually by tissue) and is a function of photon energy. Therefore, units of exposure and dose are very different. The SI unit of radiation dose is the gray (Gy), defined as 1 J absorbed per kilogram of material exposed. The traditional unit of dose is the radiation absorbed dose (rad), defined as 100 ergs per gram.

$$1 \text{ Gy} = \text{J/kg} \qquad (10.12a)$$

$$1 \text{ rad} = 100 \text{ ergs/g} \qquad (10.12b)$$

The relationship between exposure and dose in air is then:

$$1 \text{ X-unit} = \frac{1\text{ C}}{\text{kg air}} \frac{1\text{ ion}}{1.6 \times 10^{-19}\text{ C}} \times \frac{34\text{ eV}}{\text{ion in air}} 1.6 \times 10^{-19} \frac{\text{J}}{\text{eV}} \frac{\text{Gy}}{\text{J/kg}}$$

$$= 34 \text{ Gy in air}$$

$$1\text{ R} = \frac{1\text{ sc}}{\text{cm}^3} \frac{\text{ion}}{4.8 \times 10^{-10}\text{ sc}} \frac{\text{cm}^3}{0.00129\text{ g air}} \times \frac{34\text{ eV}}{\text{ion}} \frac{1.6 \times 10^{-19}\text{ J}}{\text{eV}} \frac{\text{erg}}{10^{-7}\text{ J}}$$

$$= 87.8 \frac{\text{ergs}}{\text{g}} = 0.87 \text{ rad in air}$$

The conversion between X-units and R is:

$$1 \text{ X-unit} = \frac{(34\text{ J/kg})(10^7\text{ ergs/J})(\text{kg}/10^3\text{g})}{(87.8\text{ ergs/g})/\text{R}}$$

$$= 3880 \text{ R}$$

The energy absorbed by matter is proportional to the electron density of the absorber. For example, muscle tissue (specific gravity $\simeq$ 1) is composed of 5.98×10^{22} H atoms/g, 2.75×10^{22} O atoms/g, 0.172×10^{22} N atoms/g, and 6.02×10^{21} C atoms/g. The density of electrons is then 3.28×10^{23} electrons/g. Air (density = 1.293 mg/cm^3, composition 79% N_2, 21% O_2) contains 2.1×10^{19} N_2 molecules/cm^3 and 5.6×10^{18} O_2 molecules/cm^3. The density of electrons is then 3.84×10^{20} electrons/cm^3 or 3.00×10^{23} electrons/g. The energy absorbed (J/kg tissue) corresponding to 1 X-unit (34 J/kg air) is then:

$$1 \text{ X-unit} = \frac{34\text{ J}}{\text{kg}} \frac{3.28 \times 10^{23}\ e/\text{g muscle}}{3.00 \times 10^{23}\ e/\text{g air}}$$

$$= 37 \text{ Gy in tissue}$$

Note that 1 R exposure $\simeq$ 1 rad absorbed tissue dose [(0.878)(3.28/3.00) = 0.96].

The exposure rate in air ($\dot{X}$, C/kg·sec) can be calculated from photon flux, energy, and the energy absorption coefficient and density of air from:

$$\dot{X} = \frac{(\Phi)(E)(1.6 \times 10^{-13})\mu_{e,\text{air}}}{(34)} \qquad (10.13)$$

where Φ is photon flux (photons/cm^2·sec); E is photon energy (MeV/photon); $\mu_{e,\text{air}}$ is the energy absorption coefficient for air expressed in terms of mass (cm^2/kg); 1.6×10^{-13} is the conversion from J to MeV;

and 34 is the conversion from X-units to Gy in air. The absorbed dose rate (Gy/sec) in tissue is given by:

$$\dot{D} = \frac{(\Phi)(\mathrm{E})(1.6 \times 10^{-13})(\mu_{e,\mathrm{tissue}})}{[(\mathrm{J/kg})/\mathrm{Gy}]} \qquad (10.14)$$

where $\mu_{e,\mathrm{tissue}}$ is the energy absorption coefficient for tissue ($\mathrm{cm^2/kg}$). The ratio of Eqs. (10.13) and (10.14) gives the relationship between exposure and dose such that:

$$\dot{D}\ (\mathrm{Gy/sec}) = 34 \frac{\mu_{e,\mathrm{tissue}}}{\mu_{e,\mathrm{air}}} \dot{X} \qquad (10.15a)$$

$$D\ (\mathrm{rad}) = 0.878 \frac{\mu_{e,\mathrm{tissue}}}{\mu_{e,\mathrm{air}}} \mathrm{R} \qquad (10.15b)$$

For photon energies in the range of Compton scattering (0.1–10 MeV), the dose-to-exposure ratio for tissue is approximately constant. For energies < 0.1 MeV, the photoelectric effect is the primary interaction mechanism. Since the cross section of the photoelectric effect is a function of atomic number, bone (10% Ca) absorbs more energy than soft tissues.

The specific γ-ray emission, Γ, is a unit of normalized exposure from a γ-source. Units of Γ are (C/kg)/hr at 1 m from a 1-MBq source (10^6 Bq) or R/hr at 1 m from a 1-Ci source. Γ can be calculated by determining the exposure rate $\dot{X}$ for each photon energy emitted [Eqs. (10.5) and (10.13), where f in Eq. (10.5) is the fraction of transformations that result in a photon of quantum energy under consideration]. The source strength Γ is then the sum of the exposures from each quantum energy from the γ-emitter. Table 10.8 lists specific γ-ray emission rates for a variety of radionuclides.

B. Internal radiation dosimetry

For a large energy-absorbing medium (or for radiation with limited penetration), the absorbed dose is approximately equal to the energy emitted by an internally deposited isotope (in Gy/day or rad/day). This occurs for most α- and β-radiation. The total dose is then influenced by two factors: the in situ decay of the isotope, and the biological elimination of the isotope (usually following first-order kinetics). The total dose, D, is:

$$D = \frac{D_0}{\lambda_E}(1 - \exp[-\lambda_E t]) \qquad (10.16)$$

where D_0 is the dose at time t equal to zero; and λ_E is the effective decay constant (in time^{-1}) $= \lambda_R + \lambda_B$, where R and B stand for radio-

TABLE 10.8 Gamma-Radiation Levels for 1 Ci of Some Radionuclides

Nuclide[c]	Γ^a	Nuclide[c]	Γ^a	Nuclide[c]	Γ^a
Actinium-227	2.2	Gold-198	2.3	Potassium-43	5.6
Antimony-122	2.4	Gold-199	0.9	Radium-226	8.25
Antimony-124	9.8	Hafnium-175	2.1	Radium-228	5.1
Antimony-125	2.7	Hafnium-181	3.1	Rhenium-186	0.2
Arsenic-72	10.1	Indium-114m	0.2	Rubidium-86	0.5
Arsenic-74	4.4	Iodine-124	7.2	Ruthenium-106	1.7
Arsenic-76	2.4	Iodine-125	0.7	Scandium-46	10.9
Barium-131	3.0	Iodine-126	2.5	Scandium-47	0.56
Barium-133	2.4	Iodine-130	12.2	Selenium-75	2.0
Barium-140	12.4	Iodine-131	2.2	Silver-110m	14.3
Beryllium-7	0.3	Iodine-132	11.8	Silver-111	0.2
Bromine-82	14.6	Iridium-192	4.8	Sodium-22	12.0
Cadmium-115m	0.2	Iridium-194	1.5	Sodium-24	18.4
Calcium-47	5.7	Iron-59	6.4	Strontium-85	3.0
Carbon-11	5.9	Krypton-85	0.04	Tantalum-182	6.8
Cerium-141	0.35	Lanthanum-140	11.3	Tellurium-121	3.3
Cerium-144	0.4	Lutetium-177	0.09	Tellurium-132	2.2
Cesium-134	8.7	Magnesium-28	15.7	Thulium-170	0.025
Cesium-137	3.3	Manganese-52	18.6	Tin-113	1.7
Chlorine-38[b]	8.8	Manganese-54	4.7	Tungsten-185	0.5
Chromium-51	0.16	Manganese-56	8.3	Tungsten-187	3.0
Cobalt-56	17.6	Mercury-197	0.4	Uranium-234	0.1
Cobalt-57	0.9	Mercury-203	1.3	Vanadium-48	15.6
Cobalt-58	5.5	Molybdenum-99	1.8	Xenon-133	0.1
Cobalt-60	13.2	Neodymium-147	0.8	Ytterbium-175	0.4
Copper-64	1.2	Nickel-65	3.1	Yttrium-88	14.1
Europium-152	5.8	Niobium-95	4.2	Yttrium-91	0.01
Europium-154	6.2	Osmium-191	0.6	Zinc-65	2.7
Europium-155	0.3	Palladium-109	0.03	Zirconium-95	4.1
Gallium-67	1.1	Platinum-197	0.5		
Gallium-72	11.6	Potassium-42	1.4		

[a] Γ = R·cm²/hr·mCi (Γ/10 = R·m²/hr·Ci).
[b] NBS, 1961.
[c] m indicates the isotope with the longest half-life.
SOURCE: Jaeger et al., 1968.

active and biological decay, respectively. Equation (10.16) can be expanded to the multicompartment case to track the movement of radioisotopes throughout the body.

Exposures of underground miners to high levels of radon and its daughter products have produced elevated risk of developing lung cancer (Tables 2.19, 2.20). The α-emitting, short-lived daughters are usually considered to be the principal hazard. Exposure is frequently expressed as working-level months (WLM). The WLM represents exposure for 170 hr to any combination of the short-lived daughters of radon that ultimately yield 1.3×10^5 MeV of potential energy per liter of air upon decay to ^{210}Pb. It is estimated that 1 WLM will deliver about 1 rad of α-radiation to the bronchial epithelium, the probable site of lung cancer incidence (NCRP 78, 1984).

Internal dosimetry for γ-emitters is more complex. In this case, the dose is not equal to total activity within the organ, since only a fraction of the energy originating in the tissue as photons is absorbed in that tissue. For a uniformly deposited γ-emitter, the dose at a point p from volume dV is given by:

$$dD = C\Gamma \frac{e^{-\mu r}}{r^2}\, dV \qquad (10.17)$$

where C = isotope concentration, r = distance from volume dV to point p, μ is the energy absorption coefficient, and Γ is the specific γ-emission of the internally deposited isotope. Equation (10.17) is then integrated over the entire volume of interest. Note that the solution will be a function of the geometry of the absorber modeled. Typical volumes studied are spheres and cylinders.

The Medical Internal Radiation Dose Committee of the Society of Nuclear Medicine (MIRD) has developed a method to account for the partial absorption of γ-energy in organs and tissues. The MIRD method gives the fraction of energy emitted by the source organ absorbed by a target organ (Brownell et al., 1968). The source and target organs can be the same or different.

Internal dosimetry from neutrons is computed from the energy absorbed by each element of the absorber. For fast neutrons, the dose absorbed locally is called the first collision dose, $\dot{D}_N$, and is calculated from:

$$\dot{D}_N = \Phi(E) \cdot E \cdot \sum N_i \sigma_i f_i \qquad \text{(in Gy)} \qquad (10.18)$$

where $\Phi(E)$ is the neutron flux at energy E, E is in joules, N_i is the number of atoms per kilogram of absorber of the ith element, σ_i is the scattering cross section of the ith element of energy E, and f_i is the mean fractional energy transferred to element i. For fast neutrons, the primary mechanism of energy transfer is elastic collisions.

C. Radiation protection guidelines

The toxicities of equal doses of different types of radiation are not equal. For example, neutrons and α-particles are more toxic than β- and γ-radiation when compared on an equal energy absorbed basis. In general, radiation with high linear energy transfer (LET) is more effective in damaging tissue. (High LET radiation rapidly loses energy as it travels through matter.) The relative biological effectiveness (RBE) of a form of ionizing radiation is defined as the ratio of the amount of 200-keV x-rays producing a given effect to the energy of any radiation producing the same effect. Therefore, RBE is a function of the specific type of radiation and of the species, organism, or organ exposed.

TABLE 10.9 Quality Factor and Linear Energy Transfer

LET (keV/μm in water)	QF
3.5 or less	1
3.5–7.0	1–2
7.0–23	2–5
23–53	5–10
53–175	10–20

The normalizing factors for the RBE of different types of radiation are quality factors (QF). The relationship between QF and LET is given in Table 10.9. Quality factors for common types of radiation are given in Table 10.10. The dose equivalent is the unit of absorbed dose that includes the RBE of radiation along with an additional factor for non-uniform distribution of internal radioisotopes. The SI unit of dose equivalent is the Sievert (Sv), defined as:

$$\text{Sievert} = D \cdot \text{QF} \cdot \text{DF} \tag{10.19a}$$

where D is the dose in Grays, QF is the quality factor (Table 10.10), and DF is a distribution factor (DF applies only to isotopes that concentrate in bone tissue). The cgs unit of dose equivalent is the rem, defined as:

$$\text{rem} = D(\text{rad}) \cdot \text{QF} \cdot \text{DF} \tag{10.19b}$$

The principal standards-setting organizations whose guidelines have been adopted by most federal, state, and local agencies are the International Commission on Radiological Protection (ICRP) and the National Council on Radiation Protection (NCRP). The philosophy behind radiological protection guidelines is that there is no threshold for the stochastic effects of ionizing radiation. Stochastic effects include cancer

TABLE 10.10 Quality Factor Values

Radiation	QF
Gamma-rays from radium in equilibrium (0.5-mm platinum filter)	1
x-Rays	1
Beta-rays and electrons; >0.03 MeV	1
Beta-rays and electrons; <0.03 MeV	1.7
Thermal neutrons	3
Fast neutrons	10
Protons	10
Alpha-rays	10
Heavy ions	20

and genetic alterations. The dose limits of the ICRP are based on the following three principles (ICRP, 1977):

1. No practice shall be adopted unless its introduction produces a net positive benefit.
2. All exposures shall be kept as low as reasonably achievable, economic and social factors being taken into account.
3. The dose equivalent to individuals shall not exceed the limits recommended for the appropriate circumstances by the commission.

For occupational exposures, the ICRP recommends annual dose equivalent limits of 50 rem to all tissues except the lens of the eye, 15 rem to the lens of the eye, and 5 rem for whole-body uniform radiation. For members of the general public, the ICRP recommends an annual limit of 500 mrem whole-body radiation. On average, the whole-body dose will be much less than 500 mrem. No specific limit is given for medical exposures, although it is recommended that only necessary exposures be made.

Table 10.11 lists NCRP dose limitations. These recommendations have been adopted by the ACGIH as the TLV for ionizing radiation. Note that up to one-half of the allowable annual exposure can be received in one calendar quarter. For children less than 16 years old, the annual limit to the thyroid is 1.5 rem. These recommendations are directly applicable to external radiation hazards. For inhaled or ingested radiation, dose is a function of target organ, biological half-life, character of radiation, and dose limitations. Tables 10.12*a*, 10.12*b*, and 10.12*c* list the maximum permissible body burden and concentrations in air and water for a variety of nuclides. These recommendations are based on the kinetics of radioisotopes in the body and assure that the dose equivalent will be less than 100 mrem per week in the total body, gonads, or hemopoietic system; 600 mrem per week to the thyroid; or 300 mrem per week for any other critical organ.

The Occupational Safety and Health Administration's regulations are very similar to the NCRP guidelines (OSHA, 1983). Exposure to individuals in restricted areas (an area where access is controlled by the employer for purposes of protection of individuals from exposure to radiation) in any period of one calendar quarter is limited as follows:

	rems/calendar quarter
Whole body: Head and trunk; active blood-forming organs; lens of eyes; or gonads	1.25
Hands and forearms; feet and ankles	8.75
Skin of whole body	0.50

TABLE 10.11 Dose-Limiting Recommendations

Maximum permissible dose equivalent for occupational exposure	
Combine whole-body occupational exposure	
Prospective annual limit	5 rem in any one year
Retrospective annual limit	10–15 rem in any one year
Long-term accumulation to age N years	$(N - 18) \times 5$ rem
Skin	15 rem in any one year
Hands	75 rem in any one year (25/qtr)
Forearms	30 rem in any one year (10/qtr)
Other organs, tissues, and organ systems	15 rem in any one year (5/qtr)
Fertile women (with respect to fetus)	0.5 rem in gestation period
Dose limits for the public, or occasionally exposed individuals	
Individual or occasionally exposed	0.5 rem in any one year
Students	0.1 rem in any one year
Population dose limits	
Genetic	0.17 rem average per year
Somatic	0.17 rem average per year
Emergency dose limits—life saving	
Individual (older than 45 years if possible)	100 rem
Hands and forearms	200 rem, additional (300 rem, total)
Emergency dose limits—less urgent	
Individual	25 rem
Hands and forearms	100 rem, total
Family of radioactive patients	
Individual (under age 45)	0.5 rem in any one year
Individual (over age 45)	5 rem in any one year

SOURCE: NCRP 39, 1971.

An individual in a restricted area may receive a dose above these levels as long as during any calendar quarter, the dose to the whole body is less than 3 rem, and the accumulated occupational dose is less than $5 \times (N - 18)$ rem, where N is the individual's age in years. Exposure limits for employees under 18 years of age are 10% of the limits specified above. The OSHA regulations also contain limits to the concentration of airborne materials similar to Table 10.12 (10 CFR Part 20, app. B in OSHA, 1983).

D. Administrative control

Limiting exposure time can be used to minimize ionizing radiation exposures and hazards. Based on a maximum allowable dose equivalent of 2000 mrem per calendar quarter (up to a total of 5000 mrem per year), exposures should be limited to 160 mrem per week to meet the

TABLE 10.12*a* Maximum Permissible Body Burdens and Maximum Permissible Concentrations of Radionuclides in Air and in Water for Occupational Exposure

Radionuclide and type of decay[a]		Organ of reference[b] (critical organ underlined)	Maximum permissible burden in total body q (μCi)	Maximum permissible concentrations[c] For 40-hr week $(MPC)_w$ (μCi/cm^3)	For 40-hr week $(MPC)_a$ (μCi/cm^3)	For 168-hr week[d] $(MPC)_w$ (μCi/cm^3)	For 168-hr week[d] $(MPC)_a$ (μCi/cm^3)
$^3_1$H(HTO or	(Soluble)	Body tissue	10^3	0.1	2×10^{-5}	0.03	5×10^{-6}
$^3_1H_2O$)(β^-)		Total body	2×10^6	0.2	2×10^{-5}	0.05	7×10^{-6}
($^3_1H_2$)	(Immersion)	Skin	—	—	2×10^{-3}	—	4×10^{-4}
$^{14}_6$C(CO_2)(β^-)	(Soluble)	Fat	300	0.02	4×10^{-6}	8×10^{-3}	10^{-6}
		Total body	400	0.03	5×10^{-6}	0.01	2×10^{-6}
		Bone	400	0.04	6×10^{-6}	0.01	2×10^{-6}
	(Immersion)	Total body	—	—	5×10^{-5}	—	10^{-5}
$^{32}_{15}$P (β^-)	(Soluble)	Bone	6	5×10^{-4}	7×10^{-8}	2×10^{-4}	2×10^{-8}
		Total body	30	3×10^{-3}	4×10^{-7}	9×10^{-4}	10^{-7}
		GI (LLI)	—	3×10^{-3}	6×10^{-7}	9×10^{-4}	2×10^{-7}
		Liver	50	5×10^{-3}	6×10^{-7}	2×10^{-3}	2×10^{-7}
		Brain	300	0.02	3×10^{-6}	8×10^{-3}	10^{-6}
	(Insoluble)	Lung	—	—	8×10^{-8}	—	3×10^{-8}
		GI (LLI)	—	7×10^{-4}	10^{-7}	2×10^{-4}	4×10^{-8}
$^{45}_{20}$Ca (β^-)	(Soluble)	Bone	30	3×10^{-4}	3×10^{-8}	9×10^{-5}	10^{-8}
		Total body	200	2×10^{-3}	3×10^{-7}	7×10^{-4}	9×10^{-8}
		GI (LLI)	—	0.01	3×10^{-6}	4×10^{-3}	10^{-6}
	(Insoluble)	Lung	—	—	10^{-7}	—	4×10^{-8}
		GI (LLI)	—	5×10^{-3}	9×10^{-7}	2×10^{-3}	3×10^{-7}
$^{51}_{24}$Cr (ε, γ)	(Soluble)	GI (LLI)	—	0.05	10^{-5}	0.02	4×10^{-6}
		Total body	800	0.6	10^{-5}	0.2	4×10^{-6}
		Lung	10^3	1	2×10^{-5}	0.4	8×10^{-6}
		Prostate	2×10^3	2	3×10^{-5}	0.5	10^{-5}
		Thyroid	4×10^3	3	6×10^{-5}	1	2×10^{-5}

		Kidney	8×10^{3}	6	10^{-4}	2	4×10^{-5}
	(Insoluble)	Lung	—	—	2×10^{-6}	—	8×10^{-7}
		GI (LLI)	—	0.05	8×10^{-6}	0.02	3×10^{-6}
$^{60}_{27}$Co (ε, γ)	(Soluble)	GI (LLI)	—	10^{-3}	3×10^{-7}	5×10^{-4}	10^{-7}
		Total body	10	4×10^{-3}	4×10^{-7}	10^{-3}	10^{-7}
		Pancreas	70	0.02	2×10^{-6}	7×10^{-3}	6×10^{-7}
		Liver	90	0.03	10^{-6}	9×10^{-3}	5×10^{-7}
		Spleen	200	0.05	4×10^{-6}	0.02	2×10^{-6}
		Kidney	200	0.07	6×10^{-6}	0.03	2×10^{-6}
	(Insoluble)	Lung	—	—	9×10^{-9}	—	3×10^{-9}
		GI (LLI)	—	10^{-3}	2×10^{-7}	3×10^{-4}	6×10^{-8}
$^{65}_{30}$Zn (β^{+}, ε, γ)	(Soluble)	Total body	60	3×10^{-3}	10^{-7}	10^{-3}	4×10^{-8}
		Prostate	70	4×10^{-3}	10^{-7}	10^{-3}	4×10^{-8}
		Liver	80	4×10^{-3}	10^{-7}	10^{-3}	5×10^{-8}
		Kidney	100	6×10^{-3}	2×10^{-7}	2×10^{-3}	7×10^{-8}
		GI (LLI)	—	6×10^{-3}	10^{-6}	2×10^{-3}	4×10^{-7}
		Pancreas	200	7×10^{-3}	3×10^{-7}	3×10^{-3}	9×10^{-8}
		Muscle	200	0.01	4×10^{-7}	4×10^{-3}	10^{-7}
		Ovary	300	0.01	5×10^{-7}	4×10^{-3}	2×10^{-7}
		Testis	400	0.02	6×10^{-7}	6×10^{-3}	2×10^{-7}
		Bone	700	0.04	10^{-6}	0.01	4×10^{-7}
	(Insoluble)	Lung	—	—	6×10^{-8}	—	2×10^{-8}
		GI (LLI)	—	5×10^{-3}	9×10^{-7}	2×10^{-3}	3×10^{-7}
$^{76}_{33}$As (β^{-}, γ)	(Soluble)	GI (LLI)	—	6×10^{-4}	10^{-7}	2×10^{-4}	4×10^{-8}
		Total body	20	0.4	5×10^{-6}	0.1	2×10^{-6}
		Kidney	20	0.6	8×10^{-6}	0.2	3×10^{-6}
		Liver	40	1	10^{-5}	0.4	5×10^{-6}
	(Insoluble)	GI (LLI)	—	6×10^{-4}	10^{-7}	2×10^{-4}	3×10^{-8}
		Lung	—	—	6×10^{-7}	—	2×10^{-7}
$^{89}_{38}$Sr (β^{-})	(Soluble)	Bone	4	3×10^{-4}	3×10^{-8}	10^{-4}	10^{-8}
		GI (LLI)	—	10^{-3}	3×10^{-7}	4×10^{-4}	9×10^{-8}
		Total body	40	2×10^{-3}	2×10^{-7}	7×10^{-4}	6×10^{-8}
	(Insoluble)	Lung	—	—	4×10^{-8}	—	10^{-8}
		GI (LLI)	—	8×10^{-4}	10^{-7}	3×10^{-4}	5×10^{-8}

TABLE 10.12*a* Maximum Permissible Body Burdens and Maximum Permissible Concentrations of Radionuclides in Air and in Water for Occupational Exposure (*Continued*)

Radionuclide and type of decay[a]		Organ of reference[b] (critical organ underlined)	Maximum permissible burden in total body q (μCI)	Maximum permissible concentrations[c]			
				For 40-hr week		For 168-hr week[d]	
				$(MPC)_w$ (μCi/cm^3)	$(MPC)_a$ (μCi/cm^3)	$(MPC)_w$ (μCi/cm^3)	$(MPC)_a$ (μCi/cm^3)
$^{90}_{38}$Sr (β^-)	(Soluble)	Bone	2	4×10^{-6}	3×10^{-10}	10^{-6}	10^{-10}
		Total body	20	10^{-5}	9×10^{-10}	4×10^{-6}	3×10^{-10}
		GI (LLI)	—	10^{-3}	3×10^{-7}	5×10^{-4}	10^{-7}
	(Insoluble)	Lung	—	—	5×10^{-9}	—	2×10^{-9}
		GI (LLI)	—	10^{-3}	2×10^{-7}	4×10^{-4}	6×10^{-8}
$^{95}_{40}$Zr (β^-, γ, e^-)	(Soluble)	GI (LLI)	—	2×10^{-3}	4×10^{-7}	6×10^{-4}	10^{-7}
		Total body	20	3	10^{-7}	1	4×10^{-8}
		Bone	30	4	2×10^{-7}	2	6×10^{-8}
		Kidney	30	4	2×10^{-7}	2	6×10^{-8}
		Liver	40	6	3×10^{-7}	2	9×10^{-8}
		Spleen	40	7	3×10^{-7}	2	10^{-7}
	(Insoluble)	Lung	—	—	3×10^{-8}	—	10^{-8}
		GI (LLI)	—	2×10^{-3}	3×10^{-7}	6×10^{-4}	10^{-7}
$^{95}_{41}$Nb (β^-, γ)	(Soluble)	GI (LLI)	—	3×10^{-3}	6×10^{-7}	10^{-3}	2×10^{-7}
		Total body	40	10	5×10^{-7}	4	2×10^{-7}
		Liver	60	20	7×10^{-7}	6	3×10^{-7}
		Kidney	60	20	8×10^{-7}	6	3×10^{-7}
		Bone	80	20	9×10^{-7}	7	3×10^{-7}
		Spleen	80	20	10^{-6}	7	3×10^{-7}
	(Insoluble)	Lung	—	—	10^{-7}	—	3×10^{-8}
		GI (LLI)	—	3×10^{-3}	5×10^{-7}	10^{-3}	2×10^{-7}
$^{106}_{44}$Ru (β^-, γ)	(Soluble)	GI (LLI)	—	4×10^{-4}	8×10^{-8}	10^{-4}	3×10^{-8}
		Kidney	3	0.01	10^{-7}	4×10^{-3}	5×10^{-8}
		Bone	10	0.04	5×10^{-7}	0.01	2×10^{-7}
		Total body	10	0.06	7×10^{-7}	0.02	3×10^{-7}
	(Insoluble)	Lung	—	—	6×10^{-9}	—	2×10^{-9}
		GI (LLI)	—	3×10^{-4}	6×10^{-8}	10^{-4}	2×10^{-8}

$^{131}_{53}I$ (β^-, γ, e^-)	(Soluble)	Thyroid	0.7	6×10^{-5}	9×10^{-9}	2×10^{-5}	3×10^{-9}
		Total body	50	5×10^{-3}	8×10^{-7}	2×10^{-3}	3×10^{-7}
		GI (LLI)	—	0.03	7×10^{-6}	0.01	2×10^{-6}
	(Insoluble)	GI (LLI)	—	2×10^{-3}	3×10^{-7}	6×10^{-4}	10^{-7}
		Lung	—	—	3×10^{-7}	—	10^{-7}
$^{137}_{55}Cs$ (β^-, γ, e^-)	(Soluble)	Total body	30	4×10^{-4}	6×10^{-8}	2×10^{-4}	2×10^{-8}
		Liver	40	5×10^{-4}	8×10^{-8}	2×10^{-4}	3×10^{-8}
		Spleen	50	6×10^{-4}	9×10^{-8}	2×10^{-4}	3×10^{-8}
		Muscle	50	7×10^{-4}	10^{-7}	2×10^{-4}	4×10^{-8}
		Bone	100	10^{-3}	2×10^{-7}	5×10^{-4}	7×10^{-8}
		Kidney	100	10^{-3}	2×10^{-7}	5×10^{-4}	8×10^{-8}
		Lung	300	5×10^{-3}	6×10^{-7}	2×10^{-3}	2×10^{-7}
		GI (SI)	—	0.02	5×10^{-6}	8×10^{-3}	2×10^{-6}
	(Insoluble)	Lung	—	—	10^{-8}	—	5×10^{-9}
		GI (LLI)	—	10^{-3}	2×10^{-7}	4×10^{-4}	8×10^{-8}
$^{144}_{58}Ce$ (α, β^-, γ)	(Soluble)	GI (LLI)	—	3×10^{-4}	8×10^{-8}	10^{-4}	3×10^{-8}
		Bone	5	0.2	10^{-8}	0.08	3×10^{-9}
		Liver	6	0.3	10^{-8}	0.1	4×10^{-9}
		Kidney	10	0.5	2×10^{-8}	0.2	7×10^{-9}
		Total body	20	0.7	3×10^{-8}	0.3	10^{-8}
	(Insoluble)	Lung	—	—	6×10^{-9}	—	2×10^{-9}
		GI (LLI)	—	3×10^{-4}	6×10^{-8}	10^{-4}	2×10^{-8}
$^{147}_{61}Pm$ (α, β^-)	(Soluble)	GI (LLI)	—	6×10^{-3}	10^{-6}	2×10^{-3}	5×10^{-7}
		Bone	60	1	6×10^{-8}	0.5	2×10^{-8}
		Kidney	200	4	2×10^{-7}	2	7×10^{-8}
		Total body	300	7	3×10^{-7}	2	10^{-7}
		Liver	300	8	4×10^{-7}	3	10^{-7}
	(Insoluble)	Lung	—	—	10^{-7}	—	3×10^{-8}
		GI (LLI)	—	6×10^{-3}	10^{-6}	2×10^{-3}	4×10^{-7}
$^{182}_{73}Ta$ (β^-, γ)	(Soluble)	GI (LLI)	—	10^{-3}	3×10^{-7}	4×10^{-4}	9×10^{-8}
		Liver	7	0.9	4×10^{-8}	0.3	10^{-8}
		Kidney	20	2	8×10^{-8}	0.7	3×10^{-8}
		Total body	20	2	9×10^{-8}	0.7	3×10^{-8}
		Spleen	30	4	10^{-7}	1	5×10^{-8}
		Bone	50	6	3×10^{-7}	2	9×10^{-8}
	(Insoluble)	Lung	—	—	2×10^{-8}	—	7×10^{-9}
		GI (LLI)	—	10^{-3}	2×10^{-7}	4×10^{-4}	7×10^{-8}

TABLE 10.12*a* Maximum Permissible Body Burdens and Maximum Permissible Concentrations of Radionuclides in Air and in Water for Occupational Exposure (*Continued*)

Radionuclide and type of decay[a]		Organ of reference[b] (critical organ underlined)	Maximum permissible burden in total body q (μCi)	Maximum permissible concentrations[c] For 40-hr week $(MPC)_w$ ($\mu Ci/cm^3$)	For 40-hr week $(MPC)_a$ ($\mu Ci/cm^3$)	For 168-hr week[d] $(MPC)_w$ ($\mu Ci/cm^3$)	For 168-hr week[d] $(MPC)_a$ ($\mu Ci/cm^3$)
$^{192}_{77}Ir$ (β^-, γ)	(Soluble)	GI (LLI)	—	10^{-3}	3×10^{-7}	4×10^{-4}	9×10^{-8}
		Kidney	6	4×10^{-3}	10^{-7}	10^{-3}	4×10^{-8}
		Spleen	7	4×10^{-3}	10^{-7}	10^{-3}	5×10^{-8}
		Liver	8	5×10^{-3}	2×10^{-7}	2×10^{-3}	6×10^{-8}
		Total body	20	0.01	4×10^{-7}	4×10^{-3}	10^{-7}
	(Insoluble)	Lung	—	—	3×10^{-8}	—	9×10^{-9}
		GI (LLI)	—	10^{-3}	2×10^{-7}	4×10^{-4}	6×10^{-8}
$^{198}_{79}Au$ (β^-, γ)	(Soluble)	GI (LLI)	—	2×10^{-3}	3×10^{-7}	5×10^{-4}	10^{-7}
		Kidney	20	0.07	3×10^{-6}	0.02	9×10^{-7}
		Total body	30	0.1	4×10^{-6}	0.04	2×10^{-6}
		Spleen	60	0.2	8×10^{-6}	0.07	3×10^{-6}
		Liver	80	0.3	10^{-5}	0.1	4×10^{-6}
	(Insoluble)	GI (LLI)	—	10^{-3}	2×10^{-7}	5×10^{-4}	8×10^{-8}
		Lung	—	—	6×10^{-7}	—	2×10^{-7}
$^{210}_{84}Po$ (α)	(Soluble)	Spleen	0.03	2×10^{-5}	5×10^{-10}	7×10^{-6}	2×10^{-10}
		Kidney	0.04	2×10^{-5}	5×10^{-10}	8×10^{-6}	2×10^{-10}
		Liver	0.1	7×10^{-5}	2×10^{-9}	3×10^{-5}	6×10^{-10}
		Total	0.4	2×10^{-4}	5×10^{-9}	8×10^{-5}	2×10^{-9}
		Bone	0.5	3×10^{-4}	7×10^{-9}	10^{-4}	2×10^{-9}
		GI (LLI)	—	9×10^{-4}	2×10^{-7}	3×10^{-4}	7×10^{-8}
	(Insoluble)	Lung	—	—	2×10^{-10}	—	7×10^{-11}
		GI (LLI)	—	8×10^{-4}	2×10^{-7}	3×10^{-4}	5×10^{-8}
$^{222}_{86}Rn$[e] (α, β^-, γ)		Lung	—	—	3×10^{-8}	—	10^{-8}
$^{226}_{88}Ra$ (α, β^-, γ)	(Soluble)	Bone	0.1	4×10^{-7}	3×10^{-11}	10^{-7}	10^{-11}
		Total body	0.2	6×10^{-7}	5×10^{-11}	2×10^{-7}	2×10^{-11}
		GI (LLI)	—	10^{-3}	3×10^{-7}	5×10^{-4}	10^{-7}

	(Insoluble)	Lung	—	—	5×10^{-11}	—	2×10^{-11}
		GI (LLI)	—	9×10^{-4}	2×10^{-7}	3×10^{-4}	6×10^{-8}
${}^{235}_{92}U^{f}$ (α, β^-, γ)	(Soluble)	GI (LLI)	—	8×10^{-4}	2×10^{-7}	3×10^{-4}	6×10^{-8}
		Kidney	0.03	10^{-4}	5×10^{-10}	4×10^{-5}	2×10^{-10}
		Bone	0.06	10^{-4}	6×10^{-10}	5×10^{-5}	2×10^{-10}
		Total body	0.4	4×10^{-4}	2×10^{-9}	10^{-4}	6×10^{-10}
	(Insoluble)	Lung	—	—	10^{-10}	—	4×10^{-11}
		GI (LLI)	—	8×10^{-4}	10^{-7}	3×10^{-4}	5×10^{-8}
${}^{238}_{92}U^{f}$ (α, γ, e^-)	(Soluble)	GI (LLI)	—	10^{-3}	2×10^{-7}	4×10^{-4}	8×10^{-8}
		Kidney	5×10^{-3}	2×10^{-5}	7×10^{-11}	6×10^{-6}	3×10^{-11}
		Bone	0.06	10^{-4}	6×10^{-10}	5×10^{-5}	2×10^{-10}
		Total body	0.5	4×10^{-4}	2×10^{-9}	10^{-4}	6×10^{-10}
	(Insoluble)	Lung	—	—	10^{-10}	—	5×10^{-11}
		GI (LLI)	—	10^{-3}	2×10^{-7}	4×10^{-4}	6×10^{-8}
${}^{239}_{94}Pu$ (α, γ)	(Soluble)	Bone	0.04	10^{-4}	2×10^{-12}	5×10^{-5}	6×10^{-13}
		Liver	0.4	5×10^{-4}	7×10^{-12}	2×10^{-4}	2×10^{-12}
		Kidney	0.5	7×10^{-4}	9×10^{-12}	2×10^{-4}	3×10^{-12}
		GI (LLI)	—	8×10^{-4}	2×10^{-7}	3×10^{-4}	6×10^{-8}
		Total body	0.4	10^{-3}	10^{-11}	3×10^{-4}	5×10^{-12}
	(Insoluble)	Lung	—	—	4×10^{-11}	—	10^{-11}
		GI (LLI)	—	8×10^{-4}	2×10^{-7}	3×10^{-4}	5×10^{-8}

[a]Electron (β^-), positron (β^+), electron capture (ε), gamma (γ), alpha (α), conversion electron (e^-).

[b]The abbreviations GI, S, SI, ULI, and LLI refer to gastrointestinal tract, stomach, small intestines, upper large intestine, and lower large intestine, respectively.

[c]Subscript w for water, subscript a for air.

[d]It will be noted that the MPC values for the 168-h week are not always precisely the same multiples of the MPC for the 40-hr week. Part of this is caused by rounding off the calculated values to one digit, but in some instances it is due to technical differences discussed in the ICRP report. Because of the uncertainties present in much of the biological data and because of individual variations, the differences are not considered significant. The MPC values for the 40-hr week are to be considered as basic for occupational exposure, and the values for the 168-hr week are basic for continuous exposure as in the case of the population at large.

[e]The daughter isotopes of ${}^{220}Rn$ and ${}^{222}Rn$ are assumed present to the extent that they occur in unfiltered air. For all other isotopes the daughter elements are not considered as part of the intake and if present must be considered on the basis of the rules for mixtures.

[f]ICRP Publication 6 (as amended and revised, 1962).

SOURCES: Selected from NBS, 1963; ICRP, 1962.

TABLE 10.12*b*   Maximum Permissible Concentrations of Unidentified Radionuclides in Water, $(MPCU)_w$ Values, for Continuous Occupational Exposure[a]

Limitations	$\mu Ci/cm^3$ of water[b]
If no one of the radionuclides ^{90}Sr, ^{126}I, ^{129}I, ^{131}I, ^{210}Pb, ^{210}Po, ^{211}At, ^{223}Ra, ^{224}Ra, ^{226}Ra, ^{228}Ra, ^{227}Ac, ^{230}Th, ^{231}Pa, ^{232}Th, and Th-nat is present, then the $(MPCU)_w$ is	3×10^{-5}
If no one of the radionuclides ^{90}Sr, ^{129}I, ^{210}Pb, ^{210}Po, ^{223}Ra, ^{226}Ra, ^{228}Ra, ^{231}Pa, Th-nat is present, then the $(MPCU)_w$ is	2×10^{-5}
If no one of the radionuclides ^{90}Sr, ^{129}I, ^{210}Pb, ^{226}Ra, and ^{228}Ra is present, then the $(MPCU)_w$ is	7×10^{-6}
If neither ^{226}Ra nor ^{228}Ra is present, then the $(MPCU)_w$ is	10^{-6}
If no analysis of the water is made, then the $(MPCU)_w$ is	10^{-7}

[a]Each $(MPCU)_w$ value is the smallest value of $(MPC)_w$ in Table 1 of NBS Handbook 69 for radionuclides other than those listed opposite the value. Thus these $(MPCU)_w$ values are permissible levels for continuous occupational exposure (168 hr/wk) for any radionuclide or mixture of radionuclides where the indicated isotopes are not present (i.e., where the concentration of the radionuclide in water is small compared with the $(MPC)_w$ value for this radionuclide). The $(MPCU)_w$ may be much smaller than the more exact maximum permissible concentration of the material, but the determination of this $(MPC)_w$ requires identification of the radionuclides present and the concentration of each.

[b]Use one-tenth of these values for interim application in the neighborhood of a controlled exposure area.

SOURCE: Selected from NBS, 1963.

TABLE 10.12c Maximum Permissible Concentrations of Unidentified Radionuclides in Air, $(MPCU)_a$ Values, for Continuous Occupational Exposure[a]

Limitations	$\mu Ci/cm^3$ of air[b]
If there are no α-emitting radionuclides and if no one of the β-emitting radionuclides ^{90}Sr, ^{129}I, ^{210}Pb, ^{227}Ac, ^{228}Ra, ^{230}Pa, ^{241}Pu, and ^{249}Bk is present, then the $(MPCU)_a$ is	10^{-9}
If there are no α-emitting radionuclides and if no one of the β-emitting radionuclides ^{210}Pb, ^{227}Ac, ^{228}Ra, and ^{241}Pu is present, then the $(MPCU)_a$ is	10^{-10}
If there are no α-emitting radionuclides and if the β-emitting radionuclide ^{227}Ac is not present, then the $(MPCU)_a$ is	10^{-11}
If no one of the radionuclides ^{227}Ac, ^{230}Th, ^{231}Pa, ^{232}Th, Th-nat, ^{238}Pu, ^{239}Pu, ^{240}Pu, ^{242}Pu, and ^{249}Cf is present, then the $(MPCU)_a$ is	10^{-12}
If no one of the radionuclides ^{231}Pa, Th-nat, ^{239}Pu, ^{240}Pu, ^{242}Pu, and ^{249}Cf is present, then the $(MPCU)_a$ is	7×10^{-13}
If no analysis of the air is made, then the $(MPCU)_a$ is	4×10^{-13}

[a]Each $(MPCU)_a$ value is the smallest value of $(MPC)_a$ in Table 1 of NBS Handbook 69 for radionuclides other than those listed opposite the value. Thus these $(MPCU)_a$ values are permissible levels for continuous occupational exposure (168 hr/wk) for any radionuclide or mixture of radionuclides where the indicated isotopes are not present (i.e., where the concentration of the radionuclide in air is small compared with the $(MPC)_a$ value for this radionuclide). The $(MPCU)_a$ value may be much smaller than the more exact maximum permissible concentration of the material, but the determination of the $(MPC)_a$ requires identification of the radionuclides present and the concentration of each.

[b]Use one-tenth of these values for interim application in the neighborhood of a controlled exposure area.

SOURCE: Selected from NBS 1963.

quarterly limit, or 20 mrem per day for the annual limit based on 5 days per week. For example, if the exposure rate in a radioisotope storage room is 25 mrem per hour, limiting daily exposure time to 48 min will control daily and annual exposures to 20 and 5000 mrem, the occupationally acceptable maximum dose.

Increasing the distance from the source to operator can also be used to reduce or limit exposures. For example, dose from a point source follows the inverse-square law with distance. The specific γ-emission from ^{60}Co is 1.3 R·m²/Ci·hr. For a 100-mCi source, the exposure at 1 m is:

$$(1.3\ \mathrm{R \cdot m^2/Ci \cdot hr})(0.1\ \mathrm{Ci})\left(\frac{1}{1\ \mathrm{m}}\right)^2 = 0.13\ \mathrm{R/hr\ or\ 130\ mR/hr}$$

At 2 m, the exposure falls off to:

$$(1.3\ \mathrm{R \cdot m^2/Ci \cdot hr})(0.1\ \mathrm{Ci})\left(\frac{1}{2\ \mathrm{m}}\right)^2 = 0.0325\ \mathrm{R/hr\ or\ 32.5\ mR/hr}$$

IV. Shielding

Strategies to control external radiation exposures include modification of the process to eliminate the hazard, enclosure of the hazard (shielding), and isolation of the operator from the hazard (distance and time). The concept of process alteration was discussed in Chapter 7. This discussion will assume that there are no other acceptable alternatives to the use of ionizing radiation and will focus on shielding. Isolation was discussed in Section III.D. Of principal concern are the types and amounts of radioactive materials or radiation sources and the actual levels of radiation anticipated. Consideration should be given to the necessity of shielding, ventilation, automatic or remote control, and special handling procedures.

A. Gamma-radiation

Under the assumption of good geometry, the thickness of a shield to reduce exposure from a γ-radiation source can be estimated from Eq. (10.10). This equation is based on the assumption that every photon that interacts with the shield is absorbed. The assumption does not hold for broad-beam radiation or thick shields. Under these conditions, Eq. (10.10) will underestimate the shield thickness. (A significant number of photons can be scattered, but not absorbed, by a shield.) It is therefore best to use experimental shielding data for broad-beam shielding calculations. Figures 10.3, 10.4, and 10.5 show transmission of γ-radiation through lead, concrete, and iron from commonly used radionuclides.

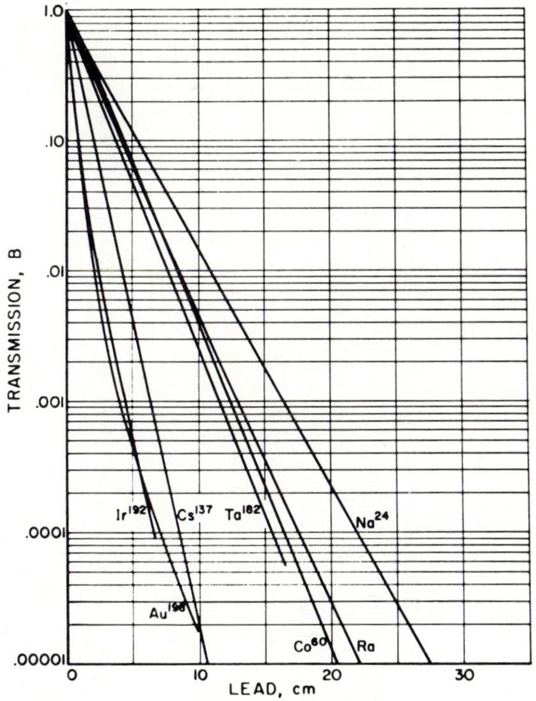

Figure 10.3 Transmission through lead of γ-rays from radium, cobalt-60, cesium-137, gold-198, iridium-192, tantalum-182, and sodium-24 (*PHS, 1970*).

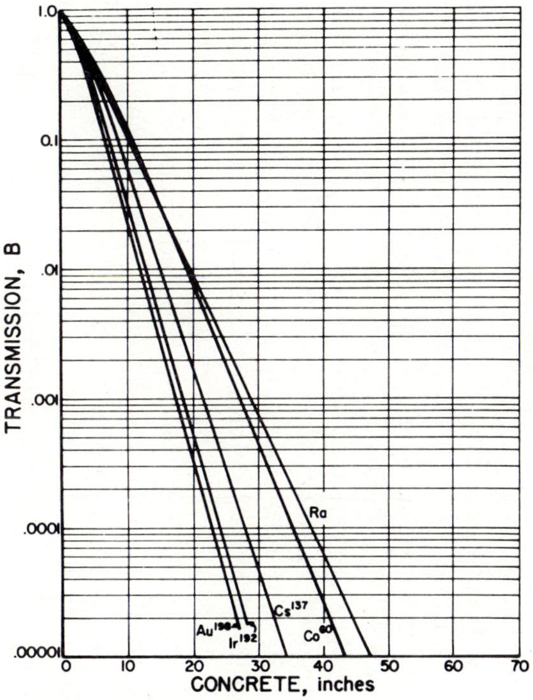

Figure 10.4 Transmission through concrete (density 2.35 $g \cdot cm^{-3}$) of γ-rays from radium, cobalt-60, cesium-137, gold-198, and iridium-192 (*PHS, 1970; NCRP 49, 1976*).

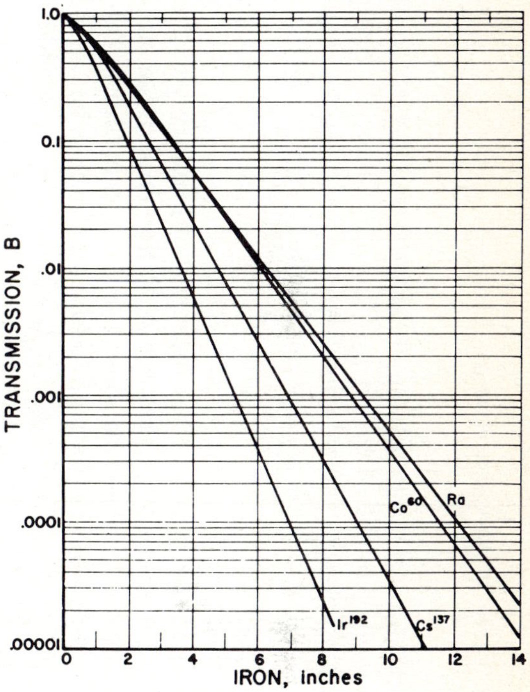

Figure 10.5 Transmission through iron of γ-rays from radium, cobalt-60, cesium-137, and iridium-192 (*PHS, 1970; NCRP 49, 1976*).

If experimental data are not available and Figures 10.3–10.5 do not apply, Eq. (10.10) can be modified with a dose build-up factor to account for scattered photon energy. The relationship among intensity reduction, attenuation, shield thickness, and dose build-up factor, B, is:

$$I = I_0 B \exp(-\mu_t t) \qquad (10.20)$$

where B is defined as the ratio of total intensity of primary and scattered radiation to the intensity of primary radiation only. Dose build-up applies to estimates of exposure, dose, or dose equivalent. Table 10.13 gives dose build-up factors for point and plane sources for various materials, energies, and values of $\mu_t t$. If the source is embedded or immediately adjacent to the shield, the point-source build-up factor is to be used. If a physically small source is far from the shield so that the radiation is nearly collimated, use the plane-source build-up factor. In the case where the source is spread over a large surface near the shield, an integration must be made over the surface (Blizard, 1958).

For the case when Eq. (10.20) is used to determine the required thickness (t) to reduce the intensity to a specified level (I), there are two unknowns, t and B. Therefore, the equation must be solved by trial and error. As a rule of thumb, the required thickness will be approximately equal to the thickness determined from Eq. (10.10) plus one half-value layer. (The half-value layer of a material is the thickness of an absorber that will reduce the photon intensity by 50%.) Half-value and tenth-value layers for x-radiation of various energies and photons from γ-emitters are given in Table 10.14. For the trial-and-error process, start with thickness t equal to the solution of Eq. (10.10) plus the half-value layer from Table 10.14. Further guesses at thickness t often are not needed.

B. x-Radiation

x-Ray shielding requires source and structural shields. The source shield is the x-ray tube housing, typically lead. Structural shielding involves the area where x-ray exposures are performed. Source shielding is intended to reduce leakage radiation to an acceptable level as defined by performance specifications of the NCRP. [Leakage radiation is the radiation field outside of the useful beam (NCRP 49, 1976).] These standards specify maximum allowable leakage radiation fields for the following three types of x-ray units:

1. Diagnostic x-ray; leakage radiation (any direction outside of useful beam) must be less than 100 mR/hr at 1 m at maximum output.
2. Therapeutic x-ray; for ≤500-kV peak voltage, leakage radiation must be less than 1 R/hr at 1 m; for >500-kV peak voltage, leakage

		$\mu_t t$ [a]						
Material[b]	MeV	1	2	4	7	10	15	20
		Dose Build-up Factor, B, for a Point Isotropic Source						
Water	0.255	3.09	7.14	23.0	72.9	166	456	982
	0.5	2.52	5.14	14.3	38.8	77.6	178	334
	1.0	2.13	3.71	7.68	16.2	27.1	50.4	82.2
	2.0	1.83	2.77	4.88	8.46	12.4	19.5	27.7
	3.0	1.69	2.42	3.91	6.23	8.63	12.8	17.0
	4.0	1.58	2.17	3.34	5.13	6.94	9.97	12.9
	6.0	1.46	1.91	2.76	3.99	5.18	7.09	8.85
	8.0	1.38	1.74	2.40	3.34	4.25	5.66	6.95
	10.0	1.33	1.63	2.19	2.97	3.72	4.90	5.98
Aluminum	0.5	2.37	4.24	9.47	21.5	38.9	80.8	141
	1.0	2.02	3.31	6.57	13.1	21.2	37.9	58.5
	2.0	1.75	2.61	4.62	8.05	11.9	18.7	26.3
	3.0	1.64	2.32	3.78	6.14	8.65	13.0	17.7
	4.0	1.53	2.08	3.22	5.01	6.88	10.1	13.4
	6.0	1.42	1.85	2.70	4.06	5.49	7.97	10.4
	8.0	1.34	1.68	2.37	3.45	4.58	6.56	8.52
	10.0	1.28	1.55	2.12	3.01	3.96	5.63	7.32
Iron	0.5	1.98	3.09	5.98	11.7	19.2	35.4	55.6
	1.0	1.87	2.89	5.39	10.2	16.2	28.3	42.7
	2.0	1.76	2.43	4.13	7.25	10.9	17.6	25.1
	3.0	1.55	2.15	3.51	5.85	8.51	13.5	19.1
	4.0	1.45	1.94	3.03	4.91	7.11	11.2	16.0
	6.0	1.34	1.72	2.58	4.14	6.02	9.89	14.7
	8.0	1.27	1.56	2.23	3.49	5.07	8.50	13.0
	10.0	1.20	1.42	1.95	2.99	4.35	7.54	12.4
Tin	0.5	1.56	2.08	3.09	4.57	6.04	8.64	—
	1.0	1.64	2.30	3.74	6.17	8.85	13.7	18.8
	2.0	1.57	2.17	3.53	5.87	8.53	13.6	19.3
	3.0	1.46	1.96	3.13	5.28	7.91	13.3	20.1
	4.0	1.38	1.81	2.82	4.82	7.41	13.2	21.2
	6.0	1.26	1.57	2.37	4.17	6.94	14.8	29.1
	8.0	1.19	1.42	2.05	3.57	6.19	15.1	34.0
	10.0	1.14	1.31	1.79	2.99	5.21	12.5	33.4

TABLE 10.13 Dose Build-up Factors *(Continued)*

		$\mu_t t$ [a]						
Material[b]	MeV	1	2	4	7	10	15	20
		Dose Build-up Factor, B, for a Point Isotropic Source						
Tungsten	0.5	1.28	1.50	1.84	2.24	2.61	3.12	—
	1.0	1.44	1.83	2.57	3.62	4.64	6.25	(7.35)[b]
	2.0	1.42	1.85	2.72	4.09	5.27	8.07	(10.6)
	3.0	1.36	1.74	2.59	4.00	5.92	9.66	14.1
	4.0	1.29	1.62	2.41	4.03	6.27	12.0	20.9
	6.0	1.20	1.43	2.07	3.60	6.29	15.7	36.3
	8.0	1.14	1.32	1.81	3.05	5.40	15.2	41.9
	10.0	1.11	1.25	1.64	2.62	4.65	14.0	39.3
Lead	0.5	1.24	1.42	1.69	2.00	2.27	2.65	(2.73)
	1.0	1.37	1.69	2.26	3.02	3.74	4.81	5.86
	2.0	1.39	1.76	2.51	3.66	4.84	6.87	9.00
	3.0	1.34	1.68	2.43	3.75	5.30	8.44	12.3
	4.0	1.27	1.56	2.25	3.61	5.44	9.80	16.3
	5.1097	1.21	1.46	2.08	3.44	5.55	11.7	23.6
	6.0	1.18	1.40	1.97	3.34	5.69	13.8	32.7
	8.0	1.14	1.30	1.74	2.89	5.07	14.1	44.6
	10.0	1.11	1.23	1.58	2.52	4.34	12.5	39.2
Uranium	0.5	1.17	1.30	1.48	1.67	1.85	2.08	—
	1.0	1.31	1.56	1.98	2.50	2.97	3.67	—
	2.0	1.33	1.64	2.23	3.09	3.95	5.36	(6.48)
	3.0	1.29	1.58	2.21	3.27	4.51	6.97	9.88
	4.0	1.24	1.50	2.09	3.21	4.66	8.01	12.7
	6.0	1.16	1.36	1.85	2.96	4.80	10.8	23.0
	8.0	1.12	1.27	1.66	2.61	4.36	11.2	28.0
	10.0	1.09	1.20	1.51	2.26	3.78	10.5	28.5

		$\mu_t t$ [a]						
Material[b]	MeV	1	2	4	7	10	15	20
		Dose Build-up Factor, B, for a Plane Monodirectional Source						
Water	0.5	2.63	4.29	9.05	20.0	35.9	74.9	
	1.0	2.26	3.39	6.27	11.5	18.0	30.8	

	2.0	1.84	2.63	4.28	6.96	9.87	14.4
	3.0	1.69	2.31	3.57	5.51	7.48	10.8
	4.0	1.58	2.10	3.12	4.63	6.19	8.54
	6.0	1.45	1.86	2.63	3.76	4.86	6.78
	8.0	1.36	1.69	2.30	3.16	4.00	5.47
Iron	0.5	2.07	2.94	4.87	8.31	12.4	20.6
	1.0	1.92	2.74	4.57	7.81	11.6	18.9
	2.0	1.69	2.35	3.76	6.11	8.78	13.7
	3.0	1.58	2.13	3.32	5.26	7.41	11.4
	4.0	1.48	1.90	2.95	4.61	6.46	9.92
	6.0	1.35	1.71	2.48	3.81	5.35	8.39
	8.0	1.27	1.55	2.17	3.27	4.58	7.33
	10.0	1.22	1.44	1.95	2.89	4.07	6.70
Tin	1.0	1.65	2.24	3.40	5.18	7.19	10.5
	2.0	1.58	2.13	3.27	5.12	7.13	11.0
	4.0	1.39	1.80	2.69	4.31	6.30	—
	6.0	1.27	1.57	2.27	3.72	5.77	11.0
	10.0	1.16	1.33	1.77	2.81	4.53	9.68
Lead	0.5	1.24	1.39	1.63	1.87	2.08	—
	1.0	1.38	1.68	2.18	2.80	3.40	4.20
	2.0	1.40	1.76	2.41	3.36	4.35	5.94
	3.0	1.36	1.71	2.42	3.55	4.82	7.18
	4.0	1.28	1.56	2.18	3.29	4.69	7.70
	6.0	1.19	1.40	1.87	2.97	4.69	9.53
	8.0	1.14	1.30	1.69	2.61	4.18	9.08
	10.0	1.11	1.24	1.54	2.27	3.54	7.70
Uranium	0.5	1.17	1.28	1.45	1.60	1.73	—
	1.0	1.30	1.53	1.90	2.32	2.70	3.60
	2.0	1.33	1.62	2.15	2.87	3.56	4.89
	3.0	1.29	1.57	2.13	3.02	3.99	5.94
	4.0	1.25	1.49	2.02	2.94	4.06	6.47
	6.0	1.18	1.37	1.82	2.74	4.12	7.79
	8.0	1.13	1.27	1.61	2.39	3.65	7.36
	10.0	1.10	1.21	1.48	2.12	3.21	6.58

[a] $\mu_t t$ = mass attenuation coefficient ($\mu_{t,m}$, $cm^2 \cdot g^{-1}$)·shield thickness (cm)·shield density (g/cm^3).
[b] For concrete use an average of aluminum and iron; e.g., B(concrete) = [B(Fe) + B(Al)]/2.

SOURCE: PHS, 1970.

TABLE 10.14 Half-value and Tenth-value Layers[a]

	Attenuation material					
	Lead (mm)		Concrete (cm)		Iron (cm)	
Peak voltage (kV)	HVL[b]	TVL[c]	HVL	TVL	HVL	TVL
50	0.06	0.17	0.43	1.5		
70	0.17	0.52	0.84	2.8		
100	0.27	0.88	1.6	5.3		
125	0.28	0.93	2.0	6.6		
150	0.30	0.99	2.24	7.4		
200	0.52	1.7	2.5	8.4		
250	0.88	2.9	2.8	9.4		
300	1.47	4.8	3.1	10.4		
400	2.5	8.3	3.3	10.9		
500	3.6	11.9	3.6	11.7		
1,000	7.9	26	4.4	14.7		
2,000	12.5	42	6.4	21		
3,000	14.5	48.5	7.4	24.5		
4,000	16	53	8.8	29.2	2.7	9.1
6,000	16.9	56	10.4	34.5	3.0	9.9
8,000	16.9	56	11.4	37.8	3.1	10.3
10,000	16.6	55	11.9	39.6	3.2	10.5
Cesium-137	6.5	21.6	4.8	15.7	1.6	5.3
Cobalt-60	12	40	6.2	20.6	2.1	6.9
Gold-198	3.3	11	4.1	13.5		
Indium-192	6.0	20	4.3	14.7	1.3	4.3
Radium-226	16.6	55	6.9	23.4	2.2	7.4

[a]Approximate values obtained at high attenuation for the indicated peak voltage values under broad-beam conditions; with low attenuation these values will be significantly less.
[b]Half-value layer.
[c]Tenth-value layer.
SOURCE: NCRP 49, 1976.

radiation must be less than 1 R/hr or less than 0.1% of useful beam exposure rate at 1 m from the tube target at any exposure rate.

3. Nonmedical x-ray; leakage radiation must be less than 1 R/hr at 1 m at maximum output.

Structural shielding encloses both the tube and space where the object is irradiated and is intended to protect people outside the area of high radiation. It is designed to reduce primary beam, leakage, and scattered radiation fields. The requirements of structural shields are determined from:

1. Maximum operating voltage of x-ray unit (kV).
2. Maximum beam current of x-ray tube (mA).

TABLE 10.15 Typical Weekly Workloads for Busy Installations

Diagnostic	Daily patient load	Weekly workload, W (mA·min): 100 kV[a] or less	125 kV[a]	150 kV[a]
Admission chest: (miniature, with photo-timing grid)	100	100	—	—
Chest: (14 × 17: 3 films per patient, no grid)	60	150	—	—
Cystoscopy	8	600	—	—
Fluoroscopy including spot filming	24	1,500	600	300
Fluoroscopy without spot filming	24	1,000	400	200
Fluoroscopy with image intensification including spot filming	24	750	300	150
General radiography	24	1,000	400	200
Special procedures	8	700	280	140
Therapy		Weekly workload, W		
Superficial (up to 150 kV[a])	32	3,000 mA·min		
Orthovoltage (200–500 kV[a])	32	20,000 mA·min		
Megavoltage (0.5 MV–10 MV)	50	100,000 R at 1 m[b]		
Cesium				
50 cm SSD[c]	16	8,000 R at 1 m[b]		
50 cm SSD	32	15,000 R at 1 m[b]		
60 cm SSD	32	24,000 R at 1 m[b]		
Cobalt				
70 cm SSD	32	30,000 R at 1 m[b]		
80 cm SSD	32	40,000 R at 1 m[b]		
100 cm SSD	32	60,000 R at 1 m[b]		

[a]Peak pulsating x-ray tube potential.
[b]R per week at 1 m = $R \cdot m^2$ per week.
[c]Source-surface distance.

SOURCE: NCRP 49, 1976.

TABLE 10.16 Use Factors for Primary Protective Barriers[a]
(To be used only if specific values for a given installation are not available)

	Radiographic installations	Therapy installations
Floor	1	1
Walls	$\frac{1}{4}$	$\frac{1}{4}$
Ceiling	—[b]	—[c]

[a] The use factor for secondary protective barriers is usually 1.
[b] The shielding requirements for the ceiling of a radiographic installation are determined by the secondary barrier requirements rather than by the use factor, which is generally extremely low.
[c] The use factor for the ceiling of a therapy installation depends on the type of equipment and techniques used, but usually is not more than $\frac{1}{4}$.
SOURCE: NCRP 49, 1976.

3. Workload factor, *W*, defined as the amount of machine use (mA·min/week). Typical workload factors are given in Table 10.15.
4. Use factor, *U*, defined as the fraction of workload during which the useful beam is pointed in the direction under consideration. Suggested use factors for installations where *U* is not known are given in Table 10.16.

TABLE 10.17 Occupancy Factors for Nonoccupationally Exposed Persons[a,b]
(For use as a guide in planning shielding where other occupancy data are not available)

Full occupancy ($T = 1$)
Work areas such as offices, laboratories, shops, wards, nurses' stations; living quarters; children's play areas; and occupied space in nearby buildings
Partial occupancy ($T = \frac{1}{4}$)
Corridors, restrooms, elevators using operators, unattended parking lots
Occasional occupancy ($T = \frac{1}{16}$)[c]
Waiting rooms, toilets, stairways, unattended elevators, janitors' closets, outside areas used only for pedestrians or vehicular traffic

[a] The occupancy factor of occupationally exposed persons, in general, may be assumed to be 1.
[b] It is advantageous in shielding design to take into account that the occupancy factor in areas adjacent to the radiation room usually is zero for any space more than 2.1 m (7 ft) above the floor, as the height of most individuals is less. It is possible, therefore, to reduce the thickness of the wall shielding above this height, provided that the radiation source is below 2.1 m (7 ft). In determining the shielding requirements for wall areas above 2.1 m (7 ft), consideration must be given to the protection of any persons occupying the floor above the areas adjacent to the radiation room. The wall areas over 2.1 m (7 ft) from the floor of the radiation room must also have sufficient shielding to adequately reduce the scattering from the ceiling of adjacent rooms toward occupants.
[c] It should be noted that the use of an occupancy factor of $\frac{1}{16}$ may result in full-time exposures in noncontrolled areas greater than 2 mR in any one hour or 100 mR in any seven consecutive days.
SOURCE: NCRP 49, 1976.

5. Occupancy factor, T, corrects for the type of occupancy in the adjacent space. Occupancy factors for nonoccupationally exposed persons are given in Table 10.17.

Areas with occupationally exposed people are considered controlled areas. Areas with nonoccupationally exposed people are uncontrolled areas. Since the maximum occupationally acceptable exposure is 5000 mrem/yr (100 mrem/week), the design benchmark for controlled areas is set at 100 mrem/week. Similarly, the design benchmark for uncontrolled areas, based on maximum nonoccupationally acceptable exposures, is 10 mrem/week. Figure 10.6 is a view of a radiation room and identifies primary and secondary barriers as well as the key distances of concern (d_{pri} = distance to nearest person in the direction of the primary beam; d_{sec} = distance to the nearest person in the direction of the leakage and scattered radiation fields; and d_{sca} = distance from the tube to scatterer in the direction of the primary beam).

The primary protective barrier attenuates x- or γ-radiation in the direction of the useful beam. All barrier calculations are on a weekly basis. The weekly exposure, X_u, from the useful beam at the point of interest (d_{pri} from source) is related to the exposure rate at 1 m, $\dot{X}_u$, by:

$$X_u = \frac{\dot{X}_u t}{(d_{pri})^2} \qquad (10.21a)$$

where t is the number of minutes of beam on-time in a week. If X_u is

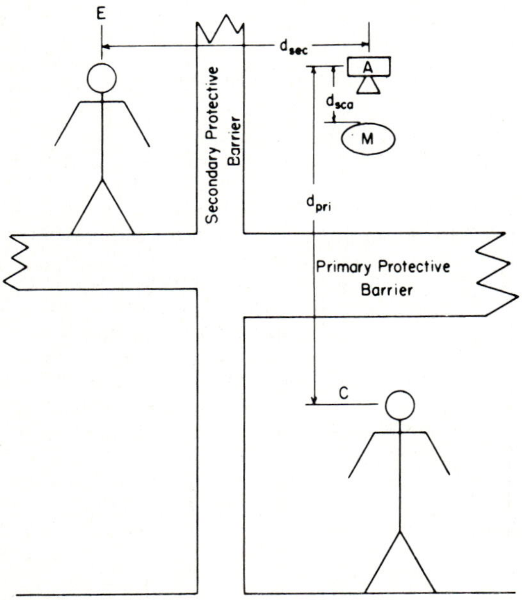

Figure 10.6 Elevation view of radiation room and its surroundings with indication of distances of interest for radiation shielding calculations (*NCRP 49, 1976*). A is the radiation source, M the patient, and C and E positions that may be occupied by persons.

greater than the permissible exposure, P, a barrier of thickness S_p is needed to give a transmission factor B_{ux} such that:

$$P = B_{ux}X_u = B_{ux}\frac{\dot{X}_u t}{(d_{\text{pri}})^2} \tag{10.21b}$$

If the normalized output $\dot{X}_n$ is defined as $\dot{X}_u/I$, where I is beam current (mA), then:

$$P = B_{ux}\dot{X}_n \frac{I\,t}{(d_{\text{pri}})^2} \tag{10.21c}$$

Since $W = I\,t$ (the product of beam current and on-time):

$$B_{ux}\dot{X}_n = \frac{P(d_{\text{pri}})^2}{W} \tag{10.21d}$$

Defining K_{ux} (where $K_{ux} = B_{ux}\dot{X}_n$) as R/mA·min for the useful beam at 1 m,

$$K_{ux} = \frac{P(d_{\text{pri}})^2}{W} \tag{10.21e}$$

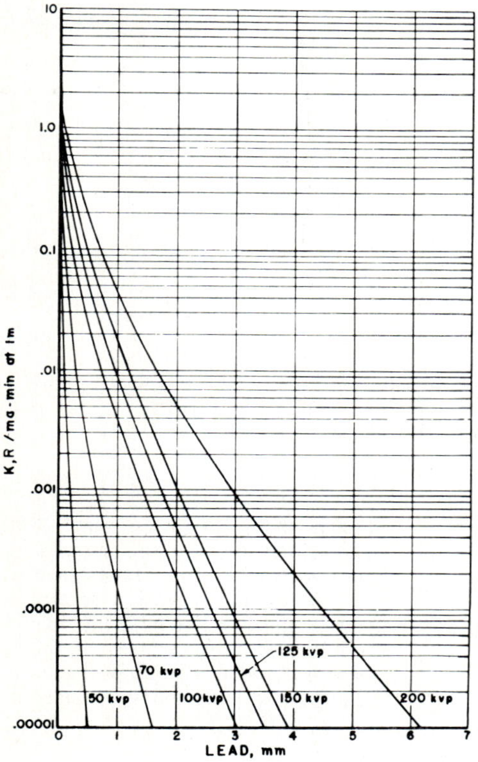

Figure 10.7a Attenuation in lead of x-rays produced by potentials of 50- to 200-kV peak (*PHS, 1970; NCRP 49, 1976*). The measurements were made with a 90° angle between the electron beam and the axis of the x-ray beam and with a pulsed waveform. The filtrations were 0.5 mm of aluminum for 50, 70, 100, and 125 kV_p, and 3 mm of aluminum for 150 and 200 kV_p (*Kelly and Trout, 1972*).

Adjusting for use and occupancy factors,

$$K_{ux} = \frac{P(d_{\text{pri}})^2}{WUT} \tag{10.21f}$$

The relationships between K_{ux} and barrier thickness, S_p, for various tube voltages and shielding materials are given in Figures 10.7–10.9. [Note that for units of K_{ux}, 0.01 R/mA·min = 3.88 × 10^{-5} (μC/kg/mA·min).] The attenuation curves are for single-phase pulsed-waveform x-ray units. The addition of one half-value layer of shielding is required for a three-phase, 12-pulse unit over a single-phase, 2-pulse unit (Trout et al., 1975).

For γ-radiation sources, W is expressed as R/per week at 1 m (from Table 10.8), and the transmission, B_{ug}, will be:

$$B_{ug} = \frac{P(d_{\text{pri}})^2}{WUT} \tag{10.22}$$

Values of B_{ug} are shown in Figures 10.3, 10.4, and 10.5.

Secondary barriers are designed to limit exposure from leakage and scattered radiation to acceptable levels. The design procedure is to calculate secondary barrier requirements for leakage and scattered radiation separately. If the two requirements are approximately equal,

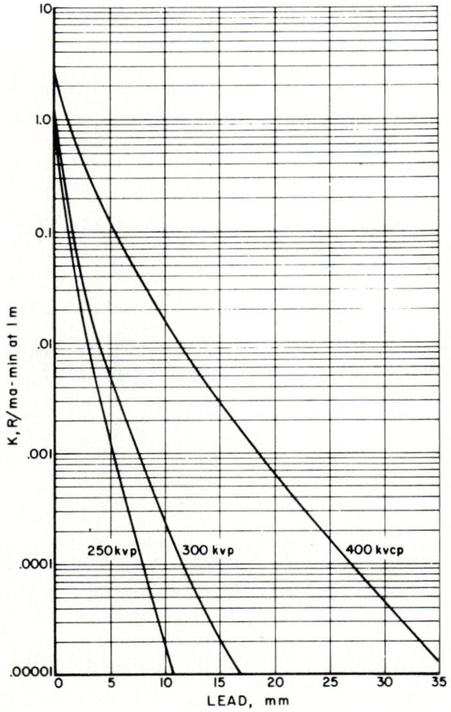

Figure 10.7b Attenuation in lead of x-rays produced by potentials of 250 to 400 kV (*PHS, 1970; NCRP 49, 1976*). The measurements were made with a 90° angle between the electron beam and the axis of the x-ray beam. The 250 and 300 kV_p curves are for a pulsed waveform and a filtration of 3 mm of aluminum (*Kelley and Trout, 1972; Trout et al., 1959*). The 400 kV_{cp} curve was obtained with a constant-potential generator and inherent filtration of approximately 3 mm of copper (*Miller and Kennedy, 1955*).

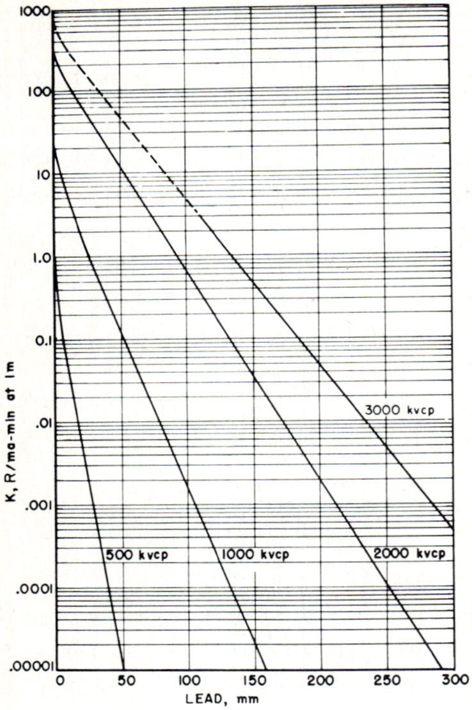

Figure 10.7c Attenuation in lead of x-rays produced by potentials of 500 to 3000 kV constant potential (*PHS, 1970; NCRP 49, 1976*). The measurements were made with a 0° angle between the electron beam and the axis of the x-ray beam and with a constant-potential generator. The 500- and 1000-kV_{cp} curves were obtained with filtration of 2.88 mm of tungsten, 2.8 mm of copper, 2.1 mm of brass, and 18.7 mm of water (*Wyckoff et al., 1948*). The 2000-kV_{cp} curve was obtained by extrapolating to broad-beam conditions (E. E. Smith) the data of *Evans et al.* (*1952*). The inherent filtration was equivalent to 6.8 mm of lead. The 3000-kV_{cp} curve has been obtained by interpolation of the 2000-kV_{cp} curve given herein, and the data of *Miller and Kennedy* (*1956*).

add one half-value layer to the greater of the two. If the difference of the two is greater than a tenth-value layer, the thicker barrier will be adequate by itself.

Leakage barriers are designed to reduce NCRP maximum leakage fields to permissible levels. For therapeutic protective housing (at >500 kV), the leakage is ≤0.1% of $\dot{X}_u$ and

$$X_L = \frac{0.001\dot{X}_u t}{(d_{\text{sec}})^2} \tag{10.23a}$$

where X_L is the weekly leakage exposure over an area of interest and t is the number of minutes of beam on-time in a week. Following a development similar to that used for Eq. (10.21) and letting $U = 1$, a barrier transmission factor B_{LX} to reduce weekly exposure X_L to permissible exposure P will be

$$B_{LX} = \frac{P}{X_L} = \frac{1000P(d_{\text{sec}})^2}{\dot{X}_n WT} \tag{10.23b}$$

Equation (10.23*b*) gives the barrier requirements for leakage fields from therapeutic x-ray > 500 kV.

For therapeutic-type protective tube housings for operating voltages below 500 kV and nonmedical x-ray, the leakage is limited to 1 R in

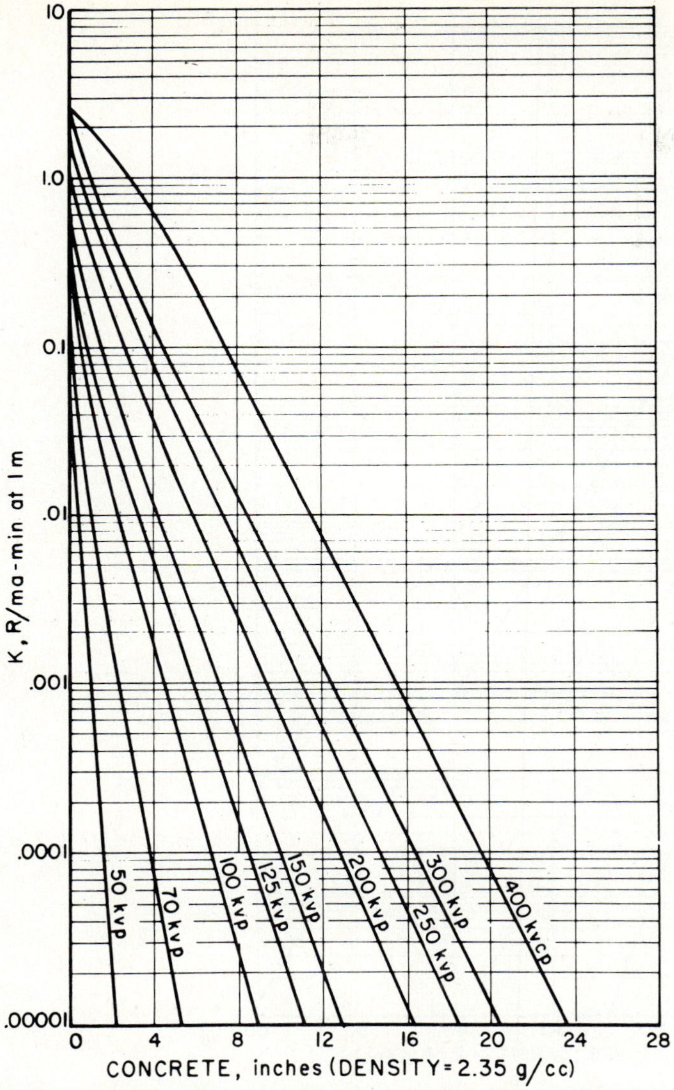

Figure 10.8*a* Attenuation in concrete of x-rays produced by potentials of 50 to 400 kV (*PHS, 1970; NCRP 49, 1976*). The measurements were made with a 90° angle between the electron beam and the axis of the x-ray beam. The curves for 50 to 300 kV_p are for a pulsed waveform. The filtrations were 1 mm of aluminum 50 kV_p, 1.5 mm of aluminum for 70 kV_p, 2 mm of aluminum for 100 kV_p, and 3 mm of aluminum for 125 to 300 kV_p (*Trout et al., 1959*). The 400-kV_{cp} curve was interpolated from data obtained with a constant potential generator and inherent filtration of approximately 3 mm of copper (*Miller and Kennedy, 1955*).

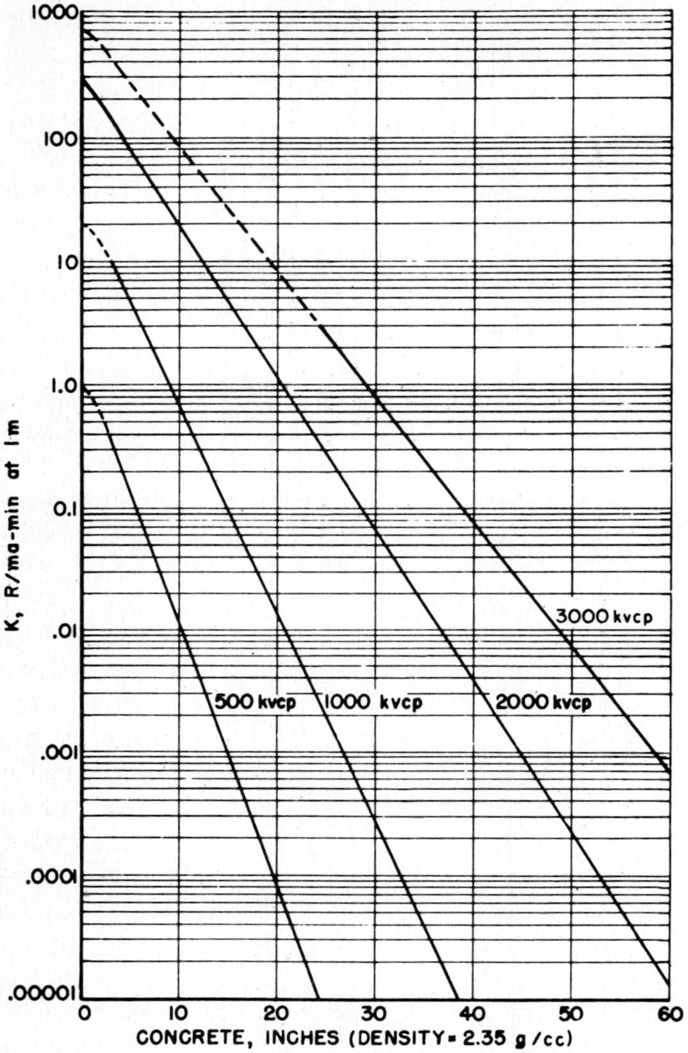

Figure 10.8*b* Attenuation in concrete of x-rays produced by potentials of 500 to 3000 kV constant potential (*PHS, 1970; NCRP 49, 1976*). The measurements were made with a 0° angle between the electron beam and the axis of the x-ray beam with a constant-potential generator. The 500- and 1000-kV$_{cp}$ curves were obtained with filtration of 2.8 mm of copper, 2.1 mm of brass, and 18.7 mm of water (*Wyckoff et al., 1948*). The 2000-kV$_{cp}$ curve was obtained by extrapolating to broad-beam conditions (E. E. Smith) the data of *Evans et al.* (*1952*). The inherent filtration was equivalent to 6.8 mm of lead. The 3000-kV$_{cp}$ curve was obtained by interpolation of the 2000-kV$_{cp}$ curve given herein, and the data of *Kirn and Kennedy* (*1954*).

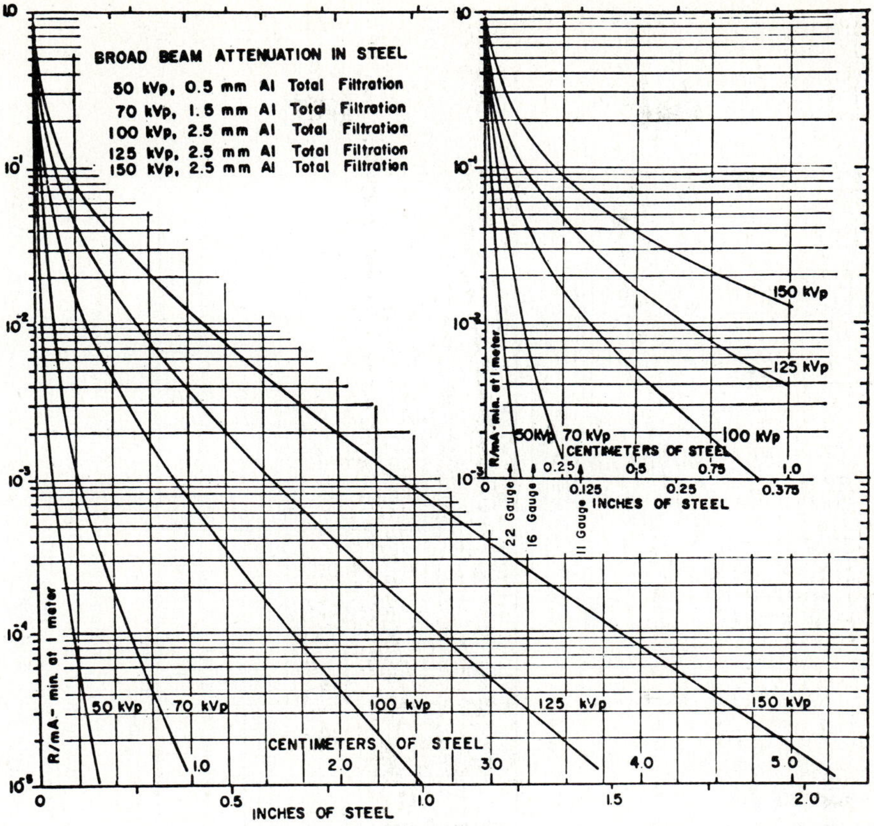

Figure 10.9a Attenuation in steel of x-rays produced by potentials of 50 to 150 kV peak (*Trout et al., 1975*). The measurements were made with a 90° angle between the electron beam and the axis of the x-ray beam. The curves are for a single-phase, pulsed waveform. The filtrations were 0.5 mm of aluminum for 50 kV_p; 1.5 mm of aluminum for 70 kV_p; and 2.5 mm of aluminum for 100, 125, and 150 kV_p.

an hour at 1 m from the source with the tube at maximum output. Therefore, at distance d_{sec}, the leakage X_L is:

$$X_L = \frac{(1 \text{ R/hr})\, t}{(d_{sec})^2 60} \tag{10.24a}$$

where t is minutes of beam on-time.

The barrier requirements for leakage fields B_{LX} for <500 kV therapeutic and nonmedical x-ray (with beam current, I) are:

$$B_{LX} = \frac{P(d_{sec})^2 60 I}{WT} \tag{10.24b}$$

For diagnostic-type tube housing with leakage field <0.1 R/hr at 1 m,

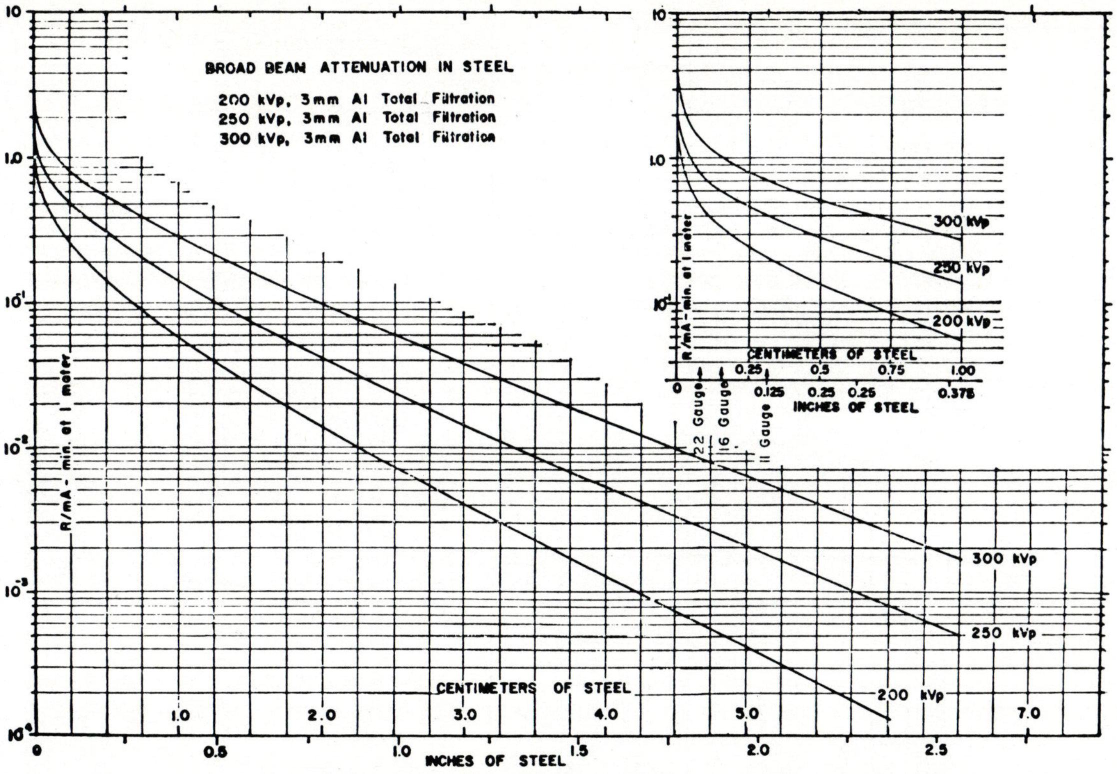

Figure 10.9*b* Attenuation in steel of x-rays produced by potentials of 200 to 300 kV peak (*Trout et al., 1975*). The measurements were made with a 90° angle between the electron beam and the axis of the x-ray beam. The curves are for single-phase, pulsed waveform. The filtrations were 3.0 mm of aluminum for 200, 250, and 300 kV_p.

the leakage barrier requirement is similar to Eq. (10.24):

$$B_{LX} = \frac{P(d_{sec})^2 600I}{WT} \tag{10.25}$$

The relationship between barrier requirement and number of half-value layers is

$$B_{LX} = \left(\frac{1}{2}\right)^n \tag{10.26}$$

where n is the number of half-value layers. Barrier thickness, S_L, is then:

$$S_L = n(\text{HVL}) \tag{10.27}$$

where HVL is the half-value layer (Table 10.14) of the shielding material.

The intensity of the scattered radiation field is a function of scattering angle, energy of the primary beam, and scattering area. Table 10.18 gives values of the ratio of scattered to incident exposure, a, as

TABLE 10.18 Ratio, *a*, of Scattered to Incident Exposure[a]

	Scattering angle (from central ray)					
Source	30	45	60	90	120	135
x-Rays						
50 kV[b]	0.0005	0.0002	0.00025	0.00035	0.0008	0.0010
70 kV[b]	0.00065	0.00035	0.00035	0.0005	0.0010	0.0013
100 kV[b]	0.0015	0.0012	0.0012	0.0013	0.0020	0.0022
125 kV[b]	0.0018	0.0015	0.0015	0.0015	0.0023	0.0025
150 kV[b]	0.0020	0.0016	0.0016	0.0016	0.0024	0.0026
200 kV[b]	0.0024	0.0020	0.0019	0.0019	0.0027	0.0028
250 kV[b]	0.0025	0.0021	0.0019	0.0019	0.0027	0.0028
300 kV[b]	0.0026	0.0022	0.0020	0.0019	0.0026	0.0028
4 MV[c]	—	0.0027	—	—	—	—
6 MV[c]	0.007	0.0018	0.0011	0.0006	—	0.0004
Gamma-rays						
^{137}Cs[d]	0.0065	0.0050	0.0041	0.0028	—	0.0019
^{60}Co[e]	0.0060	0.0036	0.0023	0.0009	—	0.0006

[a] Scattered radiation measured at 1 m from phantom when field area is 400 cm^2 at the phantom surface; incident exposure measured at center of field 1 m from the source but without phantom.

[b] Average scatter for beam centered and beam at edge of typical patient cross-section phantom; peak pulsating x-ray tube potential.

[c] Cylindrical phantom.

[d] These data were obtained from a slab placed obliquely to the central ray. A cylindrical phantom should give smaller values.

[e] F = 400 cm^2.

SOURCE: NCRP 49, 1976.

TABLE 10.19*a* Minimum Shielding Requirements for Radiographic Installations

WUT^a in mA·min			Distance in meters from source to occupied area										
100 kV[b]	125 kV[b]	150 kV[b]											
1,000	400	200	1.5	2.1	3.0	4.2	6.1	8.4	12.2				
500	200	100		1.5	2.1	3.0	4.2	6.1	8.4	12.2			
250	100	50			1.5	2.1	3.0	4.2	6.1	8.4	12.2		
125	50	25				1.5	2.1	3.0	4.2	6.1	8.4	12.2	
62.5	25	12.5					1.5	2.1	3.0	4.2	6.1	8.4	12.2
Type of area	Material		Primary protective barrier thickness[d]										
Controlled	Lead, mm		1.95	1.65	1.4	1.15	0.9	0.65	0.45	0.3	0.2	0.1	0.1
Noncontrolled	Lead, mm		2.9	2.6	2.3	2.05	1.75	1.5	1.2	0.95	0.75	0.55	0.35
Controlled	Concrete[c], cm		18	15.5	13.5	11.5	9.5	7	5.5	4	2.5	1.5	0.5
Noncontrolled	Concrete[c], cm		25	23	20.5	18.5	16.5	14	12	10	8		
			Secondary protective barrier thickness[d]										
Controlled	Lead, mm		0.55	0.45	0.35	0.3	0	0	0	0	0	0	0
Noncontrolled	Lead, mm		1.3	1.05	0.75	0.55	0.45	0.35	0.3	0.05	0	0	0
Controlled	Concrete[c], cm		5	3.5	2.5	2	0	0	0	0	0	0	0
Noncontrolled	Concrete[c], cm		11.5	9.5	7.5	5.5	4	3	2	0.5	0	0	0

[a] *W*, weekly workload in mA·min; *U*, use factor; *T*, occupancy factor.
[b] Peak pulsating x-ray tube potential.
[c] Thickness based on concrete density of 2.35 $g \cdot cm^{-3}$ (147 $lb \cdot ft^{-3}$).
[d] Barrier thickness based on 150 kV.

SOURCE: NCRP 49, 1976.

a function of energy and scattering angle for a field area F of 400 cm^2 at the scattering surface. For this geometry, and assuming that U = 1 and the scattered photons have the same energy as the useful beam, the attenuation, K_{ux}, for scattered radiation below 500 kV becomes:

$$K_{ux} = B_{sx}\dot{X}_n = \frac{P(d_{sec})^2(d_{sca})^2 400}{aWTF} \quad (10.28a)$$

where B_{sx} is the transmission factor for scattered radiation. The barrier thickness required to reduce exposure to permissible level P is determined from Figures 10.7 through 10.9. For beam energies above 500 kV, use the 500-kV curves of the figures and compensate for the extra energy of the beam with normalizing factor f such that:

$$K_{ux} = \frac{P(d_{sec})^2(d_{sca})^2 400}{aWTFf} \quad (10.28b)$$

TABLE 10.19*b* Commercial Lead Sheets

Thickness		Weight in pounds for a 1-ft² section	
Inches	Millimeter equivalent	Nominal weight	Actual weight
1/64	0.40	1	0.92
3/128	0.60	1½	1.38
1/32	0.79	2	1.85
5/128	1.00	2½	2.31
3/64	1.19	3	2.76
7/128	1.39	3½	3.22
—	1.50	—	3.48
1/16	1.58	4	3.69
5/64	1.98	5	4.60
3/32	2.38	6	5.53
—	2.5	—	5.80
—	3.0	—	6.98
1/8	3.17	8	7.38
5/32	3.97	10	9.22
3/16	4.76	12	11.06
7/32	5.55	14	12.9
1/4	6.35	16	14.75
1/3	8.47	20	19.66
2/5	10.76	24	23.60
1/2	12.70	30	29.50
2/3	16.93	40	39.33
1	25.40	60	59.00

NOTES:

1. The density of commercially rolled lead is 11.36 $g \cdot cm^{-3}$ (Lead Industries Association).
2. The commercial tolerances are ±0.005 in. for lead up to 7/128 and ±1/32 for heavier sheets.
3. It should be noted that lead sheet less than 1/32 in. thick is frequently more expensive than heavier sheet in cost of material and cost of installation.

SOURCE: NCRP 49, 1976.

where $f = 20$, 300, and 700 for operating voltages of $\leq$1000 kV, $\leq$2000 kV, and $\leq$3000 kV, respectively; $f = 1$ for operating voltages $\leq$500 kV.

Table 10.19*a* shows typical minimum shielding requirements for radiographic installations. Table 10.19*b* gives thicknesses and weights per square foot for commercial lead sheets. Table 10.20 summarizes the notation used in barrier thickness calculations. Similar methods for shielding x-rays from high-energy electron beams (>10 MeV) are also available (NCRP 51, 1977).

Low-energy x-rays are generated by x-ray analytical equipment, electron microscopes, television tubes, and high-voltage rectifiers. For these sources, structural materials or surrounding components may serve to reduce radiation dose rates significantly. The transmission of low-voltage x-rays (10, 15, 20, 30, 40, and 50 kV) through aluminum, mild steel, sheet glass, Lucite (Perspex), and wood is given in Table 10.21 (O'Riordon and Catt, 1969). The data for aluminum, steel, glass, and Lucite are given in centimeters, and the data for wood are given in grams per square centimeter. Transmission curves can be constructed from these data. The densities of the mild steel, sheet glass, and Lucite used were 7.8 g/cm^3, 2.6 g/cm^3, and 1.2 g/cm^3, respectively. The entries for wood represent the mean value of chipboard (0.71 g/cm^3), hardwood (1.03 g/cm^3), plywood (0.75 g/cm^3), and teak (0.62 g/cm^3).

C. Beta-particles

Two factors must be considered for the design of shields for high-intensity β-radiation. These are absorbing the β-particles and the bremsstrahlung generated due to the absorption of the β-particles. Therefore, a shield for high-energy β-particles should consist of: (1) a low-atomic-number layer designed to stop all the β-particles and minimize the production of bremsstrahlung; followed by (2) a high-atomic-number material (e.g., lead) to absorb a sufficient fraction of the bremsstrahlung to reduce exposures to an acceptable level.

D. Neutrons

Most materials have comparatively large cross sections for removal of thermal-energy neutrons. Once neutrons are slowed to thermal velocities, they are likely to be captured. Therefore, the range of neutrons in a given material depends on the masses and cross sections of the atomic nuclei of the absorber material. Because neutron cross sections are not predictable from other atomic constants, neutron shielding specifications are usually developed from experimental data. For example, Figure 10.10 shows the range of neutrons in water.

TABLE 10.20 Quantities and Units Used in Calculations of Barrier Thickness[a]

Quantity	Symbol	Units
Weekly design exposure rate [b]	P	R
Weekly workload:		
x-Ray equipment with current meter	W	mA·min[c]
Megavoltage and γ-sources	W	R at 1 m[d]
Use factor	U	—
Occupancy factor	T	—
Distance from radiation source:		
To person to be protected	d_{pri}, d_{sec}	m
To scatterer	d_{sca}	m
Field area	F	cm^2
Normalized output	X_n	R per mA·min at 1 m
Exposure rate at 1 m from the source of the:		
Useful beam	$\dot{X}_u$	R per min
Leakage radiation	$\dot{X}_L$	R per min
Exposure rate at 1 m from the scatter for scattered radiation	$\dot{X}_s$	R per min at 1 m
Quotient of exposure at unit distance and workload	K_{ux}	R per mA·min at 1 m
Transmission factor for:		
Useful beam:		
x-Rays	B_{ux}	—
γ-Rays	B_{ug}	—
Leakage radiation:		
x-Rays	B_{LX}	—
γ-Rays	B_{Lg}	—
Scattered radiation:		
x-Rays	B_{sx}	—
γ-Rays	B_{sg}	—
Leakage of source housing	L	R per hr at 1 m
Exposure	X_u, X_L, X_s	R
Tube current	I	mA
Weekly beam on-time	t	min
Barrier thickness:		
Primary	S_p	mm (lead) cm (concrete)
Secondary	S_s	mm (lead) cm (concrete)
Half-value layer	HVL	mm or cm
Tenth-value layer	TVL	mm or cm

[a] It is assumed that the acceleration potential is known for x-ray sources and that the energy is known for γ-ray sources.

[b] As this section deals with x- and γ-radiation only, the number of roentgens may be considered equal to the number of rem.

[c] This unit is used for x-rays produced by potentials below 4 MV, and for higher potentials when the equipment has a target-current meter.

[d] This unit is used for all γ-ray sources and for x-rays produced by potentials of 4–10 MV, when the equipment does not have a target-current meter.

TABLE 10.21 Transmission of 10- to 50-Kv x-Rays

Material	Voltage (kV)	Transmission							
		0.3	0.1	0.03	0.01	0.003	0.001	0.0001	0.00001
Aluminum (cm)	10	0.006	0.014	0.023	0.033	0.045	0.056	0.082	0.110
	15	0.011	0.026	0.047	0.072	0.108	0.144	0.224	0.314
	20	0.02	0.03	0.06	0.11	0.18	0.26	0.44	0.64
	30	—	0.04	0.10	0.22	0.40	0.60	1.16	1.78
	40	—	0.04	0.15	0.34	0.70	1.12	2.24	3.50
	40	—	0.05	0.20	0.50	1.08	1.78	3.60	
Mild steel (cm)	10	0.0013	0.003	0.0054	0.0076	0.0102	0.0126	0.0175	0.0226
	15	0.0012	0.0028	0.0048	0.0070	0.0093	0.0116	0.0166	0.0216
	20	0.0014	0.0031	0.0055	0.0081	0.0114	0.0146	0.0222	0.0302
	30	0.0015	0.0035	0.0065	0.012	0.02	0.03	0.055	0.082
	40	—	0.0038	0.008	0.016	0.033	0.054	0.109	0.176
	50	—	0.004	0.012	0.026	0.052	0.087	0.19	0.31
Sheet glass (cm)	10	—	—	—	—	—	—	—	—
	15	—	(0.03)[a]	(0.05)	(0.08)	(0.15)	(0.19)	0.30	0.40
	20	—	(0.04)	(0.10)	(0.17)	(0.25)	0.32	0.53	0.80
	30	—	(0.05)	(0.15)	0.28	0.48	0.75	1.43	2.18
	40	—	(0.07)	(0.20)	0.43	0.87	1.38	2.70	4.12
	50	(0.03)	(0.10)	(0.25)	0.62	1.30	2.15	4.15	—
Lucite (cm)	10	0.14	0.26	0.41	0.60	0.84	1.05	1.56	2.10
	15	0.22	0.48	0.88	1.35	1.95	2.60	4.10	5.70
	20	0.27	0.60	1.24	2.10	3.25	4.45	7.40	—
	30	0.30	0.70	1.80	3.50	5.90	8.40	—	—
	40	0.32	0.82	2.35	4.80	8.15	—	—	—
	50	0.34	0.90	2.85	6.10	—	—	—	—
Wood (g/cm^2)	10	0.10	0.24	0.42	0.60	0.84	1.05	1.54	2.03
	15	0.16	0.45	0.87	1.33	1.94	2.57	4.00	5.50
	20	0.22	0.60	1.20	2.02	3.15	4.40	—	—
	30	0.25	0.75	1.80	3.50	5.84	—	—	—
	40	0.26	0.85	2.33	4.77	—	—	—	—
	50	0.30	0.93	2.90	6.10	—	—	—	—

[a] Interpolation between transmission without glass and the thinnest available 24-oz sheet.

SOURCE: O'Riordan and Catt, 1969.

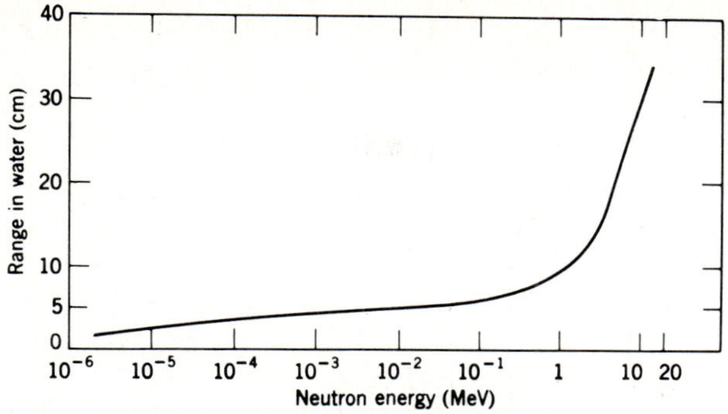

Figure 10.10 Range of neutrons in water (*Henry, 1969*). Range is taken as approximately the distance required for neutrons to slow down to indium-resonance energy of about 1.4 eV. (Values are adapted from neutron "age" calculations.)

One method of neutron shielding estimation is developed with the use of removal cross sections. For neutron energies up to 30 MeV, the removal cross section is approximately three-fourths of the total cross section. For example, the total cross sections of 4-MeV neutrons (produced in a Pu-Be generator) for hydrogen and oxygen (water) are 1.9 and 1.7 barns, respectively. Thus, the macroscopic removal cross section ($\Sigma_r = \sigma N$, where N = number of atoms/cm^3) is:

$$\Sigma_r = \frac{3}{4}\left[\left(\frac{1.9 \times 10^{-24}\ \text{cm}^2}{\text{H atom}}\right)\left(\frac{6.7 \times 10^{22}\ \text{H atoms}}{\text{cm}^3}\right) + \left(\frac{1.7 \times 10^{-24}\ \text{cm}^2}{\text{O atoms}}\right)\left(\frac{3.35 \times 10^{22}\ \text{O atoms}}{\text{cm}^3}\right)\right]$$
$$= 0.138\ \text{cm}^{-1}$$

and from Eq. (10.8), the half-value layer ($I/I_0 = 0.5$) of water for 4-MeV neutrons becomes 5.01 cm. The flux of neutrons, Φ_n, from a point source in a spherical medium is then:

$$\Phi_n = \frac{B\,S}{4\pi[n(\text{HVL})]^2 2^n}\ \text{neutrons/cm}^2\cdot\text{sec} \tag{10.29a}$$

where B is the dose build-up factor ($\simeq 5$ for hydrogenous shields > 20 cm), S is the source strength of 4-MeV neutrons (neutrons/sec), HVL is the half-value layer (cm), and n is the number of half-value layers of shield.

The following requirements must be met for the use of removal cross sections:

1. The neutron source must be monoenergetic or near monoenergetic so that the source can be described by a single energy or velocity.

2. There must be a large amount of hydrogen in the shield.
3. The shield must be effective at slowing down high-energy (>1-MeV) neutrons.

Note that these requirements were met for the example above with 4-MeV neutrons and a water shield. When these requirements can be met, Eq. (10.8) can be written as:

$$\dot{D} = N_0 C \exp[-(\Sigma t)_{\text{total}}] \tag{10.29b}$$

where $\dot{D}$ is the absorbed dose (mrad/hr); C is a conversion factor for (neutrons · $\text{cm}^{-2}\cdot\text{sec}^{-1}$) to (mrad/hr) ($\simeq 1.55 \times 10^{-2}$ for absorbed dose in tissue); N_0 is the source strength of monoenergetic neutrons (neutrons/sec); Σ_r is the macroscopic cross section; and $(\Sigma t)_{\text{total}}$ is the number of relaxation lengths separating the effective source point from the dose point. The number of relaxation lengths is computed as:

$$(\Sigma t)_{\text{total}} = \Sigma_{r_1} t_1 + \Sigma_{r_2} t_2 + \ldots \tag{10.29c}$$

where Σ_{r_i} is the macroscopic removal cross section of region i and t_i is the corresponding thickness of the material. The macroscopic removal cross section for 6–8 MeV neutrons is given by:

$$\Sigma_r = \frac{0.602\sigma_r \rho}{A} \text{cm}^{-1} \tag{10.29d}$$

where σ_r is the microscopic removal cross section (barns); ρ is the density (g/cm^3); and A is the atomic weight. Figure 10.11 can be used to obtain the removal cross section for any element or molecule for 6–8 MeV or fission neutrons. For example, the macroscopic removal cross section for $CaCO_3$, $\rho = 2.711$ and molecular weight $= 100.09$, is:

Element	A_i	$\rho_i(\text{g/cm}^3)$	$(\Sigma_r/\rho)_i$ (cm^2/g)[a]	$\Sigma_r(\text{cm}^{-1})$
Ca	40.08	$\left(\frac{40.08}{100.09}\right)(2.711) = 1.087$	0.023	0.025
C	12.01	$\left(\frac{12.01}{100.09}\right)(2.711) = 0.325$	0.050	0.016
O	16.00	$\left(\frac{(3)(16)}{100.09}\right)(2.711) = 1.299$	0.042	0.055
Macroscopic removal cross section for $CaCO_3$				0.096

[a]From Figure 10.11.

Table 10.22 gives removal cross sections for common shielding and reactor materials for fission neutrons (6–8 MeV). The removal cross sections for 14-MeV neutrons by concrete and paraffin wax are 0.052

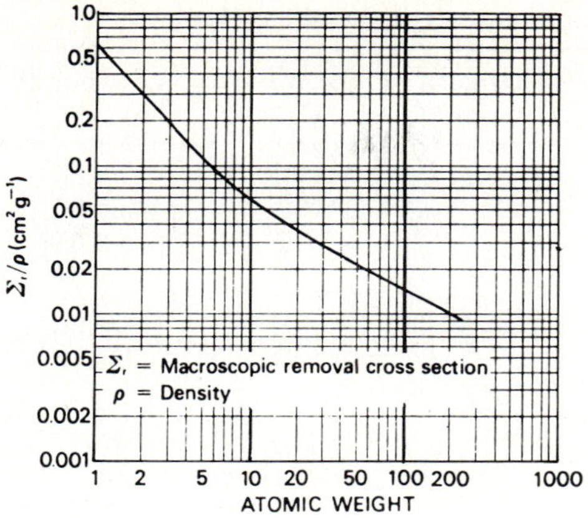

Figure 10.11 Fission neutron attenuation coefficients (*NCRP 38, 1971*).

cm^{-1} and 0.065 cm^{-1}, respectively (Bozyap and Day, 1975). The authors found it necessary to add to Eq. (10.29*b*) a dose build-up factor of 4 for the concrete shield.

The attenuation of neutron beams by slabs of various materials with the neutron beams at various angles to the slab normals have been reported (NCRP 38, 1971). Table 10.23 gives the compositions of the materials studied. The attenuation (or relative absorbed dose with and without the shield) of the dose from fluxes of 5-, 3-, and 1-MeV neutrons are shown in Figures 10.12, 10.13, and 10.14, respectively. Table 10.24 gives the neutron flux density as a function of energy that results in a dose equivalent of 100 mrem for a 40-hr exposure. (A more general relationship between flux and dose was given in Eq. 10.18.)

The absorption of neutrons can lead to induced radioactivity and the emission of γ-radiation. For low neutron energies, the γ-radiation may be more penetrating than the neutrons themselves. Table 10.25 summarizes capture cross sections for thermal-energy neutrons ($<$1 eV)

TABLE 10.22 Macroscopic Neutron Removal Cross Sections (Σ_r) and Attenuation Lengths of Some Shielding and Reactor Materials

Material	Removal cross section, Σ_r (cm^{-1})	Attenuation length (cm)
Water	0.103	9.7
Iron	0.1576	6.34
Ordinary concrete[a]	0.089	11.3
Graphite (ρ = 1.54 g·cm^{-3})	0.785	12.7

[a]The removal cross section of barytes concrete will not differ significantly from the removal cross section of ordinary concrete with similar water content and sand and mortar mix. The concrete removal cross section is, however, very sensitive to water content. The value quoted above is for 6% water by weight.

SOURCE: NCRP 38, 1971.

TABLE 10.23 Material Compositions Studied in Neutron Shielding Tests

Material	Density (g/cm^3)	Elements contained	Atoms/cm^3 ($\times 10^{-21}$)
Borated polyethylene (8% B_4C by weight)[a]	0.97	H	76.80
		C	39.20
		^{10}B	0.658
		^{11}B	2.67
Water	1.00	H	66.90
		O	33.45
Concrete	2.26	H	13.75
		O	45.87
		Al	1.743
		Si	20.15
NTS[b] soil (dry)	1.15	H	8.553
		O	22.08
		Al	2.014
		Si	9.533
NTS[b] soil (100% sat.)	1.25	H	16.87
		O	27.00
		Al	1.976
		Si	8.963

[a]Several calculations were made for pure polyethylene slabs of density 0.925 g/cm^3 up to 6 in. thick. Results differ negligibly from corresponding results for 8% borated polyethylene.

[b]Nevada test site.

SOURCE: NCRP 38, 1971.

and the maximum γ-energies encountered in neutron capture. Also shown are the average number of photons emitted per capture.

While water is an excellent medium for slowing down fast neutrons, the capture of a neutron by the hydrogen atom produces a high-energy γ (2.23 MeV, Table 10.25). The resulting γ-ray dose rate can be reduced by either increasing the γ-absorption coefficient of water by dissolving a high-atomic-number substance (e.g., $BaCl_2$); or by reducing or suppressing the rate of γ-production. Nuclei useful in suppression of capture γ-rays include ^6Li, ^{10}B, and ^{14}N. For example, the rate of production of 2.23-MeV γ from neutron capture is significantly reduced by dissolving boric acid (solubility = 63.2 g/liter) in water. The resulting concentration will be 6.17×10^{20} atoms B/ml. (The thermal cross section of boron is 3837 barns compared to a thermal cross section of hydrogen of 0.332 barns. Capture by boron results in emission of a 0.478-MeV photon compared to 2.23 MeV for hydrogen.)

Ordinary concrete is the most frequently used neutron shielding material for fixed installations (e.g., accelerators) because of its high hydrogen content and normally low Z aggregate. Materials with high cross sections for near thermal energies (e.g., B and Cd) can be used to reduce shielding thickness. Composite materials such as borated concrete are useful when incorporated into shielding barriers.

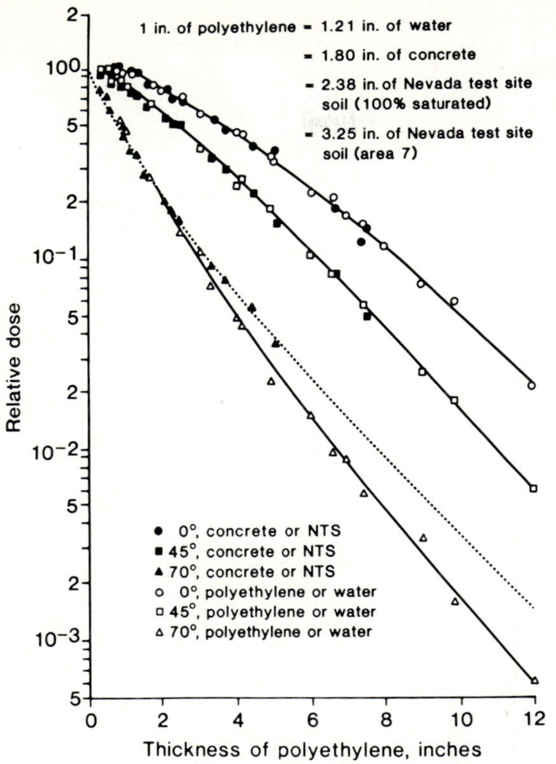

Figure 10.12 Attenuation of dose from 5-MeV incident neutrons penetrating various materials at various angles (*NCRP 38, 1971*).

Fast neutrons from low-voltage accelerators are produced by the T (d, n) ^{4}He reaction (D-T) where a beam of deuterons at energies from 100 to 400 keV is focused on a target of tritium. The maximum yield of neutrons occurs at 107 keV with a median neutron energy of 14 MeV. A 1-mA beam of 150-keV deuterons will produce approximately 10^{11} neutrons per second. The neutron dose equivalent, $\dot{D}_n$ (rem/hr), from a D-T generator at distance r (cm) from the target without shielding is given by:

$$\dot{D}_n = 1.6 \times 10^{-5}\, n/r^2 \tag{10.30a}$$

where n is the number of neutrons emitted per second (NCRP 72, 1983). The transmission factor A needed to reduce the dose equivalent rate to 2.5 mrem/hr is then:

$$A = \frac{0.0025}{\dot{D}_n} \tag{10.30b}$$

where A is a dimensionless attenuation factor.

Table 10.26 gives the thickness of concrete or water for a variety of

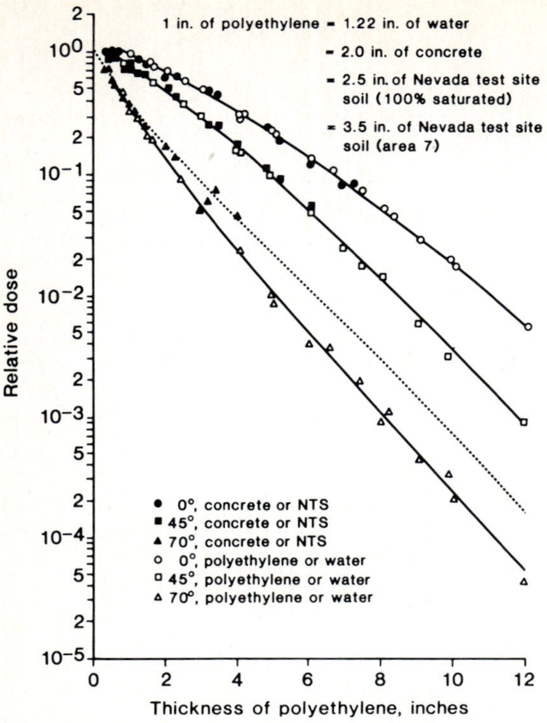

Figure 10.13 Attenuation of dose from 3-MeV incident neutrons penetrating various materials at various angles (*NCRP 38, 1971*).

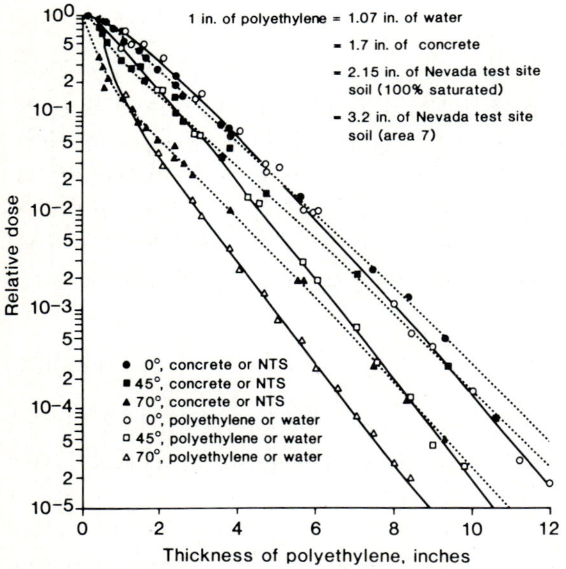

Figure 10.14 Attenuation of dose from 1-MeV incident neutrons penetrating various materials at various angles (*NCRP 38, 1971*).

TABLE 10.24 Mean Quality Factors,[a] $\overline{QF}$, and Values of Neutron Flux Density

Neutron energy (MeV)	$\overline{QF}$	Neutron flux density ($cm^{-2} \cdot sec^{-1}$)
2.5×10^{-8} (thermal)	2	680
1×10^{-7}	2	680
1×10^{-6}	2	560
1×10^{-5}	2	560
1×10^{-4}	2	580
1×10^{-3}	2	680
1×10^{-2}	2.5	700
1×10^{-1}	7.5	115
5×10^{-1}	11	27
1	11	19
2.5	9	20
5	8	16
7	7	17
10	6.5	17
14	7.5	12
20	8	11
40	7	10
60	5.5	11
1×10^{2}	4	14
2×10^{2}	3.5	13
3×10^{2}	3.5	11
4×10^{2}	3.5	10

[a]Maximum value of $\overline{QF}$ in a 30-cm phantom; the neutron flux density which, in a period of 40 hr, results in a maximum dose equivalent of 100 mrem.
SOURCE: NCRP 38, 1971.

transmission factors for D-T neutrons. One frequently encountered problem is that the required thickness of shielding from Table 10.26 is greater than the distance *r* in which the shielding must be accommodated. In this case, it is not possible to provide sufficient shielding.

For primary beam neutrons produced by constant-voltage accelerators, the shielding transmission ratio β_n of concrete ($rem \cdot cm^2$) can be calculated from the neutron fluence rate, Φ_0 (fluence rate at 1 m in neutrons$\cdot m^2 \cdot cm^{-2} \cdot sec^{-1}$) as:

$$\beta_n \leq (2.8 \times 10^{-7}) \frac{\dot{P} d^2}{\Phi_0 T} \tag{10.31}$$

where $\dot{P}$ is the maximum permissible dose (mrem/hr); d is the distance between neutron source and reference point (m); and T is the occupancy factor. The constant $2.8 \times 10^{-7} = 2.8 \times 10^{-4}$ hr/sec times 10^{-3} rem/mrem. The shielding transmission curves (β_n, dose-equivalent index transmission) for monoenergetic neutrons of thermal to 100 MeV are given in Figure 10.15. These curves include contributions from γ-pho-

TABLE 10.25 Neutron Thermal Cross Sections and Capture γ-Energies

Target	Thermal (n, γ) cross section (barns)	Highest-energy γ-ray (MeV)	Average number of photons per capture
Aluminum	0.235	7.724	2
Antimony	5.7	6.80	
Arsenic	4.3	7.30	
Barium	1.2	9.23	2.7
Beryllium	0.0095	6.814	
Bismuth	0.034	4.17	
Boron-10	3,837	0.478[a]	0.94
Cadmium	2,450	9.046	4.1
Calcium	0.44	7.83	
Carbon-12	0.0034	4.95	1.3
Chlorine	33.2	8.56	3.1
Chromium	3.1	9.716	2
Cobalt	37.0	7.486	
Copper	3.77	7.914	2.6
Fluorine	0.008	6.600	
Gadolinium	46,000	7.78	3.9
Gold	98.8	6.494	3.5
Hydrogen-1	0.332	2.230	1.0
Hydrogen-2	0.00057	6.251	1.0
Indium	196	6.38	3.3
Iron	2.53	10.16	1.7
Lead	0.17	7.38	
Lithium	70.7	7.26	
Magnesium	0.063	11.089	
Manganese	13.2	7.261	2.6
Mercury	380	7.66	3.3
Molybdenum	2.7	9.15	
Nickel	4.8	8.997	
Niobium	1.15	7.19	2.6
Nitrogen-14	0.075	10.833	
Phosphorus	0.190	7.94	
Platinum	8.8	7.920	
Potassium	2.1	9.39	
Praseodymium	11.6	5.83	
Rhodium	156	6.792	
Samarium	5,600	7.89	5.6
Scandium	24	8.85	
Selenium	12.3	10.483	
Silicon	0.160	10.599	
Silver	63	7.27	2.9
Sodium	0.534	6.41	2
Strontium	1.21	9.22	
Sulfur	0.52	8.64	
Tantalum	21	6.07	
Thallium	3.4	6.54	

TABLE 10.25 Neutron Thermal Cross Sections and Capture γ-Energies (*Continued*)

Target	Thermal (n, γ) cross section (barns)	Highest-energy γ-ray (MeV)	Average number of photons per capture
Tin	0.625	9.35	
Titanium	5.8	10.621	
Tungsten	19.2	7.42	
Vanadium	4.98	7.98	2.5
Zinc	1.10	9.51	
Zirconium	0.18	8.66	

[a]Gamma-ray from excited state of Li following (n, α) reaction.
SOURCES: NCRP 38, 1971; Blizard, 1958.

tons produced within the shield. Shielding thickness is given in terms of grams per square centimeter ($g \cdot cm^{-2}$), but is based on the atomic constituency of ordinary concrete ($\rho = 2.31$ g/cm^3). (See NCRP Reports 38, 51, and 72 for a complete discussion of neutron shielding designs.)

E. Protons and light ions

Proton ranges in aluminum, copper, lead, and uranium as a function of energy, E, are plotted in Figure 10.16. Corresponding ranges for deuterons, tritons ($^3H^+$), ^3He, and ^4He ions can be calculated from the

TABLE 10.26 Dose Equivalent Index Transmission Factors for a 14-MeV Neutron Generator in Concrete and Water[a]

Transmission factor A	Thickness (cm) Concrete	Water
0.1	55	61
0.05	64	71
0.02	76	84
0.01	84	94
5×10^{-3}	94	104
2×10^{-3}	105	118
1×10^{-3}	114	128
5×10^{-4}	123	137
2×10^{-4}	135	150
1×10^{-4}	143	161
5×10^{-5}	152	171
2×10^{-5}	165	184
1×10^{-5}	174	195
5×10^{-6}	184	—
2×10^{-6}	197	—

[a]The transmission has been multiplied by 2 to account for γ-rays.
SOURCE: NCRP 72, 1983.

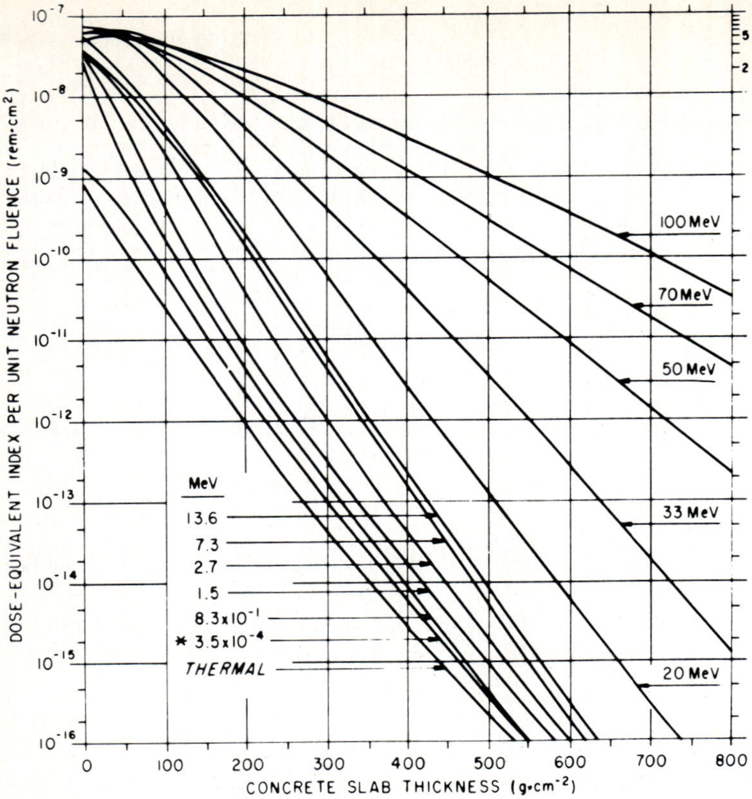

Figure 10.15 Dose equivalent index transmission through concrete of monoenergetic neutrons (*NCRP 51, 1977*). Dose equivalent index transmission per unit fluence (rem·cm²) of monoenergetic neutrons incident normally on slabs of ordinary concrete. Multicollision dose equivalent index per unit fluence is calculated to include the dose equivalent index contribution by capture γ-rays produced within the slab.

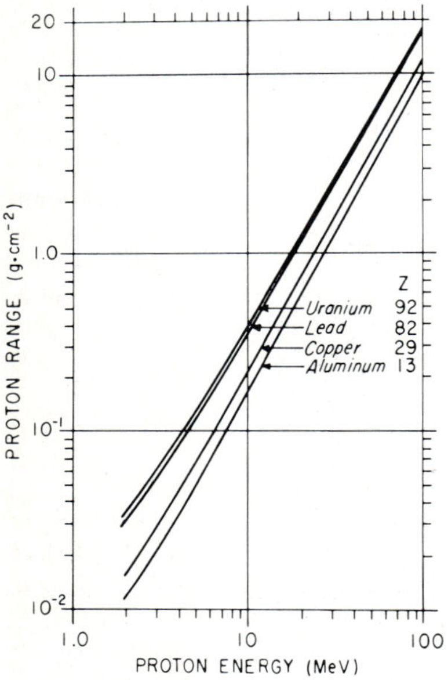

Figure 10.16 Range of protons in several materials as a function of incident proton energy (*NCRP 51, 1977*).

proton range as follows: range of deuterons = $2R_p$ (R_p = range of proton) at $E/2$; range of tritons = $3R_p$ at $E/3$; range of $^3He^+$ = $3R_p$ at $E/3$; range of $^3He^{2+}$ = $\frac{3}{4}R_p$ at $E/3$; range of $^4He^+$ = $4R_p$ at $E/4$; and range of $^4He^{2+}$ = R_p at $E/4$. Note that even though the charged particles have a finite range in matter, additional barrier thickness may be required to attenuate secondary radiations produced within the shields (including x-rays, γ-rays, and/or neutrons).

V. Example Applications

A. Beta shield

Example 10.2 Fifty milliliters of an aqueous solution containing 10 Ci of ^{90}Sr is to be stored in a laboratory. Access to the laboratory is limited to 1 hr per day, 5 days per week. The isotope will be stored 1 m from a technician's workbench. Design a shielded container to meet the NCRP dose-limiting recommendations for the technician (Table 10.11).

solution The maximum allowable dose equivalent for occupationally exposed individuals is 5000 mrem in any one year. Assuming a 5-day work week, this represents a limit of 20 mrem per day. For design purposes, assume that this exposure represents 25% of the maximum allowable exposure or a limit of 5 mrem per day. Since access is limited to no more than 1 hr per day, the shield will be designed to limit the exposure rate to 5 mrem per hour at 1 m from the source. The container will be constructed of two materials. The inner shield will consist of a polyethylene bottle designed to stop 100% of the β-particles emitted. The outer shield will be made of lead to limit exposure from the bremsstrahlung generated.

Strontium 90 has a half-life of 28 years and emits β with maximum and average energies of 0.544 and 0.200 MeV, respectively. ^{90}Sr decays to ^{90}Y with a half-life of 64 hr. Because the daughter's half-life is very much less than the parent's half-life, the activities of the isotopes will be at equilibrium. Therefore, the shield must control radioactivity from 10 Ci of ^{90}Sr and 10 Ci of ^{90}Y. From Table 10.2, ^{90}Y emits β-particles with maximum and average energies of 2.245 and 0.931 MeV, respectively. Since β-particles have a finite range in matter, a shield designed to stop 2.245-MeV β will be effective at removing 100% of the β-emissions. Figure 10.1 shows the range of β-particles in matter. For 2.245-MeV particles, the maximum range is 1.1 g/cm². From Table 10.4, the density of polyethylene is 0.95. Therefore the thickness of the polyethylene bottle to stop all β with $E < 2.245$ MeV is:

$$\frac{1.1 \text{ g/cm}^2}{0.95 \text{ g/cm}^3} = 1.16 \text{ cm}$$

To design on the conservative side, use a bottle with walls 1.25 cm thick.

The next step is to size the lead shield to control bremsstrahlung x-rays. Equation (10.4) can be used to estimate the fraction of β-particles emitted that are converted to x-ray photons. While the polyethylene bottle protects against the escape of any β-particles, the majority of β-particles are stopped by the water solvent. It is acceptable to assume, therefore, that when applying Eq. (10.4), all β-particles are stopped by water molecules. The atomic number of water (Z_{H_2O})

can be calculated from

$$Z_{H_2O} = \left(\frac{2}{18}\right)(1) + \left(\frac{16}{18}\right)(8)$$
$$= 7.22$$

and the fraction of ^{90}Sr and ^{90}Y β's converted to x-ray photons is

$$f(^{90}\text{Sr}) = (3.5 \times 10^{-4})(7.22)(0.544) \tag{10.4}$$
$$= 0.00137$$

$$f(^{90}\text{Y}) = (3.5 \times 10^{-4})(7.22)(2.245) \tag{10.4}$$
$$= 0.00567$$

Equation (10.5) is used to estimate the photon flux from a shielded β-source. For this calculation, $\overline{E}_\beta$ is the average β-energy emission rate and is calculated from the average β-energy for each isotope.

$$\overline{E}_\beta(^{90}\text{Sr}) = (10\ \text{Ci})\left(3.7 \times 10^{10}\ \frac{\text{tps}}{\text{Ci}}\right)\left(0.200\ \frac{\text{MeV}}{\text{transformation}}\right)$$
$$= 7.4 \times 10^{10}\ \frac{\text{MeV}}{\text{sec}}$$

$$\overline{E}_\beta(^{90}\text{Y}) = (10\ \text{Ci})\left(3.7 \times 10^{10}\ \frac{\text{tps}}{\text{Ci}}\right)\left(0.931\ \frac{\text{MeV}}{\text{transformation}}\right)$$
$$= 3.44 \times 10^{11}\ \frac{\text{MeV}}{\text{sec}}$$

and from Eq. (10.5):

$$\Phi(^{90}\text{Sr}) = \frac{(0.00137)(7.4 \times 10^{10}\ \text{MeV/sec})}{4\pi(100\ \text{cm})^2(0.544\ \text{MeV/photon})} \tag{10.5}$$
$$= 1483\ \text{photons/cm}^2\cdot\text{sec}$$

$$\Phi(^{90}\text{Y}) = \frac{(0.00567)(3.44 \times 10^{11}\ \text{MeV/sec})}{4\pi(100\ \text{cm})^2(2.245\ \text{MeV/photon})} \tag{10.5}$$
$$= 6914\ \text{photons/cm}^2\cdot\text{sec}$$

The exposure to bremsstrahlung photons is calculated with Eq. (10.13). From Table 10.7, the mass energy absorption coefficients for air (μ_e/ρ) for ^{90}Sr photons (0.544 MeV) and ^{90}Y photons (2.245 MeV) are 0.0297 cm^2/g and 0.0227 cm^2/g, respectively. From Eq. (10.13):

$$\dot{X}(^{90}\text{Sr}) = \frac{(1483\ \text{photons/cm}^2\cdot\text{sec})(0.544\ \text{MeV/photon})(1.6 \times 10^{-13}\ \text{J/MeV})(0.0297\ \text{cm}^2/\text{g})}{[34(\text{J/kg})/(\text{Ci/kg})](10^{-3}\ \text{kg/g})[2.577 \times 10^{-4}\ (\text{Ci/kg})/\text{R}]}$$
$$= 4.375 \times 10^{-7}\ \text{R/sec} \tag{10.13}$$

$$\dot{X}(^{90}\text{Y}) = \frac{(6914\ \text{photons/cm}^2\cdot\text{sec})(2.245\ \text{MeV/photon})(1.6 \times 10^{-13}\ \text{J/MeV})(0.0227\ \text{cm}^2/\text{g})}{[34(\text{J/kg})/(\text{Ci/kg})](10^{-3}\ \text{kg/g})[2.577 \times 10^{-4}(\text{Ci/kg})/\text{R}]}$$
$$= 6.434 \times 10^{-6}\ \text{R/sec} \tag{10.13}$$

From Eq. (10.15*b*) and Table 10.7, the tissue dose is:

$$\dot{D}(^{90}\text{Sr}) = (0.878)\frac{(0.0325)}{(0.0297)}(4.375 \times 10^{-7}\ \text{rad/sec}) \tag{10.15b}$$
$$= 4.20 \times 10^{-7}\ \text{rad/sec} = 1.51\ \text{mrad/hr}$$

$$\dot{D}(^{90}\text{Y}) = (0.878)\frac{(0.0249)}{(0.0227)}(6.434 \times 10^{-6}\ \text{rad/sec}) \tag{10.15b}$$
$$= 6.19 \times 10^{-6}\ \text{rad/sec} = 22.3\ \text{mrad/hr}$$

The total absorbed dose is (1.51 + 22.3) mrad/hr. Since the quality factor for x-radiation is 1, this represents 23.81 mrem/hr (Eq. 10.19*b*), almost five times the design benchmark of 5 mrem/hr. A lead shield is, therefore, very necessary.

The relationship between absorber thickness and intensity is given by Eq. (10.10). The linear attenuation coefficients for lead for 0.544- and 2.245-MeV x-rays are given in Table 10.6.

$$\mu_l(\text{Pb, 0.544 MeV}) = (0.145\ \text{cm}^2/\text{g})(11.3\ \text{g/cm}^3) = 1.64\ \text{cm}^{-1}$$

$$\mu_l(\text{Pb, 2.245 MeV}) = (0.0446\ \text{cm}^2/\text{g})(11.3\ \text{g/cm}^3) = 0.504\ \text{cm}^{-1}$$

The thickness (t) of the lead shield can be derived from the following:

$$I(^{90}\text{Sr}) + I(^{90}\text{Y}) = 5\ \text{mrem/hr}$$

$$I(^{90}\text{Sr}) = (1.51\ \text{mrem/hr})\ \exp[(-1.64)t] \qquad (10.10)$$

$$I(^{90}Y) = (22.3\ \text{mrad/hr})\ \exp[(-0.504)t]$$

Solving for t, 3.0 cm of lead are required to control this source. In this example, it was assumed that the bremsstrahlung has a quantum energy equal to the maximum energy of the β-particle. Since this is actually an upper limit of the photon energy, the calculated exposures are overestimated. On the other hand, dose build-up in the lead shield was not considered. This would cause the shield thickness to be underestimated. These two errors have been found to cancel (Cember, 1983). Therefore, a shield consisting of 1.25 cm polyethylene followed by 3.0 cm of lead is an appropriate solution.

B. Gamma shield

Example 10.3 A 1-Ci source of ^{137}Cs is to be stored in a spherical lead container when not in use. Determine the thickness of the container such that the dose equivalent rate at a distance of 1 m from the source will not exceed 2.5 mrem/hr. Assume the source to be a point at the center of the sphere.

solution From Table 10.8, the specific gamma-emission (Γ) for ^{137}Cs (γ = 0.622 MeV) is 3.3 R·cm²/hr·mCi. The exposure at 1 m from an unshielded source is then:

$$\dot{X} = \left(3.3\ \frac{\text{R}\cdot\text{cm}^2}{\text{hr}\cdot\text{mCi}}\right)(10^3\ \text{mCi})\left[\frac{1}{(100\ \text{cm})^2}\right] = 0.33\ \text{R/hr}$$

The absorbed dose calculated with Eq. (10.15*b*) and Table 10.7 is:

$$\dot{D} = (0.878)\ \frac{\mu_{e,\text{tissue}}/\rho_m}{\mu_{e,\text{air}}/\rho_a}\ (0.33\ \text{R/hr}) \qquad (10.15b)$$

$$= (0.878)\left(\frac{0.0324}{0.0294}\right)(0.33) \qquad (10.15b)$$

$$= 0.32\ \text{rad/hr}$$

From Eq. (10.19*b*), this represents a dose equivalent rate of 0.32 rem/hr or 320 mrem/hr.

There are three ways to determine the required shield thickness: calculate the intensity reduction with no build-up factor (Eq. 10.10) and add one half-value layer; calculate the intensity reduction with a build-up factor (Eq. 10.20); use a graphical solution (Figure 10.3).

From Table 10.6, the linear attenuation coefficient of lead for ^{137}Cs photons with quantum energy = 0.662 MeV is:

$$\mu_{t,l}(\text{Pb, 0.662 MeV}) = (0.114\ \text{cm}^2/\text{g})(11.3\ \text{g/cm}^3)$$
$$= 1.287\ \text{cm}^{-1}$$

From Eq. (10.10),

$$2.5 = 320 \exp[(-1.287\ \text{cm}^{-1})t] \qquad (10.10)$$
$$t = 3.771\ \text{cm of lead}$$

From Table 10.14, the half-value layer for ^{137}Cs is 0.65 cm of lead. Therefore, the required thickness is:

$$t = (0.65 + 3.771) = 4.421\ \text{cm of lead}$$

Solving for the shield thickness with the dose build-up factor (Eq. 10.20):

$$I = 2.5 = (320)(B) \exp[(-1.287\ \text{cm}^{-1})t] \qquad (10.20)$$

Table 10.13 gives values of B for various $\mu_t t$ products and energies. Interpolating at 0.662 MeV, B = 1.87 at $\mu_t t$ = 4 and B = 2.33 at $\mu_t t$ = 7. Since there is one equation with two unknowns, a trial-and-error solution is necessary. First, estimate a value for t. Calculate $\mu_t t$ and interpolate a value of B. With these values, calculate I. Repeat until I = 2.5.

t (cm)	$\mu_t t$	B	I (calculated)
5.0	6.435	2.25	1.14
4.5	5.791	2.15	2.07
4.4	5.663	2.125	2.36
4.35	5.598	2.115	2.50

The required shield thickness using the dose build-up factor is 4.35 cm of lead.

For the graphical method, the fraction of photons transmitted through the shield is 2.5/320 = 0.0078. From Figure 10.3, the lead shield thickness for 0.0078 transmission and ^{137}Cs photons is about 4.5 cm. Note that all three methods give comparable results.

C. x-Ray shield

Example 10.4 Determine the thickness of the primary and secondary protective barriers for a 250-kV x-ray therapy machine. Design the primary barrier to protect a controlled area 2.1 m from the target. The use and occupancy factors for this barrier are $\frac{1}{4}$ and 1, respectively. Design the secondary barrier to protect an uncontrolled area 2.1 m from the target and scatterer. The target-to-scatterer distance is 50 cm and the exposed area of the target is (20 × 20) cm^2. The wall in question adjoins an area with an occupancy factor of 1. The x-ray tube has a continuous tube current rating at 250 kV of 20 mA and a weekly workload of 20,000 mA·min.

solution The primary protective barrier is designed to protect a controlled area. The permissible exposure for a controlled area is 0.1 rem/week. Based on the notation defined in Table 10.20, $P = 0.1$ rem/week, $d_{\text{pri}} = 2.1$ m, $W = 20{,}000$ mA·min, $U = \frac{1}{4}$, and $T = 1$. From Eq. (10.21*f*):

$$K_{ux} = \frac{(0.1)(2.1)^2}{(20{,}000)(\frac{1}{4})(1)} = 8.82 \times 10^{-5} \qquad (10.21f)$$

From Figures 10.8*a* and 10.7*b*, the thickness of concrete and lead required are 15 in. and 8 mm, respectively. As shown in Table 10.19*b*, $\frac{1}{3}$-in. commercial lead sheet could be used for this installation.

The secondary barrier must be designed to stop both scattered and leakage radiation. The procedure is to evaluate these radiation fields separately. Defining the parameters for evaluating scattered radiation, $P = 0.01$ R/week (uncontrolled area), $d_{\text{sec}} = 2.1$ m, $d_{\text{sca}} = 0.5$ m, $W = 20{,}000$ mA·min, $T = 1$, $F = (20 \times 20)$ cm^2 = 400 cm^2, and f (the normalizing factor for beam energy) = 1 for operating voltages below 1000 V. Table 10.18 lists the ratio of scattered to incident exposures, a, for various x-ray voltages and scattering angles. Assuming a 90° scattering angle (worst-case assumption), $a = 0.0019$ for 250-kV x-rays. From Eq. (10.28*b*):

$$K_{ux} = \frac{(0.01)(2.1)^2(0.5)^2(400)}{(0.0019)(20{,}000)(1)(400)(1)} = 2.9 \times 10^{-4} \qquad (10.28b)$$

From Figures 10.8*a* and 10.7*b*, this represents 13.5 in. of concrete or 7 mm of lead.

Secondary barriers for leakage radiation are based on NCRP performance specifications for x-ray tube housings. For therapeutic x-ray ≤500 kV peak voltage, the leakage field must be less than 1 R/hr at 1 m. With $P = 0.01$ R/week, $I = 20$ mA, $d_{\text{sec}} = 2.1$ m, $T = 1$, and $W = 20{,}000$ mA·min, Eq. (10.24*b*) can be used to estimate the barrier requirement, and

$$B_{LX} = \frac{(0.01)(2.1)^2(60)(20)}{(20{,}000)(1)} = 2.65 \times 10^{-3} \qquad (10.24b)$$

The relationship between barrier requirement and number of half-value layers (n) is given by Eq. (10.26),

$$0.00265 = (\tfrac{1}{2})^n \qquad (10.26)$$

$$n = 8.56$$

From Table 10.14, the half-value layer of 250-kV x-rays for lead and concrete are 0.88 mm and 2.8 cm, respectively. Therefore, the barrier thickness for lead and concrete (Eq. 10.27) are 7.5 mm and 23.9 cm, respectively. Since the barrier requirements for scattered and leakage fields are approximately equal (7.0 mm vs. 7.5 mm of lead), the secondary barrier is defined as the greater of the two plus one half-value layer:

$$7.5 + 0.88 = 8.38 \text{ mm lead}$$

or $$(13.5 \text{ in.})(2.54 \text{ cm/in.}) + 2.8 \text{ cm} = 37.1 \text{ cm of concrete}$$

From Table 10.19*b*, $\frac{1}{3}$-in. commercial lead sheet would be an appropriate shield for this application.

D. Neutron shield

Example 10.5 Determine the minimum thickness of a concrete shield required for the nearest accessible point from a 0.8-MeV, 10-μA deuteron beam on a tritium target (D-T generator). The concrete shield is located 1 m from the target.

solution From Figure 10.2*b*, the thick-target neutron yield for a D-T generator at 0.8 MeV is $3 \times 10^8\ \text{sec}^{-1}\cdot\mu\text{A}^{-1}$. With a 10-$\mu$A beam current, the total number of neutrons is then $3 \times 10^9\ \text{sec}^{-1}$. From Figure 10.2*d*, the energy of D-T neutrons resulting from a 0.8-MeV beam energy is 14 MeV. Table 10.24 shows that a neutron flux of 12 neutrons /cm²·sec is equivalent to a dose equivalent of 100 mrem/40-hr period. Therefore, the dose equivalent for this problem is:

$$\dot{D}_n = \left(3 \times 10^9 \frac{\text{neutrons}}{\text{sec}}\right)\left(\frac{1}{4\pi r^2\ \text{cm}^2}\right)\frac{(0.1\ \text{rem/40 hr})}{(12\ \text{neutrons/cm}^2\cdot\text{sec})}$$

$$= \frac{4.9 \times 10^4}{r^2\ (\text{cm}^2)}\ (\text{rem/hr})$$

Alternatively, from Eq. (10.30*a*):

$$\dot{D}_n = (1.6 \times 10^{-5})\left(\frac{n}{r^2}\right) \tag{10.30a}$$

$$= \frac{4.8 \times 10^4}{r^2\ (\text{cm}^2)}\ (\text{rem/hr})$$

The transmission factor needed to reduce the dose equivalent rate to 2.5 mrem/hr (5 rem/year based on 40 hr of exposure per week, 50 weeks/year) is calculated as shown by Eq. (10.30*b*), and

$$A = (0.0025)\left(\frac{r^2}{4.9 \times 10^4}\right) \tag{10.30b}$$

Assuming that the nearest accessible point is the far face of the shield, r equals the thickness of the shield plus the target shield spacing of 1 m. The problem is solved by trial and error: (1) Estimate a value of shield thickness; (2) calculate r; (3) calculate the required thickness from the transmission factor above and Table 10.26; and (4) repeat until estimated thickness = calculated thickness.

Estimated thickness (cm)	r (cm)	A	Calculated thickness (cm)
200	300	4.6×10^{-3}	95
150	250	3.2×10^{-3}	100
110	210	2.2×10^{-3}	105

From the above calculations, a minimum of 110 cm of concrete are needed and, therefore, the nearest allowable access point to the target is 2.1 m.

The problem can also be solved by the more general relationship given by Eq. 10.31. This relationship gives a shielding transmission ratio β_n (rem·cm²) as a function of neutron fluence, maximum permissible exposure, and distance between neutron source and reference point. Given the previous solution of r = 2.1

m and solving Eq. (10.31) with $\dot{P} = 2.5$ mrem/hr:

$$\beta_n = \frac{2.8 \times 10^{-7}(\text{neutrons/sec})(\text{rem/mrem})(2.5\ \text{mrem/hr})(2.1\ \text{m})^2}{(3 \times 10^9\ \text{neutrons/sec})(10^{-4}\ \text{m}^2/\text{cm}^2)} \qquad (10.31)$$

$$= 1.03 \times 10^{-11}\ \text{rem}\cdot\text{cm}^2$$

From Figure 10.15, this dose equivalent index for 14-MeV neutrons requires a concrete slab thickness of 280 g/cm². Given the density of ordinary concrete of 2.31 g/cm³, this represents 121 cm of concrete, very close to the previously calculated 110 cm. Note that while Eq. (10.30) and Table 10.26 apply to D-T generators only, Eq. (10.31) and Figure 10.15 can be applied to any monoenergetic neutron source. Equation (10.29*b*) and the data of Bozyap and Day (1975) can be used to check the calculated concrete shield thickness. They report a removal cross section for concrete of 0.052 with a dose build-up factor of 4. From Eq. (10.29*b*) with the addition of the dose build-up factor,

$$\dot{D} = N_0 C B e^{-\Sigma t}$$

For this problem, N_0C has been calculated as

$$N_0C = \frac{4.9 \times 10^4\ (\text{rem/hr})}{r^2\ (\text{cm}^2)}$$

where r is the shield thickness (t) plus the source-to-shield distance. Substituting $r = (100 + t)$ cm, $B = 4$, $\Sigma = 0.052$, $D = 0.0025$ rem/hr, and solving for t:

$$0.0025\ \text{rem/hr} = \left[\frac{4.9 \times 10^4\ \text{rem/hr}}{(100 + t)^2\ \text{cm}^2}\right] (4) \exp[-(0.052)(t)]$$

$$t = 140\ \text{cm}$$

This is reasonably close to the 121 cm calculated from Figure 10.15 and the 110 cm from Table 10.26.

Example 10.6 Determine the minimum thickness of a shield required to control the radiation exposure from a 10-Ci Po-Be neutron generator. The shield is located 0.5 m from the source.

solution Table 10.5 shows that a ^{210}Po + Be neutron source produces 2.5×10^6 neutrons/sec per curie at average energy of 4.2 MeV. The flux density incident on the shield is, therefore,

$$\Phi_n = \left(\frac{2.5 \times 10^6\ \text{neutrons/sec}}{\text{Ci}}\right) (10\ \text{Ci}) \frac{1}{4\pi(50\ \text{cm})^2}$$

$$= 796\ \text{neutrons/cm}^2$$

From Table 10.24, the neutron flux density that results in a 40-hr exposure of 100 mrem for 4-MeV neutrons is about 15 neutrons/cm²·sec. The attenuation of the neutron flux needed to reach this dose is 15/796 = 0.019. Figures 10.12, 10.13, and 10.14 show the thickness of polyethylene, water, concrete, and soil vs. relative dose (attenuation) for 5-, 3-, and 1-MeV neutrons. Selecting the 5-MeV curves at 0° to the shield normal (both conservative assumptions), 12 in. of polyethylene would form an effective shield. As shown in the figure, 12 in. of polyethylene is equivalent to

(12)(1.21) = 14.5 in. of water

(12)(1.8) = 21.6 in. of concrete

or (12)(2.38) = 28.6 in. of saturated soil

Equation (10.29*a*) can also be used to estimate the thickness of a spherical container around a neutron source. With Φ_n = 15 neutrons/cm^2·sec, $S = 2.5 \times 10^7$ neutrons/sec, B = 5 for hydrogenous shields, the half-value layer (HVL) calculated from Eq. (10.8), and Σ (macroscopic removal cross section for water) = 0.138 cm^{-1}:

$$\frac{I}{I_0} = \frac{1}{2} = e^{-\Sigma(\mathrm{HVL})} \tag{10.8}$$

HVL = 5.02 cm

and $$15\left(\frac{\text{neutrons}}{\text{cm}^2\cdot\text{sec}}\right) = \frac{(5)(2.5 \times 10^7 \text{ neutrons/sec})}{4\pi[(n)(5.02)]^2\cdot 2^n} \tag{10.29a}$$

where n equals the number of half-value layers of the shield. Solving by trial and error, 8.5 half-value layers are needed, or

$$(8.5)(5.02) = 42.7 \text{ cm of water}$$

Although this answer is a little larger than that found with Figure 10.12, the difference is reasonable given that the spherical container is assumed to be close to the source and the previously estimated shield was located $\frac{1}{2}$ m from the source.

Neutron shields can also be evaluated using Eq. (10.29*b*). For example, the removal cross section for ordinary concrete (Table 10.22) is 0.089 cm^{-1}. With a shield of concrete 21.6 in. (55 cm) thick, the radiation absorbed dose is:

$$\dot{D} = \left(796 \frac{\text{neutrons/sec}}{\text{cm}^2\cdot\text{sec}}\right)\left\{1.55 \times 10^{-2} \frac{\text{mrad/hr}}{\text{neutrons/cm}^2\cdot\text{sec}} \exp[-(0.089)(55)]\right\}$$
$$= 0.092 \text{ mrad/hr} \tag{10.29b}$$

From Table 10.24, the quality factor for 4-MeV neutrons is about 9, and the dose equivalent (Eq. 10.19*b*) is (0.092)(9) = 0.83 mrem/hr. This suggests that Figure 10.12 gives a more conservative estimate than Eq. (10.29*b*), as Figure 10.12 was used to estimate shield thickness for a dose equivalent of 2.5 mrem/hr. However, given the difference in assumed neutron energies (Table 10.22 is based on fission neutrons in the 6–8 MeV range, Figure 10.12 is based on 5-MeV neutrons, and the Po-Be generator produces 4-MeV neutrons), the difference is reasonable.

Finally, Eq. (10.31) and Figure 10.15 can also be used to determine the shield thickness for concrete. From Eq. 10.31 (d = 50 cm):

$$\beta_n = \frac{2.8 \times 10^{-7}(\text{hr/sec})(\text{rem/mrem})(2.5 \text{ mrem/hr})(0.5 \text{ m})^2}{(2.5 \times 10^7 \text{ neutrons/sec})(10^{-4} \text{ m}^2/\text{cm}^2)}$$
$$= 7.0 \times 10^{-11} \text{ rem}\cdot\text{cm}^2$$

From Figure 10.15, this represents 225 g·cm^{-2} of concrete or 225/2.31 = 97 cm (38 in.) of concrete. This answer is about a factor of 2 more conservative than from Figure 10.12. However, the dose equivalent index in Figure 10.15 is calculated to include the dose equivalent from capture γ-rays produced within the slab. Therefore, the 97 cm predicted by Eq. 10.31 and Figure 10.15 controls the exposure to neutrons and capture gamma γ-rays, while the 55 cm predicted from Figure 10.12 limits the dose equivalent from neutrons only.

REFERENCES

Blizard, E. P. (1958). Nuclear radiation shielding. In: Etherington, H., ed., *Nuclear Engineering Handbook*. McGraw-Hill, New York, pp. 7-60 to 7-116.

Bozyap, O., and Day, L. R. (1975). Attenuation of 14 MeV neutrons in shields of concrete and wax. *Health Phys., 28,* 101–109.

Brownell, W. H., Ellett, W. H., and Reddy, A. R. (1968). Absorbed fractions for proton dosimetry. *J. Nucl. Med.,* Suppl. No. 1, MIRD Pamphlet No. 3, pp. 27–39.

Burgess, W. A. (1981). *Recognition of Health Hazards in Industry.* Wiley, New York.

Cember, H. (1983). *Introduction to Health Physics,* 2d ed. Pergamon Press, New York.

Doyle, L. E., Keyser, C. A., Leach, J. L., Schrader, G. F., and Singer, M. B. (1985). *Manfacturing Processes and Materials for Engineers,* 3d ed. Prentice-Hall, Englewood Cliffs, N.J.

Evans, W. W., Granke, R. C., Wright, F. A., and Trump, J. G. (1952). Absorption of 2 MeV constant potential roentgen rays by lead and concrete. *Radiology, 58,* 560–567.

Henry, H. F. (1969). *Fundamentals of Radiation Protection.* Wiley, New York.

ICRP (1962). *Recommendations of the International Commission on Radiological Protection,* ICRP Publication 6. Pergamon Press, Oxford.

ICRP (1977). Recommendations of the International Commission on Radiological Protection, Report 26. *Ann. ICRP,* vol. 1, no. 3.

Jaeger, R. G., Blizard, E. P., Chilton, A. B., Grotenhuis, M., Jaeger, Th. A., and Eisenlohr, H. H. (1968). *Engineering Compendium on Radiation Shielding,* vol. 1. Springer-Verlag, New York, pp. 21–30.

Kelly, J. P., and Trout, E. D. (1972). Broad beam attenuation in lead for x-rays from 50 to 300 kV. *Radiology, 104,* 171–175.

Kirn, F. S., and Kennedy, R. J. (1954). Betatron x-rays: How much concrete for shielding. *Nucleonics, 12*(6), 44–48.

Miller, W., and Kennedy, R. J. (1955). X-ray attenuation in lead, aluminum and concrete in the range 275–525 kV. *Radiology, 65,* 920–925.

Miller, W., and Kennedy, R. J. (1956). Attenuation of 86 and 176 MeV synchrotron x-rays in concrete and lead. *Rad. Res., 4,* 360–366.

NBS (1960). *Measurement of Neutron Flux and Spectra for Physical and Biological Applications,* Handbook No. 72. U.S. Department of Commerce, National Bureau of Standards, Washington, D.C.

NBS (1961). *A Manual of Radioactive Procedures,* Handbook No. 80. U.S. Department of Commerce, National Bureau of Standards, Washington, D.C.

NBS (1963). *Maximum Permissible Body Burdens and Maximum Permissible Concentrations of Radionuclides in Air and Water for Occupational Exposure,* Handbook 69. U.S. Department of Commerce, National Bureau of Standards, Washington, D.C.

NCRP 38 (1971). *Protection against Neutron Radiation.* National Council on Radiation Protection, Washington, D.C.

NCRP 39 (1971). *Basic Radiation Protection Criteria.* National Council on Radiation Protection, Washington, D.C.

NCRP 49 (1976). *Structural Shielding Design and Evaluation for Medical Use of X-rays and Gamma Rays of Energies up to 10 MeV.* National Council on Radiation Protection and Measurements, Washington, D.C.

NCRP 51 (1977). *Radiation Protection Design Guidelines for 0.1–100 MeV Particle Accelerator Facilities.* National Council on Radiation Protection, Washington, D.C.

NCRP 72 (1983). *Radiation Protection and Measurement for Low Voltage Neutron Generators.* National Council on Radiation Protection, Washington, D.C.

NCRP 78 (1984). *Evaluation of Occupational and Environmental Exposures to Radon and Radon Daughters in the United States.* National Council on Radiation Protection, Washington, D.C.

O'Riordan, M. C., and Catt, B. R. (1969). Low-energy x-ray shielding with common materials. *Health Phys., 17,* 516–518.

OSHA (1983). *General Industry Standards Part 1910.* Occupational Safety and Health Administration 2206, rev. March 11.

PHS (1970). *Radiological Health Handbook,* rev. ed. U.S. Department of Health, Education and Welfare, Public Health Service, Washington, D.C.

Summerlin, J., Jr., and Prichard, H. M. (1985). Radiological health implications of lead-

210 and polonium-210 accumulations in LPG refineries. *Amer. Ind. Hyg. Assoc. J., 46,* 202–205.

Trout, E. D., Kelley, J. P., and Herbert, G. L. (1975). X-ray attenuation in steel—50 to 300 kVp. *Health Phys., 29,* 163–169.

Trout, E. D., Kelley, J. P., and Lucas, A. C. (1959). Broad beam attenuation in concrete for 50 to 300 kVP x-rays and in lead for 300 kVp x-rays. *Radiology, 72,* 62–67.

Wyckoff, H. O., Kennedy, R. J., and Bradford, W. R. (1948). Broad and narrow beam attenuation of 500–1,400 kV x-rays in lead and concrete. *Radiology, 51,* 849–859.

Chapter

11

Nonionizing Radiation

P. A. Scheff

The photon energy necessary to ionize atomic oxygen and hydrogen is approximately 10 eV (Chapter 10). Nonionizing radiation is, therefore, defined as electromagnetic radiation with energy less than 10 eV. The frequency of a 10-eV photon is about 2.4×10^{15} Hz (Eq. 10.1). The resulting wavelength is 124 nm (Eq. 10.2). Shorter-wavelength (higher-energy) photons can dissipate energy in matter by ionizing molecules. Longer-wavelength radiation (nonionizing) primarily dissipates its energy as heat. Exposure to high levels of nonionizing radiation can, therefore, cause thermal injury to tissue.

Figure 11.1 shows the full electromagnetic spectrum, which spans over 20 orders of magnitude of photon energy. Radiofrequencies constitute the lowest energy band of the nonionizing portion of the spectrum, from about 3×10^{10} to 3×10^{2} μm. The upper portion of this band contains microwaves, with wavelengths in the range 10^6–10^3 μm. Infrared (ir) radiation extends from about 3×10^2 μm to 0.7 μm, visible from 0.7 to 0.4 μm, and ultraviolet (uv) from 0.4 to 0.04 μm. Gamma- and x-radiation are at wavelengths less than 0.1 μm.

Quantities of nonionizing radiation are given in radiometric units. Radiant *energy* is the energy emitted from a source per pulse in joules.

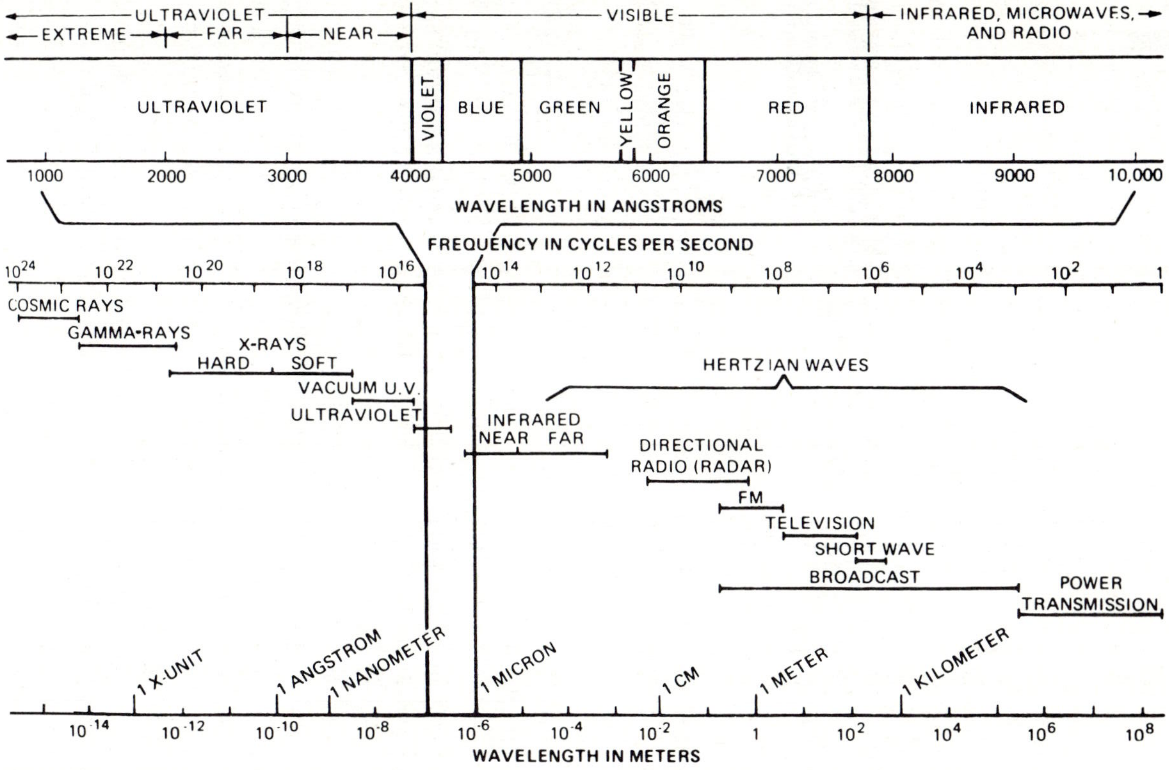

Figure 11.1 The electromagnetic spectrum (*Largent and Olishifski, 1979*). The top portion expands the spectrum between 10^{-7} and 10^{-8} m. Ionizing radiations are to the left and nonionizing radiation to the right.

Radiant *power, P,* is the total power emitted from a source in watts. Radiant *intensity, I,* is the radiant power per unit solid angle (steradian, sr) and is often expressed as W/sr. *Radiance, L,* is the intensity per unit area in $W/sr/cm^2$. Radiant *emittance* is the power emitted per unit area of the source in W/cm^2. Radiant *exposure* is the density of total radiant energy incident on a surface in J/cm^2. *Irradiance* is the power density from a laser incident on a surface in W/cm^2. For microwave radiation, power per unit area in W/cm^2 is referred to as *power density*. Visible light is measured in photometric units. These units relate to the response of the human eye to light and will be discussed in the section on visible radiation. This chapter is divided into sections on ultraviolet, visible, and infrared radiation, lasers that produce radiation in these three regions, microwaves and low-frequency radiation, and example applications.

I. Ultraviolet Radiation

Ultraviolet radiation bridges the gap between the longest-wavelength x-rays (40 nm) and the shortest wavelengths of visible light (380 nm as defined by the International Commission on Illumination devices). The uv spectrum is roughly divided into the near (320–400 nm) and far or actinic 200–315 nm regions. The extreme or vacuum ultraviolet region runs from 200 to 40 nm. Radiation in different spectral regions produces different physiological hazards. Table 11.1 shows the depth of penetration into skin of different bands of radiation between 180 and 15,000 nm. Figure 11.2 summarizes optical radiation hazards for wavelengths from 100 to 3000 nm.

The principal sources of uv are solar radiation, electronic arcs, and incandescent lamps. Table 11.2 shows values from a smoothed curve

TABLE 11.1 Penetration of Nonionizing Radiation into the Human Skin

Spectral region	Penetration (mm)
Far ultraviolet (180–290 nm)	0.01–0.1
Near ultraviolet (290–390 nm)	0.1–1.0
Visible (390–760 nm)	1–10
Near infrared (760–1,500 nm)	10–1
Far infrared (1,500–15,000 nm)	1–0.05

Microwaves—at frequencies in excess of 3,000 MHz, most of the energy received by the body converts into heat in the skin. Between 900 and 3,000 MHz, the relative heat load on skin, subcutaneous fat, and muscle tissue can vary considerably. At frequencies below 900 MHz, most of the absorbed radiant energy is converted into heat in the more deeply placed body tissues.

SOURCES: Koller, 1965; Largent and Olishifski, 1979.

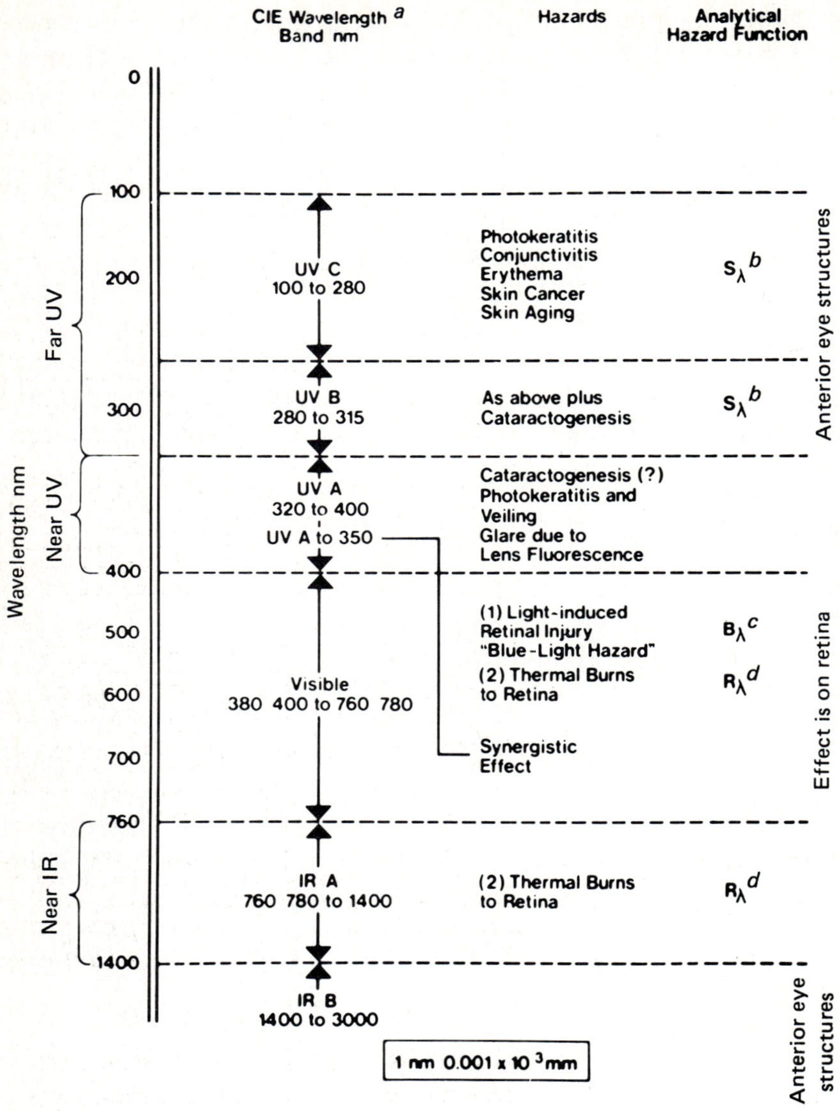

Figure 11.2 Optical radiation hazards (*Sliney et al., 1981*).

[a]International Commission on Illumination (CIE).
[b]Ultraviolet-hazard spectral weighting function, S_λ (Table 11.5).
[c]Blue-light hazard function, B_λ (Table 11.15).
[d]Burn-hazard function, R_λ (Table 11.15).

of the uv solar spectrum. Practically all of the radiation below 295 nm is absorbed in the earth's atmosphere so that none of it reaches the surface. The most efficient and useful manufactured sources of uv are arcs, which occur when a current of electricity is passed between electrodes separated by a gas or vapor. The mercury vapor lamp is the most common example. Spectral emissions from arcs are a function of the vapor pressure within the lamp and the type of envelope (usually quartz or glass). Low-pressure lamps produce a line spectrum, whereas high-pressure lamps (up to 100 atm) produce a continuous spectrum. Low-pressure mercury vapor lamps emit primarily at 253.7 nm. Other gases and vapors used include hydrogen, helium, argon, neon, krypton, xenon, cadmium, zinc, tellurium, and magnesium.

Industrial welding operations generate a significant amount of nonionizing radiation (Burgess, 1981). Shielded metal arc or stick welding (SMAW) produces radiation from 1000 μm (long-wavelength ir) down to 100 nm. Radiation in the 280–315 nm range has been associated with the condition "arc eye," where severe pain occurs 5–6 hr after exposure to the arc. Gas tungsten arc welding (GTAW) [also called tungsten inert gas (TIG)] generates uv discharges an order of magnitude greater than SMA welding (Burgess, 1981). TIG welding releases energy in wavelengths that are most damaging biologically and also are important in fixing oxygen as ozone (100–315 nm). Gas metal arc welding (GMAW) [usually referred to as metal inert gas (MIG)] also produces an order of magnitude more radiation in the 100–315 nm range than SMAW. Note that chlorinated hydrocarbons are decomposed by uv emissions from the MIG and TIG arcs, forming chlorine, hydrogen chloride, and sometimes phosgene. Plasma arc welding (PAW) generates the most intense uv spectrum of any of the inert-gas arc welding systems. Gas welding produces radiation mostly in the visible and ir wavelength ranges. Tables 11.3 and 11.4 summarize measurements of electromagnetic emissions from air-carbon arc cutting, gas welding, brazing and cutting, and single-phase direct electric arc furnaces. Note that gas welding, torch brazing, and oxygen cutting produce low levels

TABLE 11.2 Solar Spectral Irradiance

Wavelength (Å)	μW/cm^2·Å	Wavelength (Å)	μW/cm^2·Å
3100	7.6	3750	13.2
3150	8.2	3800	12.4
3200	8.5	3950	11.9
3250	10.3	4000	15.4
3300	11.6	4050	18.8
3350	11.2	4100	19.4
3400	11.1	4150	19.2
3450	11.6	4200	19.2
3500	11.8	4250	18.9
3550	11.7	4300	17.8
3600	11.6	4350	18.2
3650	13.0	4400	20.2
3700	13.3	4450	21.5

SOURCE: Koller, 1965.

TABLE 11.3 Radiation from Welding[a]

			Air Carbon Arc Cutting[b]			
			Average irradiance, (maximum) $\mu W/cm^2$			
Base material[c]	Current (A)	Number of tests	Total (200–1150 nm)	uv (200–315 nm)	uv (320–400 nm)	ir (760–1150 nm)
LCS	300	1	9,800	560	600	5,800
LCS	400	3	12,870 (13,900)	810 (1,060)	970 (1,400)	8,200 (10,300)
LCS	500	1	8,000	430	700	4,700
LCS	600	2	13,450 (13,700)	450 (610)	900 (1,100)	9,150 (9,200)
LCS	800	1	8,100	740	700	2,200
CI	230	1	10,100	920	1,000	6,000
CI	350	2	14,600 (16,100)	800 (860)	1,050 (1,600)	10,300 (14,000)
CI	600	1	16,600	990	3,100	7,600
CI	800	2	13,500 (17,300)	920	1,600	6,600 (10,600)
SS	300	1	3,700	560	—	900
SS	600	1	15,300	630	900	11,300
SS	800	2	19,400 (20,600)	1,690 (2,260)	1,800 (2,500)	10,400 (13,000)
Al	220	1	4,800	100	200	3,200
Al	400	2	8,200 (11,100)	160 (200)	450 (700)	5,450 (7,300)
Al	600	1	7,800	70	500	3,800

Gas Welding, Brazing, and Cutting[d]

Base material	Gas type[e]	Flame type[f]	Number of tests	Average irradiance, (maximum) $\mu W/cm^2$ uv (200–315 nm)	uv (320–400 nm)	ir (760–1150 nm)	Luminance[g] (cd/cm^2)
(Welding)							
LCS	OA	Oxid	4	1.70 (2.90)	2.71 (6.9)	178 (279)	0.94 (1.64)
LCS	OA	Neut	5	1.69 (5.1)	6.13 (13.0)	108 (232)	1.40 (2.74)
LCS	OA	Redu	7	1.29 (4.4)	3.53 (11.8)	108 (580)	1.36 (6.85)
Al	OA	Neut	5	0.34 (0.9)	0.77 (1.8)	7.33 (13)	1.02 (3.42)
Al	OH	Neut	1	0.03	0.04	6	0.68
LCS	OH	Neut	1	0.3	0.7	123	3.08
LCS	OM	Neut	5	1.66 (3.1)	4.18 (7.9)	97 (191)	2.46 (8.56)
Al	ONG	Neut	1	0.03	0.06	4	1.37
(Brazing)							
LCS	OA	Neut	6	0.05 (0.3)	0.19 (0.7)	8 (8)	0.10 (0.12)
SS	OA	Neut	1	0.5	1.4	6	0.14
Cu	OA	Neut	1	1.1	2.7	6	0.14
(Oxygen cutting)							
LCS	OA	Neut	2	1.13 (1.9)	2.65 (4.5)	32 (39)	1.39 (2.1)
None	OA	Neut	1	1.2	2.7	2	0.06

[a]All data taken at 1 m from source.
[b]Marshall et al., 1980.
[c]LCS = low-carbon steel; CI = cast iron; SS = stainless steel; Al = aluminum; Cu = copper.
[d]Moss and Murray, 1979.
[e]OA = oxyacetylene; OH = oxyhydrogen; OM = oxyMAPP; ONG = oxynatural gas.
[f]Oxid = oxidizing; Neut = neutral; Redu = reduction.
[g]Units of luminance commonly $candela/cm^2$; see Section II.

TABLE 11.4 Radiation from Direct Arc Furnaces

Quantity	Maximum value	Location
Luminance[a]	200 cd·cm^{-2}	4.6 m from furnace
Near uv (320–400 nm)	200 μW·cm^{-2}	4.6 m from furnace
Actinic uv (200–315 nm)	17.4 μW·cm^{-2}	3.1 m from furnace
Total irradiance (280–1100 nm)	>200 mW·cm^{-2}	4.6 m from furnace

[a]Units of luminance commonly candela/cm^2; see Section II.
SOURCE: Moss and Murray, 1981.

of uv radiation. Figure 11.3 summarizes uv irradiance as a function of arc current for a range of welding arcs. As shown in Figure 11.3, the GMAW arc (5% O_2/95% Ar) extrapolated to 800 A would have an effective uv irradiance of 10 mW/cm^2 at 1 m.

From the standpoint of optical radiation hazards, the irradiances in three wavelength bands are of interest. These are uv irradiance, E_{eff}, for the 200–315 nm band (normalized to 270 nm with the uv-hazard spectral weighting function S_λ); blue-light irradiance, E_β, for the 400–1400 nm band (normalized to 435–440 nm with the blue-light hazard

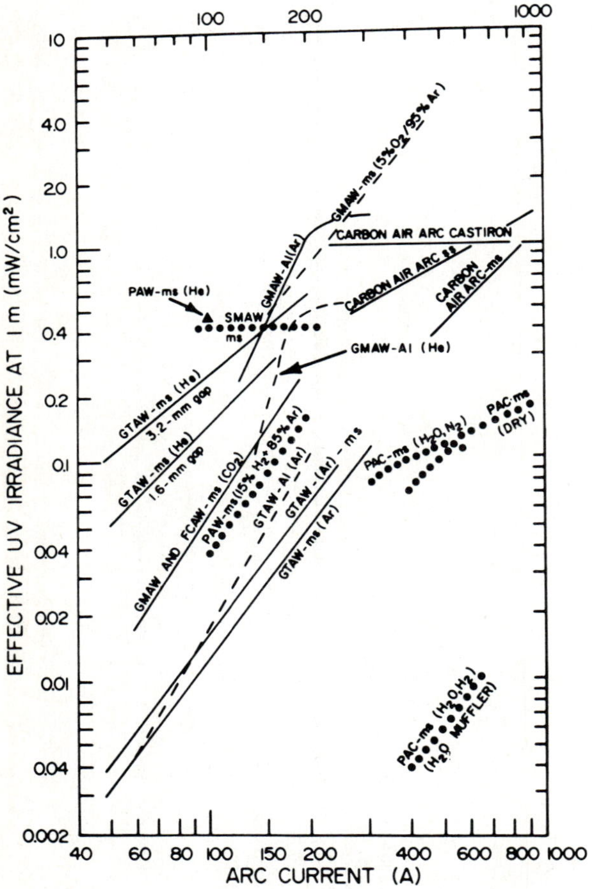

Figure 11.3 Effective uv irradiance of welding arcs (*Sliney et al., 1982*).

function B_λ); and ir irradiance, E_{ir}, for 760–2000 nm. For welding arcs, these irradiances are a function of the arc process, arc current, metals involved, and distance, r, from source to receptor location, and can be expressed as (Sliney et al., 1981):

$$E = kI^n r^{-2} \tag{11.1a}$$

where E can be E_{eff}, E_β, or E_{ir}; I is the arc current; and k and n are functions of the welding system and wavelength band involved. Values for k and n that can be used for calculating the upper limit of irradiance that can be produced under worst-case conditions by any welding process are:

$$E_{eff}(\text{W·cm}^{-2}) < 4 \times 10^{-4} I^2 r^{-2} \tag{11.1b}$$

$$E_\beta(\text{W·cm}^{-2}) < 1.1 \times 10^{-3} I^2 r^{-2} \tag{11.1c}$$

$$E_{ir}(\text{W·cm}^{-2}) < 9 \times 10^{-4} I^2 r^{-2} \tag{11.1d}$$

where I is in amperes and r is in centimeters. These steady-state relationships hold for arc currents between 30 and 700 A and are based on the most highly emissive arc processes. Values of k and n for specific welding processes and wavelength bands are also available (Sliney et al., 1981).

With respect to Eqs. (11.1), time-resolved spectral measurements of optical radiation from MIG welding arc ignition have shown that the uv irradiance can overshoot more than 10 times the steady-state average value within the first 50 msec. This was observed with a 200-A aluminum welding arc with an ER5356 electrode (American Welding Society designation) (Eriksen, 1985).

A. Exposure criteria

The threshold limit values (TLVs) for uv radiation (ACGIH, 1984) refer to radiation in the 200–400 nm region and apply for exposures of the eye and skin from arcs, gas and vapor discharges, fluorescent and incandescent sources, and solar radiation. These values do not apply to ultraviolet lasers, to exposure of photosensitive individuals or individuals concomitantly exposed to photosensitizing agents. The TLVs are intended to be used as guidelines in the control of uv exposures. For the near ultraviolet region (320–400 nm), total irradiance incident upon the unprotected skin or eye should not exceed 1 mW/cm^2 for periods greater than 10^3 sec and 1 J/cm^2 for exposure times less than 10^3 sec. For the actinic spectral region (200–315 nm), radiant exposure on the unprotected skin or eye should not exceed the values given in Table 11.5 within a 8-hr period. The effective irradiance of a broad-

TABLE 11.5 TLV and Relative Spectral Effectiveness for Actinic Ultraviolet Radiation

Wavelength (nm)	TLV (mJ/cm^2)	Relative spectral effectiveness, S_λ
200	100	0.03
210	40	0.075
220	25	0.12
230	16	0.19
240	10	0.30
250	7.0	0.43
254	6.0	0.5
260	4.6	0.65
270	3.0	1.0
280	3.4	0.88
290	4.7	0.64
300	10	0.30
305	50	0.06
310	200	0.015
315	1000	0.003

SOURCE: ACGIH, 1984.

band source is weighted against a spectral effectiveness curve (peak at 270 nm) by:

$$E_{\text{eff}} = \sum E_\lambda S_\lambda \, \Delta\lambda \tag{11.2}$$

where E_{eff} is the effective irradiance relative to a monochromatic source at 270 nm in W/cm^2, E_λ is the spectral irradiance in $W/cm^2/nm$ for wavelength band $\Delta\lambda$ (nm), and S_λ is the relative spectral effectiveness. Values of S_λ are given in Table 11.5. Permissible exposure times in seconds for actinic ultraviolet radiation may be computed by dividing $0.003\ J/cm^2$ by E_{eff} in W/cm^2. Table 11.6 summarizes this relationship. The Council on Physical Medicine of the American Medical Association has developed an exposure criterion based on erythemal thresholds for 253.7-nm radiation (Largent and Olishifski, 1979). For exposures up to 7 hr, total irradiance should not exceed $0.5 \times 10^{-6}\ W/cm^2$, and for exposures up to and exceeding 24 hr, irradiance should not exceed $0.1 \times 10^{-6}\ W/cm^2$. Note that when applying this guideline, corrections for the relative spectral effectiveness must be made for wavelengths other than 253.7 nm.

B. Shielding

In general, shielding is the most appropriate method for control of nonionizing radiation exposures above accepted guidelines. This can be implemented as shielding at the source or shielding at the operator.

TABLE 11.6 Permissible Ultraviolet Exposures

Duration of exposure per day	Effective irradiance, E_{eff} ($\mu W/cm^2$)
8 hr	0.1
4 hr	0.2
2 hr	0.4
1 hr	0.8
30 min	1.7
15 min	3.3
10 min	5
5 min	10
1 min	50
30 sec	100
10 sec	300
1 sec	3,000
0.5 sec	6,000
0.1 sec	30,000

SOURCE: ACGIH, 1984.

The intensity of a parallel beam of ultraviolet radiation passing through a nondiffusing absorbing medium falls off with increasing thickness according to the exponential relationship

$$\frac{I}{I_0} = \exp(-\alpha t) \tag{11.3}$$

where I_0 is the intensity of radiation entering the medium, I is the intensity of radiation after traversing a distance t through the medium, and α is the absorption coefficient per centimeter. The absorption coefficient is a function of wavelength. In general, solids and liquids show broad absorption and transmission bands, whereas gases and vapors may show sharp discontinuities in transmission as a function of wavelength. Most common materials are more likely to be good absorbers than transmitters of uv radiation.

Ordinary window glass is practically opaque to ultraviolet wavelengths less than 300 nm. Values of the absorption coefficient for such glass are given in Table 11.7. Figure 11.4 shows the percent transmission as a function of wavelength for two thicknesses. Ultraviolet transmission through glass is determined largely by the content of iron oxide, which absorbs strongly in this spectral region.

Appropriate protection is required when glasses or other materials are used that are designed to transmit ultraviolet radiation (Figure 11.5). Corex D No. 9700 transmission is high at 300 nm and very small at wavelengths shorter than 250 nm. This type is useful for sunlamp bulbs, as it transmits the therapeutic radiation while cutting off the

TABLE 11.7 Absorption Coefficient of Window Glass

Wavelength (nm)	α (cm^{-1})
315	11.4
320	7.9
340	1.7
360	0.54

SOURCE: Koller, 1965.

shorter wavelengths. Pyrex No. 9741, used in bactericidal lamps, transmits up to 70% of the radiation at 253.7 nm (germicidal wavelength). The transmission for Vycor No. 791 is about 85% at 253.7 nm, falling almost to zero at 184.9 nm. Comprehensive tables on the transmission of glasses are given in the *Handbook of Chemistry and Physics* (Weast, 1984).

Quartz is often used for transmission in the far region of the ultraviolet spectrum where glass becomes strongly absorbing. The transmission limit is about 180 nm (Figure 11.6). Synthetic crystals, organic liquids, and inorganic salts can also be used as filters in the uv spectral region (Koller, 1965). Didymium glass strongly absorbs in the 575–595 nm band. This material is particularly useful for reducing the intensity of the sodium spectral lines emitted from molten glass (Sliney, 1972).

The *transparency* of a material is defined as the ratio of the intensity of transmitted light to that of incident light (I/I_0). *Opacity* is the reciprocal of the transparency, and *optical density*, OD, is the base 10 logarithm of the opacity such that:

$$\text{OD} = \log\left(\frac{I_0}{I}\right) \tag{11.4}$$

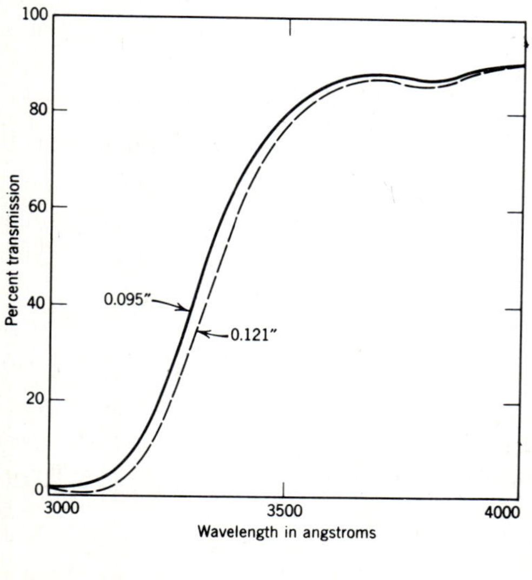

Figure 11.4 Ultraviolet transmission for two thicknesses of window glass (*Koller, 1965*).

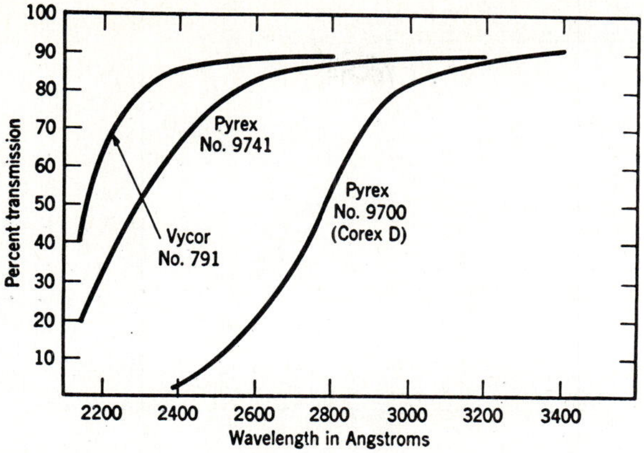

Figure 11.5 Transmission of 1-mm thickness of Pyrex and Vycor glasses (*Koller, 1965*).

Table 11.8 shows the luminous transmittance, infrared transmittance, and ultraviolet spectral transmittance for standard filter glasses with optical densities ranging from 0.214 to 5.571 (ANSI, 1979). The specification of eye-protective lenses is frequently given as a shade number. The shade numbers corresponding to optical densities are also shown in Table 11.8. Table 11.9 lists recommendations for protective lenses for welding operations.

Unfortunately, such specifications have not always been met. Numerous variations from the ANSI standard requirements were found in a comprehensive survey of 94 shade models of welding filter plates from 16 manufacturers (Campbell, 1976). Nineteen percent exceeded the maximum allowable optical density; 1% had excess ir transmittance; and 7% did not meet the ultraviolet guidelines.

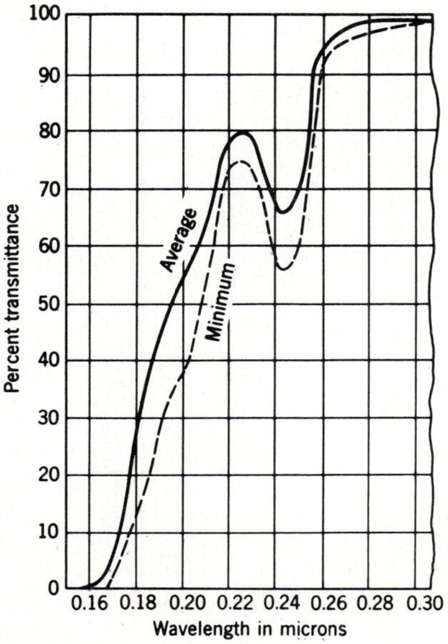

Figure 11.6 Transmission of 1-cm thickness of clear fused quartz (*Koller, 1965*).

TABLE 11.8 Transmittances and Tolerances of Various Shades of Glass Filter Lenses

The values given apply to Class I filter glass. For Class II filter lenses the transmittances and tolerances are the same with the additional requirement that the transmittance at 589.3 nm shall not exceed 15% of the luminous transmittance.

Shade	Optical density			Luminous transmittance (%)			Maximum infrared transmittance	Maximum spectral transmittance in the ultraviolet and violet (%)			
number	Minimum	Standard	Maximum	Maximum	Standard	Minimum	(%)	313 nm	334 nm	365 nm	405 nm
1.5	0.17	0.214	0.26	67	61.1	55	25	0.2	0.8	25	65
1.7	0.26	0.300	0.36	55	50.1	43	20	0.2	0.7	20	50
2.0	0.36	0.429	0.54	43	37.3	29	15	0.2	0.5	14	35
2.5	0.54	0.643	0.75	29	22.8	18.0	12	0.2	0.3	5	15
3.0	0.75	0.857	1.07	18.0	13.9	8.50	9.0	0.2	0.2	0.5	6
4.0	1.07	1.286	1.50	8.50	5.18	3.16	5.0	0.2	0.2	0.5	1.0
5.0	1.50	1.714	1.93	3.16	1.93	1.18	2.5	0.2	0.2	0.2	0.5
6.0	1.93	2.143	2.36	1.18	0.72	0.44	1.5	0.1	0.1	0.1	0.5
7.0	2.36	2.571	2.79	0.44	0.27	0.164	1.3	0.1	0.1	0.1	0.5
8.0	2.79	3.000	3.21	0.164	0.100	0.061	1.0	0.1	0.1	0.1	0.5
9.0	3.21	3.429	3.64	0.061	0.037	0.023	0.8	0.1	0.1	0.1	0.5
10.0	3.64	3.857	4.07	0.023	0.0139	0.0085	0.6	0.1	0.1	0.1	0.5
11.0	4.07	4.286	4.50	0.0085	0.0052	0.0032	0.5	0.05	0.05	0.05	0.1
12.0	4.50	4.714	4.93	0.0032	0.0019	0.0012	0.5	0.05	0.05	0.05	0.1
13.0	4.93	5.143	5.36	0.0012	0.00072	0.00044	0.4	0.05	0.05	0.05	0.1
14.0	5.36	5.571	5.79	0.00044	0.00027	0.00016	0.3	0.05	0.05	0.05	0.1

SOURCE: ANSI, 1979.

TABLE 11.9 Recommended Shade Numbers for Welding Operations

Operation	Shade number
Shielded metal arc welding: $\frac{1}{16}$-, $\frac{3}{32}$-, $\frac{1}{8}$-, and $\frac{5}{32}$-in. electrodes	10
Gas-shielded arc welding (nonferrous): $\frac{1}{16}$-, $\frac{3}{32}$-, $\frac{1}{8}$-, and $\frac{5}{32}$-in. electrodes	11
Gas-shielded arc welding (ferrous): $\frac{1}{16}$-, $\frac{3}{32}$-, $\frac{1}{8}$-, and $\frac{5}{32}$-in. electrodes	12
Shielded metal arc welding:	
$\frac{3}{16}$-, $\frac{7}{32}$-, and $\frac{1}{4}$-in. electrodes	12
$\frac{5}{16}$-, and $\frac{3}{8}$-in. electrodes	14
Gas tungsten arc welding	12
Atomic hydrogen welding	10–14
Carbon arc welding	14
Soldering	2
Torch brazing	3–4
Light cutting—up to 1 in.	3–4
Medium cutting—1 to 6 in.	4–5
Heavy cutting—6 in. and over	5–6
Gas welding (light)—up to $\frac{1}{8}$ in.	4–5
Gas welding (medium)—$\frac{1}{8}$ to $\frac{1}{2}$ in.	5–6
Gas welding (heavy)—$\frac{1}{2}$ in. and over	6–8

NOTE: In gas welding or oxygen cutting where the torch produces a high yellow light, it is desirable to use a filter or lens that absorbs the yellow or sodium line in the visible light of the operation.

SOURCES: OSHA, 1983; ANSI, 1979.

Plastic and heat-treated glass plates are less subject to fracturing than non-heat-treated shields (Campbell, 1976; ANSI, 1955). The types of eye and face protectors recommended for use in industrial operations are shown on Figure 11.7. Transparent welding curtains are available that permit supervisory personnel to observe the welder's progress and to detect accidents when outside assistance becomes necessary. Figure 11.8 shows typical spectral transmission properties of yellow, green, gray, and blue curtains. Table 11.10 shows safety characteristics of five types of commercially available welding curtains. This table gives the maximum permissible arc current allowable for various types of welding processes when personnel are more than 2 m from the arc for an 8-hr exposure period. These values are based on the TLV for chronic blue-light exposure of 1 $\mu m \cdot cm^{-2}$ for exposures $> 10^4$ sec (ACGIH, 1984). Workers closer to the welding zone will require greater protection. Additional information on a wide range of curtains is shown in Figure 11.9.

II. Visible Radiation

The visible portion of the spectrum lies between 380 and 750 nm (3.1 and 1.65 eV). The critical organ for this region is the eye. Under or-

Selection Chart for Eye and Face Protectors for Use in Industry, Schools, and Colleges

This Selection Chart offers general recommendations only. Final selection of eye and face protective devices is the responsibility of management and safety specialists. (For laser protection, refer to American National Standard for Safe Use of Lasers, ANSI Z136.1-1976.)

1. GOGGLES, Flexible Fitting, Regular Ventilation
2. GOGGLES, Flexible Fitting, Hooded Ventilation
3. GOGGLES, Cushioned Fitting, Rigid Body
*4. SPECTACLES, without Sideshields
5. SPECTACLES, Eyecup Type Sideshields
6. SPECTACLES, Semi-/Flat-Fold Sideshields
**7. WELDING GOGGLES, Eyecup Type, Tinted Lenses (Illustrated)
7A. CHIPPING GOGGLES, Eyecup Type, Clear Safety Lenses (Not Illustrated)
**8. WELDING GOGGLES, Coverspec Type, Tinted Lenses (Illustrated)
8A. CHIPPING GOGGLES, Coverspec Type, Clear Safety Lenses (Not Illustrated)
**9. WELDING GOGGLES, Coverspec Type, Tinted Plate Lens
10. FACE SHIELD, Plastic or Mesh Window (see caution note)
* 11. WELDING HELMET

*Non-sideshield spectacles are available for limited hazard use requiring only frontal protection.
**See Table A1, "Selection of Shade Numbers for Welding Filters," in Section A2 of the Appendix.

APPLICATIONS		
OPERATION	HAZARDS	PROTECTORS
ACETYLENE–BURNING ACETYLENE–CUTTING ACETYLENE–WELDING	SPARKS, HARMFUL RAYS, MOLTEN METAL, FLYING PARTICLES	7, 8, 9
CHEMICAL HANDLING	SPLASH, ACID BURNS, FUMES	2 (For severe exposure add 10)
CHIPPING	FLYING PARTICLES	1, 3, 4, 5, 6, 7A, 8A
ELECTRIC (ARC) WELDING	SPARKS, INTENSE RAYS, MOLTEN METAL	11 (In combination with 4, 5, 6, in tinted lenses, advisable)
FURNACE OPERATIONS	GLARE, HEAT, MOLTEN METAL	7, 8, 9 (For severe exposure add 10)
GRINDING–LIGHT	FLYING PARTICLES	1, 3, 5, 6 (For severe exposure add 10)
GRINDING–HEAVY	FLYING PARTICLES	1, 3, 7A, 8A (For severe exposure add 10)
LABORATORY	CHEMICAL SPLASH, GLASS BREAKAGE	2 (10 when in combination with 5, 6)
MACHINING	FLYING PARTICLES	1, 3, 5, 6 (For severe exposure add 10)
MOLTEN METALS	HEAT, GLARE, SPARKS, SPLASH	7, 8 (10 in combination with 5, 6, in tinted lenses)
SPOT WELDING	FLYING PARTICLES, SPARKS	1, 3, 4, 5, 6 (Tinted lenses advisable; for severe exposure add 10)

CAUTION:
- Face shields alone do not provide adequate protection.
- Plastic lenses are advised for protection against molten metal splash.
- Contact lenses, of themselves, do not provide eye protection in the industrial sense and shall not be worn in a hazardous environment without appropriate covering safety eyewear.

Figure 11.7 Selection chart for eye and face protectors for use in industry, schools, and colleges (*ANSI, 1979*).

dinary circumstances, visible light is not hazardous. Our discussion will focus on appropriate lighting intensities and contrasts to promote a safe and efficient workplace.

A. Units

Visible radiation is measured with a system of photometric units. The basic units of this system are defined as follows:

Lumen (lm). Represents the total light output from a source and is the SI unit of luminous flux emitted through a unit solid angle (sr) from a point source of 1 candela.

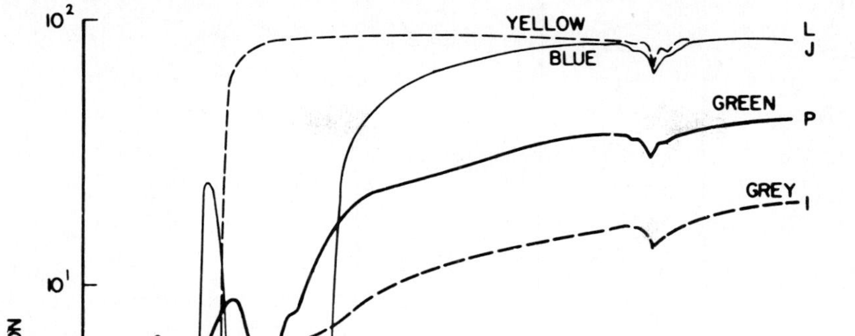

Figure 11.8 Spectral transmission properties of welding curtains (*Moss and Gawenda, 1978*).

Candela (cd). The SI unit of luminous intensity. One candela equals the intensity of a monochromatic source at 555 nm of 1/683 W/sr. (1 cd = 1 lm/sr; 1 cd $\simeq$ 1 candle power).

Lux (lx). The SI unit of illuminance; 1 lx = 1 lm/m^2, where illuminance is the density of luminous flux incident on a surface.

Footcandle (fc). The unit of illuminance when the unit of length is the foot; 1 fc = 1 lm/ft^2 (1 fc = 10.76 lx).

Nit (nt). The SI unit of luminance; 1 nt = 1 cd/m^2.

Footlambert (fL). A unit of luminance; 1 fL = $1/\pi$ cd/ft^2. One footlambert is the uniform luminance of a perfectly diffusing surface emitting or reflecting 1 lm/ft^2. The brightness in footlamberts is equal to the illumination on a surface multiplied by its reflectance (ratio of reflected to incident light).

Lambertian units are equivalent to stating the number of lumens that would leave the surface if the surface were replaced by a perfectly

TABLE 11.10 Safety Characteristics of Commercially Available Transparent Welding Curtains[a]

Type of welding curtain	Relative visibility of objects viewed through curtains; yellow = 100%[b]	Type of welding process being used (A)			
		Ferrous metal; pure gas, Ar or He	Aluminum metal; pure gas, Ar or He	Ferrous metal; mixed gas, including CO_2, O_2, H_2	SMAW or FCAW flux vapor[c]
Spectra orange	69	1300	750	550	450
Gray	56	225	125	100	75
Green	63	150	90	70	50
Yellow	100	80	50	35	30
Blue	44	40	25	20	15

[a]Maximum permissible arc current allowable for various types of welding processes. Amperage values are based on exposure of personnel located more than 2 m from the arc for an 8-hr period (Sliney et al., 1982).

[b]Subjective visibility through curtain.

[c]SMAW: shielded metal arc welding; FCAW: flux cored arc welding.

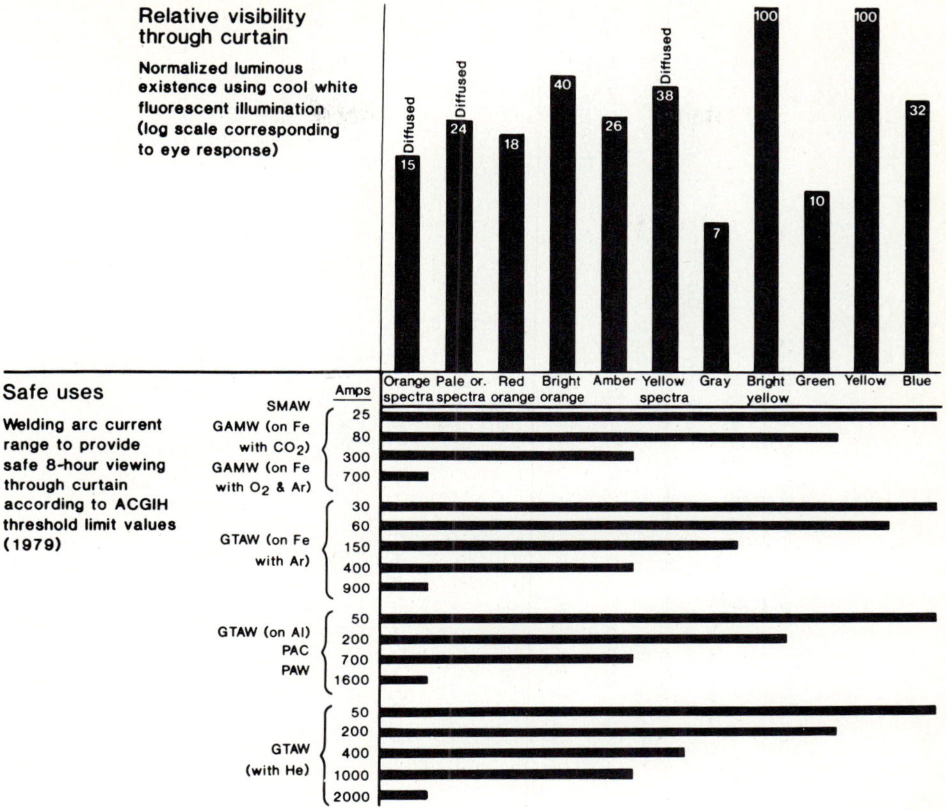

Figure 11.9 Safe use and visibility factors of transparent welding curtains (*Sliney et al., 1981*).

diffusing surface with luminance in all directions that equals the luminance of the actual surface in the direction specified. The Illuminating Engineering Society (IES) no longer recommends the use of lambertian units. The lux and the footcandle are the recommended units of luminance (IES, 1981).

B. Lighting design

Table 11.11 lists illuminance categories and values for generic types of activities indoors. Table 11.12 lists recommended illuminance categories by industrial group. Table 11.13 lists illuminance categories for measured equivalent contrast values of visual displays. These tables are intended for use in interior environments where visual performance is an important consideration. They are not intended as lighting recommendations for sensors other than the eye (e.g., television); or to achieve artistic effects; or for use when illuminances are a part of a test procedure for evaluating equipment.

TABLE 11.11 Illuminance Categories and Illuminance Values for Generic Types of Activities in Interiors

Type of activity	Illuminance category	Ranges of illuminances: Lux	Ranges of illuminances: Footcandles	Reference work plane
Public spaces with dark surroundings	A	20–30–50	2–3–5	
Simple orientation for short, temporary visits	B	50–75–100	5–7.5–10	General lighting throughout spaces
Working spaces where visual tasks are performed only occasionally	C	100–150–200	10–15–20	
Performance of visual tasks of high contrast or large size	D	200–300–500	20–30–50	
Performance of visual tasks of medium contrast or small size	E	500–750–1,000	50–75–100	Illuminance on task
Performance of visual tasks of low contrast or very small size	F	1,000–1,500–2,000	100–150–200	
Performance of visual tasks of low contrast and very small size over a prolonged period	G	2,000–3,000–5,000	200–300–500	Illuminance on task, obtained by a combination of general and local (supplementary lighting)
Performance of very prolonged and exacting visual tasks	H	5,000–7,500–10,000	500–750–1,000	
Performance of very special visual tasks of extremely low contrast and small size	I	10,000–15,000–20,000	1,000–1,500–2,000	

SOURCE: IES, 1981.

TABLE 11.12 Illumination Categories for Industrial Applications

Area/activity	Illuminance category
Assembly	
Simple	D
Moderately difficult	E
Difficult	F
Very difficult	G
Exacting	H
Bakeries	
Mixing room	D
Face of shelves	D
Inside of mixing bowl	D
Fermentation room	D
Make-up room	
Bread	D
Sweet yeast-raised products	D
Proofing room	D
Oven room	D
Fillings and other ingredients	D
Decorating and icing	
Mechanical	D
Hand	E
Scales and thermometers	D
Wrapping	D
Book binding	
Folding, assembling, pasting	D
Cutting, punching, stitching	E
Embossing and inspection	F
Breweries	
Brew house	D
Boiling and keg washing	D
Filling (bottles, cans, kegs)	D
Candy making	
Box department	D
Chocolate department	
Husking, winnowing, fat extraction, crushing and refining, feeding	D
Bean cleaning, sorting, dipping, packing, wrapping	D
Milling	E
Cream making	
Mixing, cooking, molding	D
Gum drops and jellied forms	D
Hand decorating	D
Hard candy	
Mixing, cooking, molding	D
Die cutting and sorting	E
Kiss making and wrapping	E
Canning and preserving	
Initial grading of raw material samples	D
Tomatoes	E
Color grading and cutting rooms	F
Preparation	
Preliminary sorting	
Apricots and peaches	D
Tomatoes	E
Olives	F
Cutting and pitting	E
Final sorting	E
Canning	
Continuous-belt canning	E
Sink canning	E
Hand packing	D
Olives	E

TABLE 11.12 Illumination Categories for Industrial Applications *(Continued)*

Area/activity	Illuminance category
Canning and preserving (*contd.*)	
Examination of canned samples	F
Container handling	
Inspection	F
Can unscramblers	E
Labeling and cartoning	D
Clay and concrete products	
Grinding, filter presses, kiln rooms	C
Molding, pressing, cleaning, trimming	D
Enameling	E
Color and glazing—rough work	E
Color and glazing—fine work	F
Cleaning and pressing industry	
Checking and sorting	E
Dry and wet cleaning and steaming	E
Inspection and spotting	G
Pressing	F
Repair and alteration	F
Cloth products	
Cloth inspection	I
Cutting	G
Sewing	G
Pressing	F
Clothing manufacture (men's)	
Receiving, opening, storing, shipping	D
Examining (perching)	I
Sponging, decating, winding, measuring	D
Piling and marking	E
Dairy products (*contd.*)	
Thermometer (on face)	E
Weighing room	D
Scales	E
Electrical equipment manufacturing	
Impregnating	D
Insulating: coil winding	E
Electric generating stations—interior (see also Nuclear power plants)	
Air conditioning equipment, air preheater and fan floor, ash sluicing	B
Auxiliaries, pumps, tanks, compressors, gauge area	C
Battery rooms	D
Boiler platforms	B
Burner platforms	C
Cable room	B
Coal-handling systems	B
Coal pulverizer	C
Condensers, deaerator floor, evaporator floor, heater floors	B
Control rooms	
Main control boards	D[b]
Auxiliary control panels	D[b]
Operator's station	E[b]
Maintenance and wiring areas	D
Emergency operating lighting	C
Gauge reading	D

Cutting	G
Pattern making, preparing of trimming, piping, canvas and shoulder pads	E
Fitting, bundling, shading, stitching	D
Shops	F
Inspection	G
Pressing	F
Sewing	G
Cotton gin industry	
Overhead equipment—separators, driers, grid cleaners, stick machines, conveyers, feeders, and catwalks	D
Gin stand	D
Control console	D
Lint cleaner	D
Bale press	D
Dairy products	
Fluid milk industry	
Boiler room	D
Bottle storage	D
Bottle sorting	E
Bottle washers	—[a]
Can washers	D
Cooling equipment	D
Filling: inspection	E
Gauges (on face)	E
Laboratories	E
Meter panels (on face)	E
Pasteurizers	D
Separators	D
Storage refrigerators	D
Tanks, vats	
Light interiors	C
Dark interiors	E
Hydrogen and carbon dioxide manifold area	C
Laboratory	E
Precipitators	B
Screen house	C
Soot or slag blower platform	C
Steam headers and throttles	B
Switchgear and motor control centers	D
Telephone and communication equipment rooms	D
Tunnels or galleries, piping and electrical	B
Turbine building	
Operating floor	D
Below operating floor	C
Visitor's gallery	C
Water treating area	D
Explosive manufacturing	
Hand furnaces, boiling tanks, stationary driers, stationary and gravity crystallizers	D
Mechanical furnace, generators and stills, mechanical driers, evaporators, filtration, mechanical crystallizers	D
Tanks for cooking, extractors, percolators, nitrators	D
Farms—dairy	
Milking operation area (milking parlor and stall barn)	
General	C
Cow's udder	D
Milk-handling equipment and storage area (milk house or milk room)	
General	C
Washing area	E
Bulk tank interior	E
Loading platform	C

TABLE 11.12 Illumination Categories for Industrial Applications *(Continued)*

Area/activity	Illuminance category	Area/activity	Illuminance category
Farms—dairy (*contd.*)		Glove manufacturing	
Feeding area (stall barn feed alley, pens, loose housing feed area)	C	Pressing	G
Feed storage area—forage		Knitting	F
Haymow	A	Sorting	F
Hay inspection area	C	Cutting	G
Ladders and stairs	C	Sewing and inspection	G
Silo	A	Hat manufacturing	
Silo room	C	Dyeing, stiffening, braiding, cleaning, refining	E
Feed storage area—grain and concentrate		Forming, sizing, pouncing, flanging, finishing, ironing	G
Grain bin	A	Sewing	G
Concentration storage area	B	Inspection	
Feed processing area	B	Simple	D
Livestock housing area (community, maternity, individual calf pens, and loose housing holding and resting areas)	B	Moderately difficult	E
Machine storage area (garage and machine shed)	B	Difficult	F
Farm shop area		Very difficult	G
Active storage area	B	Exacting	H
General shop area (machinery repair, rough sawing)	D	Jewelry and watch manufacturing	G
Rough bench and machine work (painting, rine storage, ordinary sheet metal work, welding, medium benchwork)	D	Laundries	
Medium bench and machine work (fine woodworking, drill press, metal lathe, grinder)	E	Washing	D
Miscellaneous areas		Flat-work ironing, weighing, marking	D
Farm office (see Reading)		Machine and press finishing, sorting	E
Restrooms (see Service spaces)		Fine hand ironing	E
Pumphouse	C	Leather manufacturing	
		Cleaning, tanning and stretching, vats	D
		Cutting, fleshing and stuffing	D
		Finishing and scarfing	E
		Leather working	
		Pressing, winding, glazing	F
		Grading, matching, cutting, scarfing, sewing	G

Flour mills	
Rolling, sifting, purifying	E
Packing	D
Product control	F
Cleaning, screens, man lifts, aisleways and walkways, bin checking	D
Forge shops	E
Foundries	
Annealing (furnaces)	D
Cleaning	D
Core making	
Fine	F
Medium	E
Grinding and chipping	F
Inspection	
Fine	G
Medium	F
Molding	
Medium	F
Large	E
Pouring	E
Sorting	E
Cupola	C
Shakeout	D
Garages—service	
Repairs	E
Active traffic areas	C
Write-up	D
Glass works	
Mix and furnace rooms, pressing and lehr, glass-blowing machines	C
Grinding, cutting, silvering	D
Fine grinding, beveling, polishing	E
Inspection, etching and decorating	F
Locker rooms	C
Machine shops	
Rough bench or machine work	D
Medium bench or machine work, ordinary automatic machines, rough grinding, medium buffing and polishing	E
Fine bench or machine work, fine automatic machines, medium grinding, fine buffing and polishing	G
Extra-fine bench or machine work, grinding, fine work	H
Materials handling	
Wrapping, packing, labeling	D
Picking stock, classifying	D
Loading, inside truck bodies and freight cars	C
Meat packing	
Slaughtering	D
Cleaning, cutting, cooking, grinding, canning, packing	D
Nuclear power plants (see also Electric generating stations)	
Auxiliary building, uncontrolled-access areas	C
Controlled-access areas	
Count room	E[2]
Laboratory	E
Health physics office	F
Medical aid room	F
Hot laundry	D
Storage room	C
Engineered safety features equipment	D
Diesel generator building	D
Fuel-handling building	
Operating floor	D
Below operating floor	C

TABLE 11.12 Illumination Categories for Industrial Applications *(Continued)*

Area/activity	Illuminance category
Paint manufacturing	
Processing	D
Mix comparison	F
Paint shops	
Dipping, simple spraying, firing	D
Rubbing, ordinary hand painting and finishing art, stencil and special spraying	D
Fine hand painting and finishing	E
Extra-fine hand painting and finishing	G
Paper box manufacturing	E
Paper manufacturing	
Beaters, grinding, calendering	D
Finishing, cutting, trimming, papermaking machines	E
Hand counting, wet end of paper machine	E
Paper machine reel, paper inspection, and laboratories	F
Rewinder	F
Plating	D
Poultry industry (see also Farm—dairy)	
Brooding, production, and laying houses	
Feeding, inspection, cleaning	C
Charts and records	D
Thermometers, thermostats, time clocks	D
Hatcheries	
General area and loading platform	C
Inside incubators	D
Dubbing station	F
Sexing	H
Printing industries (*contd.*)	
Photoengraving	
Etching, staging, blocking	D
Routing, finishing, proofing	E
Tint laying, masking	E
Sawmills	
Secondary log deck	B
Head saw (cutting area viewed by sawyer)	E
Head saw outfeed	B
Machine in-feeds (bull edger, resaws, edgers, trim, hula saws, planers)	B
Main mill floor (base lighting)	A
Sorting labels	D
Rough lumber grading	D
Finished lumber grading	F
Dry lumber warehouse (planer)	C
Dry kiln colling shed	B
Chipper infeed	B
Basement areas	
Active	A
Inactive	A
Filing room (work areas)	E
Service spaces (see also Storage rooms)	
Stairways, corridors	B
Elevators, freight and passenger	B
Toilets and wash rooms	C
Sheet metal works	
Miscellaneous machines, ordinary bench work	E
Presses, shears, stamps, spinning, medium bench work	E

Egg handling, packing, and shipping	
General cleanliness	E
Egg quality inspection	E
Loading platform, egg storage area, etc.	C
Egg processing	
General lighting	E
Fowl processing plant	
General (excluding killing and unloading area)	E
Government inspection station and grading stations	E
Unloading and killing area	C
Feed storage	
Grain, feed rations	C
Processing	C
Charts and records	D
Machine storage area (garage and machine shed)	B
Printing industries	
Type foundries	
Matrix making, dressing type	E
Font assembly-sorting	D
Casting	E
Printing plants	
Color inspection and appraisal	F
Machine composition	E
Composing room	E
Presses	E
Imposing stones	F
Proofreading	F
Electrotyping	
Molding, routing, finishing, leveling molds, trimming	E
Blocking, tinning	D
Electroplating, washing, backing	D
Punches	E
Tin plate inspection, galvanized	F
Scribing	F
Shoe manufacturing—leather	
Cutting and stitching	
Cutting tables	G
Marking, buttonholing, skiving, sorting, vamping, counting	G
Stitching, dark materials	G
Making and finishing, nailers, sole layers, welt beaters and scarfers, trimmers, welters, lasters, edge setters, sluggers, randers, wheelers, treers, cleaning, spraying, buffing, polishing, embossing	F
Shoe manufacturing—rubber	
Washing, coating, mill run compounding	D
Varnishing, vulcanizing, calendering, upper and sole cutting	D
Sole rolling, lining, making and finishing processes	E
Soap manufacturing	
Kettle houses, cutting, soap chip and powder	D
Stamping, wrapping and packing, filling and packing soap powder	D
Stairways (see Service spaces)	
Storage battery manufacturing	D
Storage rooms or warehouses	
Inactive	B
Active	
Rough, bulky items	C
Small items	D
Structural steel fabrication	E

TABLE 11.12 Illumination Categories for Industrial Applications (*Continued*)

Area/activity	Illuminance category
Sugar refining	
Grading	E
Color inspection	F
Testing	
General	D
Exacting tests, extra-fine instruments, scales, etc.	F
Textile mills	
Staple fiber preparation	
Stock dyeing, tinting	D
Sorting and grading (wool and cotton)	E[c]
Yarn manufacturing	
Opening and picking (chute feed)	D
Carding (nonwoven web formation)	D[d]
Drawing (gilling, pin drafting)	D
Combing	D[d]
Roving (slubbing, fly frame)	E
Spinning (cap spinning, twisting, texturing)	E
Yarn preparation	
Winding, quilting, twisting	E
Warping (beaming, sizing)	F[c]
Warp tie-in or drawing-in (automatic)	E
Fabric production	
Weaving, knitting, tufting	F
Inspection	G[c]
Textile mills (*contd.*)	
Finishing	
Fabric preparation (desizing, scouring, bleaching, singeing, and mercerization)	D
Fabric dyeing (printing)	D
Fabric finishing (calendaring, sanforizing, sueding, chemical treatment)	E[c]
Inspection	G[c,e]
Tobacco products	
Drying, stripping	D
Grading and sorting	F
Toilets and wash rooms (see Service spaces)	
Upholstering	F
Welding	
Orientation	D
Precision manual arc welding	H
Woodworking	
Rough sawing and bench work	D
Sizing, planing, rough sanding, medium quality machine and bench work, gluing, veneering, cooperage	D
Fine bench and machine work, fine sanding and finishing	E

[a]Special lighting such that (*a*) the luminous area is large enough to cover the surface which is being inspected and (*b*) the luminance is within the limits necessary to obtain comfortable contrast conditions.

[b]Maximum levels—controlled system.

[c]Supplementary lighting should be provided in this space to produce the higher levels required for specific seeing tasks involved.

[d]Additional lighting needs to be provided for maintenance only.

[e]Color temperature of the light source is important for color matching.

SOURCE: IES, 1981.

TABLE 11.13 Illuminance Categories for Measured Equivalent Contrast Values of Task Visual Displays

Equivalent contrast, C^a	Illuminance category[b]
>1.0	—[c]
0.75–1.0	D
0.62–0.75	E
0.50–0.62	F
0.40–0.50	G
0.30–0.40	H
<0.30	I

[a]As determined using a visibility meter. *Note:* Although specific equivalent contrasts are established scientifically, a consensus procedure has been used in establishing corresponding Illuminance categories.

[b]If task reflectance is between 5% and 20%, use next higher illuminance category: i.e., D to E, E to F, etc. If less than 5%, use two categories higher.

[c]Use 200 lux (20 footcandles).

SOURCE: IES, 1981.

The tables are to be used as follows for lighting design. First define the visual task for which the level of lighting is to be selected. Select an illuminance category from Table 11.12 if the specific task can be found, from Table 11.11 if a specific task cannot be found, or from Table 11.13 if the equivalent contrast has been determined. Using Table 11.11, determine the recommended range of illuminances for the illuminance category selected. The design illuminance will be selected from the range in Table 11.11 considering the occupants' age, reflectances, and the speed or accuracy required. For illuminance categories A through C, determine the weighting factors for room and occupant characteristics (part *a*, Table 11.14). If the sum of the factors is -2, use the lowest end of the range; if the sum of the factors is $+2$, use the highest end of the range; otherwise use the middle illuminance. For categories D through I, determine the weighting factors for age, speed, and reflectance. If the sum of the three factors is -2 or -3, use the lowest end of the range; if the sum is $+2$ or $+3$, use the highest end of the range; otherwise use the middle illuminance. Note that for categories D through I, 200 lux is the minimum acceptable general area illumination. See the IES *Lighting Handbook* for nonindustrial and outdoor categories not included in Table 11.12 (IES, 1981). ANSI Standard A11.1-1973 on industrial lighting (ANSI, 1973a) has also been developed by the Illuminating Engineering Society and contains information similar to that in Tables 11.11–11.14.

III. Infrared Radiation

The infrared region of the spectrum extends from 750 nm to approximately 0.3 cm. This range is subdivided into near ir (750–2500 nm),

TABLE 11.14 Weighting Factors to Be Considered in Selecting Specific Illuminance within Ranges of Values for Each Category

For Illuminance Categories A through C			
Room and occupant characteristics	Weighting factor		
	−1	0	+1
Occupant age	Under 40	40–55	Over 55
Room surface reflectance[a]	Greater than 70%	30 to 70%	Less than 30%

For Illuminance Categories D through I			
Task and worker characteristics	Weighting factor		
	−1	0	+1
Worker age	Under 40	40–55	Over 55
Speed and/or accuracy[b]	Not important	Important	Critical
Reflectance of task background[c]	Greater than 70%	30–70%	Less than 30%

[a]Average weighted surface reflectances, including wall, floor, and ceiling reflectances, if they encompass a large portion of the task area of visual surround. For instance, in an elevator lobby, where the ceiling height is 7.6 m (25 ft), neither the task nor the visual surround encompass the ceiling, so only the floor and wall reflectances would be considered.

[b]In determining whether speed and/or accuracy is not important, important, or critical, the following questions need to be answered: What are the time limitations? How important is it to perform the task rapidly? Will errors produce an unsafe condition or product? Will errors reduce productivity and be costly? For example, in reading for leisure there are no time limitations and it is not important to read rapidly. Errors will not be costly and will not be related to safety. Thus, speed and/or accuracy is not important. If, however, prescription notes are to be read by a pharmacist, accuracy is critical because errors could produce an unsafe condition and time is important for customer relations.

[c]The task background is that portion of the task on which the meaningful visual display is exhibited. For example, on this page the meaningful visual display includes each letter which combines with other letters to form words and phrases. The display medium, or task background, is the paper, which has a reflectance of approximately 85%.

SOURCE: IES, 1981.

intermediate (2500–5000 nm), far (5000 nm–0.03 cm), and extreme (0.03–0.3 cm). Exposures to ir radiation can occur from any surface that is at a higher temperature than the receiver. Radiation exchange is proportional to the fourth power of the absolute temperature of the two surfaces (see Eq. 12.7). The wavelength characteristics of an ir source are a function of its temperature. For example, 5%, 20%, and 40% of the radiation is below 1500 nm for source temperatures of 1000°C, 1500°C, and 2000°C, respectively (Wilkening, 1978). White, pastel-colored, and translucent products absorb energy at wavelengths longer than 2500 nm. Dark-pigmented and oxide-coated materials will absorb energy from 250 to 9000 nm.

Infrared radiation is perceived as a sensation of warmth on the skin. Radiation in the 5000 nm–0.3 cm range is completely absorbed in the surface layer of skin. Radiation in the 750–1500 nm range can cause skin burns and injure the cornea, iris, retina, and lens of the eye (Largent and Olishifski, 1979). Industrial applications of ir energy include drying and baking of paints, varnishes, enamels, adhesives, ink, and protective coatings; heating of metal parts for shrink-fit assembly; forging; thermal treating; brazing; conditioning surfaces for application of adhesives; and dehydrating of textiles, paper, leather, meat, vegetables, pottery, and sand molds. Sources of ir include heated metals, electric appliances, incandescent bulbs, furnaces, welding arcs, lasers, and plasma torches.

A. Exposure criteria

The TLV for infrared radiation is combined with that for visible light and covers the spectral region from 400 to 1400 nm (ACGIH, 1984). The guideline is based on the spectral radiance (L_λ) and total irradiance (E) of a source measured at the eye of the exposed worker. Detailed spectral data are required only if the luminance of the source is greater than 1 candela$\cdot$cm^{-2}. For sources that exceed this luminance, the integrated spectral radiance weighted against the burn-hazard function (R_λ) should not exceed

$$\sum_{400\ \text{nm}}^{1400\ \text{nm}} L_\lambda R_\lambda\, \Delta\lambda \leq \alpha^{-1} t^{0.5} \tag{11.5}$$

where L_λ is in W$\cdot$cm$^{-2}\cdot$sr$^{-1}\cdot$nm^{-1}; t is the viewing duration (or pulse duration), limited to 1 μsec to 10 s; R_λ is defined in Table 11.15; $\Delta\lambda$ is the wavelength interval (nm); and α is the angular subtense of the source in radians. If the source is oblong, α is the longest dimension that can be viewed. For protection against chronic blue-light exposure, the integrated spectral radiance should not exceed:

$$\sum_{400\ \text{nm}}^{1400\ \text{nm}} L_\lambda\, t\, B_\lambda\, \Delta\lambda \leq 100\ \text{J}\cdot\text{cm}^{-2}\cdot\text{sr}^{-1} \qquad (t \leq 10^4\ \text{sec}) \tag{11.6a}$$

$$\sum_{400\ \text{nm}}^{1400\ \text{nm}} L_\lambda B_\lambda\, \Delta\lambda \leq 10^{-2}\ \text{W}\cdot\text{cm}^{-2}\cdot\text{sr}^{-1} \qquad (t > 10^4\ \text{sec}) \tag{11.6b}$$

where B_λ is the blue-light hazard function defined in Table 11.15. For blue-light sources that exceed 10 mW$\cdot$cm$^{-2}\cdot$sr^{-1}, the permissible exposure duration t_{max} in seconds is:

$$t_{max} = 100\ \text{J}\cdot\text{cm}^{-2}\cdot\text{sr}^{-1}/L_{(blue)} \tag{11.7}$$

where $L_{(blue)}$ is the weighted product of L_λ and B_λ.

For a light source subtending an angle α less than 0.011 radian,

$$\sum_{400\ \text{nm}}^{1400\ \text{nm}} E_\lambda\, t\, B_\lambda\, \Delta\lambda \leq 10\ \text{mJ}\cdot\text{cm}^{-2} \qquad (t \leq 10^4\ \text{sec}) \tag{11.8a}$$

$$\sum_{400\ \text{nm}}^{1400\ \text{nm}} E_\lambda B_\lambda\, \Delta\lambda \leq 1\ \mu\text{W}\cdot\text{cm}^{-2} \qquad (t > 10^4\ \text{sec}) \tag{11.8b}$$

with maximum exposure duration

$$t_{\max} = 10\ \text{mJ}\cdot\text{cm}^{-2}/E_{(\text{blue})} \tag{11.9}$$

Infrared radiation for $\lambda > 770$ nm should be limited to 10 mW $\cdot$ cm^{-2}. For sources where a visual stimulus is absent, the near infrared radiance (770–1400 nm) as viewed by the eye should be limited to

$$\sum_{400\text{nm}}^{1400\text{nm}} L_\lambda\, \Delta\lambda \leq \frac{0.6}{\alpha}$$

for extended viewing conditions.

TABLE 11.15 Spectral Weighting Functions for Assessing Retinal Hazards from Broad-Band Optical Sources

Wavelength (nm)	Blue-light hazard function, B_λ	Burn-hazard function, R_λ
400	0.10	1.0
405	0.20	2.0
410	0.40	4.0
415	0.80	8.0
420	0.90	9.0
425	0.95	9.5
430	0.98	9.8
435	1.0	10.0
440	1.0	10.0
445	0.97	9.7
450	0.94	9.4
455	0.90	9.0
460	0.80	8.0
465	0.70	7.0
470	0.62	6.2
475	0.55	5.5
480	0.45	4.5
485	0.40	4.0
490	0.22	2.2
495	0.16	1.6
500–600	$10^{[(450-\lambda)/50]}$	1.0
600–700	0.001	1.0
700–1049	0.001	$10^{[(700-\lambda)/505]}$
1050–1400	0.001	0.2

SOURCE: ACGIH, 1984.

B. Exposure protection

Protective eyewear can be used to limit optical exposure to infrared radiation. Infrared transmittance as a function of optical density for ANSI standard shade numbers (ANSI, 1979) is summarized in Table 11.8. The control of radiant heat exposures is discussed further in Chapter 12.

IV. Lasers

Laser is an acronym for "light amplification by stimulated emission of radiation." Lasers can produce a monochromatic beam of light in the ultraviolet, visible, or infrared region of the electromagnetic spectrum. The light beam from a laser is coherent; that is, all of the electromagnetic waves are in phase at the same frequency. There is therefore little divergence and a high concentration of energy per area of the beam. Unlike most other sources of radiation, which follow the inverse square law, the power density of a laser is a first-order function of distance and for short distances is approximately constant. This characteristic contributes to the extreme hazard of laser radiation.

All lasers are constructed of a high-Q optical chamber, energy pump, and lasing medium, where

$$Q = \frac{\text{energy stored}}{\text{energy dissipated/unit time}} \tag{11.11}$$

Figure 11.10 shows the basic components. The optical cavity has a totally reflecting mirror on one end and a partially reflecting mirror on the end where the beam exits. The laser material is a substance that can be excited to a metastable state by the energy pump. This is a condition where the majority of atoms are at an excited state. Lasing action is stimulated by a photon emitted when an atom at the excited state spontaneously returns to a lower level or ground state. This photon starts a chain reaction by stimulating another excited atom to emit a photon. A population of photons builds in the direction parallel to

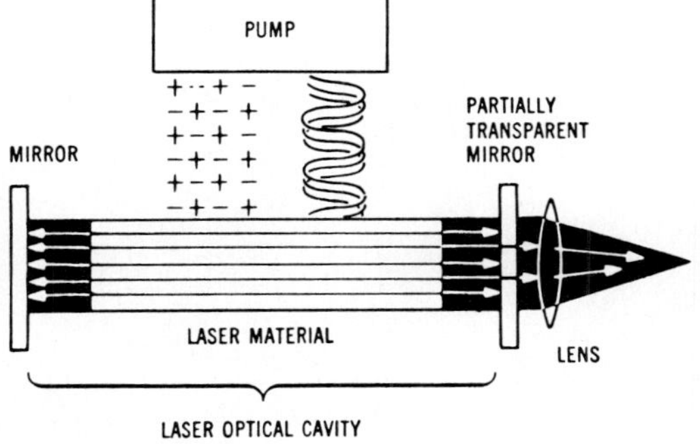

Figure 11.10 Components of a laser (*Largent and Olishifski, 1979*).

the large axis of the optical cavity between the mirrored ends. The energy that escapes through the partially reflecting mirror is all in phase, with the wavelength being a function of the difference between the metastable and ground states. These energy levels are shown schematically in Figure 11.11.

Ruby (chromium in Al_2O_3) is a common crystalline lasing medium. The metastable state of the Cr^{3+} ions in the ruby crystal is 1.79 eV. From Eqs. (10.1) and (10.2), the wavelength of the energy emitted is then:

$$\lambda = \frac{c \cdot h}{E}$$

$$= \frac{(3 \times 10^8 \text{ m/s})(10^9 \text{ nm/m})(6.625 \times 10^{-34} \text{ J}\cdot\text{sec})}{(1.79 \text{ eV})(1.602 \times 10^{-19} \text{ J/eV})}$$

$$= 693 \text{ nm}$$

Table 11.16 lists some typical lasers in common use. Helium-neon lasers can be operated in pulsed or continuous wave (cw) mode at wavelengths of 632.8, 1150, or 3340 nm. Typical output from these systems is 1–500 mW. Carbon dioxide molecular lasers typically operate in the infrared region at 10,600 nm, with peak output as high as 10,000 W. Argon ion gas lasers operate in the blue wavelengths with output at 457.8, 488, or 514.5 nm. There are also some lasers, such as Rb-Xe, H_2, Xe, and Ar-N_2, which can be operated in the ultraviolet region (<360 nm). Pulsed lasers operate in the energy range 0.1–30 J per pulse, with duration 0.1–10 msec and pulse frequency 1/30 to 10 pulses/sec. Q-switched pulse lasers produce very high power outputs for short durations ($\simeq 10^{-9}$ sec).

Lasers have a wide variety of uses, including alignment, welding, trimming, heat treating, spectrophotometry, fiber optics communications, nuclear fusion experimentation, and surgical procedures. Energy

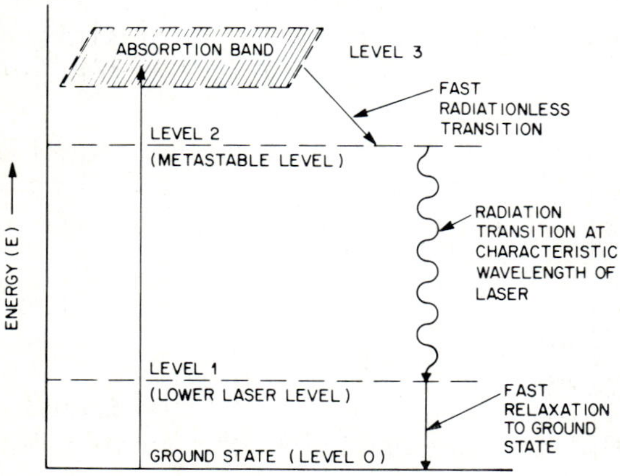

Figure 11.11 Energy levels of a laser (*Wilkening, 1978*).

TABLE 11.16 Lasers in Common Use

Laser	Wavelength (nm)	Photon energy (eV)
Ruby (Cr^{3+})	694	1.79
Nd-Y-Al-garnet	1,060	1.17
Nd-glass	1,060	1.17
Ga-As	900	1.47
Dye	360–650	3.5–1.9
He-Ne	632.8; 1,150; 3,390	1.96; 1.08; 0.37
He-Cd	325; 441.6	3.81; 2.81
CO_2	10,600	0.117
Ar	457.8; 488; 514.5	2.71; 2.54; 2.41
CO	5,500	0.225
H_2	160	7.74
H_2O	7,000–220,000	0.177–0.0056
HCN	773,000	0.0016

SOURCE: Wilkening, 1978.

densities suitable for machining metals range from 10^4 to 10^6 kW·in^{-2}. Laser beam welding is suitable for narrow and deep joints with depth:width ratios as high as 30:1 (Kalpakjian, 1984). CO_2 lasers in the 475-W to 15-kW range have been used for cutting and machining with automated presses and workstations. Carbon steel 6 mm thick can be cut at the rate of 1.0 m/min by an 800-W laser (Doyle et al., 1985). Holes from 0.25 to 1.25 mm in diameter can be cut to depths up to 15 mm in most alloys by pulsed lasers that generate 15 J-pulses (11.3 kJ/cm^2 per pulse).

A. Exposure criteria

Laser energy can damage the cornea or retina of the eye, depending on the wavelength of the radiation and absorption characteristics of the eye tissue. In visible wavelengths the energy is transparent to the cornea and damage is primarily to the retina. Lasers that operate in the uv or ir spectra produce corneal injury. Acute exposure to the skin may lead to severe burns. Table 11.17 summarizes retinal and skin damage thresholds.

The focusing action of the lens of the eye and area of the pupillary opening must be considered in developing safety standards for lasers. The diameter of the image on the retina (d_r) is given by:

$$d_r = \frac{2.44\lambda f}{d_p} \tag{11.12}$$

where λ is the wavelength of light (cm), d_p is the pupillary diameter (cm), and f is the focal length (cm) (Cember, 1983). The relationship

TABLE 11.17 Retinal and Skin Damage Thresholds

Laser	Wavelength (nm)	Duration	Level (J/cm^2)
A. Retinal damage:			
Continuous	Visible	—	6
Ruby pulse	694	200 μsec	0.85
Ruby Q-switch	694	30 nsec	0.07
B. Skin damage:			
Ruby pulse	694	0.2 msec	14–20
Argon continuous	500	6 sec	13–17
CO_2 continuous	1060	4–6 sec	4–6
Ruby Q-switch	694	10–12 nsec	0.5–1.5

SOURCE: Cember, 1983.

among retinal image size, pupillary diameter, and radiant exposure, H, is given by:

$$\frac{H(\text{retina})}{H(\text{cornea})} = \frac{d_p^2}{d_r^2} \tag{11.13}$$

Therefore, if the output from a Q-switched ruby laser (λ = 694.3 nm, 15 J/pulse, 1.6-cm beam diameter) falls on an open eye (0.7 cm pupillary diameter), the radiant exposure on the cornea will be:

$$H(\text{cornea}) = \frac{15\text{ J}}{(\pi/4)(1.6\text{ cm})^2} = 7.46\text{ J/cm}^2$$

With a focal length of 1.7 cm, the diameter of the image on the retina is:

$$d_r = (2.44)(6.943 \times 10^{-5}\text{ cm})\frac{(1.7\text{ cm})}{(0.7\text{ cm})} = 4.11 \times 10^{-4}\text{ cm} \tag{11.12}$$

with a resultant retinal exposure of:

$$H(\text{retina}) = \frac{(7.46\text{ J/cm}^2)(0.7\text{ cm})^2}{(4.11 \times 10^{-4}\text{ cm})^2}$$
$$= 2.16 \times 10^7\text{ J/cm}^2 \tag{11.13}$$

Maximum permissible exposure (MPE) criteria for lasers must take into account this enormous concentration of light.

Tables 11.18, 11.19, and 11.20 list the TLVs for direct ocular, diffuse reflection, and skin exposure for laser light from 315 nm to 1000 μm. The TLVs are averaged over a pupil diameter of 7 mm in the spectral range 400–1400 mm, 10 mm for wavelengths between 100 and 1000 μm, and 1 mm for all other laser light wavelengths. A correction factor, C_A, is given in Figure 11.12 for wavelengths between 700 and 1050

TABLE 11.18 Threshold Limit Value for Direct Ocular Exposures (Intrabeam Viewing) from a Laser Beam

Spectral region	Wavelength	Exposure time, t (sec)	TLV[a]
UVC	200–280 nm	10^{-9} to 3×10^4	3 mJ·cm^{-2}
UVB	280–302 nm	10^{-9} to 3×10^4	3 mJ·cm^{-2}
	303 nm	10^{-9} to 3×10^4	4 mJ·cm^{-2}
	304 nm	10^{-9} to 3×10^4	6 mJ·cm^{-2}
	305 nm	10^{-9} to 3×10^4	10 mJ·cm^{-2}
	306 nm	10^{-9} to 3×10^4	16 mJ·cm^{-2}
	307 nm	10^{-9} to 3×10^4	25 mJ·cm^{-2}
	308 nm	10^{-9} to 3×10^4	40 mJ·cm^{-2}
	309 nm	10^{-9} to 3×10^4	63 mJ·cm^{-2}
	310 nm	10^{-9} to 3×10^4	100 mJ·cm^{-2}
	311 nm	10^{-9} to 3×10^4	160 mJ·cm^{-2}
	312 nm	10^{-9} to 3×10^4	250 mJ·cm^{-2}
	313 nm	10^{-9} to 3×10^4	400 mJ·cm^{-2}
	314 nm	10^{-9} to 3×10^4	630 mJ·cm^{-2}
UVA	315–400 nm	10^{-9} to 10	$0.56t^{1/4}$ J·cm^{-2}
	315–400 nm	10 to 10^3	1.0 J·cm^{-2}
	315–400 nm	10^3 to 3×10^4	1.0 mW·cm^{-2}
Light	400–700 nm	10^{-9} to 1.8×10^{-5}	5×10^{-7} J·cm^{-2}
	400–700 nm	1.8×10^{-5} to 10	$1.8t^{3/4}$ mJ·cm^{-2}
	400–549 nm	10 to 10^4	10 mJ·cm^{-2}
	550–700 nm	10 to T_1[b]	$1.8t^{3/4}$ mJ·cm^{-2}
	550–700 nm	T_1 to 10^4	$10C_B$ mJ·cm^{-2}[c]
	400–700 nm	10^4 to 3×10^4	C_B μW·cm^{-2}[c]
IR-A	700–1049 nm	10^{-9} to 1.8×10^{-5}	$5C_A \times 10^{-7}$ J·cm^{-2}[d]
	700–1049 nm	1.8×10^{-5} to 10^3	$1.8C_At^{3/4}$ mJ·cm^{-2}[d]
	1050–1400 nm	10^{-9} to 10^{-4}	5×10^{-6} J·cm^{-2}
	1050–1400 nm	10^{-4} to 10^3	$9t^{3/4}$ mJ·cm^{-2}
	700–1400 nm	10^3 to 3×10^4	$320C_A$ μW·cm^{-2}[d]
IR-B & C	1.4–10^3 μm	10^{-9} to 10^{-7}	10^{-2} J·cm^2
	1.4–10^3 μm	10^{-7} to 10	$0.56(t^{1/4})$ J·cm^{-2}
	1.4–10^3 μm	10 to 3×10^4	0.1 W·cm^{-2}

[a]Not to exceed $0.56t^{1/4}$ J·cm^{-2} for ≤10 sec for 200 nm < λ < 315 nm.
[b]T_1 = 10 sec for λ = 400–549 nm; $\log T_1 = 1 + 0.02(\lambda - 550)$ for λ = 550–700 nm.
[c]C_B = 1 for λ = 400–549 nm; $\log C_B = 0.015(\lambda - 550)$ for λ = 550–700 nm.
[d]C_A: see Figure 11.12.
SOURCE: ACGIH, 1984.

nm in Tables 11.18 and 11.19; and for wavelengths between 700 and 1400 nm in Table 11.20. For multiple pulses the exposure from any single pulse is limited to the protection standard for a single comparable pulse; the average irradiance for a group of pulses is limited to the standard as given in Tables 11.18, 11.19, and 11.20 of a single pulse of the same duration as the entire pulse group; and when the pulse repetition frequency exceeds one per second, the protection stan-

TABLE 11.19 Threshold Limit Values for Viewing a Diffuse Reflection of a Laser Beam or an Extended Source Laser[a]

Spectral region	Wavelength	Exposure time, t (sec)	TLV
UV	200–400 nm	10^{-9} to 3×10^4	Same as Table 11.18
Light	400–700 nm	10^{-9} to 10	$10(t^{1/3})$ J·cm^{-2}·sr^{-1}
	400–549 nm	10 to 10^4	21 J·cm^{-2}·sr^{-1}
	550–700 nm	10 to T_1	$3.83t^{3/4}$ J·cm^{-2}·sr^{-1}
	550–700 nm	T_1 to 10^4	$21C_B$ J·cm^{-2}·sr^{-1}
	400–700 nm	10^4 to 3×10^4	$2.1C_Bt \times 10^{-3}$ W·cm^{-2}·sr^{-1}
IR-A	700–1400 nm	10^{-9} to 10	$10C_A(t^{1/3})$ J·cm^{-2}·sr^{-1}
	700–1400 nm	10 to 10^3	$3.83C_At^{3/4}$ J·cm^{-2}·sr^{-1}
	700–1400 nm	10^3 to 3×10^4	$0.64C_A$ W·cm^{-2}·sr^{-1}
IR-B & C	1.4–10^3 μm	10^{-9} to 3×10^4	Same as Table 11.18

[a]C_A, C_B, and T_1 are the same as in footnote to Table 11.18.
SOURCE: ACGIH, 1984.

dard is reduced as shown in Figure 11.13 for pulse durations less than 10^{-5} sec. For pulses of greater duration, the standard for each pulse is:

$$\text{Standard (single pulse)} = \frac{\text{standard (pulse } n\tau)}{n} \tag{11.14}$$

where n = number of pulses, τ = duration of a single pulse, and standard (pulse $n\tau$) = protection standard for one pulse having a duration equal to $(n)(\tau)$ sec.

Reflections of laser light from polished and mirrored surfaces (specular), and from rough or dull surfaces (diffuse), can also be optical hazards. The fraction of incident light reflected by a surface is the reflection coefficient. Reflection coefficients for various surfaces are given in Table 11.21. An extended source is a source that can be resolved into a geometric image on the retina. In laser hazard control,

TABLE 11.20 Threshold Limit Value for Skin Exposure from a Laser Beam[a]

Spectral region	Wavelength	Exposure time, t (sec)	TLV
UV	200–400 nm	10^{-9} to 3×10^4	Same as Table 11.18
Light & IR-A	400–1400 nm	10^{-9} to 10^{-7}	$2C_A \times 10^{-2}$ J·cm^{-2}
IR-A	400–1400 nm	10^{-7} to 10	$1.1C_A(t^{1/4})$ J·cm^{-2}
IR-A	400–1400 nm	10 to 3×10^4	$0.2C_A$W·cm^{-2}
IR-B & C	1.4–10^3 μm	10^{-9} to 3×10^4	Same as Table 11.18

[a]C_A = 1.0 for λ = 400–700 nm; see Figure 11.12 for λ = 700–1400 nm.
SOURCE: ACGIH, 1984.

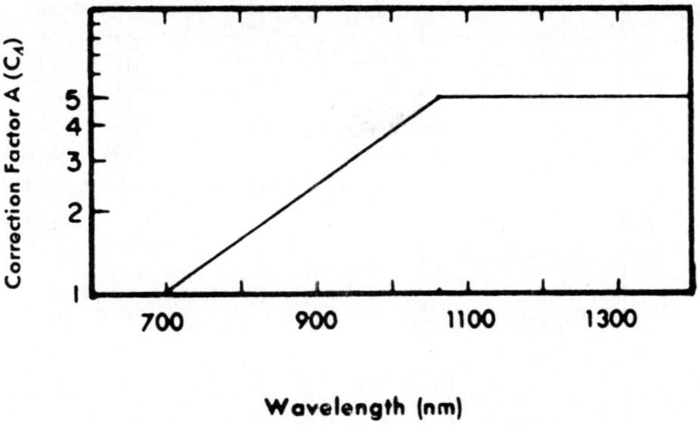

Figure 11.12 Correction factor C_A for wavelengths 0.7–1.4 μm (*ACGIH, 1984*). $C_A = 1$ for $\lambda = 0.4–0.7\ \mu$m; log $C_A = 2.0\ (\lambda - 0.7)$ for $\lambda = 0.7–1.05\ \mu$m; $C_A = 5$ for $\lambda = 1.051–1.4\ \mu$m.

an extended source is defined as a source that subtends a viewing angle greater than $\alpha_{\min}$:

$$\alpha_{\min}(\text{radians}) = \frac{d_r}{f} \tag{11.15}$$

where d_r is the diameter of the retinal image (Eq. 11.12) and f is the focal length of the eye (0.7 cm) (Cember, 1983). The geometry of an extended source is shown on Figure 11.14. If the eye is exposed to such a source, the brightness of the reflection can be calculated as the product of the irradiance and the reflection coefficient. Table 11.19 shows the TLV for reflected or extended-source laser light.

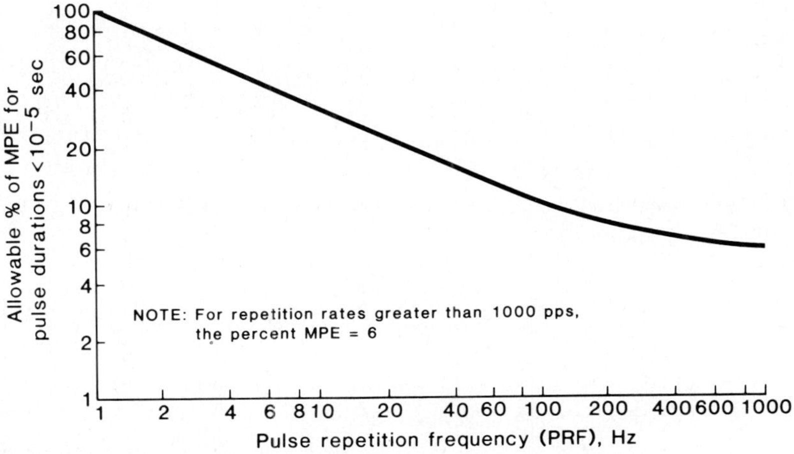

Figure 11.13 Reduction of MPE for repetitively pulsed or multiple exposures from scanning lasers, individual pulse or exposure less than 10 μsec (*ANSI, 1980*).

TABLE 11.21 Reflection Coefficients

Type	Material	Reflection coefficient
Specular	Mirrored glass	0.8–0.9
	Aluminum foil	0.84–0.87
	Rhodium	0.7–0.9
	Aluminum, polished	0.6–0.7
	Chromium	0.6–0.65
	Stainless steel	0.55–0.65
	Black structural glass	0.04–0.05
Mixed	Aluminum, oxidized	0.70–0.85
	Aluminum, brushed	0.54–0.58
	Aluminum paint	0.6–0.7
	Stainless steel, satin	0.51–0.56
Diffuse	White plaster	0.90–0.92
	White paint, flat	0.75–0.90
	Limestone	0.35–0.65
	Sandstone	0.30–0.42

SOURCE: Cember, 1983.

B. Protection guidelines

The American National Standards Institute (ANSI) has developed protection guidelines and classifications for lasers (ANSI, 1980). The Food and Drug Administration, OSHA, and the ACGIH (1976) have adopted similar guidelines. The TLVs are based on the ANSI guidelines (ACGIH, 1984). Class 1 lasers are considered to be incapable of producing damaging radiation levels and are exempt from control measures. Class 2 visible-wavelength lasers have a low hazard potential from direct viewing of the beam. Class 3 lasers are moderately hazardous. For this class, intrabeam viewing of either the direct beam or specular reflections may be hazardous. Class 4 lasers are high-power systems where direct or diffusedly reflected radiation may be hazardous and where the beam may represent a fire hazard. Table 11.22 gives the definitions of the laser hazard classifications. Tables 11.23 and 11.24 show the classification of commonly used continuous wave and pulsed lasers as a function of output power.

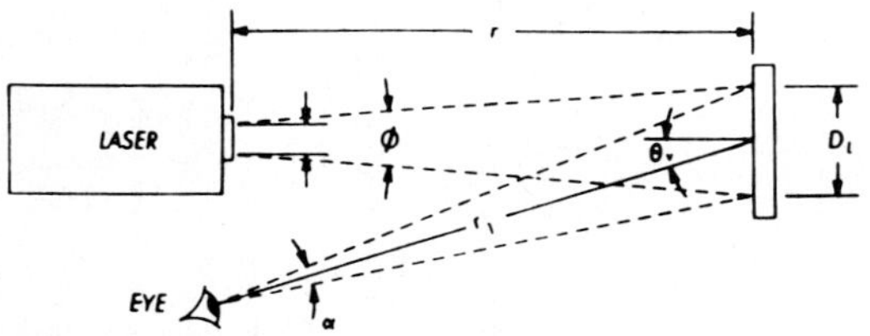

Figure 11.14 Extended-source viewing of a normally diffuse laser beam (*ANSI, 1980*).

TABLE 11.22 Laser Device Hazard Classification Definitions

Class 1: Nonrisk Laser Devices:

Any laser, or system containing a laser, that cannot emit laser radiation levels in excess of specified exempt values of accessible radiant power or energy for a given exposure duration. The exemption from hazard controls for this class applies strictly to the emitted laser radiation hazard, not to other potential hazards. The term for exempt accessible radiant power is P_{exempt}; exempt accessible radiant energy is designated Q_{exempt}. The terms P_{exempt} and Q_{exempt} apply to the longest conceivable duration of exposure (exposure time) in a "worst-case" situation or condition. Both P_{exempt} and Q_{exempt} are products of the intrabeam exposure limit (Table 11.18) and the area of the appropriate aperture. The appropriate aperture (opening) for visible wavelengths is $\pi d^2/4$ cm^2, where d is 0.7 cm, the diameter of the fully dilated pupil (worst-case exposure conditions). For cases when the laser is not a point source (e.g., laser diode, arrays), P_{exempt} and Q_{exempt} are defined as the radiant power levels and radiant energy levels that do not exceed the extended source exposure limits (Table 11.19) when the source is viewed at the expected minimum viewing distance through a 0.7-cm exit aperture after the radiation has been collected by an 8-cm entrance aperture.

Class 2: Low-Power Visible Laser Devices (Low Risk):

a. Visible (400–700 nm) continuous-wave (cw) laser devices. Devices that can emit accessible radiant power exceeding P_{exempt} for the classification duration (0.4 μW for the maximum expected exposure time greater than 0.25 sec), but not exceeding 1 mW.

b. Visible (400–700 nm) scanning laser systems. These systems may be evaluated by specifying P_{exempt} at a point 10 cm from the point of emitted radiation. All repetitively pulsed laser devices and scanning systems are considered to be low-risk if they do not exceed P_{exempt} for a 0.25-sec exposure.

Class 3: Medium-Power Laser Devices (Moderate Risk):

a. Infrared (1.4 μm–1 mm) and ultraviolet (200–400 nm) laser devices. All such devices capable of emitting a radiant power in excess of the exempt quantity (P_{exempt}) for the classification duration, but incapable of emitting (1) an average radiant power in excess of 0.5 W for the maximum exposure time greater than 0.25 sec, or (2) a radiant exposure of 10 J/cm^2 within an exposure duration of 0.25 sec or less.

b. Visible (400–700 nm) cw or repetitively pulsed laser devices. All such devices that produce a radiant power in excess of the exempt power (P_{exempt}) for a 0.25-sec exposure (1 mW for a cw laser) but cannot emit an average radiant power of 0.5 W for the maximum actual exposure time that exceeds 0.25 sec.

c. Visible and near-infrared (400–1400 nm) pulsed laser devices. All such devices that can emit a radiant energy in excess of the exempt value (Q_{exempt}) but cannot emit a radiant exposure that exceeds either 10 J/cm^2 or that required to produce a hazardous diffuse reflection (Table 11.19).

d. Near-infrared (700–1400 nm) cw laser devices or repetitively pulsed laser devices. All such devices that can emit power in excess of the exempt quantity (P_{exempt}) for the classification exposure duration but cannot emit an average power of 0.5 W or greater for periods in excess of 0.25 sec.

Class 4: High-Power Laser Devices (High Risk):

a. Ultraviolet (200–400 nm) and infrared (1.4 μm–1 mm) laser devices. All such devices that emit an average power in excess of 0.5 W for periods greater than 0.25 sec or a radiant exposure of 10 J/cm^2 within an exposure duration of $\geq$0.25 sec.

b. Visible (400–700 nm) and near-infrared (700–1400 nm) laser devices. All such devices that emit an average power of 0.5 W or more for periods $<$0.25 sec or a radiant exposure in excess of either 10 J/cm^2 or that required to exceed the hazardous diffuse reflection value (3.14 times the radiance values) given in Table 11.19.

SOURCE: Wilkening, 1978.

TABLE 11.23 Typical Laser Classification—Continuous-Wave (cw) Lasers

Wavelength range	Laser	Wavelengths	Exempt—Class 1[a]	Class 2	Class 3	Class 4
Ultraviolet (0.1–0.28 μm)	cw Neodymium: YAG (quadrupled)	266 nm only	$\leq 0.8 \times 10^{-9}$ W for 8 hr	—	> Class 1 but $\leq$ 0.5 W	> 0.5 W
Ultraviolet (0.315–0.4 μm)	Helium-cadmium Argon Krypton	325 nm only 351.1, 363.8 nm only 350.7, 356.4 nm only	$\leq 8 \times 10^{-6}$ W	—	> Class 1 but $\leq$ 0.5 W	> 0.5 W
Visible (0.4–0.7 μm)	Helium-cadmium Argon (visible) Helium-selenium	441.6 nm only 457.9, 476.5, 488, 514.5 nm, etc. 460.4–700 nm only (30 lines)	$\leq 0.4 \times 10^{-6}$ W	> Class 1 but $\leq 1 \times 10^{-3}$ W	> Class 2 but $\leq$ 0.5 W	> 0.5 W
	cw Neodymium: YAG (doubled) Helium-neon Krypton	532 nm 632.8 nm 647.1, 530.9, 676.4 nm, etc.	$\leq 0.4 \times C_B \times 10^{-6}$ W[b]	> Class 1 but $\leq 1 \times 10^{-3}$ W	> Class 2 but $\leq$ 0.5 W	> 0.5 W
Near infrared (0.7–1.4 μm)	cw Gallium-aluminum arsenide	0.85 μ (20°C)	$\leq 80 \times 10^{-6}$ W	—	> Class 1 but $\leq$ 0.5 W	> 0.5 W
	cw Gallium arsenide	0.905 μm (20°C)	$\leq 0.1 \times 10^{-3}$ W	—	> Class 1 but $\leq$ 0.5 W	> 0.5 W
	cw Neodymium: YAG	1.064 μm	$\leq 0.2 \times 10^{-3}$ W	—	> Class 1 but $\leq$ 0.5 W	> 0.5 W
	Helium-neon	1.08, 1.152 μm only	$\leq 0.2 \times 10^{-3}$ W	—	> Class 1 but $\leq$ 0.5 W	> 0.5 W
Far infrared (1.4–100 μm)	Hydrogen fluoride Carbon monoxide Carbon dioxide Helium-neon	4–6 μm 5.0–5.5 μm 10.6 μm 3.39 μm only	$\leq 0.8 \times 10^{-3}$ W	—	> Class 1 but $\leq$ 0.5 W	> 0.5 W
Far infrared (0.1–1 mm)	Water vapor Hydrogen cyanide	118 μm 337 μm	$\leq$ 0.1 W	—	> Class 1 but $\leq$ 0.5 W	> 0.5 W

[a]Assumes no mechanical or electrical design incorporated into laser system to prevent exposures from lasting to τ_{max} = 8 hr (one workday); otherwise the Class 1 AEL could be larger than tabulated.

[b]C_B is the same as in Table 11.18.

SOURCE: ANSI, 1980.

TABLE 11.24 Typical Laser Classification—Single-Pulse Lasers

Wavelength range	Laser	Wavelengths	Pulse duration	Exempt Class 1	Class 3	High-power Class 4
Ultraviolet	Neodymium: YAG Q-switched (quadrupled)	266 nm	$10–30 \times 10^{-9}$ sec (Q-sw)	No criteria	$\leq 10\ J\cdot cm^{-2}$	$> 10\ J\cdot cm^{-2}$
	Ruby (doubled)	347.1 nm	$10–30 \times 10^{-9}$ sec (Q-sw)	No criteria	$\leq 10\ J\cdot cm^{-2}$	$> 10\ J\cdot cm^{-2}$
Visible (0.4–0.7 μm)	Neodymium: YAG (doubled)	532 nm	$\sim 20 \times 10^{-9}$ sec (Q-sw)	$\leq 0.2 \times 10^{-6}$ J	> Class 1 but $\leq 74 \times 10^{-3}\ J\cdot cm^{-2}$	$> 75 \times 10^{-3}\ J\cdot cm^{-2}$
	Ruby	694.3 nm				
	Ruby (long pulse)	694.3 nm	$\sim 1 \times 10^{-3}$ sec	$\leq 4 \times 10^{-6}$ J	> Class 1 but $\leq 3.1\ J\cdot cm^{-2}$	$> 3.1\ J\cdot cm^{-2}$
	Rhodamine 6G (dye laser)	450–650 nm	$\sim 1 \times 10^{-6}$ sec	$\leq 0.2 \times 10^{-6}$ J	> Class 1 but $\leq 0.31\ J\cdot cm^{-2}$	$> 0.31\ J\cdot cm^{-2}$
Infrared	Neodymium: YAG	1.064 μm	$\sim 20 \times 10^{-9}$ sec (Q-sw)	$\leq 2 \times 10^{-6}$ J	> Class 1 but $\leq 0.16\ J\cdot cm^{-2}$	$> 0.16\ J\cdot cm^{-2}$
	Erbium: glass	1.54 μm	$\sim 10–100 \times 10^{-9}$ sec (Q-sw)	$\leq 8 \times 10^{-3}$ J	> Class 1 but $\leq 10\ J\cdot cm^{-2}$	$> 10\ J\cdot cm^{-2}$
	Carbon dioxide	10.6 μm	$1–100 \times 10^{-9}$ sec (Q-sw)	$\leq 80 \times 10^{-6}$ J	> Class 1 but $\leq 10\ J\cdot cm^{-2}$	$> 10\ J\cdot cm^{-2}$

SOURCE: ANSI, 1980.

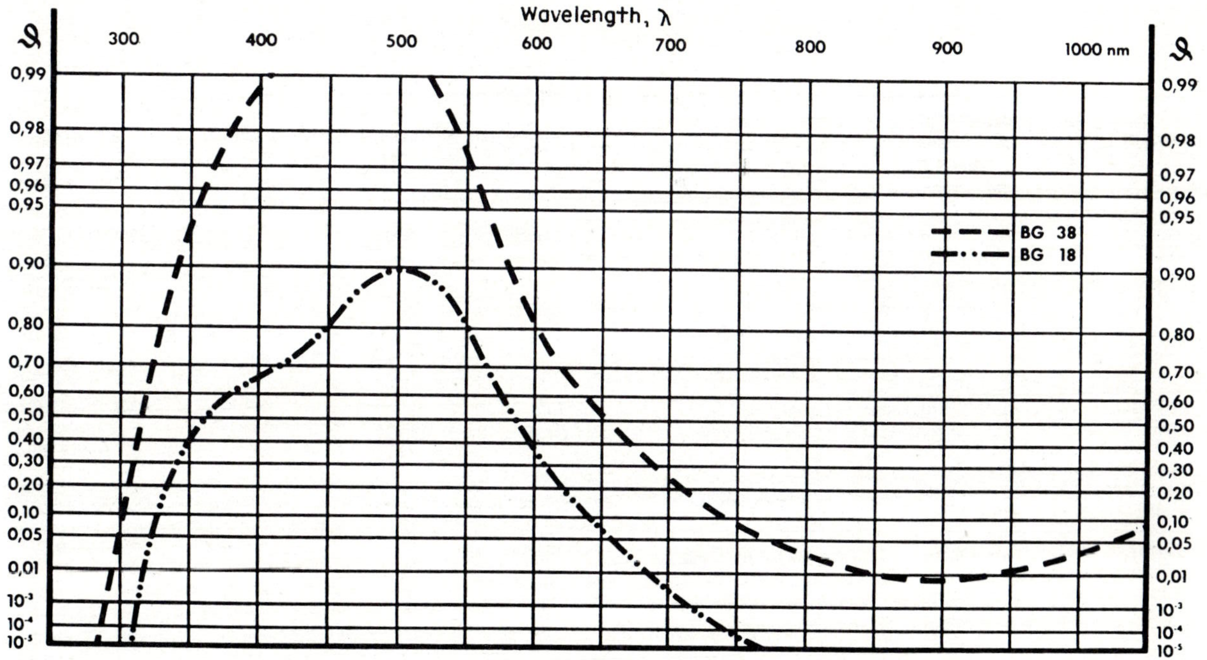

Figure 11.15 Fractional transmission of Jena BG-18 and BG-38 color glasses (*Straub, 1965*).

C. Exposure protection and source control

Protective goggles can be used to protect the eye against hazardous levels of laser light; and protective glass can be incorporated into design of the laser source. The attenuation of a protective lens is given by its optical density. The minimum optical density required is determined from Eq. (11.4) by substituting the maximum permissible exposure for I and the radiance at the unprotected eye for I_0 (ANSI, 1980). Protective lenses should be selected that provide both adequate optical density at the wavelengths of concern as well as transmit the maximum amount of ambient light. The ambient light transmitted by a lens is its luminous transmission. Table 11.8 lists luminous transmittance for standard filter lenses. Specialized protective lenses are available for various laser wavelengths. These lenses maximize both the optical density for laser light and the luminous transmittance for visible wavelengths. Jena color glass types BG18 and BG38 are very effective in controlling exposures from lasers in the pulsed ruby (1690 nm) to Nd-glass (1060 nm) spectral region (Straub, 1965). Transmission characteristics are shown in Figure 11.15. This glass has been incorporated in commercially available devices (Straub, 1965). Table 11.25 summarizes optical data for a variety of specialized laser goggles. Note that a lens for protection against laser light must be capable of absorbing the incident energy under expected operating conditions without suffering damage. Electric shock is a significant hazard associated with laser operation. All pertinent provisions of the National Electric Code should be followed.

Irradiance levels of scattered neodymium-YAG laser radiation from welding operations have been measured at levels up to 3.2 mW/cm^2 at 1 m from the source. Assuming an allowable exposure to the eye of 0.22 mW/cm^2 and an inverse square law relationship, an ocular hazard exists within 3.8 m from the operation (Rockwell and Moss, 1983). The maximum integrated near-uv irradiance (320–400 nm) for welding operations was only 6.2 μ W/cm^2, well below the 1 mW/cm^2 TLV. However, shields with optical densities greater than 6 at the 1.06-μm laser wavelength and greater than 1 in the blue region are needed to control radiation produced directly or indirectly during welding with a neodymium-YAG laser (Rockwell and Moss, 1983).

V. Microwaves and Radiofrequencies

Microwave frequencies range from about 10 to 300,000 MHz. Microwave frequency bands between 1,100 and 40,000 MHz are defined in Table 11.26. Seven industrial, scientific, and medical (ISM) frequencies have been designated by the Federal Communications Commission

TABLE 11.25 Optical Data for Laser Protection Eyewear

Catalog number	Optical density					
	Ne-N		Ar			Cu-vapor
	332 nm	337 nm	455 nm	488 nm	515 nm	511 nm
AO[a] 581						
584						
585						
588						0.3
598			17.0	13.5	9.1	
599			10.4	8.6	5.5	
680						
698						
LGS[b] NN	15	16				
HN						
A				15	11	
R						
CO_2						
NOGA						
Series A	20	20				
Series B	14	14		11	7	
FSC[c] AL-578				1	5	
AL-650[d]				0.34		

[a]American Optical Co. Safety Products, Southbridge, Mass.
[b]LAB Safety Supply, Janesville, Wisc.
[c]Fish-Schurman Corporation, New Rochelle, N.Y.
[d]Schreibis, 1968.

(13.56 ± 0.00678 MHz, 27.12 ± 0.016 MHz, 40.68 ± 0.020 MHz, 915 ± 25 MHz, 2450 ± 50 MHz, 5800 ± 75 MHz, and 22,125 ± 125 MHz). Microwave radiation can be categorized into continuous wave (cw) and pulsed modes.

The primary effect of exposure to microwaves is presently thought to be thermal (EPA, 1984). Radiation at greater than 3000 MHz is absorbed in the skin. Energy at frequencies below 3000 MHz can be absorbed in the underlying tissues. For example, the penetration depths of 2450-MHz radiation in muscle and fat are 1.67 and 8.1 cm, respectively. The eyes and testicles are two of the most sensitive organs to microwave radiation. It is generally accepted that exposure of the eye to high-intensity microwave energy may lead to cataracts. Exposures that lead to intraocular temperatures ≥45°C are believed to be cataractogenic (Largent and Olishifski, 1979; Wilkening, 1978).

Microwaves are used for communications, radar, industrial and commercial heating, freeze-drying, and wood gluing operations. Primary sources include klystrons, magnetrons, backward-wave oscillators, and semiconductor devices. Industrial microwave heating systems are summarized in Table 11.27.

Optical density						
Cu-vapor	He-Ne	Ruby	Ga-As Nd		CO_2	Luminous transmittance
578 nm	632 nm	694 nm	840 nm	1,060 nm	10,600 nm	(%)
	4.1	6.1				10
		4.4	11.2	9.5		30
		7.1	17.4	14.7		25
0.8	2.2	6.0	14.8	12.5		35
						25
						25
					15	90
				7.4		5
						70
	6					20
						45
		6				20
					12	30
			14	14		45
		6	20	20		20
			4	4		45
						12
	3.8	10	>10	>10		30

Short-wavelength microwave radiation can be transmitted through the air in sharply focused beams of energy, similar to a beam of light. Focusing of the beam is possible because the size of a typical microwave antenna greatly exceeds the wavelength of the emitted radiation. The spread of a microwave beam is measured in degrees and is called beam width. The characteristic of an antenna to concentrate microwave energy is its antenna gain (G_a), defined as

$$G_a = \frac{\text{power from antenna } (P)}{\text{power from isotropic radiator } (P_0)} \tag{11.16a}$$

TABLE 11.26 Letter Designation of Microwave Frequency Bands

Band	Frequency (MHz)
L	1,100–1,700
LS	1,700–2,600
S	2,600–3,950
C	3,950–5,850
XN	5,850–8,200
X	8,200–12,400
Ku	12,400–18,000
K	18,000–26,500
Ka	26,500–40,000

SOURCE: Largent and Olishifski, 1979.

TABLE 11.27 Summary of Conveyorized, Industrial Microwave Heating Systems

Type of process	Output power[a] (kW)	Frequency (MHz)	Maximum leakage (mW/cm)	Source of maximum leakage	Maximum leakage at conveyor slot (mW/cm^2)	Maximum potential occupational exposure at eye level (mW/cm^2)
1. Pharmaceutical drying	10	2450	190.0	Loose cleaning door	20.0	1.0
2. Food processing research	10	2450	70.0	Conveyor slot	70.0	3.0
3. Finish drying potato chips	60	915	20.0	Conveyor slot	20.0	1.0
4. Opening oysters	6	2450	25.0	Conveyor slot	25.0	1.0
5. Filament drying	2	2450	25.0	Conveyor slot	25.0	1.0
6. Film drying	5	2450	50.0	Conveyor slot	50.0	1.0
7. Film drying	30	915	2.0	Conveyor slot	2.0	1.0
8. Precooking chicken	120	2450	70.0	Waveguide feed	5.0	1.0
9. Food processing research	5	2450	1.0	Conveyor slot	1.0	1.0
10. Opening oysters	5	2450	60.0	Loose door hinge	8.0	1.0
11. Precooking onion rings	4.5	2450	<1.0	—	<1.0	1.0
12. Rapid heating of school lunches	15	2450	1.0	Conveyor slot	1.0	1.0
13. Thawing frozen food	25	915	1.0	Conveyor slot	1.0	1.0
14. Donut proofing	5	2450	30.0	Warped cleaning door	2.0	1.0
15. Macaroni drying	30	915	>200.0	Burned cleaning door	2.0	4.0

[a]Home microwave ovens generally are in the range 0.4–0.75 kW (400–750 W).

SOURCE: Largent and Olishifski, 1979.

and, for a parabolic antenna,

$$G = \frac{4\pi A}{\lambda^2} \tag{11.16b}$$

where A is the area of the antenna aperture and λ is the wavelength of the radiation. The near field of an antenna is defined as the area where energy is stored and recovered during successive oscillations and extends less than one wavelength out from the radiator. In the far field, the beam starts to spread and becomes conical in shape. The area between the near and far fields is the intermediate or Fresnel region.

The maximum power density in the near field is estimated from

$$W_{\text{near field}} = \frac{16P}{\pi D^2} = \frac{4P}{A} \tag{11.17}$$

where P is the average power output, D is the diameter of the antenna, and A is the effective area of the antenna (ANSI, 1973b). The far-field free-space power density is calculated from

$$W_{\text{far field}} = \frac{GP}{4\pi r^2} = \frac{AP}{\lambda^2 r^2} \tag{11.18}$$

where r is the distance from the antenna. The near-field equation (11.18) is valid for distances less than

$$r_1 = \frac{\pi D^2}{8\lambda} = \frac{A}{2\lambda} \tag{11.19}$$

For distances greater than r_1, Eq. (11.18) should be used. Note that these equations do not include the effect of ground reflections. Power densities may be increased by a factor of 4 when the main beam is directed toward a planar ground or reflecting surface.

A. Exposure criteria

The TLV for microwave radiation is combined with radiofrequency radiation and covers the frequency range 10 kHz to 300 GHz. The TLV is shown in Table 11.28 and is designed to limit the average whole-body specific absorption rate to 0.4 W/kg in any 6-min period. Note that the TLV is expressed in terms of electric field strength ($V^2 \cdot m^{-2}$) or magnetic field strength ($A^2 \cdot m^{-2}$). For near-field exposures, the maximum permissible power density, PD, can be calculated from:

$$\text{PD (mW} \cdot \text{cm}^{-2}) = \frac{E^2}{3770} = 37.7H^2 \tag{11.20}$$

where E is in volts per meter and H is in amperes per meter. This

TABLE 11.28 Radiofrequency/Microwave Threshold Limit Values

Frequency	Power density[a] ($mW \cdot cm^{-2}$)	Electric field strength squared[a] ($V^2 \cdot m^{-2}$)	Magnetic field strength squared[a] ($A^2 \cdot m^{-2}$)
10 kHz–3 MHz	100	377,000	2.65
3–30 MHz	$900/f^2$	$3{,}770 \times 900/f^2$	$900/(37.7 \times f^2)$
30–100 MHz	1	3,770	0.027
100–1,000 MHz	$f/100$	$3{,}770 \times f/100$	$f/(37.7 \times 100)$
1 GHz–300 GHz	10	37,700	0.265

[a] f = frequency, MHz.
SOURCE: ACGIH, 1984.

standard is very similar to those under consideration by NIOSH and the American National Standard Institute (Cember, 1983). However, the U.S. EPA has also proposed limiting whole-body specific absorption rates to 0.04 W/kg for frequencies >3 MHz; and limiting E to 87 V · m^{-1} and H to 0.23 A · m^{-1} for frequencies <3 MHz (EPA, 1986).

B. Control

Engineering controls for microwave radiation include complete enclosures with shielding materials (e.g., microwave ovens), and the re-

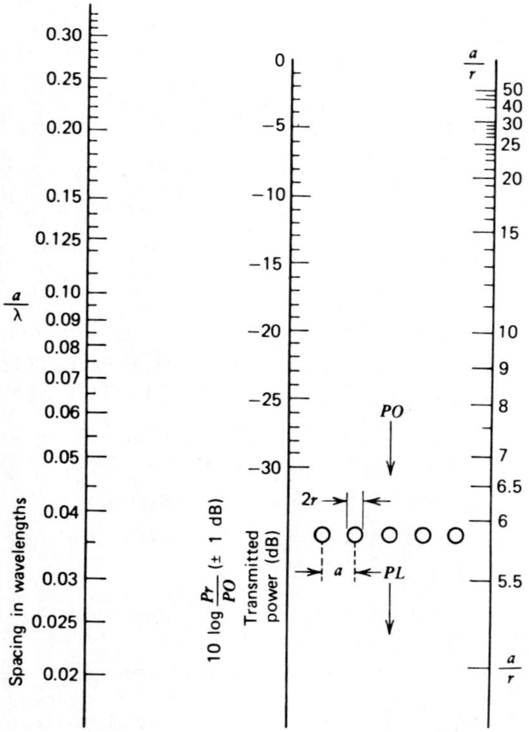

Figure 11.16 Transmission of microwaves through a grid of wires of radius r and spacing a (*Wilkening, 1978*).

TABLE 11.29 Attenuation Factors (dB) for Microwave Shielding

Material	Frequency (GHz) 1–3	3–5	5–7	7–10
60 × 60 mesh screening	20	25	22	20
32 × 32 mesh screening	18	22	22	18
16 × 16 window screen	18	20	20	22
0.25-in. mesh (hardware cloth)	18	15	12	10
Window glass	2	2	3	3.5
0.75-in. pine sheathing	2	2	2	3.5
8-in. concrete block	20	22	26	30

SOURCE: Wilkening, 1978.

striction of azimuth and elevation settings on radar antennas (Wilkening, 1978). Certain high-power microwave generators contain high-voltage (18-kV) power supplies and are therefore potential sources of x-radiation. Wire grids may serve as useful attenuation shields for microwave radiation. Figure 11.16 shows the transmission through a grid of wires as a function of wire radius, microwave frequency, and wire spacing. The attenuations provided by various materials are given in Table 11.29. For Table 11.29 and Figure 11.16, attenuation is given in units of decibels:

$$\text{Attenuation (dB)} = 10 \log \frac{P}{P_0} \tag{11.21}$$

where P is the power penetrating the shield and P_0 is the power incident in the shield.

VI. Example Applications

A. Welding

Example 11.1 Gas metal arc welding (GMAW), gas tungsten arc welding (GTAW), and plasma arc welding (PAW) produce potentially harmful exposures to uv radiation. Determine the shade numbers for welding shields to protect welders performing the following operations:

a. GMAW on steel with 5% O_2 and 95% Ar gases at 400 A of arc current

b. GTAW on steel with helium gas, 3.2 mm gap at 200 A of arc current

c. PAW on steel with 15% H_2 and 85% Ar gases at 200 A of arc current

Assume a distance of 1 m from the arc to the welder's eyes.

solution From Table 11.6, the permissible uv irradiance for an 8-hr exposure per day is 0.1 μW·cm^{-2}. From Figure 11.3, the effective uv irradiance for arc (*a*) is 3.0 mW·cm^{-2}; for arc (*b*) is 0.6 mW·cm^{-2}; and for arc (*c*) is 0.16 mW·cm^{-2}. The optical density (OD) of a shield to reduce these exposures to the TLV can be calculated with Eq. (11.4), where I_0 is the exposure 1 m from the arc (Figure

11.3) and I is the permissible uv irradiance. Therefore, the optical densities required for the three arcs are:

$$\begin{aligned} OD_a &= \log\left(\frac{3000\ \mu W\cdot cm^{-2}}{0.1\ \mu W\cdot cm^{-2}}\right) \\ &= 4.48 \\ OD_b &= \log\left(\frac{600\ \mu W\cdot cm^{-2}}{0.1\ \mu W\cdot cm^{-2}}\right) \\ &= 3.78 \\ OD_c &= \log\left(\frac{160\ \mu W\cdot cm^{-2}}{0.1\ \mu W\cdot cm^{-2}}\right) \\ &= 3.20 \end{aligned} \qquad (11.4)$$

From Table 11.8, these optical densities correspond to ANSI standard shade numbers 12, 10, and 8 or 9, respectively. These shade numbers are in good agreement with OSHA recommended shade numbers shown in Table 11.9. Figure 11.7 suggests that a welding helmet would be required for these operations.

As a check, we can determine the upper limit irradiance in three wavelength bands, E_{eff}, E_β, and E_{ir}. Equation (11.1) may be used to determine maximum irradiance as a function of arc current and distance for any welding process. The predictions from these equations and the respective TLVs are summarized below.

	Band	Wavelength (nm)	Maximum predicted irradiance ($mW\cdot cm^{-2}$)	TLV[a] ($mW\cdot cm^{-2}$)
Arc *a*[b]	eff	200–315	6.4[c]	10^{-4}
	β	400–1400	17.6[d]	10^{-3}
	ir	760–2000	14.4[e]	10
Arcs *b* and *c*[f]	eff	200–315	1.6[c]	10^{-4}
	β	400–1400	4.4[d]	10^{-3}
	ir	760–2000	3.6[e]	10

[a] ACGIH, 1984.
[b] Welding arc at 400 A and 100 cm from arc to welder's eye.
[c] From Eq. (11.1*b*).
[d] From Eq. (11.1*c*).
[e] From Eq. (11.1*d*).
[f] Welding arc at 200 A and 100 cm from arc to welder's eye.

Based on these estimates, the optical densities (Eq. 11.4) required to shield uv radiation (200–315 nm) for the three welding arcs are 4.81, 4.21, and 4.21, respectively. Required optical densities for shielding from blue-light radiation (400–1400 nm) from the three arcs are 4.25, 3.64, and 3.64, respectively. Infrared radiation appears to pose little or no hazard.

B. Single-pulse laser TLV

Example 11.2 Determine the TLV for a direct intrabeam exposure to a pulse from a 694.3-nm ruby laser having a duration 8×10^{-4} sec (0.8 msec).

solution From Table 11.18, the TLV for a ruby laser with exposure 0.8 msec is 1.8 ($t^{3/4}$) mJ·cm^{-2}. Substituting,

$$H\,(\text{TLV}) = 1.8(8 \times 10^{-4})^{3/4}\ \text{mJ·cm}^{-2}$$
$$= 8.56 \times 10^{-3}\ \text{mJ·cm}^{-2}$$

C. Repetitively pulsed laser TLV

Example 11.3 Determine the direct intrabeam TLV of a 514.5-nm argon laser for a 0.25-sec total exposure T_1 operating at 10^7 pulses per second (f = 10 MHz) with pulse width 10^{-8} sec (T = 10 ns).

solution Determination of the TLV for a repetitively pulsed user requires two analyses. First, evaluate the TLV of an individual pulse corrected for the pulse rate using Figure 11.13. Second, calculate the average irradiance TLV for the entire pulse train and exposure duration. The TLV is then the lower of the two. From Table 11.18, the TLV for a single 10-nsec pulse at 514.5 nm is 5×10^{-7} J·cm^{-2}. From Figure 11.13, the multiplicative correction factor for repetitively pulsed lasers at 10 mHz is 0.06, and the TLV per pulse is:

$$H_i \leq (0.06)(5 \times 10^{-7}\ \text{J·cm}^{-2})$$
$$= 3 \times 10^{-8}\ \text{J·cm}^{-2}$$

The TLV for the total exposure is then:

$$H\,(\text{TLV}) \leq H_i \cdot F \cdot T$$
$$= (3 \times 10^{-8}\ \text{J·cm}^{-2})(10^7)(0.25) \qquad (11.22)$$
$$= 7.5 \times 10^{-2}\ \text{J·cm}^{-2}$$

For the average irradiance calculation, the TLV of a 0.25-sec exposure is shown in Table 11.18 as 1.8($t^{3/4}$) mJ·cm^{-2}, and

$$H\,(\text{TLV average}) \leq 1.8(0.25)^{3/4}\ \text{mJ·cm}^{-2}$$
$$= 6.3 \times 10^{-1}\ \text{mJ·cm}^{-2}$$
$$= 6.3 \times 10^{-4}\ \text{J·cm}^{-2}$$

Since the TLV from the average irradiance calculation is more restrictive, the appropriate TLV is 6.3×10^{-4} J·cm^{-2}. Note that this corresponds to:

$$E = \frac{H}{T} = \frac{6.3 \times 10^{-4}\ \text{J·cm}^{-2}}{0.25\ \text{sec}} \qquad (11.23)$$
$$= 2.5 \times 10^{-3}\ \text{W·cm}^{-2}$$

D. Extended-source laser TLV

Example 11.4 Extended-source calculations apply only in the spectral region of 400–1400 nm (Table 11.19), where the source size is significantly larger than a point. Diffuse reflections are extended sources at close viewing distances. Note that Class 1 and Class 2 lasers are not capable of producing hazardous diffuse reflections, and only direct ocular intrabeam TLVs apply. Class 4 lasers are always capable of producing hazardous diffuse reflections at close viewing distances (ANSI, 1980).

Find the maximum distance (r_1) where the extended-source TLV applies for a He-Ne (633-nm) laser having an emergent beam diameter a = 4 mm, a beam divergence $\phi = 10^{-4}$ radians, and a viewing angle 20° off the central axis (θ_v). A diffuse matte target is placed 100 cm from the beam exit of the laser ($r = 100$ cm in Figure 11.14).

solution As shown in Figure 11.14, the relationship of D_L to the emergent beam divergence and diameter is:

$$\begin{aligned} D_L &= a + r\phi \\ &= 0.4 \text{ cm} + (100 \text{ cm})(10^{-4} \text{ radians}) \qquad (11.24) \\ &= 0.41 \text{ cm} \end{aligned}$$

The viewing angle α is then:

$$\alpha = \frac{D_L \cos \theta_v}{r_1} \qquad (11.25)$$

From Eqs. (11.15) and (11.12):

$$\alpha_{\min} = \frac{d_r}{f} \qquad (11.15)$$

$$\frac{d_r}{f} = \frac{2.44\lambda}{d_p} = \alpha_{\min} \qquad (11.12)$$

$$= \frac{(2.44)(633 \text{ nm})(10^{-7} \text{ cm/nm})}{0.7 \text{ cm}}$$

$$\alpha_{\min} = 2.21 \times 10^{-4} \text{ radians}$$

Substituting for $\alpha_{\min}$ and D_L in Eq. (11.25), r_1 becomes the maximum distance for extended source viewing, and

$$\begin{aligned} r_1 &= \frac{D_L \cos \theta_v}{\alpha_{\min}} \\ &= \frac{(0.41)(0.940)}{2.21 \times 10^{-4}} \\ &= 1744 \text{ cm} \end{aligned}$$

For distances up to 1744 cm, the extended-source TLV will apply (Table 11.19). Note that for distances less than 1744 cm, the light energy on the retina is independent of viewing distance.

E. Protective shielding for laser light

Example 11.5 Determine the optical density for a protective shield for conditions of Example 11.4 with a 50-mW He-Ne laser, 30-sec exposure, and distance from reflecting surface to exposed eye of 1 m. The surface is a flat, white paint.

solution From the previous problem, the source must be treated as an extended source. The brightness (B) of the reflection is given by the product of the irradiance and reflection coefficient (ρ_λ). From Table 11.21, $\rho_\lambda = 0.9$, and

$$\begin{aligned} B &= \frac{50 \text{ mW}}{(\pi/4)(0.41 \text{ cm})^2}(0.9) \\ &= 341 \text{ mW}\cdot\text{cm}^{-2} \end{aligned}$$

For a 30-sec exposure,

$$H = (0.341 \text{ W}\cdot\text{cm}^{-2})(30 \text{ sec}) = 10.2 \text{ J}\cdot\text{cm}^{-2}$$

From Table 11.19, the TLV for a diffuse reflection extended source and 30-sec exposure is:

$$\log T_1 = 1 + 0.02(\lambda\text{–}550)$$
$$= 1 + 0.02(632.8\text{–}550)$$
$$T_1 = 452 \text{ sec}$$

Therefore,

$$H\,(\text{TLV}) = 3.83T^{3/4}\ \text{J·cm}^{-2}\text{·sr}^{-1}$$
$$= 3.83(30)^{3/4}$$
$$= 49.1\ \text{J·cm}^{-2}\text{·sr}^{-1}$$

The solid angle β included in the view is given by:

$$\beta = \frac{A_L \cos\theta_v}{r_1^2}\ \text{sr} \tag{11.26}$$

where A_L is the area of reflection. Substituting into Eq. (11.26):

$$\beta = \frac{(\pi/4)(0.41\ \text{cm})^2 \cos(20°)}{(100\ \text{cm})^2}$$
$$= 1.24 \times 10^{-5}\ \text{sr}$$

The reflected radiant exposure per unit solid angle (H_r) is then:

$$H_r = \frac{H}{\beta}$$
$$= \frac{10.2\ \text{J·cm}^{-2}}{1.24 \times 10^{-5}\ \text{sr}}$$
$$= 8.22 \times 10^5\ \text{J·cm}^{-2}\text{·sr}^{-1}$$

The optical density required to reduce this exposure to the TLV is calculated from Eq. (11.4), and

$$\text{OD} = \log \frac{H_r}{H_{\text{TLV}}}$$
$$= \log \frac{8.22 \times 10^5}{49.1} \tag{11.4}$$
$$= 4.22$$

Note that this laser represents a Class 3 hazard (Table 11.23). Optical densities of laser-protective eyewear are given in Table 11.25.

F. Optical fiber manufacturing

Example 11.6 During the manufacturing of optical fibers, a gas torch (1900–2000°C) traverses a glass tube, causing it to collapse into a single rod. The fiber is subsequently drawn from the rod. Spectral radiance measurements from this process at 1970°C are summarized below (Huggins et al., 1984). Determine the optical shielding requirement for this operation.

Nonionizing Radiations at 1970°C

Wavelength band (nm)		Measured levels	TLV
Ultraviolet	(315–400)	29.9 $\mu W \cdot cm^{-2}$	1.0 $mW \cdot cm^{-2}$
Visible	(380–760)	1110 $cd \cdot cm^{-2}$	1.0 $cd \cdot cm^{-2}$
Blue light	(400–1400)	0.48 $W \cdot cm^{-2} \cdot sr^{-1}$	10.0 $mW \cdot cm^{-2} \cdot sr^{-1}$
Thermal	(400–1400)	12.72 $W \cdot cm^{-2} \cdot sr^{-1}$	31.6 $W \cdot cm^{-2} \cdot sr^{-1}$
Infrared	(>770)	3.79 $mW \cdot cm^{-2}$	10 $mW \cdot cm^{-2}$

solution The specification of optical shielding requires a comparison of measured radiation levels with the appropriate TLVs. From Section I.A, the TLV for the near ultraviolet region (320–400) is 1 $mW \cdot cm^{-2}$. From Section III.A, the TLV for visible light is 1.0 $cd \cdot cm^{-2}$. The TLV for chronic blue-light exposure is calculated from Eq. (11.6*b*). The integrated spectral radiance should not exceed 10^{-2} $W \cdot cm^{-2} \cdot sr^{-1}$ or 10 $mW \cdot cm^{-2} \cdot sr^{-1}$. The TLV for thermal radiation is based on the burn-hazard function and Eq. (11.5). Given a viewing duration of 10 sec and an angular subtense of 0.1 radians, the TLV is 31.6 $W \cdot cm^{-2} \cdot sr^{-1}$. Finally, the TLV for infrared is 10 $mW \cdot cm^{-2}$. Comparing the TLVs to the measured radiations identifies potential optical problems from visible and blue-light exposures. Thermal and infrared exposures are slightly below their respective TLVs, and ultraviolet is well below the TLV. The TLVs are listed next to the measured levels for comparison purposes.

The optical density for the shield is determined from Eq. (11.4). For visible radiation, the optical density required is:

$$\begin{aligned} \text{OD (visible)} &= \log\left(\frac{I_0}{\text{TLV}}\right) \\ &= \log\left(\frac{1110}{1}\right) \qquad (11.4) \\ &= 3.04 \end{aligned}$$

Similarly, the optical density for protection against chronic blue light is:

$$\begin{aligned} \text{OD (blue)} &= \log\left(\frac{480}{10}\right) \\ &= 1.7 \end{aligned}$$

Table 11.8 shows the transmittances of various shades as defined by ANSI (1979). This table shows that a shield with shade 9 would provide the necessary optical density while transmitting only 0.8% of the infrared (1% transmittance = optical density of 2.0). This would also provide adequate protection against chronic blue-light exposure.

G. Microwaves

Example 11.7 Personal exposures to microwave radiation can be controlled by restricting access to areas near the source. This problem defines the reach of the microwave fields and identifies the high-exposure area near a source. For a dipole antenna that radiates 1000 W of 3000-MHz microwaves from a circular parabolic reflector 0.75 m in diameter, calculate:

a. The mean power density at the aperture

b. The maximum power density in the near field
c. The distance to the far field
d. The power density at 250 m
e. The distance at which the power density will be the TLV

solution

a. The average power at the aperture, diameter 0.75 m, is:

$$W_0 = \frac{P}{A}$$
$$= \frac{1000\ \text{W}}{(\pi/4)(0.75\ \text{m})^2}$$
$$= 2263\ \text{W}\cdot\text{m}^{-2} = 226\ \text{mW}\cdot\text{cm}^{-2}$$

b. From Eq. (11.17), the maximum power in the near field is $(4)(W_0)$:

$$W_{\text{max}} = (4)(226\ \text{mW}\cdot\text{cm}^{-2})$$
$$= 905\ \text{mW}\cdot\text{cm}^{-2}$$

c. From Eq. (11.19), the distance to the far field, r_1, is (speed of light $c = f\lambda = 3 \times 10^8$ m/sec):

$$r_1 = \frac{\pi D^2}{8\lambda}$$
$$= \frac{\pi(0.75\ \text{m})^2}{8(3 \times 10^8\ \text{m/sec})/(3000 \times 10^6\ \text{Hz})} \qquad (11.19)$$
$$= 1.77\ \text{m}$$

d. From Eq. (11.18), the mean power density at 250 m (far-field conditions) is:

$$W_{\text{ff}} = \frac{AP}{\lambda^2 r^2}$$
$$= \frac{(\pi/4)(0.75)^2(1000)}{[(3 \times 10^8)/(3 \times 10^9)]^2\,(250)^2} \qquad (11.18)$$
$$= 0.71\ \text{W}\cdot\text{m}^{-2} = 0.071\ \text{mW}\cdot\text{cm}^{-2}$$

e. The TLV for 3000-MHz microwave radiation is 10 $\text{mW}\cdot\text{cm}^{-2}$ (Table 11.28). Substituting the TLV for W in Eq. (11.18) and solving for r:

$$r\ (\text{TLV}) = \left(\frac{AP}{\lambda^2 W_{\text{TLV}}}\right)^{1/2}$$
$$= \left\{\frac{(\pi/4)(0.75\ \text{m})^2(1000\ \text{W})(10^3\ \text{mW/W})}{[(3 \times 10^8\ \text{m/sec})/(3 \times 10^9\ \text{sec}^{-1})]^2\,[(10\ \text{mW})/(\text{cm}^2)]\,(10^4\ \text{cm}^2/\text{m}^2)}\right\}^{1/2}$$
$$= 21\ \text{m}$$

REFERENCES

ACGIH (1976). *A Guide for Control of Laser Hazards.* American Conference of Governmental Industrial Hygienists, Cincinnati, Ohio.

ACGIH (1984). *TLV's, Threshold Limit Values for Chemical Substances and Physical Agents in the Work Environment Adapted by ACGIH for 1984–1985.* American Conference of Governmental Industrial Hygienists, Cincinnati, Ohio.

ANSI (1955). *The Spectral-Transmissive Properties of Plastics for Use in Eye Protection.* American National Standards Institute, New York.

ANSI (1973a). *American National Standard Practice for Industrial Lighting, A11.1-1973.* American National Standards Institute, New York.

ANSI (1973b). *American National Standard Techniques and Instrumentation for the Measurement of Potentially Hazardous Electromagnetic Radiation at Microwave Frequencies, C95.3-1973.* American National Standards Institute, New York.

ANSI (1979). *American National Standard Practice for Occupational and Educational Eye and Face Protection, Z87.1-1979.* American National Standards Institute, New York.

ANSI (1980). *American National Standard for the Safe Use of Lasers, Z136.1-1980.* American National Standards Institute, New York.

Burgess, W. A. (1981). *Recognition of Health Hazards in Industry.* Wiley, New York.

Campbell, D. L. (1976). *Report on Test of Welding Filter Plates.* National Institute for Occupational Safety and Health, HEW Publication NIOSH 76-198.

Cember, H. (1983). *Introduction to Health Physics.* Pergamon Press, New York.

Doyle, L. E., Keyser, C. A., Leach, J. L., Schrader, G. F., and Singer, M. B. (1985). *Manufacturing Processes and Materials for Engineers,* 3d ed. Prentice-Hall, Englewood Cliffs, N.J.

EPA. *Biological Effects of Radiofrequency Radiation* (1984). Elder, J. A., and Cahill, D. F., eds. U.S. Environmental Protection Agency Report No. EPA 600/8-83-026F, September.

EPA (1986). Federal radiation protection guidance; Proposed alternatives for controlling public exposure to radiofrequency radiation. *Fed. Reg., 51,* 27318–27339, July 30.

Eriksen, P. (1985). Time resolved optical spectra from MIG welding arc ignitions. *Amer. Ind. Hyg. Assoc. J., 46,* 101–104.

Huggins, R. G., Testagrossa, P. A., Petersen, R. C., Philen, D. L., and Turnipseed, J. M. (1984). Nonionizing radiation aspects of optical fiber manufacturing. *Amer. Ind. Hyg. Assoc. J., 45,* 796–801.

IES (1981). *Lighting Handbook,* Kaufman, J. E., ed. Illuminating Engineering Society of North American, New York.

Kalpakjian, S. (1984). *Manufacturing Process for Engineering Materials.* Addison-Wesley, Reading, Mass.

Koller, L. R. (1965). *Ultraviolet Radiation.* Wiley, New York.

Largent, E. J., and Olishifski, J. B. (1979). Nonionizing radiation. In: *Fundamentals of Industrial Hygiene,* Olishifski, J. B., ed. National Safety Council, Chicago, pp. 311–368.

Marshall, W. J., Sliney, D. H., Hoikkala, M., and Moss, C. E. (1980). Optical radiation levels produced by air carbon arc cutting processes. *Welding J., 59*(3), 43–46, March.

Moss, C. E., and Gawenda, M. C. (1978). *Optical Radiation Transmission Levels through Transparent Welding Curtains.* National Institute for Occupational Safety and Health, DHEW Publication (NIOSH) 78-176.

Moss, C. E., and Murray, W. E. (1979). Optical radiation levels produced in gas welding, torch brazing and oxygen cutting. *Welding J., 58*(9), 37–46, September.

Moss C. E., and Murray, W. E. (1981). Optical radiation levels produced by single-phase direct arc furnaces. *Amer. Ind. Hyg. Assoc. J., 42,* 293–297.

OSHA (1983). *General Industry Standards Part 1910.* Occupational Safety and Health Administration 2206, rev. March 11.

Rockwell, R. J., and Moss, C. E. (1983). Optical radiation hazards of laser welding processes part 1: Neodymium-YAG Laser. *Amer. Ind. Hyg. Assoc. J., 44,* 572–579.

Schreibis, W. J. (1968). Laser eye protection goggles. *Amer. Ind. Hyg. Assoc. J., 29,* 504.

Sliney, D. H. (1972). Nonionizing radiation. In: *Industrial Environmental Health: The Worker and the Community.* Cralley, L. V., ed. Academic Press, New York, pp. 171–241.

Sliney, D. H., Moss, C. E., Miller, C. G., and Stephens, J. B. (1981). Semitransparent curtains for control of optical radiation hazards. *Appl. Optics, 20*(14), 2352–2366.

Sliney, D. H., Moss, C. E., Miller, C. G., and Stephens, J. B. (1982). Transparent welding curtains. *Welding J., 61*(5), 17–24, May.

Straub, H. W. (1965). Protection of the human eye from laser radiation. *Ann. N.Y. Acad. Sci., 122,* 773–776.

Weast, R. C., ed.-in-chief (1984). *Handbook of Chemistry and Physics.* Chemical Rubber Co., Cleveland, Ohio.

Wilkening, G. M. (1978). Nonionizing radiation. In: *Patty's Industrial Hygiene and Toxicology,* vol. 1, Clayton, G. D., and Clayton, F. E., eds. Wiley, New York, pp. 359–440.

Chapter 12

Heat

E. R. Hermann

Some of the health effects associated with human exposure to excessive heat in work environments have been discussed in Chapter 2. Because humans have the ability, within certain limits, to acclimatize to work at elevated temperatures, the control of heat exposures has an added dimension when compared with other types of stressor control. This capacity to adjust thermally allows people to work efficiently, albeit at reduced metabolic output, in hot environments; and, for short periods, with appropriate recovery, under temperature conditions that would otherwise be hostile to 8-hr employment. An additional consideration, even in hot industries, is the added heat stress experienced by workers during hot weather and heat wave episodes. A major concern is the control of day-to-day heat exposures in high-temperature workplaces, especially when combined with high humidity. Another important consideration is the protection of workers during periods of high ambient temperatures. Consequently, besides discussion of source isolation, insulation and shielding of hot surfaces, and the use of local exhaust ventilation, this chapter will include consideration of humidity reduction, physiological water and salt replacement, and work-rest regimens that are supportive of workers in hot environments. It is important to note that the guidelines introduced in Chapter 4 are specified for the acclimatized worker with recognition of the need for appropriate modification of work output.

To remain healthy and efficient, a person must maintain an internal body temperature near 37°C (98.6°F). When core temperatures exceed 40.6°C (105°F) with the absence of sweating, hyperthermia exists. Death may occur when the core temperature exceeds 42°C (108°F). Thermal homeostasis in the human is achieved within suitable limits by involuntary control of blood circulation from sites of metabolic heat production (large muscles, liver, and deep tissues) to the cooler body surface (skin), where heat is dissipated by evaporative cooling, radiation, convection, and conduction. People react dynamically to alter their loss or gain of heat to or from their environment so as to maintain their core temperature within the critical range.

Energy required to sustain life, maintain bodily function, and perform external work is released in body tissues by exothermic oxidation from carbohydrates, fats, and proteins. The quantity of metabolic heat produced may be estimated by simply measuring the rate of oxygen uptake by a person. Approximately 1 liter of oxygen is required to provide 5 kcal of metabolic heat output. A 70-kg man at rest uses about 0.3 liter of oxygen per minute and produces metabolic heat at the rate of 1.5 kcal/min or 90 kcal/hr.

Variation in the rate of metabolic heat production is relatively small among resting individuals, but wide differences exist for the maximal

TABLE 12.1 Oxygen Uptake, Body Heat Production, and Relative Energy Cost of Work in 70-kg Men

Work activity	Average uptake and metabolic values for work activity: Oxygen uptake (liters/min)	Average uptake and metabolic values for work activity: Body heat production, M (kcal/hr)	Maximum oxygen uptake at different workload categories (liters/min): Low 2.5 — Uptake required, percent of maximum[a]	Medium 3.0	High 3.5
Rest (seated)	0.3	90	12	10	8.5
Light machine work	0.66	200	26	22	19
Walking (3.5 mi/hr on level)	1.0	300	40	33	28
Forging	1.3	390	52	43	37
Shoveling (depends on rate, load, and lift)	1.5–2.0	450–600	60–80	50–66	43–58
Slag removal	2.3	700	92	77	66

[a]Percent of maximum oxygen uptake required at each workload category to achieve the average metabolic output for a particular work activity.

SOURCE: After Mutchler, 1978.

physical effort, and hence metabolic output, of people depending on body size, musculature, physical fitness, and age. Rates of oxygen uptake with corresponding metabolic heat production rates are shown in Table 12.1 for several types of work activity and workload categories.

I. Acclimatization

Acclimatization refers to changes in physiological mechanisms that take place over time when an individual accustomed to working in a cool or moderate environment undertakes work in a hot one. These adjustments occur progressively and reduce the strain experienced at the time of initial exposure to heat. This extended tolerance enables a person to work effectively under conditions that might have been unendurable prior to acclimatization. Work, heat, and time are needed to achieve true acclimatization. Little such tolerance for hot work is acquired unless some physical exertion is involved, but even the best physical conditioning achieved without heat stress does not confer acclimatization. By doing work in heat, however, acclimatization can readily be acquired. Two hours of hard work in heat for four or five days will result in about 80% of the maximum achievable acclimatization for that job and heat stress. Thus, after 1 or 2 weeks a healthy worker can accomplish a work output without difficulty that initially produced signs of significant strain, such as elevated body temperature and pounding heart.

The driving condition that induces acclimatization seems to be sustained elevation of body core and skin temperatures at levels greater than those that would obtain in a cool environment. Workers employed in hot industries achieve degrees of acclimatization commensurate with their average exposure. Unusual increases in work output or environmental heat factors may overload their thermoregulatory capacity and lead to signs of strain. Periodic reinforcement such as occurs daily during the work week is required for retention of heat acclimatization. Workers may lose some acclimatization during an absence of only 2 days; a long weekend without exertion in air conditioned comfort may be quite noticeable. After absences of 2 weeks or longer, the loss of acclimatization is substantial, and almost as much time at work is required to fully reestablish heat tolerance as was needed initially.

Changes in sweat composition occur as part of the acclimatization process. The principal electrolyte in sweat is sodium chloride. In unacclimatized subjects the NaCl concentration is about 3–5 g/kg (about half the concentration present in blood plasma). In heat-acclimatized subjects, sweat becomes more abundant and more dilute. The NaCl concentration is reduced to 1–2 g/kg, which has been reported as reflecting an adaptive change in hormonal balance through secretion of aldosterone, which in turn acts to conserve body salt by both the kidneys and the sweat glands (Minard, 1973).

II. Energy Balance

If heat is not to be stored or lost from the human body, all of the heat exchanges between it and its surroundings must add up to zero. Thus:

$$S = (M - W) \pm C_s \pm C_r \pm C_0 \pm E_s \pm E_r \pm R = 0 \tag{12.1}$$

where S = storage of heat in the body
M = metabolic energy production
W = energy converted into work
C_s = heat exchange by convection at the skin
C_r = heat exchange by convection of the respiratory tract
C_0 = heat exchange by conduction at the skin
E_s = heat exchange by evaporation or condensation at the skin
E_r = heat exchange by evaporation or condensation in the respiratory tract
R = heat gain or loss by radiation

It is not too unusual for body temperatures in health to vary from 37°C by +1°C or −0.5°C. In cases of very heavy work output, as in marathon racing, body temperatures in excess of 41°C (106°F) have been reported. Be this as it may, a heat storage of only 57 kcal in the "standard man" (70 kg) gives a temperature elevation of about 1°C.

The foregoing heat balance equation may be simplified by incorporating the physical work term W into an adjusted metabolic rate M' and lumping the relatively minor exchanges due to C_0, C_r, and E_r with respective terms for C_s and E_s. We can now write:

$$S = M' \pm C \pm E \pm R = 0 \tag{12.2}$$

where M' = metabolic heat adjusted for work output
C = convective and conductive heat exchanges
E = evaporative heat exchanges
R = radiant heat exchanges

These factors may be estimated as functions of air speed, dry- and wet-bulb temperatures, and the globe temperature (discussed in Section II.B). The physical units used in measuring environmental heat exchanges and metabolic human rates of heat production have been various. For the sake of consistency with fuel intake (caloric value of foods) and because time sense regarding duration of heat exposure is most conveniently measured in minutes, kilocalories per minute will be used in expressing the current developmental stage of heat exchange equations. Obviously, resultant estimation in kilocalories per hour or kilocalories per day may be obtained by multiplying by 60 or 1440, respectively.

A. Convection/conduction

Convective heat exchanges are primarily a function of the temperature gradient between the skin, t_s, and ambient air, t_a, and the speed, V, with which air moves past the given skin area, A. In equation form this is:

$$C = A(a + bV^d)(t_a - t_s) \tag{12.3}$$

where a, b, and d are empirical constants. The constant a is a coefficient of natural convection, where the only air movement is that furnished by thermal currents. It may be omitted when bV^d is large. The constant b is the coefficient of forced convection. Empirical development of this equation has led to assignment of values to a, b, and d so that:

$$C = A(0.0325 + 0.1066V^{0.67})(t_a - t_s) \tag{12.4}$$

where C = convective-conductive heat exchange (kcal/min)
A = surface area of body (m^2)
V = air speed (m/sec)
t_s = skin temperature (°C)
t_a = air temperature (°C)

The surface area of the body is often assumed to be 1.8 m^2. An empirical equation for obtaining the surface area of a human being from height and weight measurements is given by:

$$A = 0.7184 \times 10^{-2} W^{0.425} H^{0.725} \tag{12.5}$$

where A is in square meters, W is in kilograms, and H is in centimeters.

The amount of body area that is used in making the convective heat exchange calculation may be somewhat reduced by the clothing worn by the individual engaged in hot work. However, the usual uniform for hot work consists of socks, safety shoes and glasses (if needed), and very light or short pants. The worker's skin is fully wetted with sweat. In chemically hazardous environments, the wearing of full protective clothing may be dictated, in which case convective cooling of a worker approaches zero unless a flow of cool dry air is supplied through the suit.

B. Radiation

Radiation, R, is primarily a function of the difference between the mean radiant temperature of the solid surroundings and the worker's skin temperature. Skin temperature, t_s, under hot work conditions is usually not widely variable; however, the common assumption that it always approximates 35°C may introduce error. The temperature of solid surroundings or mean wall temperature, t_w, may be highly variable, depending on the type of work conditions. In fact, individual wall temperatures in a single work environment during a single day or even only a few minutes may be widely different. The *globe thermometer* developed by Vernon (1930) in his early researches has been used effectively to integrate surrounding wall temperatures into a mean value. The equation for this estimation of t_w is:

$$t_w = [(t_g + t_k)^4 + (0.248 \times 10^9 V^{0.5})(t_g - t_a)]^{0.25} - t_k \tag{12.6}$$

where t_g = globe temperature (°C)
t_a = air temperature (°C)
t_k = 273.17 K
V = air speed at the globe (m/sec)

The globe thermometer consists of a thermometer, thermocouple, or thermistor with its sensing element at the center of a blackened hollow sphere, 0.15 m in diameter (Graves, 1974). If equilibrium conditions between convective loss (gain) and radiant heat gain (loss) by the globe thermometer are attained, valid measurements of t_a, t_g, and V can be used through Eq. (12.6) to provide a good estimate of the mean wall temperature at the globe (which serves as a physical model for a human worker).

Knowing t_w is necessary to determine the rate of heat exchange by radiation between the person and the environment. Most of the experimental work on such exchanges has been with nude or nearly nude people. As indicated earlier, near nudity is the usual situation in real-life hot work. The equation for R is:

$$R = ekFA(T_w^4 - T_s^4) \tag{12.7}$$

where e = emissivity
k = universal radiation constant
F = fraction of surface area involved in the radiant exchange
A = surface area of the body (see Eq. 12.5)
T_w = absolute temperature of the solid surroundings
T_s = absolute temperature of the skin

Emissivity of radiant heat of the skin surface and of enclosing walls are both essentially unity. Except in unusual circumstances, F is also close to unity. Within the temperature range of occupational interest, the difference between the fourth powers of the absolute temperatures

T_w and T_s is equal to the temperature difference multiplied by a constant. Combining all constants gives:

$$R = k_r A(t_w - t_s) \tag{12.8}$$

where k_r = combined coefficient of heat exchange by radiation
t_w = wall or solid surroundings temperature (°C)
t_s = skin temperature (°C)
A = surface area of skin (m^2)

For R in kilocalories per minute, k_r becomes 0.0728 and, using the units above for the other variables, a working expression for R is:

$$R = 0.0728A(t_w - t_s) \tag{12.9}$$

C. Evaporative cooling

Under the condition that the skin is completely wet with sweat, a maximum rate of heat loss provided by evaporation, E_{max}, may be calculated from an equation similar to Eq. (12.3) used for convective heat exchange:

$$E_{max} = A(f + gV^h)(P_a - P_s) \tag{12.10}$$

where f, g, and h are constants similar to a, b, and d; P_a refers to the partial pressure of water at the ambient air humidity; and P_s refers to vapor pressure of water at skin temperature. In Eq. (12.10), the constant f, attributable to diffusional evaporation with little air movement, is usually combined with the predominant factor gV^h into a coefficient of evaporative heat exchange, k_e. Thus we have:

$$E_{max} = k_e AV^h(P_a - P_s) \tag{12.11}$$

A working equation for E_{max} in kilocalories per minute may be written with h assigned a value of 0.63 and k_e = 0.198. Thus:

$$E_{max} = 0.198AV^{0.63}(P_a - P_s) \tag{12.12}$$

where A = surface area of skin (m^2)
V = air speed (m/sec)
P_a = partial pressure of water at humidity of ambient air (mmHg)
P_s = vapor pressure of water at the temperature of the skin (mmHg)

It should be noted that a value of skin temperature is necessary for solving Eq. (12.12) for evaporative heat exchange as well as for Eq. (12.9) and Eq. (12.4) for radiation and convective exchanges. Despite the common assumption of 35°C, the temperature of the skin is *not* constant in heat stress with time or location on the body. Under most heat stress situations, the mean skin temperature will vary between 30°C (86°F) and 37°C (98.6°F) (Peterson, 1977). Actual measurement of mean skin temperature or a better method of estimating it is needed.

D. Affordable metabolism and work

Depending on size, activity, living environment, and other factors, the standard man may typically expend energy at the rate of 2200–3000 kcal/day, which averages only 1.5–2.1 kcal/min. Persons engaged in athletic endeavor and heavy manual labor may have a food input and metabolic output considerably higher. Basal metabolism in the standard man is about 1.0 kcal/min. During extremely heavy work, for short periods of time, a man may have a metabolic rate of 15.0 kcal/min or even higher. An average metabolic rate of 9.0 kcal/min, however, corresponds to very heavy work activity excluding basal metabolism. Thus, for a manual laborer engaged in heavy work 400 min of a day at 9.0 kcal/min plus 1440 min/day of basal metabolism at 1.0 kcal/min and 1040 min/day of light activity equivalent to 1.0 kcal/min above basal metabolism, it may readily be calculated that the total daily metabolic output when working is some 6080 kcal/day. Food input (fuel rate) must match this rate of energy output (power) if there is to be neither loss nor gain of weight. It is obvious why hard workers, athletes, and others who have a high metabolic output rapidly gain weight when they adopt sedentary work habits without drastically reducing their habits of food intake.

With regard to heat stress, if the algebraic sum of radiation R, convection/conduction C, and evaporative cooling E is not sufficient to offset the metabolic heat output of a worker, heat will be stored in the body and his or her temperature will rise. A motivated worker who is unaccustomed to this situation may continue to work at a high rate, with the inevitable induction of heat strain and even heat injury! The seasoned worker's response to overheating is to simply reduce work activity so that $M' + R + C + E = 0$, a natural as well as logical solution. If the sum of $R + C + E$ is not less than -1.0 kcal/min, the worker cannot even afford basal metabolic activity and must either alter the environment or move out of it. To afford moderate to heavy work activity it is essential that the sum of $R + C + E$ be -6.0 to -10.0 kcal/min and preferably somewhat lower.

In a hot, wet industry, dehumidification may be an effective means of reducing heat stress to an acceptable level; in a hot, dry industry, the control method is usually by reduction of radiant heat. Frequently the amount of humidity is temporal. The importance of considering humidity as a major factor in reducing heat stress is exemplified in the following problem.

Example 12.1 The extruder and finishing building of a butyl rubber manufacturing plant has an ambient air temperature of 30°C, a globe temperature of 60°C, air speed of 1.5 m/sec, and relative humidity ranging from 20% to 90%. If the heat exchange area of a worker's body is 1.8 m², what is the maximum metabolic energy (adjusted for work output) that he can afford while working in this building if he is not to suffer a rise in body temperature?

solution Solve for extremes of relative humidity at 20% and 90%. The average wall temperature of the surround from Eq. (12.6) is:

$$t_w = [(60 + 273.17)^4 + 0.248 \times 10^9 \times (1.5)^{0.5}(60 - 30)]^{0.25} - 273.17 \qquad (12.6)$$
$$= 109.5°C$$

a. Assume that skin temperature is 34°C when relative humidity is 20%. Then convection-conduction heat exchange from Eq. (12.4) is:

$$C = 1.8[0.0325 + 0.1066 \times (1.5)^{0.67}](30 - 34) \qquad (12.4)$$
$$= -1.24 \text{ kcal/min}$$

Thermal exchange by radiation (Eq. 12.7) is:

$$R = 0.0728 \times 1.8(109.5 - 34) = 9.89 \text{ kcal/min} \qquad (12.9)$$

The maximum heat loss provided by evaporation of sweat (Eq. 12.12), when P_s = 39.9 mmHg at 34°C and P_a (at saturation) = 31.8 mmHg at 30°C, is:

$$E_{max} = 0.198 \times 1.8 \times (1.5)^{0.63}(31.8 \times 0.20 - 39.9) \qquad (12.12)$$
$$= -15.43 \text{ kcal/min}$$

Since storage of heat is to be zero (no rise in body core temperature),

$$S = M' + C + R + E = 0 \qquad (12.2)$$

and

$$M' = 1.24 - 9.89 + 15.43 = 6.8 \text{ kcal/min}$$

which on a continuous basis permits moderate work activity with occasional short periods of heavy work output without developing heat strains.

b. Assume that skin temperature is 36°C when relative humidity is 90% (less evaporative cooling, greater blood flow through the skin).

$$C = 1.8[0.0325 + 0.1066 \times (1.5)^{0.67}](30 - 36) \qquad (12.4)$$
$$= -1.86 \text{ kcal/min}$$
$$R = 0.0728 \times 1.8(109.5 - 36) = 9.63 \text{ kcal/min} \qquad (12.9)$$
$$E_{max} = 0.198 \times 1.8 \times (1.5)^{0.63}(31.8 \times 0.90 - 44.6) \qquad (12.12)$$
$$= -7.35 \text{ kcal/min}$$

$$M' + C + R + E = 0$$
$$M' - 1.86 + 9.63 - 7.35 = 0$$

Therefore,

$$M' = -0.42 \text{ kcal/min}$$

a situation indicating that the worker cannot even afford basal metabolism without developing heat strains.

The reader is left to work out how much relative humidity would permit sitting metabolism with no physical work activity.

The reactive physiological response to heat stress is heat strain. The circulatory system and sweating mechanism carry the burden of balancing heat gain and heat loss so as to prevent elevation of body temperature. With respect to hot environments, several indices of heat strain have been identified and studied. Although many physiological parameters could qualify as heat strains by being altered in response to heat stress, the more common ones are sweat rate, sweat evaporation rate, skin temperature, heart rate, and core temperature. No single one of these provides complete information concerning other heat strains. Also, correlation among various measures of strain is far from perfect. The best measure of overall heat strain is probably sweat rate.

Sweating occurs over a wide range of thermal exposure in quantities just sufficient to provide enough evaporative cooling E to balance the net heat load $M' + R + C$ in Eq. (12.2). In a typical human being, some 2.5 million sweat glands may secrete more than 3 kg of sweat in 1 hr under conditions of maximum heat stress. Thoroughly acclimatized men can maintain sweat rates of 1.0–1.5 kg/hr for several hours (Mutchler, 1978). Approximately 580 kcal of body heat may be eliminated by evaporation of 1 liter of sweat at skin temperatures ranging from 27 to 37°C. The efficiency of cooling by sweat depends on the rate of actual evaporation E from the skin surface (Eq. 12.12).

When evaporation of sweat is restricted, the skin temperature t_s rises above that resulting from less humid conditions at the same air temperature t_a. Input data from the entire body is gathered, integrated, and fed into the heat control center (hypothalamus). (The hypothalamus reacts to maintain a person's core temperature at a constant level regardless of metabolic rate, within limits.) This gland then responds by recruiting more sweat glands, thereby increasing the fraction of wetted body surface. Sweat production in excess of that which can be evaporated from the skin is essentially wasted as a coolant, because the liquid perspiration drips off the body. Under even more humid conditions (higher P_a) with less evaporation, body heat is not dissipated, thus raising the core temperature and the temperature of the skin t_s. In response the sweat rate will be increased. If the water vapor pressure gradient between the new skin temperature and the air temperature is sufficient to provide adequate evaporative cooling, a new heat balance may be achieved with little rise in body temperature. If, however, the humidity is so high that free evaporation of sweat from the wetted skin will not take place, sweat will be wasted and a substantial increase in body temperature will arise. Persistence of such conditions with time produces a serious thermoregulatory strain. Thus, sweat rate during free evaporation varies in proportion to heat load ($M' + R + C$). When sweat production exceeds that needed to just wet the body surface exposed to evaporation, the sweat rate exceeds the evaporative capacity of the system and increases as a function of the core temperature and the skin temperature.

Sweat rate is therefore not only an index of *heat stress* over the entire range of compensation but also an index of *heat strain* in the zone of time-limited compensation. In the latter case sweat rate rises in parallel to deep body temperature and heart rate. Sweat rate, a heat strain, may be used to serve as an index of time-weighted average of heat stress. It is measured by the difference in body weight over a given time period, corrected for weight gained by water and food intake and weight lost in urine, feces, and respiration (Mutchler, 1978).

E. Heat stress index

One method for determining allowable heat stress based on the wet-bulb globe temperature index and workload is given in Chapter 4. Another approach is determination of the heat stress index (HSI). If there is to be no gain or loss of heat from the human body under hot work conditions, the required cooling (E_{req}) to be provided by evaporation of sweat must equal the algebraic sum of the heat load factors, $M' + R + C$. Thus:

$$E_{req} = M' + R + C \tag{12.13}$$

The HSI is the ratio of E_{req} and the *maximum* amount of cooling *afforded* by the given environment, E_{max} (Belding and Hatch, 1955). Expressed on a scale of 100:

$$\text{HSI} = 100\left(\frac{E_{req}}{E_{max}}\right) \tag{12.14}$$

The values of E_{req}, E_{max}, M', R, and C are all in kilocalories per minute and are applicable to seminude workers wearing only shorts and footwear. For workers wearing shirt and trousers, the heat exchange values of R, C, and E_{max} should be multiplied by a factor of 0.60. In other words, the heat exchange rates are reduced by about 40%. Various coefficients have been assigned to these heat exchange equations by different authors. The refinements presented here, though current, should not be regarded as absolute. Care should be taken that all factors are measured accurately, especially skin temperature and the vapor pressure associated with it. Interpretations of numerical values obtained by the Belding and Hatch HSI are shown in Table 12.2.

The heat stress index provides an engineering analysis of occupational heat exposure, and two factors entering into it may be used to estimate the amount of time required to raise the body temperature of a standard man 1.11°C (2°F) when the heat load $M' + R + C = E_{req}$ is greater than the evaporative cooling afforded by the actual environment E_{max}. Because a permissible temperature elevation of 1.11°C corresponds to a heat storage of 63 kcal (250 Btu or 263,700 J), the allowable exposure time (AET) may be equated as follows:

$$\text{AET} = \frac{63}{E_{req} - E_{max}} \tag{12.15}$$

in which AET is in minutes and E_{req} and E_{max} are in kilocalories per

TABLE 12.2 Evaluation of Values in Belding and Hatch Heat Stress Index

Index of heat stress (HSI)	Physiological and hygienic implications of 8-hr exposures to various heat stresses
−20 −10	*Mild cold strain.* This condition frequently exists in areas where people recover from exposure to heat.
0	*No thermal strain.*
+10 20 30	*Mild to moderate heat strain.* For a job that involves higher intellectual functions, dexterity, or alertness, subtle to substantial decrements in performance may be expected. In performance of heavy physical work, little decrement is expected unless ability of individuals to perform such work under no thermal stress is marginal.
40 50 60	*Severe heat strain,* involving a threat to health unless people are physically fit. Break-in period required for workers not previously acclimatized. Some decrement in performance of physical work is to be expected. Medical selection of personnel desirable because these conditions are unsuitable for those with cardiovascular or respiratory impairment or with chronic dermatitis. These working conditions are also unsuitable for activities requiring sustained mental effort.
70 80 90	*Very severe heat strain.* Only a small percentage of the population may be expected to qualify for this work. Personnel should be selected (*a*) by medical examination and (*b*) by trial on the job (after acclimatization). Special measures are needed to assure adequate water and salt intake. Amelioration of working conditions by any feasible means is highly desirable and may be expected to decrease the health hazard while increasing efficiency on the job. Slight "indisposition" that in most jobs would be insufficient to affect performance may render workers unfit for this exposure.
100	*Maximum strain* tolerated daily by fit, acclimatized young men.

SOURCE: Belding and Hatch, 1955.

minute. Obviously the useful range of this equation is from about 5 to 500 min and the numerator, 63 kcal, should be adjusted for individuals whose weight is much greater or smaller than 70 kg. Actually the use of heat gain in a 70-kg "standard man" may not be reasonable, inasmuch as the U.S. male typically now weighs about 80 kg and his female counterpart weighs about 60 kg. Thus the numerator in the AET equation might be 72 for males and 54 for females neglecting several physiological factors other than mass. Similarly, if environmental data are available for the rest area, which presumably is much cooler than the work area, the minimum recovery time (MRT) in minutes for "cooling out" during a work-rest regimen may be estimated from:

$$\text{MRT} = \frac{63}{E_{\text{max}} \text{ at rest} - E_{\text{req}} \text{ at rest}} \qquad (12.16)$$

Again the 63 kcal in the numerator is for a 70-kg "standard man," and this number might be replaced by 72 and 54, respectively, for U.S. men and women.

III. Engineering Control

Ideally consideration of control measures should accompany preliminary planning of new plants or facilities in which it is anticipated that workers will encounter heat stress. The health engineer or industrial hygienist should establish procedures with top management so that he or she is consulted as soon as any consideration is being given to instituting a new manufacturing process or product line. It is generally much cheaper and more effective to control occupational health and safety problems through proper planning and design rather than waiting to rectify problems after they occur. Several factors relating to worker activity and the process environment are important:

1. Physical demands and bodily heat production
2. Number and duration of heat exposures
3. Clothing requirements with respect to safety hazards other than heat
4. Environmental conditions of temperature, humidity, and air speed in relation to heat exchange components
5. Thermal condition of the rest area if heat stress is so severe that a work-rest regimen is required

An outline of alternatives that may be employed in controlling heat stress is shown in Table 12.3 (Brief and Confer, 1973).

TABLE 12.3 Alternative Measures for Controlling Heat Stress

Item	Possible actions
1. Heat components	Reduce by
M'	Mechanization of some or all tasks Sharing workload with others (particularly during peak heat periods) Increasing rest time
R	Reduce by Insulating furnace walls Using reflective screens Wearing reflective aprons or clothing Covering exposed parts of body
C	If air temperature is above skin temperature, reduce C by Lowering air temperature Lowering air speed Wearing clothing If air temperature is below skin temperature, reduce C by Lowering air temperature Increasing air speed Removing clothing
E_{max}	Increase by Increasing air speed Decreasing humidity
2. Work schedule	Duration Shorten duration of each exposure Use more frequent rest periods Recovery Use nearby air conditioned space for rest area Adjust V in rest area for effective cooling Other Allow worker to self-limit exposure on basis of signs and symptoms of heat strain Provide cool, potable water containing 0.1% salt
3. Clothing	For extreme conditions, use cooled clothing (by vortex tube or other means) Wear type of clothing to obtain $E_{max} > E_{req}$ with minimum sweating

SOURCE: Brief and Confer, 1973.

A large fraction of the total heat load is generally contributed by the metabolic heat M'. Furthermore, reduction of this factor is limited by basal metabolism at one end and maximum work output at the other. Even a sedentary worker of medium size who is pushing buttons will produce excess heat at the rate of 1.7 kcal/min, whereas one who is manually transferring heavy materials at a steady pace will tend to limit his metabolic rate to about 5.0 or 6.0 kcal/min (TCE, 1971). Thus control measures such as partial mechanization may reduce the metabolic rate for steady work by only about 4.3 kcal/min. However, if only observation of a process is required much of the time, mechanization may make it possible to isolate the worker from the heat, perhaps in an air conditioned cab or booth.

High-energy tasks such as shoveling, which result in heat production at rates up to 8.0 or 10.0 kcal/min, require that rest be taken one-half to two-thirds of the time simply because of excessive physical demands. Thus the hourly contribution to heat load may be cut to 300 kcal (5 kcal/min). Mechanization of such work is desirable and often is cost-effective.

Many of the critical exposures to heat in industry are at irregular intervals, such as emergency repairs of hot equipment or relining furnaces. In these cases the level of heat stress and physical effort are high and unpredictable. Such exposures usually force a progressive rise in body temperature. Ideally, body temperature and heart rate should be monitored and used as criteria for limiting such exposures on an ad hoc basis; and new low-cost instrumentation is becoming readily available for such use. In past practice, however, tolerance limits have been based on the experience of the worker as well as of the supervisor.

As has been presented, heat stress is affected by clothing. Conventional work clothes interfere substantially with heat loss by convection, radiation, and evaporation. A man doing moderately hard physical work (5.0 kcal/min) and wearing only shorts, socks, and shoes has a comfort temperature of about 20°C. In conventional work clothes, his ambient comfort temperature might well be 13°C. If the ambient temperature was 20°C, the energy cost of wearing conventional work clothes would be equivalent to an added sweat rate of 240 ml/hr.

Special personal protection equipment and clothing are required for very hot work. Oversized, insulated gauntlets are used for handling hot materials. Infrared-reflecting face shields are necessary for performing visual tasks in high radiant heat. Insulated suits that provide full-body protection may be obtained for the manual work required in such hot jobs as relining furnaces. Textiles used in manufacturing cloth for these suits are now made from highly heat-resistant synthetic fibers instead of asbestos with its related health hazards. The protection afforded by such suits is enhanced by aluminizing the surface and sometimes by interlining with foil between insulating cloth layers.

In instances where reduction of overall temperature and humidity to an acceptable level is technically or financially impractical and not enough skilled personnel are available to complete a job using a work-rest regimen, air conditioned suits may be employed. These are usually made of pervious material supplied with a generous flow of cooled air. Cooling of the air may be accomplished by one of several means. One method is to force the air through a coil of conductive metal tubing immersed in an ice bath or routed through a small freezer. The flow of air delivered to the suit through a flexible tube may range from 15 to 200 ft^3/min, and the temperature should be about 20°C. Air cooling may also be provided by using a vortex tube that separates compressed air at 100 lb/in.^{2} and about 38°C into a cold air stream and a hot air stream. The air flow requirement for a one-person vortex tube is about 25 ft^3/min, of which some 75% is converted into 5°C air for distribution in the ventilated suit and 25% is exhausted as air at a high temperature (up to 130°C). Both equipment and power costs are appreciable for such devices, and the hot air exhaust of the vortex tube may exacerbate heat problems in a confined space unless temperatures already exceed 130°C.

Self-contained sources of conditioned air have also been developed. A liquid refrigerant, sealed in a finned container, is cooled in a deep freeze. The container is placed in a backpack, and a small rechargeable battery-driven fan circulates air across the fins into the suit. A single cooling and charging may provide tolerable conditions for several minutes up to 1 hr while relining a furnace wall.

Cold vests containing frozen water or carbon dioxide (Dry Ice) have also been employed in somewhat less severe hot work such as gold mining. When using these vests the worker's skin must be protected from "freezer burn" by appropriate clothing on the inside, and the vest must be insulated from external heat exchange and abrasion on the outside. Utilization of the optimal cooling or heat reduction system for a particular job or type of work provides a challenge to both creative technical engineering and engineering judgment. Good engineering control of heat stress requires thorough consideration of human factors, job conditions, work requirements, the physics of heat transfer, available technology, economics, and the effects of nuisance factors.

An engineering solution to the problem of heat stress entails the same types of control strategy applied to other environmental stressors such as airborne gases, vapors, fumes, and dusts. As with these, alternative engineering control methods include substitution, control at the source, localized reduction, isolation, and general ventilation.

The most fundamental approach to engineering control of heat is to limit it at its point of generation. If this can be done at the time the operation or process is developed, it may eliminate a later need to change the operation or substitute a process component that is less wasteful of heat. One example of heat reduction at the source is the selection of induction heating rather than direct-fired furnaces for certain forging operations.

A. Control at the source

The options for control of heat at the source are selection of the most heat-efficient process for making the desired product, isolation, reduction in emissivity, insulation, radiation shielding, and local exhaust ventilation. Process alteration has been discussed in Chapter 7. Regarding source control, it is important that the health engineer or industrial hygienist be consulted early in the initial planning stages of developing, constructing, and adopting any new manufacturing line or process.

1. Isolation. If considered in the early planning and design stages, a very practical method for limiting exposure of workers to excessive heat is to locate hot processing equipment and operations away from spaces occupied by workers. Such operations may be partitioned and separated from the rest of the facility, located in a separate building, or located outdoors with minimal shelter. For example, industrial furnaces and boilers are nearly always segregated from other operations in the same facility.

2. Reduced emission from hot surfaces. The amount of thermal radiation absorbed or emitted by a surface depends on the absorptivity and emissivity of the surface. Consider two small bodies of areas A_1 and A_2, placed in an evacuated enclosure (no conduction or convection). Then, when the system reaches thermal equilibrium, the bodies will emit radiation at the rates A_1W_1 and A_2W_2, where W is the total emissive power, energy per unit time per unit area of the surface, emitted through the hemisphere above each element of surface. Let the energy impinging on a unit area of any small body in the enclosure due to radiation from the wall be I. If the bodies have absorptivities (fraction of incident radiation that is absorbed) of α_1 and α_2, then energy balances on the bodies will be

$$IA_1\alpha_1 = A_1W_1 \quad \text{and} \quad IA_2\alpha_2 = A_2W_2 \qquad (12.17)$$

from which

$$\frac{W_1}{\alpha_1} = \frac{W_2}{\alpha_2} = \frac{W_x}{\alpha_x} \qquad (12.18)$$

(where x indicates characteristics of any body). This generalization, that the ratio of the emissive power of a surface to its absorptivity is the same for all bodies, is known as Kirchoff's law. Since α cannot exceed unity, Kirchoff's law places an upper limit on W called W_B (for black body). Any surface having this upper limiting emissive power is designated a perfect radiator. Since such a surface must have an absorptivity of unity and therefore a reflectivity of zero, the perfect radiator is commonly referred to as a black body. The ratio of the emissive power of an actual surface to that of a black body is called the emissivity. Kirchoff's law may be restated that at thermal equilibrium the emissivity and absorptivity of a body are the same.

Emissivity and hence the amount of heat radiated from a hot source can be controlled by treatment of the surface. Reducing the emissivity of a hot source not only helps alleviate heat stress problems but conserves energy. Accomplishing both is usually desirable. The emissivity of a hot source can be lowered by polishing, painting, or sheathing. The reflectance and emissivity of material surfaces are complementary. These factors are listed in Table 12.4 for a variety of shielding materials. The heat emission properties of a coating or shielding material must be considered in conjunction with its aging, cost, and even insulative qualities. Conservation of heat energy so as to avoid elevated temperatures in work areas is illustrated by the following example.

Example 12.2 A hot chemical process vessel has an iron outer surface area of 12 m^2. Under operating conditions the average temperature of the vessel surface is 190°C. The work environment surrounding the vessel is at 30°C. At what rate will radiant heat energy be conserved if an aluminum foil sheath is installed on the vessel surface?

solution Modify Eq. (12.7) to show the difference in emissivity of two surfaces. Thus:

$$\Delta R = (e_1 - e_2)kFA(T_w^4 - T_s^4) \qquad (12.19)$$

TABLE 12.4 Emissivity and Reflectance Factors for Common Shielding Materials

Material surface	Emissivity of radiant heat (%)	Reflectance of radiant heat (%)
Bright aluminum	5	95
Polished aluminum	8	92
Polished tin	8	92
Bright zinc	10	90
Oxidized aluminum	16	84
Varnished aluminum	20	80
Varnished tin	20	80
Oxidized zinc	27	73
Aluminum paint, new, clean	35	65
Sheet iron, smooth	55	45
Aluminum paint, dull, dirty	60	40
Sheet iron, oxidized	65	35
Steel and iron	80–90	20–10
Brick	80–96	20–4
Black lacquer	90	10
Wood	92–96	8–4
Asbestos board	92–96	8–4
Oxide paints, all colors	94	6
Flat black lacquer	97	3
Perfect black body	100	0

where: e_1 = 0.85 for iron (Table 12.4)

e_2 = 0.16 for oxidized aluminum (only 0.05 when new foil first applied) (Table 12.4)

$k = 0.081 \times 10^{-8}$ kcal/m²·min·K⁴

$F = 1.00$

$A = 12$ m², area of radiant flux

$T_w = 190 + 273 = 463$ K, temperature of vessel surfaces

$T_s = 30 + 273 = 303$ K, temperature of ambient surfaces in work area

$$\Delta R = (0.85 - 0.16)0.081 \times 10^{-8} \times 1.00 \times 12[(463)^4 - (303)^4]$$
$$= 252 \text{ kcal/min conserved} \quad \text{or} \quad 60 \times 10^3 \text{ Btu/hr} \qquad (12.19)$$

3. Insulation. Thermal insulation reduces heat transfer from hot equipment to the work environment by reducing thermal conductivity. Materials or combinations of materials that have air- or gas-filled pockets or void spaces that retard the transfer of heat with reasonable effectiveness are poor thermal conductors and hence good thermal insulators. Such materials may be particulate, fibrous, or cellular, singly or in combination, with or without binders, or they may be assembled, such as multiple heat-reflecting surfaces that incorporate air- or gas-filled voids.

The heat transfer through a simple or complex material may be expressed in terms of its thermal conductivity (for unit thickness) or conductance (for a specific thickness). Good thermal insulators have low values of thermal conductivity or conductance.

Thermal insulation is produced from a great number of substances in various forms, sizes, shapes, and thicknesses. These insulative materials are available as blankets, felts, batts, blocks, cements, loose fill, and foams that may be flexible, rigid, semirigid, preformed, or foamed in place. They are often combined with foils or thin sheets of heat-reflective metals such as aluminum, nickel, and stainless steel. Most commonly used materials fall within the following categories:

Fibrous or cellular (mineral): alumina, asbestos, glass, perlite, rock, silica, slag, or vermiculite

Fibrous or cellular (organic): cane, cotton, wood and wood bark (cork)

Cellular organic plastics: elastomer, polyisocyanate, polyisocyanurate, polystyrene, and polyvinyl acetate

Cements: insulating and/or finishing

Thermal conductivity values may be obtained from engineering handbooks and manufacturers' literature (e.g., Perry et al., 1984; ASHRAE, 1985). A good overview of typical published values is given in Figure 12.1 Depending on the type of insulation, the thermal conductivity (K factor) can vary with age, manufacturer, moisture content, and temperature. The mean temperature on the abscissa of Figure 12.1 is the arithmetic mean of the operating temperature on both sides of the insualting material. The rate at which heat is lost from a system is calculated from:

$$Q = \frac{KS\theta}{t} \tag{12.20}$$

where Q = heat transfer rate, typically in Btu/hr or kcal/hr
K = thermal conductivity, in (Btu·in.)/(hr·ft^2·°F), (kcal·m)/(hr·m^2·°C), or (kcal·cm)/min·m^2·°C)
S = surface area, in ft^2 or m^2
θ = temperature difference across insulation, in °F or °C
t = thickness of insulation, in in., ft, or m

Units of measurement must be consistent with those used in the thermal conductivity factor K. Multiply K values given in (Btu·in.)/(hr·ft^2·°F) by 0.1240 to obtain (kcal·m)/(hr·m^2·°C) or by 0.2067 to obtain (kcal·cm)/(min·m^2·°C). One Btu/hr = 4.2×10^{-3} kcal/min.

Actual system heat loss to the environment will normally exceed calculated values because of projections, seams, expansion-contraction openings, moisture, construction imperfections, and material damage. Heat and fire resistance of materials varies widely, and appropriate insulation must be used for the temperatures involved. Autoignition can occur if combustible fluids are absorbed by insulations having a wicking action. Under actual fire conditions, calcium silicate, cellular glass, glass fiber, and mineral wool are all fire-resistant but do not perform equally well. A steel jacket provides protection, but aluminum does not. Additional information on application of thermal insulation can be obtained from Perry et al. (1984). However, some idea of the amount of heat that may be conserved by a simple application of an insulating material is provided by the following example.

Example 12.3 A high-pressure steam boiler operating at 1543 lb/in.^{2} absolute pressure is to be insulated so that the external jacket temperature does not exceed 120°F in a furnace room at about 80°F. It is found that 2.5 in. of mineral wool insulation will hold up at the temperatures given and will meet performance specifications. The boiler has a surface area of 70 ft^2. At what rate would heat be lost from the insulated boiler in comparison to the same boiler without insulation? The boiler shell is 1.375-in.-thick steel with a thermal conductivity K of 9.0 (Btu)/[(hr·ft^2)(°F/ft)].

solution

a. Heat loss by conduction-convection from uninsulated boiler: For consistency, convert the conductivity of boiler shell to (Btu·in.)/(hr·ft^2·°F):

$$9.0 \times 12 = 108 \text{ (Btu·in.)/(hr·ft}^2\text{·°F)}$$

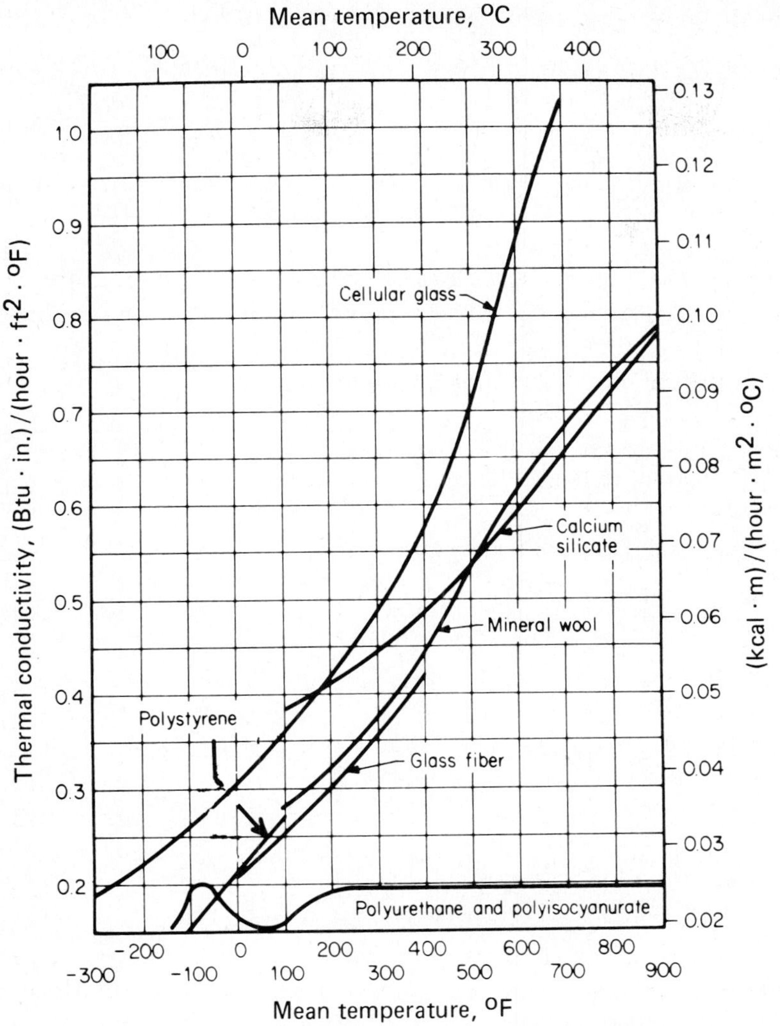

Figure 12.1 Thermal conductivity of several insulating materials (*Perry et al., 1984*).

Temperature of steam at 1543 lb/in.^{2} = 600°F [from steam tables, e.g., in Perry et al. (1984)].

Heat transfer (from Eq. 12.20):

$$Q = 108 \times \frac{70(600 - 80)}{1.375}$$

$$= 2.86 \times 10^6 \text{ Btu/hr} \quad \text{or} \quad 12 \times 10^3 \text{ kcal/min} \tag{12.20}$$

b. Heat loss from insulated boiler: The insulative property of the steel shell is negligible in comparison to 2.5 in. mineral wool. Assume that the temperature at the inside surface of the insulation is 600°F, and at the outside surface is 120°F (given). The mean is 360°F. Thermal conductivity for mineral wool at this temperature is 0.42 (Btu·in.)/(h·ft^2·°F) (Figure 12.4).

Heat transfer to room by conduction-convection (Eq. 12.20):

$$Q = 0.42 \times \frac{70(600 - 80)}{2.5}$$

$$= 6.1 \times 10^3 \text{ Btu/hr} \quad \text{or} \quad 26 \text{ kcal/min} \qquad (12.20)$$

About 470 times as much heat would be lost from the uninsulated boiler as from the insulated boiler by means of conduction and subsequent convection.

In addition to heat losses by conduction-convection, heat may be transferred to the surrounding environment by radiation. A comparison between the two modes of heat loss may be made using the uninsulated boiler from the preceding problem.

Example 12.4 Consider the same boiler and operating conditions as in Example 12.3. Find the thermal radiation to the boiler house if the uninsulated boiler shell has a reflectance of 10%.

solution Determine data for all factors in Eq. (12.7):

$$e = 1 - \text{reflectance} = 1 - 0.10 = 0.90$$

$$k = 0.081 \times 10^{-8} \text{ kcal/m}^2\cdot\text{min}\cdot\text{K}^4 \text{ or } 0.17 \times 10^{-8} \text{ Btu/ft}^2\cdot\text{hr}\cdot{}^\circ\text{R}^4$$

$$T_w \text{ (boiler wall)} = 1060^\circ\text{R}$$

$$T_s = 540^\circ\text{R}$$

$$F = 1.00$$

$$A = 70 \text{ ft}^2$$

Thus,

$$R = 0.90 \times 0.17 \times 10^{-8} \times 1 \times 70[(1060)^4 - (540)^4]$$

$$= 126 \times 10^3 \text{ Btu/hr} \quad \text{or} \quad 530 \text{ kcal/min} \qquad (12.7)$$

which is very small compared to conductive-convective heat losses from the uninsulated boiler, but large when compared to the insulated boiler. The use of reflective foils as discussed in the following section may further reduce heat losses.

4. Radiation shielding. The characteristics of thermal radiation are quite different from those of thermal conduction and convection, which depend on molecular contact. The radiant form of heat energy passes through space without heating the matter occupying that space unless the matter is capable of absorbing it. Absorption by air is negligible. Consequently, air is not heated significantly for other than climatological considerations. In a work environment, shielding of workers from radiant heat involves placing a barrier between the worker and the source of radiant energy. Radiation shielding may involve both reflection and/or absorption. For certain uses such shielding must be transparent or flexible.

Reflective shields may be constructed from foils or sheets of aluminum, stainless steel, or other bright-surfaced metallic materials. Aluminum offers the advantage of 85–95% reflectance (Table 12.4) and as a foil may economically be combined with insulative materials. The successful use of reflective shielding requires an understanding of certain principles. The bright reflecting surface should be exposed directly to the incident radiant energy; it should not be imbedded in other materials. Shield surfaces should not be painted; they should be kept clean. When shields are mounted on the wall of a furnace or oven, they should be spaced a few inches away from the wall by brackets. The space between the shield and the furnace wall creates a chimney effect, removing heat from the work area. Corrugated shield sheeting should be installed so that the corrugations are vertical, not horizontal, in order to minimize turbulence in the "chimney" flow and reduce collection of foreign matter on the surfaces.

5. Composite shielding and insulation. Predicted radiant emissions for a single-foil shield and various reflectances, placed in 21°C and 43°C environments, are shown in Figures 12.2 and 12.3, respectively (Takata and Gilligan, 1978). Interpolation may be used to obtain values at intermediate environmental temperatures.

Five basic heat shield sections are shown in Figure 12.4. Predicted surface temperatures and radiant heat fluxes for a multiple-foil shield arrangement in which two foils are separated by a 6.4-mm air gap are given in Figures 12.5 and 12.6 for environmental temperatures of 21°C and 43°C. Reflective shielding can be combined with insulation to provide an even more effective barrier to the transfer of thermal radiation into unwanted spaces. The results of calculations for a heat barrier consisting of 5 cm of insulation having an overall thermal conductance of 3.9×10^{-5} cal/cm²·sec·°C are shown in Figures 12.7 and 12.8. It should be noted that the addition of insulation becomes quite advantageous as reflectance decreases. The latter situation is one that takes place naturally on an originally bright reflective barrier due to soiling and oxidation. The foregoing information may be applied quantitatively in the engineering of reflective-insulative thermal barriers.

Example 12.5 The intensity of thermal radiation onto a shield placed around a pottery kiln was found to be 300 kcal/m²·min. The surrounding environment has an air temperature of 27°C. The shield is a composite of 5-cm conductive insulation having a conductance of 3.9×10^{-5} cal/cm²·°C, with reflecting foils attached to both front and rear surfaces. Emissivity of each foil is 0.90. Predict the radiant emission rate from the cool side of the shield and the temperature at the front and rear surfaces.

solution Read estimated values from Figures 12.7 and 12.8 and make any necessary interpolations.

$$I_{IN} = 300 \text{ kcal/m}^2\cdot\text{min} = 0.5 \text{ cal/cm}^2\cdot\text{sec}$$

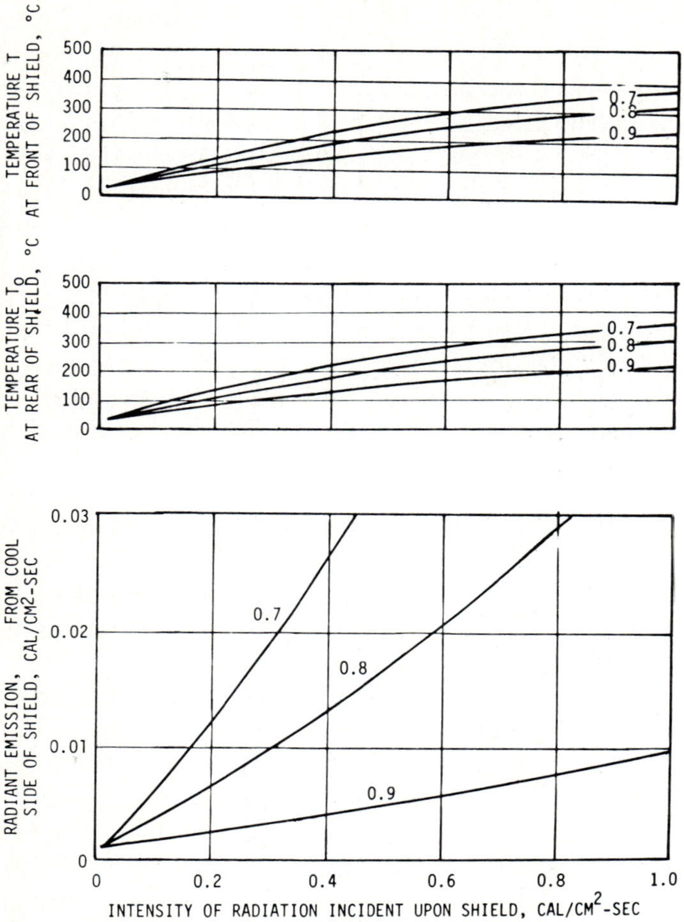

Figure 12.2 Predicted surface temperatures and radiant heat fluxes for single-foil shields in 21°C environment (*Takata and Gilligan, 1978*). Reflectances of exterior shield surfaces are given on curves.

From the ordinate of Figure 12.7, for 21°C:

I_E = 0.002 cal/cm²·sec

T_0 = 80°C

T = 220°C

From the ordinate of Figure 12.8 for 43°C:

I_E = 0.0025 cal/cm²·sec

T_0 = 100°C

T = 250°C

Assuming that the results obtained for radiant emission I_E and temperatures T_0

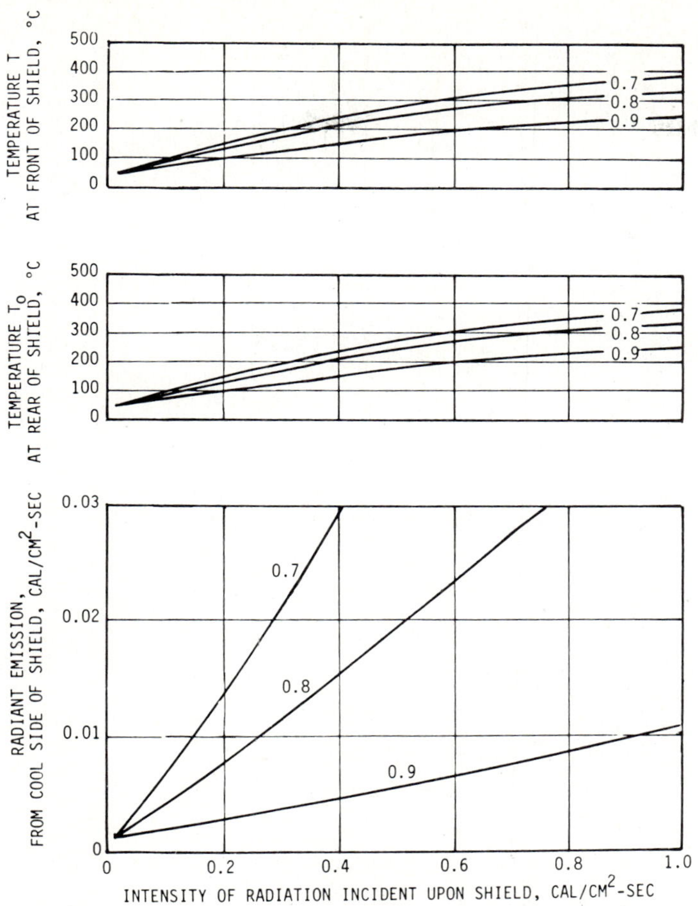

Figure 12.3 Predicted surface temperatures and radiant heat fluxes for single-foil shields in 43°C environment (*Takata and Gilligan, 1978*). Reflectances of exterior shield surfaces are given on curves.

and T are essentially linear between the 21°C and 43°C environments, an arithmetic interpolation for a 27°C environment gives:

$$I_E = 0.002 + 0.0005 \times \tfrac{6}{22}$$
$$= 0.00214 \text{ cal/cm}^2\text{·sec}$$

which converts to 1.28 kcal/m²·min, a 99.6% reduction relative to the incident radiation I_{IN}.

$$T_0 = 80 + 20 \times \tfrac{6}{22} = 86°\text{C}$$

and $$T = 220 + 50 \times \tfrac{6}{22} = 234°\text{C}$$

The reader should compare the results with those expected from single- and double-foil shields as given in Figures 12.5, 12.6, 12.7, and

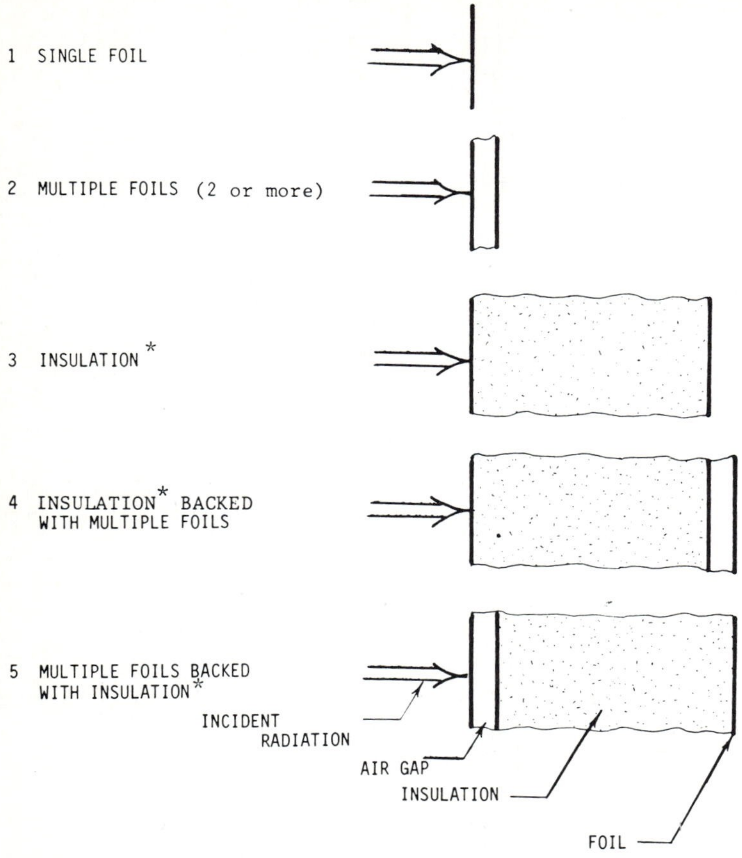

Figure 12.4 Heat shield configurations. (*Foils are attached to both surfaces of insulation.)

12.8. For actual installations the engineering economics of amortizing the costs of installing and maintaining shielding will have to be compared with the costs of waste heat plus installation, maintenance, and operation of cooling and ventilating equipment if needed.

Infrared radiation is readily absorbed by dull, dark-colored surfaces. Absorptive shielding is typically constructed of two or three metal sheets separated by air spaces. Heat may be removed by air convection (chimney effect) or by running water between the metal plates to remove heat by conduction. Radiation from the cool side of the shield can be minimized by coating it with aluminum foil or other material of low emissivity. Consideration should be given to energy conservation by utilization of heat transferred to air or water by the absorbing shields. The economies of accomplishing this are often negative, but coupling the heat-wasting process with one requiring warm air or warm water may in some instances prove cost-effective. [The engi-

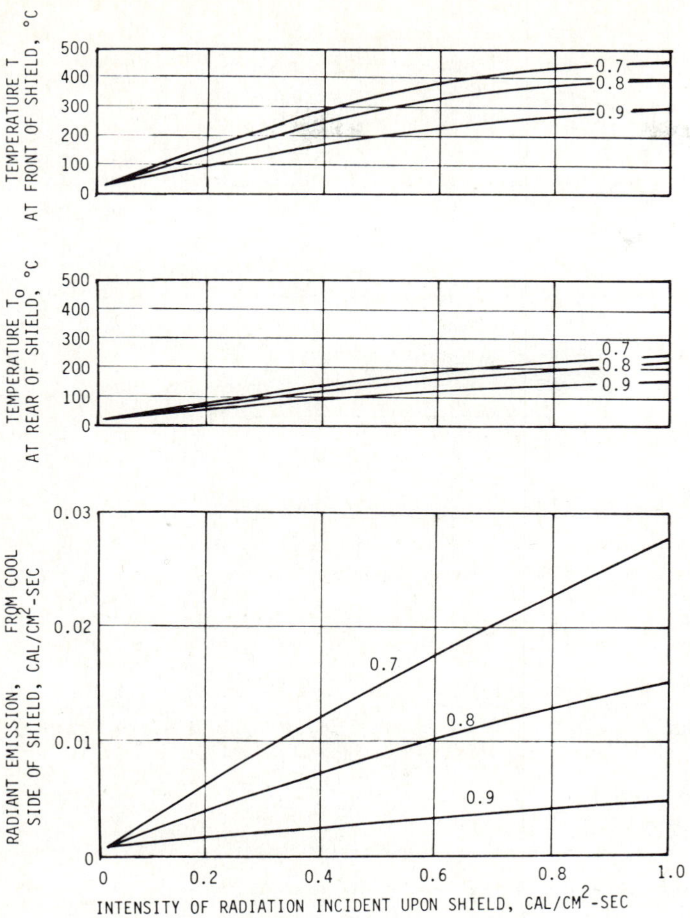

Figure 12.5 Predicted surface temperatures and radiant heat fluxes for multiple foil shield in 21°C environment (*Takata and Gilligan, 1978*). Multiple-foil shield: two foils separated by 0.64-cm ($\frac{1}{4}$-in.) air gap, reflectances of inner surfaces = 0.9. Reflectances of exterior shield surfaces are given on curves.

neering design of heat transfer equipment is covered in Sections 9, 10, and 11 of Perry et al. (1984).]

Transparent heat shielding is usually glass or metallic mesh. Both reduce transmission of thermal radiation through absorption or reflection. Reflection is more effective. Consequently, glass coated with gold or silver specular film, or special infrared reflecting glass, is to be preferred. The transmittance of visible light and a portion of the near-infrared spectrum through metal-coated glass is shown in Figure 12.9. Transmittance of infrared wavelengths greater than 900 nm and extending through the far infrared (30 μm to 1 mm) is relatively small, a few percent down to essentially zero. The radiation that is not trans-

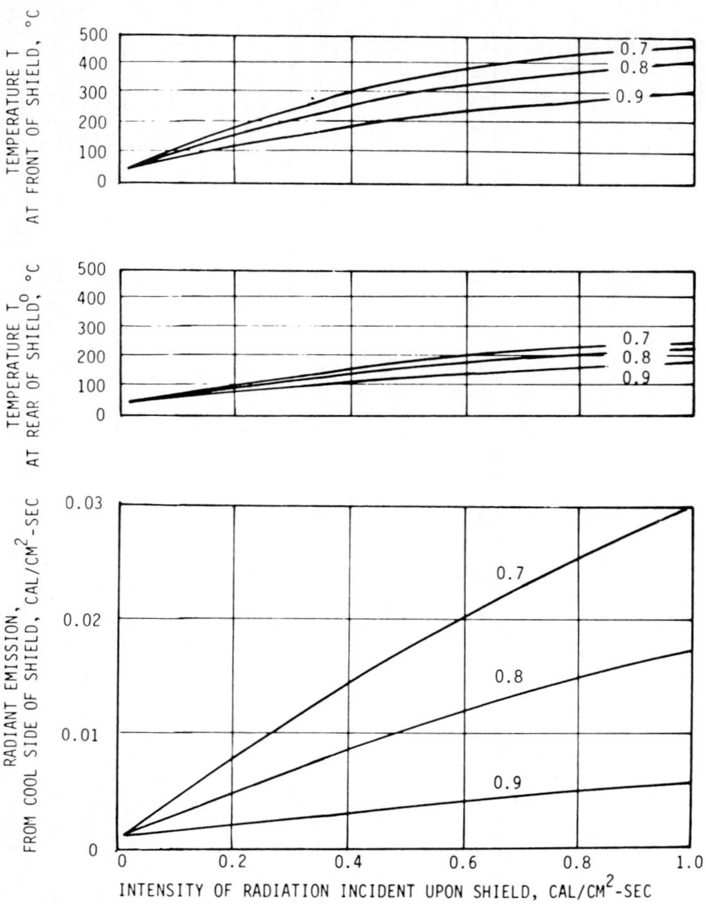

Figure 12.6 Predicted surface temperatures and radiant heat fluxes for multiple-foil shield in 43°C environment (*Takata and Gilligan, 1978*). Multiple-foil shield: two foils separated by 0.64-cm ($\frac{1}{4}$-in.) air gap, reflectances of inner surfaces = 0.9. Reflectances of exterior shield surfaces are given on curves.

mitted through such glass is reflected or absorbed. Thus such shielding may on an overall basis be 80–90% effective in preventing transmission of the incident radiant heat. Transmission of the absorbed radiant heat by reflective glass shielding is kept low by limiting the surface area of the window. It should be noted from Figure 12.9 that transmittance of electromagnetic radiation in the visible light range of 400 to 700 nm is quite good for gold films on glass and certainly acceptable for silver films on glass for most viewing tasks at ordinary light intensities. Operating booths, equipment cabs, and furnace viewing ports are examples of the uses of such windows. Some hot manufacturing operations, however, require that air or solid materials pass through an

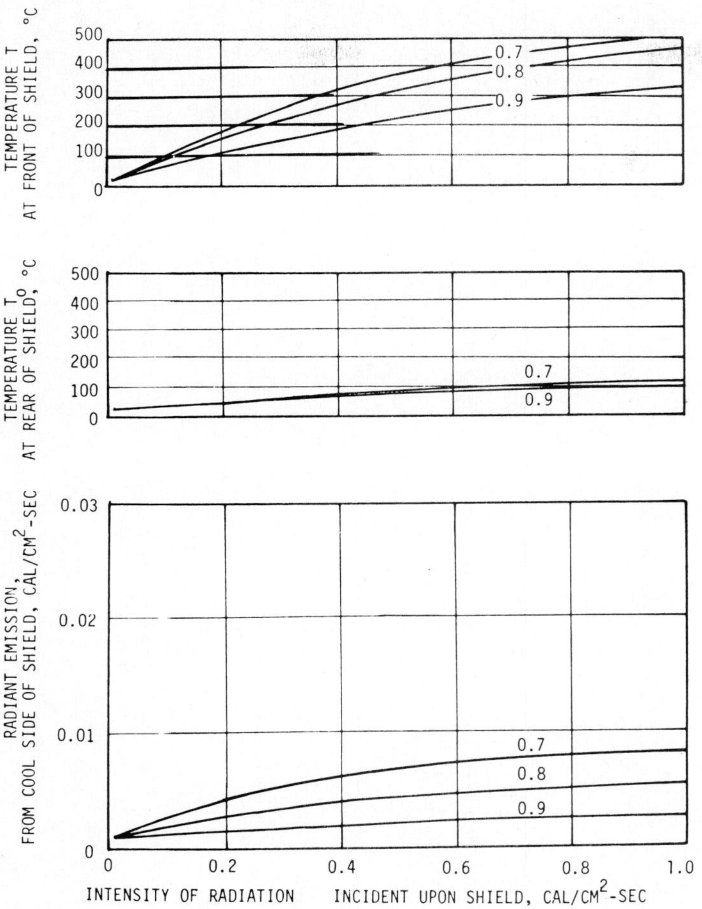

Figure 12.7 Predicted surface temperatures and radiant heat fluxes for an insulation shield (~5 cm thick) in a 21°C environment (*Takata and Gilligan, 1978*). Insulation shield: insulation with thermal conductance of 3.9×10^{-5} cal/cm²·sec·°C with reflecting foils attached to front and rear surfaces. Reflectances of exterior shield surfaces are given on curves.

opening. In these cases windows of chain or wire mesh help reduce the amount of radiant heat reaching an operator. The metal portions of such windows are essentially "black body" and may absorb as much as 50% of the incident radiant heat depending on the ratio of air openings to total window area. Subsequent reemission of the heat is reduced because the mesh is at a much lower temperature than the radiant heat source, and cooler air is nearly always drawn through the window toward the hot emission source so as to cool the wire mesh or chain by convection.

Fabrics may be aluminized by special processes and may reflect up

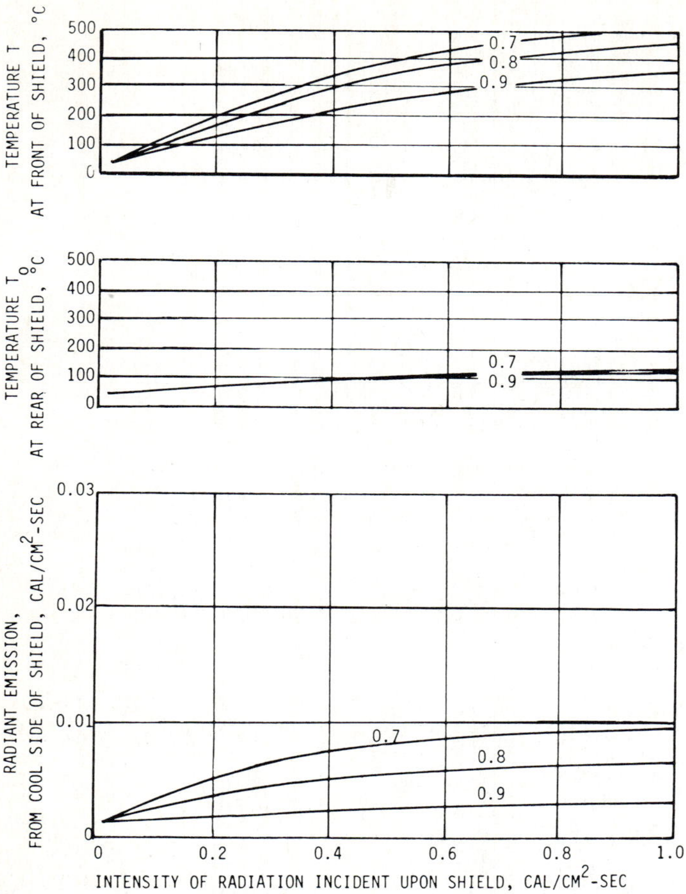

Figure 12.8 Predicted surface temperatures and radiant heat fluxes for insulation shield (~5 cm thick) in a 43°C environment (*Takata and Gilligan, 1978*). Insulation shield: insulation with thermal conductance of 3.9×10^{-5} cal/cm²·sec·°C with reflecting foils attached to front and rear surfaces. Reflectances of exterior shield surfaces are given on curves.

to 90% of incident radiant heat. Such flexible shielding may be fashioned into various articles of protective clothing, ranging from small disposable aprons to relatively high-cost firefighting suits and even spacesuits for astronauts.

6. Local exhaust ventilation. Removal of hot air surrounding furnaces and other hot equipment is commonly accomplished by installing canopy hoods employing natural draft or mechanical exhaust ventilation. Such arrangements are limited to removing convective heat. Thermal radiation, which is often considerably greater than convective heat, is not removed by exhaust hooding, and appropriate radiation shielding

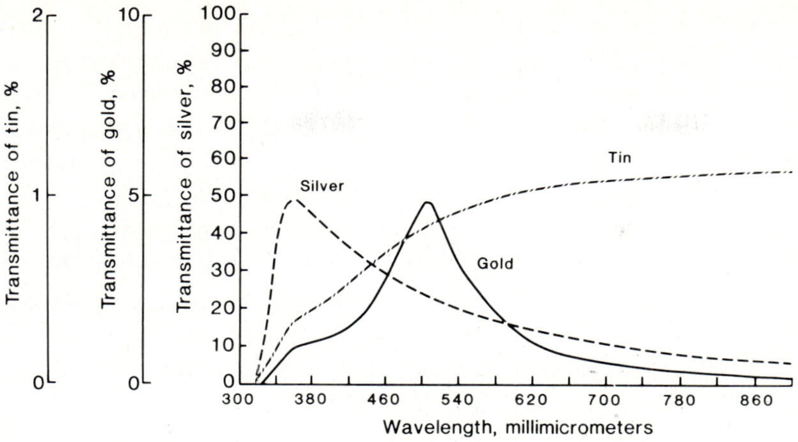

Figure 12.9 Spectral transmittances of metal films on glass (*Stair, 1948*). The metal films are of undetermined thickness on clear glass, and only illustrate what may be obtained through the use of metals having varying degrees of relative spectral absorption. In general the "white" metals, e.g., aluminum, platinum, chromium, cadmium, rhodium, would be expected to show but little selective spectral adsorption, with total density depending on the amount of metal deposited on the glass. (Note: 1 nm = 1 millimicrometer.)

must also be used. Local exhaust ventilation is most applicable to removing water vapor and convective heat from driers, quenching baths, cookers, and other hot, wet processes.

B. Cooling at work stations

In some situations it is impractical to provide adequate heat reduction at the source or throughout the general work area. In these cases spot cooling may be used. Cool air with lowered relative humidity, preferably between 20% and 60%, is supplied in sufficient quantity to surround the worker. This localized ventilation may serve three functions: (1) It displaces hot air surrounding the worker; (2) it provides increased air speed; and (3) it increases evaporative cooling. Thus the convection-conduction component of the heat load C may be utilized to offset the radiant heat load R, and evaporative cooling E is increased; that is, E becomes more negative in the heat balance $M' + R + C + E = 0$ (Eq. 12.2). It must be realized that the use of high air speed to offset radiation may be counterproductive if the supplied air adds to the convective load when it is warmer than skin temperature and fails to provide sufficient evaporative cooling because the water vapor content (humidity) is not low enough. This latter set of conditions defines the problem of increased heat load that commonly occurs in a hot work environment where simple cooling fans are used with a supply of warm humid air. Workers frequently respond by making such fans inoperable.

The air from a localized cooling system should be supplied so that minimum mixing takes place with ambient hot air. Discharge openings for the cool air ducts should be located above the work station, so that air flow is assisted by gravity. These may be directed straight down if the work activity is quite localized or angled so as to optimize the air flow along the most common path of the worker's movements.

In situations of high heat, workers may be stationed inside an insulated, locally cooled observation-control booth or relief room, which may serve the multiple purpose of also providing protection from excessive noise, dusts, fumes, vapors, or gases. The increasing use of air conditioned tractors, crane cabs, and trucks provides examples of this concept.

C. General ventilation

Gravity-induced general ventilation is most commonly employed for dissipating heat from hot industrial processes. Cool air is permitted to enter the building through strategically placed wall openings and is discharged after heating through roof openings. For a building H feet high located where the barometric pressure is P in inches of Hg, containing air at an absolute temperature T_1 (°R) surrounded by outside air at an absolute temperature T_2 (°R), the theoretical intensity of the draft, d, in inches of water pressure is:

$$d = 0.256HP\left(\frac{1}{T_1} - \frac{1}{T_2}\right) \tag{12.21}$$

The effect of friction is neglected in this formula and is obviously not negligible for a sizable building. To obtain the actual draft for a chimney or stack, the losses that occur with flow must be deducted from

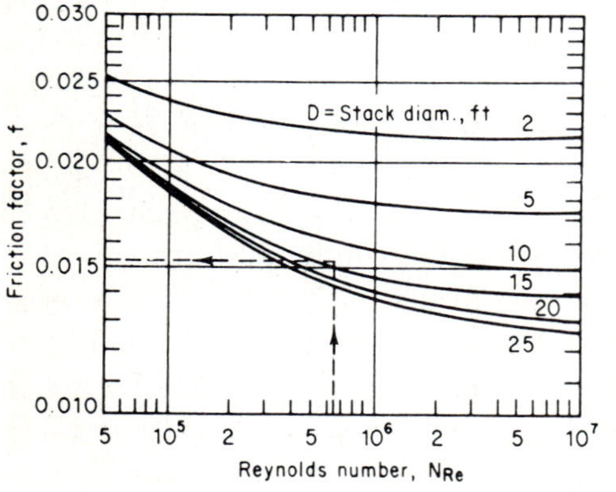

Figure 12.10 Friction factors for stacks (*Babcock & Wilcox Co., in Perry et al., 1984*).

the theoretical draft. These losses include both friction and exit velocity. They may be established from:

$$\Delta d = \frac{v^2}{2g_c}\left(1 + \frac{fH}{D}\right)\left(\frac{\rho_g}{5.2}\right) \tag{12.22}$$

where Δd = stack flow loss (in.H_2O)
v = stack exit velocity (ft/sec)
g_c = gravity constant, 32.17 (lb_{mass}/lb_{force})(ft/sec^2)
f = friction factor from Figure 12.10
H = stack height above entry (ft)
D = stack diameter (ft)
ρ_g = average density of gases in the stack (lb/ft^3)

The Reynolds number N_{Re} for Figure 12.10 may be approximated from:

$$N_{Re} = \frac{24{,}000W}{D(t_1 + 715)} \tag{12.23}$$

where t_1 = average stack temperature (°F)
W = gas flow rate (lb/hr)
D = stack diameter (ft)

Despite the fact that most heat loss from hot sources is by radiation, such heat is absorbed and reradiated by walls and any intervening structures. All such surfaces eventually become secondary sources of convective heat, which must be removed by ventilation or lost through the walls. General ventilation is an essential part of overall heat control. However, it generally fails to offset direct radiant heat exposures and is frequently inadequate for overall heat removal because of an insufficient supply of cool make-up air. Improper location of inlets without enough intake area results in poor distribution of too little air to achieve desired effects. The following example illustrates a few aspects of gravity-induced general ventilation.

Example 12.6

a. A large foundry building depends on gravity-induced general ventilation for exhausting much of the waste heat and other contaminants from the workspace. The foundry is located in Denver, and the mean height of the loft windows is 35 ft above that of ground-level doors. What is the theoretical intensity of the draft, *d*, if the temperature inside the building is 83°F when the temperature outside is 25°F? How much force would be exerted by this air pressure on a large door 8 ft wide and 10 ft high? Friction and exit velocity losses are negligible.

b. Assume that the loft windows are shut tight (negligible leakage) and gravity-induced ventilation is entirely through a 5-ft-diameter round stack, 35 ft high, with the same temperature difference as before. Find the intensity of the draft under these conditions.

solution

a. The barometric pressure at Denver may be obtained from altitude/pressure tables or from the empirical relation:

$$A = 25{,}000 \ln \left(\frac{29.92}{P}\right)$$

where A = altitude (ft)
P = barometric pressure (in.Hg)

Assume that the foundry is at 5280 ft (Denver is 1 mile above sea level). Then:

$$P = \frac{29.92}{\exp(5280/25{,}000)}$$
$$= 24.22 \text{ in.}$$

From Eq. (12.21):

$$d = 0.256 \times 35 \times 24.22[(25 + 460)^{-1} - (83 + 460)^{-1}]$$
$$= 0.0478 \text{ in.H}_2\text{O} \quad \text{or} \quad 0.249 \text{ lb/ft}^2$$

Because the door has an area of 8 ft × 10 ft, the force F on the door is

$$F = 0.249 \times 8 \times 10 = 19.9 \text{ lb}$$

which is definitely noticeable during manual operation and may even present a safety problem.

b. Let the draft under the new conditions be $d' = d - \Delta d$. A trial-and-error solution is required. A first estimate in this reiterative solution might be that $d' = d/2$ and is equal to the velocity pressure in Eq. (5.15). Thus:

$$v = 4005\sqrt{0.0239} = 619 \text{ ft/min} \tag{5.15}$$

The unknown factors for determining Δd may now be estimated:

$$\rho_g = 0.0808 \times \frac{460}{83 + 460}$$
$$= 0.0684 \text{ lb/ft}^3 \text{ for air at 83°F in the stack}$$

$$W = \rho_g\, av = 0.0684 \left(\frac{\pi}{4}\right)(5)^2(619)(60)$$
$$= 49{,}880 \text{ lb/hr, air flow in stack of area } a$$

$$N_{\text{Re}} = 24{,}000 \times \frac{49{,}880}{5(83 + 715)}$$
$$= 3 \times 10^5 \tag{12.23}$$

from which $f = 0.0187$ (from Figure 12.10).

With these estimates,

$$\Delta d = \frac{(619/60)^2}{2 \times 32.17}\left(1 + \frac{0.0187 \times 35}{5}\right)\left(\frac{0.0684}{5.2}\right)$$
$$= 0.0246 \text{ in.H}_2\text{O} \tag{12.22}$$

$$d' = d - \Delta d = 0.0478 - 0.0246 = 0.0232 \text{ in.H}_2\text{O}$$
(compared to 0.0239 in.H_2O assumed)

Second iteration: A better estimation of d' should lie about halfway between the first assumed value and the value calculated in the first iteration. Try $d' = (0.0239 + 0.0232)/2 = 0.0235$ in.H_2O, for which:

$$v = 4005\sqrt{0.0236} = 615 \text{ ft/min} \tag{5.15}$$

$$W = 0.0684 \times 19.6 \times 615 \times 60 = 49{,}490 \text{ lb/hr}$$

$$N_{\text{Re}} = 24 \times 10^3 \times \frac{49{,}490}{5(83 + 715)} = 2.97 \times 10^5 \tag{12.23}$$

$$f = 0.0187$$

$$\Delta d = \frac{(615/60)^2}{2 \times 32.17}\left(1 + \frac{0.0187 \times 35}{5}\right)\left(\frac{0.0684}{5.2}\right)$$

$$= 0.0243 \text{ in.H}_2\text{O} \tag{12.22}$$

$$d' = 0.0478 - 0.0243 = 0.0235 \text{ in.H}_2\text{O, which is close enough}$$

The preceding example illustrates the reduction of gravity-induced pressure differential that arises from localizing general ventilation to one large stack. The overall ventilation is, of course, also reduced, and removal of contaminants from distant sources is impaired. However, removal of contaminants from a nearby major source is greatly enhanced by localization, and dispersal to the general work atmosphere may be largely prevented. The engineer should design the system accordingly, based on both hygienic and operational requirements.

1. Make-up air supply. Air that is exhausted from a building must be replaced with air from outside. Depending on climate, this may be done without additional cooling. However, it is frequently most practical to design new facilities for general ventilation so as to include the capability of cooling outside air prior to distribution. In dry climates, evaporative cooling is particularly economical in lowering the dry-bulb temperature. An evaporative cooler provides direct contact between incoming air and water by using wetted packing or sprays. Although the dry-bulb temperature of the air may be reduced considerably, the process is adiabatic and there is no appreciable amount of heat transferred from or to the air. Sensible heat in the air stream is converted to latent heat in that relative humidity is increased at the expense of temperature. Even though water vapor is added to the incoming air, the effect is not necessarily negative, because the new dry-bulb temperature may be sufficiently below skin temperature to cool personnel by convection.

In addition to evaporative cooling, three types of chilled-coil heat exchange systems are in common use for reducing the temperature of fresh air supplies. These are once-through water-cooled coils, recirculated cooled water systems, and refrigerated coils. The simplest of these is the water-cooled coil, in which water passes directly through the coil and absorbs heat and condenses moisture from the air being supplied to the workspace. This system is most practical where a plentiful, low-cost supply of cool water is available. Generally the water enters the coil at less than 11°C and is wasted or passed into another process.

When sufficient water at a satisfactory temperature is not available on a continuous basis, a recirculating cooled water system may be employed. The water must be cooled artificially by mechanical refrigeration or by melting ice. Air may be cooled directly by passing it over finned coils which are refrigerated by the mechanical compression-liquefaction-vaporization-expansion cycle of a liquid refrigerant. This method is used in small, localized air cooling units as well as for large, ducted, central systems.

2. Distribution and removal of make-up air. Hot bodies naturally induce upward convection. In a general ventilation system combined with radiation shielding, make-up air should enter near the floor level, pass across the workers, and then pass onto the hot equipment. The workers benefit by receiving the cooler make-up air prior to its mixing with warm building air or being heated by convection over hot processes or equipment. The air, of course, rises and is discharged through vent openings in the roof. Appropriate planning and design consideration should be given to locations of air inlets, work stations, hot equipment, and exhaust vents whenever possible. When general ventilation is used to remove heated air, the basic strategy is to locate exhaust openings above the heat sources and as close to them as practical. Unfortunately, many hot industries are characterized by large, flat buildings, requiring exhaust fans and forced air supplies if satisfactory ventilation is to be achieved.

D. Moisture control

In contrast to hot-dry work situations, there are certain industries that may be characterized as warm-moist. To be tolerated, temperatures in these industries must be considerably lower due to the limitations placed on evaporation of sweat. In some of these work settings, high relative humidity is unavoidable due to the presence of inexhaustible sources of humidity, e.g., in coal or gold mines. In others, such as munitions plants, textile mills, and flour mills, high relative humidity is purposely maintained so as to prevent dust diseases, improve product quality, or reduce problems arising from build-up of static electricity. Whenever feasible, dehumidification clearly helps to reduce heat stress by increasing the maximum available cooling capacity of sweat E_{max} (Eq. 12.10).

Moisture control entails limiting the release of water vapor at its points of origin as well as employing dehumidification. Simple, direct measures of control are effective. These include covering normally open hot water tanks, repairing leaky valves and joints in steam piping, and covering hot water drains. Dehumidification may be accomplished by refrigeration, absorption, or condensation. For the localized cooling of relief areas and operating booths, refrigeration is most widely used to condition the air. Even though it may be exhorbitant or clearly impractical to air condition some operations such as the rolling mills in a steel plant, the use of air conditioned control booths and work-rest regimens for certain jobs is quite practical.

IV. Administration of a Heat Stress Control Program

Heat stress is a complex problem that requires a multifaceted approach for its solution and control. It involves physical and chemical factors in the environment, physiological and psychological human factors, and a variety of job factors. The definition and implementation of control measures must depend on appropriate applications of engineering, industrial hygiene, and occupational medicine. As an ongoing problem involving personnel, money, machines, regulation, and productivity, enlightened participation of industrial management is essential. In addition to engineering control of heat exposures, each of the following elements should also be included:

1. Industrial hygiene monitoring and supervision
2. Medical examination and selection
3. Employee education and training
4. Forecasting of episodes
5. Work-rest regimen

A. Industrial hygiene monitoring and supervision

Knowledge of the levels of heat stress occurring in work situations throughout an industry is essential to its control. Hot areas and hot jobs should be characterized throughout an industrial facility in relation to heat stress indices, recommended standards, and current health regulations. Despite shortcomings, at least a WBGT survey should be made for each hot area and hot job as a minimum. In addition, an engineering evaluation of heat stress in critical locations should be made using all of the human factors, environmental measurements and concepts that are required for calculating the heat stress index

(HSI), allowable exposure time (AET), and minimum recovery time (MRT).

Because heat stress is only seasonally related to certain jobs or reaches heavy or severe proportions only on occasion, an "action level" based on one of the environmental heat stress indices may be desirable. The NIOSH Criteria Document (NIOSH, 1972) recommends that a WBGT profile be established for each workplace for winter and summer seasons to aid in deciding when hot work practices should be initiated. It is suggested here that a value of 40 in the HSI be used as an action level. Current forms of the heat exchange equations should be used, environmental measurements should be accurate, the metabolic heat generation should be corrected for physical work, and consideration should be given to clothing and skin area available for sweat evaporation and convection. Once initial profiles have been established, monitoring may be needed only during the warm months of the year, starting early in the season.

In characterizing the thermal exposure of a person engaged in hot work, the Standards Advisory Committee (SACHS, 1974) recommended use of the hottest 2-hr period of the work shift in which regular work is performed. Where seasonal and diurnal heat loads affect the work environment, this period usually occurs in the summer between 2 and 5 P.M.

Because a program of control and management of heat stress involves many disciplines and organizational divisions within a company—e.g., health, safety, engineering, personnel, maintenance, operations, and training—it should be supervised or coordinated by an industrial hygienist who has had both formal education and some practical experience in all of these areas.

B. Medical examination and selection

Medical examinations of personnel who are to be assigned to hot jobs are important in selecting individuals who will be able to cope with heat stress without developing undue strain. In the examination, special emphasis should be placed on the cardiovascular, renal, hepatic, and respiratory systems, and the skin. The examination should include a complete medical history of the worker, with specific emphasis on prior heat-related disorders or illnesses. Young, physically fit workers who are not overweight are more resistant to developing heat strains and illnesses than older, out-of-condition workers. One medical recommendation (NIOSH, 1972) is that all employees exposed to hot environmental conditions should be given a periodic physical examination every 2 years if under age 45, and every year if 45 years of age or older. All components of the preplacement medical examination should be included. In addition, a person who has had first-aid training in recognizing the signs and symptoms of heat disorders and illnesses should be available during working hours. The examining physician should also look for evidence of intolerance to heat, either on or off the job. A recorded history of successful acclimatization and extended work on prior hot jobs is generally the best basis for predicting future performance under heat stress. Care should be taken, however, that no significant change in the worker's health status has occurred since the previous employment. Where heat stress is increased by hot weather, it is desirable that the periodic examination of exposed personnel be scheduled in the summer. Absenteeism due to illness or repeated accidents among individuals or groups of workers should alert the occupational physician to possible heat intolerance or the presence of a chemical stressor such as carbon monoxide that will aggravate the problem of heat exposure. Nutritional status of employees should be noted and advice relative to weight adjustment provided.

C. Employee education and training

Successful application of basic health and safety procedures depends in large measure on the understanding and performance of affected workers. Education increases knowledge and understanding, and training improves desired performance. Excessive heat exposure can result in primary heat illnesses, accidents, and increased susceptibility to the toxic effects of certain chemicals and physical agents. Employees subject to high levels of heat exposure and their first-line supervisors should be well informed and trained regarding: (1) heat acclimatization, (2) water intake for replacement, (3) salt replacement to maintain electrolyte balance, (4) recognition of the acute effects of excessive heat stress, (5) instruction on the potential effects of combining heat stress with alcohol or other drug use, or with exposure to chemical or physical agents peculiar to a particular work environment, and (6) the use of appropriate clothing.

Any worker, however healthy, physically conditioned, and motivated, when exposed to high heat stress for the first time, will develop heat strains such as elevated body temperature, pounding heart, excessive sweating, or psychological aberrations. On each succeeding day of heat exposure, however, his ability to work at the same level of heat stress improves and the strains diminish. Usually within a period of 2 weeks, a worker will begin to perform without difficulty. This process of acclimatization may be hastened by a formal training program. One proposed procedure recommends that unacclimatized workers be trained over a period of 6 days (NIOSH, 1972). The acclimatization schedule begins on the first day in the hot environment with work performance at only 50% of the anticipated ultimate workload and time, followed by daily 10% increments building up to the total work and time exposure on the sixth day. Because acclimatization is readily lost, acclimatized workers who return from nine or more consecutive calendar days of leave should be reacclimatized in a 4-day regimen. The reacclimatization schedule should begin with 50% of the work performance on the first day, followed by daily 20% increments building up to 100% workload in the hot environment on the fourth day. Previously acclimatized employees who return from four consecutive days of illness should have medical approval to return to the job, and should undergo a 4-day reacclimatization regimen as outlined for employees who return after leaves of nine or more consecutive days.

Along with acclimatization training, workers should be educated regarding the necessity for replacement of salt and water. This involves drinking small (200–300 ml) draughts of cool, salted (0.1% NaCl) water at 15- 20-min intervals or in sufficient amount to almost equal their sweat loss. Muscle cramps and low urinary chloride concentrations are clear indications of salt deficiency.

D. Forecasting of episodes

In some hot industries heat stress may be independent of climate, in many it tends to be a seasonal phenomenon, and in still others it develops only during the dog days of summer. Unexpected heat waves early in the summer are notable for creating acute heat strains. Thus, good administrative practice would include alertness to weather service forecasts predicting hot spells. As needed, workloads could be redistributed, and performance of the hottest jobs could be scheduled for the coolest portion of a day or work shift. Mutchler et al. (1976) have presented a systematic technique for relating the WBGT at any worksite to external weather conditions, current or predicted. Similar modeling can also be done for other heat stress indices. Correlation of climatic episodes with heat stress index profiles, which have been developed in advance for selected work stations, provides needed lead time for making administrative changes.

E. Work-rest regimen

The authors of the NIOSH Criteria Document on occupational exposure to hot environments recommended in 1972 that a work-rest regimen be implemented to reduce the peaks of physiological strain and to improve recovery during rest periods (NIOSH, 1972). In 1974 the Standards Advisory Committee on Heat Stress (SACHS, 1974) recommended that the duration of work periods, the frequency and length of rest pauses, and the pace and tempo of work should be adjusted to avoid heat strain. The 1985–1986 TLV listing on heat stress (ACGIH, 1985) contains work-rest regimens referenced to permissible WBGT values in degrees celsius and workloads designated as light, moderate, or heavy (see Chapter 4). Obviously, the duration of work and rest periods depends on the severity of the heat stress including its length of time. Rest periods should be taken before excessive fatigue develops and should be long enough to reduce the pulse rate below 100 beats/min. In actual situations work-rest regimens are often adjusted to the tolerance of individual workers or to work performance requirements. If all of the factors entering into calculation of E_{max} and E_{req} are measured accurately and good heat stress evaluations are made, the allowable exposure time, AET, and minimum recovery time, MRT, may be calculated from Eqs. (12.15) and (12.16) and an appropriate work-rest regimen prescribed for any given hot job. Such a work-rest regimen, however, should not be enforced if it is not tolerated by individual workers employed on that specific job. The industrial hygienist should be consulted and necessary adjustments made.

Brouha (1967) has stated that the possibility of rest in cool surroundings considerably reduces the total cost of work in heat. Although definitive studies are lacking regarding the optimal temperature, humidity, and air movement for the recovery room, a temperature around 25°C with about 40% relative humidity and moderate air movement has seemed acceptable in practice. The cooled rest or recovery room for work-rest regimens should be located near the hot workplace so as to afford optimal utilization of AET and MRT during the work cycle. Regardless of whether the work-rest regimen is derived from E_{req} or E_{max} to give AET and MRT periods, or from the ACGIH-TLV adaptation of WBGT, it is generally more desirable to employ shorter work-rest cycles than long ones, with definite consideration being given to worker responses.

REFERENCES

ACGIH (1985). *TLV's, Threshold Limit Values for Chemical Substances and Physical Agents in the Work Environment Adapted by ACGIH for 1985–1986*. American Conference of Governmental Industrial Hygienists, Cincinnati, Ohio.

ASHRAE (1985). *ASHRAE Handbook—1985 Fundamentals*. American Society of Heating, Refrigerating and Air Conditioning Engineers, Atlanta, Ga.

Belding, H. S., and Hatch, T. F. (1955). Index for evaluating heat stress in terms of resulting physiological strain. *Heat. Pip. Air Condition. 27,* 129–135.

Brief, R. S., and Confer, R. G. (1973). Alternative measures for controlling heat stress. *Med. Bull., 33,* 229–253.

Brouha, L. A. (1967). *Physiology in Industry—Evaluation of Industrial Stresses by the Physiological Reactions of the Workers,* 2d ed. Pergamon Press, Elmsford, N.Y.

Minard, D. (1973). Physiology of heat stress. In: *The Industrial Environment—Its Eval-*

uation & Control. U.S. Department of Health, Education, and Welfare, Public Health Service, U.S. Government Printing Office, Washington, D.C., pp. 399–412.

Mutchler, J. E. (1978). Heat stress: Its effects, measurement and control. In: Clayton, G. D., and Clayton, F. E., eds., *Patty's Industrial Hygiene and Toxicology, vol. I, General Principles,* 3d. ed. Wiley, New York, pp. 927–992.

Mutchler, J. E., Malzahn, D. D., Vecchio, J. L., and Soule, R. D. (1976). An improved method for monitoring heat stress levels in the workplace. *Amer. Ind. Hyg. Assoc. J., 37,* 151–164.

NIOSH (1972). *Criteria for a Recommended Standard Occupational Exposure to Hot Environments.* U.S. Department of Health, Education and Welfare, National Institute of Occupational Safety and Health, HSM 72-10269, Cincinnati, Ohio.

Perry, R. H., Green, D. W., and Maloney, J. O. (1984). *Perry's Chemical Engineering Handbook,* 6th ed. McGraw-Hill, New York.

Peterson, J. E. (1977). *Industrial Health.* Prentice-Hall, Englewood Cliffs, N.J.

SACHS (Standards Advisory Committee on Heat Stress) (1974). *Recommendations for a Standard for Work in Hot Environments,* Draft No. 5. U.S. Department of Labor, Washington, D.C.

Stair, R. (1948). *Spectral-transmissive Properties and Use of Eye-protective Glasses.* National Bureau of Standards Circular 471, U.S. Government Printing Office, Washington, D.C.

Takata, A., and Gilligan, J. (1978). *Heat Shielding Materials and Design Guidelines.* Contract 210-88-0064, National Institute for Occupational Safety and Health, U.S. Department of Health, Education and Welfare, Cincinnati, Ohio.

TCE (Technical Committee on Ergonomics) (1971). Ergonomics guide to assessment of metabolic and cardiac costs of physical work. *Amer. Ind. Hyg. Assoc. J., 32,* 560–564.

Vernon, H. M. (1930). The measurement of radiant heat in relation to human comfort. *Proc. Physiol. Soc.* (*J. Physiol., 70*), 15–17, July 5.

Chapter

13

Noise and Vibration

E. R. Hermann

Mechanical vibration is defined as a periodic motion of the particles of an elastic body or medium in alternately opposite directions from their positions at rest. Those vibrations that affect the sense or organs of hearing we call sound. The human ear may respond to sound waves in a frequency range from as low as 1 or 2 hertz (Hz) to as high as 30,000 Hz. The nominal acoustic range of human hearing is given as 20–20,000 Hz. Typical minimum audible sound levels for young persons are shown in Figure 13.1. The permanent hearing loss that can occur because of occupational exposures has been discussed in Chapter 2. Various existing design guidelines are summarized in Chapter 4. Although most of our discussion will focus on noise control, many of the techniques are equally applicable to reduction of vibration to mitigate other undesirable physiological endpoints (e.g., whole-body vibration).

The root-mean-square (rms) sound pressure, P_{rms}, is a common measure of sound. It is obtained by squaring the value of sound pressure disturbance at each instant of time. The squared values are then added and averaged over the sample time. The rms sound pressure is the

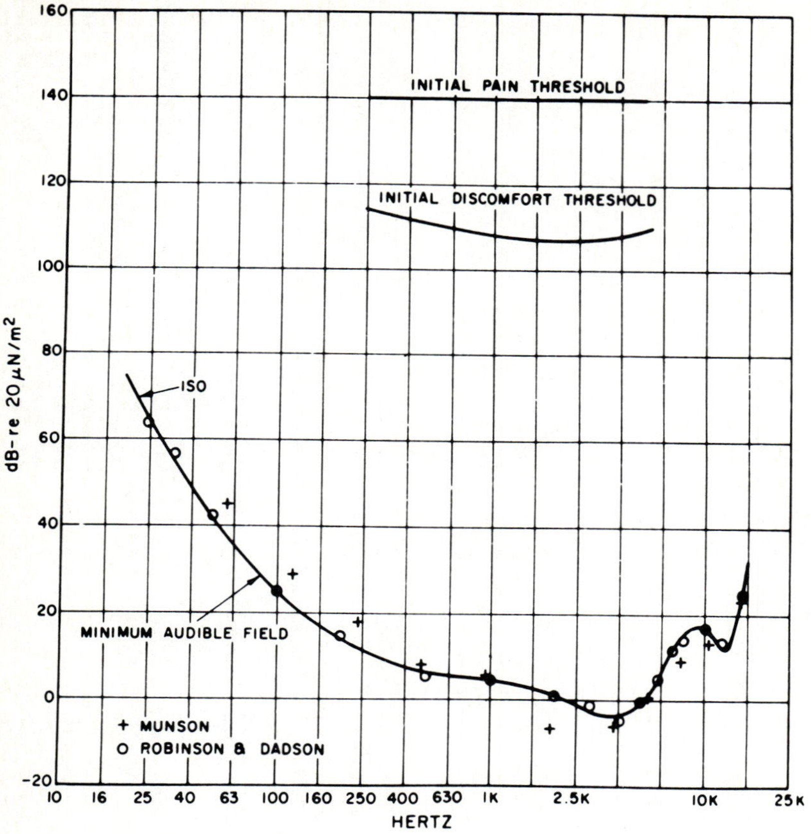

Figure 13.1 Thresholds of hearing and tolerance. (+ *Munson, as reported in Sivian and White, 1933;* ○ *Robinson and Dadson, as reported in 1956; ISO, 1967; figure from Peterson and Gross, 1974.*)

square root of this time average. Since squaring converts all negative values to positive ones, the rms sound pressure is a useful nonzero measure of the magnitude of a sound wave. The units of sound pressure normally used are microbars, dynes per square centimeter, newtons per square meter, or pascals. Relations among these units are:

$$1 \text{ microbar} = 1 \text{ dyne/cm}^2 = 0.1 \text{ N/m}^2 = 0.1 \text{ Pa}$$

In working with periodic motion, wavelength is measured between two analogous parts on two successive parts of the wave. Lambda is the usual symbol for wavelength, and the distance is usually expressed in feet or meters. The frequency of sound vibrations is measured in hertz, Hz (cycles per second) and is designated by f. The time in which analogous pressure points on successive waves of sound pass a given point defines the speed of sound in a particular medium. This is given by:

$$c = f\lambda \tag{13.1}$$

where c = speed of sound (ft/sec, m/sec)
f = frequency of vibration (Hz = cycles/sec)
λ = wavelength (ft, m)

The speed with which a sound travels through a medium depends on the mass and elastic properties of the material. In air at 72°F (22°C), the speed of sound is about 1130 ft/sec (344 m/sec).

In a homogeneous medium, the speed of sound is independent of frequency. At a given frequency in various media, the wavelength varies directly with the square root of the density and inversely with the square root of the compressibility of the transmitting medium. Because the speed of sound is proportional to wavelength, it is similarly affected. Thus, for example, the speed of sound is about 4700 ft/sec in water and 16,500 ft/sec in steel.

Intensity of sound, I, relates to an energy flux, and the units used are usually watts per square meter (W/m^2). In practical applications, sound intensities apply most frequently to the energy flux at a point of reception such as the human ear. Sound intensities are usually reported with respect to a relative scale such that

$$L_I = 10 \log \left(\frac{I}{I_0}\right) \tag{13.2}$$

where L_I, the logarithmic level, is in decibels (dB); and I_0, the reference intensity, is usually assigned a value of 10^{-12} W/m^2. One decibel is the minimum difference in loudness that is usually perceptible to the human ear. It should be noted that base-10 logarithms are generally used in the field of sound measurement, and that all decibel scales are actually unitless as far as real physical values are concerned. Real physical values, however, can readily be obtained from a measurement of decibel level L_I if a definite value has been assigned to I_0. Thus:

$$I = I_0 \cdot 10^{L_I/10}$$

It is clear from this expression that for each multiple change in intensity by one decade order of magnitude, the number of decibels is changed by plus or minus 10; or for each change in intensity by a factor of 2, the number of decibels is changed by 3.0103, say ±3.

We now may develop a logarithmic relative scale for sound pressure level using real physical units of sound pressure variation. This may most simply be done by using an equation from fundamental physics of sound relating pressure variation to intensity, namely:

$$P^2 = I\rho c \tag{13.3}$$

in which c, P, and I are as previously defined, and ρ is the density of the transmitting medium. For our immediate purpose it is most important to note that P^2 is directly proportional to I. Therefore, by substitution we get:

$$L_p = 10 \log \left(\frac{P^2}{P_0^2}\right) = 10 \log \left(\frac{P}{P_0}\right)^2 \tag{13.4}$$

$$= 20 \log \left(\frac{P}{P_0}\right) \tag{13.5}$$

where L_p = sound pressure level (dB)
P = sound pressure variation, rms
P_0 = reference sound pressure variation, rms (usually 20 μN/m^2 or 20 μPa)

Note that, for a dynamic range of $P = 1P_0$ to 10^6P_0, the sound pressure level reduces to a range from 0 dB to 120 dB.

Sound may also be expressed in terms of power output from a source, and because intensity I is equal to power W divided by area a, we have $W = Ia$. The common unit for the power of a sound is watts. The consistent reference power for use in the sound power level equation,

$$L_w = 10 \log \left(\frac{W}{W_0}\right) \tag{13.6}$$

is $W_0 = 10^{-12}$ W. Some sound pressure levels at a receiver located less than 2 ft from various sound sources of various power outputs are shown in Table 13.1.

When a vibration is a continuing periodic function, it is said to be in a steady state. Noise, however, may be in the form of multiple steady-state vibrations. For occupational health purposes, noise exposures that persist for the total duration of a workday are labeled continuous, even though it is obvious that they are interrupted at the close of each workday or at least reduced to a much lower and hopefully insignificant level. When environmental noise levels fluctuate markedly or the exposures are discontinuous, they are considered to be intermittent. In making evaluations of noise exposures for industrial hygiene and engineering control purposes, it must be borne in mind that human noise exposures are a summation of occupational, recreational, and avocational experiences and are highly variable. For many workers, however, their industrial noise exposures are dominant.

TABLE 13.1 Levels of Some Common Sounds

Sound power, W (W)	Sound power level, L_w (dB) (re 10^{-12} W)	Sound pressure, P (N/m^2) (Pascal)	Sound pressure level, L_p (dB) (re 20 $\mu N/m^2$) (μPascal)	Sound source
3,000,000.	200	100,000 (1 bar)	194	Saturn rocket
	185	20,000.0	180	
	175		170	
30,000.	165	2,000.0	160	Ram jet
	155		150	Turbo jet
300.	145	200.0	140	Propeller aircraft
			135	Threshold of pain
	135		130	Pipe organ
3.0	125	20.0	120	Riveter, chipper
	115		110	Punch press
0.03	105	2.0	100	Passing truck
	95		90	Factory
0.0003	85	0.2	80	Noisy office
	75		70	
0.000003	65	0.02	60	Conversational speech
	55		50	Private office
0.00000003	45	0.002	40	Average residence
	35		30	Recording studio
0.0000000003	25	0.0002	20	Whisper
	15		10	Threshold of good hearing
0.000000000003	5	0.00002	0	Threshold of excellent youthful hearing

I. Problem Analysis

A. Propagation of sound

The propagation of sound waves through air ranges from a simple free-field situation to one in which there is a complex pattern of reflections, absorptions, diffractions, and transmissions by or through barriers. To describe the sound field, we will initially consider the limiting case of a free field. Later we will discuss another limiting case at the other end of the sound spectrum, the reverberant field.

In a free field, the sound from a nondirectional point source radiates equally in all directions in the form of a spherical wave. Such waves are thought of as waves propagating radially away from a small "pulsating" sphere. The intensity of the wave follows the usual inverse-square law for energy propagation, and the intensity drops to one-fourth its value each time the distance is doubled. Since the sound pressure is proportional to the square root of the intensity, the sound pressure drops by one-half its value for each doubling of distance. It must be emphasized that this decrease for each doubling of distance holds only for the region defined as the far field (i.e., beyond about two wavelengths). Closer to the sound source is the near field, and a different mathematical treatment is required to describe the sound field in this region. The intensity of sound over a unit area on a spherical surface at a distance r from a point source suspended in a free field is:

$$I = \frac{W}{S} = \frac{W}{4\pi r^2} = \frac{P^2}{\rho c} \qquad (13.7)$$

where I = intensity (W/m^2)
W = total sound power radiated by the source (W)
P = rms sound pressure variation (N/m^2)
ρ = density of the medium (kg/m^3)
c = velocity of sound (m/sec)
S = surface area (m^2)

The product of the density and velocity of sound, ρc, is called the characteristic impedance of the medium through which the sound is traveling and is the constant of proportionality that equates the sound pressure squared to the sound intensity.

Sound power is the amount of energy per unit time that radiates from a source in the form of an acoustic wave. The sound power level of a source is independent of its surroundings, whereas sound pressure level represents the pressure in a sound field at a particular point and is a function of the power generated by the source, the distance from the source, and the "acoustic characteristics" of the room. No available instrument measures the power level of a source directly. However, power levels can be computed from sound pressure measurements if the measurement and environmental media conditions can be described adequately.

Fortunately, there is a unique relationship between the intensity and the pressure of a spherical sound wave. Recall from Eqs. (13.3) and (13.7) that:

$$I = \frac{W}{S}$$

and $$I = \frac{P^2}{\rho c}$$

The simplifying assumption can be made that all noise sources such as aircraft engines, motorcycles, gear boxes, turbines, horns, and rock band amplifiers emit sound uniformly from a spherical surface area S_0 of 1.0 m^2. Then:

$$\log I = \log W - \log S$$

or, in decibel level form,

$$10 \log \frac{I}{I_0} = 10 \log \frac{W}{W_0} - 10 \log \frac{S}{S_0} \qquad (13.8a)$$

or $$L_I = L_w - 10 \log S \qquad (13.8b)$$

where the subscripts I and W indicate logarithmic levels of intensity and power, respectively.

For spherical radiation in a free field and $S = 4\pi r^2$, substitution in Eq. (13.8*b*) gives:

$$\begin{aligned} L_I &= L_w - 10 \log(4\pi r^2) \\ &= L_w - 10 \log r^2 - 10 \log(4\pi) \\ &= L_w - 20 \log r - 11 \qquad (\text{dB re } 10^{-12}\ \text{W/m}^2) \end{aligned} \qquad (13.9)$$

where r *must* be in meters. If r is to be expressed in feet, the source surface area S_0 becomes 10.76 ft^2 and the derived constant, $10 \log 4\pi/S_0$, becomes 0.7. Thus for free-field radiation we may write:

Spherical radiation (dB re 10^{-12} W/m^2)

$$L_I = L_w - 20 \log r - 11 \qquad (r \text{ in meters}) \qquad (13.9)$$

$$L_I = L_w - 20 \log r - 0.7 \qquad (r \text{ in feet}) \qquad (13.9a)$$

Similarly, if sound is radiated from a source on the ground, assuming no absorption by the ground, all of the power is radiated into a hemisphere and the area becomes $2\pi r^2$ and we may obtain:

Hemispherical radiation (dB re 10^{-12} W/m^2)

$$L_I = L_w - 20 \log r - 8 \qquad (r \text{ in meters}) \tag{13.10}$$

$$L_I = L_w - 20 \log r + 2.3 \qquad (r \text{ in feet}) \tag{13.10a}$$

Root-mean-square values of sound pressure variations are sensed by a microphone and measured by a sound level meter that indicates sound pressure level directly. Therefore the relationship between sound power level and sound pressure level is actually more useful in practice. Using the expression $W/S = P^2/\rho c$ and multiplying through this equation with needed terms, we may state:

$$\frac{W}{S} \cdot \frac{1}{W_0} = \frac{P^2}{\rho c} \cdot \frac{1}{W_0} \cdot \frac{P_0^2}{P_0^2} = \frac{P^2}{P_0^2} \cdot \frac{P_0^2}{\rho c W_0}$$

In decibel form this may be rewritten as:

$$10 \log \frac{W}{W_0} - 10 \log S = 10 \log \frac{P^2}{P_0^2} + 10 \log \frac{P_0^2}{\rho c W_0}$$

The last term in this equation proximates zero for most applications because ρc, the characteristic impedance of air at 20°C and 750 mmHg atmospheric pressure, is about 408 rayls (kg·m^{-2}·sec^{-1}), and

$$10 \log \left(\frac{P_0^2}{\rho c W_0}\right) = 10 \log \left(\frac{(2 \times 10^{-5})^2}{408 \times 10^{-12}}\right) = -0.1$$

which is inconsequential in nearly all decibel measurements. Even for fairly wide variations in atmospheric temperature, pressure, and humidity, the value of $10 \log(P_0^2/\rho c W_0)$ will be less than 0.25 dB and may be neglected. Consequently, the relationship between logarithmic levels with respect to sound power and sound pressure becomes

$$L_w \simeq L_p + 10 \log S$$

or

$$L_p \simeq L_w - 10 \log S \qquad (\text{dB re } 2 \times 10^{-5} \text{ Pa}) \tag{13.11}$$

where S is in square meters. Notice that this expression for L_p is the same as for the one previously developed for L_I. Thus, in a free field, $L_p = L_I$.

From the foregoing development (e.g., Eq. 13.8*a*), it can be seen that for two identical sound sources, the sound intensity or sound pressure at a receiver will be twice that contributed by one source, with a resultant increase of 3 dB in sound pressure level. (Substitute $2W$ for W

in Eq. 13.8*a*). It should also be noted that the resultant sound intensity level or sound pressure level at a receiver that is a distance r from a source radiating under hemispherical conditions is 3 dB greater than for spherical conditions. This is analogous to placing a lightbulb in front of a plane mirror. All light emitted into the hemisphere containing the mirror is reflected into the hemisphere with which we are concerned.

Another example of sound propagation as a one-dimensional free progressive wave can be obtained by considering sound emitted from a source that is a long straight line, such as an elevated steam line with little interfering support. If the surface of the pipeline is pulsating uniformly in a radial direction, let us consider the sound intensity received at various distances r_1, or r_2, normal to the line source (Figure 13.2). Laplace's law states that the intensity of a radiant field at a point is directly proportional to the strength of the source and inversely proportional to the square of the distance between source and point. Figure 13.3 shows the geometric relationships between the strength of a uniform line source, W, (power/length), and the intensity, dI, at the point. Therefore,

$$ds = \frac{C\,d\theta}{\sin\theta} \quad \text{and} \quad C = \frac{r}{\sin\theta}$$

giving

$$ds = \frac{r\,d\theta}{\sin^2\theta} \quad \text{and also} \quad C^2 = \frac{r^2}{\sin^2\theta}$$

Therefore,

$$dI = \frac{W\,(r\,d\theta/\sin^2\theta)\sin\theta}{r^2/\sin^2\theta} = \frac{W\sin\theta\,d\theta}{r}$$

and

$$I = \frac{W}{r}\int_0^{\pi}\sin\theta\,d\theta = \frac{2W}{r}\left|-\cos\right|_0^{\pi/2}$$

$$I = \frac{2W}{r} \tag{13.12}$$

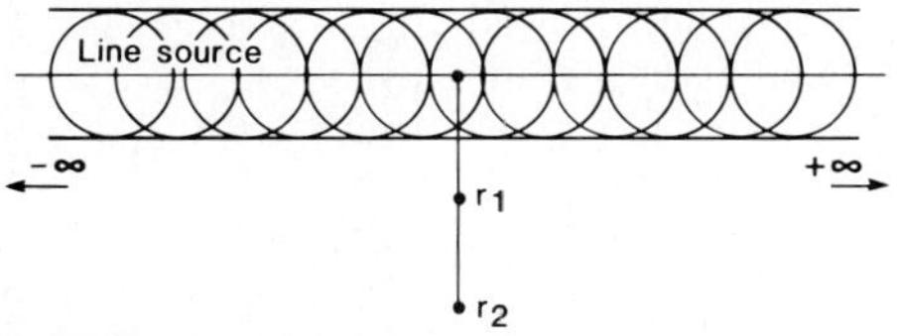

Figure 13.2 Cylindrical propagation of sound from a line source in a free field.

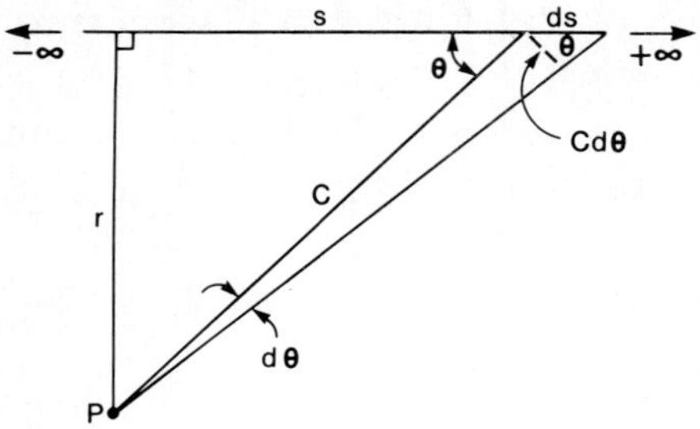

Figure 13.3 Illustration of development of intensity of a radiant field in accordance with Laplace's law.

Thus it may be seen that sound intensity for an ideal infinitely long line source in a free field follows an inverse linear relation $I \propto 1/r$ or, for two different distances,

$$L_1 - L_2 = 10 \log \left(\frac{r_2}{r_1}\right) \tag{13.13}$$

in decibel form. This may be contrasted with the inverse square law $I \propto 1/r^2$, which holds for free-field radiation from a point source, or

$$L_1 - L_2 = 20 \log \left(\frac{r_2}{r_1}\right) \tag{13.14}$$

in decibel form.

The foregoing ideal situations may be used in approximating the radiant attenuation of many sources. Obviously, however, a variety of other situations exist that we would like to estimate. One of these is the line source of quite finite length; another would be large panels providing plane sources of sound emission, such as large window areas of factories. Methods of approximating the sound pressure level L_p (or L_I) are given in Figure 13.4. Although such predictive models are not

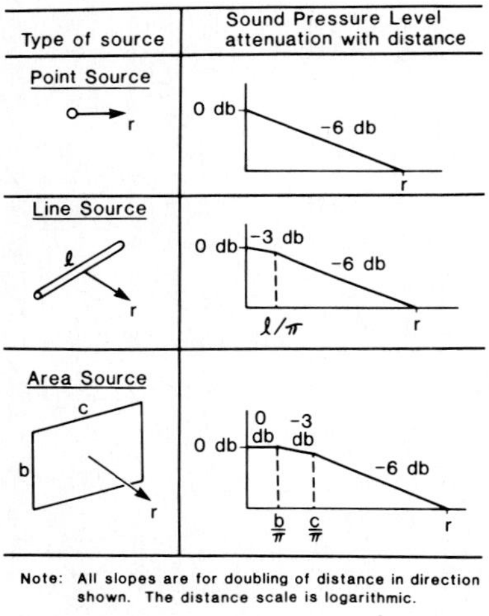

Figure 13.4 Sound attenuation as a function of distance from three different types of noise source: a point source, a line source of finite length, and a flat area.

absolute substitutes for actual measurements under field conditions, they generally work very well.

B. Adding and subtracting sound

A common problem encountered in acoustics, hearing conservation, and noise control is the evaluation of sound levels when multiple sources are added together or one or more is removed as a contributor to the total sound. It must be kept in mind that sound levels are expressed as a log X; they are unitless numbers. To add or subtract sounds we must convert their numerical levels in decibels back to *real* units of power, pressure, or intensity, perform the necessary addition or subtraction, and then convert to levels in decibels as desired. The principle involved and its application are best illustrated by examples.

Example 13.1 Two sources are radiating in a free field. One source has a sound power level of 123 dB and the other source has a sound power level of 117 dB, re 10^{-12} W. What is the sound power level when the two sources are combined?

solution

$$L_w = 10 \log \frac{W}{W_0} \quad \text{or} \quad \frac{W}{W_0} = 10^{L_w/10}$$

$$\frac{W_1}{W_0} = 10^{L_{w_1}/10} = 10^{123/10} = 1.995 \times 10^{12}$$

$$\frac{W_2}{W_0} = 10^{L_{w_2}/10} = 10^{117/10} = 5.012 \times 10^{11}$$

Therefore,

$$\frac{W_1}{W_0} + \frac{W_2}{W_0} = 2.496 \times 10^{12}$$

and

$$\begin{aligned} L_w\,(\text{total}) &= 10 \log \left(\frac{W_1}{W_0} + \frac{W_2}{W_0}\right) \\ &= 10 \log(2.496 \times 10^{12}) \\ &= 10 \times 12.40 \\ L_w\,(\text{total}) &= 124 \text{ dB re } 10^{-12} \text{ W} \end{aligned}$$

The same process may be applied to sound intensity level or sound pressure level received at a point. Note, however, that $I \propto P^2$ and $(P/P_0)^2$ must be employed in dealing with sound pressure levels.

Example 13.2 Suppose that the sound pressure level of each of three individual sound sources is measured at a point of reception such that the first source alone contributes an L_p of 86 dB, the second source running alone is received as 84 dB, and, with only the third source, the L_p is 89 dB (re 20 μPa). What will the sound pressure level be at this point with all three sources running?

solution

$$L_p = 10 \log \left(\frac{P}{P_0}\right)^2 \quad \text{or} \quad \left(\frac{P}{P_0}\right)^2 = 10^{L_p/10}$$

$$\left(\frac{P_{\text{total}}}{P_0}\right)^2 = 10^{L_{p_1}/10} + 10^{L_{p_2}/10} + 10^{L_{p_3}/10}$$

$$= 10^{8.6} + 10^{8.4} + 10^{8.9}$$

$$= (3.981 + 2.512 + 7.943) \times 10^8$$

$$= 1.4436 \times 10^9$$

$$L_p \text{ (total)} = 10 \log \left(\frac{P_{\text{total}}}{P_0}\right)^2$$

$$= 10 \log 1.4436 \times 10^9$$

$$= 10 \times 9.16$$

$$L_p \text{ (total)} = 91.6 \text{ dB}$$

In general, then, the procedure for adding sound levels may be summarized as follows:

$$L_\Sigma = 10 \log \left(\sum_{i=1}^{n} 10^{L_i/10} \right) \qquad (13.15)$$

where L may be sound power level, sound intensity level, or sound pressure level.

Example 13.3 In addition to point and line sources of sound, flat areas may also emit sound. For example, sound may be transmitted through a bank of open factory windows 12.57 ft high × 157.1 ft long. If the sound pressure level 3 ft from the windows is 90 dB, what is the sound pressure level 80 ft from the windows?

solution Refer to the area source in Figure 13.4. The sound pressure level is constant to b/π. The attenuation due to distance between b/π and c/π from the windows is -3 dB per doubling of distance $= -10 \log(r_2/r_1)$. At distances greater than c/π, the attenuation is -6 dB per doubling of distance $= -20 \log(r_3/r_2)$.

$$r_1 = \frac{b}{\pi} = \frac{12.57}{\pi} = 4 \text{ ft}$$

$$r_2 = \frac{c}{\pi} = \frac{157.1}{\pi} = 50 \text{ ft}$$

$$r_3 = 80 \text{ ft}$$

$$L_{p_{r1}} = 90 \text{ dB} \qquad \text{(as well as } L_{p0} \text{ or } L_{p3} \text{ at 0 and 3 ft, respectively)}$$

$$L_{p80} = L_{p_{r1}} - 10 \log \left(\frac{r_2}{r_1}\right) - 20 \log \left(\frac{r_3}{r_2}\right)$$

$$= 90 - 10 \log \left(\frac{50}{4}\right) - 20 \log \left(\frac{80}{50}\right)$$

$$= 90 - 11 - 4 = 75 \text{ dB}$$

Adequate approximations can frequently be obtained by using the simplified schedule in Table 13.2 for adding sound levels. The principle of combining sound levels should also be used to correct for ambient

TABLE 13.2 Small-Difference Adjustments with Two Decibel Levels

Difference in decibel values	Add to higher value
0 or 1 dB	3 dB
2 or 3 dB	2 dB
4 to 9 dB	1 dB
10 dB or more	0 dB

Examples: 83 dB + 82 dB ≃ 86 dB
83 dB + 80 dB ≃ 85 dB
83 dB + 78 dB ≃ 84 dB
83 dB + 73 dB ≃ 83 dB

background noise when measuring the sound level contributed by a particular source. In this case a subtraction is involved. To determine how much the background sound level contributes to the total sound level at a given point with a given source in operation, one should turn off the sound source and measure the background sound level. This measurement should be made with the same analysis system that was used when the sound source was operating. If the difference between the background sound level and the total sound level is greater than 10 dB, the background sound does not contribute significantly to a measurement of the source sound. However, if the difference is less than 10 dB, the total sound (or noise) level must be corrected to obtain the actual rather than the apparent sound level of the source (Table 13.3). Although such corrections are frequently applied to overall sound level readings, when apparatus noise is analyzed, the background sound level in each frequency band should also be analyzed to determine if the correction for the sound level in each band is necessary. The spectrum of the background sound is usually different from that of the source noise to be measured, and the corrections in each band will be different. For example, if the reading is 90 dB with the source emitting

TABLE 13.3 Correction for Background Sound to Be Applied Separately to Each Band

dB difference between total and background sound levels	dB to be subtracted from total sound level to get actual sound level of source
8–10	0.5
6–8	1.0
4.5–6	1.5
4–4.5	2.0
3.5	2.5
3	3.0

TABLE 13.4 Ideal Limits for Octave Bands Designated by Center Frequencies

Frequency of lower band limit (Hz)	Center frequency of band (Hz)	Frequency of upper band limit (Hz)
22	31.5	44
44	63	88
88	125	177
177	250	354
354	500	707
707	1,000	1,414
1,414	2,000	2,828
2,828	4,000	5,657
5,657	8,000	11,314
11,314	16,000	22,627

and 83 dB with the source silent, the correct reading for the sound source is 89 dB. If the difference is less than 3 dB, one cannot determine with adequate accuracy the sound level contributed by the noise source alone.

C. Calculation of an A-weighted sound level

All sound pressure measurements are frequency dependent. To simplify our descriptions of sound, it is convenient to divide the frequency scale into octaves or one-third octave bands. For practical purposes the pure tones on audiometers (used to test hearing) and the octave bands utilized by noise-measuring devices are ordinarily set at the center of the frequency bands given in Table 13.4.

Workplace noise measurements are often expressed in terms of a frequency-weighted scale (see also Chapter 2). The most common of these scales is the A-weighted network. To convert a set of octave-band sound pressure levels into an equivalent A-weighted sound level, A-scale correction factors are applied for the nine standard octave center

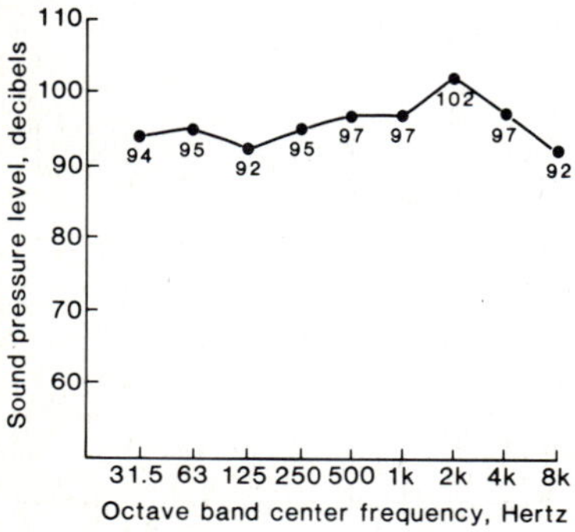

Figure 13.5 Whole-octave band analysis of sound contributed by a boiler feed pump to a point 3 ft away.

TABLE 13.5 Calculation of an A-Weighted Sound Level Contributed by a Boiler Feed Pump to a Point 3 ft Away

Octave band center frequencies (Hz)	Boiler feed pump sound pressure levels (dB) at 3 ft	A-scale correction factors (dB)	Corrected band values (dB)	Decibel addition	A-weighted sound level (dBA)
31.5	94	−39	55		
				69	
63	95	−26	69		
				77	
125	92	−16	76		
				87	
250	95	−9	86		
500	97	−3	94		105
				99	
1K	97	0	97		
				104	
2K	102	+1	103		
				105	
4K	97	+1	98		
				99	
8K	92	−1	91		

or: $L_A = 10 \log(10^{5.5} + 10^{6.9} + 10^{7.6} + 10^{8.6} + 10^{9.4} + 10^{9.7} + 10^{10.3} + 10^{9.8} + 10^{9.1}) = 105$ dBA

frequencies (neglecting 16,000 Hz in Table 13.4), and the corrected values are combined by decibel addition. The A-scale correction factors to be used are the values of the A-weighting network at the center of each particular octave band and are shown in Table 13.5. The resulting value is the A-weighted sound level (dBA). An example of this type of calculation is given for a boiler feed pump (Figure 13.5) in Table 13.5 using small difference adjustments from Table 13.2. The calculated result using Eq. (13.15) for combining sound levels is also given. The reader should compare the tabular estimation with one made employ-

TABLE 13.6 Sources of Noise and Methods of Noise Reduction in Chemical Process Plants

Equipment	Source of noise	Method of noise reduction
Heaters	Combustion at burners	Acoustic plenum[a] (10 Bwg plate)
		Seals around control rods and over sight holes
	Inspiration of premix air at burners	Inspirating intake silencer
	Draft fans	Intake silencer or acoustic plenum
	Ducts	Lagging
Motors	Forward-curved cooling air fan	Intake silencer
		Vaneaxial fan
	Cooling system	Absorbent duct liners
	Mechanical and electrical	Enclosure
Airfin coolers	Fan	Decrease rpm (increasing pitch)
		Tip and hub seals
		Increase number of blades[b]
		Decrease static pressure drop[b]
		Add more fin tubes[b]
	Speed changer	Belts in place of gears
	Motors	Quiet motor
		Slower motor
	Fan shroud	Streamline airflow
		Stiffening and damping (reducing vibration)
Compressors	Discharge piping and expansion joint	Inline silencer and/or lagging
	Antisurge bypass	Use quiet valves and enlarge and streamline piping[b]
		Lag valves and piping
		Inline silencers
	Intake piping and suction drum	Lagging
	Air intake	Silencer
	Discharge to air	Silencer

ing the exponential equation and also note the negligible difference in estimation entailed by ignoring sound contributions from octave bands more than 10 dB smaller than the sound from the octave band contributing the greatest sound pressure level (SPL). For most acoustic applications, the octave frequency band readings are the most useful. It is always possible to reduce an octave band analysis to an A-weighted sound level; but it is not possible to reconstruct the octave band readings from the single A-weighted reading.

II. Noise Control

The general frame of reference in all noise-reduction problems consists of (1) a sound source, (2) the sound wave path, and (3) a sound receiver. The best way to achieve noise reduction is to reduce the amount of waste energy that occurs at the source as mechanical vibration. Mod-

TABLE 13.6 Sources of Noise and Methods of Noise Reduction in Chemical Process Plants (*Continued*)

Equipment	Source of noise	Method of noise reduction
Compressors (*cont'd.*)		
	Timing gears (axial)	Enclosure (or constrained damping on case)
		Silencers on discharge and lagging
	Speed changers	Enclosure (or constrained damping on case)
Engines	Exhaust	Silencer (muffler)
	Air intake	Silencer
	Cooling fan	Enclose intake or discharge or both
		Use quieter fan
Miscellaneous	Turbine steam discharge	Silencer
	Air and steam vents	Silencer
		Use quiet valve
	Eductors	Lagging
	Piping	Limit velocities
		Avoid abrupt changes in size and direction
	Valves	Lagging
		Limit pressure drops and velocities
		Limit mass flow
		Use constant velocity or other quiet valve
		Divide pressure drop
		Size adequately for total flow
		Size for control range
	Pumps	Enclosure

[a]If oil-fired, provide for drainage of oil leaks and inspection.
[b]Usually limited to replacement or new facilities.
SOURCE: Jensen et al., 1978.

ifications of equipment and machinery may include better quality control, specification of closer tolerances on moving parts, better balancing of rotating parts, and sometimes even complete redesign of devices or substitution of basic methods for achieving a particular step in the processing of materials. Usually these modifications are not within the capability of the user and must be left to the equipment manufacturer. The user, however, should exercise the option of specifying the amount and characteristics of noise that will be acceptable from a piece of equipment or machinery. This of course is written into the purchase contract and involves guarantees on the part of the manufacturer and the installer. Such guarantees should extend over a significant length of time, because most moving machinery becomes noisier with age.

The receiver is at the other end of the path. Two methods of reducing noise exposure are available at the receiving end. One is to remove the affected person from the sound field or limit his or her time of exposure; the other is to have the person wear earmuffs or earplugs. The latter procedure is actually a control of the path of noise, but since it directly involves the receiver it is considered a receiver application.

The middle course of action is modification of sound along its path from source to receiver. In the following sections we will consider and provide examples of noise control at the source, along its path of propagation, and at the receiver. Applications of some of these control methods in the chemical process industry are illustrated in Table 13.6.

A. Source radiation

Although the sound power of a source is independent of the environment in which it is located, sound pressure levels around the source are not, and sound pressure is what ears and microphones respond to. If a machine could be suspended in midair, far away from any reflecting objects, including the ground, it would be in a free field. Sound pressure levels around the machine would be due only to the machine, and if it were a point source, sound pressure levels would decrease 6 dB every time the distance from the machine were doubled, that is, 0.5 to 1 m, 1 m to 2 m, etc. The sound attenuation would follow an inverse-square law.

In practice, neither of these conditions exists. Industrial machinery cannot be suspended in midair, and machines are not point sources. Even if a machine could be suspended in a true free field, the inverse-square law may not apply to measurements close to its surface. Noise is produced by vibration of various parts of the machine, and not all parts vibrate in phase with each other. Some parts move out of phase while others move in. This is true even on relatively small areas of the same part. Higher frequencies can be produced by very complex "vibration mode shapes." There is partial cancellation of sound waves at some locations, and some reinforcement at others. The result is that sound pressure level does not decrease 6 dB each time the distance from the source is doubled, even though the inverse-square law says it should. At a distance of about twice the largest dimension of the machine, these near-field effects seem to disappear, and the inverse-square law becomes valid (see Figure 13.4). However, when predicting employee noise exposure, we cannot rely on a decrease of 6 dB per doubling of distance, and the larger the machine, the less the drop. If the machine is mounted on a hard, smooth floor, but still with no other reflecting surface anywhere near it—no wall, ceiling, or other object, in any direction—it is considered to be in a free field over a reflecting plane.

Measured sound pressure levels will be 3 dB higher than they would be if the machine was hanging in midair, because the sound power that was radiated downward when it was hanging in midair now strikes the reflecting floor and is radiated upward. In the free-field case, the total sound power was radiated into a hypothetical sphere surrounding the machine. When the machine is placed on the floor, all the sound power is radiated into a hypothetical hemisphere over the machine. That is, the hemisphere now receives twice the sound power that it did in the first case. Twice the sound power represents a 3-dB increase, and thus the measured sound pressure levels will be 3 dB greater.

If the floor is not perfectly reflecting, not all the sound power striking it will be reflected; some will be absorbed by the floor. Most industrial floors, such as concrete, reflect about 98% of the sound striking them. If all the walls, the floor, and the ceiling are hard reflecting surfaces, all the sound is reflected, again and again. The sound level measured at any point in the room is the sum of that radiated directly by the machine, plus all the reflected components coming from many different directions. The sound pressure level does not decrease according to the inverse-square law but becomes uniform within a somewhat reverberant room. Practically all industrial machinery installations are in environments that are neither a true free field nor a true reverberant field, but somewhere in between the two. This is called a semireverberant field, and it is the condition of greatest practical importance.

Noise measured around a machine in a semireverberant location is due to the sum of two components: the noise radiated directly by the machine; and the noise reflected from walls, floor, and ceiling. Close to the machine, most of the noise is due to the directly radiated component. Close to the walls, the reflected component may be predominant. The total may be calculated mathematically if the sound power level of the machine, distance from the machine, and room constant are known. Room constant depends, in turn, on the area of the floor, ceiling, and walls, and the type of material used in their construction. Materials that reflect most sound and absorb only a little are said to have a low "absorption coefficient." Materials that absorb sound and reflect very little have high absorption coefficients. Whenever noise measurements are made indoors, these concepts must be kept in mind.

In actual practice, noise sources do not radiate sound uniformly in all directions. It is also true that at low frequencies, many sources of noise are nearly nondirectional. However, as the frequency increases, directivity increases. Directivity can best be illustrated by noise radiating out of an opening, such as a forced-draft fan inlet. For cases such as this, directivity correction factors can be applied. As can be seen in Figure 13.6, the absolute magnitude of correction factors for directivity increases with frequency for a given angle between the opening and receiver. In addition, as the angle between opening centerline and receiver increases, the magnitude of the directivity correction factor, plus or minus, also increases. It should also be mentioned that if a given source has a given directivity in free space, the actual direc-

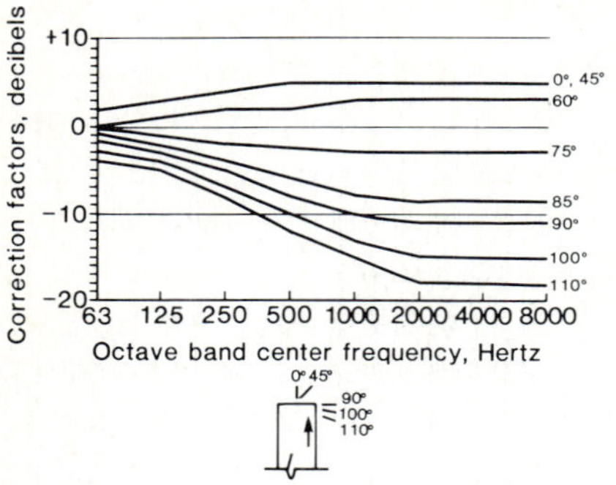

Figure 13.6 Directivity corrections in relation to frequency and angle between centerline of sound source opening and receiver.

tivity within a room may vary depending on the location of the source in the room due to reflections from nearby surfaces. This will be shown in the section dealing with sound in large rooms.

B. Barrier design

Barriers are frequently used to reduce sound propagation. The design of an effective barrier involves a proper application of the basic physics of transmission loss through masses, standing waves behind reflectors, absorption at surfaces, and sound diffraction around obstacles.

Diffraction refers to the bending of sound around obstacles. If the wavelength of the sound is small (i.e., high frequency) in comparison with the size of the object, the sound will be reflected or scattered in many directions and the obstacle will cast a shadow.

Transmission loss is generally not a major consideration in the design of barriers. As a rule of thumb, if a barrier of sufficient height to shield the noise is constructed of a material that can support itself and simultaneously resist the wind load imposed upon it, it will meet the transmission loss requirements. Some suggest a minimum mass of 4 lb/ft^2.

The placement of a barrier near a sound will create standing waves of sound between the source and the wall if discrete tones are present. When the tone is reflected by a wall, the original wave and the reflection combine to form standing waves. Improper location of a barrier can result in a sound level "build-up" because of these standing waves on the interior side of the barrier and reduce its effectiveness. In order to limit this increased sound level, the source-wall distance should be an odd multiple of the quarter-wavelength of the predominant noise frequency. Lining the inside of the barrier with sound-absorbing material also increases attenuation by reducing the reflected sound.

The main factors to consider in barrier design are the dimensions of the barrier, height and width. The diffraction of sound around a barrier is particularly troublesome at low frequencies. Sound diffraction at a barrier produces "spill-over" of sound at the edges. Usually, significant reduction is obtained only when the barrier extends one wavelength

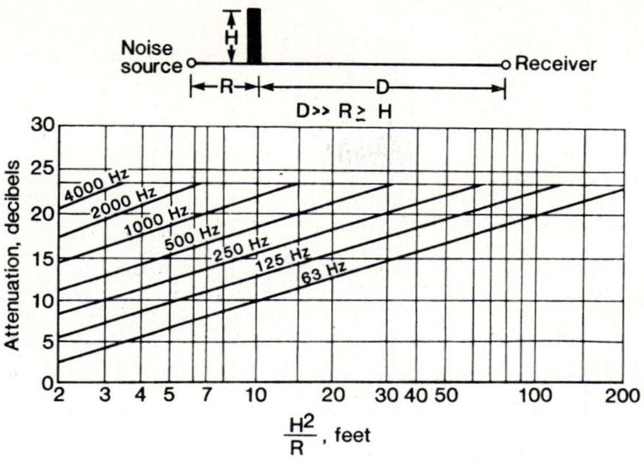

Figure 13.7 Sound attenuation by barriers relative to frequency and as a function of barrier height and distance from source. (Maximum attenuation value is 24 dB.)

above and to either side of the line of sight between the noise source center and the observer. The size of such extensions is obviously quite large for low-frequency sound.

Values for barrier attenuation are provided in Figure 13.7. Note that barrier attenuation increases with the ratio (H^2/R). Note also that the high frequencies are attenuated more than the low frequencies for a given configuration. An upper limit of barrier attenuation equal to 24 dB is also designated in the figure. This limit is due to atmospheric irregularities and turbulence and has been empirically validated in practice.

Example 13.4 A source of sound is emitting 94 dB in the octave band centered on 31.5 Hz. In the octave bands identified by center frequencies of 63, 125, 250, 500, 1k, 2k, 4k, 8k, and 16k Hz, the sound emissions are 95, 92, 95, 97, 97, 102, 97, 92, and 70 dB, respectively. If a barrier having a mass of 12 lb/ft^2 that is 8.5 ft high and is located 20 ft from the source on level ground, what would the sound pressure level be at a receiver located 200 ft from the source on the other side of the barrier? What would the A-weighted sound level be?

solution Referring to Figure 13.7, the conditions that $D >> R \geq H$ are met. Transmission through the barrier is insignificant (12 lb/ft^2 > 4 lb/ft^2).

Attenuation attributable to the barrier in each octave band when $H^2/R = 8.5^2/20 = 3.61$ is: 2 dB for 31.5 Hz, 5 dB for 63 Hz, 8 dB for 125 Hz, 11 dB for 250 Hz, 13 dB for 500 Hz, 17 dB for 1 kHz, 21 dB for 2 kHz, 24 dB for 4 kHz, 24 dB for 8 kHz, and 24 dB for 16 kHz (from Figure 13.7).

Let: L_Σ = summation of sound level components
L_p = sound pressure level at a point of reception
L_{SB} = sound level after attenuation by a barrier but not including attenuation due to distance
L_P = A-weighted sound pressure level at a point of reception

As a function of the barrier alone,

$$\begin{aligned} L_\Sigma &= 10 \log \left(\sum_{i=1}^{n} 10^{L_i/10} \right) \\ &= 10 \log(10^{9.2} + 10^{9.0} + 10^{8.4} + 10^{8.4} + 10^{8.4} \\ &\quad + 10^{8.0} + 10^{8.1} + 10^{7.3} + 10^{6.8} + 10^{5.6}) \\ &= 95.6 \text{ dB} \end{aligned} \tag{13.15}$$

which, when attenuated by hemispherical radiation at a distance of 200 ft, becomes

$$\begin{aligned} L_p &= L_{SB} - 20 \log r + 2.3 \\ &= 95.6 - 20 \log 200 + 2.3 \\ &= 51.9 \text{ dB} \end{aligned} \qquad (13.10a)$$

The A-weighted level of the sound if attenuated by the barrier would be

$$10 \log(10^{5.3} + 10^{6.4} + 10^{6.8} + 10^{7.5} + 10^{8.1} + 10^{8.0} + 10^{7.4} + 10^{6.7}) = 86.6 \text{ dBA}$$

which, when attenuated by hemispherical radiation at a distance of 20 ft, becomes

$$L_{P_A} = 86.6 - 20 \log 200 + 2.3 = 42.9 \text{ dBA}$$

C. Boundary surface treatment

In addition to barrier dimensions, the noise-reflecting and -absorbing characteristics of the barrier surface are also significant design considerations. Some of these characteristics are depicted schematically in Figures 13.8, 13.9, and 13.10.

1. A good sound-absorbing material is a poor barrier (Figure 13.8). Because sound-absorbing materials have soft, porous structures, they offer low resistance to a sound wave and permit passage of sound energy through to the other side with little attenuation. If air can pass through the material, so can sound.
2. A solid partition has a high transmission loss, and most of the incident sound is reflected back toward the source (Figure 13.9).

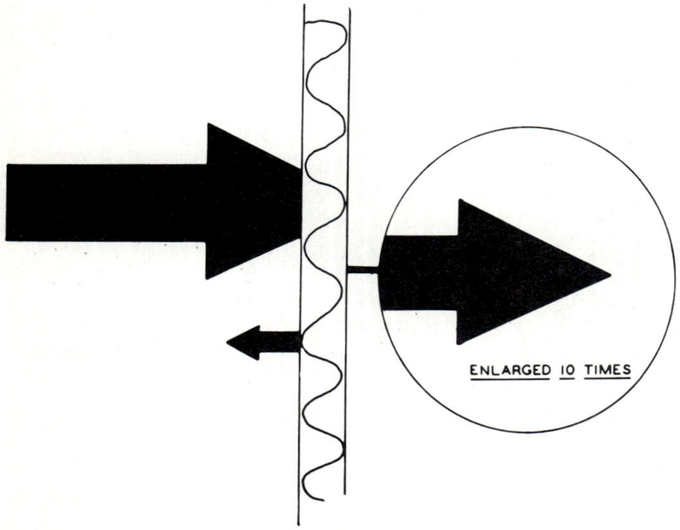

Figure 13.8 A graphical representation of the transmission of sound through a sound-absorbing material.

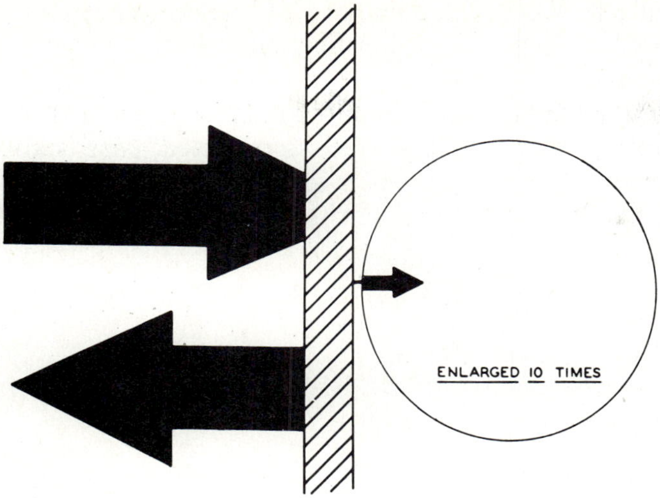

Figure 13.9 A graphical representation of the transmission of sound through a solid partition.

Good sound barrier materials are hard and heavy. They generally possess considerable mass, and as such offer a high resistance to the passage of sound.

3. A good sound barrier material is usually a poor absorber, and good absorbent material is usually a poor barrier. Thus, the better acoustical treatment is almost always a combination of materials (Figure 13.10). The solid partition acts to reduce sound transmission, while

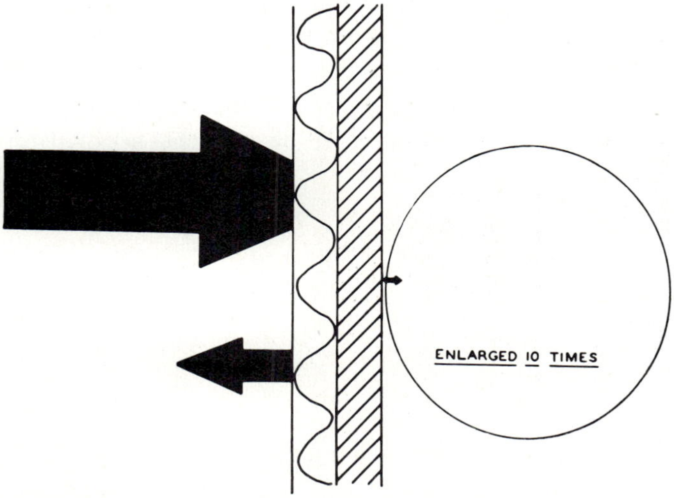

Figure 13.10 A graphical representation of the transmission of sound through a combination of sound-absorbing material and a solid partition.

the absorbent layer converts sound energy to heat, thereby reducing the amount of sound available for both reflection and transmission.

1. Sound absorption coefficients. From the foregoing, it is obvious that part of the sound impinging upon any surface is reflected, part is absorbed, and part is transmitted. If I_0 is the original intensity of sound at a boundary surface, we could determine the intensity of sound absorbed, I_A, from:

$$I_0 = I_{\text{Reflected}} + I_{\text{Absorbed}} + I_{\text{Transmitted}}$$

and the absorption coefficient of the partition under consideration may be designated α. Thus,

$$\alpha = \frac{I_A}{I_0} = \frac{I_0 - I_R - I_T}{I_0} \tag{13.16}$$

Unfortunately, the absorption coefficient of most materials is not the same for all frequencies. The absorption coefficients at six frequencies by octaves ranging from 125 Hz to 4000 Hz are given in Table 13.7 for a variety of construction materials. Table 13.8 is for carpeted surfaces. The values shown are approximate. More precise sound absorption coefficient values depend on specific construction, density of material, type of finish, size of panel, type of mounting, angle of incidence, and other factors.

2. Noise-reduction coefficients. The noise-reduction coefficient (NRC) of a material is a single acoustical index number that is the arithmetic mean of the absorption coefficients of the material at frequencies of 250, 500, 1000, and 2000 Hz. This average is expressed to the nearest multiple of 0.05. For example, if a loop pile tufted carpet having a pile weight of 20 oz/yd^2 is laid on a hair pad of 40 oz/yd^2, the absorption coefficients from Table 13.8 are:

Frequency (Hz)	*Absorption coefficient*
125	0.10
250	0.19
500	0.35
1000	0.79
2000	0.69
4000	0.79

Then the NRC for this material is $(0.19 + 0.35 + 0.79 + 0.69)/4 = 0.50$.

In addition to decorative and ergonomic advantages, the carpeting of concrete floors with and without pad underlayments has very noticeable acoustical effects. Noise-reduction coefficients achieved with various types of carpets are shown in Tables 13.9 and 13.10. The use of underlayment significantly increases the NRCs, especially if the carpet backing is permeable and uncoated. The type of fiber (i.e., nylon, wool, or acrylic) has no significant effect on absorption.

3. Sound absorption in large rooms. The absorption of sound by a surface is the product of the absorption coefficient and the area of the surface. The unit of sound absorption is the sabin. By way of definition, the absorption of sound by 1 ft^2 of perfectly absorbing surface ($\alpha = 1.00$) is 1 sabin. The average absorption, $\overline{\alpha}$, of an enclosure or room may be determined by the sum of absorptions of each area comprising the room. Thus:

$$\overline{\alpha} = \frac{\sum \alpha_i S_i}{\sum S_i} \tag{13.17}$$

where α_i = random incidence absorption coefficient of the ith surface; and S_i = area of that surface in square feet.

The total absorption of the room is given by the numerator:

$$R = \sum \alpha_i S_i = S\overline{\alpha} \tag{13.18}$$

where S is the total surface area and R is called the room constant.

The effect that room absorption and distance from a noise source have on sound pressure level is shown in Figure 13.11, where the sound pressure level, L_p, relative to the sound power level, L_w, of a noise source is plotted vs. distance, r, between the two (see p. 690). A family of curves for several values of room constant has been drawn, and practical values of $L_p - L_w$ in decibels may be obtained from the diagram for distances up to 100 m and room constants ranging from zero to infinity. If the room does not affect the sound field ($\overline{\alpha} = 1$), then the relationship between L_p and L_w reduces to the free-field case (Eq. 13.11). If the room constant is near zero (almost perfect reflectance), the sound pressure level does not differ anywhere in the room, and an ideal reverberant field exists. This information is very important when considering acoustical treatment of a room.

In many cases, the operator of a piece of machinery is probably affected more by the direct field of the noise source rather than the reverberant field. Consequently, absorption treatment on the walls and ceiling will not reduce the noise of this machine at the operator's position. It can be seen in Figure 13.11 that the sound pressure level can

TABLE 13.7 Sound Absorption Coefficients

Material	Sound absorption coefficients					
	125 Hz	250 Hz	500 Hz	1000 Hz	2000 Hz	4000 Hz
Ballast or other crushed stone:						
3.18-cm ($1\frac{1}{4}$-in.) screened ballast 15.2 cm (6 in.) deep	0.19	0.23	0.43	0.37	0.58	0.62
3.18-cm ($1\frac{1}{4}$-in.) 30.5 cm (12 in.) deep	0.27	0.58	0.48	0.54	0.73	0.63
3.18-cm ($1\frac{1}{4}$-in.) 45.7 cm (18 in.) deep	0.41	0.53	0.64	0.84	0.91	0.63
0.64-cm ($\frac{1}{4}$-in.) or less granite aggregate 15.2 cm (6 in.) deep	0.22	0.64	0.70	0.79	0.88	0.72
Brick, unglazed	0.03	0.03	0.03	0.04	0.05	0.07
Brick, unglazed, painted	0.01	0.01	0.02	0.02	0.02	0.03
Carpet, heavy, on concrete	0.02	0.06	0.14	0.37	0.60	0.65
Same on 1350 g/m^2 (40 oz/yd^2) hairfelt or foam rubber	0.08	0.24	0.57	0.69	0.71	0.73
Same, with impermeable latex backing on 1350 g/m^2 (40 oz/yd^2) hairfelt or foam rubber	0.08	0.27	0.39	0.34	0.48	0.63
Concrete block, coarse	0.36	0.44	0.31	0.29	0.39	0.25
Concrete block, painted	0.10	0.05	0.06	0.07	0.09	0.08
Drapes:						
Light velour 338 g/m^2 (10 oz/yd^2) hung straight, in contact with wall	0.03	0.04	0.11	0.17	0.24	0.35
Medium velour, 475 g/m^2 (14 oz/yd^2) draped to half area	0.07	0.31	0.49	0.75	0.70	0.60
Heavy velour, 610 g/m^2 (18 oz/yd^2) draped to half area	0.14	0.35	0.55	0.72	0.70	0.65
Fiber-glass boards and blankets:						
2.54-cm (1-in.) glass wool 24 kg to 48 kg/m^3 (1.5 to 3.0 lb/ft^3)	0.08	0.25	0.65	0.85	0.80	0.75
5.1-cm (2-in.) glass wool 24 kg to 48 kg/m^3 (1.5 to 3.0 lb/ft^3)	0.17	0.55	0.80	0.90	0.85	0.80

2.54-cm (1-in.) glass wool, 2.54-cm (1-in) air space	0.15	0.55	0.80	0.90	0.85	0.80
5.1-cm (2-in.) glass-fiber panels with plastic sheet wrapping and perforated metal facing, as installed	0.33	0.79	0.99	0.91	0.76	0.64
Floors:						
Concrete or terrazzo	0.01	0.01	0.015	0.02	0.02	0.02
Linoleum, asphalt, rubber, or cork tile on concrete	0.02	0.03	0.03	0.03	0.03	0.02
Wood	0.15	0.11	0.10	0.07	0.06	0.07
Wood parquet in asphalt on concrete	0.04	0.04	0.07	0.06	0.06	0.07
Glass:						
Large panes of heavy plate glass	0.18	0.06	0.04	0.03	0.02	0.02
Ordinary window glass	0.35	0.25	0.18	0.12	0.07	0.04
Gypsum board, 1.27 cm ($\frac{1}{2}$-in.), nailed to 5. 1 cm × 10.2 cm (2 in. × 4 in.) studs 41 cm (16 in.) c.-to-c.	0.29	0.10	0.05	0.04	0.07	0.09
Marble or glazed tile	0.01	0.01	0.01	0.01	0.02	0.02
Mineral spray-on materials:						
1.27-cm ($\frac{1}{2}$-in.) mineral fiber	0.05	0.15	0.45	0.70	0.80	0.80
1.9-cm ($\frac{3}{4}$-in.) mineral fiber	0.10	0.30	0.60	0.90	0.90	0.95
2.5-cm (1-in.) mineral fiber	0.16	0.45	0.70	0.90	0.90	0.85
1.27-cm ($\frac{1}{2}$-in.) mineral fiber on metal lath, 2.54-cm (1-in.) air space	0.25	0.50	0.80	0.90	0.90	0.85
Plaster, gypsum or lime, smooth finish on tile or brick	0.013	0.015	0.02	0.03	0.04	0.05
Plaster, gypsum or lime, rough finish on lath	0.14	0.10	0.06	0.05	0.04	0.03
Same, with smooth finish	0.14	0.10	0.06	0.04	0.04	0.03
Plywood paneling, 1 cm ($\frac{3}{8}$-in.) thick	0.28	0.22	0.17	0.09	0.10	0.11
Water surface, as in a swimming pool	0.008	0.008	0.013	0.015	0.020	0.025

SOURCE: Sabine and Moulder, 1979.

TABLE 13.8 Typical Sound Absorption Coefficients for Carpet

Construction	125 Hz	250 Hz	500 Hz	1000 Hz	2000 Hz	4000 Hz	NRC
Carpet (no pad):							
Woven wool loop, 1.2 kg/m^2 (35 oz/yd^2) 2.4 mm ($\frac{3}{32}$ in.) pile ht	0.10	0.16	0.11	0.30	0.50	0.47	0.25
Woven wool loop, 1.4 kg/m^2 (40 oz/yd^2) 6.4 mm ($\frac{1}{4}$ in.) pile ht	0.15	0.17	0.12	0.32	0.52	0.57	0.30
Woven wool loop, 2.3 kg/m^2 (66 oz/yd^2) 9.5 mm ($\frac{3}{8}$ in.) pile ht	0.17	0.18	0.21	0.50	0.63	0.83	0.40
Loop pile tufted carpet, 1.4 kg/m^2 (40 oz/yd^2) 9.5 mm ($\frac{6}{16}$ in.) pile ht:							
Hair pad, 1.4 kg/m^2 (40 oz/yd^2)	0.03	0.25	0.55	0.70	0.62	0.84	0.45
Hair pad, 3.0 kg/m^2 (86 oz/yd^2)	0.10	0.40	0.62	0.70	0.63	0.88	0.60
Hair and jute pad, 3.0 kg/m^2 (86 oz/yd^2)	0.20	0.50	0.68	0.72	0.65	0.90	0.65
Loop pile tufted carpet, 0.7 kg/m^2 (20 oz/yd^2):							
Without pad	0.04	0.08	0.17	0.33	0.59	0.75	0.30
With 1.4 kg/m^2 (40 oz/yd^2) hair pad	0.10	0.19	0.35	0.79	0.69	0.79	0.50

SOURCE: American Carpet and Rug Institute, Sound conditioning with carpet, in Sabine and Moulder, 1979.

TABLE 13.9 Sound Absorption of Carpet on Bare Concrete

Construction	Pile weight	Pile height	Surface	Fiber	NRC
Woven	1.2 kg/m² (35 oz/yd²)	4 mm (0.175 in.)	Loop	Wool	0.30
Woven	1.2 kg/m² (35 oz/yd²)	4 mm (0.175 in.)	Cut	Wool	0.35
Tufted	1.1 kg/m² (32 oz/yd²)	14 mm (0.56 in.)	Cut	Nylon	0.50
Tufted	1.3 kg/m² (36.5 oz/yd²)	11 mm (0.44 in.)	Cut	Acrylic	0.50
Tufted	1.5 kg/m² (43 oz/yd²)	13 mm (0.50 in.)	Cut	Wool	0.55
Woven	1.5 kg/m² (44 oz/yd²)	6 mm (0.25 in.)	Loop	Wool	0.30
Woven	2.3 kg/m² (66 oz/yd²)	10 mm (0.375 in.)	Loop	Wool	0.40
Woven	3.1 kg/m² (88 oz/yd²)	13 mm (0.50 in.)	Loop	Wool	0.40
Tufted	0.5 kg/m² (15 oz/yd²)	6 mm (0.25 in.)	Loop	Nylon	0.25
Tufted	1.4 kg/m² (40 oz/yd²)	6 mm (0.25 in.)	Loop	Wool	0.35
Tufted	2.1 kg/m² (60 oz/yd²)	6 mm (0.25 in.)	Loop	Wool	0.30

SOURCE: Sabine and Moulder, 1979.

TABLE 13.10 Sound Absorption of Various Carpets on a 1.4-kg/m² (40-oz/yd²) Hair Pad (Underlayment)

Construction	Pile weight	Pile height	Surface	Fiber	NRC
Woven coated	1.5 kg/m² (44 oz/yd²)	6 mm (0.25 in.)	Loop	Wool	0.40
Woven uncoated	1.5 kg/m² (44 oz/yd²)	6 mm (0.25 in.)	Loop	Wool	0.60
Knitted	1.4 kg/m² (40 oz/yd²)	5–10 mm (0.2–0.4 in.)	Loop	Wool	0.65
Tufted	0.5 kg/m² (15 oz/yd²)	6 mm (0.25 in.)	Loop	Nylon	0.65
Tufted	1.5 kg/m² (43 oz/yd²)	6 mm (0.25 in.)	Loop	Acrylic	0.50
Tufted	1.4 kg/m² (40 oz/yd²)	10 mm (0.39 in.)	Loop	Wool	0.60
Tufted	1.1 kg/m² (32 oz/yd²)	14 mm (0.56 in.)	Cut	Nylon	0.70
Tufted	1.5 kg/m² (43 oz/yd²)	13 mm (0.50 in.)	Cut	Wool	0.70

SOURCE: Sabine and Moulder, 1979.

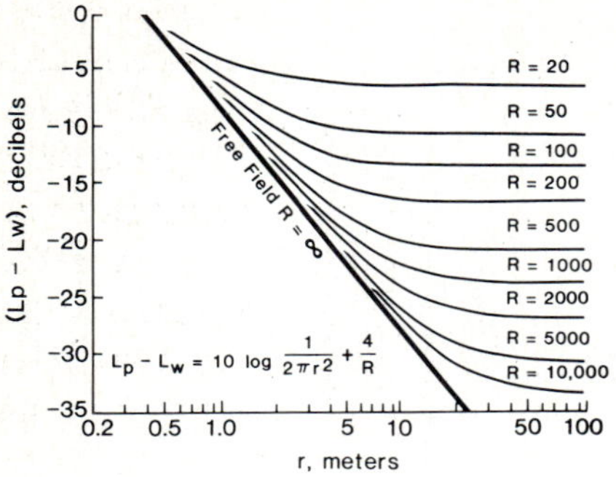

Figure 13.11 Relative sound pressure level vs. distance from the source for semireverberant fields; r = distance from acoustic center, R = room constant.

never be below the straight line corresponding to the free-field $1/r^2$ decrease.

Employees a little farther from the machine will probably be in a region that can be called a semireverberant field, that is, where the sound pressure level is made up of some combination of the direct and reflected sound. Figure 13.11 can be used as a quick guide to determine if sound absorption treatment reduces the noise level at a given location.

Suppose that it is desired to reduce the sound pressure level at a particular operator's position that is 8 m from the sound source with a sound power level of L_w. If the room constant is determined to be about 1000, any absorbent material on the walls will have a negligible effect on the sound pressure level at this position. If, however, the room constant is significantly below 1000, an absorption treatment of the surfaces (for example, R after treatment $= R' = 1000$) can have an appreciable effect. The effect would be equal to $(L_p - L_w)$, when $R < 1000$, minus $(L_p' - L_w)$, when $R' = 1000$.

Thus, in practice, one must first determine when absorption treatment will be useful. It is important to look at the equations on which Figure 13.11 is based. At a point some distance r from a sound source in a large room, the sound pressure level is equal to the sum of the contributions of the direct field and the reverberant field. This is expressed mathematically in the following general form, using mks units:

$$L_p = L_w + 10 \log\left(\frac{Q}{4\pi r^2} + \frac{4}{R}\right) \quad (\text{dB re } 2 \times 10^{-5} \text{ Pa}) \qquad (13.19)$$

where L_w = sound power level of source (dB re 10^{-12} W)
r^2 = squared distance from the source (m^2)
R = room constant (m^2)
Q = directivity factor of the source

Equation (13.19) describes the sound field at all points in a room regardless of the relative magnitudes of the direct and reverberant sound fields. If English units are used, 10 dB must be added to the right side of the above equation.

The directivity factor Q in Eq. (13.19) deserves further discussion. For a nondirectional source in a free field, $Q = 1$. If a machine is mounted on a hard floor in the center of a large room, $Q = 2$. Close to where a wall meets the floor, there are two reflecting surfaces instead of one, and $Q = 4$. In the corner, where two walls meet the floor, there are three reflecting surfaces, and $Q = 8$. Another way to look at the problem is to consider that all the sound power that would be radiated into a hypothetical sphere surrounding the machine, if it were in a free field, will be directed into one-eighth of the sphere when the machine is backed into a corner.

Clearly, the larger R becomes, the lower the sound pressure level will be. To determine the decrease in sound pressure level when absorption is added to a room, which increases the room constant R, one could calculate the value of L_p from Eq. (13.19) for R before treatment and again for R', the room constant after treatment; or by the equivalent relation:

$$\text{Reduction in dB} = NR = (L_p - L_w) - (L_p' - L_w)$$

$$L_p - L_p' = 10 \log\left(\frac{(Q/S) + (4/R)}{(Q/S) + (4/R')}\right)$$

where L_p' is the sound pressure level after acoustic treatment. The use of this equation is illustrated by the following example.

Example 13.5 The dimensions of a room are 50 ft (length), 25 ft (width), and 12 ft (height). The absorption in the room is shown below using typical absorption coefficients (Table 13.7).

Room component	S_i (ft^2)	α_i @ 500 Hz	$S_i\alpha_i$ (sabins)
Floor	1250	0.072	90.0
Ceiling, plaster	1250	0.06	75.0
Side walls	1200	0.054	64.8
End walls	600	0.082	49.2
			279.0

Given these initial room conditions:

a. Can absorption treatment be effective in reducing the 500-Hz octave band sound level of a newly installed machine at a position 8 m from an observer?

b. How much reduction of the sound level from this source will be achieved using absorption treatment?

solution

a. The room constant can be calculated by:

$$R = \sum S_i\alpha_i = 279.0 \text{ ft}^2 = 26 \text{ m}^2$$

From Figure 13.11, a room constant corresponding to 26 m^2 is well above the free-field curve at $r = 8$ m. Absorption can thus be effective for this position. Note, however, that for a given room constant, absorption treatment will have less of an effect as the distance from the machine, r, decreases. Also, the use of absorber materials can never reduce the sound to a level below that of the free-field radiation. At best, a sound absorber can reduce the reflections to zero, which is the same as removing the surface entirely (i.e., no surface = no reflection = perfect absorber).

b. Now suppose that an acoustical wall treatment with an absorption coefficient of 0.95 at 500 Hz and a type of mineral-fiber ceiling panels with an absorption coefficient of 0.80 at 500 Hz are selected for absorption treatment (Table 13.7). The following room components are used to calculate the effects of absorption treatment:

Room component	S_i (ft^2)	α_i @ 500 Hz	$S_i\alpha_i$ (sabins)
Floor	1250	0.072	90.0
Treated ceiling	1250	0.80	1000.0
Treated side walls	1200	0.95	1140.0
Treated end walls	600	0.95	570.0
			2800 (260 m^2)

The reduction in sound level for applications of the absorption materials are calculated below with a directivity constant $Q = 2$:

$$\text{Reduction in dB} = 10 \log\left(\frac{(2/4\pi r^2) + (4/R)}{(2/4\pi r^2) + (4/R')}\right)$$

Substituting $r = 8$ m, $R = 26$ m^2, $R' = 260$ m^2 yields a reduction = 9.42 dB. This agrees with an estimate obtained from Figure 13.11. Note that this procedure can be applied to partial absorption treatment of the room as well.

In summary, the noise measured at a machine operator's location is due to a combination of several factors:

1. The direct noise of the machine itself, usually stated in terms of free field (over a reflecting plane)
2. The dimensions of the room where the machine is installed
3. The kind of material used on the floor, ceiling, and walls
4. The location of the machine in the room with respect to the various walls
5. The number of machines in the room and their spacing
6. The ambient noise when no machine is operating

D. Source control

Several noise control efforts that do not involve process design changes can be effective in reducing noise generation at the source. These alternatives are not necessarily simple or cheap, but they should be considered before exploring more complex solutions. The controls are: proper maintenance; changing operating procedures; replacing equipment; relocating the equipment; and operating machines at proper speed. If the foregoing controls are ineffective or seem noncompetitive with other control measures, vibration control at the source may be achieved by: reducing impacts in the source, balancing rotating or reciprocating components, smoothing mechanical and inertial fluctuations, reducing vibrations at connections, employing isolators, or damping the mechanical energy of vibration.

Vibration due to *impact forces* may be reduced by modifying the interaction of components that strike each other to produce the impact, provided that such modification does not impair their functioning. For example, in a stamping machine, a reduced force that is applied more slowly and that acts over a longer period might be substituted for a more rapid, but more severe, stamping force. Where feasible, impact forces may be reduced by introducing cushioning, e.g., in the form of a soft metal (such as lead, which may need to be replaced from time to time), rubber, or plastic.

Forces that result from the *lack of balance* of rotating or reciprocating components can be reduced by balancing or by replacing such components with better-balanced ones. Balancing of rotating systems requires specialized equipment as described by Muster and Stadelbauer (1976). Such balancing of reciprocating systems in the field generally is difficult and impractical for noise control purposes. For many types of machines, balance quality grades have been established by International Standard ISO 1940–1973 (E) (ISO, 1973). These grades give permissible unbalance magnitudes and permit estimation of the amount of unbalance in a given machine or specification of the balance quality required.

The unbalanced force in a given machine acts at the rotational frequency when n denotes the rotational speed:

$$f_r = \frac{n}{60} \qquad \text{(Hz)} \tag{13.20}$$

in revolutions per minute. For a machine whose imbalance is known, one may calculate the magnitude, F, of the unbalanced force from

$$F = m_u \epsilon_u \omega^2 \tag{13.21}$$

where $\omega = 2\pi f_r$, m_u represents the unbalanced mass of the rotor, and

ϵ_u is its eccentricity (the distance of the rotor's center of gravity from the axis of rotation). Consistent units must be used. For example, to obtain F in pounds, use m_u in slugs and ϵ_u in feet; to obtain F in newtons, use m_u in kilograms and ϵ_u in meters.

Nonconstant friction forces usually are caused by poor lubrication or by unfavorable combinations of sliding interface materials, geometric arrangements, and sliding velocities. These forces sometimes can be eliminated by providing improved lubrication or by changing interface materials and/or the geometry of the sliding surfaces.

Unsteady fluid forces that result from fluid-moving devices such as compressors, pumps, fans, or blowers may be reduced by changing or modifying these devices or by inserting components into the fluid path that reduce these forces. Unsteady forces resulting from turbulent flows or vortices may be reduced by the application of turbulence or vortex suppression devices. Where unsteady forces arise due to flutter (an unstable interaction between a structural component and a fluid flow), reduction of these forces generally requires that this unstable condition be eliminated, e.g., by changing the structural configuration exposed to the flow or by changing the flow speed or direction.

Fluctuating magnetic forces, such as occur in electrical motors and transformers, usually cannot be changed without modifying those devices significantly.

Unsteady mechanical interaction forces, such as occur in ball and roller bearings or between meshing gears, usually can be reduced by component replacement or design changes. The vibrations due to bearings may be reduced by using high-quality bearings, which are made to closer tolerances. The vibration and noise due to gears may be reduced by using helical or spiral gear sets instead of spur or bevel gears, by using higher-quality gears (made to closer tolerances) or gears with finer pitches and shorter teeth, by improving the alignment between mating gears, or by improving gear lubrication.

E. Vibration reduction at connections

Where an item to be protected from vibrations is attached to a vibrating structure, as shown schematically in Figure 13.12*a*, its vibration may be reduced: (1) by reducing the motion of the vibrating structure, and/or (2) by providing interconnections that are so "soft" (resilient) that little of that structure's motion is transmitted to the item, as indicated schematically in Figure 13.12*b*. The term "isolation" usually is employed to designate the process of reducing the vibration of the item to be protected by means of connections that transmit relatively little vibratory motion; the resilient connectors usually are called vibration

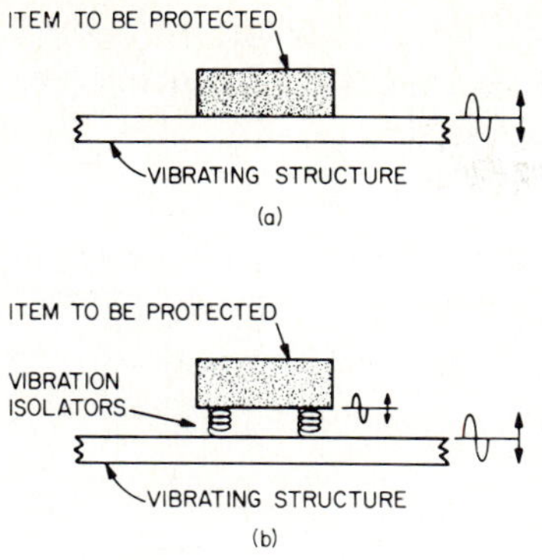

Figure 13.12 Schematic diagram of item connected to vibrating structure: (*a*) without vibration isolators, (*b*) with vibration isolators (*Ungar and Cohen, 1979*).

isolators. A common example of such an isolator is the cloth connection between a centrifugal-fan air blower and a main ventilation duct.

To reduce the vibration of an item that is attached to a vibrating structure, one or more of the following means may be employed:

1. *Reduction of the vibration of the vibrating structure at the connection location.* Depending on the structure and the frequencies involved, this reduction may be achieved by adding stiffeners, mass, or damping, as shown schematically in Figure 13.13. For example, reduction of vibration of a sheet metal structure (at frequencies above about 30 Hz) usually may be accomplished most simply by the addition of mass; the added mass generally should be at least twice the mass of the local vibrating structure.
2. *Insertion of soft connectors (vibration isolators),* as shown schematically in Figures 13.12 and 13.13. These connectors should be as soft as possible, consistent with practical limits on deflection due to static and shock loads. The vibration isolators must be soft enough so that the natural frequency, f_n, of the item to be protected, as

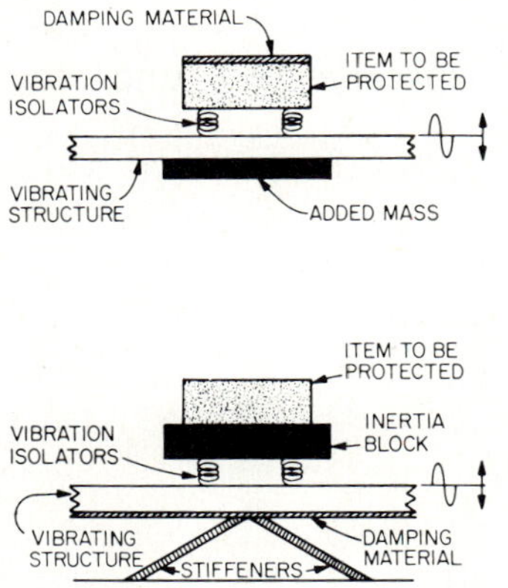

Figure 13.13 Schematic diagram of means for reducing vibrations of item connected to vibrating structure (*Ungar and Cohen, 1979*).

attached to the isolators, is smaller by at least a factor of 2 than the lowest excitation frequency of concern. (Vibration isolators are discussed in more detail later in this chapter.)

3. *An increase in the rigidity of the item to be protected,* or at least of its mounting base, so that this item deforms less as the result of vibrations. Increasing the rigidity may require local redesign of the item's structure, the addition of stiffeners, and/or the addition of damping (Figure 13.13).
4. *An increase in the mass of the item to be protected* by mounting it on a rigid platform or inside a rigid enclosure (Figure 13.13). The added mass, called an "inertia block," results in a lowering of the natural frequency f_n of the item to be protected, as mounted on given vibration isolators. Increasing the mass by a given factor has the same effect as reducing the isolator stiffness by the same factor, usually, however, without introducing problems arising from large deflections due to transient or shock forces.
5. *Use of two-stage systems of vibration isolation* with a mass between them, as shown schematically in Figure 13.14. Two-stage isolation can be extremely effective for reducing vibration at high excitation frequencies, but may increase the vibration at low frequencies.

To avoid this detrimental effect at low frequencies, the mass m_i between the two isolation systems must satisfy

$$m_i > \frac{k}{20f^2} \tag{13.22}$$

where m_i is expressed in kilograms, k is expressed in newtons per meter, and f in hertz; or

$$m_i > \frac{20k}{f^2} \tag{13.23}$$

where m_i is expressed in pounds mass, k is expressed in pounds per

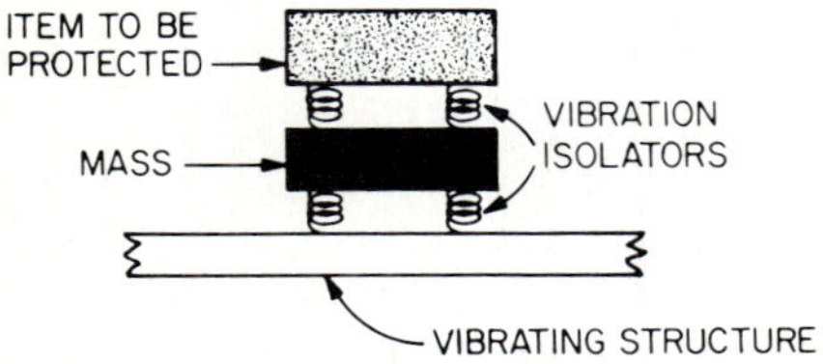

Figure 13.14 Schematic diagram of two-stage isolation. The two vibration-isolation systems have the same stiffness (*Ungar and Cohen, 1979*).

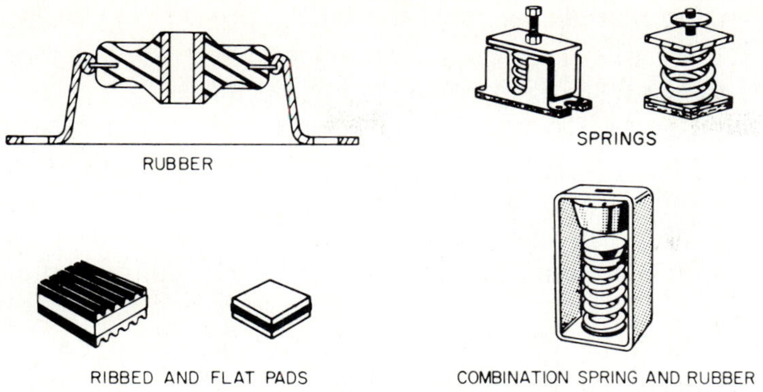

Figure 13.15 Some common isolator types (*Ungar and Cohen, 1979*).

inch, and f in hertz. Here k denotes the stiffness of the isolation system, and f the lowest excitation frequency of concern.

F. Vibrational isolation

A vibration isolator is a resilient support that tends to isolate a system from steady vibration. Isolators generally are made of steel springs, molded rubber or Neoprene components, pads or slabs (which may be flat, ribbed, or waffle-patterned) of resilient materials, or combinations of these, usually together with housings and attachment means suitable for specific applications. Some common types of vibration isolators are shown in Figure 13.15.

For any given application, vibration isolators must be (1) soft enough to provide the desired isolation, (2) able to carry the imposed load, and (3) compatible with the environment to which they are exposed (e.g., temperature, humidity, corrosive atmospheres, fuel vapors). To provide the required isolation, an isolator must have a stiffness that is less than the stuctural stiffness at the point of attachment; and it must provide a natural frequency that is considerably lower than the lowest excitation frequency of interest.

If a mass, m, subject to an excitation at a frequency of f hertz, is to be mounted on vibration isolators, then effective isolation can be obtained only if the total stiffness k of the isolators is selected so that

$$k < 10mf^2 \tag{13.24}$$

where k is expressed in newtons per meter and m is expressed in kilograms; or

$$k < \frac{mf^2}{40} \tag{13.25}$$

where k is expressed in pounds per inch and m is expressed in pounds mass.

Expressions (13.24) and (13.25) pertain to a rigid mass that is constrained to vibrate along a straight line. This situation represents a reasonable approximation to many practical problems. For example, in isolating a vibrating machine from a floor that radiates noise, isolation of essentially only the vertical motion of the machine may be required, because only such motion results in significant sound radiation by the floor.

If the isolator in this type of system is such that the isolator's deflection is proportional to the force on it (the case for steel springs over most of their operating range, but true for rubber elements only for small deflections), then the natural frequency, f_n, of the system is related to its static deflection (i.e., the amount the isolator deflects due to the weight it supports) by

$$f_n = \frac{5}{\sqrt{\delta_1}} = \frac{3.13}{\sqrt{\delta_2}} \quad \text{(Hz)} \tag{13.26}$$

where δ_1 denotes the static deflection in centimeters and δ_2 the static deflection in inches. For isolators with rubber elements, the natural frequency generally is somewhat higher than the value given by the foregoing equation (up to about a factor of 2).

For excitation frequencies that are below the natural frequency, an isolator provides no vibration reduction at all. For excitation frequencies that are approximately equal to the natural frequency, the vibration is amplified, often to the point of destruction, unless amplitude-limiting means are built into the system. Useful isolation (vibration reduction) results only for excitation frequencies that are higher than the natural frequency.

Table 13.11 may be used to evaluate qualitatively the effect of a vibration isolator if the exciting frequency is known, as well as the natural frequency of the mass isolator system (e.g., from its static deflection, according to Eq. 13.26). In complex systems, where several excitation frequencies may be present, each must be considered; a more detailed analysis than that using this simplified approach should usually be employed for complex problems. In general, however, motion of the item to be isolated in the horizontal direction, as well as rocking motion, may occur in conjunction with its vertical vibration. In such situations, isolators should be selected that are soft enough in all directions to have the natural frequencies associated with all motions that fall below the lowest excitation frequency of interest.

In order to reduce rocking effects, the individual vibration isolators

TABLE 13.11 Approximate Degree of Vibration Isolation Based on Ratio of Driving Frequency f to the Natural Frequency f_n

Ratio, f/f_n, of driving frequency of source to natural frequency of the isolated system	Degree of vibration isolation
Below 1.4	Amplification
1.4–3	Negligible
3–6	Low
6–10	Fair
Above 10	High

comprising a system for isolation of predominantly vertical vibrations (e.g., for a machine resting on a floor) should be selected and arranged so that the stiffness of each isolator is proportional to the deadweight load it carries. For nonvertical vibrations, the same rule of thumb may be applied by considering the entire arrangement as if it were rotated so that the vibration direction were vertical. Because specification and installation of isolators of different stiffnesses often cause problems, the use of an "inertia block," described below, is recommended where possible to make the weight distribution of the item to be protected nearly uniform. In problems where considerable design and analysis effort is warranted, the procedures given in Crede and Ruzika (1976) should be applied.

Where vibration isolators are used, care must be taken to eliminate alternate vibration transmission paths ("bridging"), such as may occur via piping, electrical conduits, shafting, or unisolated fasteners. Where it is necessary to have a bolt extend through a vibration isolation pad, this bolt must be provided with soft elastic (rubber) washers and sleeves, so as to avoid metal-to-metal short-circuit paths for vibration transmission (e.g., Figure 13.16). Figure 13.17 illustrates the use of flexible connectors and conduits to avoid the bridging of isolation. Further information concerning flexible connectors is given later in this section.

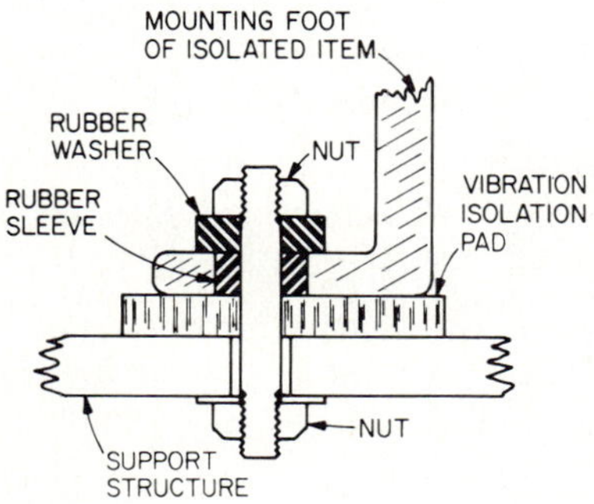

Figure 13.16 Typical installation of elastic washer and sleeves to avoid vibration transmission via bolt extending through isolation pad. (NOTE: Alternatively or additionally, elastic washer and sleeve may be used on the portion of the bolt in contact with the support structure.) (*Ungar and Cohen, 1979.*)

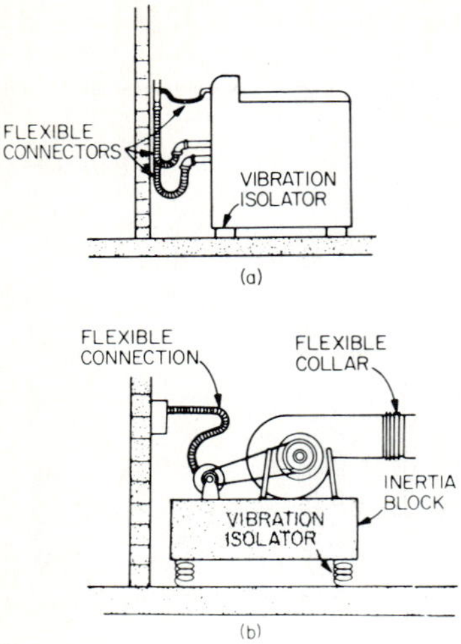

Figure 13.17 Typical uses of flexible connectors and conduits to avoid "bridging" of isolation. (*a*) Fluid-using (e.g., washing) machine isolated from floor and fixed piping. (*b*) Motor and fan mounted on a vibration-isolated inertia block. (*Ungar and Cohen, 1979.*)

1. Steel springs. Steel springs are generally used for vibration isolators that must carry heavy loads and/or must provide static deflections greater than 5 cm (2 in.), that is, natural frequencies below 2 Hz; or where temperature or other environmental conditions make other materials unsuitable. Because steel springs have little inherent damping and can have their own resonances in the audible frequency range,

TABLE 13.12 Minimum Coil Outside Diameters Recommended for Free-Standing Helical Compression Springs of Steel to Avoid Sideward Buckling

Tabulated Numbers Are Minimum Coil Diameters (cm)

		Load (kg)		
		Up to 350	350–1150	1150–2700
Static deflection (cm)	Up to 3	7.0	10.0	12.5
	3–5	12.5	12.5	18.0
	5–7.5	15.0	18.0	21.5
	7.5–10	21.5	21.5	25.5
	10 and up	25.5	25.5	30.5

Tabulated Numbers Are Minimum Coil Diameters (in.)

		Load (lb)		
		Up to 750	750–2500	2500–6000
Static deflection (in.)	Up to 1.25	2.75	4.00	5.00
	1.25–2.00	5.00	5.00	7.00
	2.00–3.00	6.00	7.00	8.50
	3.00–4.00	8.50	8.50	10.00
	4.00 and up	10.00	10.00	12.00

SOURCE: After Eberhart, 1967.

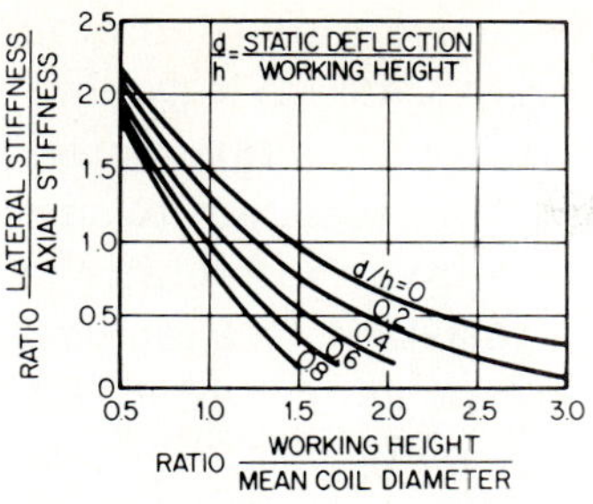

Figure 13.18 Ratio of lateral to axial stiffness of helical compression springs, as a function of ratio of working height to mean coil diameter, for several ratios of static deflection to working height (*Crede, 1976*).

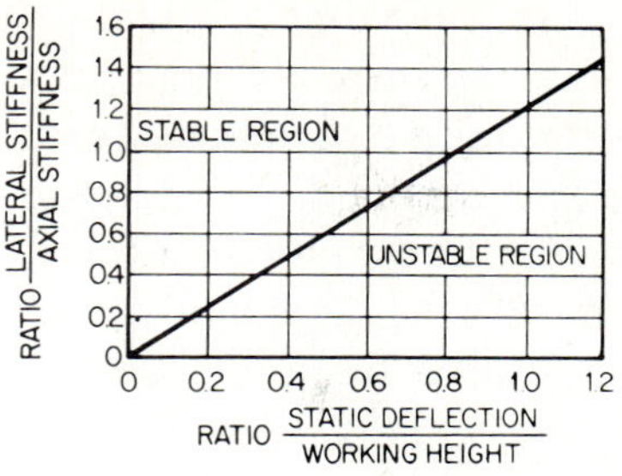

Figure 13.19 Separation between stable and unstable regions of a system supported solely on helical springs loaded in compression (*Crede, 1976*).

they should be used in conjunction with wire mesh inserts and in series with elastomeric or fibrous material pads (to provide damping and to interrupt all-metal vibration transmission paths).

Steel springs (or similar springs of other metals) usually are helically coiled and are fabricated from wire or bar. Springs also are made from flat or curved bars, sheet metal strips, and spring washers. Helical and flat leaf springs have stiffnesses that vary little with load in their design range; their features and design are discussed in Church (1976). Selection of springs involves consideration of their load-carrying capacities (in terms of stresses or excessive deflections) and, where such springs carry vertical loads in compression, of their lateral stability.

Free-standing helical compression springs must have large enough coil diameters so that the springs do not buckle sideward under load. Minimum coil diameters recommended to avoid this buckling are given in Table 13.12. Where space limitations do not permit the use of such large diameters, a stabilizing housing for the spring must be used. The lateral stiffness of a helical spring can be determined from its axial stiffness by use of Figure 13.18. The stability of a helical spring under load can be evaluated by use of Figure 13.19.

2. Rubber or Neoprene vibration isolators. Vibration isolators incorporating molded rubber or Neoprene (i.e., "elastomeric") parts are available in a wide variety of forms, designed for various combinations of stiffnesses in the various directions. Vibration isolators of such materials can be used for isolating exciting frequencies as low as 10 Hz, particularly if the isolator is designed so that the elastomer is loaded essentially in shear. Elastomeric isolators can be relatively light, strong, and inexpensive, but their stiffnesses can vary considerably with temperature. Natural rubber is strong, incorporates considerable damping, and is relatively inexpensive, but has limited resistance to deterioration when exposed to hydrocarbons, ozone, and high temperatures. Neoprene and some other synthetic rubbers are more resistant to such deterioration and are well suited for applications involving relatively high temperatures. Silicone rubbers are very resistant to chemicals and are available for applications at temperatures as low as −75°C (−100°F) and as high as 200°C (400°F). All elastomers possess plastic creep and deform slowly but continuously under high loads, particularly at high temperatures.

The stiffness of a rubber or Neoprene isolator or pad depends on its size and shape, on how its various surfaces are confined or bonded, on the direction, frequency, and magnitude of the loading, and on the elastic modulus of the material (Frye, 1976). The latter varies with frequency and temperature, but is usually characterized by a "durometer" number (determined from a measurement of the indentation produced by a cone pressed into the material), measured at room temperature—the higher the number, the stiffer the material. For rubbers used in isolators, the number is generally between 30 and 70. These values are suitable only for comparing similar materials (e.g., Neoprene elastomers of like composition). Two different types of rubber with the same durometer number may exhibit dynamic stiffnesses that differ considerably from each other.

Because rubber stiffens with increasing frequency and with increasing static load, the static deflection required to specify a rubber isolator typically should be between two and three times that for an ideal constant-stiffness isolator (i.e., that for an isolator whose deflection is proportional to the force applied to it).

3. Isolation pads. *Sponge rubber* pads are commercially available in many varieties and stiffnesses. The stiffness of such a material usually increases rapidly with increasing load and increasing frequency. Open-cell sponge rubber tends to take on liquids. Therefore, a closed-cell material should be used in the presence of combustible or other fluids.

Fiber glass vibration isolation board is available in a variety of densities and fiber diameters. Fiber glass boards typically consist of a multitude of matted fine fibers, usually with a small amount of binder. The stiffness of this material generally increases with increasing density, fiber diameter, and static deflection. Because they contain a multitude of capillaries, fiber glass products tend to absorb moisture and other contaminants. Some manufacturers supply pads and blocks with sealed surfaces; alternatively, one may enclose fiber glass pads in plastic (e.g., PVC "bags") to protect them from the atmosphere. When a vibratory load is applied to a previously unused fiber glass board or pad, it tends to compress and stiffen during the first few (10 to 100) cycles of loading; thereafter, no further stiffening or thickness reductions tend to occur. Some commercially available materials are "precycled" and have guaranteed stiffness characteristics.

Although fiber glass pads are usually specified in terms of their density, the stiffness of pads supplied by different manufacturers may differ greatly, even for pads of nominally the same density. The stiffnesses of materials supplied primarily for use as thermal insulation may differ markedly from those of similar pads supplied for vibration isolation. Figure 13.20 shows the load-vs.-deflection characteristics of fiber glass supplied by one manufacturer for vibration isolation uses. Where pads of fiber glass or other fibrous materials are used between large areas of closely spaced surfaces, the entrapped air may increase the effective stiffness of the isolation system significantly, unless efficient venting paths are provided (Ungar, 1975).

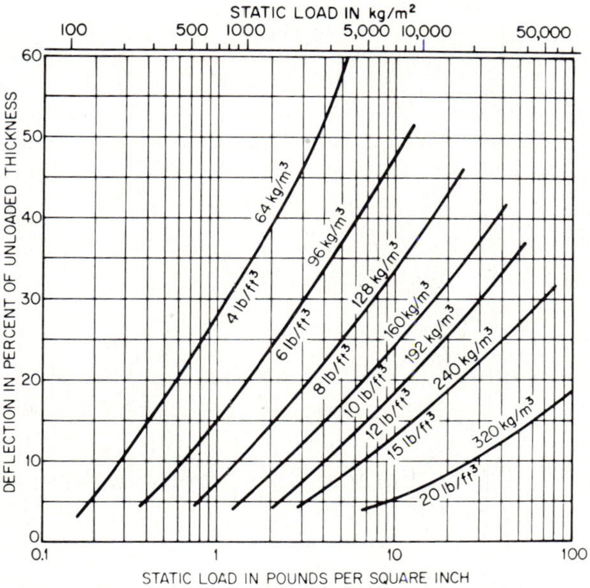

Figure 13.20 Static load-vs.-deflection curves for fiber glass floor isolation board (molded Aerocor supplied by Birma Products). The floor isolation boards were loaded and unloaded 10 times before data were taken. Curves represent averaged and smoothed data (for the densities indicated), which typically exhibited ±30% variation from the average. Note that pads supplied by other manufacturers may have significantly different characteristics. (*Ungar and Cohen, 1979.*)

Cork is used primarily in the form of composition board, consisting of ground cork plus a binder. This composition board generally is relatively insensitive to moisture and petroleum products.

Example 13.6 Very forceful impacts are used to form the heads on steel wire nails in an automatic nail-making machine (Crocker et al., 1971). In addition to direct impact noise, the resulting vibration may be transmitted through the machine to a concrete floor, which also radiates considerable noise. An octave band analysis of the sound level at the machine operator is shown in Figure 13.21. This may be converted to 103.5 dBA by Eq. 13.15 using the A-scale correction factors shown in Table 13.5. Ten machines, each operating at 300 strokes/min, contributed to the sound field in this problem.

solution A transparent enclosure around the point of impact constructed of polycarbonate or Lexan plastic would provide sufficient transmission loss at all frequencies to reduce the sound below 90 dBA if it could be made airtight. Unfortunately, openings to permit introduction of wire feed stock, movement of machine components, and removal of product destroy much of the effective noise reduction attainable with such shielding. The results actually obtained with such an enclosure are shown in Figure 13.21. Calculation of transmission losses by the transparent shielding material is not reliable in comparison to minimizing enclosure openings, keeping the shielding in place and yet making it easily removed and replaced for machine maintenance and operation.

Reduction of the direct-line noise levels from 103.5 dBA to 95 dBA is not satisfactory for 8-hr/day, 40-hr/week exposures according to OSHA regulations. Consideration must be given to sound radiated by the floor.

Although the noise generated by this machine is from individual impacts (rise time about 1 msec and duration about 10 msec), the frequency from each machine is 5 Hz (period = 0.2 sec or 200 msec) and from multiple machines is considerably

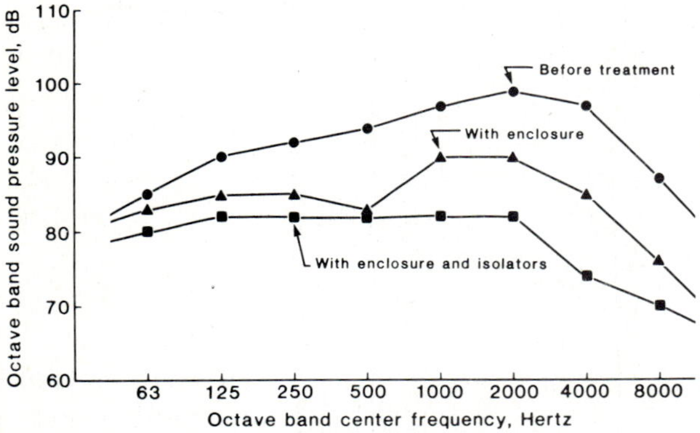

Figure 13.21 Operator position sound pressure levels, before and after treatment of nail-making machine (*after Jensen et al., 1978*).

	dBA level
Before treatment	103.5
After treatment:	
Enclosure only	95
Enclosure and isolators	90

greater. The sound is considered to affect the human ear as continuous sound and should be regulated accordingly. Note that analysis and noise control efforts are carried out singly on each machine, but final evaluation of human exposures must be done in concert.

As a design consideration, the natural period of an isolator plus machine should be much greater than the shock pulse duration of 10 msec, but should be less than the time between impacts, 200 msec. If an elastomer-type isolator is used, having only 0.1 in. of static deflection under machine load, the natural period may be estimated from the reciprocal of Eq. (13.26).

$$\frac{1}{f_n} = \frac{\sqrt{\delta_2}}{3.13} = \frac{\sqrt{0.1}}{3.13} = 0.1 \text{ sec} \qquad (13.26)$$

or 100 msec, which is <200 msec.

Combination of enclosure and isolator treatment of each nail-making machine gives the aggregate results shown in Figure 13.21 (Jensen et al., 1978). Although solutions to problems such as this one are highly empirical, a good understanding of underlying principles enables the design engineer to more readily achieve desired results.

4. Means of avoiding bridging. When providing vibration isolation for any mechanical system or component, care must be taken to eliminate all vibration transmission paths that can bypass (or "bridge") the primary isolators. As illustrated in Figure 13.17, flexible sections should be included in all tubing, conduit, ducts, shafts, and electrical cabling connecting between isolated and unisolated components.

Flexible sections of rubber or plastic hose usually are preferable for isolation in piping systems, but sections of flexible metal hose or tubing may be required if the temperature or pressure is too high, if chemically active fluids must be transmitted, or if prescribed by codes. The strip-wound flexible metallic hose illustrated in Figure 13.22 derives its flexibility from sliding action at its interlocked joints and obtains its pressure-tight qualities from packing (usually of asbestos) included in the joint grooves. This type of hose is suitable for air, oil, water, and steam, but is generally unsatisfactory for handling gases at high pressure or dangerous, highly volatile substances.

Corrugated tubing, as illustrated in Figure 13.22*a* and 13.22*b*, is commonly made of steel, stainless steel, or bronze, but is also available in other alloys. Depending on materials and construction methods, manufacturers supply such tubing in diameters from about 5 mm to 50 cm ($\frac{1}{8}$ in. to 20 in.), usually with integral fittings at its ends. Seamless or welded corrugated tubing is not susceptible to leakage and therefore is satisfactory for containing volatile substances and gases under pressure. For high-pressure applications, corrugated tubing normally is enclosed within a flexible metal-braid sleeve that is attached to the end fittings. This sleeve prevents axial elongation of the tubing and permits it to carry greater pressures without bursting.

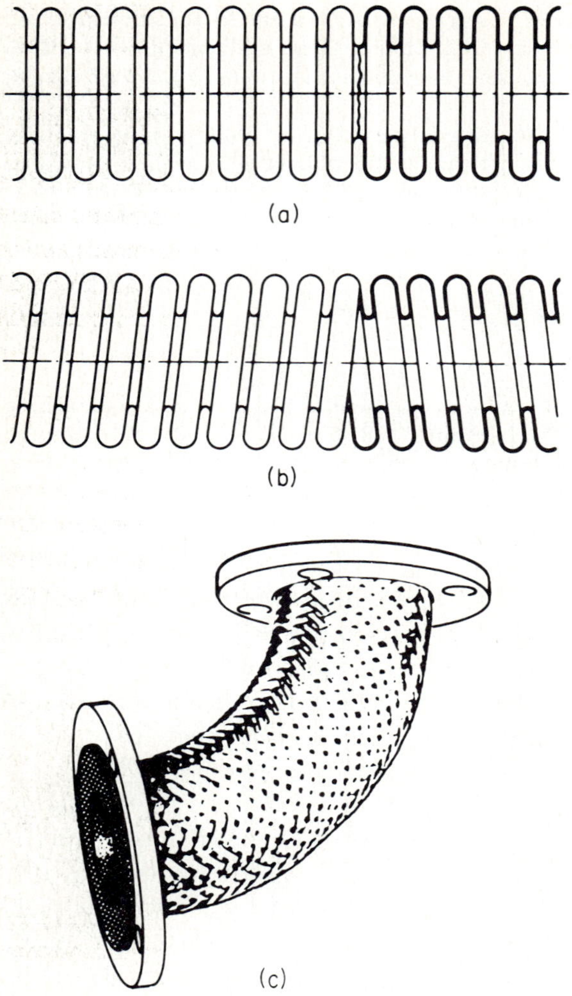

Figure 13.22 Cutaway views of flexible connectors: (*a*, *b*) flexible seamless tubing of spiral type, (*c*) flexible pipe connector. (*Crede, 1976*.)

Expansion joints made of rubber, plastics (e.g., Teflon), or metal bellows should not be used instead of the longer flexible connections discussed above. Such expansion joints require tie-rod assemblies, because the joints themselves cannot withstand the axial loads imposed on them by internal pressures; and such tie-rod assemblies serve as bridging vibration transmission paths.

The following general guidelines should be observed in the installation of flexible conduits:

1. Attach the flexible conduit to the items it interconnects at locations where there is relatively little vibratory movement and where the

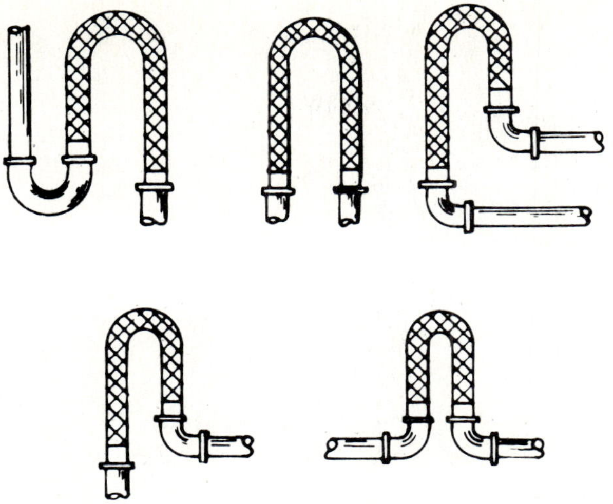

Figure 13.23 Recommended configurations of flexible connector installations (*Crede, 1976*).

structures of these items are most massive. This tends to reduce the vibration transmitted along a given conduit.

2. Include a large bend or loop in the flexible conduit, if one end of the conduit experiences large displacements relative to the other or if the conduit must have very low stiffness. A loop should be so arranged that the vibratory movement is substantially distributed over its entire length. Excessive wear resulting from stress concentrations at the end connections may shorten the life of the conduit, if the movement occurs only near its end. Figure 13.23 shows several recommended arrangements that ensure distribution of movement over the length of the flexible conduit.

5. Inertia blocks. An inertia block usually is a concrete block that serves as a base for mechanical equipment, such as fans or pumps, and is mounted on resilient supports. The concrete may be poured with steel channel and other inserts, so as to allow attachment of machines and vibration isolators (see Figure 13.24). Inertia blocks (1) reduce motion; (2) lower the center of gravity of the isolated system, increasing its stability; (3) minimize the effects of unequal weight distribution, if the concrete mass is appropriately distributed; and (4) reduce the effects of reaction forces such as occur when vibration isolation is applied to fluid machinery. An inertia block should weigh at least as much as the item being isolated; preferably it should weigh twice as much. Because reciprocating engines and compressors may have perhaps as much as one-third of their total weight oscillating, such a machine generally requires an inertia block of three to five times its weight. In

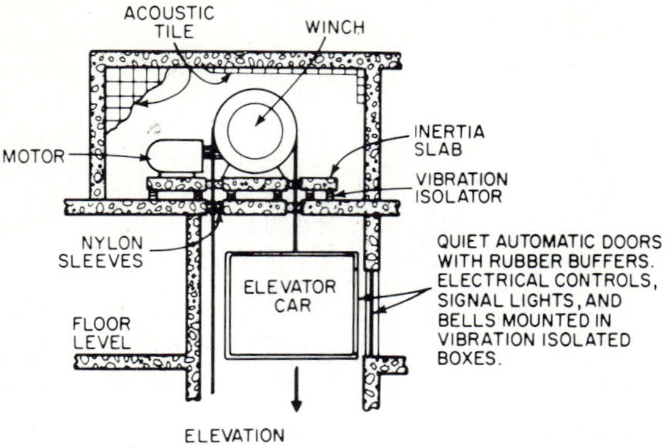

Figure 13.24 Measures to reduce elevator noise (*Vér, 1979*).

installations consisting of several pieces of equipment requiring vibration isolation, such as a motor driving a fan or generator (see Figures 13.17*b* and 13.24), it is advisable to mount all components on a common inertia block.

G. Damping

"Damping" is the dissipation of the mechanical energy associated with vibration, i.e., the conversion of this energy into heat. A "damping treatment" is any material or combination of materials that is applied to a structural component to increase its damping. If a large, relatively thin metal structure is excited by vibration or airborne sound, damping treatments applied to such a structure and to adjacent connecting supports and structures may result in significant noise reduction. Depending on the specific conditions, the reduction in radiated or transmitted noise levels may range from negligible amounts to as much as 10 to 20 dB. Added damping generally is most effective when applied to thin structures that are excited near their resonances, either by random forces or by impacts.

Although all materials have some inherent damping, most conventional structural materials (such as steel) have so little that they require addition of a damping treatment for noise control purposes. A damping treatment generally is useful for noise reduction if it is applied to a structure that:

1. Is forced into vibration at a natural frequency
2. Is set into vibration by random forces or forces made up of many frequency components
3. Is set into vibration by impacts or other transient forces

4. Transmits vibration to noise-radiating surfaces
5. Radiates broad-band sound

If vibration-isolated structures have resonances in the frequency range of interest, then application of damping treatment to such structures can improve the effectiveness of the isolation at frequencies near those associated with the natural frequencies of the structure.

Damping treatments generally may be ineffective for cases other than those cited above. For example, the addition of a damping treatment has no effect on the steady vibration of a structure due to forces that act at frequencies that do not correspond to natural frequencies of the structure; damping added to a panel does not affect its low-frequency sound transmission.

1. Damping mechanisms and treatments. The damping mechanisms of greatest interest for noise control may be categorized as mechanical hysteresis, friction, energy transport, and frequency conversion.

Mechanical hysteresis, also called "internal friction" or "material damping," results from dissipative processes in materials at the molecular or metallurgical grain level. Although all materials have some damping, conventional structural materials have relatively little. In order to obtain well-damped structural components, one may apply a damping treatment consisting of a high-hysteresis material to a component made of a conventional material, such as steel.

Friction occurs between (dry or lubricated) surfaces that move relative to each other, thereby converting mechanical vibration into heat and providing damping—for example, friction between adjacent structural components or between fiber glass insulation and metal ductwork to which it is fastened. Large relative motions are required to achieve significant damping by this mechanism. Damping due to friction depends on interface pressures and clearances, as well as on the viscosity of the fluid in the interface gap. All of these parameters are difficult to control.

Energy transport, the flow of mechanical energy away from the structure under consideration, has the same effect as local energy dissipation. Such transport may occur as a result of wave propagation along an extended structure, radiation of sound from the structure to a fluid in contact with it, or fluid pumping produced by structural vibrations. The latter effect occurs, for example, if a fibrous blanket is placed near a vibrating panel but not in contact with it. In this arrangement, which often is convenient for damping of sheet metal panels, the air between a panel and blanket is pumped through the orifices in the blanket and thus transports energy away from the panel.

Frequency conversion occurs when vibratory energy at frequencies of interest is converted to other (usually higher) frequencies (where it usually is converted to heat more readily), so that this energy is no longer available at the frequencies of concern. Such energy conversion occurs, for example, when granular materials such as sand, gravel, or lead shot are placed in contact with vibrating surfaces, so that "rattling" occurs between these surfaces and between adjacent grains. Increased damping of large structures, such as floors or walls of buildings, often may be obtained most economically by the use of boxes that are partially filled with sand, attached to or built into these structures at locations where major motions occur. The damping here depends markedly on amplitude and may be small if the motions are not intense enough to produce rattling. The weight, grain size, moisture content, and bed depth of the sand affect the damping in ways that cannot usually be predicted reliably.

2. Characterizing damping. Several measures are commonly used to quantify the damping of a structure or of a material sample. Most of these measures are based on: (1) how rapidly vibrations decay when structures are displaced from equilibrium and allowed to vibrate in absence of externally applied forces, or (2) characteristics of structural responses at or near resonances. The most common measures of damping include the three defined below (for more complete information see Table 37.1 of Plunkett, 1976; and Cremer et al., 1973).

Loss factor (η): a dimensionless quantity defined by:

$$\eta = \frac{D}{2\pi W_0} \tag{13.27}$$

where D = energy dissipated per cycle of vibration
W_0 = average total (kinetic plus potential) energy of the vibrating system

Damping ratio (ζ), (or fraction of critical damping): a dimensionless quantity, given by:

$$\zeta = \frac{\Delta}{\Delta_c} = \frac{\eta}{2} \tag{13.28}$$

where Δ = viscous damping coefficient pertaining to the viscous damper of a simple mass-spring-damper system with vibration characteristics like those of the structure of interest
Δ_c = critical damping coefficient (i.e., the smallest value of Δ for which the mass-spring-damper system does not oscillate when deflected and released)

Percent critical damping: a dimensionless quantity equal to 100 times the damping ratio.

3. Specifying damping. For a structure forced to vibrate at a natural frequency, the vibration amplitude is inversely proportional to the loss factor, η. In this case the vibration reduction ratio, r, due to increasing the loss factor from η_0 to η_1 is given by

$$r = \frac{\text{amplitude with initial loss factor } \eta_0}{\text{amplitude with increased loss factor } \eta_1} = \frac{\eta_1}{\eta_0} \tag{13.29}$$

and, correspondingly,

$$\text{Vibration reduction} = 20 \log\left(\frac{\eta_1}{\eta_0}\right) \tag{13.30}$$

For a structure that vibrates as the result of random excitation, the *root-mean-square* amplitude is inversely proportional to the loss factor. Consequently,

$$r = \frac{\text{rms amplitude with initial loss factor } \eta_0}{\text{rms amplitude with increased loss factor } \eta_1} = \left(\frac{\eta_1}{\eta_0}\right)^{0.5} \tag{13.31}$$

and $$\text{Vibration reduction} = 10 \log\left(\frac{\eta_1}{\eta_0}\right) \tag{13.32}$$

Note that an increase in damping (i.e., an increase in loss factor) produces double the decibel reduction in vibration of a structure forced to vibrate at a natural frequency compared to one vibrating as a result of random excitation. Further, an increase in damping does little to reduce vibration that is nonrandom and is not at a natural frequency.

The effect of increased damping on recurrent transient vibrations, such as those due to repetitive impacts, depends in a complex manner on the repetition rate of the excitation and on the damping that is already present. Each transient excitation "pulse" produces a vibration whose amplitude decreases with time, with the decrease being more rapid for greater damping. If the repetition time is long compared with the time required for vibrations to decay significantly (i.e., if the damping present is already relatively high), so that the vibration due to one pulse has decayed substantially before the next pulse is applied (see lower part of Figure 13.25), the vibration consists essentially of a se-

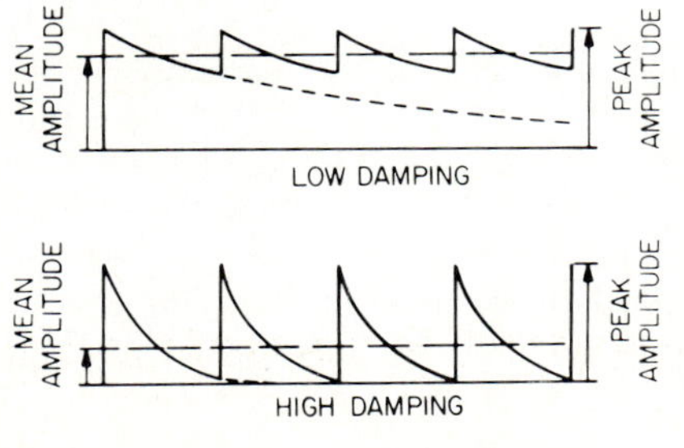

Figure 13.25 Effect of damping in reducing the average vibration amplitude of a structure subject to repetitive impacts. Dashed line indicates decay of response to a single impact (*Ungar and Cohen, 1979*).

quence of independent decaying vibrations. Here the rate at which each such independent vibration decays increases with increasing damping. On the other hand, if the repetition time interval is relatively short (i.e., if the damping is relatively low), considerable vibration energy due to one pulse still persists when the next one is applied (see upper part of Figure 13.25). The vibration then consists of a nearly continuous "ringing." The average vibration amplitude decreases with increasing damping.

The effect of damping on sound transmission through walls or panels is complicated, and usually is significant only in a very limited range of frequencies in the vicinity of the so-called coincidence frequency. In all practical cases, an applied damping treatment can provide a significant benefit only if this treatment increases the damping by a considerable factor (typically, at least 3). Thus, if a structure already is damped reasonably well (e.g., by supports around its edges or by contact with adjacent structures), enough damping usually cannot be added to produce a noticeable effect. On the other hand, even a small amount of damping added to a very lightly damped structure (such as a freely hanging gong) can have a dramatic effect.

Since added damping rarely produces adverse effects in terms of noise and vibration control, it is common practice to apply as much damping as possible within the constraints of cost, weight, or space. The point of diminishing returns often can be determined best by a simple trial-and-error process. This begins with the application of a considerable amount of damping treatment (perhaps in an impractical configuration) to determine whether damping can overcome the problem at all; continuing with lesser amounts of treatment; and, where stringent constraints exist, experimenting to determine the minimal amounts and optimum placement of the damping treatment.

Example 13.7 In the manufacture of folding cardboard cartons, printed sheets are cut on presses that have automatic strippers to remove waste from between the cartons. The cardboard scrap is removed through a horizontal air-vane conveyor to the intake of a centrifugal fan, which sends the material to a baler in a remote location (Figure 13.26). Operation of the cutting presses alone is well below the 90-dBA limit. However, the scrap removal system results in sound levels at the pressman station reaching 95 dBA on an almost continuous basis (Jensen et al., 1978).

solution Investigation of the system revealed that noise generation was largely due to pieces of cardboard scrap striking the sides of the intake conveyor under the press stripper, the sides of the intake hood to the fan, and the fan and outlet ducts. These all are sheet metal panels requiring damping. Damping of all such surfaces to a distance 10 ft above the floor was provided using $\frac{1}{32}$-in.-thick, 2-lb/ft^2 lead sheeting attached with a resin glue. Sound transmission was also reduced by the limp mass of the lead sheet. Noise levels were reduced to 88–90 dBA at the pressman's platform.

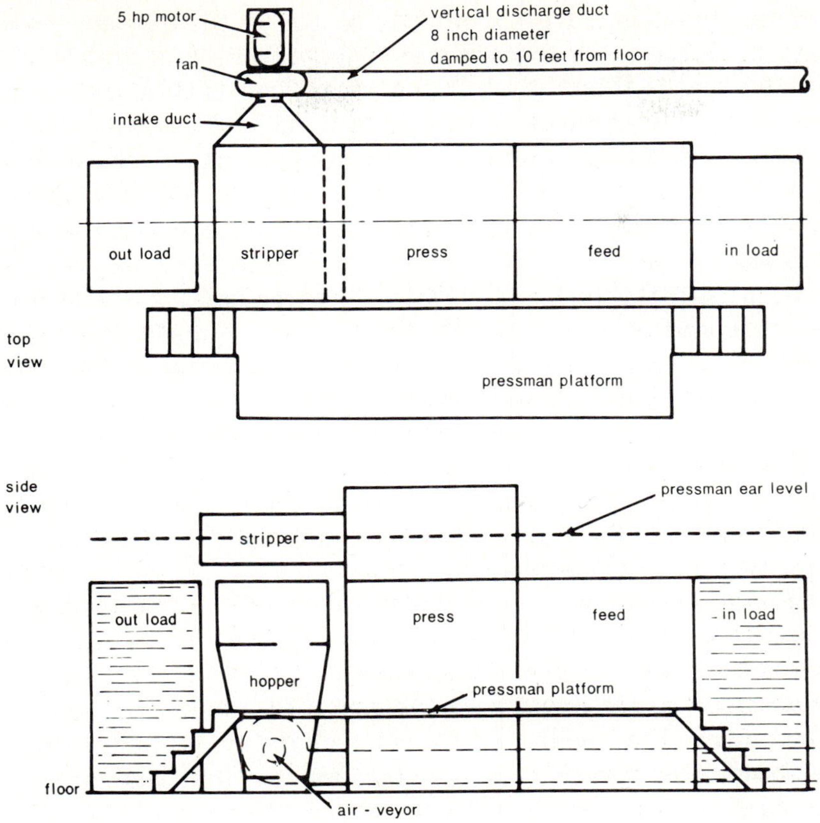

Figure 13.26 Scrap-handling system for cutting press (*Jensen et al., 1978*).

For much higher vibration and sound levels, duct treatment may require lagging, which is a spring-absorber-mass combination of materials. An example would be 1–3 in. of resilient acoustic absorbing material (glass fiber or polyurethane foam) with a heavy cover sound barrier of sheet lead or lead-loaded vinyl sheeting over the sound-radiating surfaces (Jensen et al., 1978).

4. Damping treatment using viscoelastic layers. A viscoelastic material is a material that can store a considerable amount of strain energy when deformed, but also can dissipate a significant fraction of the stored energy because it flows somewhat like a very viscous liquid. All rubbers and plastics exhibit this combination of elastic and viscous properties to some extent, and thus are viscoelastic.

The properties of viscoelastic materials tend to vary considerably with frequency and temperature, although they generally are relatively independent of strain amplitude, particularly at the small amplitudes of interest for noise control. Consequently, it is important that materials for viscoelastic treatments be selected for the frequency and temperature ranges to be encountered. The stiffness of a typical viscoelastic material is relatively low and nearly constant at high temperatures and low frequencies, increases by a factor of the order of 100 in an intermediate range of temperatures and frequencies [extending over 20 to 40°C (68 to 104°F) and over several decades of frequency], and is relatively high and nearly constant at low temperatures and high frequencies. The loss factor of such a material tends to be relatively low in both the high-temperature/low-frequency and the low-temperature/high-frequency regions, and to have its maximum value in the "transition" region where the rapid stiffness change occurs. The values of the stiffnesses and loss factors, as well as the frequencies and temperatures corresponding to the aforementioned regions, vary considerably from material to material. Data on specific commercially available viscoelastic damping materials are generally provided by their manufacturers. The July issues of *Sound and Vibration* contain names and addresses of suppliers.

The damping of a conventional structural panel may be increased by bonding to that panel a layer of viscoelastic (damping) material. In selecting such a damping layer and deciding on its placement, remember that a damping treatment can have a significant effect only: (1) if it is placed where it is made to deform considerably as the structure vibrates; (2) if it is "matched to the structure" so that the damping treatment can take on a considerable fraction of the structure's energy; and (3) if it is capable of dissipating much of the energy it takes on (i.e., it has a high loss factor). For example, consider a rectangular sheet metal panel that is bolted at its edges to a relatively massive machine housing. If a strip of damping material is placed along one of the edges of the panel, this strip does not bend or stretch significantly as the panel deflects in vibration, and therefore provides little damping. In contrast, if the same strip is placed along one of the panel's diagonals, it stretches and bends considerably as the panel vibrates, and thus dissipates a greater fraction of the panel's energy. In this position, the strip provides significantly more damping than it does if it is located along a panel edge. If the aforementioned strip is very soft, it can store relatively little of the panel's vibratory energy. Even though such a strip may participate considerably in the panel's vibration, it provides little damping, because it (like any component) can dissipate only a fraction of the energy it takes on. If the strip is very stiff, it tends to restrict the motion of the panel and again can take on and dissipate little energy. A strip of intermediate stiffness that restricts the panel's motion only slightly and shares considerably in its vibratory energy can contribute a significant amount of damping.

However, in the absence of definite information, or if all parts of a structure are expected to vibrate about the same, it is common practice to cover an entire panel with damping treatment, thereby ensuring coverage of the parts that have considerable vibratory motion.

If a layer of viscoelastic material is bonded to a structural panel without an additional covering ("constraining" layer), the viscoelastic layer stores energy and provides damping primarily due to this layer's stretching, which results from bending of the panel. The damping produced by such an "unconstrained" or "free" layer increases with increasing layer thickness. This damping typically varies as between the first and second power of thickness, with the greater rate of variation corresponding to thinner layers and softer damping materials.

If a layer of a viscoelastic material is bonded to a structural panel and a second "constraining" layer of a structural material is bonded atop the viscoelastic layer, then the viscoelastic layer stores energy and provides damping. The effect is due primarily to the shear deformation that is induced in this layer as this three-component sandwich configuration bends. The damping produced by such a "constrained" viscoelastic layer at a given frequency is small for small and large thicknesses of this layer and has a maximum for an intermediate thickness. The optimum thickness at which maximum damping occurs depends in a relatively complex manner on the stiffness and thicknesses of the structural panel and the covering layer, on the shear stiffness of the viscoelastic layer, and on frequency.

Constrained layer treatments generally can provide more damping than free layer treatments, but generally must be selected carefully in order to realize this advantage. [Ungar (1971) provides a corresponding design procedure.]

The damping due to a free layer treatment varies with frequency and temperature only as the result of the frequency and temperature dependence of the properties of the viscoelastic material. But the damping due to a constrained layer treatment exhibits an additional frequency and temperature dependence that varies with the thicknesses and stiffnesses of the three component layers. One may obtain an effective free layer treatment simply by using a reasonable thickness (typically two to four times the structural panel thickness) of a viscoelastic material that exhibits high damping and reasonable stiffness in the frequency and temperature range of interest. In order to obtain an effective constrained layer treatment, a material must be employed that exhibits a high loss factor in the frequency and temperature range of concern, and a material thickness selected so that the aforementioned damping maximum occurs in the range of interest. Thus, an arbitrarily selected free layer treatment using any reasonable viscoelastic material generally will work quite well; whereas an arbitrarily selected constrained layer treatment may be relatively ineffective.

For efficient damping performance, the viscoelastic material must be bonded well and continuously to the structural panel and, where one is used, to the constraining layer. If the viscoelastic material is not self-adhesive, a thin layer of a stiff adhesive, such as a rigid epoxy, should be used.

H. Silencers

Silencers are devices intended to be used in piping or ducted systems to reduce the amount of sound that is transmitted from one section of the system to the next, or to the surrounding area. Silencers reduce the noise while permitting the passage of air or other fluid that is flowing in the system. Silencers can be broken down into two fundamental groups, absorptive and reactive. Absorptive silencers contain either fibrous or porous materials and depend on the absorptive properties of these materials to reduce noise. An example application was given in Chapter 1. Reactive silencers depend on the reflection or expansion of the sound waves, with corresponding self-destruction as the basic noise reduction mechanism.

1. Absorptive silencers. Probably the simplest absorptive silencer is the lined duct. Lining the inside walls of a duct with sound-absorbing material can provide a silencer of almost any desired length. The lining of duct bends is most effective. The acoustical performance of lined bends is due primarily to absorption of acoustic energy by the lining and reflection of axial waves back toward the source. Typical treatment materials are 1- or 2-in. acoustical-grade fibrous or porous materials with a perforated sheet metal protective facing. In installations where liquids or finely ground powders are present, the absorbing material may be bagged in 1- to 2-mil-thick plastic (PVC, Mylar, etc.). This treatment can be specified to suppliers at a modest increase in cost.

Another simple form of absorptive silencer is parallel baffles. Basically, these baffles consist of an aerodynamically streamlined entrance and exit with perforated walls backed by highly absorbent acoustical material (usually fibrous glass or mineral wool). The acoustical performance of parallel baffles depends on the thickness of the absorbing materials, spacing of baffles, and length of baffles. Attenuation also increases with length, but not in a linear fashion; i.e., doubling the length does not double attenuation. This nonlinearity is due principally to the rapid absorption of high-order transverse modes in the first few feet of the silencer, leaving a plane wave propagation that has a grazing incidence (90° from normal) to the absorbing treatment. Parallel baffles are available commercially from a large number of manufacturers in modular units so that they can be assembled together to fit almost any duct size or attenuation requirements. Because there is little flow resistance in parallel baffles, they can be readily applied to installations where pressure losses are critical, such as air conditioning systems, forced-draft fans, and gas turbines.

2. Reactive silencers. The simplest form of reactive silencer is the single expansion chamber (Figure 13.27). At each change in the cross-sectional area of the duct, the impedance change causes a reflection of the sound waves. This reflection when added to the incoming sound results in destructive interference and subsequent noise reduction. By considering the reflection and expansion of sound waves as they flow through the discontinuities at 1 and 2, the transmission loss (TL) can be shown to be:

$$\text{TL(dB)} = 10 \log\left[1 + \frac{1}{4}\left(m - \frac{1}{m}\right)^2 \sin^2 kl\right] \tag{13.33}$$

where $m = \dfrac{\text{muffler cross-sectional area}}{\text{duct cross-sectional area}} = \dfrac{A_2}{A_1}$

$k = \text{wave number} = \dfrac{2\pi}{\lambda} = \dfrac{2\pi f}{c} \ (\text{ft}^{-1})$

$$l = \text{muffler length (ft)} \tag{13.34}$$

By examining this equation, one sees that the magnitude of the attenuation depends on the value of m and on the frequency (through the argument of the sine function, kl). Note that peak attenuation increases with the factor m and that the attenuation vanishes at values of the argument $kl = n\pi$ ($n = 0, 1, 2, 3, \ldots$). Maximum attenuation occurs at values of:

$$kl = \frac{n\pi}{2} \qquad (n = 1, 3, 5, \ldots) \tag{13.35}$$

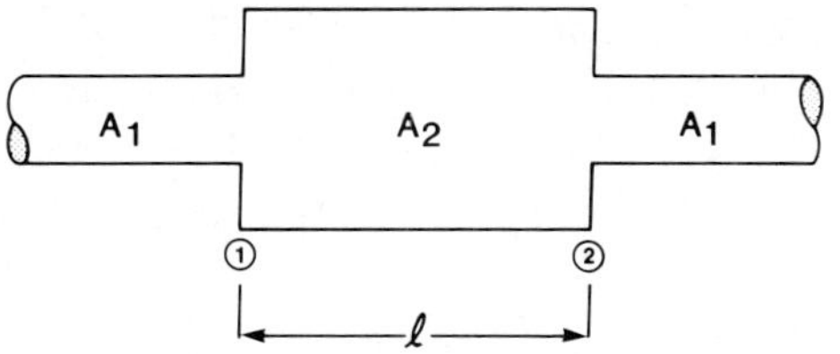

Figure 13.27 Reactive silencing by means of a single expansion chamber.

Solving Eq. (13.34) for the muffler chamber length, and substituting $k = 2\pi/\lambda$, gives:

$$l = \frac{n\lambda}{4} \qquad (n = 1, 3, 5, \ldots)$$

Therefore, peak attenuation occurs at frequencies that correspond to a chamber length:

$$l = \frac{\lambda}{4}, \frac{3\lambda}{4}, \frac{5\lambda}{4}, \ldots$$

Because of this relationship, these silencers are often called quarter-wave mufflers. In the same way, the attenuation vanishes at frequencies corresponding to chamber length:

$$l = \frac{\lambda}{2}, \lambda, \frac{3\lambda}{2}, \ldots$$

The spectral character of these silencers is shown in Figure 13.28. In order to provide a wider range of attenuation, several chambers can be put in series. Generally, the critical length of each chamber is varied so that where the attenuation vanishes for one chamber, another is at, or near, its peak.

Example 13.8 A circular duct of 2 in. inside diameter opens into an expansion chamber 24 in. long having a cross-sectional area of 50 in^2. The exhaust pipe on this reactive silencer is also 2 in. in diameter. What is the transmission loss through this muffler of 1000-Hz sound and of 500-Hz sound? Assume that the exhaust gas is dry air at 150°C.

solution The wave number k is obtained from Eq. (13.32). The velocity of sound in dry air is approximately

$$V = 331.5 + 0.607t$$

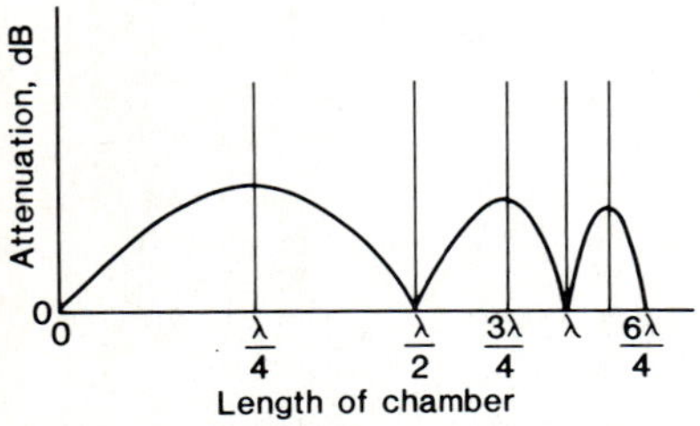

Figure 13.28 Schematic of sound attenuation in quarter-wave mufflers (not drawn to scale).

for t in degrees Celsius and V in meters per second. Therefore,

$$V = 331.5 + (0.607)(150) = 422.6 \text{ m/sec or } 1386 \text{ ft/sec}$$

and $$kl \text{ at } 1000 \text{ Hz} = 2\pi \times 1000 \times \frac{2}{1386} = 9.07 \text{ radians} \qquad (13.35)$$

Thus: $$\text{TL} = 10 \log\left[1 + \frac{1}{4}\left(m - \frac{1}{m}\right)^2 \sin^2 kl\right] \qquad (13.33)$$

$$= 10 \log\left[1 + \frac{1}{4}\left(\frac{50}{3.14} - \frac{1}{15.9}\right)^2 \sin^2 9.07\right]$$

$$= 10 \log[1 + 62.9 \times \sin^2 520°]$$

$$= 9 \text{ db at } 1000 \text{ Hz}$$

At 500 Hz, $kl = 4.53$ radians and:

$$\text{TL} = 10 \log[1 + 62.9 \times \sin^2 260°] = 18 \text{ dB at } 500 \text{ Hz} \qquad (13.33)$$

The reader is left to determine the frequencies that are maximally and minimally attenuated. It is obvious from the example given that the temperature of exhaust gas affects attenuation as a function of frequency. In addition, the composition of exhaust gas, especially regarding CO_2 and water vapor, affects the velocity of sound and hence the frequencies attenuated by a particular muffler design.

From the above example it should be apparent that multiple chambers in series may be used to provide silencing of broad-band noise.

3. Helmholtz resonator. An effective but limited type of sound absorber is the side-branch or cavity resonator (Helmholtz resonator) illustrated in Figure 13.29. When sound waves strike the neck of the resonator, they set the air within it into motion. The air in the neck behaves like a plug that moves in and out, compressing the air in the enclosed volume. In so doing, acoustic energy is converted to heat due to friction along the walls of the neck. These losses can be increased by placing light, porous material over the mouth of each neck or by lining the

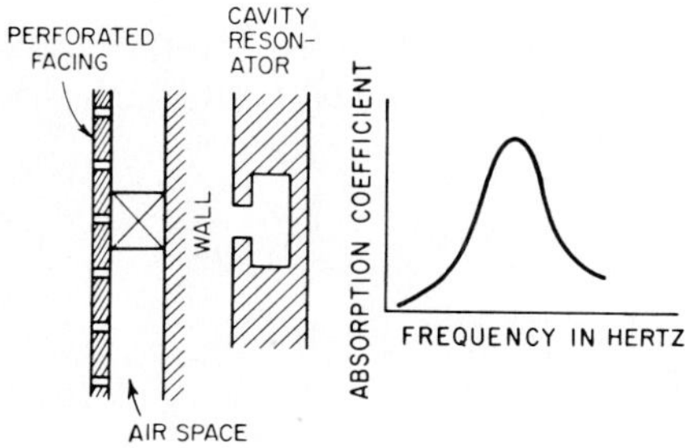

Figure 13.29 Sound-absorption characteristics of a "resonance-type" absorber; the perforated facing, spaced from a wall, acts as a Helmholtz (cavity) resonator (*Sabine and Moulder, 1979*).

enclosed volume with absorptive material. Maximum sound absorption occurs at the resonant frequency of the mass of air in the neck and the "spring" furnished by compression of air in the cavity; little sound is absorbed at other frequencies. As indicated by the spectral characteristic curve in Figure 13.29, such a resonator is effective only over a narrow frequency band.

If the volume of the resonator is V, the neck has a cross-sectional area of S, and the volume of the neck is v, the resonance frequency, f, is given by:

$$f = \frac{2160S}{(vV)^{0.5}} \tag{13.36a}$$

where all dimensions are in inches. If all dimensions are in centimeters, the expression is:

$$f = \frac{5500S}{(vV)^{0.5}} \tag{13.36b}$$

Although the Helmholtz resonator may appear to be of limited value, the principle is widely used. Generally the size is such that these devices are mostly used for frequencies below about 400 Hz, and their sharp resonance-type absorption makes them useful when very selective absorption is required. Perforated panel absorbers can be thought of as a large number of individual Helmholtz resonators. When a perforated panel is spaced away from a solid backing, the holes in the panel constitute the necks of the individual resonators, which share the common volume behind the panel as their cavities. There are also slotted concrete blocks available that use the slots as the necks and the inside spaces as the cavities of Helmholtz resonators to absorb the sound. To broaden the absorption characteristic, the cavities can be filled with a porous sound-absorbing material. These blocks are frequently used to absorb the strong, 120-Hz hum encountered when dealing with electrical transformer noise.

Example 13.9 If a perforated wall-facing is 0.50 in. thick and has circular holes of 0.125 and 0.25 in. in diameter, what frequencies of sound does it best absorb if the effective cavity volume for each hole is 4 in.^{3}?

solution

$$\begin{aligned} f &= \frac{2160S}{(vV)^{0.5}} \\ &= \frac{(2160)(\pi/4)(0.25)^2}{[(\pi/4)(0.25)^2(0.50)(4)]^{0.5}} \qquad (13.36a) \\ &= 339 \text{ Hz for 0.25-in. holes and 169 Hz for 0.125-in. holes} \end{aligned}$$

I. Enclosures

In many cases the purpose of an acoustic enclosure is to keep noise from getting inside. Examples are soundproof booths for machine operators and audiometric test booths for testing the hearing of employees. More often, however, an enclosure (box) is placed around a noise source to keep the noise from getting outside. In predicting the noise reduction for this case, there are some subtleties that warrant further discussion.

Noise reduction (NR) is the difference between the sound pressure levels inside and outside the box. The real question concerns the reduction in noise level at a point outside the box, measured before and after the installation. An interesting phenomenon occurs in a complete enclosure. If a box is built around a sound source, the intensity inside will increase until the amount of power absorbed by the walls is equal to the power emitted by the source. This phenomenon is referred to as "sound build-up." The theories underlying enclosure design can best be illustrated by an example.

Example 13.10 Suppose that a machine is mounted on the concrete floor of a large room and that the sound power level of this sound source is 105 dB in the 500-Hz band. First, calculate the L_p at a distance of 2 m, assuming a point source, free-field conditions, and hemispherical radiation.

$$\begin{aligned} L_p &= L_w - 20 \log r - 8 \\ &= 105 - 20 \log(2) - 8 \\ &= 105 - 14 \\ &= 91 \text{ dB} \end{aligned} \tag{13.10}$$

Now suppose that we enclose the source by placing a box over it as shown in Figure 13.30. The box is made of $\frac{1}{16}$-in. steel, having a transmission loss of TL = 29 dB and $\alpha = 0.02$ in the 500-Hz region. Assuming that the enclosure is sufficiently large (dimensions many times the wavelength of any sound wave being considered), the sound field within the enclosure can be described, as a first approximation, by Eq. (13.19):

$$L_p = L_w + 10 \log\left(\frac{Q}{4\pi r^2} + \frac{4}{R}\right) \tag{13.19}$$

For this problem we will use the distance to the nearest wall, $r = 2$ m. Since

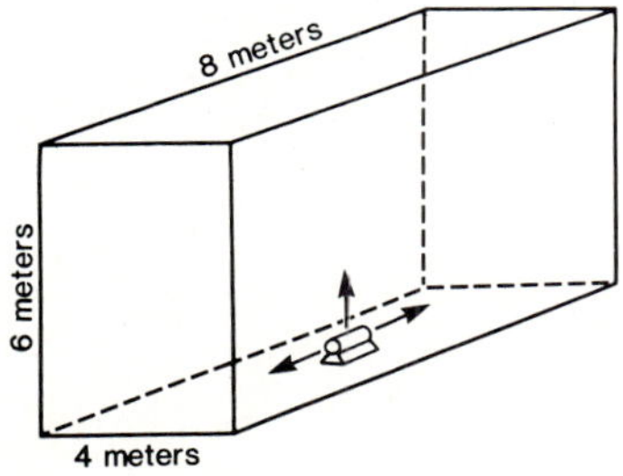

Figure 13.30 Source enclosure.

radiation is assumed to be hemispherical, $Q = 2$. Also, the room constant for the box must be estimated from Eq. (13.18). From Table 13.7, in the 500-Hz region, α (concrete) $\simeq 0.02 \simeq \alpha$ (steel). The area of the box including the concrete floor is $S_c + S_s = 208 \text{ m}^2$. Because α is approximately the same for concrete and steel at the given frequency,

$$R = S\overline{\alpha} = (0.02)(208) = 4.16 \text{ m}^2 \tag{13.18}$$

Entering these values into Eq. (13.19):

$$\begin{aligned} L_p &= 105 + 10 \log\left[\frac{2}{4\pi(2)^2} + \frac{4}{4.16}\right] \\ &= 105 + 10 \log(0.04 + 0.96) \\ &= 105 \text{ dB} \end{aligned} \tag{13.19}$$

Thus, the sound pressure level at the closest inside wall of the enclosure is 105 dB. Note that this is 14 dB greater than the level that existed without the box (free field). Also note the relative magnitudes of the direct-field contribution (0.04) and the reverberant-field contribution (0.96). The sound pressure level within the box is essentially independent of r and is dominated by the reverberation term. Obviously, the space inside the box is almost a perfect reverberant field.

The noise reduction of the enclosure may be calculated by the following relation:

$$\text{NR} = 10 \log\left(1 + \frac{\overline{\alpha}}{\overline{\tau}}\right) \tag{13.37}$$

where $\overline{\alpha}$ = average absorption coefficient
$\overline{\tau}$ = average transmission coefficient

which are calculated from

$$\overline{\alpha} = \frac{\alpha_s S_s + \alpha_c S_c}{S_s + S_c} = 0.02$$

$$\begin{aligned} \overline{\tau} &= \frac{\tau_s S_s + \tau_c S_c}{S_s + S_c} \\ &= \frac{(10^{-2.9})(176) + (0)(32)}{176 + 32} = 1.07 \times 10^{-3} \end{aligned}$$

where subscripts s and c refer to steel and concrete, respectively. Thus:

$$\begin{aligned} \text{NR} &= 10 \log\left(1 + \frac{\overline{\alpha}}{\overline{\tau}}\right) \\ &= 10 \log\left(1 + \frac{0.02}{1.07 \times 10^{-3}}\right) \\ &= 13 \text{ dB} \end{aligned}$$

Using the definition of noise reduction:

$$\text{NR} = L_p \text{ (inside)} - L_p \text{ (outside)} \tag{13.38}$$

or

$$\begin{aligned} L_p \text{ (outside)} &= L_p \text{ (inside)} - \text{NR} \\ &= 105 - 13 \\ &= 92 \text{ dB} \end{aligned}$$

The free-field sound pressure level was previously calculated to be 91 dB at 2 m. The effective noise reduction provided by the enclosure has been estimated to be −1 dB, a slight gain rather than a reduction.

In this case,

$$\text{Effective NR} = 91 - 92 = -1 \text{ dB}$$

Several assumptions were made in this example so that any numerical answer must have certain tolerance limits. However, the effective noise reduction computed above, actually a gain of 1 dB, dramatically illustrates the phenomenon of "sound build-up" in a complete enclosure. The solution to this problem is to add absorption to the inside walls of the enclosure. If the job is good enough, the level at the inside of the wall can approach the free-field level.

Let us now line the walls and ceiling of this enclosure with 1-in.-thick glass fiber insulation having an absorption coefficient $\alpha = 0.65$ in the 500-Hz region (Table 13.7). The average absorption coefficient then becomes:

$$\bar{\alpha} = \frac{(0.65)(176) + (0.02)(32)}{208} = 0.55$$

Enclosure noise reduction becomes:

$$\begin{aligned} \text{NR} &= 10 \log\left(1 + \frac{0.55}{1.07 \times 10^{-3}}\right) \\ &= 27 \text{ dB} \end{aligned} \tag{13.37}$$

The room constant, R', is

$$R' = (0.02)(32) + (0.65)(176) = 115 \text{ m}^2$$

and the sound pressure level just at the enclosure wall will be

$$\begin{aligned} L_p &= 105 + 10 \log\left[\frac{2}{4\pi(2)^2} + \frac{4}{115}\right] \\ &= 105 + 10 \log(0.04 + 0.035) \\ &= 105 - 11.3 \\ &= 94 \text{ dB} \end{aligned} \tag{13.19}$$

Note in the above calculation that increasing the absorption increases the room constant, and decreases the magnitude of the reverberant contribution. (The direct-field and reverberant terms are now of the same order of magnitude.)

The most important information to be gleaned from the above calculation is that the sound pressure level at the inside wall has been reduced from 105 dB to 94 dB with the addition of absorptive lining. Note also that 94 dB is close to the free-field sound pressure level of 91 dB.

$$\begin{aligned} L_p \text{ (outside)} &= L_p \text{ (inside)} - \text{NR} \\ &= 94 - 27 \\ &= 67 \text{ dB} \end{aligned} \tag{13.38}$$

$$\text{Effective NR} = 91 - 67 = 24 \text{ dB}$$

In most practical situations, sound pressure build-up inside a complete enclosure can cause an effective *decrease* in noise reduction up to about 10 dB. It is for this reason that enclosures are normally lined with a sound-absorbent material in order to increase the room constant of the enclosure.

Another factor that must be considered is resonance. If the dimensions of the box result in resonance due to one of the modes of the sound, the box can be driven to high levels of vibration and become a new radiation of these components of sound. When this occurs, the sound pressure level outside the box can be *higher* at certain frequencies than it was even before the box was installed. This effect is significantly reduced by the use of absorptive lining on the interior surfaces of the enclosures, and by damping and stiffening the panels.

REFERENCES

Church, A. H. (1976). Mechanical springs. In: Harris, C. M., and Crede, C. E., eds., *Shock and Vibration Handbook.* McGraw-Hill, New York, chap. 34.

Crede, C. E. (1976). Application and design of isolators. In: Harris, C. M., and Crede, C. E., eds., *Shock and Vibration Handbook.* McGraw-Hill, New York, chap. 32.

Crede, C. E., and Ruzicka, J. E. (1976). Theory of Vibration Isolation. In: Harris, C. M., and Crede, C. E., eds., *Shock and Vibration Handbook.* McGraw-Hill, New York, chap. 30.

Cremer, L., Heckl, M., and Ungar E. E. (1973). *Structureborne Sound.* Springer-Verlag, Berlin.

Crocker, M. J., Hamilton, J. F., and Price, A. J. (1971). Vibration isolation for machine noise reduction. *Sound Vibrat., 5,* 30–34.

Eberhart, L. L. (1967). Specification of products for vibration control. *Sound Vibrat., 1,* 8–13, May.

Frye, W. A. (1976). Rubber springs. In: Harris, C. M., and Crede, C. E., eds., *Shock and Vibration Handbook.* McGraw-Hill, New York, chap. 35.

ISO (1967). *Recommendation R226 (regarding auditory sensitivity).* International Organization for Standardization.

ISO (1973). *Balance Quality Standards.* International Organization for Standardization, 1940–1973 (E).

Jensen, P., Jokel, C. R., and Miller, L. H. (1978). *Industrial Noise Control Manual.* DHEW (NIOSH) Publication No. 79-117, U.S. Government Printing Office, Washington, D.C.

Muster, D., and Stadelbauer, D. G. (1976). Balancing of rotating machinery. In: Harris, C. M., and Crede, C. E., eds., *Shock and Vibration Handbook.* McGraw-Hill, New York, chap. 39.

Peterson, A. P. G., and Gross, E. E. (1974). *Handbook of Noise Measurement,* 7th ed. General Radio Company, Concord, Mass.

Plunkett, R. (1976). Vibration control by applied damping treatments. In: Harris, C. M., and Crede, C. E., eds., *Shock and Vibration Handbook.* McGraw-Hill, New York, chap. 37.

Robinson, D. W., and Dadson, R. S. (1956). A re-determination of the equal loudness relations for pure tones. *Brit. J. Appl. Phys., 7,* 166–181.

Sabine, H. J., and Moulder, R. (1979). Sound absorptive materials. In: Harris, C. M., ed., *Handbook of Noise Control,* 2d ed. McGraw-Hill, New York, chap. 21.

Sivian, L. J., and White, S. D. (1933). On minimum audible sound fields. *J. Acous. Soc. Amer., 4,* 228–321.

Ungar, E. E. (1971). Damping of panels. In: Beranek, L. L., ed., *Noise and Vibration Control.* McGraw-Hill, New York, chap. 14.

Ungar, E. E. (1975). Design of floated slabs to avoid stiffness effects of entrapped air. *Noise Control Eng., 5,* 12, July–August.

Ungar, E. E., and Cohen, R. (1979). Vibration control techniques. In: Harris, C. M., ed., *Handbook of Noise Control,* 2d ed. McGraw-Hill, New York, chap. 20.

Vér, I. L. (1979). Structureborne sound insulation. In: Harris, C. M., ed., *Handbook of Noise Control,* 2d ed. McGraw-Hill, New York, chap. 23.

Index

ABOUT THE AUTHORS

Dr. Wadden is presently Professor of Environmental and Occupational Health Sciences at the University of Illinois School of Public Health. He has degrees in chemical engineering from Iowa State and North Carolina State and received his Ph.D. in chemical and environmental engineering from Northwestern in 1972. Before beginning his academic career he was a practicing engineer for eight years in industry and in government. He is a Registered Professional Engineer (Illinois), a Certified Industrial Hygienist, and a Diplomate of the American Academy of Environmental Engineers and has twice been a Senior International Fellow with the Fogarty Center, National Institutes of Health. He teaches a variety of courses in the control and characterization of hazards which occur in occupational and environmental settings.

Dr. Scheff received his M.S. in environmental health sciences and his Ph.D. in environmental and occupational health sciences from the University of Illinois at Chicago. He is presently Associate Professor in the Pritzker Department of Environmental Engineering, Illinois Institute of Technology, where, since 1983, he has taught courses on engineering control of industrial hazards, indoor air pollution, air pollution control, and air monitoring. At IIT, he has integrated industrial hygiene into the environmental engineering curriculum and currently advises industrial hygienists in the school's graduate program. Dr. Scheff is certified by the American Board of Industrial Hygiene.